To Julia, Stephen, and Nicholas,
with love and appreciation
for their support and encouragement.

Library of Congress Cataloging in Publication Data:

Clevett, Kenneth J.
Process analyzer technology.

"A Wiley-Interscience publication."
1. Chemical process control. 2. Instrumental analysis. I. Title.
TP155.75.C574 1986 660.2′81 85-26302
ISBN 0-471-88316-6

Printed in the United States of America

10 9 8 7 6 5 4 3 2 1

Process Analyzer Technology

Kenneth J. Clevett

BSc., F.Inst.P., F.Inst.M.C.

Exxon Chemical Company
Florham Park, New Jersey

A Wiley-Interscience Publication

JOHN WILEY & SONS

New York • Chichester • Brisbane • Toronto • Singapore

Preface

Process stream analysis, the use of analytical instrumentation for continuous on-line measurement of the quality of industrial process streams, is an important part of the wide field of process instrumentation. The technique first became important in the process industries in the late 1950s and early 1960s. Since then, the technology has expanded enormously and continues to do so.

Process analyzers, while having a chemical basis in the theoretical sense, embody the practical electronic aspects of process instrumentation. Consequently, the analyzer engineer must have an in-depth knowledge of both chemistry and instrumentation, a rather unusual combination.

Successes that can be attributed to the use of process analyzers include savings in production, product giveaway, operating manpower, and energy conservation. In fact, many of today's complex processes could not operate at optimum efficiency without analyzers. Although the theoretical aspects of analytical techniques used in the plant laboratory are the subject of many excellent textbooks, comparatively few cover the subject of on-line process analyzers in detail, especially from the practical application standpoint.

Handbook of Process Stream Analysis, first published in 1973 tried to close this rather large gap and make life a little easier for the field analyzer engineer. Since 1973, we have seen many significant changes in this area of technology. The advent of the microprocessor in the early 1970s and its subsequent commercialization created a major revolution in the instrument world, leading to the programmable controllers and digital instrumentation systems used so extensively today. Likewise, the microprocessor has had a major impact on the design of process analyzers, providing for the first time the capability for carrying out multianalyzer control, internal self-diagnostics, report writing, and digital data transmission. The extensive use today of computers for advanced process control and optimization has markedly increased the importance of process analyzer technology, since it is essential that process stream quality data be available on a timely basis for use in process control algorithms.

This major revision, which includes the addition of several new chapters and reorganization of others, surveys the process analyzer scene today, at the same time maintaining an emphasis on practical aspects of the subject. The title of the book has been changed for two reasons. First, in view of the significant increase in the volume of material, it is no longer a "handbook" but a "reference book." Second, and more important, I feel strongly that the subject has now progressed in stature sufficiently to justify calling it a technology in its own right.

Although process gas chromatographs (GCs) are still the most extensively used of the chromatographic techniques for on-line analysis, liquid chromatography is gradually becoming accepted as an on-line analytical technique. In addition, capillary gas chromatography (GC^2) is now finding extensive use in on-line applications. For this reason, Chapter 1 is now titled "Chromatography" and includes information on these new areas, as well as a deeper treatment of column engineering, from a practical sense.

Spectroscopy has probably seen as much change as any other analytical technique in recent years. It is a wide and varied subject, each individual technique almost a specialist technology in its own right. During the past decade, we have also seen major developments in on-line spectrometric analyzers; techniques such as mass spectrometry and x-ray fluorescence, hitherto restricted to the laboratory, have now migrated into the on-line process field. Such development is still proceeding today and will almost certainly continue in the future. Chapter 10 has thus been expanded to reflect this situation.

The subjects of air quality and water quality monitoring have become extremely important from an industrial/environmental viewpoint, with interest and concern being shown by academia, industry, and government worldwide. A separate chapter has been devoted to each of these subjects, the complexity and variety of water quality monitoring being shown by the length of the chapter. The subject of pH measurement has been included here, since its primary application is in water quality monitoring.

In recent years there has been an increased awareness of the fundamental importance of the sample handling system and its vital contribution to the overall performance of the total analyzer system. Chapter 20 therefore deals with the subject in far greater detail than before, with an emphasis on the practical aspects. Again, because of their important contributions to overall performance, the subjects of design and installation, calibration, maintenance, and performance evaluation of analyzer systems are dealt with in new chapters. A separate chapter is also devoted to analyzer data systems in view of its importance in today's world of computer control.

Combustion efficiency monitoring and control has also become an important aspect of process plant operations, from energy conservation, environmental, and operating efficiency standpoints; a separate chapter is devoted to the subject.

Those involved in process analyzer technology today agree that the main justification for installing on-line process analyzers is their contribution to efficient control and optimization of industrial processes. Therefore, no book on process analyzer technology would be complete without mentioning this very important subject. Rather than trying to cover the thousands of industrial applications in which analyzers are used, in Chapter 24 I have dealt briefly with major applications in the petroleum and petrochemical industries, with which I am intimately familiar. Many of the concepts outlined here will apply equally well to analyzer applications in other industries. The subject of pH control, unique in the process control field, is also dealt with in Chapter 24.

The list of manufacturers' names and addresses is missing from this revision for two reasons. First, the list would be much longer than ten years ago. Second, and more important, documentation in the process analyzer area has improved so much in the past few years that I considered it unjustified. I recommend two sources of information to the reader:

- The Instrument Society of America (ISA) *Directory of Instrumentation,* which names parent companies, and U.S. and international representatives. No analyzer engineer should be without this book.
- *Anadata,* published by Xertex Corporation, United Controls Division, which lists commercially available process analyzers in specification form, by measurement parameter. Again a useful document, but a full-time occupation for Xertex!

Once again I am indebted and wish to express my thanks and appreciation to the many process analyzer manufacturers and analyzer systems specialists in the United States, the United Kingdom, and Europe for their interest in and support for this project and for their permission to publish illustrations, photographs, and specification data concerning their equipment. Without their help it would not have been possible to write this book. Unfortunately, it would also have been impossible to include every process analyzer commercially available today; some arbitrary selection had to be made. Therefore I wish to express my regrets and apologies to any manufacturer whose equipment is not covered in any great detail, or not at all. I have attempted, at the end of each chapter where appropriate, to cover in a summary table all process analyzers of a specific type. It is a fact of life, based on the rapidity with which technology is advancing today, that these tables may be incomplete.

I also wish to acknowledge the Exxon Chemical Company for the inclusion of material in Chapters 1 and 20. This material has been used successfully over the past few years, as part of a training course for Exxon's analyzer engineers and maintenance technicians.

Finally, and by no means least, a word of thanks to microprocessor technology. With the help of a personal computer and an excellent word processing software package, many hours of hard labor were saved in preparing the written script in a form that would not have been possible on a conventional electronic typewriter.

KENNETH J. CLEVETT

Florham Park, New Jersey
April 1986

Contents

1

Chromatography

1.1 INTRODUCTION AND HISTORICAL BACKGROUND

Chromatography, a physical process by which a mixture of chemical compounds may be separated, has rapidly become one of the most successful analytical techniques, both in the laboratory and for on-line process analysis.

The chromatographic process works in a discontinuous manner, similar to a batch distillation. A small sample is taken and the individual components of the mixture are retained on the column to different extents, as if they were distilled off one by one. Because of its nature, the separation normally takes from 1 to 10 min. As the components emerge from the process, they are individually measured and reported.

Note that this is a *physical* process; no chemical change is involved. In practice we are usually talking about gases simply dissolving in liquids or being attracted to the surface of solid materials.

The invention of chromatography is attributed to the work of the Russian biochemist Tswett [1], who was interested in the green coloring matter found in plants. In 1903 he reported the separation of different plant pigments, which were visible as colored bands when a solution of chlorophyll was washed by a suitable solvent through a tube containing an adsorbent such as powdered chalk. In a paper published in 1906 [1], Tswett named this technique "chromatography" (literally, color writing).

Not a lot more was heard about chromatography until a technique known as partition chromatography was introduced by Martin and Synge [2] in 1941, using a liquid moving phase. The method was further developed by Martin and his co-workers to a special form of the technique known as paper chromatography. For this highly successful contribution to the field of biological and medical research, Martin and Synge received the Nobel Prize in 1952.

The possibility of using a gaseous mobile phase instead of a liquid phase was mentioned in the 1941 paper of Martin and Synge, but there was no follow-up of this suggestion. Eventually, James and Martin [3,4] started to elaborate on it in 1949 and the results were presented at the Analytical Chemistry Congress at Oxford, England, in 1952. One of the characteristic features of the method described was the very small samples used for estimations.

The simplicity and analytical power of the method were instantly recognized. Because of its promise, the technique received much attention and its development was very rapid. Since 1952, growth in both the theoretical and practical aspects of the technique has been tremendous. Not only has it provided a simple solution to many complex routine laboratory analyses, but, as we shall see later, it has provided an efficient method by which industrial processes may be monitored and controlled on line.

Chromatography is now recognized as one of the most important tools of analytical and industrial chemistry, with the great advantage that separation and quantitative estimation of components in a sample may be carried out rapidly in a single operation.

1.2 TYPES OF CHROMATOGRAPHY

The basis of chromatography is that a discrete sample of the mixture to be analyzed is transported through a static medium by a moving carrier. The various types of chromatography are classified by the nature of the carrier (or *mobile phase*) and the nature of the static medium (or *stationary phase*). There are therefore four practical possibilities:

Mobile Phase	Stationary Phase	
Liquid	Liquid	LLC
Liquid	Solid	LSC
Gas	Liquid	GLC
Gas	Solid	GSC

Tswett's early work is an example of liquid-solid chromatography.

In this chapter we are primarily concerned with gas-liquid, gas-solid and liquid-liquid chromatography. The last of these is currently in the same phase of development as GLC was in the early 1960s as far as on-line process analysis is concerned. The technique is beginning to make itself felt in the specialty chemicals industry now that component reliability is increasing. It remains to be seen whether LLC will become as popular as GLC.

1.3 THE BASIC GAS-LIQUID CHROMATOGRAPH

The basic components of a simple gas-liquid chromatograph are shown in Fig. 1.1. The *carrier gas,* which is normally nitrogen, helium, or hydrogen, continuously flows through the *column,* where the separation occurs. Samples, which can be either gas (typical volume 0.5 ml) or liquid (typical volume 1μl), are periodically injected into the carrier gas by a specially designed *sample injection valve*. Liquid samples must vaporize immediately on injection and pass through the system in the vapor phase.

After separation, the components emerge from the column and pass to the *detector,* which produces a signal proportional to the instantaneous concentration of sample components in the carrier gas. When this signal is graphed against the time from sample injection, the characteristic *chromatogram* record is obtained.

Since the sample volume, columns and detector are temperature-dependent in operation, they are installed in a temperature-controlled enclosure known as the *oven*.

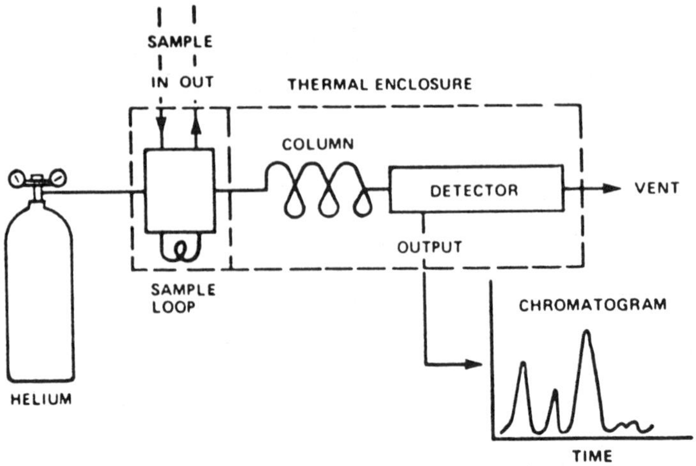

FIGURE 1.1 The basic gas-liquid chromatograph.

1.4 BASIC THEORY OF CHROMATOGRAPHY

The following discussion of the chromatographic process is a simplification, but it gives an insight into what actually occurs inside the chromatographic column. Although gas-liquid chromatography is considered initially, the general principles apply also to the other types previously described.

To understand gas-liquid chromatography, we need to know how gases and liquids interact. Consider what happens when a glass of water is gently heated. Long before the water boils, small bubbles of air appear and cling to the sides of the glass. This illustrates the basic fact that *gases dissolve in liquids*. Clearly, gas solubility decreases as the temperature increases.

Pressure also has an effect on gas solubility, as may be seen by opening a bottle of champagne or a can of beer. More important, the amount of carbon dioxide dissolved far exceeds the amount of air that can be dissolved in water. This second important fact, that *different gases dissolve to different extents in the same liquid,* is the fundamental basis of gas chromatography.

To explain the mechanism, we start by considering how a single gas behaves in a model of the chromatographic process. Imagine a liquid and some air in a completely closed vessel, as shown in Fig. 1.2*a* and then think what would happen if some carbon dioxide (or other gas) were added to the air. The carbon dioxide starts to dissolve, but soon reaches a point where the tendency for more to dissolve is exactly balanced by the tendency for some of the already dissolved gas to come out of solution. This equilibrium condition is shown in Fig. 1.2*b*.

The ratio of the amount of carbon dioxide in each phase at the equilibrium

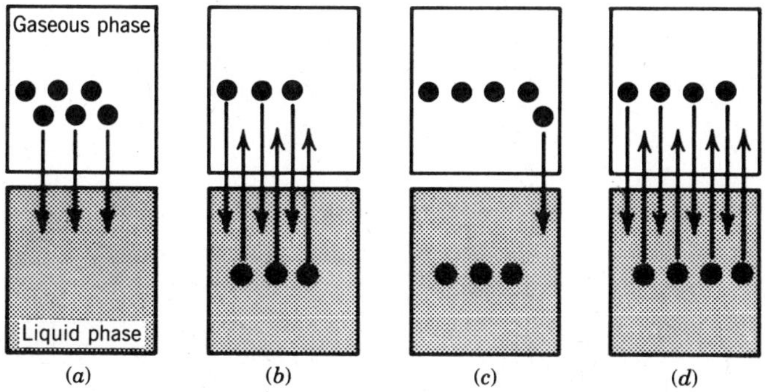

FIGURE 1.2 Static model of the chromatographic process. (*a*) CO_2 molecules introduced; (*b*) dynamic equilibrium; (*c*) more CO_2 introduced; (*d*) new equilibrium. (Courtesy of the Foxboro Company)

point is known as the partition coefficient (K), where

$$K = \frac{\text{conc. of gas in the liquid phase}}{\text{conc. of gas in the gas phase.}}$$

The actual value of K is an indicator of how soluble the gas is in the particular liquid phase. In Fig. 1.2 it is assumed that $K = 1$, that is, half the gas is dissolved and half remains in the gas phase.

From the preceding discussion we know that the solubility will be constant if the pressure and temperature remain constant. Therefore, if more carbon dioxide is added to the system (within certain limits), half of it will dissolve to re-establish the equilibrium shown in Fig. 1.2*d*.

1.4.1 A Model of the Chromatographic Process

So far, we have considered only the static equilibrium of a gas-liquid system. However, in a gas-liquid chromatograph the carrier gas is moving continuously over the stationary liquid phase. How can we study what is happening when the situation in the column is continuously changing? Admittedly this is not easy; even complex theory does not completely explain the process. However, we may achieve a good understanding by the simple expedient of breaking down the movement into a series of separate steps, rather like taking a motion picture and playing it back frame by frame. We can then look closely at what is happening at each stage of the action.

In Fig. 1.3 it is assumed that the column is divided into a series of closed compartments, similar to those considered in Fig. 1.2. For the moment we will restrict our interest to what happens if a pure gas, such as carbon dioxide, is injected into the carrier gas. The flowing carrier gas transports the sample into the first "compartment." At this point we will imagine that the

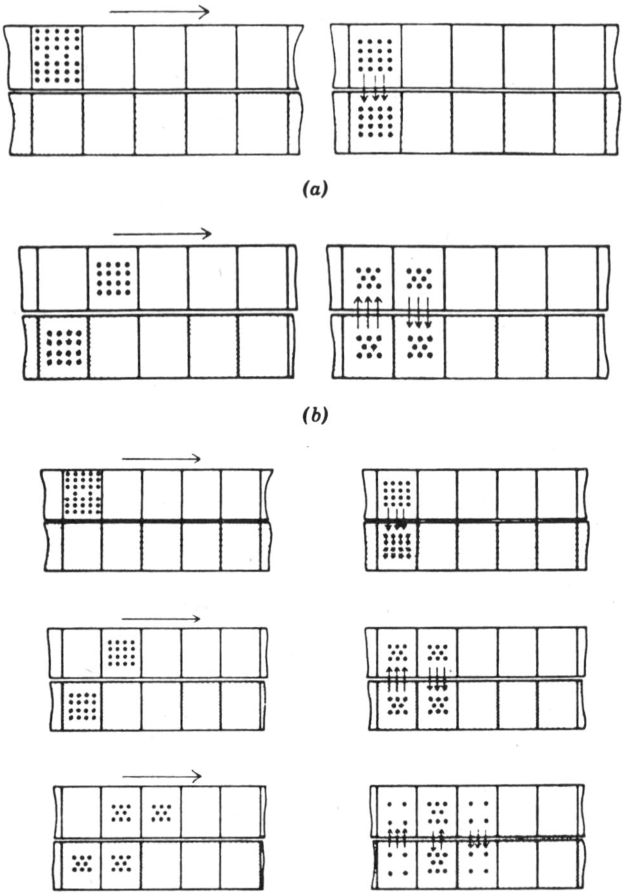

(a) (b) (c)

FIGURE 1.3 Gas-liquid equilibria. (*a*) First equilibrium; (*b*) second equilibrium; (*c*) third equilibrium. (Courtesy of the Foxboro Company)

carrier gas flow is stopped and the compartments sealed. We now have a situation similar to the previous assumption; equilibrium will soon be reached with half the sample dissolved in the liquid phase and half remaining in the gas phase, as shown in Fig. 1.3*a*. Once equilibrium is attained, assume that the carrier gas flow is re-established, allowed to move a distance equivalent to one compartment, and stopped again for examination. The sample in the gas phase is carried along with the carrier into the second compartment, disturbing the equilibrium and leading to the formation of two new equilibria as shown in Fig. 1.3*b*.

If the process is allowed to continue in this start-stop fashion, during each movement the leading edge of the component band will encounter a fresh liquid phase and divide by two, while the trailing edge of the band, dissolved

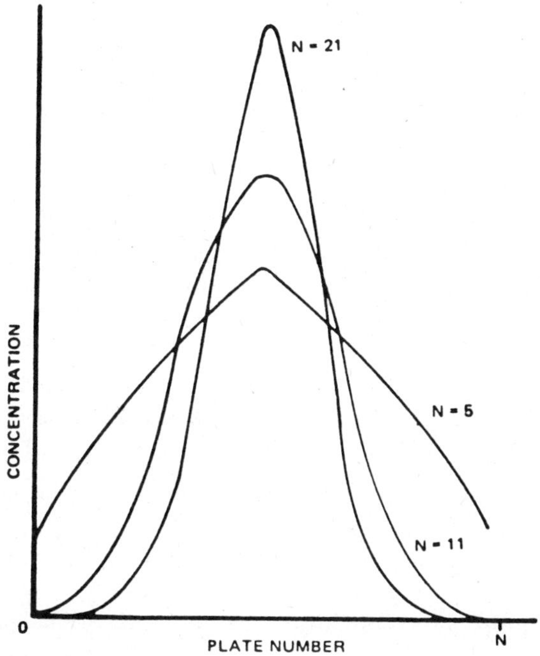

FIGURE 1.4 Graphical presentation of 5, 11, and 21 equilibria.

in the liquid, will be flushed by clean carrier gas and similarly divide by two. Intermediate compartments will reach equilibrium by partial readjustment.

Fig. 1.3*c* shows the development of the first five equilibria and these results are presented graphically in Fig. 1.4, together with the results that would have been obtained if the process had been continued for 11 and 21 steps. Note that in each case the sample is distributed through all the compartments, with the maximum amount at the center. For large numbers of equilibria, however, the amount of sample in the outside compartments is negligible and the distribution at the center begins to resemble the characteristic shape of a chromatogram peak.

1.4.2 Relationship to Theory

The description just given is very close to what actually happens in a chromatographic column. The curves model the distribution of a component along the column after the passage of the carrier gas. Clearly, the width of a peak in this model is dependent on the number of compartments into which the column is divided. Had we continued the process through 100 compartments, the peak would have been quite narrow.

The idea of separate compartments is, of course, purely imaginary, but it does explain chromatographic behavior reasonably well. It is possible to

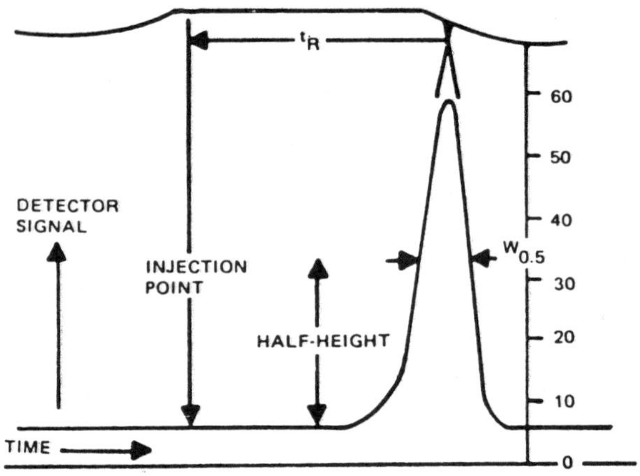

FIGURE 1.5 Data required for calculation of column efficiency.

calculate the number of theoretical compartments from the actual chromatogram peaks obtained; the measurements necessary to do this are shown in Fig. 1.5. The compartments are known as "theoretical plates," analogous to the theoretical plates of a distillation column. The number of theoretical plates (N) is calculated from the chromatogram measurements by the expression

$$N = 5.54(t_R/W_{0.5})^2$$

where t_R = retention time from injection and $W_{0.5}$ = peak width at half height.

Although this is only a discontinuous model of a continuous process, it is nevertheless a powerful and well used one. Most of the advanced theory is concerned with how other operating factors affect the number of theoretical plates. This is because the theory that helps us to design columns with more "plates" will lead to narrower peaks and more efficient separation of components. For this reason, N is used as a measure of column efficiency.

1.4.3 Achieving Separation

The preceding discussion has shown how a peak is formed as the component is carried through the column. The peak may be narrow or wide according to how many times it has reached equilibrium during its passage. However, this idea of theoretical plates does not explain how separation is obtained. Note that in Fig. 1.4, where component distribution is shown for 5, 11, and 21 plates, the peak *position* does not change. This is because an assumption was made that K = 1. For more soluble gases (K greater than 1), the peak would be left of center, whereas for less soluble gases (K less than 1), the peak would be right of center, as shown in Fig. 1.6.

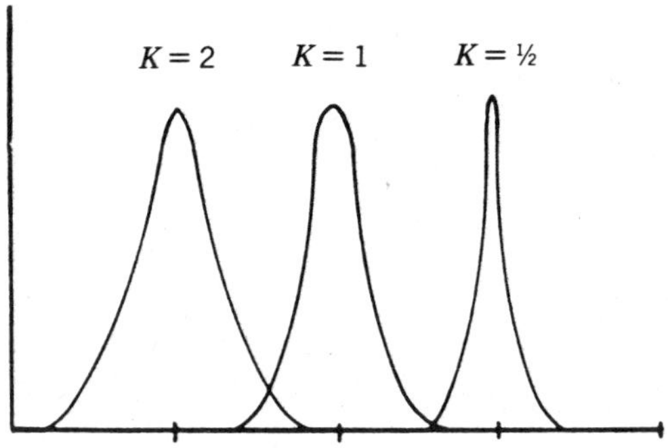

FIGURE 1.6 Principle of chromatographic separation.

This result is to be expected. More soluble gases will be retained longer on the column than those that are less soluble. Thus, it is simply the solubility of the gas in the liquid that determines peak position in the chromatogram and not some mysterious chemical process.

Resolution as used in chromatography is defined as the distance between two peaks in the chromatogram and describes how column efficiency determines their width. Both of these factors are important in obtaining complete resolution and will be discussed in more detail later.

1.4.4 Gas-Solid Chromatography

The previous discussion dealt with gas-liquid systems. In gas-solid chromatography there is neither liquid nor solution. Instead, the carrier gas is in contact with a solid of high surface area. Sample gases are attracted to such surfaces and reach an equilibrium very much like that shown in Fig. 1.2.

Thus, the mechanism is essentially the same, with the "degree of attraction" to the surface replacing solubility. Gas-solid interactions are very strong and the technique is generally only useful for the common inorganic gases and light hydrocarbons. In addition, these solids irreversibly retain water and other substances, which leads to changes in their retention behavior. For these reasons, gas-solid chromatography is used in process gas chromatographs only when no other way of effecting the desired separation can be found.

1.5 LIQUID CHROMATOGRAPHY

Liquid chromatography has been known even longer than gas chromatography, but its use has largely been confined to the laboratory until quite recently. As mentioned previously, process liquid chromatography is now at about the same stage as process gas chromatography was 20 years ago.

Liquid chromatography is, however, a very well established laboratory analytical technique.

1.5.1 Subcategories of Liquid Chromatography

There are basically four subcategories of liquid chromatography:

(a) Liquid-Liquid Chromatography

This is partitioning chromatography, where a separation is achieved by partitioning of a sample between a liquid carrier and a liquid stationary phase that coats a solid packing material. For stability, the liquid stationary phase is chemically bonded to the packing material. This eliminates the problem of washing off the stationary phase with the liquid carrier.

LLC can further be subdivided into normal-phase and reversed-phase partitioning systems. Normal phase means that the carrier is less polar than the stationary phase, while reversed-phase systems use a carrier that is more polar than the stationary phase. The latter has become very popular and is often referred to as "reversed-phase LC." Most laboratory-type LC separations today involve reverse phase partitioning, principally because the technique separates many organic compounds with water as a major component in the carrier solvent. This is attractive both in laboratory and on-line process applications. The technique is most useful in the separation of nonpolar hydrocarbons, particularly those that differ only in their carbon number.

(b) Liquid-Solid Chromatography

This technique, also known as adsorption chromatography, involves the competition between sample components and solvent molecules for the active adsorption sites of the solid packing material, the most popular of which is silica. It is most useful for organic compounds with an intermediate molecular weight. It is not suitable for high molecular weight organics or for ionic compounds, since the latter tend to "tail" appreciably. The technique is attractive for separating isomeric mixtures of similar compounds, such as aromatic isomers that differ only in the position of the functional group.

(c) Ion-Exchange Chromatography

This technique almost exclusively involves the separation of ionic compounds in an aqueous medium. The typical packing used is a highly permeable and porous ionic resin. However, these are gradually being replaced by the new bonded-phase ion-exchange resins that give more efficient separation, shorter analysis time and a more stable column system. The greatest potential for ion-exchange chromatography is in the separation of inorganic compounds, although the demand for such separations has so far been slow to develop.

(d) Size Exclusion Chromatography

This technique is also known as steric exclusion, liquid-exclusion, gel filtration, and gel permeation chromatography. They refer to essentially the same method by which components are separated according to their molecular size.

Molecules that are smaller than the average size of the packing material will spend a significant amount of time within the pores, while larger molecules will spend less time within the pores. This results in a molecular size or weight distribution chromatogram where the larger molecules elute first, while the smaller ones have longer elution times. The technique is ideal for analysis of many polymer streams and is therefore the single most important application for on-line liquid chromatographs.

1.6 GAS CHROMATOGRAPH HARDWARE DESIGN

In this section, the individual components of the process gas-liquid chromatograph as outlined in Section 1.3 are discussed in more detail, in particular the basic concepts used in design.

1.6.1 Thermal Enclosures

Chromatographic operations are usually sensitive to changes in temperature, particularly the gas sample injection valve, the column system, and most of all the thermal conductivity detector. To eliminate this problem, the sample valve, column system and detector are mounted in a constant temperature enclosure, often called the "oven," controlled to ±0.05%. Constancy is the objective here, not accuracy, and may have to be maintained over a very wide ambient temperature range, perhaps −30°C to 50°C, since the process GC may be installed in an unsheltered environment, unlike the laboratory GC.

Process GC ovens are often rated to operate up to 225°C, but the lowest practical temperature is always selected to extend the life of columns and valves. Normally, 130°C is the maximum, the minimum being about 10°C above ambient to ensure good control. Some work has been done with vortex coolers, but the additional complexity is not normally justified. The choice of oven operating temperature is often dictated by sample injection considerations (discussed in Section 1.6.3). Some manufacturers can provide a separate temperature-controlled compartment for the liquid sample injection valve, if required.

(a) Thermal Mass Oven Designs

Any oven design must meet the requirements of reliable on-line operation in hazardous, often corrosive environments, while allowing reasonable access to valves, columns, and detectors for maintenance. Early designs evolved from explosion-proof electrical enclosures, sometimes with a large number of bolt fixings that made access difficult and required the power to be switched off before work could begin.

The basic idea of the thermal mass approach is to keep the temperature-sensitive components in contact with a substantial mass of metal. The main advantages are that low temperature gradients are achieved throughout the enclosure and good control is possible using a simple thermostat. Offsetting these virtues is the long stabilization time required after any work has been carried out on the instrument. With the development of reliable and inexpensive electronic controllers, the advantage of using a simple thermostat has also been lost and the relatively expensive casting and machining requirements have generally made thermal mass enclosures uneconomic. In most modern process GCs, thermal mass ovens have been replaced by various air-bath arrangements.

(b) Air-Bath Oven Designs

This method of heating by hot air circulation has been used since the very first process GCs were built. Initially, stability was poor, extreme temperature gradients were experienced, and it was difficult to make the design safe for hazardous environments because of the use of electric fans and electric wire heaters with high surface temperatures. Most of these problems have been overcome, and modern ovens usually employ an electric cartridge or steam-powered heater housed in an explosion-proof cartridge.

A typical design using electrical energy is shown in Fig. 1.7. The oven is

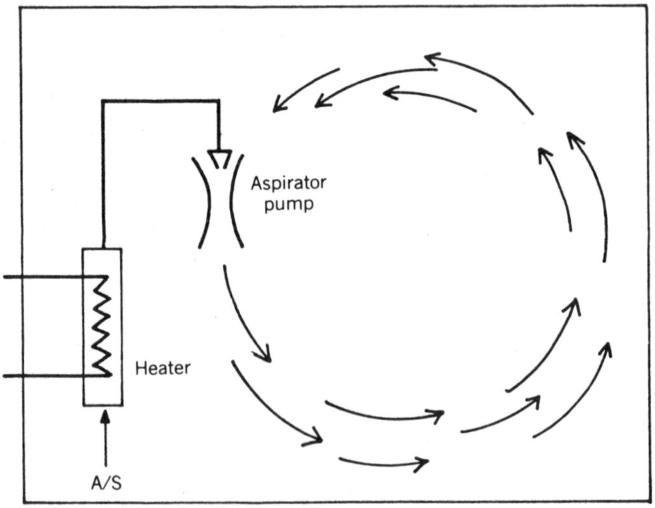

FIGURE 1.7 Typical air-bath oven design.

thermally insulated. Compressed air passes through a heater, is mixed with oven air, and is circulated by the flow eductors. Temperature is controlled by electronic regulation of the heater element, and safety thermostats provide protection in case of controller malfunction. Steam-heated ovens are not common since they cannot operate at an adequately high temperature for many applications. For moderate-temperature analysis, however, they offer a low-cost and safe installation.

(c) Programmed Temperature Control

For laboratory GC work, temperature programming is used quite frequently, particularly for wide boiling range component analysis. With this technique, the oven temperature is gradually increased in a controlled manner after sample injection, to speed up the elution of the high boiling components. The technique is not often used in process GCs, although several manufacturers offer this option. It is possible that with increasing reliability of electronic sequencing and control circuity, temperature programmed GCs may become more attractive.

(d) Safety of GC Ovens

It is essential to recognize and understand the safety hazards inherent in the situation where flammable samples are piped into a heated enclosure such as a GC oven. There are two sources of explosive gas concentration—the surrounding atmosphere itself and leakage of sample within the enclosure. Therefore it is of the utmost importance that the analyzer does not generate a source of ignition under normal or fault conditions. Potential sources of ignition include electrical arcing, high temperature caused by heater runaway, and hot surfaces during normal operation.

Protection against ignition from electrical power sources is usually by conventional conduit rated for the gases constituting the hazard. The U.S. National Electrical Code (NEC) rates the gases by group. Other national codes have similar, though not identical, ratings. It is important to realize that the hazard caused by the sample may be different from that of the surrounding area. Provided that the correct classification is adopted and the equipment is built according to the code, no problems should occur. If conduit sealing is required, this often must be done after site installation. It should not be forgotten.

The potential for overheating due to electrical failure is always present in electrically heated ovens. Two sensors are used to monitor the situation. One senses the heater element temperature; the other senses the general oven temperature. The latter may be the temperature control sensor or a separate, dedicated sensor. Typically, oven temperature measurement is provided with preset limits, first for high alarm, second for auto shutdown. In addition, a pressure switch cuts off electrical power if the heater air supply falls below a preset limit.

The last and much neglected cause of ignition is a hot surface. The ability of a hot surface to cause ignition depends upon the type of gas present. The NEC code classifies gases by "T" ratings. Note that the ignition temperature of gases are in a different order from their NEC group ratings.

Hot surfaces exist in any GC oven. Inspection of the NEC "T" ratings will reveal that some gases ignite at 180°C. Clearly, it is not permissible to run any oven above this temperature if such gases may be present. The true situation is in fact worse, since, for heaters to work efficiently, their temperature must be well above the desired oven control temperature. In some designs, this differential may be hundreds of degrees, making "T" rating conformance impossible. One technique used to overcome this is to mix very hot heater air with cool air before discharging it into the oven. This segregates the hot surfaces from the potentially flammable oven air. One problem with this is diffusion of oven air into the hot heater air should the air supply fail. Another problem may be the thermal capacity of the heater, causing contact surfaces to overheat after air and power are off. Each individual design must be carefully tested under normal and shutdown conditions to assure safety.

1.6.2 The Carrier Gas System

The purpose of the carrier gas system is to provide a stable transport and detection medium for the components of the sample. Any change in carrier flowrate will have a detrimental effect upon detector signal and component retention times. Most process GCs rely on constant carrier flowrate to maintain detector zero and span settings and to properly locate the components of interest on a fixed time base.

The choice of carrier depends to a large extent on the type of detector selected. Thermal conductivity detectors, for instance, usually require helium or hydrogen to give maximum differential thermal conductivity between the carrier gas and eluting components. Of secondary importance is the effect of carrier gas on column efficiency and pressure drop. These effects will be dealt with in more detail later. For the moment we will concentrate on problems common to all carrier gases.

First, carrier gases must be pure and dry. Because the requirement for purity is not well understood, the degree of purity purchased is based on a "reasonable price" rather than on technical considerations. Of primary importance is the presence in the carrier of components to be measured. For example, if a 10 ppm methane sample is to be analyzed using a carrier containing 1 ppm methane, a measurement of 9 ppm will be obtained, giving a 10% error. If a 1 ppm methane sample is analyzed, the chromatograph will give no signal at all. However, the presence of methane in the carrier gas will have no effect on the measurement of 1 ppm ethane. In other words, impurities in the carrier gas cause zero elevations for the components present.

Another problem is the effect of carrier gas purity on baseline. If the supply of carrier were infinite and the purity constant, there would be no problem; in the real world, carrier gas cylinders may vary in their impurity levels, resulting in baseline shifts outside the capabilities of signal processing circuits. Generally, the total impurity level should result in baseline elevation of less than about 5% of the most sensitive measurement range. When using a flame ionization detector, the presence of argon in the nitrogen carrier can be tolerated, but the hydrocarbon content is critical. It is also best to minimize the amount of oxygen and water present, since these will gradually destroy some column materials. Water is detrimental to active solid packings and is undesirable in most other separations.

When designing carrier gas systems, thought must be given to the maintenance aspects. Cylinders may be dedicated to individual GCs or small groups of GCs and mounted, for instance, outside the shelters in which the GCs are located; this is common practice. Alternatively, cylinders may be manifolded together at a central location and piped to individual analyzer locations. When large numbers of gas chromatographs are in use, the central cylinder approach can lead to significant savings in cylinder-handling labor.

To ensure continuity of supply, it is best to use two cylinder manifolds, as shown in Fig. 1.8, each fitted with a pressure regulator, one set at a slightly higher pressure than the other. With this system the cylinder with the higher pressure is used first, with the other switching in automatically when the pressure begins to fall. A pressure switch set midway between the two regulator settings will give an alarm signal when the first set of cylinders is empty.

It is essential that all carrier gas piping and fittings between cylinders and GCs be chemically cleaned with a dry solvent such as hexane or acetone and then dried with clean nitrogen or helium purge. This is best done prior to

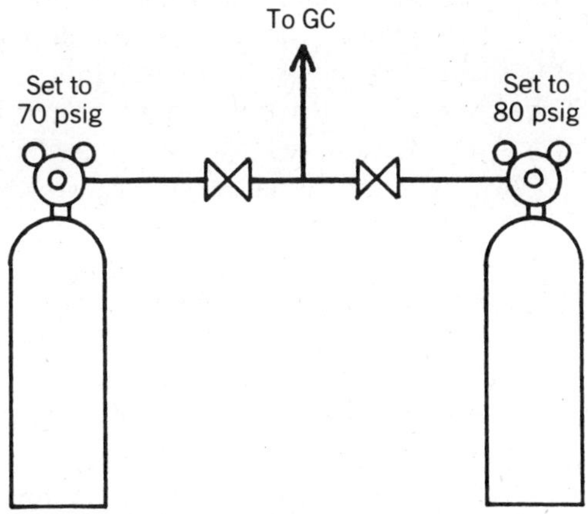

FIGURE 1.8 Carrier gas manifold system.

installation. When analyzing for less than 1000 ppm of any component or using an active solid column, a drier tube containing molecular sieves should always be fitted in the carrier gas line close to the GC. A dual-tube system with shut-off valves will allow one drier to be changed while the other is in use. The spent drier may be reactivated by purging with pure helium or nitrogen while heating to about 300°C.

Another important step in pre-startup checking is to leak test the carrier gas system. This can be a tedious business but will pay dividends later. Helium and hydrogen leak much faster than air, so even a small leak can have a disastrous effect on carrier gas consumption.

Carrier gas is usually supplied to the GC at about 15 psig above column head pressure to provide optimum conditions for the internal pressure regulator. In modern GCs, carrier flow regulation is usually achieved by a precision pressure regulator mounted inside the temperature controlled enclosure. Flow regulators, which rely on maintaining a constant differential pressure across a restriction, are necessary only if temperature programming is being used; otherwise, maintaining a constant pressure across the columns will provide better flow control. Flow regulators should never be used in conjunction with backflush or other column switching arrangements involving pressure changes in the columns. It is important that when pressure changes are introduced by column switching, the column head pressure is brought back to normal pressure as rapidly as possible. Flow regulators, by their very nature, prevent carrier gas from quickly flowing into the column to make up the deficiency. Consequently, it may take more than one minute for the detector flow to stabilize after switching.

Column head pressure is typically 15 to 60 psig and monitored by a pressure gauge. Flow is more important, but more difficult to measure. Typically, flow is set between 20 and 80 cm^3/min, in accordance with application data, and is sometimes monitored by a rotameter. It is not good practice to install the rotameter at the detector vent since it might affect detector performance. Therefore it is normally mounted after the pressure regulator. Unfortunately this reduces the reading to the point where it acts more as a leak detector than a viable measurement of carrier flowrate.

Detector vent flowrate is best measured using a soap film flowmeter, a device by which the flow is measured by timing the rate at which soap bubbles travel up a calibrated tube. It is less accurate with hydrogen as this gas readily diffuses through the bubbles.

1.6.3 Sample Injection Systems

The function of the sample injection system is to deliver to the head of the column a precisely metered volume of a representative sample of the process stream to be analyzed.

GC detectors respond to the absolute quantity of components in the sample; thus it is important to hold the sample volume constant for reproducible results. To ensure good GC performance and repeatability, the whole of the sample volume must be injected into the column in a time period negligible compared with the narrowest peak width expected. Normally referred to as "plug injection," this also implies close-coupling the sample valve to the column, with minimum dead volume in between.

Sample injection valves are mechanical devices and must be carefully designed to give adequate operating life at the fastest analysis repetition rate. Note that a 1-min analysis cycle will require over 0.5 million valve operations per year and that the severe process environment may reduce the performance of a valve below the "operations lifetime" quoted for ideal conditions.

(a) Types of Sample Injection Valves

The Rotary Valve. One of the first sample injection valves used in process GCs was the rotary type, a typical example of which is shown in Fig. 1.9. In the rest condition, process sample flows through the metered volume, while carrier flows into the column. The valve rotates through 60 degrees when energized, causing the carrier gas to sweep the metered volume of sample into the column. After a few seconds, the valve is de-energized to refill the sample volume ready for the next injection.

A rotary valve for liquid injection is shown in Fig. 1.10. This valve rotates through 90° to operate. The metered volume in this case is a small slot machined into the rotor. All sample valves have a nominal delivery volume; it is repeatability that matters, not volumetric accuracy.

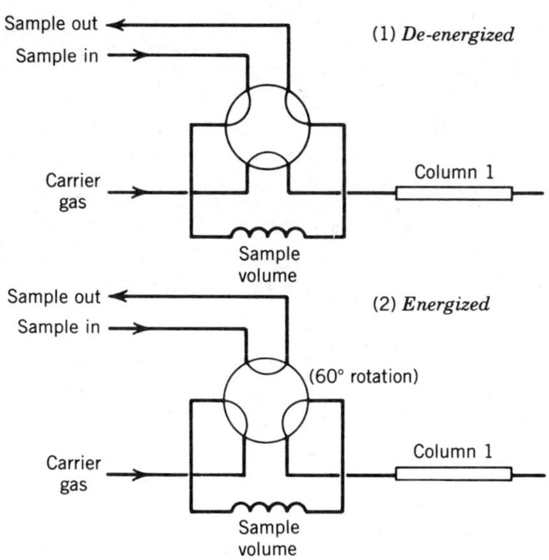

FIGURE 1.9 Rotary valve (six-port) for gas injection.

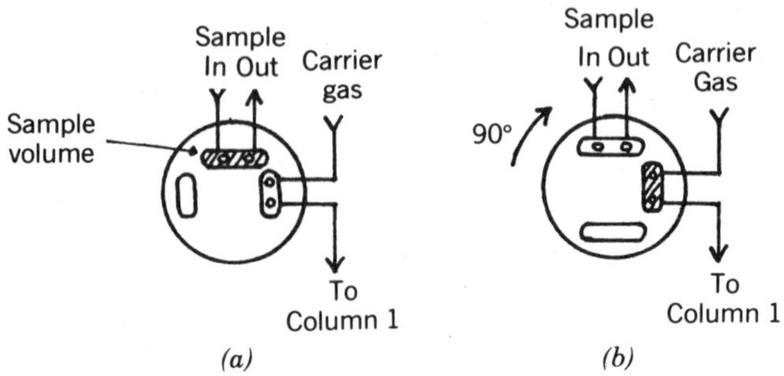

FIGURE 1.10 Rotary valve for liquid injection. (*a*) Sample fill position; (*b*) sample inject position.

The Slide Valve. This type of sample injection valve is undoubtedly the most popular one in use today; a typical example is shown in Fig. 1.11. The principle of operation is to physically transfer a metered volume of the process fluid from the sample flow into the carrier gas flow by means of a moving plate or slide.

Slides for liquid samples may have a "straight-through" configuration or a "cavity" type sample volume. The two alternatives are shown in Fig. 1.12. The "straight-through" geometry is demonstrably faster than the cavity type, but both are sufficiently fast for classical process GLC. Many valve designs surround the sample volume with PTFE (Teflon), but this tends to slow vaporization of the sample because of the good insulating properties of PTFE.

The Piston or Spool Valve. A different type of linear valve is the piston or spool valve. The valve uses elastomer "O"-ring seals to divide the piston

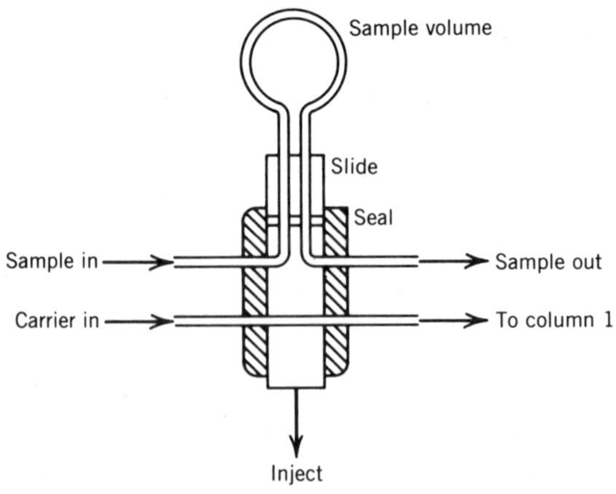

FIGURE 1.11 Typical slide valve for gases.

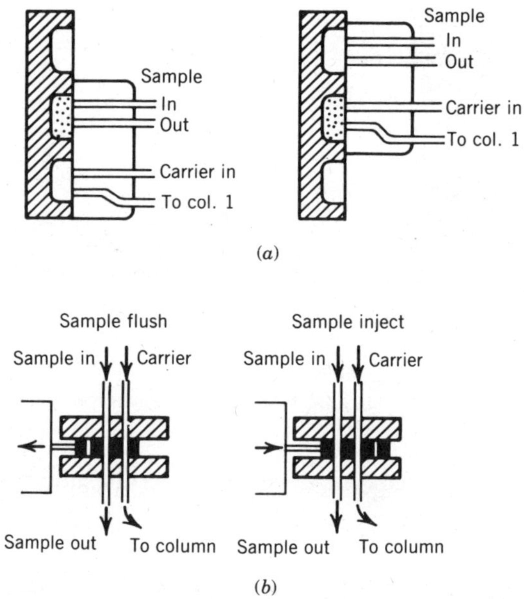

FIGURE 1.12 Typical slide valve for liquids. (*a*) Cavity-type geometry; (*b*) straight-through geometry.

into annular sections that connect various ports, as shown in Fig. 1.13. Actuation of the valve changes the port connections to inject the sample. This type of valve must be carefully designed to reduce "dead" volumes and is primarily used for gas samples. It overcomes the major difficulty with slide and rotary valves—obtaining a good gas-tight sliding seal—and is often used when a large sample volume is required.

The Diaphragm Valve. This valve, which does not contain any sliding surfaces in contact with the sample, uses an entirely different design approach; an example is shown in Fig. 1.14. The valve consists of six ports and six plungers, arranged in a circular pattern. The plungers are operated in

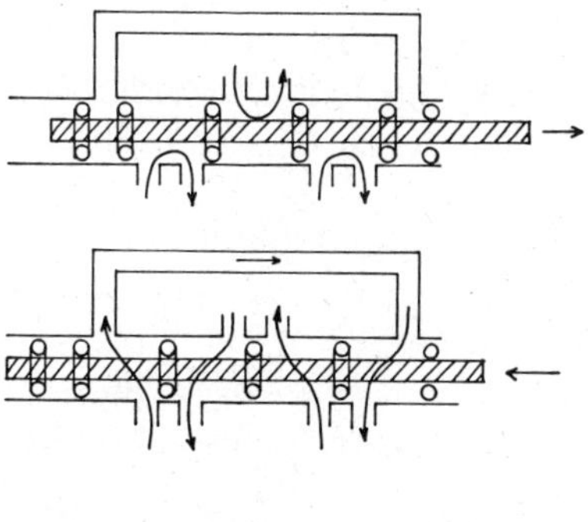

FIGURE 1.13 Piston/spool type valve.

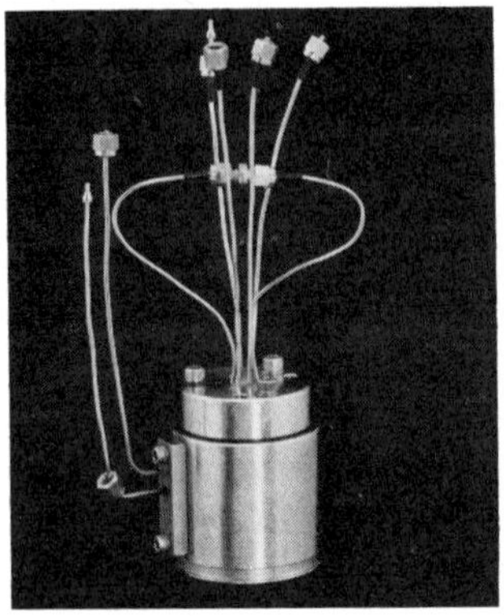

FIGURE 1.14 Typical diaphragm valve. (Courtesy of Applied Automation Inc.)

alternate sets of three and, in view of their small size, produce very high sealing pressure on the diaphragm. During valve actuation, a mechanical interlock ensures normally closed passages are opened. Operation of the valve is claimed to be very fast and thus suitable for high-speed GLC because of the very small distances moved. Potential problems are diaphragm leakage and occlusion of minute quantities of liquid samples in the diaphragm material or at the points where the diaphragm is pinched.

Valve Operations. The valves mentioned may be remotely actuated by pneumatic, solenoid, or electric motor actuators. The pneumatic variety are the most common and require from 15 to 60 psig to operate, depending upon the design. Normally, three-way explosion-proof solenoid valves are used to control the air to the operator cylinder. Most valves use a spring return, although this requires double power in the operator. A few use double-acting operators.

(b) Materials of Construction

Selection of materials of construction for sample injection valves is most important. They must not be liable to corrosion or other forms of attack by any chemical substance in the sample, nor should they catalyze chemical changes within the sample itself. In view of the possibility of sample leakage, the same criteria apply to external materials of construction also. It must be appreciated that sample valves are precision devices, often employing optically flat surfaces and very small passages. Therefore, a material suitable for process plant construction may not last longer than a few days in a sample valve.

In most cases, stainless steel and PTFE are suitable, but there are exceptions. For instance, stainless steel is subject to attack by chlorine and hydrogen chloride. If these compounds are present, then hastelloy, tantalum, or all-PTFE construction might be suitable alternatives. PTFE is excellent for chemical resistance, but is soft and tends to extrude under pressure. It is also

subject to penetration and swelling by some liquids, notably styrene, which can block the gas passages and may change the sample volume. Many valves use glass-filled PTFE, which offers better dimensional stability but not much relief from attack by organic liquids. One valve uses two metal valve plates with a thin PTFE membrane between to act as a seal and lubricant. Fused alumina ceramic is a new material for slide valves. It is chemically inert, very hard and can be polished almost perfectly flat. The main problem has been to get a good ceramic-to-ceramic seal without too much frictional resistance to movement.

(c) Gas Sample Injection

The sample size required is quite small: 0.3 to 1.0 cm^3 is normal for $\frac{1}{8}$-in. columns and up to 5 cm^3 for $\frac{1}{4}$-in. columns. Larger volumes up to about 25 cm^3 may be taken for trace analysis. The size is adjustable by changing the dimensions of the external sample loop. Occasionally, where high-sensitivity detectors and capillary columns are used, very small gas samples may be necessary. In these situations, a valve designed for liquid samples is probably the best choice.

Gas samples are usually taken at atmospheric pressure. Since the flowing sample generates a small back-pressure, "atmospheric balancing" may be used to bring the sample to exactly atmospheric pressure. Needless to say, carrier gas must not leak into the metered volume, but this is often an unrecognized source. Sometimes it is preferable to sample at process pressure in a few applications, especially when trying to sample from a process under moderate vacuum. This can easily be accomplished by having a vacuum pump and flow control on the sample valve vent line. Results will vary with process pressure, and volume calibration is not possible. Often, the ratio of two component concentrations is required and, since ratioing cancels out any errors, this is a perfectly valid method. It is certainly far simpler than trying to control the sample volume at a specified subambient pressure.

(d) Liquid Sample Injection

For liquid samples, the objective is to inject about the same molar quantity of sample as in gas injection, followed by rapid vaporization of the sample. This requires the precise metering of very small volumes of liquid. Liquid sample volumes usually lie between 0.5 and about 10 microliters (μl). Liquid sampling is more difficult; therefore, process streams that can be easily maintained in the vapor state are better vaporized prior to injection. However, samples with final boiling points above the preferred column temperature must be injected in the liquid phase.

When injecting liquids, it is important that sample pressure is high enough to prevent formation of gas bubbles in the liquid when the sample is heated to analyzer temperature, since the smallest bubble will have a drastic effect on the volume of sample injected. Liquid sample valves are therefore rated

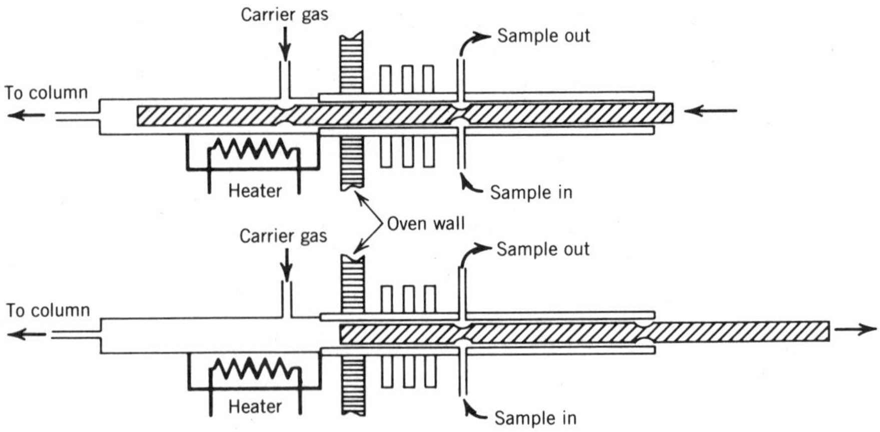

FIGURE 1.15 Dual zone liquid injection.

at up to 300 psig sample pressure. However, it is inadvisable to attempt liquid sample injection above about 135 psig, since all valves have a propensity to fail when used repeatedly under such adverse conditions.

Incoming process liquid must be preheated to sample valve temperature before entering the valve. It is often thought that the sample valve temperature should be above the atmospheric boiling point of the sample so that it will flash-vaporize when injected. In actual practice, however, good injections may be obtained at valve temperatures 100°C *below* the atmospheric boiling point. Of course, lower temperatures act in our favor too, as they increase separation, prolong the life of columns and valves, and prevent samples from polymerizing or decomposing. This gives us some latitude in selecting oven temperature to suit both sample vaporization and column performance.

For high boiling liquids that are very difficult to handle and cannot be injected at column temperature, the additional complication of dual-zone operation cannot be avoided. Special valves have been designed to sample the process at low temperature and inject the sample through the oven wall into the hot column zone. This technique is illustrated in Fig. 1.15. The sample volume is an annular groove in the valve piston. When energized, the operator drives the piston through the seal, injecting the sample into the hot carrier gas. A thermal shield prevents excessive heat transfer between the two temperature zones.

1.6.4 Types of Columns

Three distinct types of columns are used in process gas chromatographs today: (a) liquid-phase columns, (b) active solid columns, and (c) columns using synthetic packings.

(a) Liquid Phase Columns

Separation in these columns occurs as a result of components interacting with the liquid by "partition" (partially dissolving in it). There is an almost infinite choice of liquid phases and a wealth of excellent reference literature on this subject. Any non-volatile liquid may be used. In practice, however, a limited list of reliable, repeatable, and stable liquids has been developed for process GC use. The liquid must be non-volatile and is held in place by being very thinly coated over a solid support.

Supports are generally of two types. The "pink" variety, a more refined descendant of the classical ground firebrick, is excellent for hydrocarbon (non-polar) separations. The "white" variety is basically a diatomaceous earth (the aggregated skeletal remains of microscopic sea creatures) and is best for separating more polar chemical compounds.

Both types of support are available in various mesh sizes and chemical treatments. For columns of $\frac{1}{8}$ in. diameter, 80–100 or 100–120 mesh is normal, the larger the mesh size the smaller the particle size. Smaller particle sizes generally provide more efficient separations but require higher carrier gas pressure to achieve the same carrier flowrate.

The purpose of chemical treatment of the support is to improve its inertness. There are two types, acid-washed (AW) and silicone-treated (DMCS). These help to reduce "tailing" of certain polar components but are not always beneficial.

(b) Active Solid Columns

These columns are packed with a granulated solid of large surface area. Separation occurs as a result of the varying attraction of gas molecules to the surface of the solid (adsorption). Active solid columns are suitable for separating only a few gases. They have a propensity to permanently adsorb heavier molecules (particularly water) and therefore their use has declined in favor of the porous polymers.

The only solid now in regular use is molecular sieves, which is valuable for its ability to separate oxygen, argon, nitrogen, hydrogen, and methane. To protect the column from deactivating molecules, a pre-column is necessary, usually Poropak-T.

(c) Synthetic Packings

The most popular of these valuable new developments is the Poropak series, which consist of polystyrene beads. Their properties overlap active solids and liquid phases and make them particularly suitable for light hydrocarbon and permanent gas analysis. New columns must be conditioned at about 200°C and at low carrier flow. Each type must be considered individually for an application. There is a wealth of good reference literature on the subject available from manufacturers of these packings.

1.6.5 Detectors for Gas Chromatography

The detector responds to components emerging from the columns and provides a signal (usually electrical) that can be processed to yield component concentration data. In practice, all detectors in use operate on the differential principle, that is, they show a constant response to carrier gas, generate a relatively large signal whenever a component elutes from the column, and return to a steady "baseline" level after the component has passed through completely.

In process gas chromatography, the requirements for long-term stability in a harsh environment and generalized sensitivity to a wide range of chemical substances have limited the choice to the thermal conductivity detector (TCD), the flame ionization detector (FID), and the flame photometric detector (FPD). Photo-ionization detectors (PID) are used primarily in environmental monitoring.

Any detector used in a process GC must respond rapidly and preferably linearly to changes in component concentration. The response should not be too sensitive to changes in operating parameters such as carrier gas flow and temperature. Calibration at weekly intervals must be sufficient to maintain good accuracy. The baseline response must be stable during one analysis and must not vary outside the limits of the electrical compensation system for long periods of time (months). The physical construction should be simple, inexpensive, and capable of withstanding vibration and long-term exposure to the process plant environment. Installed, it must conform to the industrial electrical and work safety codes applicable at the user location.

(a) The Thermal Conductivity Detector

The TCD is the most frequently used detector for process GCs. The main reasons for its present popularity are probably historic and economic, rather than it's being the logical technical choice. It has two advantages: it is sensitive to almost all sample components encountered except the carrier gas, and it is very simple and inexpensive, requiring only simple electronic circuitry.

There are, however, some serious disadvantages. The thermal elements are liable to burn-out and subject to oxidation effects that change their characteristics. They are shock and vibration sensitive and may become embrittled after long service. The detector is very sensitive to temperature, resulting in baseline and calibration drift, so very good oven temperature control is essential. The linear dynamic range is restricted to about three orders of magnitude, which, although adequate for older fixed-range instruments, limits its usefulness with modern auto-ranging circuits.

In view of these serious shortcomings, one may well ask why it is so popular. Before about 1960, it was the only detector available. It was the subject of intense research and development at that time because of the rapid growth of GLC, and although its response is poorly explained by theory, its practical design is well understood. Above all, it is simple and inexpensive. Detector cell geometry has been the subject of considerable development. Early work produced the straight-through, semi-diffusion and diffusion designs. These are compromises between fast response (straight-through) and high stability (diffusion). The diffusion design exposes the element to the thermal effect of a sample component without direct physical contact and is used in the measurement of corrosive gases such as chlorine. Examples are shown in Fig. 1.16.

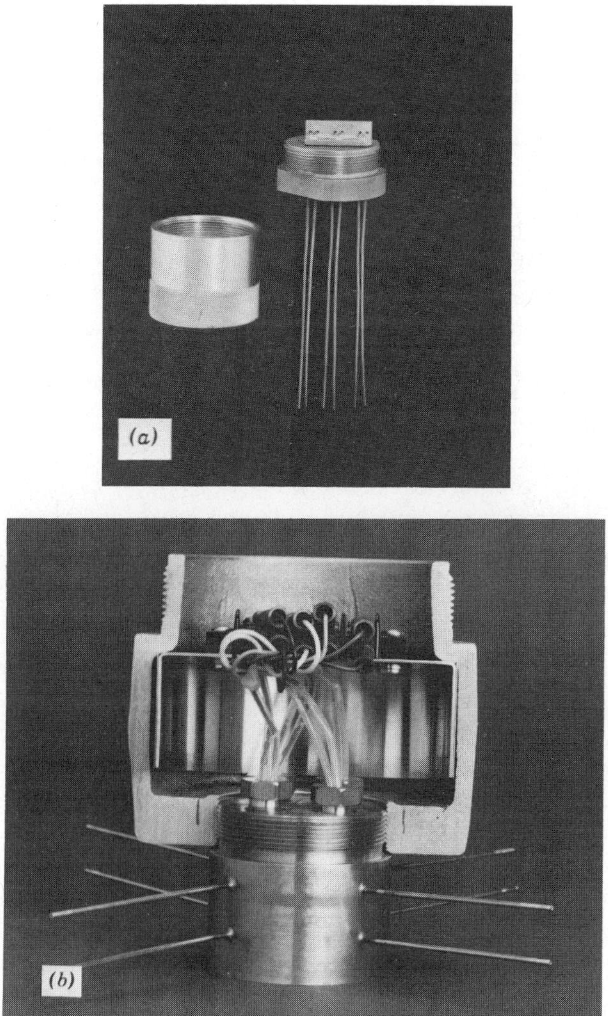

FIGURE 1.16 Thermal conductivity detector designs (*a*. Courtesy of Applied Automation Inc.). (*b*. Courtesy of The Foxboro Company)

More recently, development has concentrated on reducing the internal volume of the detector to improve the response and make it compatible with the smaller columns. Detectors with internal volumes of about 40 μl are now common and may be used with highly efficient columns of small internal diameter. Generally, it is best to restrict TCDs to measurements of 0–1000 ppm or higher; for low ppm measurements, other types are more suitable.

TC Detector Elements. The thermally responsive elements in a TCD may consist of coiled resistive wire or heat-sensitive semiconductor (thermistor). The element chosen must exhibit good response and reliability with varying carrier gas impurities and sample component concentrations.

Early TCDs used tungsten wire elements and hydrogen or helium carrier gas. With constant voltage supply the passage of a sample component reduced heat loss and caused the wire temperature to increase. The corresponding increase in resistance was the basis for the output signal. The use of a constant current power supply increased the temperature rise and the sensitivity but resulted in increased electrical noise and propensity to failure.

A significant improvement in detector life and stability has been obtained by controlling the filament current to maintain constant wire resistance; the temperature of the element does not vary. This technique uses the energy supply as the measured variable. Advantages are better life expectancy (even when carrier gas runs out) and better response time, since thermal lags in heating and cooling the element are eliminated.

During the 1960s, thermistors showed considerable promise as the basis for low-volume, high-sensitivity GC detectors. However, many practical problems have reduced these advantages. Thermistors attain their highest sensitivity at low temperatures, are not stable over long periods of time (especially in hydrogen service), and have non-linear output characteristics. Consequently, wire elements have maintained their general use. For process applications, corrosion-resistant materials often must be used. Various types are available, including platinum, nickel, platinum-iridium and tungsten-rhenium. Another approach has been to coat the element. Gold-plated wires are successful, but other elements such as glass-coated thermistors and PTFE-coated wires tend to increase response time.

When analyzing samples containing chlorine, hydrogen chloride, or other corrosive substances, it is wise to prevent the corrosive material from entering the detector by appropriate use of column switching. If the corrosive gases must be measured directly, a diffusion cell geometry will reduce direct contact with the thermal elements and prolong their life expectancy.

The configuration of carrier gas flows through the TCD deserves special attention. Reference elements should be identical to the measurement elements and be exposed to exactly the same conditions. TCDs for process GLC are usually four-element types, but the two-element version is sometimes seen. Innovative use of all four elements may be made if each is separately accessible. For instance, an eight-port detector may be used to measure hydrogen in nitrogen carrier gas on one side and methane and carbon dioxide in helium carrier gas on the other. Essentially, this blends two detectors into one and allows the use of common signal processing circuits, while avoiding the anomalous response to hydrogen in helium carrier gas. A similar approach may be used to measure oxygen and nitrogen on one side and carbon dioxide on the other, both using helium carrier gas. These configurations realize significant savings over alternative designs and add further justification for the continuing use of the TCD.

(b) The Flame Ionization Detector

In contrast to the TCD, the FID, developed by McWilliam and Dewar, is very simple. The basic detector, shown in Fig. 1.17, consists of a hydrogen flame into which is fed the carrier gas and components eluting from the column. At the high temperature of the flame, compounds containing carbon-hydrogen bonds break down into positive and negative ions. A high electric potential is applied across the flame, causing the negative ions to discharge electrons to the positive plate, from where they flow round the circuit to the negative plate to be absorbed by the positive ions. This electron flow constitutes an electrical current proportional to the amount of carbon in the flame. The current is measured by a high-impedance electrometer circuit and transmitted as a component concentration signal.

The FID detector is very sensitive and linear over about six decades of response. It is possible to achieve a measurement range of 0–1 ppm for

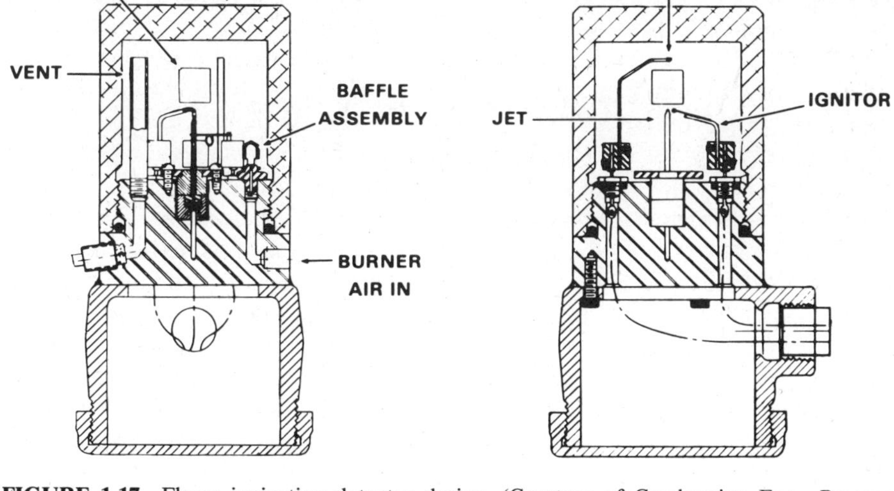

FIGURE 1.17 Flame ionization detector design. (Courtesy of Combustion Eng., Process Analytics)

hydrocarbons. However, the FID does not respond to compounds that do not contain carbon, nor does it respond to carbon oxides or sulfides. This lack of response to inorganic molecules has limited the general acceptance of the FID.

When either detector is suitable, the FID will usually provide superior results. In fact, there are many cases where, for the sake of overall standardization, user companies are specifying the FID for all applications, both percentage and ppm measurements. The FID's inherent sensitivity and linearity provide a dynamic range that can be exploited by microcomputer data processing to provide accurate measurements from 10 ppm up to 100%. Its high sensitivity and small internal volume allow the use of very small sample volumes and narrow-bore columns, which provide fast, highly efficient separations.

The following points should be borne in mind regarding the design of the FID. The burner jet should be cooled by a finned heat sink to prevent it getting too hot. If this happens, the components can be "carbonized" before they reach the flame, and the resulting carbon deposits in the mouth of the jet will cause noisy signals. If the jet is used as an electrode, it should be positive as this provides a better signal. The best geometry for the collector is in the form of a cylinder. This gives maximum ion collection and therefore highest signal level. The positive and negative electrodes should be separated by about 6 mm. The detector should be tuned by adjusting hydrogen flow for maximum response, thereby optimizing the flame temperature for the production of ions.

The detector should be operated at an oven temperature high enough to prevent water produced by combustion to condense in the cell. The detector should have a bottom vent, so that any water condensed when the oven is cold can drain away. The drain (vent) from the detector should be winterized in cold climates to prevent freeze-up and consequent blockage.

(c) The Flame Photometric Detector

This type of detector is used to measure low level concentrations of sulfur compounds after elution from the GC column. A typical example is shown in Fig. 1.18. When sulfur-containing compounds are burned in a hydrogen flame, elemental sulfur (S_2) is produced, which chemiluminesces in the region above the flame. This emitted light is gathered and transmitted through a narrow band-pass filter (394 nm) onto a photomultiplier tube, which produces an output proportional to the square of the sulfur concentration. In the example shown in Fig. 1.18, the gathering and transmission of the light is performed by a fiber-optic bundle. The output current is amplified and converted to a voltage by an electrometer circuit. The hydrogen/air flow ratio is set at approximately 2 : 1 to provide a reducing atmosphere.

A flame-out sensor and igniter coil are located in the vent region above the flame and emission region. The effluent from the GC column is mixed near

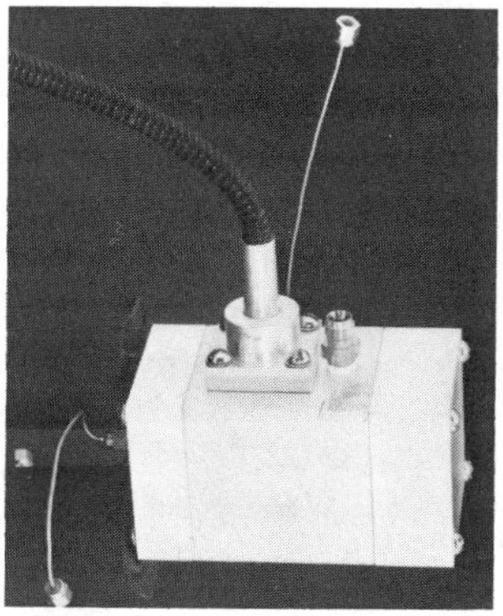

FIGURE 1.18 Typical flame photometric detector. (Courtesy of Applied Automation Inc.)

the burner tip with a constant flow of hydrogen (if hydrogen is not being used as the carrier gas) and passed into the burner tip. Air is supplied to furnish oxygen for combustion and to purge the area near the fiber-optic bundle to prevent optical transmission loss due to hydrogen diffusion.

1.7 COLUMN ENGINEERING FOR PROCESS GLC

In Section 1.4 the basic theory of process GLC was explained and it was shown how separation on a chromatographic column is achieved. This section deals with the fundamental concepts of what is known as "column engineering" and outlines the various parameters that must be adjusted to obtain maximum column efficiency. To do this we must first understand the meaning of peak interference, resolution, liquid phase efficiency, and column efficiency and how they are interelated.

1.7.1 Peak Interference and Resolution

Consider the case of two "ideal" peaks shown in Fig. 1.19. We assume that the peaks are triangular in shape. If the peaks have equal width, they are perfectly separated when peak separation (S) = peak width (W). Resolution (R) is defined as R = peak separation/peak width. Therefore, for perfect separation of ideal peaks, $R = 1.0$. If the peaks are not of equal width, then R is still 1.0 provided the average peak width (W_{av}) is used in the calculation. The expression $R = S/W_{av}$ works for peaks with the same or different widths. Of course, in the real world of chromatograms, the peaks are not triangular, but resolution can still be calculated from the above formula. Peak width in this case is estimated by drawing the tangents to the sides of the peak as shown in Fig. 1.20.

As peaks get closer together (their separation decreases), they will eventually interfere and it will be difficult to distinguish them on the chromato-

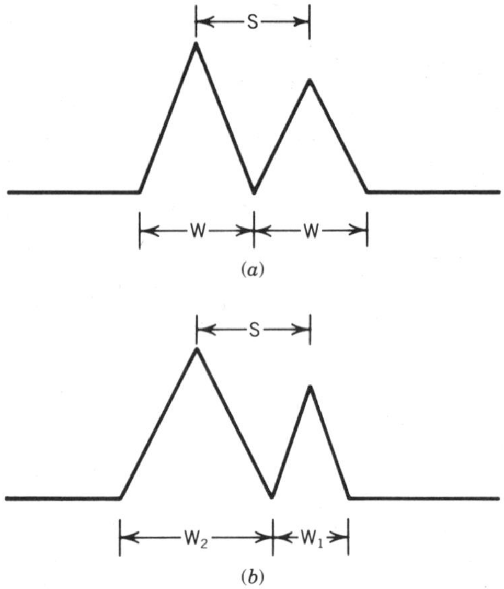

FIGURE 1.19 Ideal peak separation. (*a*) Peaks of equal width; (*b*) peaks of unequal width.

gram. An example is shown in Fig. 1.21. It must be remembered that the GC detector will see the combined effect of the two peaks; the chromatogram output will show a worse separation than that actually occurring on the column itself. The art in reading the chromatogram is to be able to identify peaks that are incompletely separated and spot extra peaks. In some cases, they are impossible to see except for a flat top or a slight mis-shape on the edge of a peak.

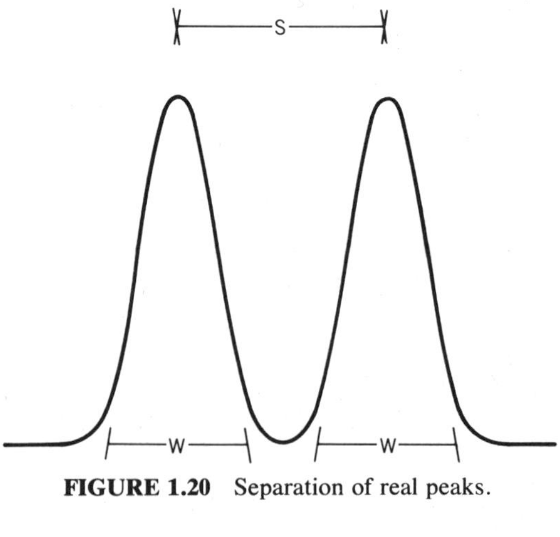

FIGURE 1.20 Separation of real peaks.

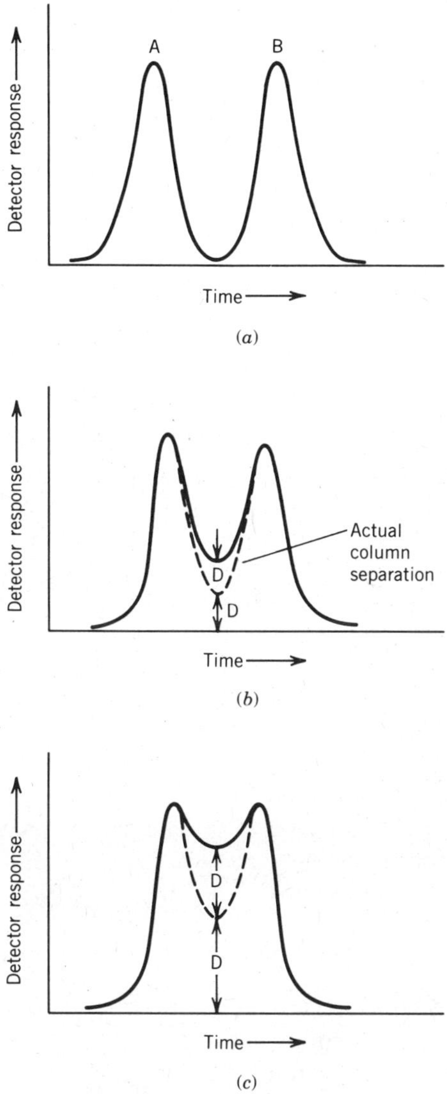

FIGURE 1.21 Peak interference. (*a*) Complete separation; (*b*) incomplete separation; (*c*) poor separation.

Given that resolution = peak separation/average peak width, in order to improve resolution we can increase peak separation, reduce peak width, or do both. The separation on a chromatographic column is entirely due to the liquid phase efficiency, whereas the peak width is entirely due to the column efficiency. Thus we have two main parameters for maximizing resolution.

1.7.2 Liquid Phase Efficiency

Factors that affect the separation between peaks can be deduced from the model of the chromatographic process described earlier. They are:

1. The identity of the two peaks, that is, their solubility in the liquid phase.
2. The liquid phase itself.
3. The temperature of the column.
4. The pressure of the column.
5. The flow rate through the column.
6. The amount of liquid phase (%) on the column.

To find out exactly what the liquid phase is doing in terms of component separation, we must learn to read the chromatogram. Consider the example shown in Fig. 1.22. The two components A and B elute at 8 and 10 min respectively, referred to the sample injection time. The air peak elutes at 1 min. Remember that the component has only two options—it moves at carrier gas velocity (column dead time) or it dissolves in the liquid (time in the liquid). Thus the air peak time is equal to the column dead time for the particular carrier gas flowrate. The time in front of the air peak has nothing to do with the liquid phase, but the liquid phase is *totally* responsible for the time after the air peak.

Therefore, from Fig. 1.22, it can be seen that component A is in the liquid phase for 7 min and component B for 9 min; the column dead time is 1 min and peaks A and B are therefore in the gas phase for 1 min. The relative solubility (α) is defined as

$$\alpha = \frac{\text{solubility of B}}{\text{solubility of A}}$$

For Fig. 1.22, $\alpha = 9/7 = 1.286$.

The time spent by the component in the liquid phase may be calculated from the expression K = peak time after air peak/air peak time. For component A, $K = 87\%$ and for component B, $K = 90\%$.

The two basic rules for maximizing liquid phase efficiency are therefore:

1. Choose a liquid that provides the greatest difference in solubility between the two peaks required to be separated, that is, maximize α. This is the fundamental separating ability of the liquid phase. It should be as high as possible.
2. Keep the peaks in the liquid phase between 50 and 90% of the time. This ensures that the liquid phase is being used to its maximum possible effect.

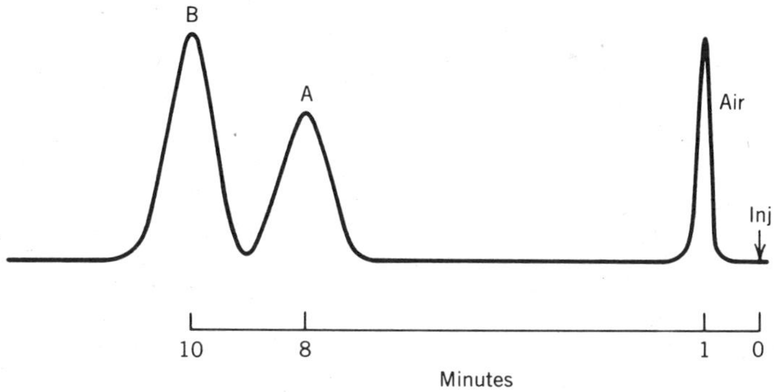

FIGURE 1.22 Liquid phase efficiency.

Having optimized liquid phase efficiency, only then is it worth increasing the separation by adding more column or back-pressuring the column. A common mistake is to put in more column before liquid phase efficiency is even thought about. The column temperature is one parameter that can be easily adjusted to meet the requirement of rule 2 just given. When the temperature required is too high, the percent of liquid phase on the column must be reduced, a much more difficult operation. Column temperature and percent liquid loading work together with regard to liquid phase efficiency, whereas column length and carrier flowrate work in a similar fashion regarding column efficiency.

1.7.3 Column Efficiency

Having optimized liquid phase efficiency, it is often found that further attempts to increase peak separation, usually by increasing column length, result in wider peaks. Also, some recipes for reducing peak width result in less peak separation. Nevertheless, the column efficiency is improved if relatively more separation is gained than the loss in peak width, or peak width is reduced relatively more than the loss in separation. Therefore, an efficient column may be defined as one that produces peaks that are narrow relative to their *retention time* (peak time from sample injection), that is:

$$\text{column efficiency } (N) = \text{retention time/peak width}$$

$$\text{or } N = \frac{T_R}{W}$$

This formula requires a slight calibration, according to detailed theory, to bring us back to the idea of *theoretical plates*. The number of theoretical plates is a measure of the efficiency of the chromatographic separation and is calculated from the expression

$$\text{number of plates } (N) = 16\left(\frac{T_R}{W}\right)^2$$

Once the liquid phase efficiency has been optimized, the more "plates" the better the resolution.

The number of plates may be measured from the chromatogram, as shown in Fig. 1.23. First, draw the tangents to the sides of the peak, then draw the baseline under the peak. Measure the peak width between tangents. Then measure the retention time from the sample injection point in the same units as peak width (e.g., seconds, mm, squares). Then calculate N from the formula.

The preceding obviously indicates that chromatography is a practical science. It is important to understand the basic theory, but what matters is the final result. Therefore it is essential to use the chromatogram to decide what is best. All separations are unique, so it is necessary to work on the specific separation required. Factors that must be considered are:

1. Relative retention achievable.
2. Concentration ranges to be accommodated.
3. Tailing and skewing of peaks.
4. Amount of sample injection required.
5. Actual hardware to be used.
6. Analysis time requirements.

What is theoretically ideal may not be so in practice.

The first reaction to a problem separation is to add more column. This often works, but it may result in unacceptable analysis time, interference

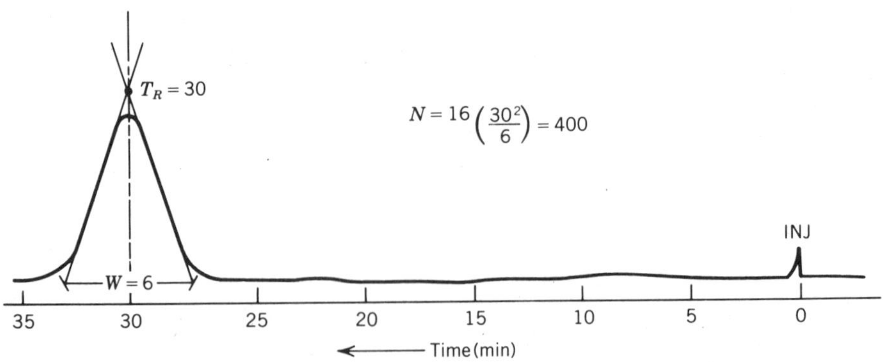

FIGURE 1.23 Calculation of N from a chromatogram.

with the separation of other components, too high carrier pressure with consequent valve leakage, peak widths increasing as much as separation, and therefore no improvement in resolution.

To check column efficiency without considering the effect of changing column length, the concept of "plate height" is used, where the plate height (H) = column length (mm)/number of plates (N). Note that for the highest column efficiency, the plate height should be minimized. For packed columns, H is typically 0.3 − 1.0 mm, and a good packed column might have the following parameters:

$$L = 1.0 \text{ m} \qquad N = 2000 \qquad H = 0.5 \text{ mm}$$

H is the length of column theoretically needed to achieve one single gas-liquid equilibrium, or it is equivalent to the number of "compartments" or "frames" in the original model described in Section 1.4.

The choice of carrier gas flowrate has an important bearing on column efficiency. It is easy to adjust, has a direct effect, helps to give an understanding of the effect of changing the column length, determines the analysis time, and brings in the effects of all the other variables. Above all, there is a wealth of theory to explain it. Van Deemter [5] expressed the dependence of plate height (H) on the carrier gas linear velocity (u) with the equation

$$H = 2\lambda\, dp + \frac{2\gamma\, Dg}{u} + \frac{8u}{\pi^2}\left(\frac{K'}{(1 + K')^2}\right)\left(\frac{2d_f^2}{D_L}\right)$$

where λ = column packing density
dp = average particle size of column packing
γ = tortuosity factor of channels in packed column
Dg = molecular diffusity of sample in gaseous phase
K = distribution coefficient and $K' = KF_L/F_g$
F_g = volume fraction of gas in column
F_L = volume fraction of liquid in column
d_f = statistical average film thickness of stationary liquid
D_L = molecular diffusivity of sample in liquid phase

This equation may be expressed more simply as

$$H = \frac{B}{u} + Cu$$

A graph of carrier gas velocity versus plate height in accordance with the Van Deemter equation is shown in Fig. 1.24. A, which is fixed, expresses how peak width is increased due to multiple gas paths in the column:

$$A = \text{particle size} \times \text{packing factor}$$

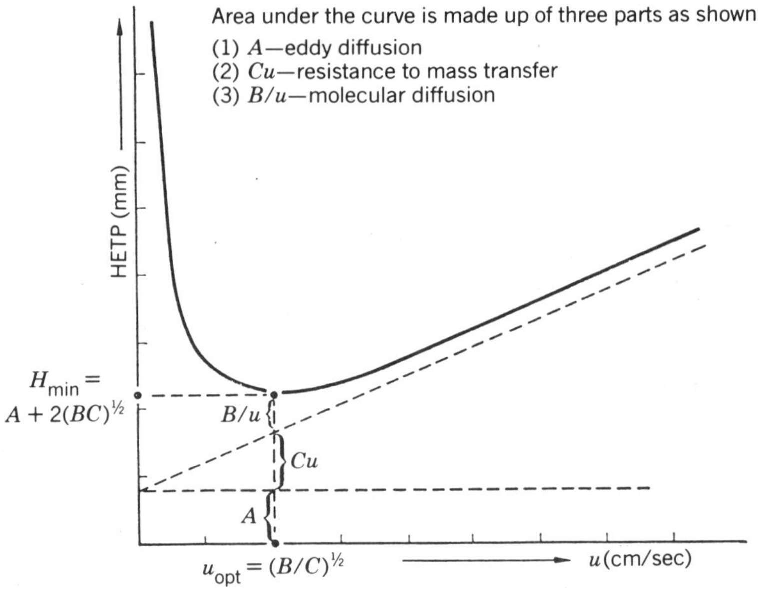

FIGURE 1.24 Graphical presentation of the Van Deemter equation.

A can be kept small by reducing particle size without causing excessive pressure drop; 100–120 mesh is optimum for most columns. Also, the packing factor can be kept low by using close-cut mesh and avoiding crushing, fines, and voids. Obtaining uniform packing is more difficult for smaller particle sizes; thus, there is an optimum particle size. For $\frac{1}{8}$-in. columns, 100–120 mesh is usually best.

The *B* term, which is dominant at low carrier flowrates, allows for the diffusion of the component molecules in the gas phase:

$$B = \frac{\text{diffusion in gas phase}}{\text{gas velocity}}$$

To keep this term small, a dense carrier gas such as nitrogen or argon, should be used, rather than hydrogen or helium, but the choice must be compatible with the detector. The carrier should be run at high pressure and high velocity. The column should be run at low temperature to minimize diffusion.

The first part of the *C* term is another diffusion term, describing how quickly equilibrium is reached in the gas phase. If velocity is in the high region of the curve, the operating conditions may be optimized for maximum diffusion rate by using a low-density carrier gas and reduced pressure. Also, a low viscosity carrier will minimize pressure drop (hydrogen is often best).

The second part of the *C* term is the most important, as it expresses the speed of equilibrium in the liquid phase. Two factors are important—the liquid film thickness and the diffusion of the peak in the liquid. To minimize this term, a lower percent liquid load or a support with a higher surface area should be used. The diffusion rate can be increased by using a lower viscosity liquid, and a higher temperature may also help (although this will also increase the *B* term).

Note that the Van Deemter curve shows a minimum, and it would appear that this is the best place to operate since it gives the minimum plate height. However, it is always better to operate at higher carrier gas velocity because at the "optimum" there is no leeway for change in carrier velocity. A small decrease will result in operation on the steep part of the curve, that is, the plate height will increase quite rapidly and column efficiency will be lost.

With small sample injection volumes, the peak width should not change with sample volume, but there is a maximum. This will depend upon many column factors, but the most important is column diameter. For $\frac{1}{8}$-in. columns the maximum is about 0.1 cm^3, and for $\frac{1}{4}$-in. columns it is about 1.0 cm^3. Larger samples can be used, provided that the volume of individual components do not exceed these limits. In practice, a smaller sample will almost always improve a difficult resolution.

Narrow columns are more efficient, but other factors must be considered:

1. Sample volume must be large enough to obtain the required sensitivity.
2. Small columns are more difficult to pack.
3. Small columns require small particle size, which results in higher pressure drop.
4. Optimum velocity is obtained at low flowrate; this makes dead volume more critical.

For process GLC work, $\frac{1}{8}$-in. diameter columns are usually the best.

1.7.4 Column Switching Systems

First, it should be appreciated that column switching systems are not inherently desirable, since they add complexity to the overall chromatograph design. If it could be guaranteed that the process stream is a simple mixture (e.g., ethane, propane, and butane), then a single silicone oil column may well provide a complete separation without the need of extra columns, column switching valves, and the necessary timing mechanism. However, in virtually every process there is always the chance that other components will be present. Therefore the GC system must be designed to guard against this possibility. In addition, many separations cannot be carried out by a single injection onto one column.

Column switching systems allow us to apply different separating columns to different groups of components, all in the same analysis. Much of the "art" of chromatography lies in the skillful design and maintenance of column switching systems. Unfortunately, many systems work very badly. They are designed with timing that is too critical for reliable operation, and maintenance personnel do not understand how they are supposed to work or how to optimize the settings. Often this is the result of poor system documentation by the manufacturer.

(a) Requirements for an Effective System

The ideal column design would separate the required components from all other components, but this specificity is often only available at the price of increased complexity and reduced reliability. The compromise solution is to protect the separation from the effect of other components likely to occur in the process stream. It is therefore most important that the specification for a process chromatograph include process sample composition, not only for the normal operating condition, but for upset conditions as well. This allows the designer to take account of every component and possible interferences in column engineering work.

Once the full stream specification has been established, the next problem is to decide what components need to be measured. It is the user who makes this decision and therefore defines the complexity of the column system. The column designer will do his or her best to achieve the specified analysis, usually making alternative recommendations only if the separation is particularly difficult. In the past, users were concerned about making enough measurements to justify the cost of the analyzer. From experience, many have learned to balance the value of each measurement against its contribution to the overall complexity of the chromatograph system.

Almost any single component, with one or two exceptions, can be easily separated from all other components using a simple column system. Two components will only occasionally cause difficulties, but separation of three or four components will often require a more complex column design unless they are all members of the same homologous series. To separate more than four components will nearly always require a complex multiple column arrangement. Therefore, one may reach the conclusion that in order to obtain reliable separations the number of measured components must be minimized.

For advanced control applications, we are usually interested in a single key component, so the column system may not need to be very complex. However, there are applications where the complexity of multi-component analysis cannot be avoided. An example is product specification point monitoring. In such a case, a simple backflush or heart-cut system will probably not be adequate. Rather than developing a complex multiple-column configuration, it may be wise to consider using a second sample injection into a different set of columns. In this way, each column system can be optimized individually and the GC will be more reliable and easier to maintain. The sample injections may be simultaneous from separate sample valves or sequential from the same sample valve.

(b) Types of Switching Systems

Four column switching systems are commonly used in process GLC work:

1. Backflush.
2. Heart-cut.
3. Dual column selector.
4. Trap and bypass.

These are used either singly or in combination. The two most important are backflush and heart cut. If one understands the principles behind these two methods, an understanding of other two will follow automatically.

Backflush. One of the most popular of column switching systems is known as *backflush*. The reasons for using the backflush technique are:

1. To reduce the total analysis time by flushing unwanted heavier components to vent.
2. To remove components that would interfere with the required separation.
3. To remove components that would damage one of the columns.
4. To regroup a number of components for measurement.
5. To maintain good performance by ensuring that the columns do not accumulate impurities from the sample.

The basic idea behind backflush is that if a component takes T sec to flow forward through the column, it will take exactly T sec to return to the front of the column on backflush, as shown in Fig. 1.25*a* for two components A and B. Therefore, under ideal conditions, the flow during backflush (B/F) is identical to the forward flow, the B/F peak is exactly recombined after equal time, and the B/F peak is exactly half eluted after equal time. Therefore, the backflush must occur during the first half of the cycle (ideally 30–40% of the cycle), *never* during the last half of the cycle. Obviously, if this were done, components would remain on the backflush column and would interfere with the next analysis.

Unfortunately, the situation is never ideal and it is impossible for the conditions experienced by every peak to be the same during B/F as they were during forward flow. The deviation from ideal is caused by pressure drop effects in the column, and produces a pressure surge and consequent baseline upset when the column switch occurs. The result is shown in Fig. 1.25*b*, where the two peaks fail to recombine completely. If the purpose of B/F is to remove the heavier components, the lack of total recombination can help in setting up the timing, since the first peak to be backflushed will be the last to elute from the column. If this peak can be identified on the chromatogram, one is assured that all other components have eluted and the

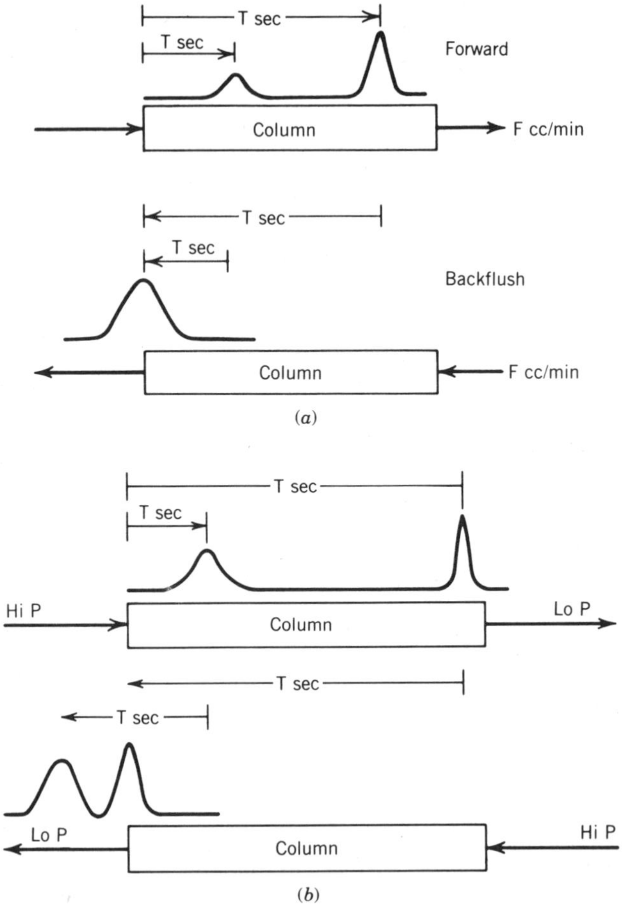

FIGURE 1.25 (*a*) Ideal backflush. (*b*) Real backflush.

column is clear for the next analysis. This is an example of high pressure drop B/F.

If the purpose of B/F is to regroup peaks for measurement, then a low-pressure drop across the B/F column is required, and this will also reduce the baseline upset that occurs at backflush. When designing backflush systems, one should consider the following sources of pressure drop in columns:

1. Column length.
2. Column diameter.
3. Packing "mesh" size.
4. Carrier gas flowrate.
5. Ratio of B/F column length to main column length.
6. Column exit pressure.

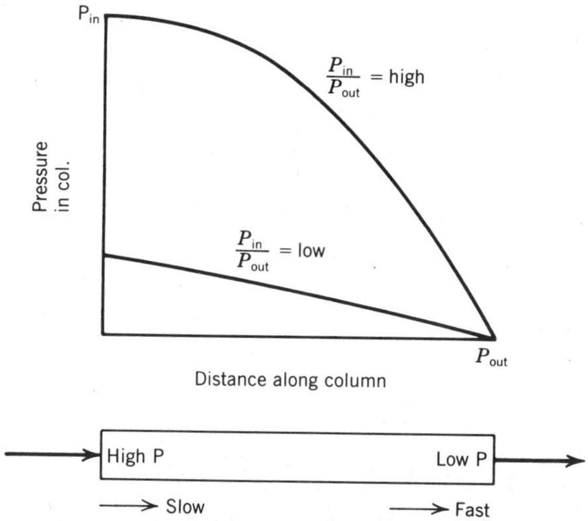

FIGURE 1.26 Effect of column pressure drop.

The problem with column pressure drop is that when it is high, the relationship between column pressure and distance along the column is nonlinear. This effect is illustrated in Fig. 1.26, assuming a column inlet pressure of, say, 75 psia and a column outlet pressure of, say, 15 psia. As a rule of thumb, the backflush column should be twice as effective as the main column. If the two columns have the same packing, the backflush column should be one-third of the total column length.

Figure 1.27 shows an example of a poor backflush system design. Note that the two columns are of the same type and are the same length; whenever this occurs, there are always problems. First, time is wasted at the end of the cycle waiting for the B/F peaks to clear before the next sample injection can take place. Second, the measured peaks may be mixed up with the B/F pressure surge, which can be a major problem. If the B/F column must be the length shown, the main column should be made longer. This provides better separation of the measured peaks and moves these peaks after the B/F

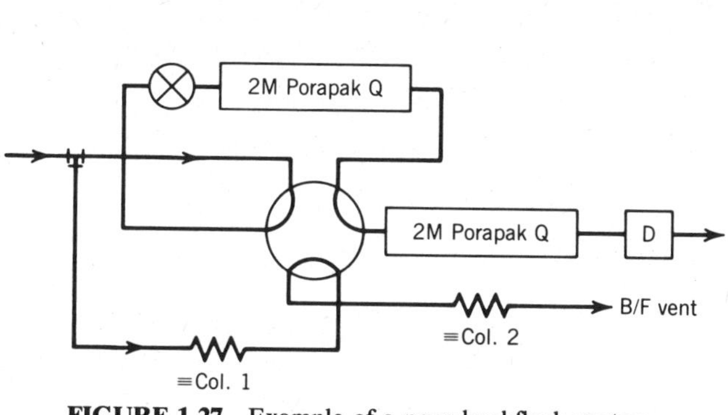

FIGURE 1.27 Example of a poor backflush system.

point. Of course, one has to take into account the increase in analysis time and the pressure drop along the column. Ideally one should aim to complete the B/F at the same time as the analysis. If the B/F column can be made shorter, it should be reduced to about one-third of the total column length. This will result in an unchanged chromatogram and therefore no increase in analysis time. The following are examples of some of the backflush systems commonly used in process GLC work.

Six-Port Backflush to Vent. In this example, shown in Fig. 1.28*a*, a six-port column switching valve is used. Note that there is independent operation of the sample inject and B/F valves. This system is almost always used for liquids. The sample injected valve may be inside or outside the B/F circuit. In the system shown, the B/F is done at high pressure. If a low-pressure B/F is required, a restrictor is fitted after column 1.

Ten-Port Sample Inject and Backflush. This system, shown in Fig. 1.28*b*, is used for vapor samples only. It is very commonly used since it is inexpensive (only one switching valve is used for both sample inject and B/F). The main problem with this system is that sample inject occurs on top of the B/F pressure surge, which makes it less flexible for troubleshooting. Also low-pressure B/F cannot be done.

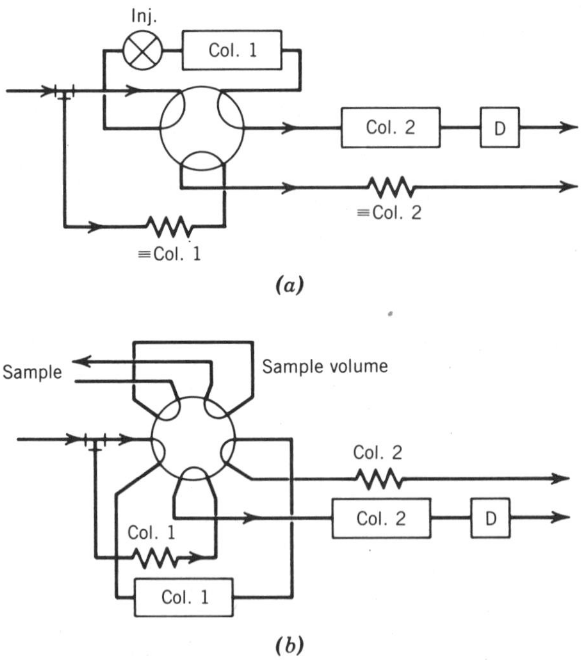

FIGURE 1.28 Typical backflush systems. (*a*) six-port backflush to vent; (*b*) 10-port inject and backflush.

Pressure Balance Backflush. This system has the advantage that no precision valves are used; in principle, inter-column volumes may be almost eliminated. The B/F is done at low pressure, which is good for removing very strongly retained components. The major problem is that the system requires very careful setting up. It relies on a constant mid-point pressure; therefore, a small change in flow resistance will upset operation. If the bleed stops, this will cause a serious dead-leg in the column connection. The pressure regulator must be capable of changing from a trickle flow to about twice the normal carrier flow at constant pressure. This system is relatively uncommon in the United States but more widely used in Europe.

Tail-Cut by Backflush. This method is frequently used to separate a minor component that appears on the tail of a major component after elution from column 1. It is effective only if the tail is fully formed in column 1 and if column 2 strongly retains the component of interest. For peaks less than about 1000 ppm, this method is not satisfactory, and heart-cut should be used. Given the right conditions, however, the tail-cut B/F system provides an elegant solution.

Basic Rules for Backflush Systems. From the preceding discussion and examples, it is possible to derive some basic rules to be used in the design of B/F systems, as follows:

1. Use pressure regulators rather then flow controllers, since the latter will not achieve sufficiently fast return to baseline stability after the B/F pressure surge.
2. Backflush in the first half of the analysis cycle, preferably at about 30–40%.
3. Don't use a long B/F column. If the same packing as column 2 is used, it should be about one-third of the total column length.
4. Make sure that carrier head pressure is kept high. Use a restrictor or dummy column if necessary.
5. Make sure that the system is easy to set up. Ensure that the first B/F peak appears on the chromatogram if not backflushed.
6. Make the B/F column do some of the separating work; don't waste separation.
7. It is bad practice to B/F fragile columns such as molecular sieves. The packing will fracture, creating an increase in flow resistance.
8. Put the sample inject valve in the B/F circuit, especially with liquid inject, so that inject valve bleed does not ruin the separation.
9. Don't have the B/F flowrate much higher than the forward flowrate unless it is unavoidable.

Heart-Cut Systems. Heart-cut is another very popular and effective column switching system, the purposes of which are:

1. To remove most of a major component.
2. To reduce the amount of a major component reaching the second column down to a concentration level that does not tail.
3. To increase the amount of the measured component relative to the major component to 1 in 20 or better.
4. To remove other components that may interfere with the analysis.

In gas-liquid chromatography, one must assume that all peaks have "tails." The tail will be revealed merely by increasing the sensitivity of the response, as shown in Fig. 1.29. For good peaks, the tail is only a problem for a size ratio of 100 : 1 or more. Note that a tail is normal, but a triangular peak is not; it is essential that one learns to recognize the difference.

As already mentioned, the backflush tail-cut technique is very effective for this type of separation, but the main problem is that the timing is critical; switch too early and part of the peak is lost; switch too late and a bad baseline is obtained. Also, 1000 ppm is about the limit for B/F tail cut, whereas it is simple for heart-cut.

A typical heart-cut valve system is shown in Fig. 1.30. Note that the valve is switched on and off to transfer peak B and the remnant of the tail of peak A to column 2, which then carries out the separation of peak B and the remnant. The separation shown in the top diagram is entirely due to column 1, whereas column 2 is entirely responsible for the separation of peak B and the remnant. Note also the typical shape of the separated remnant. Its sides are vertical at the point of heart-cut but are shaped like a peak after passing through column 2. However, the curved shape of the top of the remnant is the same as the original baseline. Therefore, reference to this shape will identify the place on the chromatogram where the remnant came from. The heart-cut system has the advantages that there is no flow or pressure surge on switching and all ports are at the same pressure, so no leakage should occur.

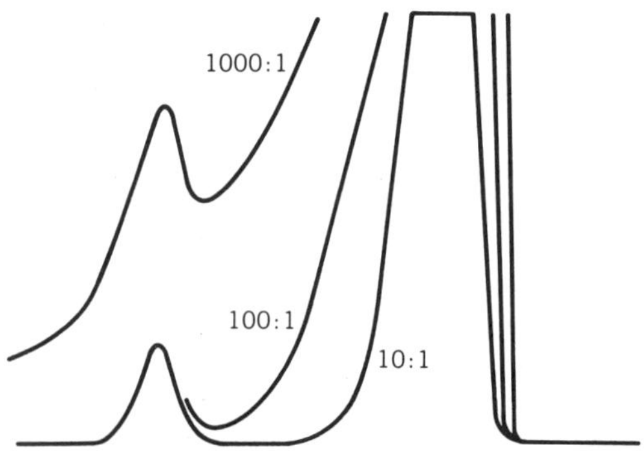

FIGURE 1.29 Normal peak tailing.

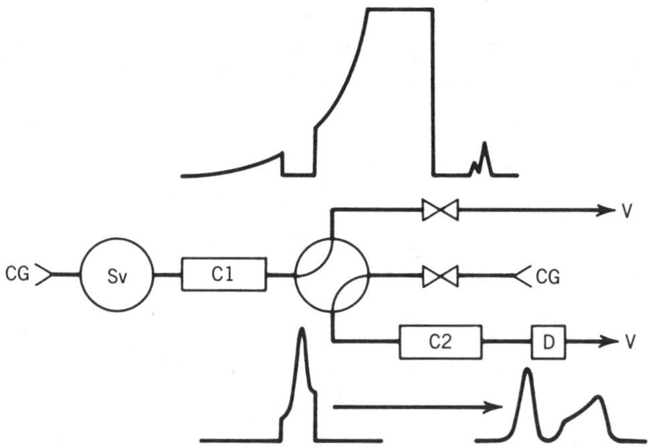

FIGURE 1.30 A typical heart cut valve system.

An example of a more complex heart-cut system is shown in Fig. 1.31. In this case, the function is to separate two components using one cut. This system is far more difficult to set up. Minimum separation of components Y and Z is required on column 1, otherwise a wide heart-cut would be necessary. In addition, column 2 must be more powerful, to move component Y out past the wedge of X, as well as to achieve a good separation of Y and Z. If column 1 separation of Y and Z is too wide, column 1 length must be reduced and one must accept more of the major component being cut onto column 2.

It is possible to carry out multiple heart-cuts; this system is often used for cutting in an early component in front of the major. It is usually not a problem, but one needs to watch out for the tail of the previous major. Two or more cuts into a major tail are usually critical and are best avoided if possible, since the cut times tend to interact. However, complex heart-cut

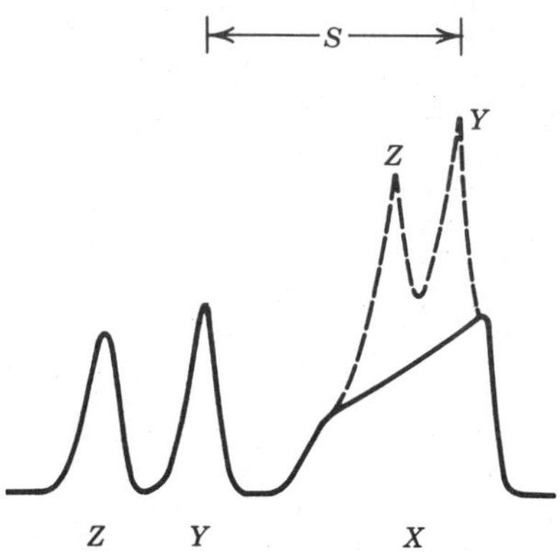

FIGURE 1.31 A more complex heart cut system.

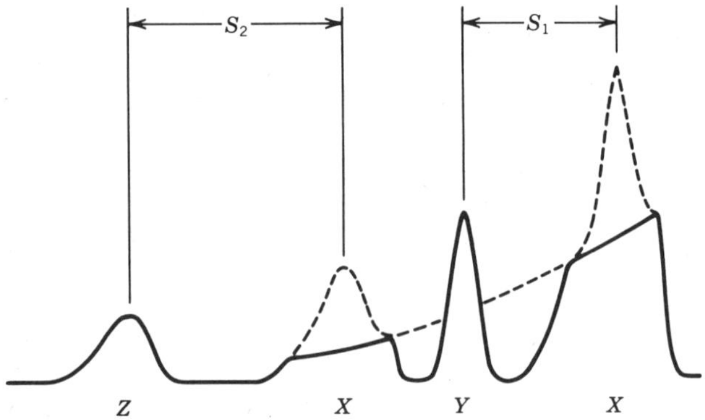

FIGURE 1.32 A two-component, two-cut system.

systems are commonly used in process GLC. An example of a two-component, two-cut system is illustrated in Fig. 1.32, showing the measured component peaks, major remnants, and the original peak positions.

In a good heart-cut, peaks do not tail in column 2. Therefore, every part of every remnant will move at the same speed in column 2 (the retention time of that component). If, however, there are two major components cut onto column 2, there will be two remnants, which will travel at different speeds on column 2. This can make life a little difficult in reading the chromatogram. If a large remnant (20 : 1 or more) is cut, it will start another tail on column 2. In this case, the front time of the cut should be adjusted, or alternatively, column 1 may not be powerful enough.

Basic Rules for Heart-Cut Systems. Once again, we can derive some basic rules for heart-cut systems, based on the preceding discussion:

1. Don't waste time with an excessively long first column. Usually, about 20–40% of column 2 length will suffice.
2. Use the simplest system for the required separation. Use a tail-cut backflush if it will do the job.
3. Always use heart-cut if the measured component concentration is less than about 1000 ppm.
4. When there are many impurities in the samples, don't use the same packing in column 2 as in column 1 unless you know for certain that no interferences are possible.
5. Don't be concerned about carrier gas control. Pressure or flow regulation will work equally well for heart-cut.
6. Keep the heart-cut system as simple as possible. Don't use complex or multiple cuts if at all possible.
7. Allow a little leeway on valve timing; don't cut too close, otherwise it will be very difficult to set up and small changes can have drastic effects.

8. Don't use a backflush to terminate the cut. Switch the heart-cut valve off first, then backflush.
9. Finally, remember that "bumps" on the chromatogram may not be switching upsets at all; they may be remnants. Also, the remnant from a narrow cut may look very much like a peak. One should learn to recognize these shapes.

1.8 ON-LINE PROCESS GAS CHROMATOGRAPHY

Thus far, this chapter has covered the history of chromatography as an analytical technique, the basic theory, the use of the technique as a powerful analytical tool in the laboratory, and its further development for on-line process analysis. The basic components of the process gas chromatograph have been described, as well as the more important column engineering techniques used specifically in on-line process instruments. Some mention has also been made of liquid chromatography, including the various types and their application in laboratory analyses.

The remainder of the chapter is devoted to a discussion of the major advances in technology that have taken place in this field in recent years and a description of some of the equipment commercially available today for on-line chromatographic analysis on an industry-wide basis.

1.8.1 Recent Advances in Technology

(a) Process Chromatograph Computer Systems

One of the most significant advances in technology in recent years has been the advent of the microprocessor. This device has had a remarkable effect on the instrument world in general, one of the results being the development of digital instrumentation systems such as those marketed by Honeywell, Foxboro, Fisher Controls, Taylor, and many others. Its presence has also been felt very strongly in the process analyzer field, as will be evident in the chapters following.

One of the first uses of the microprocessor in the process analyzer field was in the area of on-line process gas chromatography. As we saw earlier in this chapter, the GC is a rather unique device, in that it is a discontinuous analyzer (i.e., it has a finite cycle time) and requires a means of programming its operation to carry out functions such as sample injection, column switching, auto zero, peak detection, etc. In addition, its basic output is a chromatogram, which is unlike that of conventional instrumentation such as flow, pressure, level, and temperature transmitters.

The first on-line GCs introduced in the late 1950s had programmers with motor-driven cams and microswitches to carry out the required switching functions. Component peaks were displayed on a recorder using what was known as the "bargraph" format. This was quite difficult for the process operator to read, especially if a large number of components were being measured, and was quite different from the trend-type recordings provided for other process variables. As electronics technology advanced in the 1960s, peak picking circuits were introduced to provide trend outputs for components.

The next major step forward was the introduction of solid-state circuit technology into GC programmers. This eliminated the timing inaccuracies associated with the cam-type programmer and also led to vastly improved techniques for peak detection and gating, auto zero, and trend output generation. By the early 1970s, the majority of process GC manufacturers were using solid-state programmers.

The introduction of the microprocessor in the early 1970s created a major revolution in the instrument world. The traditional control room instrument panel was replaced by an operator's console, where a digital electronic control system replaced the traditional analog electronic or pneumatic panel instruments, the operator's interface with the process was now a CRT and keyboard, and a high-speed printer provided status and alarm reports for the benefit of the process operator and for management information purposes. Thus began a completely different philosophy regarding process operation and control.

It had long been recognized that one of the problems with analyzer systems, and specifically process GCs, was the requirement that every instrument needed a dedicated programmer, which, if part of a computer control scheme, had to be interfaced individually with the process control computer. The introduction of microprocessor-based chromatograph systems provided many advantages over the solid-state programmers:

- Multiple GCs can be controlled from one programmer.
- GC programming functions can be carried out more precisely.
- The microprocessor has the capability to perform extensive internal diagnostic routines to monitor system performance.
- The programmer can be interfaced to a printer to provide analysis reports, alarm reports, system status reports, and so on.
- The multi-analyzer system can be connected by a single data link to a host computer, thus simplifying the communication requirements.
- Analog signals can be provided, if necessary, for direct transmission to the process instrumentation system.
- The power of the microprocessor can be used to perform special calculation routines, such as BTU, specific gravity, time-weighted averages, and so on. It can also carry out more sophisticated peak resolution routines such as perpendicular drop, tangent skim, and so on.

Microprocessor-based chromatograph systems are now fully established in industry and are continuing to develop in sophistication and complexity. The latest systems, as described later in this section, are very powerful and flexible and address the question of reliability and redundancy. In view of the important part data acquisition and presentation play in today's world of instrumentation and process control, the subject of analyzer data systems is covered separately and in more detail in Chapter 19.

(b) Capillary Chromatography

Another important development that has taken place in the field of chromatography is in the area of column technology. Almost all of the work done in this field has been initiated in the laboratory, but the results have been used to upgrade also the performance of on-line process GCs.

The first process GCs used conventional packed columns of 0.25-in. diameter, with somewhat limited separating ability. As column technology advanced, these columns were replaced by columns of smaller diameter (e.g., 0.187 in. and 0.125 in.), which resulted in better peak resolution. However, there were still limitations on resolution and some component separations were particularly difficult to achieve with available column technology.

Capillary columns have been used in laboratory chromatographs for many years. They are capable of providing excellent component resolution, reduced analysis time, and carrying out separations previously unobtainable using conventional packed columns. The major constraint on the use of capillary columns in process GCs has been mainly due to the large internal volume of components such as sample injection and column switching valves. Nevertheless, Siemens has used fused silica capillary columns in its process chromatographs for several years, particularly for ambient air monitoring applications. To avoid the internal volume problem, Siemens uses the pressure balance column switching technique, a technique not commonly used by other process chromatograph manufacturers.

However, recent advances in chromatographic valve design has resulted in very low volume components, acceptable for use with capillary columns. For instance, Applied Automation (AA) is using semi-capillary SCOT (support coated open tubular) columns in its Optichrom 2100 chromatograph. The use of these columns has resulted in much improved component resolution, particularly in previously difficult separations such as the xylene isomers. Recently, both AA and Combustion Engineering have started to successfully apply fused silica capillary column technology in their process GCs.

Although little experience with this technology is available, it is expected that capillary columns will eventually replace conventional packed columns in process chromatographs. Although these very small diameter columns are expensive and must be handled very carefully, their phenomenal performance far outweighs their other limitations.

1.8.2 Commercially Available Equipment

In view of the important part gas chromatography has played as an on-line analytical technique for the past 25 years, there is a very wide and varied range of process chromatographs available today for industrial applications. This section only provides information on the latest equipment available from a few of the major manufacturers, with a more complete summary given in Table 1.1 on page 80.

Likewise, considerable experience has been gained over the years by the process GC manufacturers in the area of applications engineering. This has led them to compile extensive libraries of column configuration data for specific applications covering a wide variety of industries such as petroleum, petrochemical, natural gas, iron and steel, utilities, heavy chemicals, pharmaceuticals, food and beverage, and pulp and paper. One can safely say that at present almost all of the conventional applications of process GCs in industry have been covered. We are now involved in looking at the more difficult applications, using the latest column technology.

The potential industrial applications of the process GC are very extensive; therefore no attempt has been made to categorize them here. For the analyzer engineer with an interest in a particular application, it is recommended that contact be made with a process GC manufacturer, who will almost certainly be able to advise whether the application has been tackled before and, if not, whether the application is possible using available column technology. The majority of process GC manufacturers are willing to run tests on customer samples to evaluate the feasibility of a particular application; this is always a sound policy where uncertainty exists.

(a) Amscor Inc. (United States)

The MS-IV Process Analyzer System is the latest in the Amscor range, utilizing state-of-the-art microprocessor technology (Intel SBC) and including a central programmer that communicates with up to eight process analyzers (gas chromatographs or continuous analyzers of various types).

The central programmer, shown in Fig. 1.33, is the heart of the system and the user interface. A built-in CRT provides the operator with clear, menu-driven set-up instructions, analyzer status, and analyzer results. A membrane-type keyboard gives user access to status, trend and fault modes, fault messages and alarms, as well as control of individual analyzer cycles. Program changes are made via a technician's keyboard, which is provided as a separate unit, accessible by pulling out the programmer chassis and lifting the flip-up panel. This keyboard is dedicated to entry of analyzer set-up instructions and manual control of a selected analyzer. For additional security, a password lockout code may be entered.

The MS-IV Gas Chromatograph operates under the control of its own microprocessor and communicates with the central programmer over a digital link. Single or dual detectors may be utilized (thermal conductivity or flame ionization). The GC is capable of analyzing up to 64 streams and 99 components per stream using up to eight separate timetables.

Oven temperature may be configured for P & I control of up to four different temperature zones, with setpoints entered from the programmer

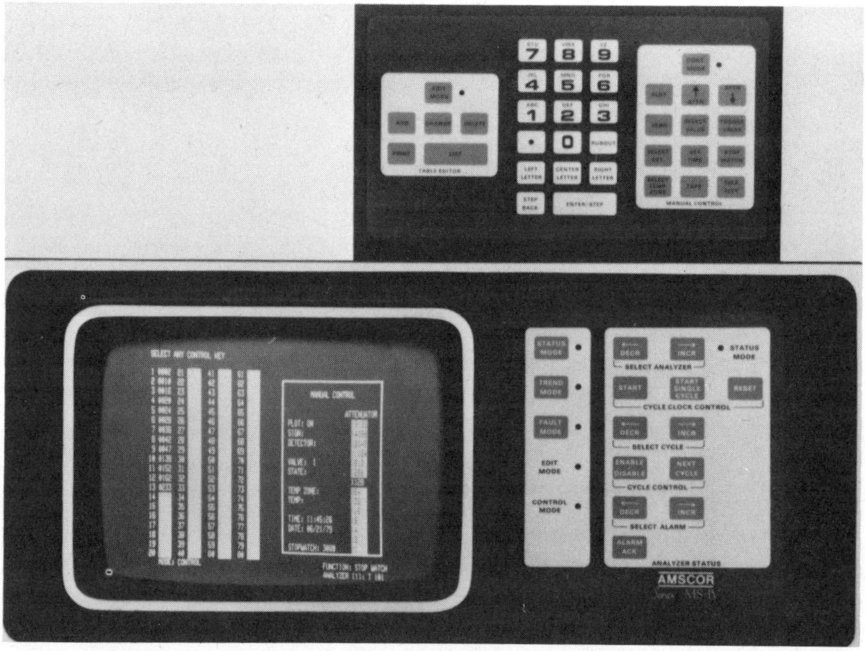

FIGURE 1.33 MS-IV central programmer. (Courtesy of Amscor Inc.)

and displayed on the CRT. The oven can accommodate up to four sliding plate valves and two detectors and can be functionally split into two sections for dual zone applications. Flow controls for all gases, instrument air, and purge air are located in an electrically non-hazardous section of the chromatograph. All of the various utilities are monitored for loss of pressure and reported to the programmer.

The chromatograph is suitable for installation in NEC Class 1, Group B, C, and D, Division 2 hazardous locations, and for Division 1 using ISA Type X purge. Built-in diagnostics are provided, comprising software routines that monitor system performance and report any malfunctions as faults displayed on the programmer CRT and printer. The gas chromatograph controller (GCC) is designed to interface with most process GCs, allowing upgrading of existing GCs into the MS-IV system.

Calculation methods for GC component concentration include area percent, linear area percent, and external standard methods. The peak handling software allows user selection of four peak type definitions: area allocation, perpendicular drop, tangent skim, and peak grouping (for backflush-to-measure applications). Other special calculation options include ratioing, BTU, specific gravity, data averaging, and other user specified routines.

Continuous analyzers (e.g., oxygen, infrared, etc.) are interfaced to the central programmer via a microprocessor-based continuous analyzer controller (CAC), which may be slaved to the GCs to provide time-synchronized data from master and slave units. The CAC also has alarms, diagnostics, auto stream selection, range change, and auto-zero/span calibration facilities, and is suitable for installation in Division 2 locations. Input signals from continuous analyzers can be voltage (0–100 mV, 0–1 V, or 0–10 V dc) or current (4–20 mA dc).

The MS-IV system generates short-form and long-form analysis reports, fault and alarm reports, and analyzer tables (set-up and analysis execution programs). These are available on the CRT and may be hard-copied on a serial or parallel printer connected to the system. Host computer communications utilize Amscor standard protocol (or user specified protocol as an option) via RS232, RS422, or current loop. A 24-channel analog output board is available for trend or bargraph recording. A built-in recorder jack allows the user to reconstruct and plot any analyzer chromatogram or oven zone temperature.

(b) Applied Automation Inc. (United States)

Applied Automation manufactures both analog and microprocessor-based chromatograph systems, although the analog systems are now being phased out. The analog Optichrom 102 system has been used successfully in industry for many years. The Optichrom 2100 microprocessor-based chromatograph system, developed in the mid-1970s, has gradually replaced the earlier Optichrom 2C system, which was one of the first mini-computer-based chromatograph systems to be produced and is still operating successfully in many locations. Recently, AA introduced its new Optichrom Advance System, which is a truly distributed analyzer system and an outgrowth of the 2100 system.

The Optichrom 102 Process Chromatograph System shown in Fig. 1.34 has an air-purged designed analysis unit suitable for bench mounting in NEC Class 1, Group B, C, and D, Division 1 hazardous locations or, when supplied with a sample conditioning system, as a floor mounted unit. The electronics section and oven section are separated and each has a sheet metal, lift-off enclosure for convenient maintenance access. The oven temperature is controlled by a three-mode controller to within ±0.02°F over the range

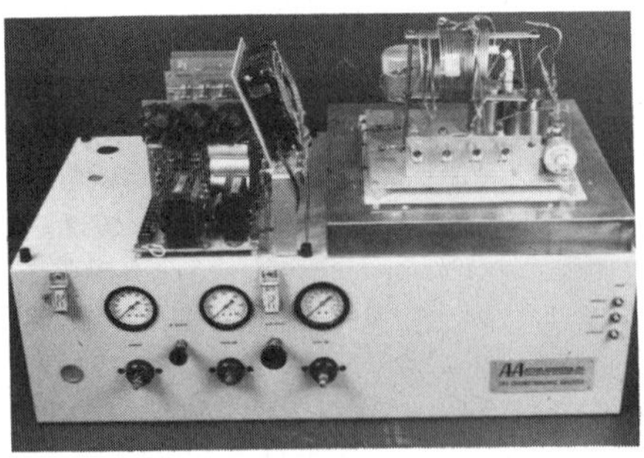

FIGURE 1.34 Optichrom model 102 chromatograph. (Courtesy of Applied Automation Inc.)

60–400°F under steady ambient conditions and can accommodate conventional packed columns or capillary columns. Thermistor TC, filament TC, and flame ionization detectors are available. Plug-in integrated circuit modules are used for ease of maintenance. Power supply, relays, detector amplifier, and other electronics are mounted in the analyzer unit to allow direct computer communications without a programmer.

The Model 102 programmer, which is suitable for Division 2 locations, utilizes a 10 × 10 timebase diode matrix board for programming analyzer functions. Time resolution is adjustable for $\frac{1}{4}$, $\frac{1}{2}$, 1, 2, or 4 sec increments, giving maximum cycle times from 4 to 68 min. Auto zero, three attenuator functions, cycle reset, and one relay function are all standard and functions can be programmed to occur simultaneously. Check points and function tests are available as maintenance aids. Up to six stored analog outputs (voltage or current) are available for remote recording. Stream switching capability is available, with or without auto attenuator switching, for up to a maximum of four streams. A separate enclosure is used to house components for expansion up to 16 streams. In addition, special calculating modules are available for ratio, BTU, specific gravity, and octane number (calculated from total aromatics).

The Optichrom 2100 Process Chromatograph System is microprocessor based, the programmer being able to control up to four chromatographs. The chromatograph itself, shown in Fig. 1.35*a,* is designed for surface mounting and consists of three sections—the electronics section (top); the carrier gas, instrument air, and utility air controls (center); and the chromatographic oven (bottom). If a sample conditioning system is included, it will be

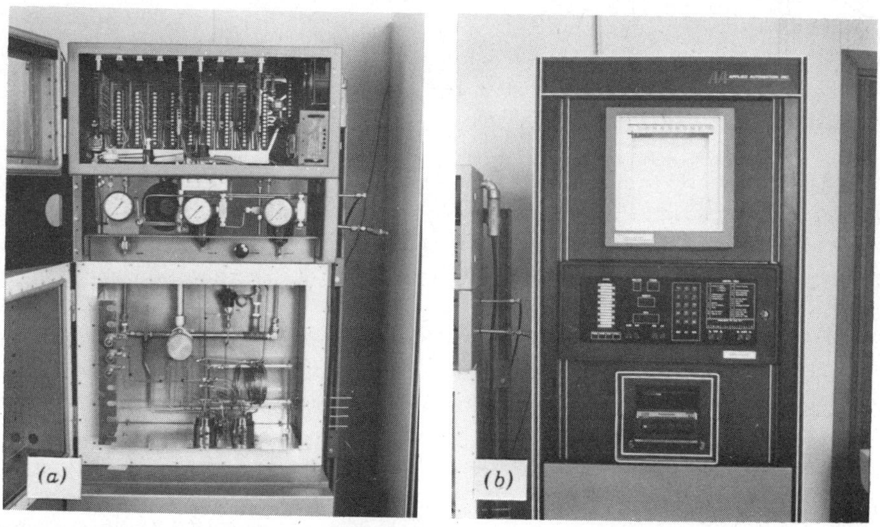

FIGURE 1.35 Optichrom Model 2100 system. (Courtesy of Applied Automation Inc.) (*a*) Chromatograph; (*b*) programmer (center).

mounted in an enclosure at the base of the unit. This design allows front access to all components, electronics, oven, and valves for maintenance.

The same oven components are available as for the Model 102 GC. Four detector options are available—thermistor TC, filament TC, flame ionization (FID), and flame photometric (FPD). Conventional 0.125-in. columns or capillary columns can be accommodated. All analog-to-digital conversion of analysis information is carried out at the analyzer unit, allowing serial data transmission to the 2100 programmer via a single twisted pair of wires. Optical isolation is standard and IS barriers can be provided at the programmer to meet IS requirements. The electronics section is purged to meet NEC Class 1, Group D, Division 1 installation. Manual switches are accessible from outside this section for operating analyzer valve functions. Oven temperature is controlled to within ±0.02°F over the range 140–350°F.

The model 2100 programmer shown in Fig. 1.35*b* (center) is capable of supporting up to a maximum of four GCs. It is a microprocessor-based instrument, providing extensive data reduction and simple programming via a push-button keyboard and digital display. A cassette loading facility is provided for easy downloading of GC operating programs. A variety of outputs are available, including chromatogram, bargraph, trend, digital printer, and an RS232C serial data link to a host computer. Four levels of alarm are provided and visually displayed; up to 80 alarm conditions can be monitored as standard. Up to 16 streams can be monitored per analyzer and up to 50 components and 99 factors per stream. Time-of-day and cycle time clocks can be incorporated. Normal data reduction functions of integration and scaling are a standard part of the system. In addition, special calculation packages are available. Special interface packages can be provided to allow the Model 102 chromatograph to be used in conjunction with the 2100 programmer.

The Optichrom ADVANCE Chromatograph System, the latest in the range of AA chromatograph computer systems, is a microprocessor-based, truly distributed system and has grown out of the very successful Optichrom 2100 System.

The system is designed so that each analyzer can operate on a stand-alone basis, independent of any central controller or programmer, utilizing the data hiway concept. The system can be expanded without hardware or program modifications to existing equipment. However, the option of a central operator's station is also available.

The Advance analyzer shown in Fig. 1.36 is suitable for wall mounting in NEC Class 1, Group D, Division 1 hazardous locations, and consists of the chromatographic oven and electronics section, which require front access only. Complete sample conditioning systems can also be provided, in which case the analyzer is built into a free-standing rack as shown, with the sample conditioning system mounted in the bottom compartment. The oven is of air-bath design, with the temperature setpoint software controlled and hardware limited by an override thermistor to meet the NEC Code for Division 1. In

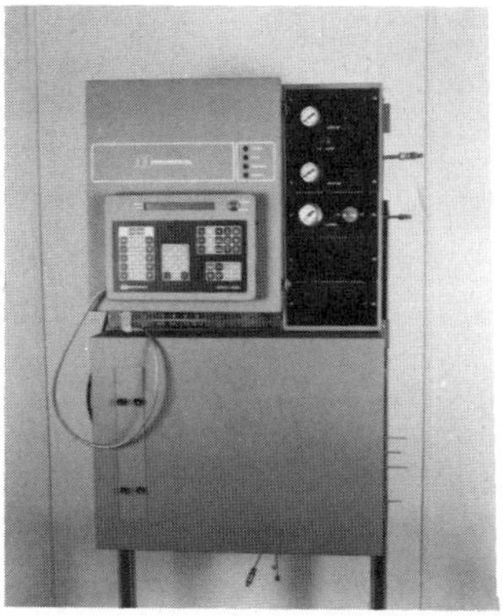

FIGURE 1.36 Optichrom advance analyzer. (Courtesy of Applied Automation Inc.)

addition, oven-temperature shutdown is provided. Valves, tubing, flow restrictors, columns, and so on are mounted on a removable plate, allowing the entire valve and column system to be removed easily from the oven for maintenance or modification. Thermistor TC and flame ionization detectors are available, together with an integral intercolumn detector (ITC). The analyzer is suitable for use with either packed columns or fused silica capillary columns, depending upon the particular application.

The analyzer has a dedicated microprocessor and memory with 30-day battery back-up. The controller section electronics is built into the removable door for easy servicing or replacement. I/O boards can be plugged directly into the electronics section or into a remotely mounted I/O Unit. Visual indicators are provided on the front panel of the electronics section for indicating purge status, fault, warning, and normal operation. The analyzer electronics has the capability of storing up to five chromatograms for future call-up or reference and for bringing in external analog signals.

A portable service panel (PSP) provides complete access to all analyzers functions and allows all maintenance, operating, and programming functions to be carried out at each analyzer. The PSP is suitable for Division 1 and magnetically fastens to and plugs directly into the electronics section of the analyzer. It has two lines of LCD readout display and a keyboard that is functionally labeled for easy use. The permanent keyboard is for operational mode, with an overlay provided for applications work. One panel key is dedicated for selecting the mode. A keylock is also provided to restrict the PSPs use to displaying data.

An important part of the Advance system is the data hiway network, shown schematically in Fig. 1.37. Each data hiway will accommodate up to 32 functional units, examples of which are analyzers, printers, and I/O card racks. Data transmission rate is 14.2 kbaud. For large systems, up to eight data hiways can be tied together into a high-speed network, with a data transmission rate of 1 Mbaud. The system will then accommodate up to 256 functional units, thus allowing over 200 analyzers to be tied into one net-

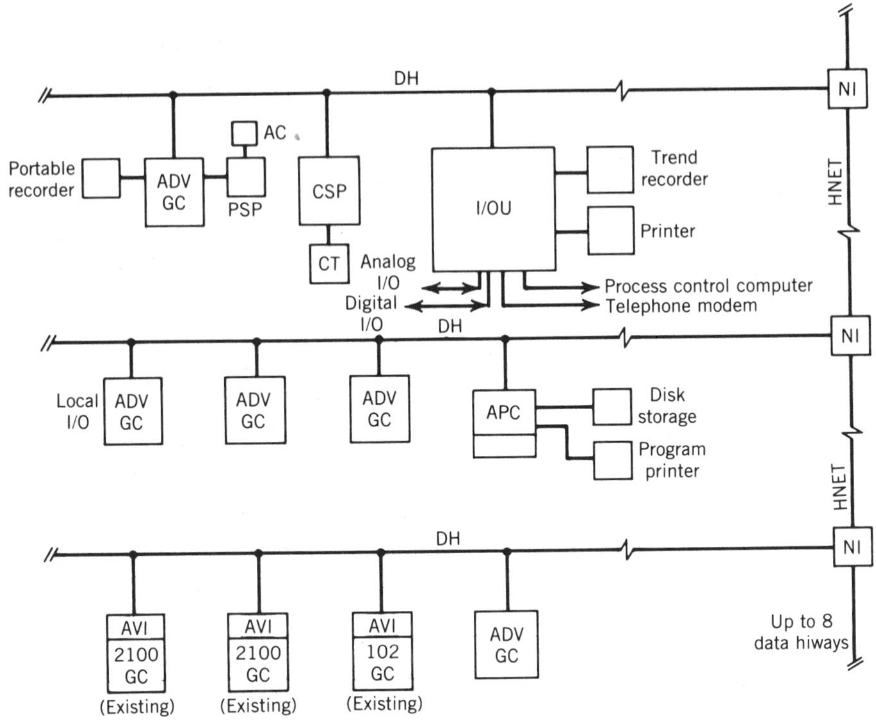

FIGURE 1.37 Optichrom advance data hiway network. KEY: PSP, portable service panel; CSP, central service panel; APC, application personal computer; AC, application cartridge; CT, cassette tape; DH, data hiway; HNET, high-speed network; NI, network interface; AVI, advance interface; I/OU, input/output unit. (Courtesy of Applied Automation Inc.)

work. Where applicable, all units tied to the hiway can communicate with all other units. Thus, a PSP plugged into one analyzer can communicate with all other analyzers on the hiway. However, failure of one unit has no effect on any other unit. Hiway redundancy is provided by using two sets of cables, with both signal paths active.

Special interfaces are available to allow existing Optichrom 102 and 2100 analyzers to be tied into the network. These interfaces contain the same microprocessor and similar software as that used in the new Advance analyzers, thus allowing existing AA installations to be upgraded to the same performance as the Advance analyzers and allowing removal of 102 and 2100 programmers.

The central service panel (CSP) is also available, and connects directly to the data hiway. This device is functionally identical to the PSP and is designed for wall mounting in Division 2 locations. A single CSP has access to any analyzer on the network, providing a central operations and maintenance station for the entire system.

The application personal computer (APC), available as an option, is an IBM Model PC or XT that ties directly to the data hiway. Although the APC is not required for a completely functional analyzer system, it enhances the capabilities of the system by providing data storage, CRT display, application programming, and off-line data processing utilizing many of the programs available for the IBM PC. APC features include:

1. Interactive creation or editing of application programs from the computer keyboard.
2. Uploading and downloading of programs to and from each analyzer.
3. Optional mass storage peripherals such as floppy and hard disks.
4. Display of program tables and stored chromatograms.
5. Graphic display capability.
6. Access to any analyzer on the network.
7. Direct listing on a peripheral printer.

For long-term analyzer program storage, three options are available. An applications cartridge, which plugs directly into the PSP and is suitable for

FIGURE 1.38 Optichrom advance application personal computer. (Courtesy of Applied Automation Inc.)

Division 1, has the capacity to store a single analyzer program. A cassette tape, using a commercial autocassette recorder, plugs directly into the CSP and can be used for loading and saving programs. When the APC is included, additional floppy disk or hard disk storage is available.

The APC is available in two versions. The small configuration uses the IBM Model PC with 128 kbyte memory, 9-in. monochrome CRT, single floppy disk, and Epson printer. The large configuration uses the IBM Model XT with 256 kbyte memory, 12-in. color CRT, single floppy disk, 10 mbyte hard disk, Epson printer, and Microsoft mouse. A typical example of the APC is shown in Fig. 1.38.

(c) Beckman Industrial Inc. (U.S.A.)

Beckman has been in the process chromatograph business since its inception in the late 1950s. The Series D and E chromatographs, coupled with the Models 320, 520 and 620 programmers, have been used successfully in industry for many years; a considerable number are still in operation.

The mid 1970s saw the introduction of the Model 6700 Process Chromatograph, which was quickly updated to the Model 6710. This unit is still in production and is shown in Fig. 1.39. The analyzer unit consists of three sections—the chromatographic oven, a gas flow control section, and an electronics section that houses the oven temperature controller and detector amplifier. The analyzer unit is designed for NEC Class 1, Groups C and D, Division 1 hazardous locations, with stand-alone capability for direct computer control or control from the Model 6710 programmer. Thermal conductivity or flame ionization detectors may be used with common cable interconnection to the programmer. Sliding plate valves are used for sample injection and column switching functions. The oven is of air-bath design and temperature is maintained within ±0.05°C over the range 55–225°C by a proportional temperature controller.

The 6710 programmer is of modular design and uses digital timing for precise resolution of all GC timing functions, with auto-zero before each component. Peak resolution may be by peak height or peak area integration. Plug-in circuit board options are available for dual alarm, trend, memory, integration, and so on, with capability for future expansion. Front panel LED indicators are provided for all functions, allowing verification of analyzer performance. The programmer is capable of manual or automatic stream selection for up to five or 10 sample streams, with front panel indication of all active streams. Up to six analog outputs are available for trend recording.

The Model 6750 Process Chromatograph System is the latest in the Beckman range and represents a new generation of gas chromatographs utilizing microprocessor technology. Up to six analyzers may be controlled simultaneously from a single data processor. Each analyzer and the data processor

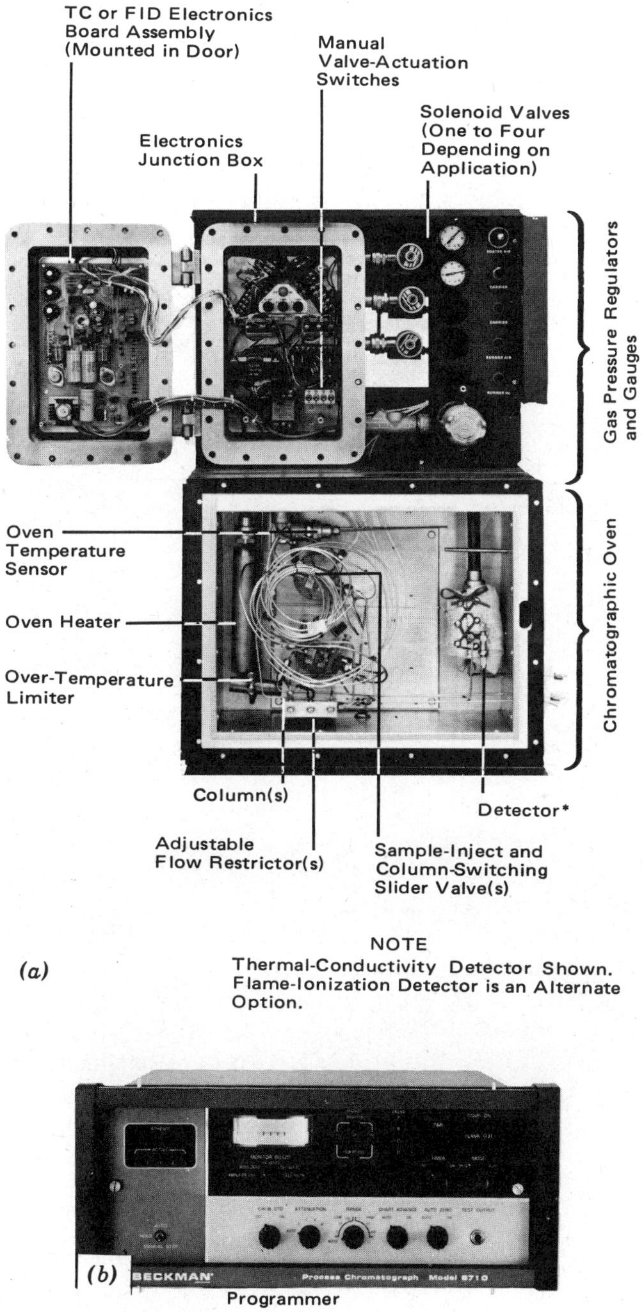

FIGURE 1.39 Model 6710 chromatograph. (*a*) Analyzer unit; (*b*) programmer. (Courtesy of Beckman Industrial Inc.)

contains a microprocessor and a full complement of memory and input/output interface logic.

The analyzer microprocessor directs all analyzer functions, monitors critical and non-critical conditions of the analysis and support systems, controls stream switching and calibration functions, and provides alarm notification to the data processor.

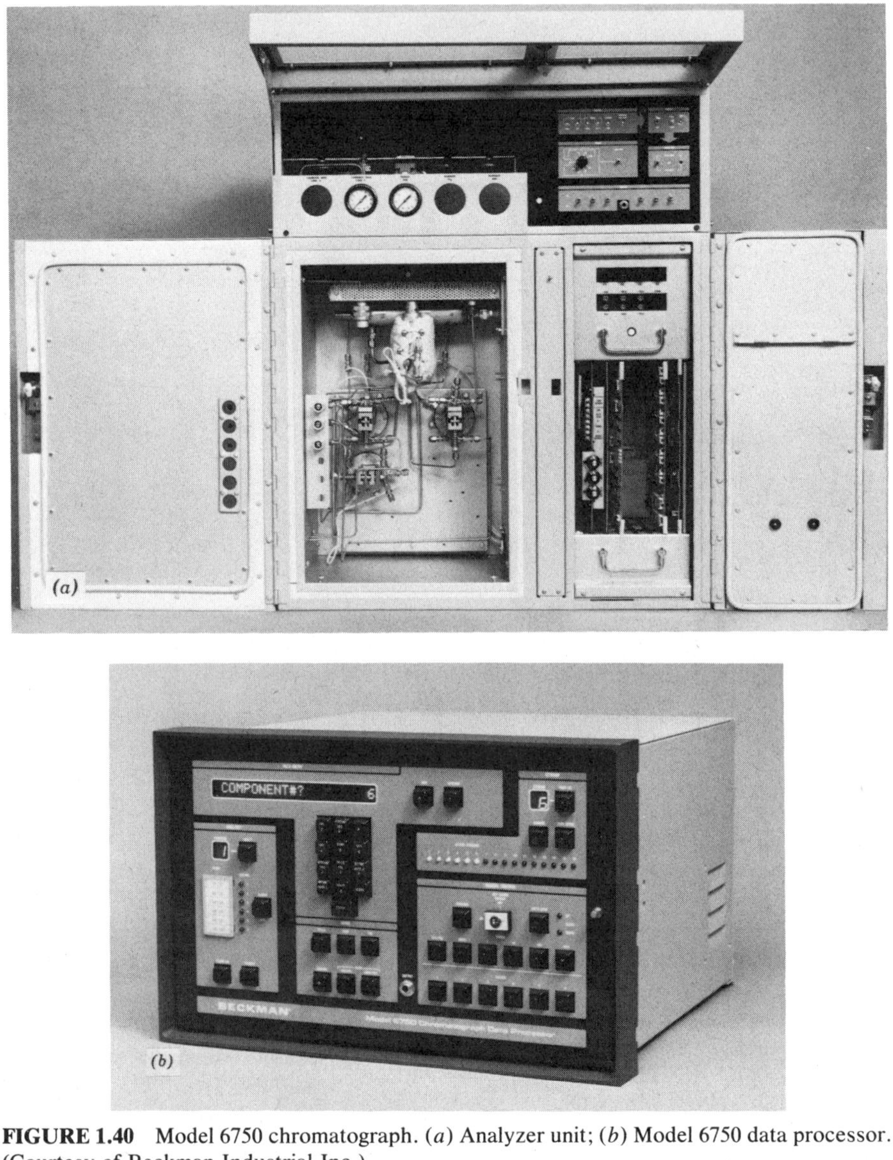

FIGURE 1.40 Model 6750 chromatograph. (*a*) Analyzer unit; (*b*) Model 6750 data processor. (Courtesy of Beckman Industrial Inc.)

A typical analyzer unit is shown in Fig. 1.40*a* and consists of five sections—the chromatographic oven, sample handling (if integrally mounted), pneumatic control, manual control, and electronics sections. The electronics section houses the microprocessor, memory, general purpose logic, signal conditioning, and detector boards. The oven is of air-bath design and can accommodate up to six valves. Thermal conductivity, flame ionization, and flame photometric detectors are available. Communication between analyzer unit and data processor is by a five-wire bidirectional ASCII data link, capable of operating over distances up to 3000 ft. The intrinsically safe manual control section allows complete manual stand-alone chromatograph operation and provides status indication, range switching, zeroing facilities, and switches for valve operation.

The data processor shown in Fig. 1.40*b* provides communication with up to six analyzers, system status monitoring, data logging and report generation, and remote manual control of a selected analyzer. The front panel has five functional sections. The data entry and display section has a 20-character alphanumeric display and keyboard that uses a conversational mode for table set-up and editing. The analyzer control section allows communication with a particular analyzer to obtain manual control or status. The stream control section includes all the functions of a conventional 10-stream selector. The mode control section controls the auto/manual status of the current selected analyzer and allows initiation of the calibration routine. The manual control section allows manual operation of analyzer functions during set-up and maintenance, including valve control, detector range, and auto zero. A key lock is provided to prevent operation of controls or modifying functions when enabled. The processor is designed for panel mounting in NEC Class 1, Group B, C, and D, Division 2 or non-hazardous locations.

A wide variety of user-selectable outputs are provided. Up to 26 isolated current outputs are available for trend recording, as well as up to six bargraph outputs (one per analyzer). A jack connector is provided for chromatogram recording on a user-selected analyzer. Interfaces are provided for host computer communication, printer, and a cassette deck. The cassette loading capability allows loading of applications programs and tables so that the system can be readily restored in the event of a long-term power failure. A variety of different report formats are available, including stream table, analysis, input data, status, and alarm reports. Special calculation routines are also available for BTU, specific gravity, summation, difference, multiplication, ratio, square root, and so on.

(d) Combustion Engineering, Process Analytics (United States)

The Process Analytics arm of Combustion Engineering (formerly Bendix EPID) has a long history in on-line process chromatography dating back to the Greenbrier process GCs of the late 1950's. The cam-type programmers of the 1960s, such as the Chromamatic and Chromatrend series, were gradually replaced with digital programmers in the early 1970s. In addition, there were many enhancements of chromatograph design based on advances in column, valve, and materials technology.

The latest in the digital programmer line is the Model 7000 Process Chromatograph System, which consists of the Model 002 analyzer and Model 7000 digital programmer. However, the accent today appears to be on microprocessor-based chromatograph systems, of which CE offers a wide variety, from single, stand-alone GCs up to large systems capable of supporting many GC's and continuous analyzers.

The Model 9020 Process GC System consists of the Model 002 analyzer and the Model 9000 microprocessor controller, which can be mounted integral with or remotely from the analyzer. The Model 002 Analyzer shown in Fig. 1.41 is the backbone of all CE chromatograph systems and is designed

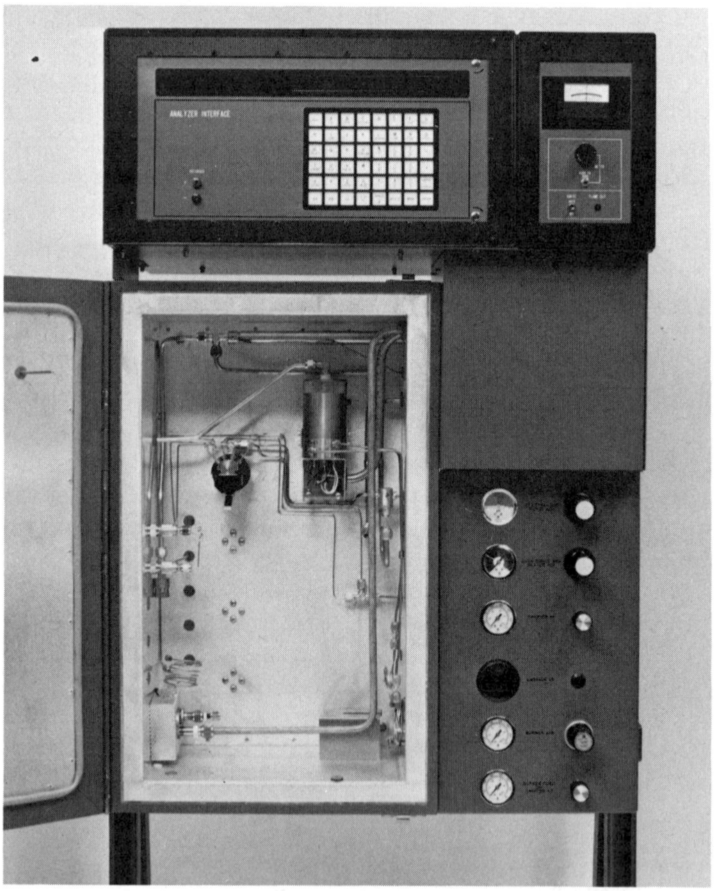

FIGURE 1.41 Model 002 process chromatograph. (Courtesy of Combustion Eng., Process Analytics)

for wall mounting in NEC Class 1, Group B, C, and D, Division 1 or 2 hazardous locations using ISA purging techniques. Oven temperature control is capable of maintaining ±0.1°C over an operating range of 50–220°C with stable ambient conditions. Sliding plate valves are used for vapor sample injection and column switching functions. The liquid sample injection valve is designed for "in-wall" installation, where the liquid sample is kept outside the analyzer oven and only the injected volume is transferred into the oven. Standard detectors include constant-temperature thermal conductivity, flame ionization, and flame photometric types. Other types such as electron capture, gas density, catalytic, and thermistor detectors are also available.

The front panel of the Model 9000 controller has a 32-character alphanumeric display and a 6 × 8 keyboard providing conversational mode communication. Featured in the front panel design is a bidirectional link between the microprocessor and the front panel/keyboard display allowing location of the front panel at distances up to 6000 ft from the controller. The front panel has a key lock to protect data entry and an output connector for a service recorder.

Software features include automatic calibration, area normalization, analyzer status/alarm display, and baseline correction. Programs are available for BTU, specific gravity, and time-weighted average calculations. Peak height or area may be used as a basis for concentration calculations for up to 12 components, selectable from the keyboard. Up to seven process streams can be accommodated as standard, expandable to 20. Background display of real-time analyzer operation is provided, including time-of-day, cycle time, analysis cycle, alarm status, component number, stream number, and key component concentration. Analog outputs (voltage or current) are available as bargraph, chromatogram, or trend, selectable from the keyboard. Serial transmission is available for interface to a host computer and printer.

The 5000 Family of Multi-Analyzer Computer Systems is available in three versions. The *MicroMac 5000-1,* shown in Fig. 1.42, is capable of controlling and processing data from up to eight GCs. The communication network is of buss design, where a common set of lines are used to communicate with all GCs on the loop over distances up to 4000 ft. Two pairs of twisted 18-gauge wires are used, one pair dedicated to data transmission, the other to data reception. A field communications module (dataconverter) is mounted as an integral part of each GC and has the following features:

- Auto-ranging amplifier with analog-to-digital converter.
- Serialization and transmission of data.
- Reception of digital output data from the controller.
- Manual field valve switching for the GC, sampling system, and calibration valves (up to a total of 16).
- Remote analyzer start, stop, abort, and report.
- Recorder jack and zero/attenuation controls for local chromatogram generation.

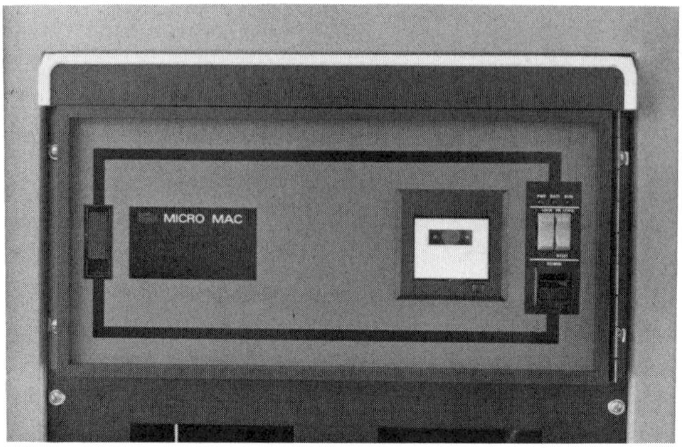

FIGURE 1.42 Micro-Mac Model 5000-1. (Courtesy of Combustion Eng., Process Analytics)

Optional analyzer status sensors may be installed to be read by the dataconverter at the start of each analysis cycle. Individual dataconverter packages can also be supplied for interfacing existing analyzers into the system.

The MicroMac is designed for panel or 19-in. relay rack mounting and is programmed entirely in RAM, allowing new software features to be implemented by cassette loading of a new executive program without hardware changes. An uninterruptable power supply is provided to protect memory loss on short-term power failure (less than three minutes duration). Standard analog outputs include 15 trends (assignable to any component in any stream for any analyzer) and a high-resolution signal for chromatogram reconstruction at the controller. The operator interface is a Perkin Elmer Bantam video display and standard ASCII keyboard. Printer and host computer interfaces are also provided.

The MicroMac Model 5000-2, shown in Fig. 1.43, is configured to handle up to 16 GCs. The single or dual floppy disk-based controller offers complete system load in 60 sec and complete diagnostics for system testing. It is available in a 30-in. floor mount cabinet, with CRT display/keyboard as the basic I/O device. Printer and host computer interfaces are also provided. Analog trend output capability is expandable to 128 outputs, user assignable.

The Model 5500 Chromatograph System, a typical example of which is shown in Fig. 1.44, is a minicomputer-based, dual floppy disk system capable of controlling up to 32 multi-stream, multi-component GCs. Three serial outputs are available for printer, CRT, and host computer communication. The unit is mounted in a 72-in. cabinet. Options include expansion to 64 trend outputs, back-up system with manual or semi-automatic changeover, and built-in recorder for chromatogram presentation.

The 5000 family of controllers uses a continuous peak detection technique based on changes in baseline slope and peak naming. Start and stop of peak search is set over wide time periods and only interrupted to exclude un-

FIGURE 1.43 Micro-Mac Model 5000-2. (Courtesy of Combustion Eng., Process Analytics)

FIGURE 1.44 Model 5500 chromatograph system. (Courtesy of Combustion Eng., Process Analytics)

wanted baseline excursions. Peaks are matched with a time/name pair after peak search is complete. Capability is provided for determining unknown peaks and peak matching by reference peaks. Peak resolution is program selectable and includes perpendicular drop, area reallocation, tangent skimming, and baseline correction techniques. Peak measurement can be by area integration or peak height, user selectable. Two peak summation techniques are available, forced integration and summation by peak name.

Techniques for calibration include normalization, external standard, and internal standard. Storage of characterized peaks, including uncorrected peak areas, peak start, crest, and end amplitudes and times is also a feature of the system. An auto-calibration feature is available and can be scheduled on the basis of hours or days within a particular month. Special calculation procedures can be provided for BTU, specific gravity, and time-weighted average. Host data link packages can be provided for the majority of process control computers, using a priority interrupt-based protocol that provides link establishment, complete data transfer, and data validation. A variety of different reports can be generated on operator demand or on a preprogrammed time basis. A status change report is generated automatically each time the status of an analyzer changes.

The latest addition to the CE range of chromatograph computer systems is the Distributed Analyzer Control System (DACS). This system has all the features of the 5000 family and is also capable of integrating continuous analyzers (e.g., infrared, moisture, pH, conductivity, etc.) with chromatographic analyzers to form a complete, distributed analyzer control system with the ability to provide extensive, automatic verification of compositional data.

The building blocks of the system are two types of field-mounted controller, one for new and existing chromatographs (the GCC) and one for continuous analyzers (the CAC). The GCC is dedicated to a single GC and is capable of supporting multi-component analysis of up to 20 sample streams, four analytical methods, 16 analog trend outputs, bargraph output, and a variety of special-purpose calculations. It is similar to the 9000 controller and is mounted as an integral part of the chromatograph as shown in Fig. 1.45.

The CAC is configured to scan the outputs of up to eight continuous analyzers at a rate of one sample per second. Independent function tables, assignable to each analyzer, allow adjustable data averaging cycles and trend updates every 5–4000 sec, as well as zero and span routines. The CAC also has the additional feature of allowing a lock on data to the previous "good" value in the event of a data alarm or out-of-service condition.

The third component of the DACS is the central data manager (CDM) (shown in Fig. 1.46), a microcomputer with integral single or dual floppy disk storage, which is built around the Intel Multibus and uses standard, commercially available board products. This unit provides a central database of analytical information and a point from which the maintenance technician can address the complete process analyzer system. Interfaces are provided at the CDM for a CRT/keyboard (the operator interface), a printer for report

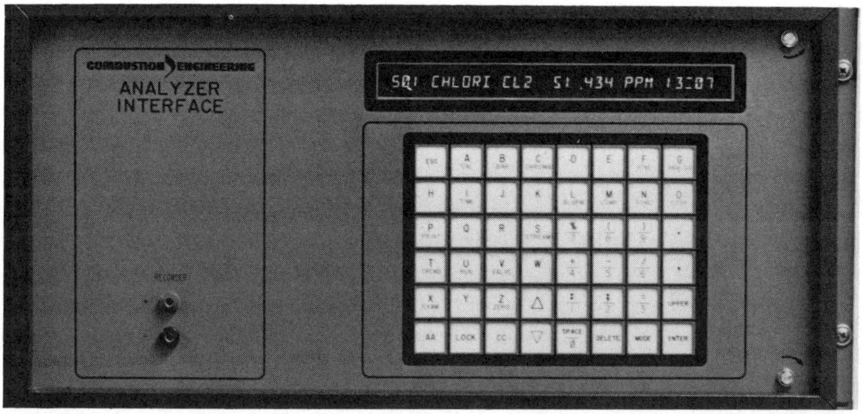

FIGURE 1.45 DACS chromatograph controller. (Courtesy of Combustion Eng., Process Analytics)

generation, a digital plotter for chromatogram generation, and a serial data link to a process control computer.

The DACS may be configured in three ways as illustrated in Fig. 1.47. System I is adaptable to small process plants or applications where analysis requirements are limited, and uses remote, stand-alone controllers providing analog trend outputs and remote printer interface. System II is a distributed

FIGURE 1.46 DACS central data manager. (Courtesy of Combustion Eng., Process Analytics)

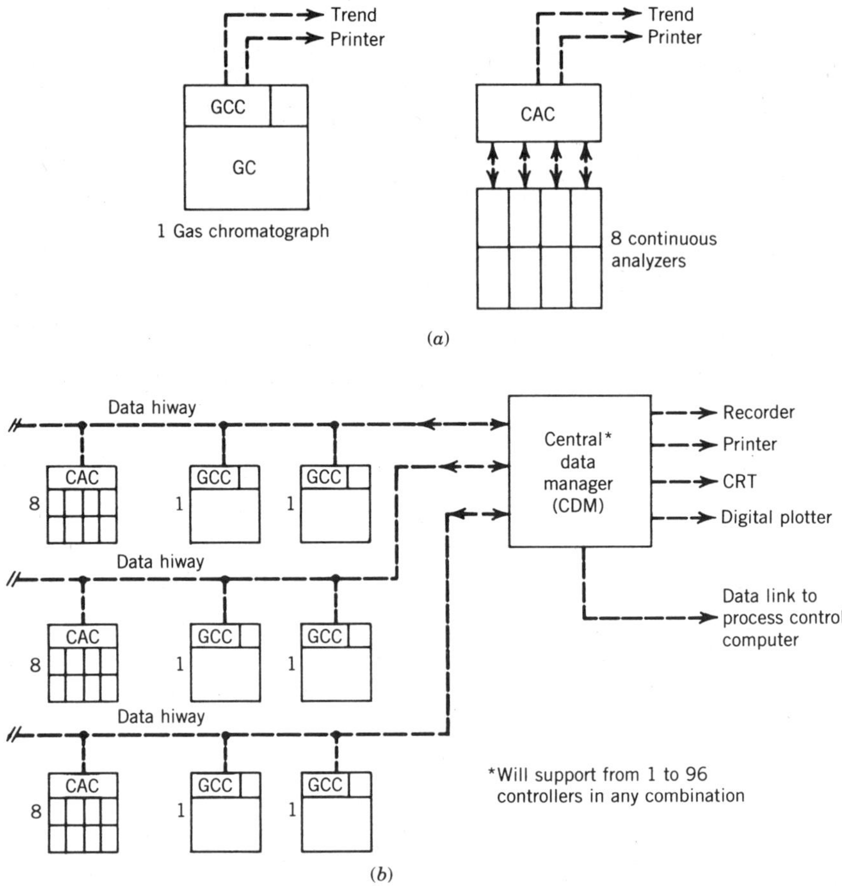

FIGURE 1.47 DACS configurations. (*a*) System I; (*b*) System II; (Courtesy of CE, Process Analytics)

system utilizing one CDM and is capable of supporting up to 96 field-mounted controllers for GCs or continuous analyzers in any combination. All controllers are linked via a data hiway to the CDM. System III is a distributed system with fault tolerance and utilizes two CDMs for complete system back-up and to maintain operation in multiple failure modes. All field mounted controllers are linked to the CDMs via two data hiways (primary and back-up). In addition, each CDM is serial data linked to the process control computer. A third data hiway is reserved for transmission of real-time chromatograms to a digital plotter.

(e) Foxboro Analytical (United States)

Foxboro has been in the process chromatograph business for over 20 years. When Foxboro acquired the microprocessor-based chromatograph line of

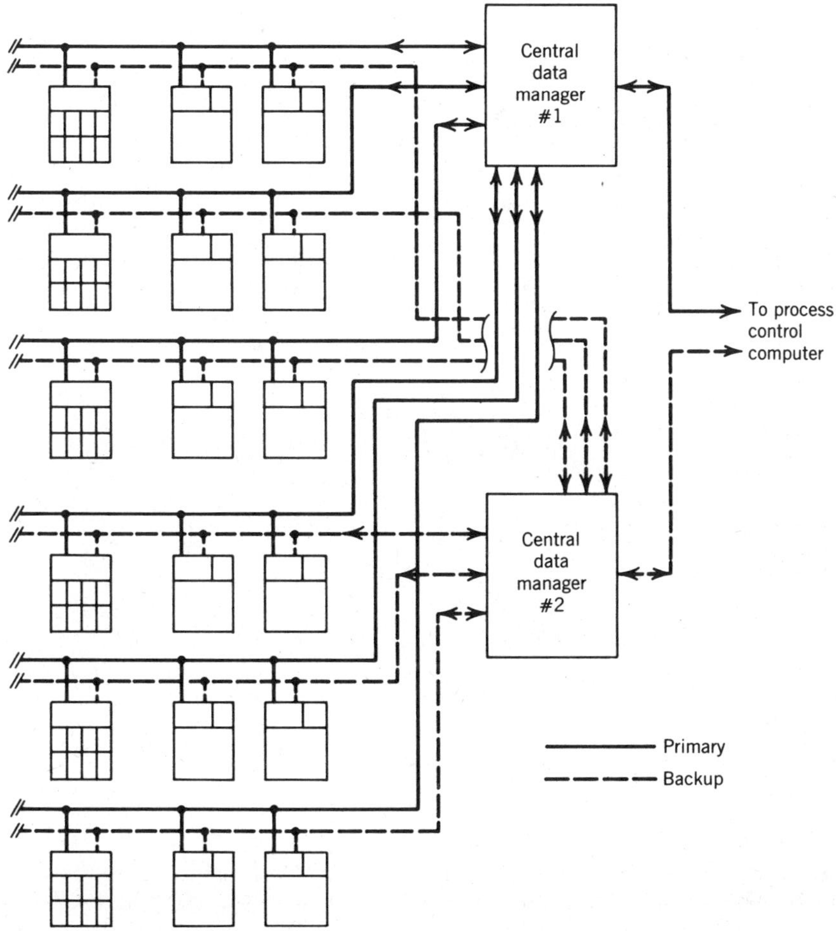

FIGURE 1.47 (*c*) System III. (Courtesy of CE, Process Analytics)

Arcas Inc., its own line of process chromatographs was discontinued, except for the very successful Model 91PCT Pneumatic Composition Transmitter (to be discussed later). Foxboro's current line of process GC systems includes the Model 91PCT and the Model 931A/935 Microprocessor-Based Chromatograph System.

The Model 91PCT Series Pneumatic Composition Transmitter, as its name implies, is an all-pneumatic instrument, the operation of which is based on gas chromatographic principles. It utilizes pneumatic components developed by Foxboro over many years of experience in this area.

Simplicity is the key factor in design; the instrument is designed for straightforward analyses and for a maximum of two-component measure-

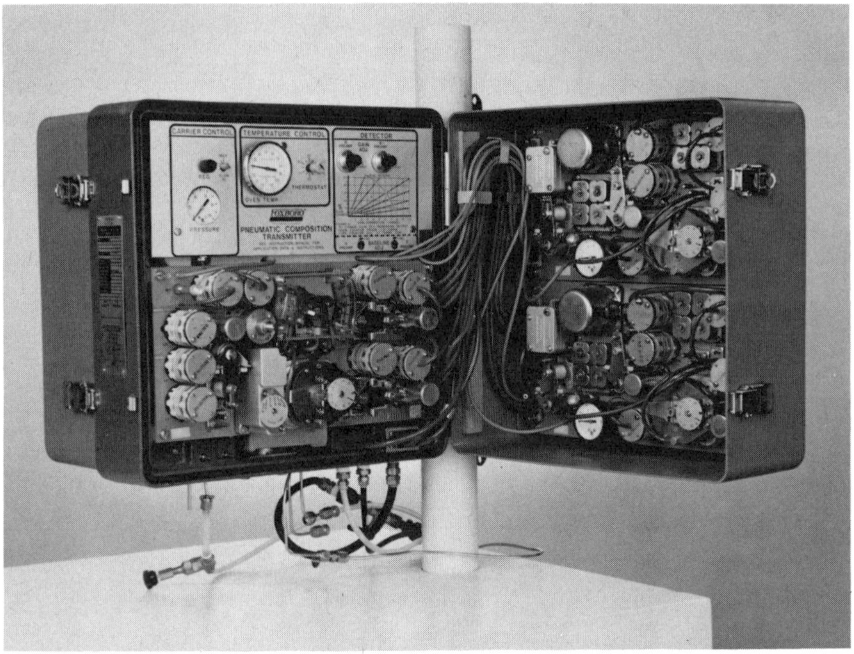

FIGURE 1.48 Model 91-PCT (pneumatic composition transmitter). (Courtesy of The Foxboro Company)

ment. The analyzer, shown in Fig. 1.48, is housed in a single lightweight NEMA 3 enclosure suitable for pipe mounting close to the sample point. Components are mounted on pneumatic circuit boards for easy access and maintenance. Being completely pneumatic, the instrument is inherently safe from an electrical hazard standpoint. The oven temperature thermostat may be set within the range 122–212°F. Steam is the standard heating medium for the oven, but electrical heating is available as an option (this, of course, makes the analyzer no longer wholly pneumatic). Helium and hydrogen are normally used as carrier gases, but the final choice will depend upon the application.

Sample injection valves are designed for vapor injection only at a maximum pressure of 30 psig. An alternative valve design is available for corrosive sample applications. Column configurations are backflush, foreflush, and tailcut. Maximum column length is 20 ft for single-column applications and 40 ft for dual column applications. A maximum of two sample components can be measured and a trend 3–15 psig output is provided for each component. Concentration ranges are adjustable between the limits 0–1 and 0–100%. Applications include measurement of hydrocarbons in the C_1 to C_6 range as well as ethylene oxide, water vapor, hydrogen, oxygen, nitrogen, carbon monoxide, and carbon dioxide.

The latest in the Foxboro line of microprocessor-based chromatograph systems is the Model 931A/935 Chromatograph System, which consists of the Model 931A Chromatograph and the Model 935 Data Collector. The Model 931A chromatograph shown in Fig. 1.49*a* is designed for wall mounting in NEC Class 1, Groups C and D, Division 2 hazardous locations and for Division 1 locations as an option by the use of ISA Type Y purge. It contains its own microprocessor controller, located in the upper electronics section, which is dedicated to one GC and provides control of the chromatographic oven, detector signal integration, sample handling system control, monitors alarm sensors, concentration levels, and data transfer validity.

The oven is of air-bath design. Valves, columns, and flow controllers are mounted on a single assembly inside the oven. This assembly can be removed from the oven as a single unit for ease of servicing and checking. Three types of detector are available as standard—the filament TC, flame ionization, and photometric detector. Peak compositions are available as both analog and digital signals at the GC.

Peak identification is by forced gate or slope detection techniques. Peak area resolution is by gate, baseline, vertical drop, or peak skimming. Baseline correction techniques used for peak area include perpendicular drop,

FIGURE 1.49 (*a*) Model 931A chromatograph. (Courtesy of The Foxboro Company)

FIGURE 1.49 (*b*) Model 935 data collector. (Courtesy of The Foxboro Company)

prior baseline, peak skimming, and integration. The controller is capable of monitoring both software and hardware generated alarms. An extensive self-diagnostics package is provided for monitoring system performance. The controller has a data entry keyboard which permits user-friendly set-up, operation, and display. A service jack and manual attenuator are provided for chromatogram recording. Eight 4–20 mA dc trend outputs are available for remote recording. Also available is a host computer communication link via an optically-isolated 20 mA current loop with transfer rates selectable from 300, 600, 1200, 2400, 4800, and 9600 baud.

The Model 935 Data Collector shown in Fig. 1.49*b*, is a microprocessor-based unit that provides centralized analysis data collection and reporting for up to eight 931A GCs as standard (with expandability up to 12) and capability of uploading and downloading information. The unit includes a CRT/keyboard operator interface and interfaces to a printer and host computer. Extensive internal diagnostics monitor total system performance. A reconstructed chromatogram is available from each GC as a 4–20 mA dc signal complete with event markers. Communication with the Model 931A GC is via a four-wire, full duplex link, either as a 20 mA current loop or an

RS-232C, with baud rate selectable from 600, 1200, 2400, 4800, or 9600. Up to 16 4–20 mA dc trend outputs are available, together with 32 alarm outputs.

1.9 ON-LINE PROCESS LIQUID CHROMATOGRAPHY

1.9.1 Historical Development

In Section 1.5, the subject of liquid chromatography was introduced and the various subcategories of LC briefly described. It is appropriate at this point to discuss the evolution of modern LC and its relationship to the LC techniques used by the scientists who pioneered this field.

Over the past 20 years there has been a phenomenal development in the technique now known as high performance liquid chromatogrpahy (HPLC). Although it is very difficult to compare the evolution of the various analytical techniques used in the laboratory, the growth of HPLC is considered by many scientists to be unparalleled in the field of analytical chemistry, even compared with that of gas chromatography.

The term HPLC, as applied to the present-day technique, should not be construed to imply that earlier LC did not exhibit high performance. On the contrary, all the pioneering scientists in Germany, Hungary, England, and the United States carried out separations hitherto impossible. Therefore, even by present-day standards, the separations can certainly be classed as "high performance" LC. In fact, the HP in the name originally stood for "high pressure," since the change from atmospheric conditions and gravity flow to pumping systems and high pressures seemed to be the most significant difference between the old and the new techniques. However, it was soon recognized that high pressure was not the main characteristic of the new technique. The name was changed to high performance liquid chromatography by C. Horvath in 1970 and was almost immediately accepted on an international basis to describe modern liquid chromatography.

There are four basic areas where present-day LC performance is superior to that obtained by the pioneering scientists of LC:

- Speed. Separations that took one or two hours to perform 40 years ago can now be done in minutes.
- Column performance. Present-day LC is simpler, more precise and reproducible, mainly due to improvements in column technology.
- Sample size. Whereas classical LC was a semi-preparative laboratory method, HPLC is a micro-technique, requiring only very small samples.
- Theoretical basis. While classical LC was essentially empirical in nature, the development of HPLC was based on theoretical principles that led to improvements in the technique.

1.9.2 Liquid Chromatography versus Gas Chromatography

The on-line process GC was the first instrument to provide the capability of measuring process stream composition by separation and measurement of individual components. Process LC expands this capability to include component separations previously considered impossible by GC or other analytical techniques. Therefore process LC should be looked upon as complementary to rather than competing with process GC because it provides additional capability in areas where process GC cannot provide an analysis.

It should be emphasized that there is a considerable overlap between the two chromatographic techniques, since many components can be separated and measured using either technique. However, in view of the more advanced development of process GC at the present, it is the preferred method, being more cost effective and requiring less effort from a maintenance and operational standpoint. In addition, although experience in process GC is very helpful, it is not sufficient for maintenance of a process LC. Some form of specialized training is essential.

Some of the limitations of process GC may be summarized as follows. Normally GC ovens are operated at or below 175°C, since at higher temperatures the life span of sample injection and column switching valves is drastically reduced, as is detector sensitivity. In addition, at higher temperatures there may be problems in meeting the requirements for safe installation in hazardous locations. Many hydrocarbons and petrochemicals requiring high temperatures for analysis by GC can be analyzed by LC at relatively low temperatures. Some compounds tend to decompose or polymerize on heating and therefore cannot be analyzed by GC. It is not possible to analyze for ionic species using GC. All of these compounds may be analyzed using LC. In addition, LC has the capability of measuring the molecular weight distribution of polymeric materials, a measurement hitherto impossible except by empirical techniques.

1.9.3 Laboratory LC versus Process LC

As mentioned earlier, LC is now an internationally recognized and powerful analytical technique for use in the laboratory. Process LC, however, is still in a relatively early stage of development. In fact, the situation is rather similar to that of process GC in the early 1960s. As with the development of almost all on-line process analyzers, the reasons for this are best illustrated by looking at the differences between laboratory and process LC.

First, the laboratory instrument is designed for manual operation in a safe, stable environment, whereas the process instrument must be designed to operate automatically and unattended in a relatively harsh environment with wide variations in ambient temperature. It must be weather-resistant and meet the requirements for installation in locations classified as hazardous for electrical equipment.

Second, the laboratory instrument is designed for versatility (frequent changes in operating parameters and programming). Conversely, the process instrument is designed for a dedicated analysis and must provide high stability and reliable operation.

Third, the emphasis in process LC is to provide the fastest possible analysis utilizing multi-column and multi-valve technology. In addition, more emphasis is placed on solvent selectivity and optimization than in laboratory LC.

Fourth, the process LC requires a sampling system to transport the process sample to the LC location and condition it so that it is compatible with the requirements of the instrument. In contrast, the laboratory instrument requires no complicated sampling facilities.

Another difference is in the life expectancy of the columns. The life span of a laboratory column is reasonably estimated at one thousand injections. By contrast, a process LC might be expected to complete a thousand injections in under a week. Therefore at least a 10-fold increase in life expectancy is desirable for a process LC column.

One significant factor that must be considered is the lower level of expertise in the technology of process LC maintenance personnel compared with laboratory personnel. This is to be expected, since process LC is a young technique and very little field experience is available at present. Adequate training of field maintenance forces in process LC techniques is therefore absolutely essential.

Finally, the cost of the process LC is much higher than the laboratory LC, since in most cases the cost of the process LC includes both hardware and applications engineering, whereas the laboratory LC is often sold as hardware only. The sample handling facility required by the process LC is usually a very significant cost item and may be as complex and costly as the analyzer itself. Good sampling system design is usually critical to the success of the process LC installation.

The potential purchaser of a process LC should bear the above differences in mind and avoid the temptation of using a laboratory-type instrument where a process instrument is required.

1.9.4 Commercially Available Equipment

Although several process chromatograph manufacturers have shown an interest in process LC and a number of prototypes have been produced, to the author's knowledge only one company, Applied Automation Inc., currently manufactures on-line liquid chromatographs for industrial process applications. AA installed its first prototype process LC in 1974 and the first publications covering on-line experience appeared in 1976 and 1977. AA now reports over 50 units on order or in operation in the United States. Many of these installations are in applications involving the analysis of the molecular weight distribution of polymers utilizing the technique of size-exclusion chromatography (SEC), often known as gel permeation chromatography (GPC).

(a) Applied Automation Inc. (United States)

For process LC applications, AA manufactures the Optichrom 2100 LC analyzer/programmer combination. The microprocessor-based 2100 programmer, described earlier, controls the analytical cycle, provides internal self-diagnostics for the complete system, carries out the necessary data reduction, and may be interfaced to a printer, recorder, and host computer for data presentation. AA is also planning to interface the LC analyzer to its new Optichrom Advance Chromatograph System, scheduled for 1985.

The LC analyzer, which is shown in Fig. 1.50 and is similar to the Optichrom 2100 gas chromatograph, is suitable for wall mounting in NEC Class 1, Groups B, C, and D, Division 1 hazardous locations. Since molecular weight distribution (using GPC) appears to be the most important on-line process application of LC to date, the following equipment description is limited to this particular application only. Process GPC requires the integration of a number of sophisticated hardware components to obtain good results, whereas the column separations are relatively simple. In contrast, other forms of process LC require greater column engineering effort to obtain suitable component separations, whereas the required hardware components are not usually so sophisticated. In addition, for process GPC a microprocessor-based programmer or computer is essential to provide the extensive data reduction necessary.

A block diagram of a typical process LC system for molecular weight distribution measurement is shown in Fig. 1.51. The main components of the GPC are:

- Carrier (solvent) storage and flow control system.
- Sample preparation (dilution) system.
- Sample injection system.
- Columns and column switching system.
- Detector(s).

Proper carrier (solvent) handling is an issue that cannot be over-emphasized. Most LC solvents are inflammable and often toxic. They must be handled with caution and operating procedures established. The solvent reservoir is designed to store from four to eight weeks supply of carrier for the LC and is usually located adjacent to the analyzer. Solvent is drawn from the reservoir through a filter by a carrier pump. A shutdown system, consisting of a restrictor and low pressure switch, provides pump protection in the event of a solvent line rupture.

Several high-pressure pumping systems have been tried for process LC, most of which were inadequate for precise control. The pneumatic amplifier

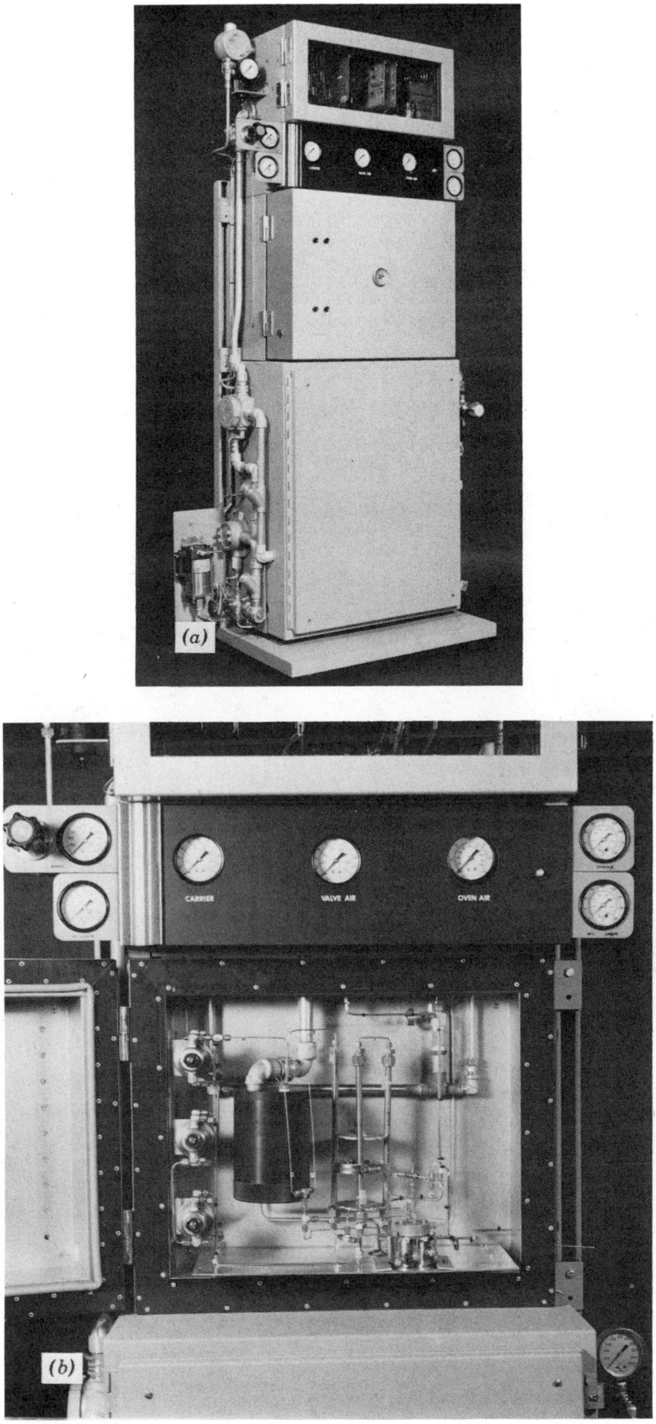

FIGURE 1.50 Optichrom LC analyzer. (*a*) Analyzer unit; (*b*) interior view of chromatographic oven. (Courtesy of Applied Automation Inc.)

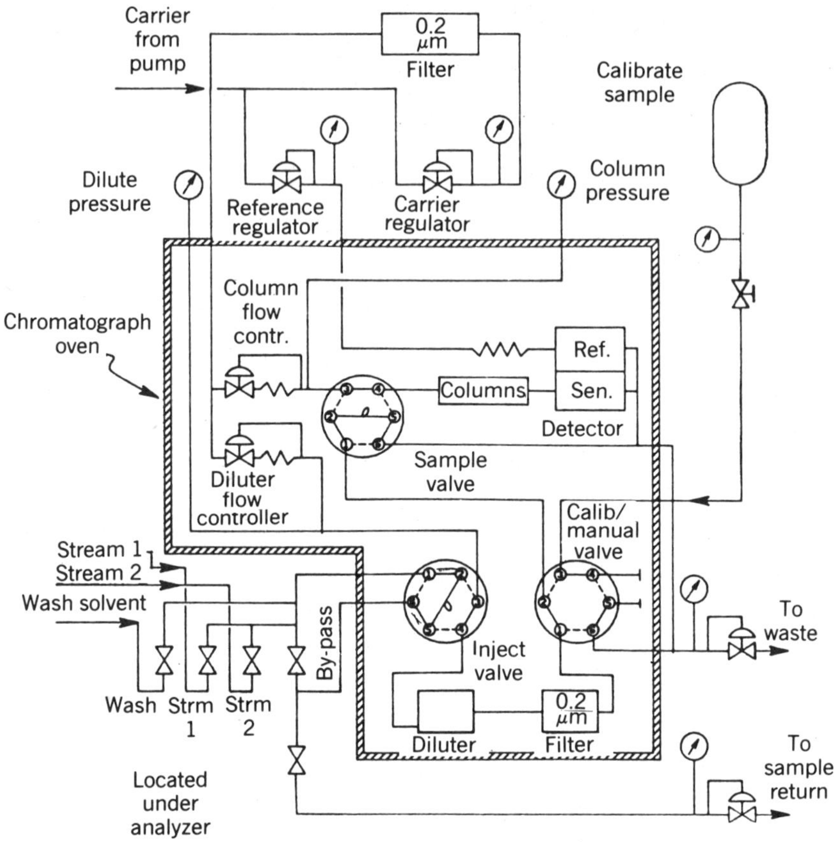

FIGURE 1.51 Optichrom LC for mol. wt. distribution. (Courtesy of Applied Automation Inc.)

pump appears to be the most satisfactory and is currently in use. No doubt as technology advances in the future, more efficient and reliable solvent pumps will become available. The pressure regulator is located downstream of the shutdown system. For most process LC applications, a 0–2000 psig output regulator has been found satisfactory, although in some cases the pump discharge is maintained as high as 3000 psig and the regulator output set at about 1000 psig. This arrangement provides a very stable pressure source essentially independent of carrier pump fluctuations.

The next component, the carrier flow controller, is one of the most important in process GPC. In polymer analysis, an error of only a few seconds in elution time can represent an error of thousands of molecular weight units. Therefore very precise flow control is essential for molecular weight distribution analysis. One flow control device that has been used successfully is the Veriflow Corp. Model LC221S. Since carrier viscosity is a function of temperature, the flow controller must be placed in a constant temperature zone (usually the oven). Maximum allowable flowrate change for GPC is about 2%.

An important component of the GPC system concerns sample preparation. In addition to the customary sample conditioning, the process sample requires dilution prior to injection. Normally the sample dilution system is mounted as an integral part of the sample conditioning system. However, for certain applications involving high-viscosity polymers, it must be carried out adjacent to the sample point. Sometimes, sample dilution requires the use of a separate solvent, which necessitates the provision of separate storage, pumping, and flow control systems for the dilution solvent, together with a safety protection and shutdown system for the pump.

The AA Model 20 diaphragm plunger valve is preferred for sample injection and column switching functions because it has low dead volume, a short activation time of 150 msec, and very few moving parts that move about 10 thousandths of an inch. As with process GC, several different column configurations can be used for process LC separations, including forward column stepping, reverse column stepping, backflush, heartcut, and column backwashing (peculiar to LC). Separation in process LC must be optimized by the use of selected solvents and column switching techniques, the major constraint being maximum allowable analysis time. Oven temperature is controlled to within ±0.02°F over the range 140–300°F under steady ambient conditions. For ambient changes from 0–122°F, control is to within ±0.3°F.

AA offer three detectors for process LC. The most popular is the optical absorbance detector, which has a UV-near IR wavelength range of 214–1100 nm, incrementally selectable. Reference and sample flows are through individual fluid cells designed to focus the radiation onto a dual photodiode. A typical optical path length for the detector is 1 cm, with a cell volume of 18 μl and a pressure rating of 500 psig.

The second most popular detector is the refractive index detector, which is of the deflection type, employing a single beam of light that passes twice through the detection and reference cells. The detector features an explosion-proof housing and temperature compensation electronics.

A more recent introduction is the dielectric constant detector, which measures small changes in the dielectric constant of a flowing liquid stream. The detector consists of two adjacent, identical, parallel-plate capacitors, with the sample and carrier acting as the dielectric material between the plates. Each cell (ref. and meas.) forms part of a parallel inductance/capacitance resonant circuit, where the frequency of each cell oscillator is determined by the capacitance of the cell. The output frequency of the two oscillators is fed to a mixer that senses the frequency difference.

Software for Molecular Weight Distribution. Commonly used parameters for polymer characterization are the number average molecular weight (M_n)

TABLE 1.1 On-Line Process Chromatographs

Manufacturer	Country of Origin	Model Number	Detector Types[a]	Column Types	Switching Valves	Micro-based	No. of Analyzers	Streams/ Analyzers	Com-ponents/ Stream	Oven Control	Continuous Analyzer Control	Remarks
1. Amscor Inc.	USA	MS-IV	TCD FID	Packed $\frac{1}{8}$ in.	Series 4 Slider type	Yes	8	64	99	Forced hot air Up to 4 Zones P&I Temp. program available	Yes	
2. Applied Automation Inc.	USA	Optichrom 102	TCD FID	Packed $\frac{1}{8}$ in. Semicapillary (SCOT)	Diaphragm type Model 10 Model 20	No, but computer compatible w/o controller	1	4	6	PID 60–400°F Forced hot air	No	Can be interfaced to Optichrom advance system
		Optichrom 2100	TCD FID FPD	Packed $\frac{1}{8}$ in. Semicapillary Fused silica	Diaphragm type as above	Yes	4	16	50	Forced hot air 140–350°F PID	No	Can be interfaced to Optichrom advance system
		Optichrom Advance	TCD FID FPD PID	Packed $\frac{1}{8}$ in. Semicapillary Fused silica	Diaphragm type	Yes Distributed system	240+ (256 devices total)	32	99	Forced hot air PID 140–350°F	Yes	Distributed analyzer control system
		Optichrom LC (liquid chrom)	RI DC	Packed	Diaphragm type Model 20	Yes	1			Forced hot air PID 140–350°F	No	To be integrated into Optichrom advance system
3. Beckman Industrial Inc.	USA	Model 6710	TDC FID	Packed $\frac{1}{8}$ in.	Slider type (6 or 10 port)	No Computer compatible w/o controller	1	5 or 10	6	P only 55–225°C Forced hot air	No	
		Model 6750	TCD FID FPD	Packed $\frac{1}{8}$ in.	Slider type (6 or 10 port)	Yes	6	6	16	P only 55–225°C Forced hot air	No	

4. Carlo Erba sPA	Italy	Fractomatic Series 1000	TCD FID ECD	Packed $\frac{1}{8}$ in.	Diaphragm type bimatic	No	1	2, 6, or 12	14	Forced hot air	No	
5. Combustion Engineering Process Analytics	USA	002-9000 (9020)	TCD FID FPD ECD	Packed $\frac{1}{8}$ in. Fused silica (capillary)	C-6 linear sliding plate (4, 8, or 10 port) Liquid sample valve	Yes	1	20	7	P only 50–220°C Forced hot air	No	
		Micromac 5000-1				Yes	8 (Std)	20	7		Yes (13 max per Dataverter)	Analyzer computer systems utilizing Model 002 GCs
		Micromac 5000-2				Yes	16	20	7		Yes (13 max per Dataverter)	
		5500-1				Yes	32	20	7		Yes (13 max per Dataverter)	
		DACS				Yes	96 controllers (any mix)	20	7		Yes (8 max per Controller)	
6. Daniel Industries Inc.	USA	500	TCD FID	Packed $\frac{1}{8}$ or $\frac{1}{16}$ in.	Diaphragm type	Yes (Model 2250)	1	5	16		Yes (2 max)	
		1000	TCD FID FPD	Packed $\frac{1}{8}$ or $\frac{1}{16}$ in.	Diaphragm type	Yes (Model 2250)	1	5	16		Yes (2 max)	
7. Foxboro Analytical	USA	931A 935	TCD FID FPD	Packed $\frac{1}{8}$ in.	Rotary style slider type	Yes Yes	1 12 (max)	16 16	10 10		Yes	Data collector
8. Hone-Addison Ltd.	Great Britain	Series 604	TCD FID	Packed $\frac{1}{8}$ in.	4, 6, 8, or 10 port Piston actuated slider type	No	1	1, 2, 6, (expandable to 30)		Forced hot air Zero crossing control 120–320°F	No	

TABLE 1.1 (*Continued*)

Manufacturer	Country of Origin	Model Number	Detector Types	Column Types	Switching Valves	Micro-based	No. of Analyzer	Streams/ Analyzers	Com-ponents/ Stream	Oven Control	Continuous Analyzer Control	Remarks
9. Servomex Ltd.	Great Britain	Series 400	TCD	Packed $\frac{1}{8}$ in.	Sliding plate type (6 or 10 port)	No	1	6 (max)	4 or 12	Forced hot air P + D Control 300°F max Hi-temp version 570°F max	No	
10. Siemens AG	West Germany	P101	TCD FID FPD	Packed $\frac{1}{8}$ in. or fused silica (capillary)	Slider type for liquid sampling	Yes	1	1 to 15	8	140–400°F Wall centered heating	No	"Valveless" column switching
		P180	TCD FID	Packed $\frac{1}{8}$ or $\frac{1}{4}$ in.	Diaphragm or rotary types	Yes	1	10	10	Forced ventilation Zero crossing control 104–320°F	No	Used with Model 2/4 or 10/10 programmers
		P200	TCD FID	Packed $\frac{1}{8}$ or $\frac{1}{4}$ in.	Diaphragm or rotary types	No	1	10	10	Isothermal controller two-step 104–320°F	No	Used with Model 2/4 or 10/10 programmers
		P300	TCD FID	Packed $\frac{1}{8}$ or $\frac{1}{4}$ in. or capillary	Diaphragm or rotary types	No	1	10	15	Isothermal controller two-step 104–320°F	No	"Valveless" column switching also available
11. Yokogawa Ltd.	Japan	GC 8	TCD FID	Packed $\frac{1}{8}$ or $\frac{1}{16}$ in.	?	Yes	1	12	24	On-off 120–270°F	No	
		Signal station				Yes	6 (max)					

[a] Detector Types—TCD: Thermal Conductivity: FID: Flame Ionization FPD: Flame Photometric ECD: Electron Capture RI: Refracture Index DC: Dielectric Constant.

and the weight average molecular weight (M_w), where

$$M_n = \Sigma N_i M_i$$

$$M_w = \frac{\Sigma N_i M_i^2}{\Sigma N_i M_i}$$

and N_i is the number of species with a molecular weight M_i. The ratio of the two averages (M_w/M_n) is also of interest and is known as the polydispersity of the polymer.

To calculate N_i and M_i, the software program splits the polymer peak into a large number of slices (60–80) and measures slice width and slice area. Computations are then carried out for M_w, M_n, and the ratio M_w/M_n.

REFERENCES

1. Tswett, M. *Ber. Deut. Bot. Gesellsch.*, **24,** 318, 384 (1906).
2. Martin, A. J. P., and Synge, R. L. M. *Biochem. J. (London)*, **35,** 1358 (1941).
3. James, A. T., and Martin, A. J. P. *Analyst,* **77,** 915 (1952).
4. James, A. T., and Martin, A. J. P. *Biochem. J. (London)*, **50,** 679 (1952).
5. Van Deemter, J. J., Zeiderweg, F. J., and Klinkenberg, A. *Chem. Eng. Sci.*, **5,** 271 (1956).

BIBLIOGRAPHY

Ayers, B. O. Chromatographic analyzer in process instrumentation. In *Gas Chromatography,* pp. 249–267. Academic Press, New York (1958).

Bailey, S. J. On-line analyzers prevent plant operating pitfalls. *Contr. Eng.* (Aug 1983).

Bossart, C. J., Heller, H., Maier, H. J., and Zinn, T. L. Development of programmers in gas chromatography. *Instr. & Contr. Syst.* (March 1965).

Clemons, J. M., and Leaseburge, E. Separation and measurement of PPM levels of thiophene in benzene with a microprocessor-controlled process gas chromatograph. *Proc. ISA Conf.*, Philadelphia, PA (Oct 1978).

Clemons, J., and Leaseburge, E. Capillary columns in process gas chromatography. *ISA Symp.*, Boston, MA (May 1985).

Clevett, K. J. On-line application of vapour phase chromatography to process plant. *Instr. Rev.* (Aug, Sept. 1966).

Desty, D. H. (Ed.). *Vapor Phase Chromatography*. Academic Press, New York (1959).

Fickey, E. W. Carrier gases for gas chromatography. *Trans. Soc. of Instr. Tech.* (Dec 1965).

Frost and Sullivan Inc. Industrial on-stream process analyzers, *Report No. 669* (1979).

Fuller, E. N., Porter, G. T., and Roof, L. B. Size exclusion chromatography moves to the plant for on-line polymer analysis. *J. Chrom. Sci.*, **20** (March 1982).

Golay, M. J. E. *Anal. Chem.*, **29,** 928 (1957).

Golay, M. J. E. *Nature* **180,** 435 (1957).

Golay, M. J. E. In Desty, D. H. (Ed.), *Gas Chromatography*. Butterworths, London (1958).

Houser, E. A., and Villalobos, R. Analysis of catalytic cracker regenerator flue gas. *Anal. Instr.*, **8,** *16th ISA-AID Symp.*, Pittsburgh, PA (May 1970).

Jutila, M. Guide to selecting gas and liquid chromatographs. *InTech* (Aug 1980).

Keulimans, A. I. M. *Gas Chromatography*. Reinhold, New York (1959).

Layne, T. G. Process liquid chromatography: Proving its worth. *InTech* (Sept 1984).

Manka, D. P. Chromatographic analysis of chemicals derived from coal carburization. *Adv. in Instr. ISA Intl. Conf.*, Philadelphia, PA (Oct 1982).

Martin, F. D. Developments in process gas chromatography. *InTech* (Jan 1970).

McCoy, R. D. Adding capabilities to process chromatographs with microprocessor-based programmers. *Proc. ISA Conf.*, Houston, TX (May 1978).

McCoy, R. D. *Anal. Instr.*, **16,** 155–162 (1978).

Mowery, R. A. Jr. Process liquid chromatography and its applications in industry, *Proc. ISA Conf.*, Houston, TX (Oct 1980).

Mowery, R. A. Jr. On-line process liquid chromatography. *Chem. Eng.* (May 1981).

Mowery, R. A. Jr. *Automated Stream Analysis for Process Control,* Vol 1, pp. 119–187. Academic Press, New York (1982).

Ogelsby, M. W., and Hobbs, J. W. Chromatographic analyzers for distillation control. *Oil & Gas J.* (Jan. 10, 1966).

Parris, N. A. *J. Chromatogr.*, **157,** 161–170 (1978).

Rice, R. B., Carver, D. H., Simmons, J. R., and Parsons, J. R. Chromatographic analyzer system in a butadiene plant. *Proc. 34th. Ann. Symp. on Instr. for the Proc, Ind.*, Texas A & M (Jan 1978).

Snyder, L. R., and Kirkland, J. J. *Introduction to Modern Liquid Chromatography,* 2nd Ed. Wiley, New York (1974).

Verdin, A. Principles and applications of process chromatographs. *Chem. Engr.* (April 1963).

Villalobos, R. How wet carrier gas cuts analyzer performance. *Instr. Tech.* (Oct 1967).

Villalobos, R. Process gas chromatography. *Anal. Chem.*, **47,** 983A (Sept 1975).

Villalobos, R. Calibration and data reduction in microprocessor-based chromatographs. *Trans. Inst. Meas. & Control,* **1,** No. 4 (Oct–Dec 1979).

Villalobos, R., and Turner, G. Peak height versus area integration. *Anal. Instr. Symp. ISA,* San Francisco, CA (1976).

Villalobos, R., and Annino, R. The application of capillary columns in process gas chromatography. *ISA Symp.*, Boston, MA (May 1985).

Villalobos, R., Rowland, R., and Roggenkamp, R. An air purification and methanator accessory for process chromatographs with hydrogen flame ionization detector. *22nd Ann. ISA-AID Symp.*, San Francisco, CA (May 1976).

Villalobos, R., Porter, R., LeBlanc, R., and Hearn, R. New concepts in microprocessor-based process chromatographs. *33rd Ann. Symp.*, Texas A & M Univ. (Jan 1978).

Wightman, R. E., Karasek, F. W., and Fourroux, M. M. Automatic control with high-speed gas chromatography. *ISA Journal* (May 1960).

2

Viscosity Measurement

2.1 BACKGROUND AND DEFINITIONS

Viscosity is an important physical property of fluids. Knowledge of it is essential in the design of process equipment such as pumps, piping, and flow metering systems. Measurement and control of viscosity play a fundamental role in many industrial processes, particularly in the petroleum, petrochemical, paper, textile, and printing industries. The qualitative investigation of flow processes calls for the measurement of the fluid viscosity, or, more generally, the rheological properties of the fluid. The flow behavior of a fluid is often a decisive factor in the performance of the fluid under the process operating conditions and therefore has a major role in the success of the process itself. Changes in viscosity may often result from molecular changes; therefore measurement of viscosity may enable chemists to estimate the degree of polymerization, homogenization, crystal concentration, degree of purity, and so on.

Viscosity is a measure of the *internal friction* of a fluid. Perhaps a simple way to understand viscosity is to start by defining the related parameter *friction* and its effect on *solid* matter.

When two solid bodies are in contact, with one surface sliding over the other, the force that opposes relative motion of the two bodies is friction. The effect of friction can be easily demonstrated by placing a solid body on an inclined plane and gradually increasing the angle of inclination of the plane, as shown in Fig. 2.1. Assuming the body has a mass M, the weight Mg of the body may be resolved into two components, $Mg \cos \theta$ at right angles to the plane and $Mg \sin \theta$ parallel to the plane; that is:

$$f = Mg \sin \theta$$
$$F = Mg \cos \theta$$

The coefficient of friction (μ) is defined as the ratio f/F. Therefore,

$$\mu = \frac{Mg \sin \theta}{Mg \cos \theta} = \tan \theta$$

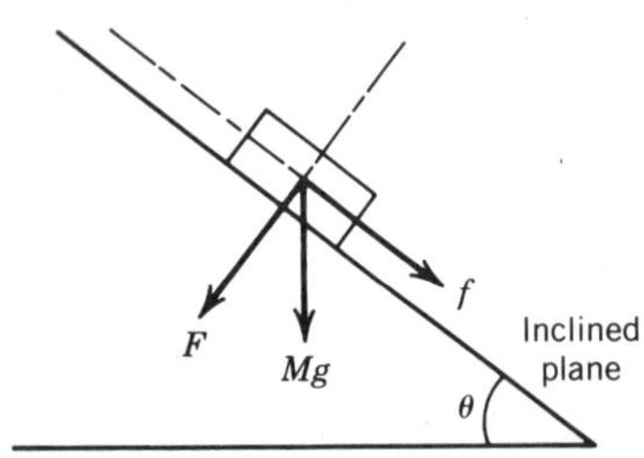

FIGURE 2.1 Coefficient of friction (solids).

This value of μ is called the *coefficient of static friction*. If the experiment is repeated, with the body given a slight push, the angle θ' can be found at which the body will continue to slide without accelerating. This value of μ is called the *coefficient of sliding friction*. The true nature of friction between solid bodies is rather difficult to explain. Although it is certain that surface roughness must be a contributing factor, the effect of molecular layers of certain substances in reducing friction between sliding surfaces seems to point to friction being a molecular phenomenon, closely associated to adhesion.

Friction in fluids is a little easier to explain. Since a fluid cannot support a shearing stress, however small it is, various layers of the fluid will move relative to each other and therefore a velocity gradient will exist at right angles to the direction of flow.

Newton's definition of viscosity can be explained by reference to Fig. 2.2, which shows two parallel planes *AB* and *CD* with area *A* cm^2, separated by a perpendicular distance of Y cm, moving in a fluid. The upper layer of fluid has a velocity which is greater than the lower layer and will tend to drag the lower layer forward. At steady flow conditions, there is no acceleration and therefore an equal and opposite drag will be exerted on the upper layer by the slower moving lower layer. Assuming that the velocities of the upper and lower layers are V_1 and V_2 respectively, the shear rate (D) is equal to $(V_1 - V_2)/Y$. The force per unit area of the layer required to act on surface A to maintain this shear rate is known as the shear stress (T), where $T = F/A$.

Newton postulated that shear stress is proportional to shear rate, and defined the ratio T/D as the coefficient of viscosity or, more simply, the viscosity of the fluid (η). He assumed this to be constant for each fluid (the law of constancy).

In the cgs system, viscosity is measured in poise (P), where 1 P = 1 dyne · sec · cm^{-2} and is the tangential force per square centimeter required to maintain a velocity difference of 1 cm/sec between layers 1 cm apart. In the SI system, the unit of measurement is the newton · sec · m^{-2} or pascal · sec. Normally, industry uses the centipoise (cP), one hundredth of the poise. The expression T/D defines the absolute or dynamic viscosity. The ratio of the absolute viscosity to density is the kinematic viscosity (ν), where $\nu = \eta/\rho$. This is sometimes called the viscosity/density ratio. In the cgs system, the unit of kinematic viscosity is the stokes (St) and the usual industrial unit is the centistokes (cSt). The reciprocal of absolute viscosity is known as fluidity.

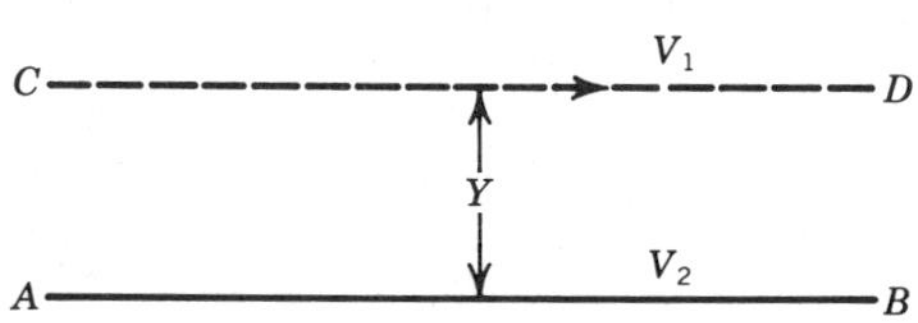

FIGURE 2.2 Newton's definition of viscosity.

Another term often used in describing the flow properties of a fluid is its *consistency*. However, consistency and viscosity do not mean the same thing, except in the case of simple Newtonian fluids. The consistency of a fluid refers to a point measurement at a given set of conditions (e.g., temperature, rate of flow). If this single measurement is sufficient to define the shape of the shear stress/shear rate curve, then the material is Newtonian and its consistency and viscosity are the same. Thus the term consistency is related to what is termed non-Newtonian behavior, since the viscosity characteristics of these materials cannot be defined by a single measurement. In the process industries, consistency is most often used to describe, for example, the solid content of a slurry or suspension such as pulp stock (paper industry) or batter mixtures (food industry). In these cases, the consistency is a measure of the degree of firmness of the fluid.

2.2 NON-NEWTONIAN FLUIDS

There are many fluids that conform to Newton's law of constancy first defined. Water, hydrocarbon liquids, all gases, and many others are Newtonian fluids at room temperature. For these substances, the rate of shear at which their viscosities are measured is unimportant and, provided that the temperature and pressure are given, the viscosity is a constant for the measured substance. If the shear stress of any Newtonian fluid is plotted against the shear rate, the result, called a *rheogram,* is a straight line through the origin, the slope of which depends upon the viscosity of the fluid. If viscosity is plotted against shear stress, the result is a straight line parallel to the shear stress axis (see Fig. 2.3).

Many substances, however, show deviations from Newton's law of constancy and exhibit ratios of shear stress/shear rate that are not constant. In most cases, the ratio decreases with increase in shear stress (see Fig. 2.4*b* and *d*), although there are substances that show increased viscosity with increasing shear stress (see Fig. 2.5*b*). It follows therefore that if such a substance is subjected to a low shear rate, using a "single-point" method, a

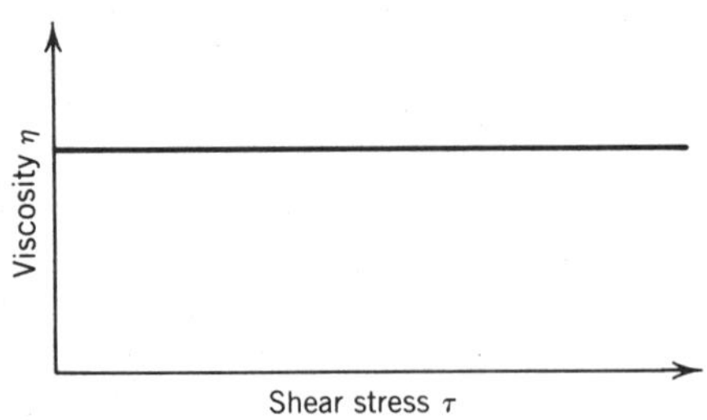

FIGURE 2.3 Rheogram for Newtonian fluids.

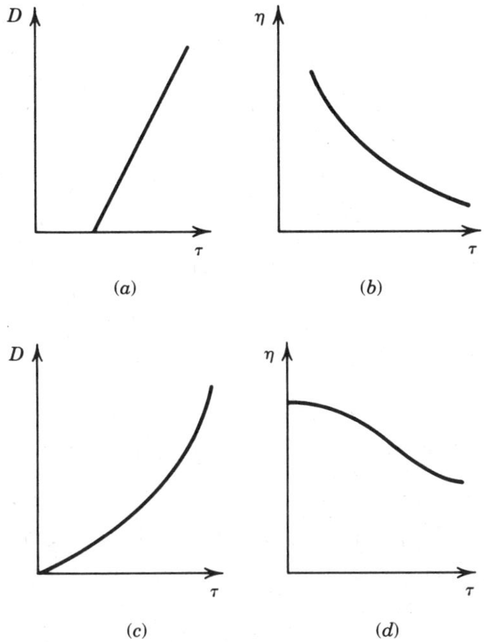

FIGURE 2.4 Rheograms for non-Newtonian fluids. (*a, b*) Plastic flow behavior; (*c, d*) pseudo-plastic flow behavior.

low value of viscosity will be obtained. The correct assessment of rheological behavior can be highlighted by reference to the design of a centrifugal pump to handle liquid with the characteristic as shown in Fig. 2.5*b*. If the pump is designed on the basis of a low shear rate viscosity measurement, at the higher shear rates and forces met under operating conditions the pump will be inadequate and this could result in a costly failure. This can be avoided by careful attention to the rheological behavior of the fluid.

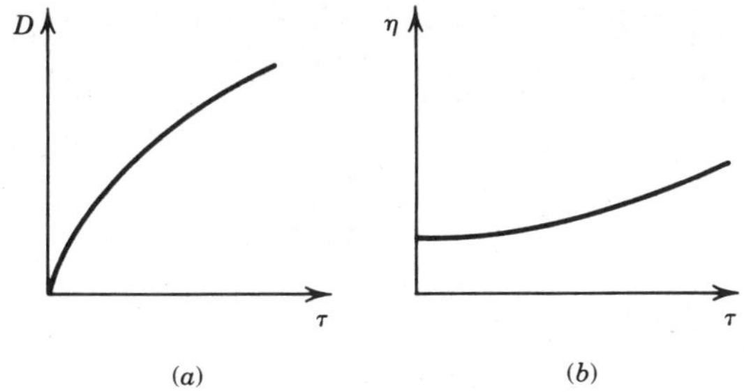

FIGURE 2.5 Rheograms for non-Newtonian fluids. (*a, b*) Dilatant flow behavior.

There are a number of variations in non-Newtonian flow characteristics which may be classified as follows:

2.2.1. Plastic Flow Behavior

Shown in Fig. 2.4*a* and *b,* this usually occurs with crowded, dispersed substances, the disperse part of which is wetted. Examples are printing inks, paints, vaseline, mayonnaise and toothpaste. These substances have a shear stress below which the substance no longer flows and becomes a solid body. These substances also tend to exhibit thixotropic behavior (see Section 2.2.4).

2.2.2. Pseudo-Plastic Flow Behavior

Shown in Fig. 2.4*c* and *d,* this occurs with long molecular chains in solvents and is sometimes called "shear thinning." Examples are glues, lacquers, and doped lubricants. These substances also tend to be visco-elastic (see Section 2.2.6).

2.2.3. Dilatant Flow Behavior

Shown in Fig. 2.5*a* and *b,* this mostly occurs with tightly-packed dispersed substances that are unwetted or lightly wetted. Examples are starch in water and oil-based paint sediment.

2.2.4. Thixotropic Flow Behavior

Thixotropy is a time-dependent effect that is completely reversible. The viscosity of the substance is seen to decrease during shearing at a constant shear rate and again increases when the substance is at rest. The rheogram of a thixotropic substance exhibits a hysteresis loop, as shown in Fig. 2.6.

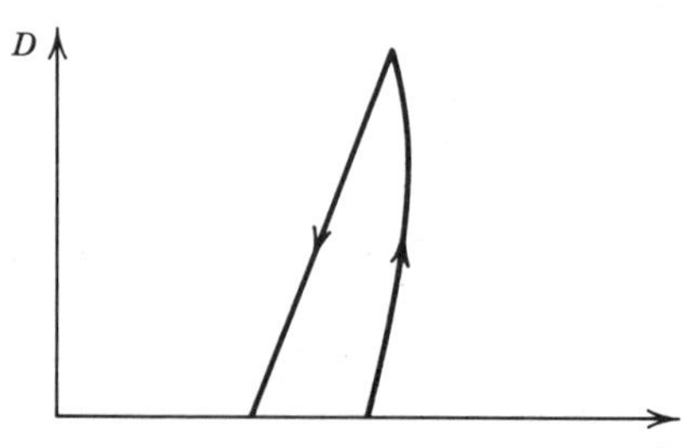

FIGURE 2.6 Rheogram for a thixotropic fluid.

2.2.5. Rheopectic Flow Behavior

Rheopectic substances exhibit a gradual increase in viscosity under shear, the reverse of thixotropy. Such substances are rare, whereas thixotropic substances are fairly common. Vanadium oxide suspension is an example of a rheopectic substance.

2.2.6 Visco-Elastic Flow Behavior

This type of flow behavior is said to exist in substances that exhibit both elastic and viscous properties. If pure shear stress is applied, forces normal to the shear plane are also present. With a body rotating in such a substance these forces cause the body to move radially inward and upward.

2.3 THE EFFECT OF TEMPERATURE ON VISCOSITY

So far, we have discussed the variation of viscosity with shear rate. However, viscosity also varies with temperature, density, and composition of mixtures. The best known parameter relating viscosity with temperature is the *viscosity index*. The viscosity of liquids decreases with increase in temperature, whereas for gases there is a small increase in viscosity with increase in temperature. In the case of liquids, various equations relating viscosity with temperature have been put forward, but these are mainly empirical and have limited applicability. Perhaps the most successful attempt was made by Andrade [2], who derived the expression

$$\eta v^{1/3} = Ae^{B/RTv}$$

where A and B are constants and v is the specific volume of the liquid at absolute temperature T. This relationship is found to hold in practice for many pure liquids, although in the case of water there is considerable discrepancy. With most liquids, the viscosity-temperature curve is exponential, and an extremely small temperature change could cause a large change in viscosity. In some materials the viscosity changes by as much as 50% per °C. It is therefore essential that viscosity measurements be made at a constant temperature. The exponential shape of the curve makes temperature compensation rather difficult, except over very narrow ranges. Often the simplest way to eliminate temperature effects is to carry out the measurement at a closely-controlled temperature.

2.4 CHARACTERISTICS OF FLUID FLOW

Since rheological characteristics of a process fluid have such a marked effect on flow behavior, an understanding of the nature of fluid flow is important to the analyzer engineer, particularly in the design of sample handling systems and in the measurement of viscosity itself. There are two entirely different types of fluid flow, which are dependent upon the viscosity, the velocity, and other properties of the fluid. This can be amply demonstrated by injecting small streams of dye into a clear liquid (e.g., water) flowing in a glass pipe and observing the behavior of these colored filaments at various liquid velocities.

At low velocities, termed *streamlined* (or *laminar*) flow conditions, the velocity of the liquid at any place is constant and the flow path in the liquid is steady in form, that is, the filaments move in straight lines. As the velocity is increased, a point is reached, called the "critical velocity," where the filaments of colored fluid will waver and start to break into diffused patterns. At higher velocities, termed *turbulent* flow conditions, the filaments are dispersed at random throughout the main body of the liquid.

The work of Osborne Reynolds showed that the nature of fluid flow in pipe (laminar or turbulent) is dependent upon the pipe diameter, fluid density, fluid viscosity, and the flow velocity. He defined a dimensionless variable, the Reynolds Number (R_e) by the expression:

$$R_e = \frac{dv\rho}{\eta}$$

where d = pipe inside diameter (in.)
v = fluid velocity (ft/sec)
ρ = fluid density (lb/ft^3)
η = fluid viscosity (cP)

Fluid flow is considered laminar at values of R_e less than 2000 and turbulent if R_e is greater than 4000. Between these two values is the *critical zone* where the flow is unpredictable.

2.5 CLASSICAL METHODS OF VISCOSITY MEASUREMENT

Background on the classical techniques used for the measurement of viscosity will give the reader an insight into the development of methods of measurement used in the laboratory and how these methods have been adapted for use in on-line process viscometers. All of the techniques described are used in one form or another in industry today. Therefore it is important that the basis of these techniques is understood.

2.5.1 Poiseuille's Method

This method, developed by the French physicist Poiseuille in 1842, involves measurement, under laminar flow conditions, of the volume of liquid, under

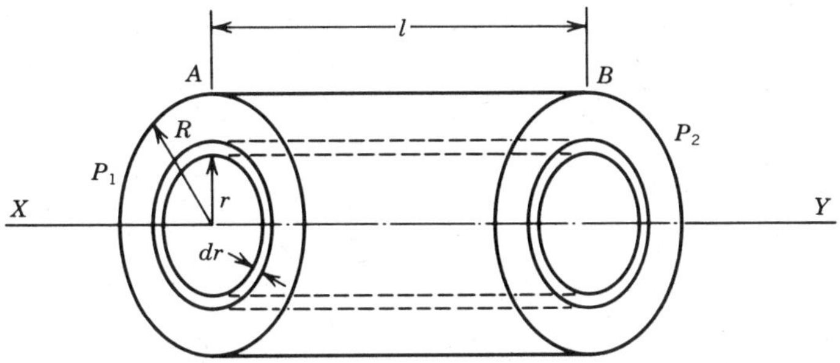

FIGURE 2.7 Poiseuille's method.

constant head conditions, flowing through a narrow-bore tube in a given time. A greatly enlarged section of the tube is shown in Fig. 2.7. The flow of liquid is from face A to face B and everywhere parallel to the axis XY. The velocity varies over the section, being a maximum at the axis and zero at the tube wall. If r is the radius of any cylinder of the liquid, the velocity gradient at the surface of this cylinder of liquid is dv/dr and the surface area is $2\pi rl$. The viscous drag on the cylinder surface is therefore $2\pi rl\eta dv/dr$. Each part of the liquid has a constant velocity, which varies from the axis to the tube wall.

Therefore the viscous drag is equal to the resultant force due to pressure difference between the two ends of the cylinder. There is no radial flow and therefore the pressure is constant over any section. Let the pressure be P_1 over Section A and P_2 over Section B. The force over Section A is $\pi r^2 P_1$ and over Section B, $\pi r^2 P_2$. Therefore,

$$2\pi rl \cdot \eta \cdot dv/dr = -\pi r^2(P_1 - P_2)$$

or

$$dv/dr = -\frac{r(P_1 - P_2)}{2l\eta}$$

the negative sign indicating opposing forces. Integrating with respect to r:

$$v = \frac{-(P_1 - P_2)}{2l\eta}\int_0^r r \cdot dr \quad \text{or} \quad v = \frac{-(P_1 - P_2)r^2}{4l\eta} + C$$

Now $v = 0$ when $r = R$, therefore,

$$C = \frac{(P_1 - P_2)R^2}{4l\eta}$$

and

$$v = \frac{(P_1 - P_2)(R^2 - r^2)}{4l\eta}$$

All of these quantities can be measured except η; therefore η can be calculated. However, it is very difficult to measure the value of v. A more convenient approach would be to measure the volume V of liquid crossing any section of the tube in unit time.

For a thin cylindrical surface of radius r and thickness dr, the volume passing in unit time is given by:

$$dV = 2\pi r \cdot dr \cdot v$$

$$= 2\pi r \frac{(P_1 - P_2)(R^2 - r^2)}{4l\eta} \cdot dr$$

Integrating with respect to r:

$$V = \frac{\pi(P_1 - P_2)}{2l\eta} \int_0^R r(R^2 - r^2) \cdot dr$$

or

$$V = \frac{\pi(P_1 - P_2)}{2l\eta} \left[\frac{R^4}{2} - \frac{R^4}{4}\right]$$

Therefore,

$$V = \frac{\pi(P_1 - P_2)}{8l\eta} \cdot R^4$$

and

$$\eta = \frac{\pi(P_1 - P_2) \cdot R^4}{8lV}$$

Since the fourth power of the small quantity R (the radius of the narrow-bore tube) enters into the calculation, this measurement must be made with the utmost accuracy.

2.5.2 U-Tube Method

This method is particularly useful when only small quantities of the sample liquid are available and compares the viscosity of the sample with that of a standard liquid such as water. A typical U-tube viscometer is shown in Fig. 2.8 and is essentially a more refined version of Poiseuille's method described earlier.

The viscometer is placed in a constant-temperature bath. The sample liquid is poured through inlet A into vessel E, where it is allowed to stabilize

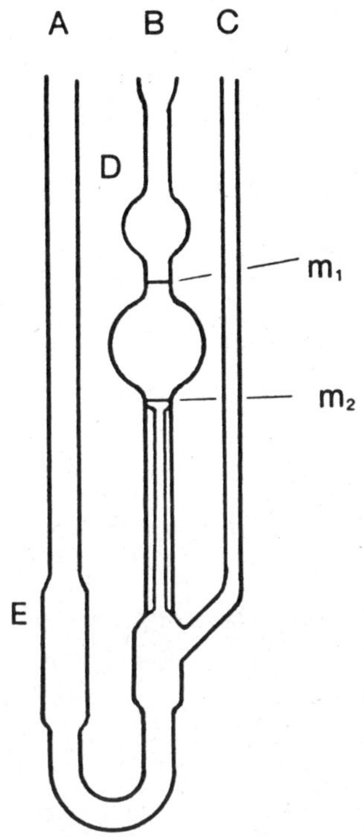

FIGURE 2.8 U-tube (capillary) viscometer (Ostwald).

at the bath temperature. It is then sucked through the capillary tube F to the upper edge of bulb D with C closed. A, B, and C are then opened and a measurement is made of the time taken (t_1) for the liquid level to drop from level L_1 to level L_2. The same test is done using the standard liquid and the time (t_2) is noted.

For each liquid, the volume Q is the same and therefore flow conditions are the same for both liquids, the only difference being the viscosities and densities of the two liquids. Flowrate is proportional to the pressure difference (density) and inversely proportional to the viscosity. Therefore,

$$\frac{dQ_1}{dt_1} = \frac{k\rho_1}{\eta_1}$$

for the sample liquid, where k varies with time but may be assigned the mean value $\bar{k}$.

Integrating with respect to time:

$$Q_1 = \int_0^{t_1} \bar{k}\frac{\rho_1}{\eta_1} \cdot dt = \bar{k}\frac{\rho_1}{\eta_1} \cdot t_1$$

For the standard liquid:

$$Q_2 = \bar{k}\frac{\rho_2}{\eta_2}\cdot t_2$$

Now $Q_1 = Q_2$; therefore,

$$\bar{k}\frac{\rho_1}{\eta_1}\cdot t_1 = \bar{k}\frac{\rho_2}{\eta_2}\cdot t_2$$

Since the viscosity and density of the standard liquid are known accurately, and letting $K = \eta_2/\rho_2$,

$$\eta_1 = \frac{K\rho_1 t_1}{t_2}$$

This method allows the direct measurement of the kinematic viscosity of liquids. Since the kinematic viscosity $(\nu) = \eta/\rho$, then

$$\nu_1 = K \cdot t_1/t_2.$$

The U-tube normally has a calibration factor that takes account of the viscosity and density of the standard liquid and provides a direct reading in terms of kinematic viscosity units. This method is described in more detail in Section 2.6.

2.5.3 Rotating Cylinder Method

In the rotating cylinder method, also known as the Couette principle, the liquid under test fills the annular space between two concentric cylinders. The outer cylinder is fixed, whilst the inner cylinder is rotated evenly about its vertical axis. The torque required to rotate the inner cylinder at a constant angular velocity is a measure of the fluid viscosity. The system is shown schematically in Fig. 2.9.

Assume that cylinder A, radius R_1, is fixed, and cylinder B, radius R_2, is rotated at constant angular velocity. The sample liquid in the annular space between A and B will eventually settle into steady rotation with the angular velocity increasing from zero at A to a maximum of Ω at B. Let the angular velocity at a distance r from the axis be ω. The velocity gradient corresponding to the slippage of adjacent liquid layers is $r \cdot d\omega/dr$ and results in a viscous drag of $\eta r \cdot d\omega/dr$ per unit area. The area represented by the dotted line in Fig. 2.9 is $2\pi rL$ assuming a cylinder of length L.

The total force due to viscosity is $2\pi L\eta \cdot r^2 \cdot d\omega/dr$, and gives rise to a torque τ, where

$$\tau = 2\pi L\eta \cdot r^2 \cdot d\omega/dr \cdot r$$

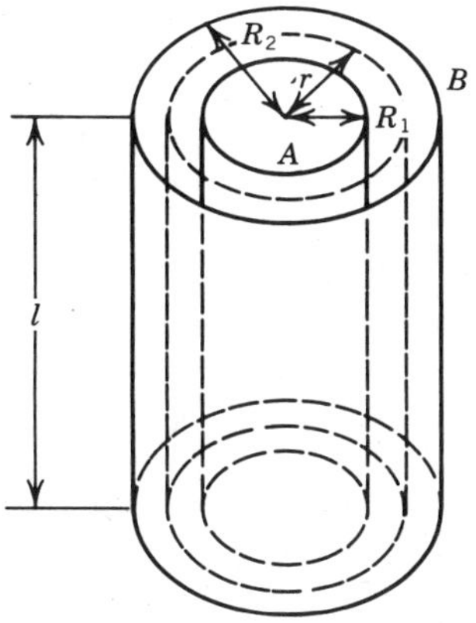

FIGURE 2.9 Rotating cylinder method (Couette).

or

$$\tau = 2\pi L\eta \cdot r^3 \cdot d\omega/dr$$

Therefore,

$$d\omega = \frac{\tau}{2\pi L\eta} \cdot \frac{dr}{r^3}$$

Integrating with respect to r, we get

$$\int_0^{\Omega} d\omega = \frac{\tau}{2\pi L\eta} \int_{R_1}^{R_2} \frac{dr}{r^3}$$

and

$$\Omega = \frac{\tau}{4\pi L\eta} \left(\frac{1}{R_1^2} - \frac{1}{R_2^2}\right)$$

The shear rate is constant throughout the annular space; therefore this method is very useful for studying the effect of variations in the shear rate on fluid viscosity (rheological investigations). There are several varieties of rotational viscometer available commercially, based on the principle outlined here. These include the coaxial cylinder, the cone-and-plate, the concentric-disc, and the parallel-disc types. Originally, these types of viscometer were used mainly in the laboratory, but in recent years industrial viscometers for in-line measurement have appeared and will be discussed later in this chapter.

2.5.4 Stokes' Falling Sphere Method

A body falling through a fluid experiences a resistance to its motion that is related to the viscosity of the fluid and the velocity of the body. For a spherical body travelling at a low velocity in a fluid of which the outer boundary is far removed from the sphere, the retarding force was calculated by Stokes to be equal to $6\pi\eta rv$, where r is the radius of the sphere and v its velocity. When the sphere reaches its limiting velocity, that is, when there is no acceleration, the resulting downward force is equal to the retarding force due to the fluid viscosity.

The downward force on the sphere due to gravity is equal to $\frac{4}{3}\pi r^3 gD$, where D is the density of the sphere and g is the gravitational constant. The upward force on the sphere due to buoyancy is equal to $\frac{4}{3}\pi r^3 gd$, where d is the density of the fluid. Therefore the resultant downward force is equal to $\frac{4}{3}\pi r^3 g(D - d)$. Since this force is equal to the viscous drag on the sphere,

$$\frac{4}{3}\pi r^3 g(D - d) = 6\pi\eta rv$$

Therefore,

$$\eta = \frac{2r^2 g(D - d)}{9v}$$

When using this method, care must be taken to ensure that the sphere reaches its limiting velocity. The method is useful for measuring the viscosity of very viscous liquids, since, under these conditions, the limiting velocity is reached very quickly.

2.6 LABORATORY METHODS OF VISCOSITY MEASUREMENT

There are many different methods for viscosity measurement in the laboratory, most of them based on the classical techniques described. These methods are adopted as standard specifications by the American Society for Testing Materials (ASTM), the British Standards Institution (BSI), the Institute of Petroleum (IP), the International Standards Organization (ISO), and the Deutscher Normenausschaus (DIN). D.C-H Cheng [1] has categorized the methods in an excellent series of articles for *Measurement and Control.* Two are used extensively throughout industry as standard laboratory methods.

2.6.1 Capillary Viscometers

The U-tube method is recognized as a standard for industrial routine estimations, particularly in the petroleum industry, mainly because of its simplicity. ASTM method D445-79, for example, lays down the procedure for viscosity estimations on both transparent and opaque petroleum products. The method involves measurement of the time taken for a fixed volume of liquid to flow under gravity through a calibrated glass capillary viscometer under a reproducible driving head and at a closely controlled temperature. The method applies to Newtonian fluids only. Standard specifications and operating instructions for glass capillary viscometers are covered in ASTM standard D446-74.

There are many capillary viscometers. They fall under the categories of Ostwald and modified Ostwald, suspended-level, and reverse-flow types (a typical Ostwald capillary viscometer is shown in Fig. 2.8). Some versions have several bulbs at various heights. With them, it is possible to determine deviations from Newtonian behavior. Since the use of the capillary viscometer in the laboratory involves the measurement of time, often these viscometers are calibrated in "time" units. Examples are the Saybolt-second unit (United States), the Redwood-second unit (UK), and the Engler-degree (Germany). Conversion from these units to kinematic viscosity units (cSt) can be made with the tables covered in ASTM standards.

2.6.2 Rotational Viscometers

Rotational viscometers have been used for many years as a means of carrying out rheological research on various substances. The flow behavior may be ascertained from a plot of angular speed against torque (shear stress against shear rate), as described in Section 2.2. Two companies are well known in this field, Contraves AG (Switzerland) and Brookfield Engineering Laboratories (United States).

Contraves manufactures the range of Rheomat rotational viscometers. Measuring systems include concentric cylinder and cone/plate types. The Rheomat-115 includes microprocessor-based programming electronics and provision for x-y recording, covering a viscosity range of 1 to 6×10^8 cP. The Low Shear 30 is designed for the determination of visco-elastic properties within the low shear range. The type LS measuring system covers the viscosity range 9.3×10^{-4} to 6×10^6 cP. The Rheomat 30 has a 30-way speed control module with a range of 0.048 to 350 min^{-1} and torque measurement reproducibility of $\pm 1\%$ of selected range. Viscosity measurement range is 10 to 6×10^8 cP.

The Rheoscan 30 is a universal programming unit for the Rheomat 30 and Low Shear 30. It can be used in conjunction with an x-y recorder for automatic curve plotting. Combining the different functions and program variables allows more than 100 measuring programs to be chosen. The Rheoanalyzer is a special microprocessor-based unit that interfaces with a rheometer such as the Rheomat 30 to allow computer control of the operation, computation, and documentation/printout of results.

Brookfield Engineering Laboratories manufactures the range of Synchro-Lectric rotational viscometers and accessories. The LV models are designed to measure materials in the low viscosity range. The LVF has four speeds (60, 30, 12, and 6 rpm) and 16 ranges (min 0–100 cP to max 0–10^5 cP). The

LVT has eight speeds (from 60 to 0.3 rpm) and 32 ranges (min 0–100 cP to max 0–2 × 10^6 cP). The RV models are for medium viscosity ranges, with a lower limit of effective measurement of near 85 cP. There are three models, the RVF, RVF-100, and RVT, with speeds from 100 to 0.5 rpm and covering viscosity ranges from 0–100 cP to 0–8 × 10^6 cP. The HA and HB series are for very high viscosity materials such as tars, asphalts, and roofing compounds, with speeds from 100 to 0.5 rpm and covering viscosity ranges from 0–200 cP to 0–64 × 10^6 cP. UL- and CSA-approved explosion-proof construction can be provided for any viscometer model.

2.6.3 Other Methods

(a) *Efflux Cup Viscometers*

The efflux cup viscometer operates like the capillary viscometer—the efflux time of a specified volume of the test sample is measured through a fixed orifice at the bottom of the cup and is representative of the viscosity of the sample. For Newtonian fluids the efflux time may be converted to kinematic units by means of charts or formulas. The efflux cup is mounted in a temperature-controlled bath for high measurement accuracy. Several efflux cup viscometers are available—for examples, the Saybolt Cup (shown in Fig. 2.10), the Ford Cup, and the Zahn Cup.

The Saybolt viscometer is the standard for testing petroleum products and is covered under ASTM method D88-56. There are three types of orifice: universal, furol, and asphalt. The cup-orifice combination is selected to give an efflux time of between 20 and 100 sec. The efflux time of the universal orifice is designated as Saybolt-second units and is the time required for 60

Catalog Number	Description
74981	Saybolt tube, brass
74983	Saybolt tube, stainless steel
74990	Orifice, universal
74991	Orifice, Road Oil, with wrench
74992	Orifice, Furol

74981 74983

74990 74991 74992

FIGURE 2.10 Efflux cup viscometer (Saybolt).

ml of sample to flow through an orifice of 0.176 cm diameter and 1.225 cm long.

The Ford cup viscometer is used for low viscosity liquids that do not deviate significantly from Newtonian behavior. The procedure and apparatus is covered in ASTM method D1200-58 for paints, varnishes, lacquers, and related liquid materials. Three different orifices are available.

The Zahn cup viscometer is widely used by paint manufacturers to standardize their products during manufacturing stages because of its convenience and rapid use. It is also used for measuring the viscosity of varnishes and lacquers. Five different orifices are available.

(b) Falling Sphere Viscometers

These devices operate in accordance with Stokes' Law as defined in Section 2.5.4. A typical example of a falling sphere viscometer is shown in Fig. 2.11.

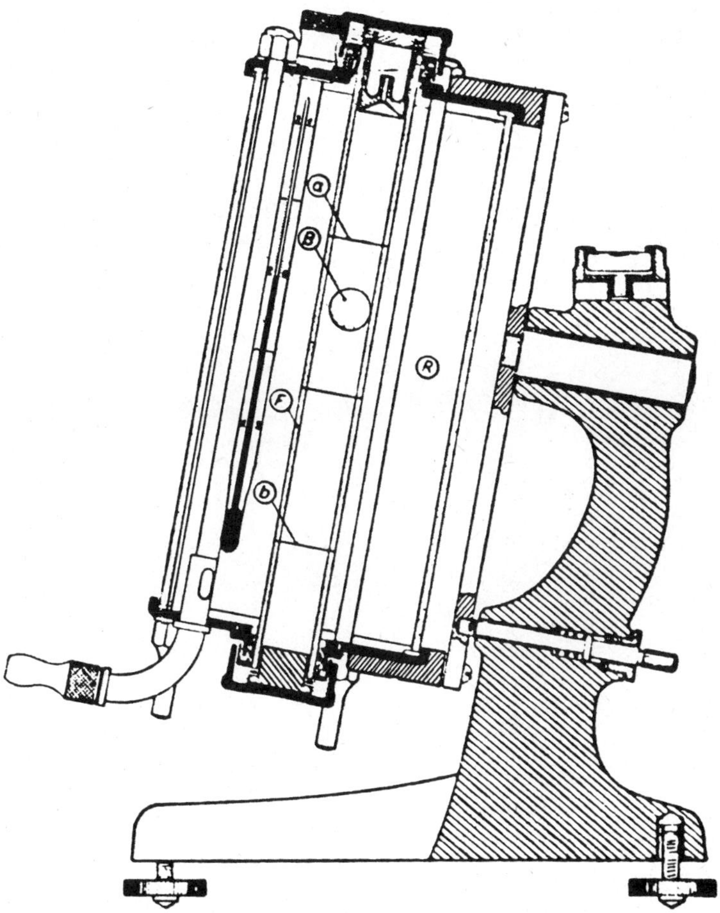

FIGURE 2.11 Falling sphere viscometer.

The sample under test is located in an inclined, thermally controlled glass tube with two calibration lines. A ball is allowed to sink through the liquid and the time required for the ball to cover the distance between the two marks is related to the sample viscosity. The viscosity of opaque substances cannot be measured with this device.

(c) Capillary Extrusion Viscometers

The capillary viscometers described in Section 2.6.1 are limited to low-viscosity Newtonian fluids because of the small driving force available. Several manufacturers have developed capillary-extrusion viscometers to study the flow behavior of non-Newtonian high-viscosity fluids. One type uses compressed nitrogen to drive the sample through a capillary, as shown in Fig. 2.12. The sample is placed in a thermostatically-controlled reservoir with a free floating follower plug riding on top of the sample to distribute the pressure evenly over the fluid surface. The flowrate is calculated from the weight extruded and the elapsed time of each extrusion. The driving force is measured by a pressure gauge. The viscosity of the sample, the rate of shear, and the stress are calculated from the observed data using formulas provided by the manufacturer.

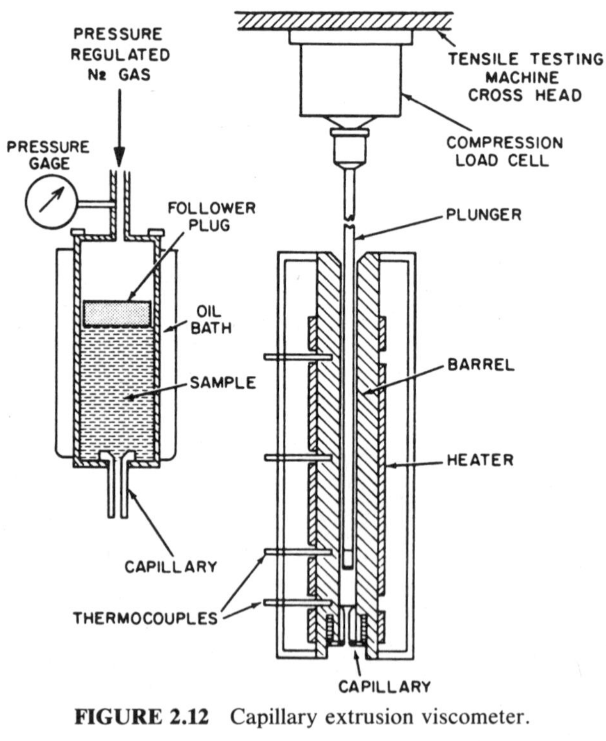

FIGURE 2.12 Capillary extrusion viscometer.

2.7 PROCESS VISCOMETERS

On-line process viscosity analyzers commercially available may be conveniently classified into four categories, based on the measurement principle used:

- Capillary (based on the Poiseuille method).
- Vibrational.
- Rotational (based on torque measurement).
- Falling piston.

2.7.1 Capillary-Type Viscometers

The first viscometers of this type, manufactured by Hallikainen in the United States, appeared in the late 1950s. They were also manufactured for the European market, under licence by GEC-Elliot in England. Since that time, there have been several significant changes. When Hallikainen went out of business in the mid 1960s, its line of process analyzers was taken over by Totco, while GEC-Elliot continued to market the Hallikainen line in Europe. Recently, Totco was bought by GCA Precision Scientific (PSG), who continue to market the updated version of the Hallikainen Model 1077 Viscometer.

Meanwhile, GEC-Elliot became part of Fisher Controls International and continued to market its viscometers under this name. Recently, Fluid Data Inc. purchased the complete Fisher Controls range of process analyzers and is planning to market them in the United States and Europe under the name "Hallikainen Instruments."

Apart from the viscometers, the product range includes analyzers for measurement of moisture, boiling point, pour point, vapor pressure, and color. The wheel has come full circle and the name Hallikainen still remains.

In the United States there appear to be two sources of the capillary-type viscometer, GCA Precision Scientific and Fluid Data Inc., while in Europe, the new Fluid Data organization in England is the only source.

(a) GCA Precision Scientific Group (United States)

PSG manufactures two process viscometers, the Model 41070 Series, which is the updated version of the original Hallikainen 1077, and the Model 44860. Both operate on the Hagan-Poiseuille principle, which states that the pressure drop of a Newtonian fluid flowing through a capillary under conditions of constant flow and constant temperature, will vary in proportion to changes in the absolute viscosity of the fluid. The two units are packaged in a similar fashion. However, the Model 44860 has additional design features based on extensive operating and maintenance experience with this type of equipment.

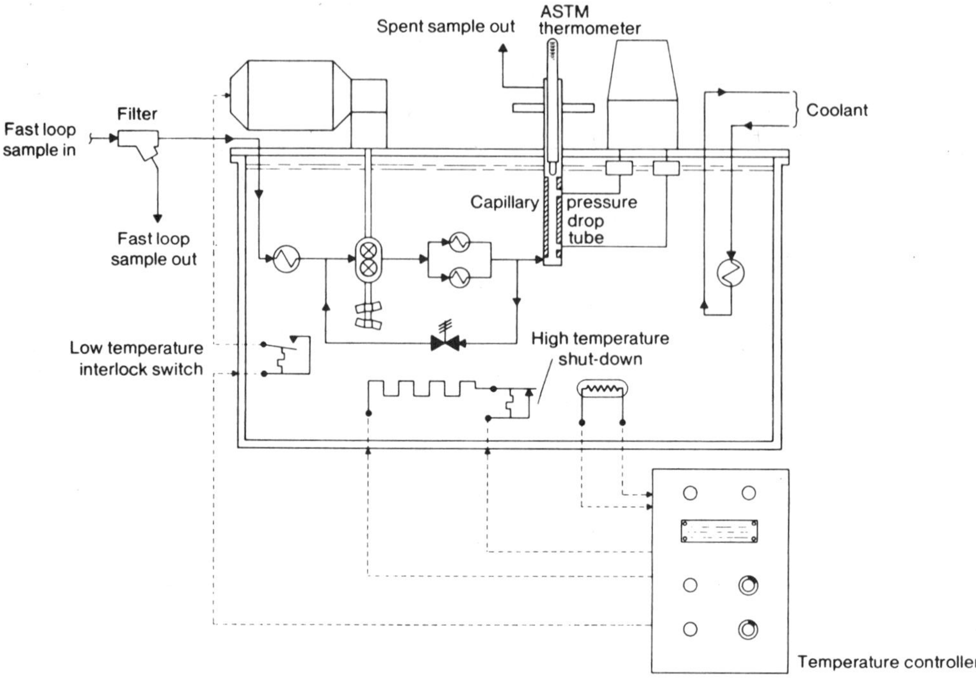

FIGURE 2.13 Model 41070 process viscometer. (Courtesy of GCA Corporation)

The Model 41070 Series Viscometer is shown schematically in Fig. 2.13. Process sample from the fast loop filter (usually 60-mesh) enters the constant temperature bath and flows through the primary heat exchanger coil before passing to a precision, close-tolerance gear pump which establishes a constant flowrate of 64 ml/min (50 ml/min on 50-Hz operation). Before entering the measuring capillary, the sample is brought to the precise bath temperature by a secondary, parallel-path heat exchanger. Flow continues through a 60-mesh filter screen, through the capillary restriction, past an ASTM thermometer, and out of the analyzer.

Bath temperature is measured by an RTD, which forms one arm of a balanced bridge in the controller circuit that regulates the power to a heater. Cooling coils work in conjunction with the heater to maintain constant bath temperature. The metering pump motor also drives a pair of impellers to provide bath agitation and minimize temperature gradients. Safety features include a relief valve, installed ahead of the capillary for pump protection, and a thermostat for bath high-temperature cut-off, which de-energizes the heater.

When the sample inlet temperature is more than 200°F higher than the bath temperature, an external sample cooler is required. For samples with high particulate loading, a self-cleaning filter with a 38–50 micron screen is recommended in the fast loop take-off, to protect the close-tolerance gear pump internals.

The differential pressure across the capillary is measured with a conventional force-balance DP transmitter with the option of either pneumatic 3–15 psig output (Model 41077) or electronic 4–20 mA dc output (Model 41078). For high viscosity applications, the transmitter can be provided with mechanical seals (Models 41251 and 41252) or the impulse lines steam traced. The analyzer may also be supplied calibrated but without a transmitter (Model 41079).

Viscometer measurement is in spans up to 2500 cP, with a repeatability quoted as ±1% of calibrated span. Bath temperature limits are 100–275°F. Analyzer dead time is 48 sec and measuring system response is 36 sec. The analyzer is mounted on a free-standing four-legged structure and is designed for operation in NEC Class 1, Group D, Division 1 hazardous locations.

The Model 44860 Continuous Viscosimeter, shown schematically in Fig. 2.14, is designed for applications such as fuel oil blending, lube oil produc-

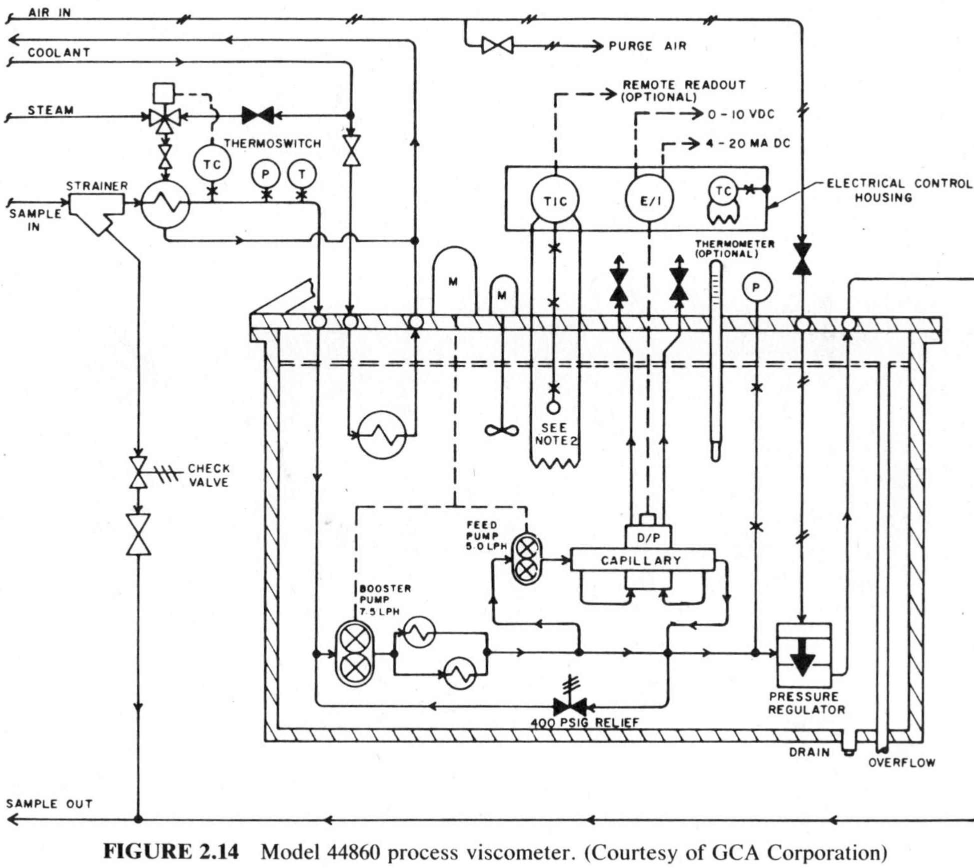

FIGURE 2.14 Model 44860 process viscometer. (Courtesy of GCA Corporation)

tion, and residual fuel oil dilution. The analyzer has an integral sampling system consisting of a fast loop strainer, check valve, and block valve, and a fast-acting temperature control system, to protect the analyzer from major changes in process stream temperature. The booster pump/back pressure regulator combination ensures that the analyzer is immune to changes in back pressure and supply pressure, by establishing analyzer operating pressure at about two atmospheres above the sample return pressure. The metering pump is driven by the same motor that drives the booster pump and has about two-thirds the capacity of the booster pump. The excess flow from the booster pump bypasses the metering pump/capillary tube assembly and establishes a condition of constant pressure across the assembly, ensuring constant flow across the capillary.

The differential pressure sensor is submerged in the bath at the same elevation as the capillary and requires no chemical seals. The sensor is a variable-reluctance differential transformer, the output of which is transduced to a conventional 0–10 V dc or 4–20 mA dc output for remote recording. The differential pressure sensor, capillary range tube, and both differential legs of the measurement system can be solvent-flushed in place without disassembly.

All the components submerged in the bath are suspended from the bath cover, giving easy access for maintenance. Air, sample, and coolant lines are installed on the axis of the cover hinges. The analyzer is supplied with eight capillary range tubes as standard, allowing operation at spans from 2 to 4000 cP, with a repeatability quoted as $\pm 0.5\%$ of full scale. The analyzer is designed for installation in NEC Class 1, Group D, Division 1 hazardous locations.

(b) Fluid Data Inc. (United States)

Fluid Data manufactures three Hallikainen-type Continuous Viscometers—the basic Model 1077, the Model 491 with diaphragm seals, and the Model 421 with multi-range capability. All instruments are suitable for installation in NEC Class 1, Group D, Division 1 hazardous locations. The Model 1077 is shown in Fig. 2.15. The analyzer operates in the same manner as the PSG Model 41070. Maximum measurement range is 0–2500 cP, with a minimum of 0–10 cP. The differential pressure transmitter may be electronic 4–20 mA dc or pneumatic 3–15 psig. Repeatability is quoted as $\pm 0.5\%$ of full scale for ranges up to 0–1000 cP.

(c) Servomex Ltd (England)

In conjunction with British Petroleum, Servomex has developed a viscometer for the measurement of equiviscous temperature. This is the temperature at which a fluid has a particular viscosity and can be used to derive bitumen penetration and viscosity values over a very wide range using one wide-bore measurement capillary. The development of this analyzer came out of BP's

FIGURE 2.15 Hallikainen Model 1077 viscometer. (Courtesy of Fluid Data Inc.)

need for rapid and continuous determination of bitumen penetration. The laboratory test is simple and relatively inexpensive. The equipment consists of a weighted needle that is allowed to sink vertically into the bitumen sample, which is held at a constant temperature of 25°C for a predetermined time. The amount of penetration, in tenths of millimeters, is called the penetration (or Pen Number). The major problem is that it takes several hours for the sample to cool down to a constant temperature of 25°C.

Work with conventional viscometers was unsuccessful because of two problems. First, the upper temperature constraint of the constant temperature bath limited the range of bitumens that could easily be pumped. Second, the nonlinear relationship of penetration with viscosity limited the penetration ranges that could be covered with one capillary.

The Equiviscous Temperature (EVT) Analyzer, shown schematically in

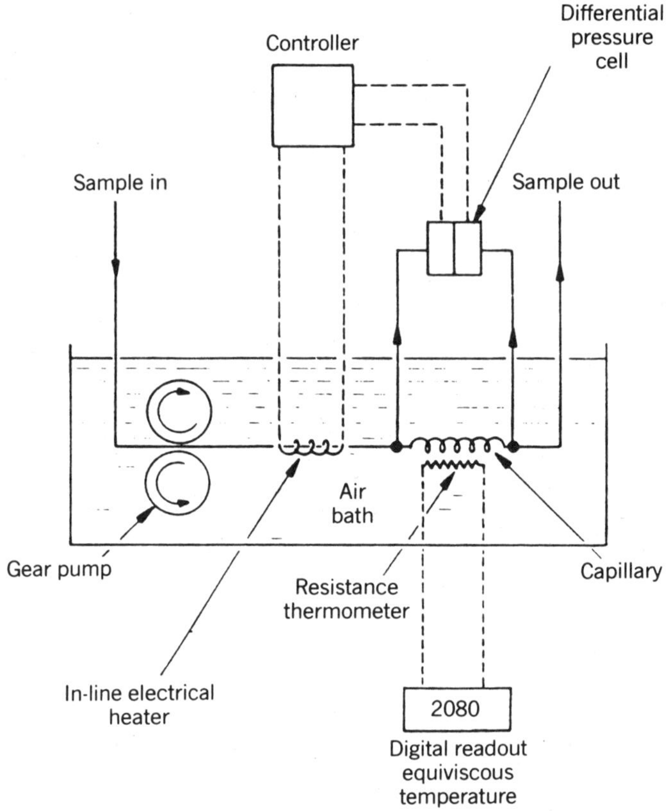

FIGURE 2.16 Equiviscous temperature (EVT) analyzer. (Courtesy of Servomex Ltd.)

Fig. 2.16, consists of an air bath controlled at about 120°C (maximum 150°C), a gear pump driven by a constant speed motor, a 3-mm bore measurement tube (the capillary), a differential pressure (DP) sensor, and an in-line heater between the pump and capillary. The DP sensor provides an input to a controller that controls the temperature of the in-line heater heating the sample, thus changing the viscosity until the differential pressure reaches a preset value.

The temperature at which the viscosity is equal to the preset value is called the *equiviscous temperature* (*EVT*). The sample temperature is measured by a platinum resistance element and displayed on the digital panel meter. A 4–20 mA dc output is also available for remote recording. Three temperature ranges are available: 150–250°C, 0–300°C, and 50–150°C. Repeatability is quoted as 0.5°C for the 150–250°C range. Sample flowrate, as set by the gear pump and motor, is 110 ml/min nominal. The viscosity measuring capillary range is 10–65 or 10–145 cP and viscosity set point adjustment is selected by a two-position toggle switch (10–35 and 35–65 cP or 10–65 and 65–145 cP respectively).

Two versions of the analyzer are available. The Type 397A is suitable for installation in Zone 1 hazardous locations, the Type 398A is suitable for safe area installation only. Two types of sample head can be provided. The Type 396A is for manual loaded samples only. The Type 395A is designed to enable the operator to switch from manual loaded to continuous automatic on-line sampling and vice versa.

2.7.2 Vibrational-Type Viscometers

(a) Automation Products Inc. (United States)

The Dynatrol 900 Series Viscometers are designed to measure the viscosity of process liquids within the range 1–100,000 cP. The analyzer consists of two parts, a probe unit immersed in the process liquid and a converter that produces a 4–20 mA dc signal proportional to viscosity. The probe unit is shown in Fig. 2.17.

In the detector part of the probe unit, mechanical vibration of the driver armature at 120 Hz, via drive coil excitation, is transferred along the attached spring rod, through a welded node point, to the probe itself. The amplitude of vibration is dependent upon the viscosity of the process liquid. Probe vibration is transferred via a second welded node point to the pickup end of the detector. Vibration of the pickup armature in the field of a permanent magnet induces a 120 Hz ac voltage in the pickup coil. The output from the detector feeds a solid-state converter, which produces a 4–20 mA dc output for remote recording. The converter also provides regulated power to the detector, and temperature compensation circuitry can be supplied as an option. Span and zero controls are provided on the front panel. Various configurations are available depending upon the need for temperature compensation and the required measurement range. The converter is available

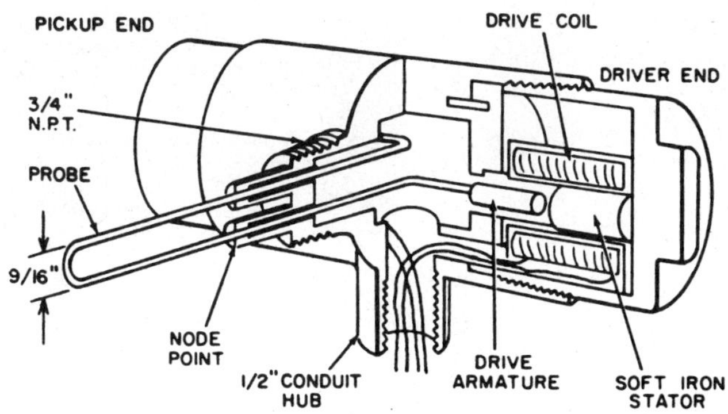

FIGURE 2.17 Dynatrol 900 Series viscometer. (Courtesy of Automation Products Inc.)

either as a general purpose panel-mounted device or suitable for installation in NEC Class 1, Group D, Division 2 hazardous locations.

(b) CE Process Analytics (United States)

The Ultraviscoson Model 1800 Viscometer is designed to accurately measure the viscosity of a process liquid by deriving the mathematical product of viscosity and density. Measurement range is 0–50,000 cP $\times$ gm/ml, with attenuation factors at $\times 1$, $\times 10$, $\times 100$, and $\times 1000$.

The viscometer shown in Fig. 2.18 consists of a probe, which is inserted into the process liquid to be measured, and an electronics unit that is essentially a small computer. The probe acts as a transducer, measuring the dampening effect or "viscous drag" of the liquid. The actual sensing element is a thin strip of metal, one end of which is a magnetostrictive alloy, the other end (exposed to the process liquid) is stainless steel. The electronics unit produces a short pulse of current to a coil inside the probe and around the thin blade sensing element. The blade is made to vibrate at its natural frequency, determined by the length of the strip. The portion of the blade emerging from the barrel is $\frac{1}{4}$ wavelength at 28 Hz.

FIGURE 2.18 Ultraviscoson Model 1800 viscometer. (Courtesy of Combustion Eng., Process Analytics)

Blade motion varies from almost zero at the diaphragm entrance to maximum amplitude at the blade tip. Vibration is longitudinal and less than 0.25 microns. Once the electrical drive pulse sets the blade in motion, no other pulse will be generated until the vibration amplitude is dampened to a predetermined level. The viscosity of the process liquid will determine when a drive pulse is to be generated; the higher the viscosity, the greater the dampening, resulting in a more frequent drive pulse being generated.

Probes are available to accommodate a wide variety of liquids at temperatures up to 200°C and pressures up to 750 psig. Special designs are offered for particular applications and for various types of fittings and mounting arrangements. All probes have a replaceable tip design. Zero and span controls allow each probe to be calibrated to a liquid of known viscosity.

The electronics unit is suitable for installation in non-hazardous locations only, although purge facilities can be provided. The sensors are explosion-proof for NEC Class 1, Group D, Division 1 locations. Connection is by a single conductor coaxial cable up to 500 ft long. The reference temperature compensator is capable of providing compensation for variations in sample temperature of ±25°F.

(c) Nametre Company (United States)

Nametre manufactures a range of laboratory and in-line process viscometers that operate on the oscillating sphere principle. The sensing element is a 1.26-in. diameter stainless steel sphere immersed in the sample liquid. Viscosity is determined by measuring the power required to maintain torsional vibrations of the sphere at a constant amplitude. A block diagram of the operating system is shown in Fig. 2.19.

A variable-gain amplifier drives a magnetic coil that causes the spherical torsional member to oscillate at one of its natural resonant frequencies. The amplitude of the oscillations is sensed as an ac voltage by an amplitude monitoring circuit. The voltage is rectified and referenced to a dc voltage equivalent to that obtained when the sphere oscillates in air. The resulting error signal is used to control the gain of the amplifier in order to maintain the amplitude of mechanical oscillation precisely at 25 microns over a wide range of viscous loss. The output signal from the transducer, which is proportional to the product viscosity × density, is obtained from the voltage developed across a resistor in series with the magnetic coil. This signal is relatively insensitive to electrical noise and to small variations in the electrical parameters of the measurement system. The signal is rectified and filtered and the dc signal is squared to provide an output that varies linearly with viscosity.

The viscometer has provisions for entering density values so that the viscosity may be displayed in absolute units (cP), kinematic units (cSt), or even in Saybolt units (by using a multiplying factor with the density value). The dynamic range of the instrument is 0.1–100,000 cP × g/cm^3, with six

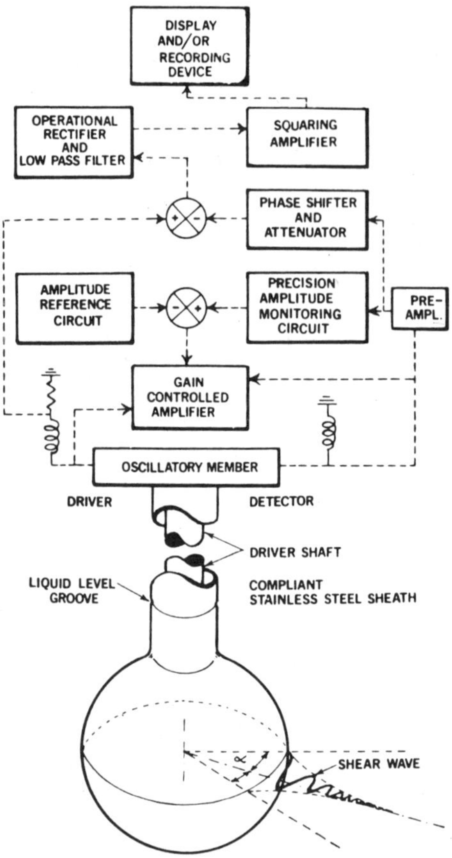

FIGURE 2.19 Nametre oscillating sphere viscometers: principle. (Courtesy of the Nametre Company)

linear decades of 0.1–1, 1–10, 10–100, 100–1000, 1000–10,000, and 10,000–100,000 cP, selectable by pushbutton range switches. Overrange is 90% of full scale of selected range. The response time to 99% of full scale is 5 seconds for the 0.1–1 and 1–10 ranges and 2 seconds for higher ranges. Voltage output is standard, but current outputs (1–5, 4–20 or 10–50 mA dc) are available as well as BCD output. The electronics unit is designed for installation in non-hazardous locations only and may be mounted up to 100 ft from the sensing unit.

The Model 7.006C4P in-line viscometer shown in Fig. 2.20 consists of two component parts, the sensing unit and the electronics unit. The 316 stainless steel hermetically sealed sensor is an integral part of a 6-in., 316 stainless steel mating flange as shown. This flange is available in 150, 300, 600, 900, and 1500 lb rating. The instrument is also available with a temperature probe to allow the electronics unit to display sample temperature and viscosity. Viscosity sensors other than the spherical type are also available. For exam-

FIGURE 2.20 Model 7.006C4P in-line viscometer. (Courtesy of Nametre Company)

ple, thin cylinders allow viscosity measurement to hundreds of millions of centipoise. However, the basic operating principle remains the same.

The transducer assembly has not yet been certified as an explosion-proof unit by FM or UL. However, it is being used in hazardous locations by suitable installation with Fisher Controls AC-305 Safety Barriers.

2.7.3 Rotational-Type Viscometers

(a) Brookfield Engineering Laboratories Inc. (United States)

Brookfield manufactures a wide range of rotational viscometers for both laboratory and on-line process use. The well-known Synchro-Lectric viscometers have been used for many years as standards for viscosity measurement in the laboratory.

The Viscosel Model VTA120 Rotational Viscometer is designed to measure the viscosity of process liquids over the range 0–2000 cP and senses the viscous drag imposed on a disc or cylinder rotated at constant speed through a torsional element. The instrument, shown in Fig. 2.21, incorporates the sensing, readout, and control in one package. The system is pneumatic in principle and designed to operate on a 20 psig clean air supply. The rotational element is driven by a constant-speed air motor and the torque of the

FIGURE 2.21 Model VTA120 rotational viscometer. (Courtesy of Viscosel Corporation)

element is converted to a pneumatic signal, which is transmitted to an indicator controller. A 3–15 psig output is also provided.

The sample chamber, which is available in 316 stainless steel or plastic, has a design maximum flowrate of 2 gal/min. Correct operation of the measurement system can be verified by carrying out three checks. First, with the spindle rotating in air, a zero reading should be obtained. Second, exerting a momentary drag on the spindle, the indicator should respond with an upscale swing, returning to zero when the drag is removed. Third, with the spindle stalled, a full-scale reading should be obtained. A special friction clutch within the instrument protects the measuring system from damage due to overloading.

In view of its fully pneumatic design, the instrument is suitable for operation in hazardous locations without additional protection. Three measuring elements are available—a disc cone, a cylindrical cone, and a hollow cylindrical element. Together with various spindle extensions, they allow viscosity measurement over a wide variety of ranges depending on the particular application. The instrument is not equipped with temperature compensation. Therefore, measurement is made at the sample temperature or some form of sample temperature control is necessary. Repeatability is quoted as ±1% of full scale.

The Model TT100 In-line Process Viscometer, the sensing unit of which is shown in Fig. 2.22, is designed for use in fully flooded systems under pressure or vacuum. In the sensing unit, viscous drag causes a linear and very small angular deflection measured by an electromagnetic signal generator—a precision rotary differential transformer (Microsyn). This well-proven device was developed during World War II and is used extensively in gyros for inertial navigation in planes and missiles, as a signal source for precision pressure and DP transmitters, and in digital vacuum/pressure calibrators.

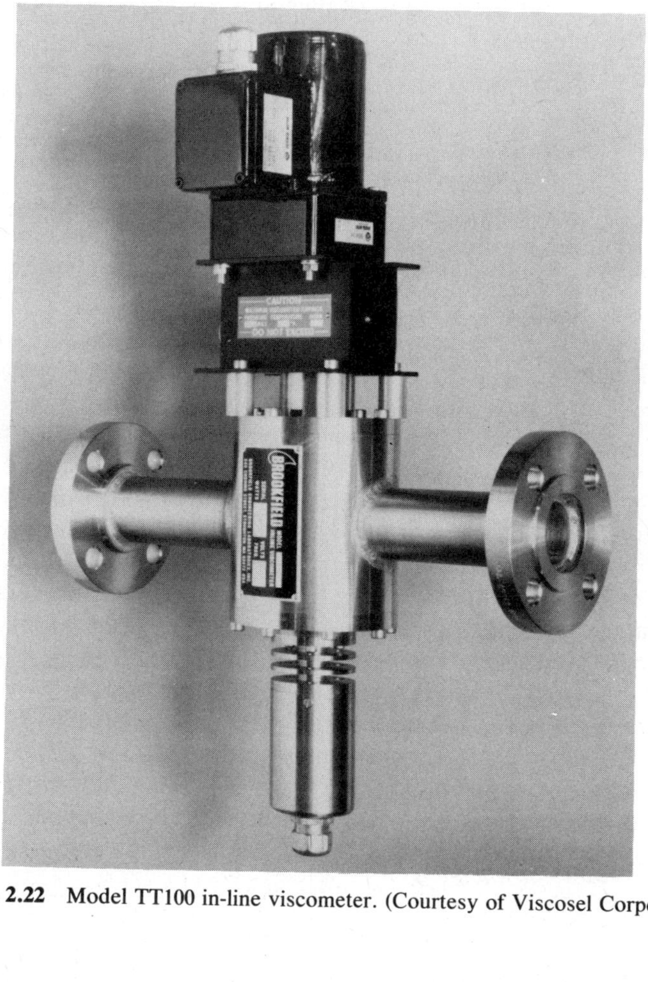

FIGURE 2.22 Model TT100 in-line viscometer. (Courtesy of Viscosel Corporation)

The analyzer electronics unit provides square wave excitation to the Microsyn primary and demodulates the secondary output to a dc voltage or 4–20 mA dc signal, which is a linear function of sample viscosity.

Instrument range is 10–50,000 cP, with a repeatability of ±0.5% of span. Shear rate range is 10–1000 sec^{-1}. The standard sensing unit is constructed from 18-8 stainless steel and is suitable for sample pressures up to 200 psig and temperatures from −40 to 200°F. A unit for pressures up to 500 psig and temperatures up to 500°F is also available as an option. Sample flowrate should be within the range 0.5 to 20 gal/min, depending upon sample viscosity. Screwed or flanged sample connections can be provided. FM-approved explosion-proof construction is available for the sensing unit, suitable for installation in NEC Class 1, Groups C and D, Divisions 1 and 2 hazardous locations.

The complete instrument consists of three component parts—the sensing unit just described, the transducer electronics, and the motor capacitor/switch circuitry, both of which are mounted in individual Hoffman-type NEMA 1 or 4 enclosures. The electronics unit is connected to the viscometer by a 10-ft cable.

(b) Contraves AG (Switzerland)

Tekmar Company is the U.S. representative for Contraves AG (Switzerland). Apart from the wide range of laboratory rheometers described earlier, Contraves also manufactures a range of on-line process viscometers operating on the rotational principle. These viscometers consist of a measuring cell, torque measuring head, and remote control unit. During 1984, Contraves introduced a new fully encapsulated, explosion-proof torque head enclosure and a separate control unit, which are used with a variety of different measuring cells to meet specific customer application requirements. The control unit contains the rotational speed selector. Five rotational speeds are now available instead of the three available with the older units. The five speeds are distributed over a wider range, such that one measuring bob covers a measuring range in a ratio of 200 : 1 instead of 100 : 1 in the older units. The control unit also incorporates on/off switches for power supply and motor drive and can provide a current (0–20 or 4–20 mA dc) output for remote recording. This updated range of viscometers is known as the Covimat range, the Covimat 101 being single-speed and the Covimat 105 being five-speed units. The overall measuring range of the series is 1 to 10^7 cP.

The Type TO Viscometer has a measuring cell designed for direct immersion in an open vessel at atmospheric pressure. The immersion depth (flange of casing to lower end of measuring cell) may be 200, 300, or 500 mm, with the proviso that the upper surface of the measuring bob be at least 20 mm below the liquid surface. The measuring cell is rated for 65°C maximum.

The Type DD Viscometer has a flow-through measuring cell suitable for maximum operating conditions of 140 psig, 90°C, and 5 l/min flowrate. This unit has been used successfully in gelatin production and the control of paint viscosity for robot sprayers.

The Type DC40-D Viscometer is suitable for sample pressures up to 1300 psig and temperatures up to 300°C. The measuring cell is of double-walled construction, incorporating a heating jacket, and has flanged connections for installation directly in piping systems. Sample temperature is measured by a platinum resistance thermometer. This instrument has been used successfully in the plastics fiber industry for viscosity control prior to spinning.

The Type DC40-E Viscometer is suitable for wide-range viscosity applications and is used in the chemical, pharmaceutical, and food industries (e.g., viscosity of cake icing, tomato sauce, and tomato paste). The measuring cell is suitable for maximum sample conditions of 355 psig, 250°C, and flowrates up to 40 l/min. A typical example, showing the new explosion-proof torque head enclosure and separate control unit, is illustrated in Fig. 2.23.

The Type DC43 Viscometer, shown schematically in Fig. 2.24, has a measuring cell designed for maximum operating conditions of 1770 psig, 350°C, and flowrates up to 40 l/min. It has been used successfully in plastics polymerization processes.

The Type DC44 Viscometer is similar in design to the DC43, except that the measuring cell is suitable for maximum operating conditions of 2350 psig and 350°C.

The Covicontrol Viscosity Control System, shown in Fig. 2.25, is designed for viscosity control of paints, colors, varnishes, glues, and a wide

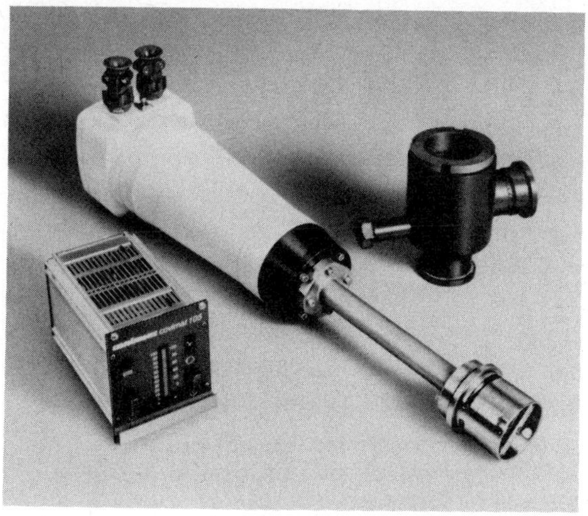

FIGURE 2.23 Model DC-40E viscometer. (Courtesy of Tekmar Company)

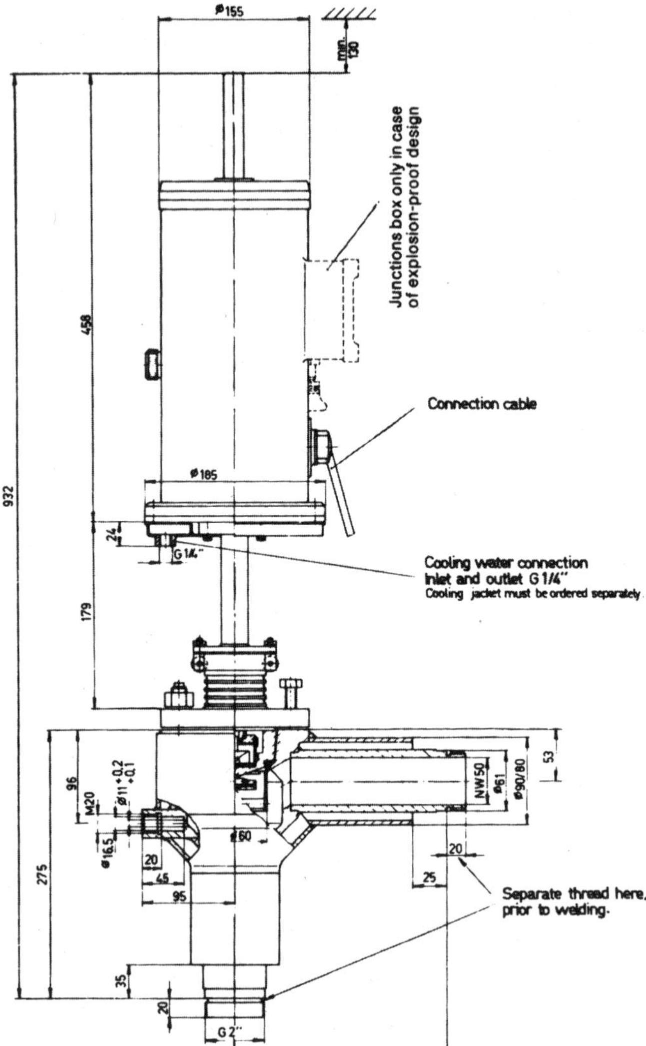

FIGURE 2.24 Contraves Model DC-43 viscometer. (Courtesy of Tekmar Company)

variety of coating materials. This unit has not been updated, since the torque head enclosure is already fully encapsulated. The available measuring bob series covers a measuring range from 1 to 1650 cP, utilizing immersion type or flow-through measuring cells (three different types are available). The controller is a double acting proportional device, with a dead zone adjustable from ±0.8% to ±3.2% of set value. Also incorporated are adjustable feed delay time and upper and lower alarm limits.

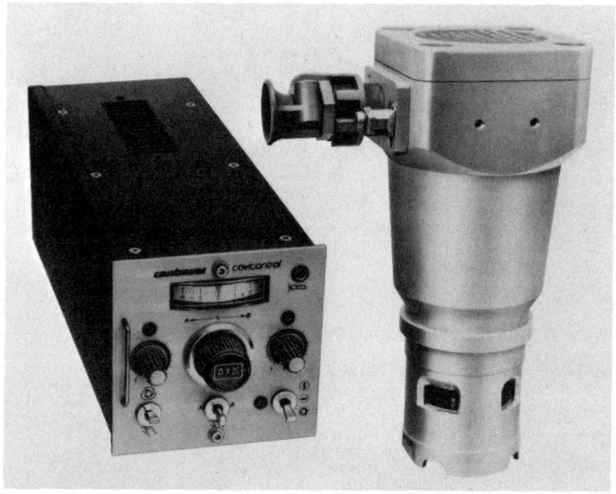

FIGURE 2.25 Covicontrol viscosity control system. (Courtesy of Tekmar Company)

(c) Eurcontrol Kalle AB (Sweden)

The Model VISC 21.5 In-line Viscosity Transmitter is available with either electronic (E) or pneumatic (P) transmission. The analysis unit consists of a flow-through sample chamber, drive motor, and assembly (with Crane-type drive shaft mechanical seal) and the transducer assembly. The measuring system consists of a rotating disc and stationary disc. The rotating disc is driven at a constant speed by an electric motor via a drive shaft. The disc has radial slots to ensure that up-to-date sample is continuously drawn into the space between the discs. The drive shaft length is adjustable to allow range change (variation of gap between the discs).

The torque exerted on the stationary disc is proportional to the fluid viscosity and is transmitted through the measuring shaft to the position sensing system. In the case of the pneumatic unit, this is a force-balance flapper-nozzle system, which provides a 3–15 psig output for remote recording, proportional to viscosity, and produces a feedback signal to create an equal and opposite torque that balances the measuring torque.

In the electronic viscosity transmitter shown in Fig. 2.26 the torque on the stationary disc is transmitted through the measuring shaft to a lever system (force-balance principle). The position of a magnet attached to the end of the lever system is sensed by a pick-up head and associated electronics. A proportional voltage is transmitted to a remote electronics unit, which produces a feedback current to the feedback coils located in the transducer assembly. This creates a torque equal and opposite to the measuring torque that restores the system to balance. A 0–20 or 4–20 mA dc output, proportional to viscosity, is provided for remote recording.

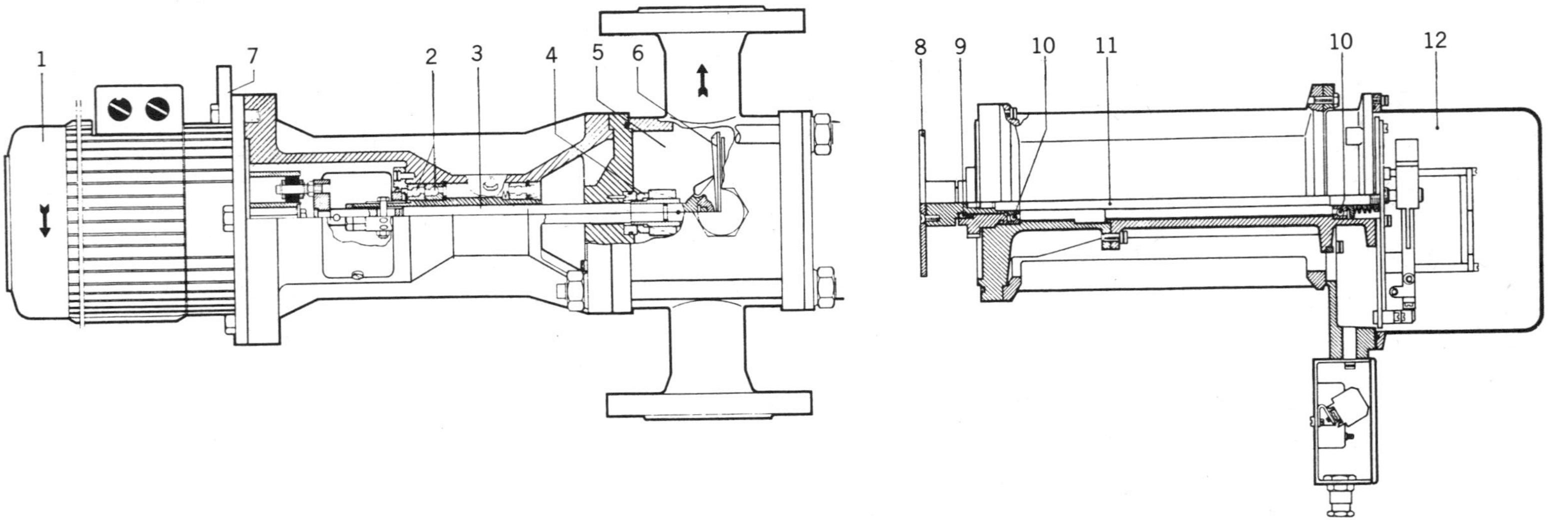

FIGURE 2.26 Model VISC 21.5 in-line viscometer. KEY: 1, electric motor; 2, angular contact ball bearings; 3, drive shaft; 4, mechanical seal; 5, measuring vessel; 6, rotating disc; 7, support bracket; 8, measuring disc (stationary); 9, O-ring seal; 10, angular contact ball bearing; 11, measuring shaft; 12, transducer cover. (Courtesy of Eurcontrol Kalle)

The measuring chamber is constructed of stainless steel as standard and is available with 2, 3, or 6-in. flanges for installation in pipe systems. Maximum sample flowrate is dependent upon chamber size:

2 in.	3700 USG/hr
3 in.	9510 USG/hr
6 in.	33,290 USG/hr

The rotating disc system is self-cleaning and relatively insensitive to particulates and to pressure and flow changes. Sample pressures up to 1200 psig and temperatures up to 330°F can be accommodated.

The pneumatic transmitter has a dynamic viscosity range of 1500 cP maximum, with a minimum span of 25 cP. The electronic transmitter has a range of 700 cP maximum, with a minimum span of 25 cP.

(d) C.W. Brabender Instruments Inc. (United States)

The Convimeter In-line Viscometer shown in Fig. 2.27 is designed for on-line continuous measurement (and control, if required) of liquids or pastes over a wide viscosity range. The viscometer is suitable for viscosity monitoring in pipelines, tanks, or mixing vessels and consists of two parts, the analysis unit and a remote mounted control unit.

The sensing device is immersed in the process stream and rotates at a constant speed in gyratory motion, imparted by a drive motor. This motion is obtained by a bend in the lower part of the drive shaft, ending inside the

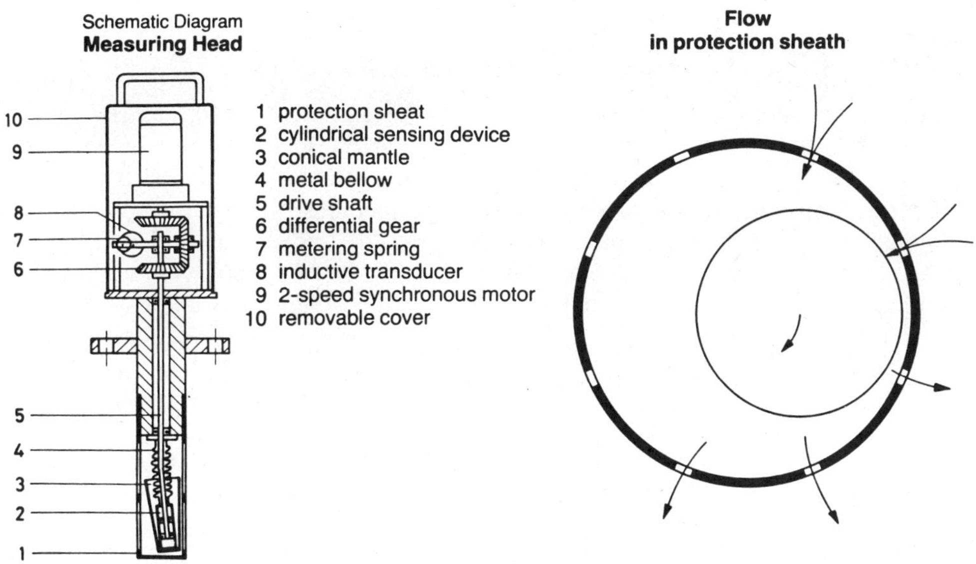

FIGURE 2.27 Convimeter in-line viscometer. (Courtesy of Brabender Instruments Inc.)

sensing device, as shown in Fig. 2.27. A flexible metal bellows surrounding the drive shaft provides a seal that is frictionless and insensitive to pressure and temperature changes. Torque measurement is by a differential gearbox with helical metering spring and inductive transducer. The drive motor can be switched to either of two speeds from the control unit. This, together with various designs of sensor and two helical metering springs (Types II and III), allows the instrument to measure a wide range of viscosities from approximately 5 to 10^7 cP.

The measuring head is available in explosion-proof construction, with PTB approval, and is suitable for sample pressures up to 150 psig and temperatures up to 300°C. A special version is also available for pressures up to 750 psig. Temperature compensation is available as an option, utilizing a resistance element. For detection of low viscosities, the measuring head type L is used, with a minimum range of 0–35 cP. By modifying the drive shaft and sensing device, the Convimeter NW50 can be adapted to measure very high viscosities. Two types are available, the H1 (highest range 0–10^6 cP) and H2 (two ranges, 0–3 $\times$ 10^6 cP and 0–1.5 $\times$ 10^6 cP). The standard measuring head type NW80 is suitable for viscosity ranges from 150–250,000 cP.

The control unit is suitable for rack or panel mounting in non-hazardous locations only. Controls are provided for motor speed, zero, sensitivity, reference temperature (adjustable from 0–300°C), and temperature coefficient. Either analog or digital display of viscosity can be provided at the control unit. A current output (0–20 or 4–20 mA dc) is provided for remote recording. Measurement repeatability is quoted as $\pm 1\%$ of full scale.

2.7.4 Falling Piston Type Viscometers

(a) Norcross Corporation (United States)

The wide range of in-line, falling-piston viscometers manufactured by Norcross have been used extensively in a variety of industries for about thirty years. The principle of measurement is an adaptation of Stokes' method, described earlier. The product line includes instruments for open tanks or vessels operating at atmospheric pressure, closed vessels operating above or below atmospheric pressure and for in-line or sidestream applications. They are available as transmitters in conjunction with indicators or recorders, with or without control, and/or high and low alarm facilities.

The basic sensing element is a piston and tube assembly, shown schematically in Fig. 2.28, which is immersed in the process liquid. The piston is periodically raised by a lifting mechanism, an action that draws sample liquid in through the tube openings. The liquid then passes down through the clearance between the piston and the cylinder bore and into the space formed below the piston when it is lifted. The piston is then allowed to fall

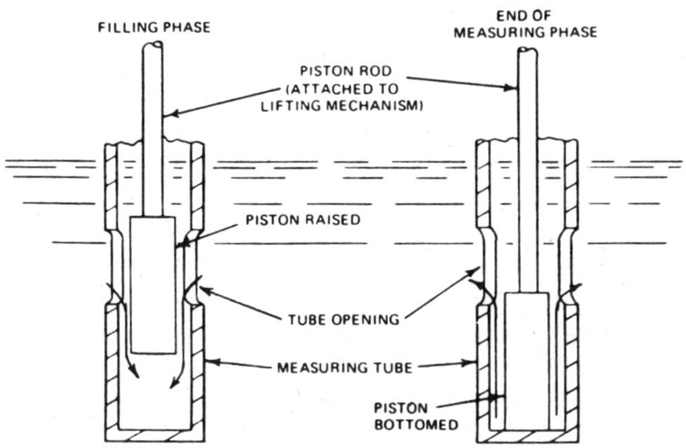

FIGURE 2.28 Norcross falling piston viscometers: principle. (Courtesy of Norcross Corporation)

under gravity and the sample is expelled through the same path by which it entered the tube. The time taken for the piston to fall from its start position to a predetermined level is a measure of the liquid viscosity, with the clearance between the piston and the tube bore forming the measuring orifice.

Time measurements are transmitted as an electrical signal to a remote recorder or controller. The flushing action created by the up-and-down motion of the piston provides effective self-cleaning during instrument operation. All wetted parts are stainless steel (303 and 304 are standard).

Models M8, M8BO, M10, M15, and M16 are designed for applications where the instrument is to be mounted directly on a vessel. The piston lifting mechanism is a pneumatically operated cylinder. In some models, a two-way air cylinder is fitted and air is used to drive the piston in the downward direction as well as to lift it.

Models M8 and M8BO are designed for atmospheric pressure applications and are suitable for sample temperatures up to 700°F. The viscometer has an overall range of $0.1–10^5$ cP and may be supplied with any intermediate range to suit customer requirements.

The Model M10 is designed for top of vessel (flange) mounting in applications from full vacuum up to 300 psig, with measuring tube length from 1–6 ft and temperatures up to 700°F. Overall viscosity range is $0.1–10^6$ cP.

The Model M15 is for small atmospheric vessels from 4 in. to 12 in. deep and temperatures up to 250°F.

The Model M20 shown in Fig. 2.29 is designed for flange mounting in any direction on the vessel, for pressures up to 600 psig and temperatures up to 500°F, with measuring tube 7–14 in. long. Overall viscosity range is $1000–10^6$ cP. Model M20 may also be used as an in-line device.

The Models M16, M20, M24, M28, M30, M31, and M32 are in-line visco-

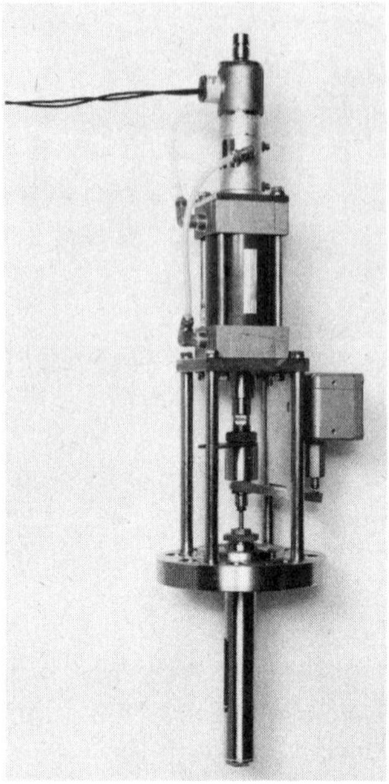

FIGURE 2.29 Model M20 viscometer. (Courtesy of Norcross Corporation)

meters, designed for installation directly in piping systems or in a bypass. Viscosity ranges, together with pressure and temperature limitations are as follows:

Model	Viscosity Range (cP)	Temperature (°F)	Pressure (psig)
M16	$0.1–10^4$	400	150
M24	0.1–9000	500	600
M28	$0.1–10^5$	1000	7000
M30	0.1–9000	400	1500
M31	0.1–9000	500	1500
M32	0.1–9000	500	1500

The Model 36 In-Line Viscometer, shown schematically in Fig. 2.30, is designed for process applications at pressures up to 500 psig and temperatures up to 300°F. Interchangeable pistons are used for measuring viscosities up to 6500 cP; a single piston covers a 10 : 1 range. The analyzer consists of the measuring unit and a remote mounted receiver/recorder.

Sample flow through the measuring unit is controlled by a solenoid valve, which is actuated by the receiver. When the valve is opened, the sample flow raises the piston. When the valve is closed, the sample flows around the bypass and the piston is allowed to fall under gravity, the time of fall being a measure of the sample viscosity. The time starts when the valve is closed

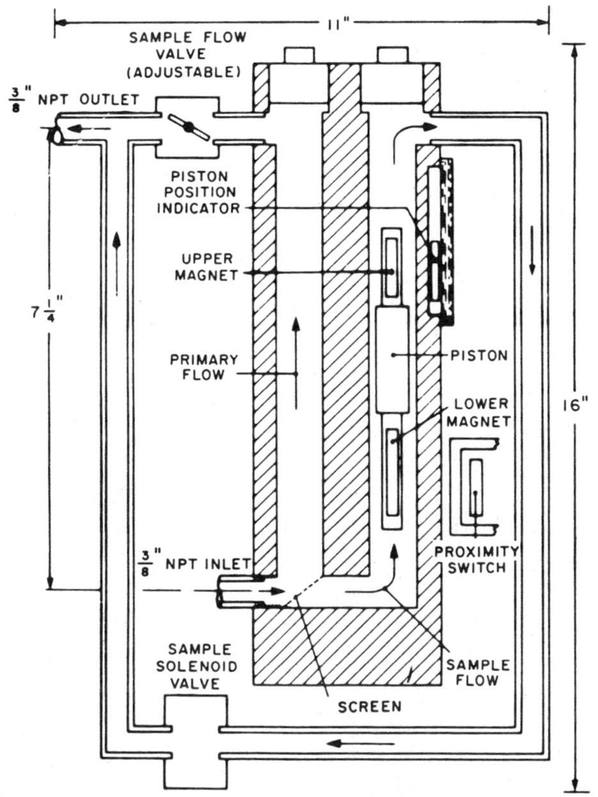

FIGURE 2.30 Model M36 in-line viscometer. (Courtesy of Norcross Corporation)

and ends when the lower magnet in the piston assembly actuates the proximity switch. The sample flowrate is adjusted (within the limits 1–10 gal/min) so that the piston can be raised in about 3 sec. A piston position indicator is provided to facilitate this adjustment. The measuring unit incorporates a self-cleaning screen, located as shown in Fig. 2.30. Foreign material collected on this screen is washed away by the bypass sample flow when the solenoid valve is closed.

The proximity switch enclosure and solenoid valve are UL-approved for NEC Class 1, Group D, Division 1 hazardous locations. Wetted parts are of aluminum, brass, and stainless steel. All stainless steel construction is also available.

Norcross also manufactures a wide variety of receiver instruments for use with its range of falling-piston viscometers. Models RL, RS, and IS are the standard recorders and indicators. Various options are available, including on-off electric control, high and low alarms, pneumatic transmitter (3–15 psig), electronic transmitter (4–20 mA dc), and (for the RL Series recorder) a temperature pen.

The latest addition to the Norcross product line is the MP Series Microprocessor-Based Digital Viscosity Receiver. The functions of the receiver include control of the cyclic operation of the piston assembly, digital display of viscosity, control action, and diagnostic verification of system operation. Electronic (4–20 mA dc) or pneumatic (3–15 psig) outputs are available for remote recording. An RS232 serial port is also available for transmission of viscosity data to a central computer. Temperature compensation facilities can also be provided.

2.7.5 Special-Purpose Instruments

The needs of the chemical process industries (CPI) has generated many special-purpose viscometers, both for laboratory research and for on-line process use. Perhaps the most unique and complex are those encountered in the thermoplastics industry for measurement of the viscosity (or related parameter) of highly viscous, non-Newtonian polymeric materials.

As with all non-Newtonian fluids, the melt viscosity of a polymer is not constant at any given temperature, but varies greatly with shear rate and shear stress. Therefore it is desirable to measure the flow characteristics at both low and high shear rates. Three laboratory devices are used for the measurement of polymer melt viscosity:

1. The ASTM Extrusion Plastometer, designed by DuPont and suitable for measurement at low shear rates.
2. The Canadian Industries Ltd. (CIL) high shear viscometer.
3. The Instron Capillary Rheometer, a constant shear rate device designed by Monsanto.

The ASTM and CIL devices measure the flowrate of molten polymer passing through an orifice of specified length and diameter under prescribed conditions of temperature and pressure. The Instron device uses a tensile testing machine to drive a piston at constant speed in a temperature controlled reservoir, forcing the polymer sample through a capillary tube.

The majority of early on-stream viscosity instruments were designed to correlate some end effect with the ASTM Standard Melt Index Test (D1703-62), with its varied test conditions. This test maintains a constant shear stress (pressure) and measures the rate of flow through an orifice of specified L/D ratio. Melt Index is defined as the number of grams extruded through this device in 10 min.

(a) Seismograph Service Corporation (United States)

The Seiscor Division of SSC has specialized in the development and manufacture of capillary rheometers for the thermoplastics industry for about twenty years. The latest Seiscor analyzers and accessories are described here.

The AK-2000 Series Continuous Melt Rheometers are designed for on-line monitoring of the melt viscosity of thermoplastic polymers. These instruments provide direct readout of ASTM Melt Index, CIL Flow Index, or other required test parameter and do not require a special correlation to agree with laboratory methods.

The instrument consists of two component parts, the control chassis, which may be panel mounted or supplied as a console, and the die sampling unit. The control unit contains temperature control and calibration systems. The die sampling unit contains the flow-producing and measuring, pressure-sensing, and temperature-stabilizing equipment, and is suitable for operating pressures up to 10,000 psig. Single or dual cavity systems are available, the latter providing capability for automatic and alternate measurement at high and low shear rates, with variable time adjustment. The die sampling unit is designed to operate in conjunction with the polymer processing equipment and may be mounted directly onto the process extruder.

A schematic diagram of the die sampling unit is shown in Fig. 2.31. Polymer melt from the process extruder passes to the unit by extruder pressure and enters a precision metering pump, which has a positive displacement of 0.297 ml/rev and is driven by a variable speed motor. The polymer flows through a passage with sufficient residence time for it to stabilize at the test temperature. It then flows, via a selector valve, into one of two cavities containing a pressure transducer and finally through the capillary. The pressure transducer signal is the input to a controller, the output of which controls the speed of the dc motor driving the metering pump. Polymer is thus metered at the rate necessary to maintain the specified test pressure at the capillary. The output of the tachometer on the metering pump is linearly proportional to this polymer metering rate and is recorded as either the ASTM Melt Index or the CIL Flow Index.

By virtue of the die sampling unit design, actual process operating conditions do not affect the test conditions established for the analysis. The main advantage of the Melt Rheometer is its ability to handle any two test conditions specified as product quality parameters, within the temperature range 190–300 ± 0.2°C and at pressures up to 10,000 psig. Readout is provided directly in laboratory units of measurement. Viscosity ranges from 400–1.8×10^8 cP may be accommodated. Measurement repeatability is quoted as $\pm 3\%$ of range.

The Model AK-4100 Four-Point Capillary Rheometer is designed for continuous on-line simulation of various capillary rheology and can cover up to four shear stress per shear rate conditions. The polymer melt from the process extruder is subjected to four shear rate conditions through a capillary and the resulting shear stresses recorded. From this information, a flow profile of the polymer may be derived. A schematic diagram of the rheometer is shown in Fig. 2.32. The instrument consists of a dual-cavity die sampling unit and a console-mounted control unit.

Rheometer operation is controlled at a high and low shear range. For each shear range, the polymer melt is diverted into separated cavities having the

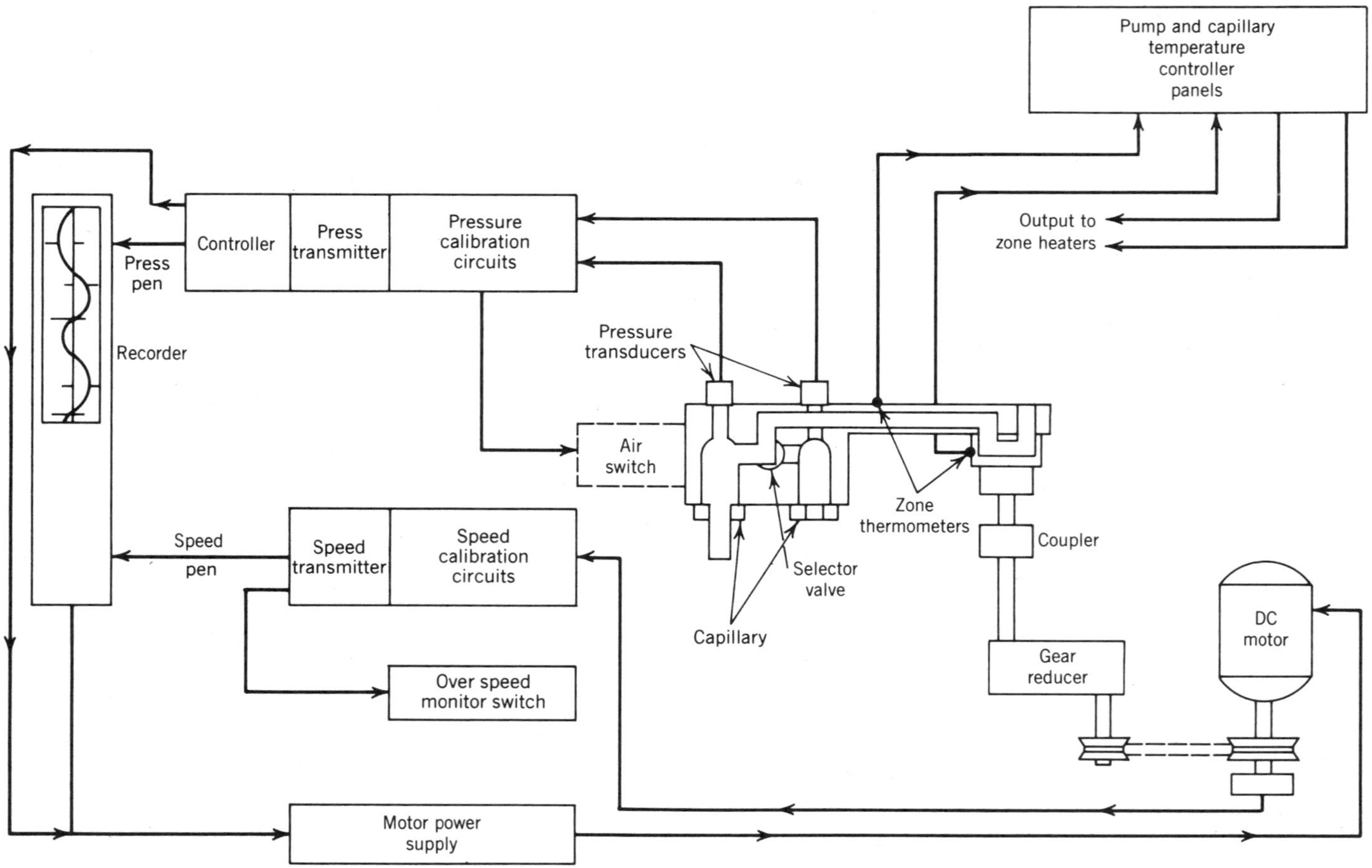

FIGURE 2.31 AK-2000 Series melt rheometer. (Courtesy of Seismograph Service Corporation)

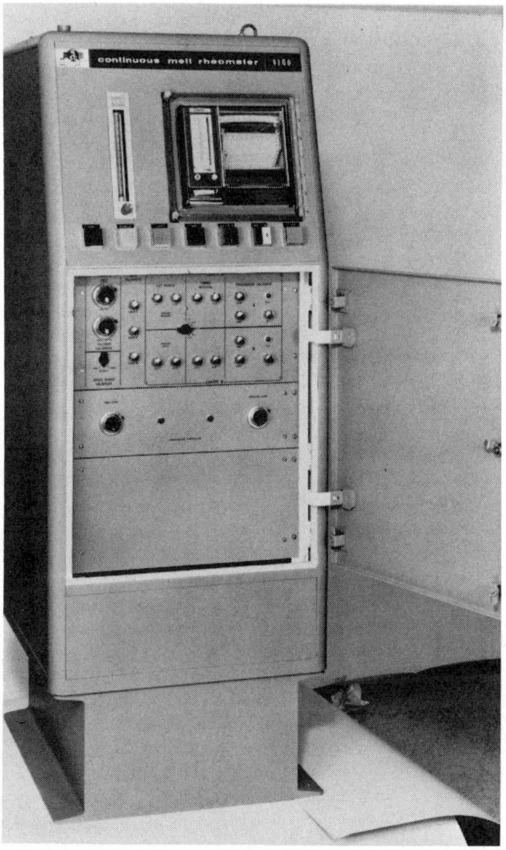

FIGURE 2.32 AK-4100 four-point rheometer. (Courtesy of Seismograph Service Corporation)

same size capillaries with different L/D ratios, but different pressure range transducers, corresponding to high and low ranges. Combined with the two speed ranges, this gives a total of four shear rate per shear stress points. Cycling between these four points is achieved automatically by a program timer circuit that selects two shear rate conditions for each shear range. Pump speeds and cavity pressures are recorded by a two-pen recorder/controller. The time for a complete analysis is 20 min, with 4 min for each shear rate condition.

The controller has a remote set point station, which is adjusted by the timer for the correct metering pump speed. The programmer also switches the speed and pressure calibration circuits, controls the three-way valve for diverting polymer flow, changes rotation of the dc motor, and provides an identification marker on the recorder chart at the start of the high and low shear rate condition.

Seiscor also manufactures two sampling systems for use with its rheometers. These were developed to allow the analysis point to be moved from the polymer extruder to an upstream point closer to the process reactor.

The Dry Particle Sampling System is designed to extract a sample of dry polymer, condition it to the molten state, and extrude it directly into the continuous rheometer for analysis.

The Polymer Solution Sampling System extracts a solution sample and supplies a volatiles-free melt to the rheometer. Design is based on the flash-drying principle.

REFERENCES

1. Cheng, D. C-H. *Meas. & Contr.* (March, Apr, May 1981).

2. Andrade, E. N. da C. *Viscosity and Plasticity*. Chemical Publ. Co., Columbus (1951).

BIBLIOGRAPHY

Contraves A. G. *Measurement of Rheological Properties,* Bull. No. 6704-685, Switzerland (from original paper by Heidiger, M.).

Dinsdale, A., and Moore, F. *Viscosity And Its Measurement*. Chapman & Hall, London (1962).

Ferry, J. D. *Viscoelastic Properties of Polymers,* 2nd Ed. Wiley, New York (1970).

Ferry, J. D. *Meas. & Contr.* (Sept 1977).

Ferry, J. D. Oscillation viscometry: Effects of shear rate and frequency. *Meas. & Contr.* (Sept–Oct 1977).

Hallikainen, K. E. Viscometry. *Insts. & Contr. Sys.* (Jan 1970).

Hammond, M. T. A new concept in viscometry. *Meas. & Contr.* (Aug 1977).

Holzmuller, W. Physik Der Kunstoffe. Academie-Verlag, Berlin (1961).

Leegwater, M. G. Effects of molecular weight distribution on HDPE container molding. *SPE J.,* **25** (Nov 1969).

Matusik, F. J., and Scarna, P. C. Latest instrument makes on-line viscosity control of slurries possible. *Contr. Eng.* (Dec 1981).

Matusik, F. J., Fitzgerald, J. V., and Oppliger, H. R. Oscillation viscometry. *Meas. & Data* (May–June 1975).

McKennell, R. The influence of viscometer design on non-Newtonian measurement. *Anal. Chem.* (Oct 1960).

Mill, C. C. *Rheology of Disperse Systems*. Pergamon, London (1962).

Odler, I., Becker, T., and Weiss, B. Rheological properties of cement pastes. *Il Cemento* (March 1978).

Reid, R. K., and McGinnis, F. H. Reactor sampling for continuous polymer viscosity analysis. *Trans. ISA,* **7,** No. 4 (1968).

Sherman, P. *Rheology of Emulsions*. Pergamon, London (1962).

Sherman, P. *Industrial Rheology*. Academic Press, New York (1970).

Van Wazer, J. R., Lyons, J. W., Kim, K. Y., and Colwell, R. E. *Viscosity and Flow Measurement*. Wiley, New York (1963).

Viscometer survey. *Insts. & Contr. Sys.* (Jan 1971).

Wehmeyer, K., Eakins, J., and Powers, J. Computer-controlled viscometer provides coating thickness control of approximately 30% at IBM. *Chem. Proc.* (June 1984).

Welty, R. O. An on-stream melt viscosity analyzer. *21st Ann. Tech. Conf. Plast. Eng.* (March 1965).

Whorlow, R. W. *Rheological Techniques.* Horwood, Chichester (1980).

Wohl, M. H. Instruments for viscometry. *Chem. Eng.* (March 1968).

3

Boiling Point Measurement

3.1 INTRODUCTION AND HISTORICAL BACKGROUND

Distillation forms the backbone of many industrial processes and is of particular importance in the petroleum, petrochemical, and chemical industries. The process of distillation, as used in some form for separation and purification, is very old and thought to date back to 1000 B.C. or even earlier. Some isolated recordings appear in the writings of early historians in the fourth century A.D. Distillation of whiskey is said to have originated in Ireland in the 12th century. The resulting product was called *Usquebagh* (meaning "water of life" in Gaelic), from which the word "whiskey" was derived. A commercial, continuous "still" was built by Aeneas Coffey in Dublin in 1832 for distillation of alcohol and was similar in characteristics to present-day units.

Although born in antiquity, the process of distillation had developed quite far by the mid-19th century, but the major advances came with the discovery of crude oil and the need to develop an efficient method of separating and purifying petroleum products. The first oil well was drilled by Colonel Drake in Pennsylvania in 1859; a year later the first refinery was built in Titusville. By 1873 hundreds of small refineries (and a few large ones) were built and running.

So the technique of distillation has developed over a period of centuries, but most of this development (in the theoretical aspects and adaptation to continuous operation) has taken place quite recently and continues with the introduction of new industrial processes.

Before proceeding to the subject of process on-line distillation, or boiling point analyzers, it is appropriate to first establish what is meant by "distillation," how the process is used in modern industrial processes, and how the process analyzer fits into this picture.

3.2 DEFINITIONS

Distillation may be defined as a process for the separation of a homogeneous liquid mixture, in which the liquid is partially vaporized and the vapor thus formed is subsequently condensed. The condensed vapor is enriched in the lower-boiling components of the mixture and the residual liquid in the higher boiling components. The process therefore makes use of the *difference in volatilities (or vapor pressures)* of the various components in the mixture. When the vapor and liquid phases are in equilibrium, the maximum relative difference in concentration of the components in the two phases occurs. Distillation is sometimes used loosely to describe certain processes where chemical reactions are involved, such as the "destructive distillation" of coal or wood.

There are many types of distillation processes in use today. These may be classified according to the number of components to be separated:

Binary (two components)	Ternary (three components)
Multi-component	Complex

In complex distillation, individual components cannot be classified as pure compounds. Distillation processes may be classified according to the type of separation, namely equilibrium flash vaporization or fractionation.

They may also be classed as batch or continuous and as pressure, vacuum, or steam distillation. One may also use various combinations, depending upon the particular process requirements.

3.3 DISTILLATION PROCESSES

3.3.1 Equilibrium Flash Vaporization

The equilibrium flash vaporization (EFV) process is simple in operation and easy to understand. As shown in Fig. 3.1, residue and vapor are continuously withdrawn from the boiling vessel. It is assumed that all components exist in both liquid and vapor phases so that an equilibrium is established. It is also assumed that the temperature and pressure of both phases are the same while they are in contact. In actual practice, true equilibrium is never reached, although it can be closely approximated.

Assume a feed rate F, a vapor rate V, and a liquid rate L at equilibrium conditions, and consider a more volatile component i, where

Z_i is the mole fraction of i in the feed

X_i is the mole fraction of i in the liquid

Y_i is the mole fraction of i in the vapor

For overall material balance: $F = V + L$. For material balance of the more volatile component,

$$Z_i \cdot F = Y_i \cdot V + X_i \cdot L$$

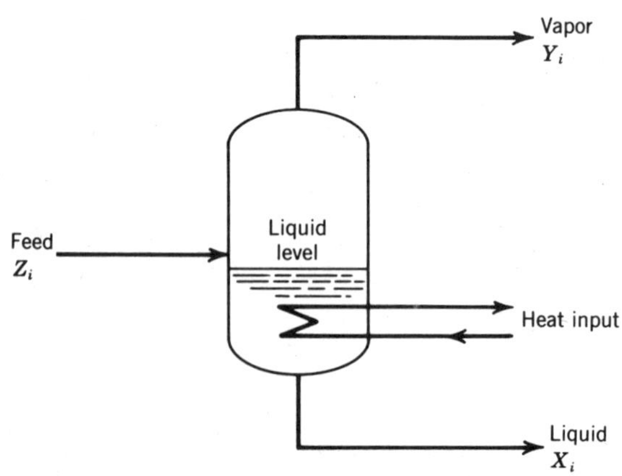

FIGURE 3.1 The equilibrium flash vaporization process.

Combining and rearranging, we get

$$Z_i(V + L) = Y_i \cdot V + X_i \cdot L$$

and therefore,

$$\frac{L}{V} = \frac{Y_i - Z_i}{Z_i - X_i}$$

and

$$\frac{V}{F} = \frac{Z_i - X_i}{Y_i - X_i}$$

where L/V is the liquid vapor ratio and V/F is the fraction vaporized at operating temperature T_o and operating pressure P. These are two important measurement parameters in the process, which forms the basis of all flash calculations carried out by chemical engineers involved in process equipment design.

The separation of the feed components obtained by the EFV process is somewhat limited; therefore it can only be used for separations under certain conditions.

3.3.2 Continuous Fractional Distillation

In most modern industrial processes, a mixture is required to be separated into products that meet certain specifications based, for example, on purity and boiling point. The product may be sold directly to a customer or used as feedstock to a subsequent refining process. In most cases a fairly tight specification that is not possible using the EFV process described must be met. Such considerations led to the process known as continuous fractional distillation.

The equilibrium in a system that consists of several phases is governed by the Phase Rule of Willard Gibbs, published in 1878 and expressed as

$$P + F = C + 2$$

where P is the number of phases, F the number of degrees of freedom, and C the number of components. The number of "degrees of freedom" is the "variability" of the system or the number of independent ways in which the temperature, pressure, and composition may be varied.

For a pure substance, with liquid and vapor in equilibrium, $F = 1$; therefore, if the pressure is specified, the temperature is automatically fixed. In other words, pure liquids vaporize at one fixed temperature, when the pressure is specified. For a liquid consisting of a number of pure components, there is a temperature range between the boiling points of the various components. The liquid may thus be characterized by initial, intermediate, and final boiling points at a specified pressure.

To demonstrate the basis of continuous fractionation, consider the case of a binary mixture consisting of benzene and toluene, and refer to the vapor pressure/composition curve shown in Fig. 3.2. The lower curve shows the composition of the liquid boiling at the given temperature and the upper curve the composition of the vapor in equilibrium with the liquid. The horizontal lines AB, CD, EF, and so on, connect points showing the composition of liquid and vapor in equilibrium.

If distillation of a liquid of composition A is carried out sufficiently slowly for the liquid/vapor equilibrium to be unappreciably disturbed, a distillate of composition BC will be obtained initially. If this fraction is redistilled, it yields at first a distillate of composition DE; redistillation of this gives an initial distillate of composition FG, and so on. Thus, starting with a large quantity of a mixture of any composition, a small amount of almost pure benzene can be obtained by a series of successive distillations (EFV stages). The process was illustrated by reference to a binary mixture, but it applies equally well to multi-component mixtures, although in these cases fractions with a specific boiling range will be obtained.

In practice, for multi-component separations, this apparently very laborious operation can be carried out simply and efficiently in a distillation (or fractionating) column, in which a large number of EFV stages are arranged in series so that the products from each stage are fed to adjacent stages. A typical example of a cross-flow distillation column is shown in Fig. 3.3.

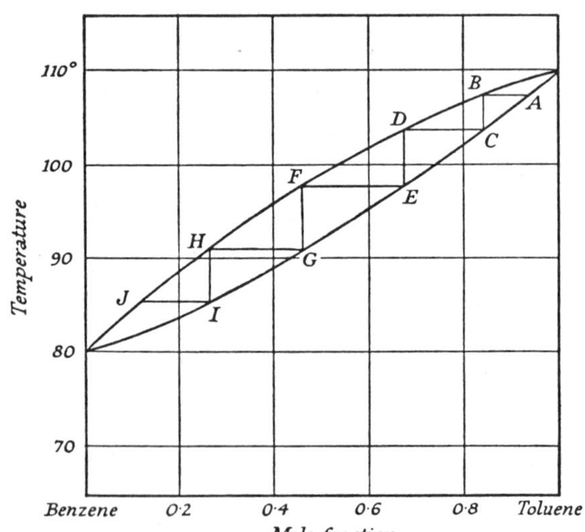

FIGURE 3.2 Vapor pressure/composition curve, benzene-toluene.

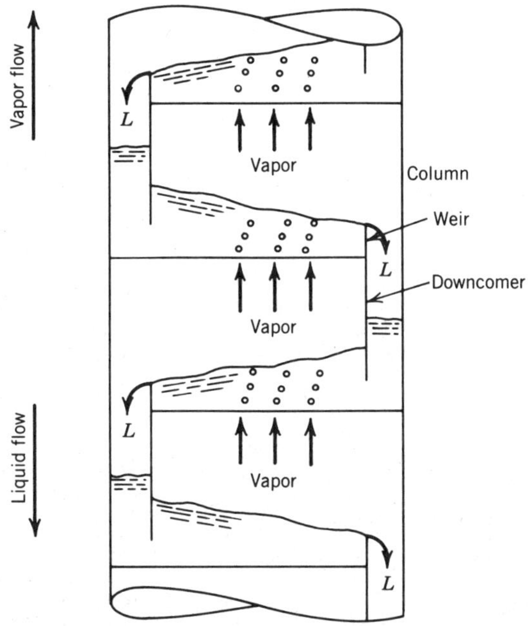

FIGURE 3.3 Typical cross-flow distillation column internals.

Vapor from one stage passes to the stage above and liquid to the stage below. In turn, each stage receives liquid from the stage above and vapor from the stage below. The concentration of the lower boiling components is increased in the vapor phase in the direction of vapor flow (up the column) and is decreased in the liquid phase in the direction of liquid flow (down the column). The reverse is true for the higher boiling components.

Therefore there is a temperature gradient in the system, reaching a minimum as the final vapor is withdrawn from the top of the column and a maximum as the final liquid is withdrawn from the bottom of the column. Heat energy is supplied to the process at the last stage from which liquid is withdrawn (known as the reboiler), by the feed to the column, or both. The final vapor is condensed, and part of the condensate returned to the top of the column to remove all or part of the latent heat contained in the vapor in the last vapor stage.

This liquid is known as *reflux,* and the *reflux ratio* is defined as the number of moles of liquid returned to the column per mole of liquid withdrawn as distillate. A pressure gradient is induced across the column to provide the necessary counter-flow of liquid and vapor. The column contains trays or plates, each of which acts as a stage for intimately contacting liquid and vapor to create equilibrium conditions.

Such fractionating columns are widely used in industry to separate multicomponent mixtures, the number of stages in the column depending upon

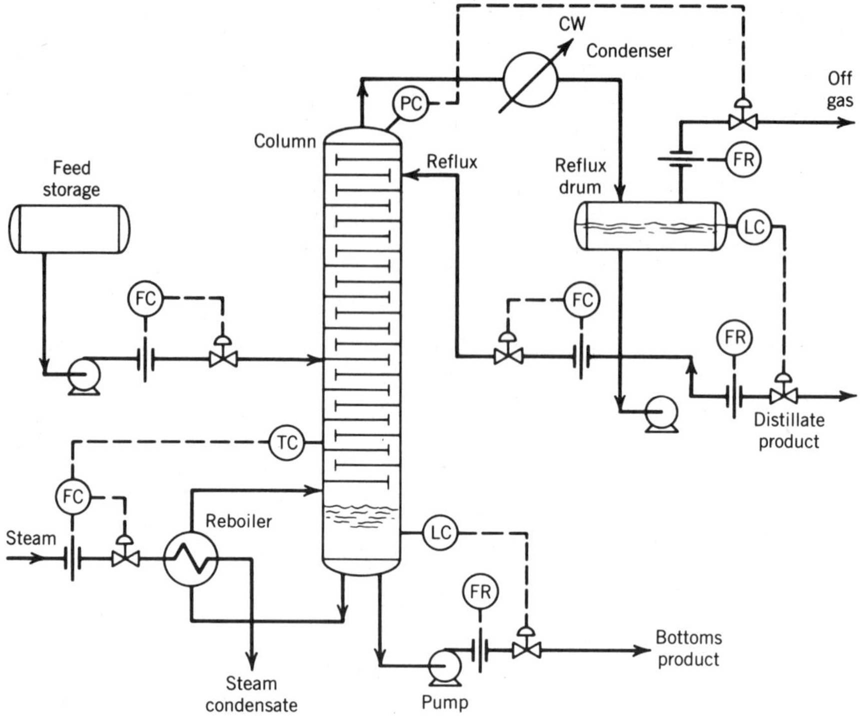

FIGURE 3.4 Typical fractionating column and controls.

the degree of separation required. An example of a fractionating column with its associated controls is shown in Fig. 3.4. Note that there are several ways of controlling a fractionating column; the example given is typical of one method.

3.3.3 Complex System Fractionation

In complex system fractionation, it is not possible to use pure component properties in column design calculations because of the large number of components and their isomers present in the feed. This poses a problem for the design engineer. An empirical technique is used instead, based on the "pseudo-component" concept. In this method the mixture is considered to be composed of a large number of fractions, each of which is characterized by an average boiling point, average density, average molecular weight, and so on. From these data it is possible to design a suitable fractionating column. Although such designs are only approximate, the equipment designed in this manner is generally more than satisfactory.

Some of the more important complex systems are the petroleum-based materials such as crude petroleum, natural gas, natural gasoline, and their associated products. Crude distillation is the primary process in any oil

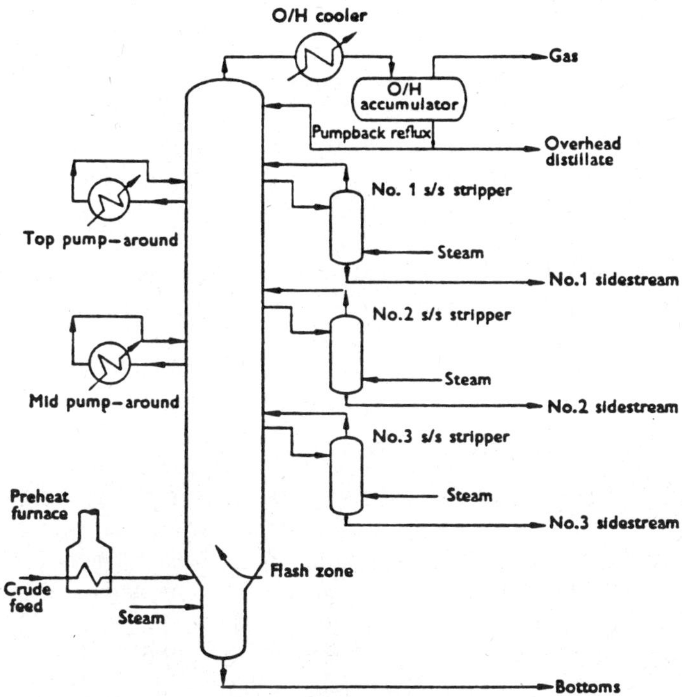

FIGURE 3.5 Typical crude pipestill.

refinery, since this process provides the feedstock for all downstream refinery units, including petrochemical plants.

The crude distillation process is carried out in a complex fractionating column known as a *pipestill,* a typical example of which is shown in Fig. 3.5. In this example the crude feed is separated into six fractions:

Overhead gas	Naphtha
Three distillate sidestreams	Residue or bottoms

Modern pipestills normally have from two to five sidestreams, depending upon refinery requirements. Each sidestream is steam-stripped to improve the separation between streams. The crude feed enters the column at the *flash zone,* where steam stripping is also used to improve distillate yield. Each raw cut from the pipestill is further processed in downstream units to produce end-products to meet customer specifications. Usually, the crude distillation process has built-in flexibility to enable a range of products to be cut on one fractionating column. The following summary will give the reader some idea of the flexibility:

Overhead gas is usually compressed and used as fuel gas for furnaces, boilers, turbines, and so on.

Overhead distillate (naphtha) may be used as feedstock for LPG production, steam cracking units, or gasoline blending.

First sidestream may be used for kerosine, turbo-jet fuel or gasoline production.

Second sidestream (light gas oil) may be used for diesel fuel production.

Third sidestream (heavy gas oil) may be used for heavy diesel fuel production, as blendstock for light fuel oil, or as a feedstock for catalytic cracking plants.

Residue (or bottoms) may be used for the production of various grades of fuel oil or may undergo vacuum distillation to provide feedstock for lubricating oil production or for catalytic cracking.

Vacuum distillation columns usually operate at 50–100 mm Hg, making it possible to produce distillates that boil at temperatures as high as 600°C at atmospheric pressure.

3.3.4 The Role of the Continuous Boiling Point Analyzer

In the atmospheric pipestill illustrated in Fig. 3.5, each fraction withdrawn from the column, except the overhead gas, can be characterized by an initial and final boiling point. In addition, there is a close relationship between the boiling ranges of the various products, which is a function of the tightness of control on column operation. Even a brief look at the primary distillation process indicates the tremendous complexity of refinery operations and the extent of interrelation between the various processes necessary to meet customer demand and tight specifications on final products. Therefore, precise control of the pipestill plays a very important role in overall refinery operations, since it is the initial step in the process.

Although conventional flow, level, pressure, and temperature control systems are the basic regulatory mechanisms for column control, some measurement of product quality is essential, if only for a monitoring-only check on process operation. Until the advent of the continuous boiling point analyzer, such measurements were carried out in the laboratory, where the standard ASTM test methods are used and have satisfied the requirements of industry for many years. However, the routine laboratory test route suffers from inherent delays in sampling, testing, and getting the results to the plant operations personnel. The greatest problem occurs when pipestill operation is changed to make a different range of products. During this period, frequent testing is required to ensure that on-grade product is being made.

This is precisely the situation in which the continuous boiling point analyzer is most useful. It provides a continuous record of product quality and gives timely indication of process upset or off-grade conditions, enabling operating efficiency to be increased substantially. Today, when most pipestill operations are under computer control, on-line quality data forms an essential input to an advanced "interacting" control system that takes automatic action to maintain product quality specifications.

3.4 STANDARD LABORATORY METHODS OF MEASUREMENT

Since the process boiling point analyzer results must be correlated finally against the standard laboratory methods of analysis, an understanding of the methods and their precision are essential to the analyzer engineer.

The standard laboratory test methods, as laid down by the ASTM, have been used on a multi-industry basis for many years. ASTM method D86-78 has a long history and covers the atmospheric distillation of a wide range of petroleum products. It was adopted as a joint ASTM/IP (Institute of Petroleum) standard in 1964. The specification for the equipment to be used for the test is covered in ASTM standard E133.

A 100-ml sample of the process stream is distilled in a flask at atmospheric pressure under conditions that approximate to a single-plate batch distillation with empirical heat input and distillation rate. The vapor is condensed and collected in a graduated cylinder. The vapor temperature is measured by an ASTM thermometer located in the neck of the flask. Observations are made of thermometer readings and volume of condensate over the full boiling range of the sample. The thermometer readings are corrected to 760 mm Hg barometric pressure and volumes of condensate are corrected for the fraction of sample not recovered. Readings are then recalculated from the "percent recovered" basis to the "percent evaporated" basis to give meaningful results. The test, done manually as described here, may take as long as 15 min to complete, not including the calculation procedure.

At present, the test is rarely done manually: it is usually carried out using an automatic distillation apparatus, such as the Gallenkamp ASDA Mark 3 or 4 or the Precision Scientific ADA III. This equipment uses the specified distillation flask as laid down in ASTM E133, but vapor temperature is measured using a thermocouple. The results of the test are recorded as a graph of temperature against condensate volume. The ASTM D86 procedure was revised in 1960 to make allowance for the use of automatic distillation equipment.

Another standard test of considerable importance in the petroleum industry is ASTM method D1160-77. This method covers the distillation of petroleum products under reduced pressure and is applicable to those products that are partially or completely vaporized at 399°C (750°F) and pressures down to 1 mm Hg. In this case, the equipment specifications are laid down in ASTM D1160 itself.

A 200-ml sample of the process stream is distilled under conditions approximating a single-plate batch distillation at some predetermined and accurately controlled pressure between 1 mm Hg and atmospheric pressure. The pressure-sensing device must be capable of an accuracy of ± 0.05 mm Hg at 5 mm Hg and below, and pressure control must be within $\pm 1.0\%$ of the absolute pressure at 10 mm Hg and within ± 0.1 mm Hg at lower pressure. The sample should be moisture-free. Vapor temperature is measured by a thermocouple connected to a potentiometer, the complete system being capable of detecting changes of ±0.5°C over the complete range of operation (0–371°C or 32–700°F).

There are a number of other ASTM standard distillation test methods applicable to petroleum products; they are specific to one type of product:

ASTM D216-77	Natural gasoline
D285-62	Crude petroleum
D850-79	Aromatic hydrocarbons and related products
D1078-78	Volatile organic liquids
D1088-53	Butadiene

The ASTM methods described form the basis for most distillation testing carried out in oil refinery laboratories. Of these, D86 and D1160 are the tests with which we are concerned when considering correlation of process boiling point analyzers.

3.4.1 Other Related Laboratory Methods

The technique of temperature-programmed gas-liquid chromatography has been used in the laboratory for several years. One specific use is in the estimation of boiling point distribution. There are two relevant ASTM standard test methods for the Petroleum Industry:

- Method D2887-73, boiling range distribution of petroleum products (boiling range 100–1000°F).
- Method D3710-78, boiling range distribution of gasoline and gasoline fractions (FBP 500°F).

The sample is introduced into a gas chromatographic column, which separates hydrocarbons in boiling point order. The oven temperature is raised at a reproducible rate and the area under the chromatogram is recorded throughout the run. Boiling point temperatures are assigned to the time axis from a calibration curve, obtained under the same conditions by running a known mixture of hydrocarbons covering the sample boiling range.

The IBP is defined as the point at which a cumulative area count of 0.5% of the total area under the chromatogram is obtained. The FBP is similarly defined as 99.5% of total area. The cumulative area at each selected interval between IBP and FBP is divided by the total area to obtain the percent of sample recovered at each interval.

The use of this technique provides an insight into the composition of the sample. In addition, the distillation curve obtained by Method D86 can be predicted by the data obtained by GC stimulated distillation. In the case of gasolines, the technique provides the data necessary to calculate the vapor pressure of gasoline, traditionally determined by Method D323 (see Chapter 6).

3.5 PROCESS BOILING POINT ANALYZERS

Process analyzers for boiling point measurement may be classified arbitrarily into the following categories:

- Initial boiling point.
- Intermediate boiling point.
- Final boiling point.
- Vacuum distillation.
- Percent recovered at a given temperature.
- Multipoint distillation curve analyzers.

There is considerable overlap in these arbitrary categories, as will become evident in the following description of commercially available equipment. In some cases, initial and final boiling point analyzers may be adjusted to give intermediate boiling points. Some analyzers may be modified to give percent recovered readings at a selected temperature. Others provide a complete distillation curve for the process sample and record several points on this curve. Some analyzers follow the ASTM D86 procedure, others utilize a completely different technique to simulate the ASTM test.

3.5.1 Carlo Erba S.p.A. (Italy)

The Carlo Erba Model PD155 Process Distillator is designed to provide boiling point data on petroleum products and is capable of recording initial, final, and five intermediate boiling points on a sample. Multistream capability is also available. The instrument consists of three component parts—the analyzer unit, the programmer, and a recorder.

The analyzer unit, shown schematically in Fig. 3.6, is housed in an explosion-proof enclosure mounted on a free-standing rack for convenient field installation. The unit contains all equipment necessary to carry out a distillation of the process sample and includes the following components:

1. A stainless steel distillation flask with built-in heater.
2. A specially designed pneumatically operated sampling valve, which injects a reproducible 40 ml volume of sample into the flask.
3. A condenser unit for condensing flask vapors during the heating cycle.
4. A burette-type receiver, specially calibrated, for collecting distillate from the condenser unit.
5. A number of photoelectric detectors, mounted at various heights on the receiver unit, their locations depending upon the various boiling points to be measured.
6. A thermocouple located in the neck of the distillation flask for measuring the vapor temperature.

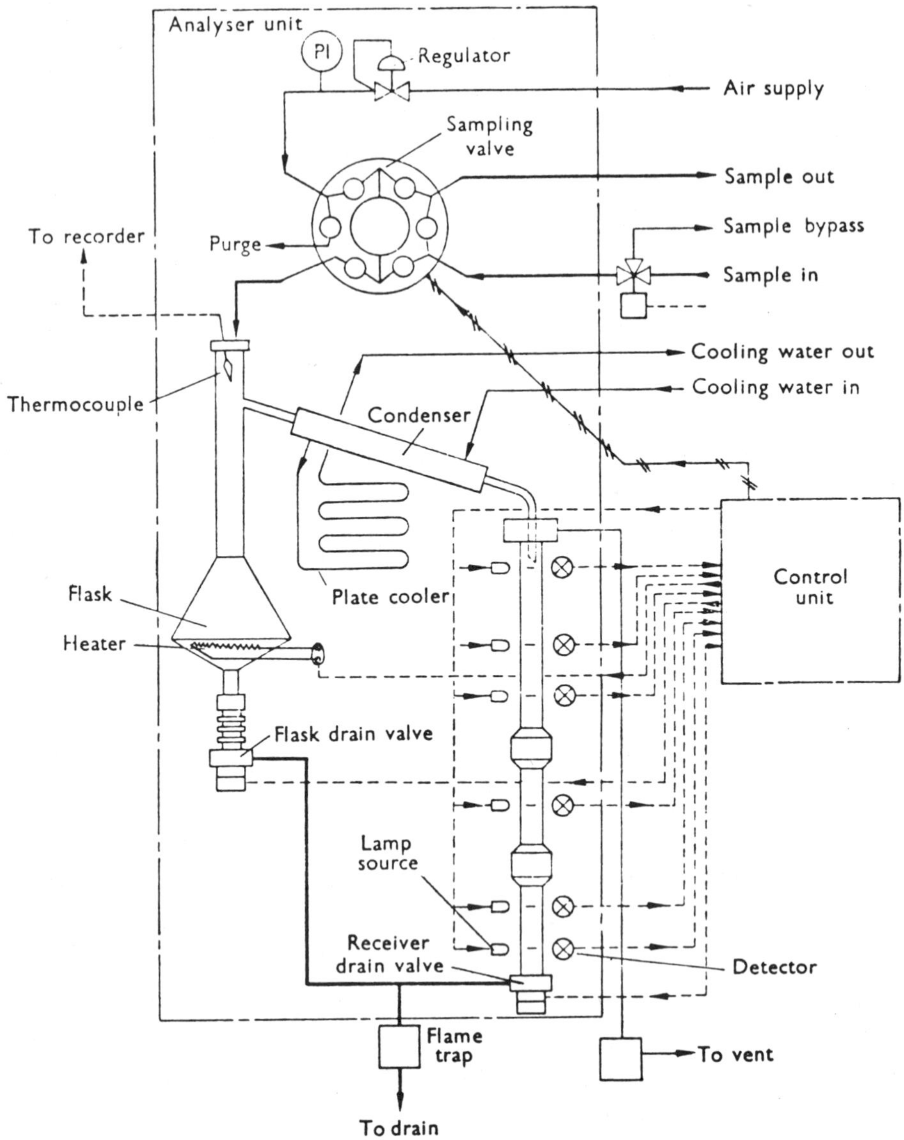

FIGURE 3.6 Model PD155 process distillator. (Courtesy of Carlo Erba spA)

The programmer (control) unit provides all the necessary functions for remotely controlling the analyzer unit operation and is suitable for panel or rack mounting in non-hazardous locations up to 1000 ft from the analyzer unit. Programming the analyzer cycle is accomplished by a series of cams driven by a synchronous timer motor. The unit also contains the logic circuitry for the photoelectric detectors, safety devices, a stabilized dc power supply, and a power selector circuit for multi-stream operation (when required). The unit controls the introduction of sample into the distillation flask, heater start-up, the washing and cooling cycle, and optional stream selection.

Analyzer operation is as follows. At the start of the cycle, the sample valve is switched to allow precisely 40 ml of process sample into the flask, power is supplied to the flask heater, and IBP detector is energized. The flask thermocouple measures the vapor temperature and transmits a signal to a remote recorder, usually mounted adjacent to the control unit. When the first drop of condensate passes into the receiver, the IBP detector trips and the first intermediate level detector is energized. This sequence continues until the last intermediate level detector is tripped and the FBP detector is energized. The FBP detector is tripped when the last drop of liquid boils off, causing a large temperature change in the base of the flask. When this occurs, the heater is switched off, the receiver drain valve is opened, and the washing/cooling cycle initiated in readiness for the next distillation cycle.

The *samplematic* sampling valve is a diaphragm-type injection valve, operated by solenoid three-way valves that change the arrangement of the connecting passages in the sample valve by directing control air to the relevant points. It is basically the same type of valve as used in the Carlo Erba process chromatographs for sample injection. Minimum air pressure is 60 psig; a safety switch is provided to shut off power to the flask heater and stop the analyzer cycle if air pressure falls below 20 psig.

The flask heater has various connections to allow flexibility in setting up the analyzer for the application. The desired heating rate is generally set at the factory, based on the boiling range of the sample to be measured.

The front panel of the control unit incorporates a semi-graphic of the analyzer unit, with pilot lamps for indicating the status of the analyzer cycle:

"sampling"	= sample injection
"heater"	= heater on
"drain"	= washing and drain cycle on
"off-normal"	= analyzer out of service
"%vol. EPB"	= detecting devices in operation

Also provided are a power switch, a cycle start pushbutton, and a sample pump start/stop switch (when a sample pump is used).

The recorder may be any potentiometric type with thermocouple input. The recorder input circuit is shorted out and the chart advance motor energized by the control unit, whenever a photoelectric detector circuit is tripped. The result is a bargraph trace, the top of the bars being the temperature of the various boiling points being measured.

The analyzer has an operating range of 50–300°C and the following distillation points are usually measured: IBP, 5%, 10%, 50%, 90%, 95–98%, and EBP. Readings may be correlated with ASTM D86 distillation data.

3.5.2 Fluid Data Inc. (United States and England)

Fluid Data manufactures the Model 1463D Boiling Point Analyzer, which is an updated version of the original Hallikainen Model 1463.

This instrument, shown schematically in Fig. 3.7, is designed to measure the percent recovered boiling point temperature of hydrocarbon liquids from the 5% to the 97% point, within the temperature range 150–700°F. Results may be correlated with ASTM D86 distillation data. The analyzer has the capability of selecting any boiling point without the need to add or change any system elements.

Process sample is metered and pumped to the boiler assembly by one head of a duplex pump at a normal flowrate of 25 ml/min. Before entering the boiler, sample is preheated by a tube-in-tube heat exchanger and further heated by a preheat coil within the boiler. The sample then expands through a relief valve into a flash cup, which is vented to atmosphere through a flame arrestor. The residual liquid is directed onto an evaporator tube where, as the sample flows down the tube, the desired percent overheads are boiled off. Liquid drops off the end of the evaporator tube into a bottoms cup, where its temperature is measured by a thermocouple.

Sample liquid in the bottoms cup is pumped out through the tube-in-tube heat exchanger by the second head of the duplex pump, at a predetermined rate set by the percent boiling point to be measured, and then drained via a flame arrestor. The sample level in the bottoms cup is identical to that in the adjoining float housing. Automatic regulation is provided by a level sensor and control circuit that regulates power to the evaporator heater in order to boil off the desired volume of liquid equal to the difference between sample inflow and outflow. The circuit is controlled by the output of a photocell in the level sensor. A beam and flag assembly is pivoted about a diaphragm mounted in the sensor. The end of the beam protruding from the sensor is fitted with a ball float that senses the sample level in the float housing and thus varies the position of the flag balanced between the photocell and the light source.

If insufficient heat is generated by the heater, the liquid level (and the float) will rise, activating the beam and flag assembly. This action increases the photocell resistance, increasing power to the heater until a stable liquid level is reached. If excessive heat is generated by the heater, the liquid level in the float cup will decrease, reversing the action.

The analyzer is mounted on a free-standing steel framework and is suitable for installation in NEC Class 1, Group D, Division 1 hazardous locations. Separate enclosures of flame-proof design are provided for the boiler assembly, level controller, and electronics section. The thermocouple for sample temperature measurement is connected to a temperature transmitter, also mounted on the analyzer frame, which provides a 4–20 mA dc output for remote recording. The analyzer time constant for a step change at the sample pump inlet is typically 60 sec at the mid-range temperature.

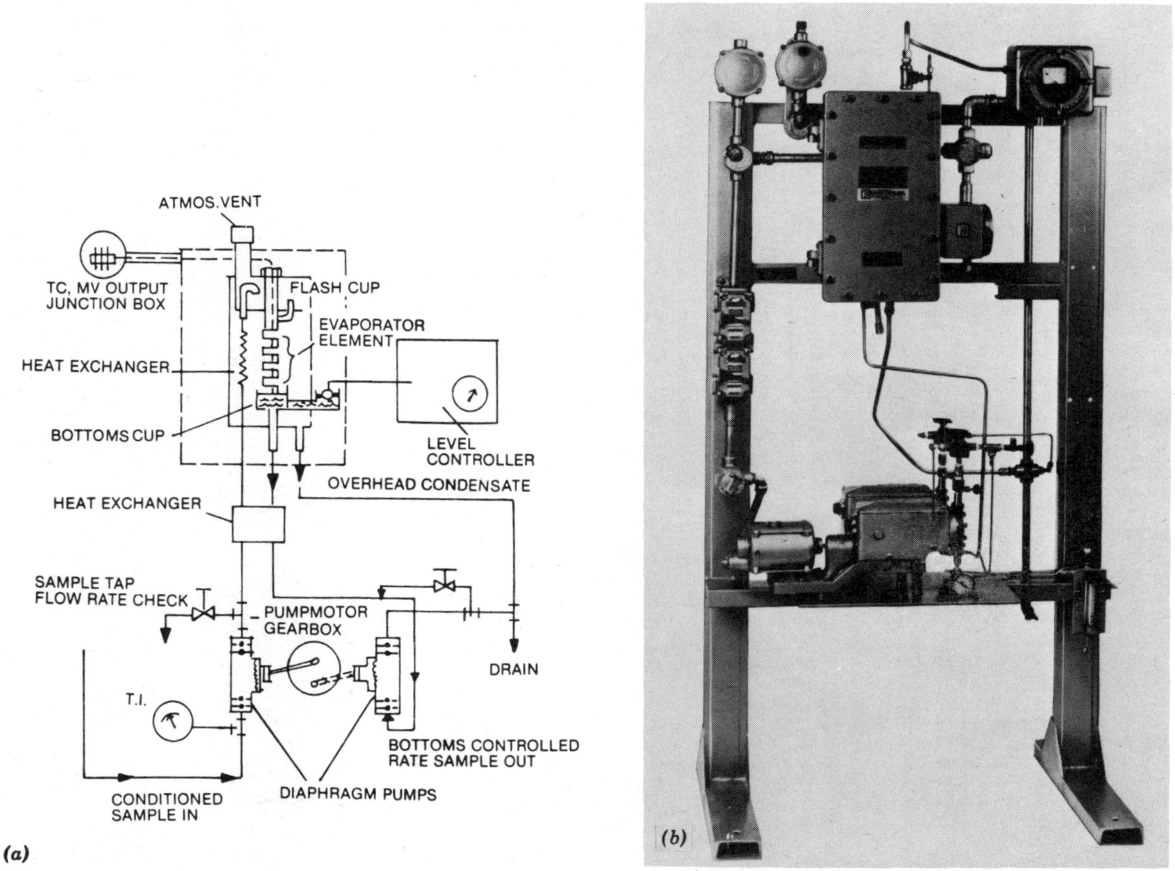

FIGURE 3.7 Fluid Data Hallikainen Model 1463D. (*a*) Operating principle; (*b*) equipment layout. (Courtesy of Fluid Data Inc.)

A recent introduction by Fluid Data is the Model 1463M Distillation Analyzer. This instrument is based on the proven Hallikainen design of the Model 1463D using diaphragm pumps for sample metering and the simple low-power evaporator element.

The design is enhanced by the use of microprocessor electronics for operation, control, and checking functions to measure boiling point temperature for any given percent recovered from 5% to 95%, within the temperature range 100–650°F. Alternatively, it is possible to measure percent recovered at a given temperature. Both modes of operation are microprocessor controlled utilizing the principle of liquid level measurement in the bottoms cup, as in the Model 1463D.

The instrument has a digital display, transducer level control, and modulating capability of the bottoms diaphragm pump. The microprocessor provides system self-diagnostics, can be programmed for multipoint operation, is computer compatible, and correlates data to ASTM D 86/IP 123 for boiling point applications. Instrument response is quoted as 1–2 minutes nominal, and the equipment is suitable for installation in NEC Class 1, Group D, Division 1 hazardous locations.

3.5.3 GCA/Precision Scientific Group (United States)

GCA/PSG manufactures five boiling point analyzers, three of which are from the line originally made by Totco. Two are recent introductions by GCA, based on extensive development work and field testing.

(a) Initial Boiling Point Analyzer, Model 41269

This instrument, shown in Fig. 3.8, is designed to measure any preselected point on the ASTM distillation curve from IBP to approximately the 30% point of a hydrocarbon liquid, covering a boiling point temperature range of 100–650°F. Repeatability is quoted as ±1% of sample boiling range and readings may be correlated with ASTM D86 distillation data.

The conditioned process sample is preheated and enters the top of the distillation column through a throttling valve. A float in the boiling pot responds to liquid level changes, controls the throttling valve position and thus the volume of sample entering the analyzer. The sample flows through the float guide tube to the feed zone of the column and into the boiling pot, where heat is supplied by an electrical heater. Heater selection within the range 400–3200 watts is determined by factory analysis of the customer's sample distillation requirements.

As sample distillation progresses, the sample feed is split into vapor overheads and liquid bottom fractions. Overhead vapor exits from the top of the column; bottoms liquid exits from the boiling pot and is combined with the overhead vapor. A check valve in the bottoms outlet line prevents overhead vapor from flowing back into the boiling pot. A calibrated orifice in the overheads line, together with the pressure drop in the bottoms line, combine

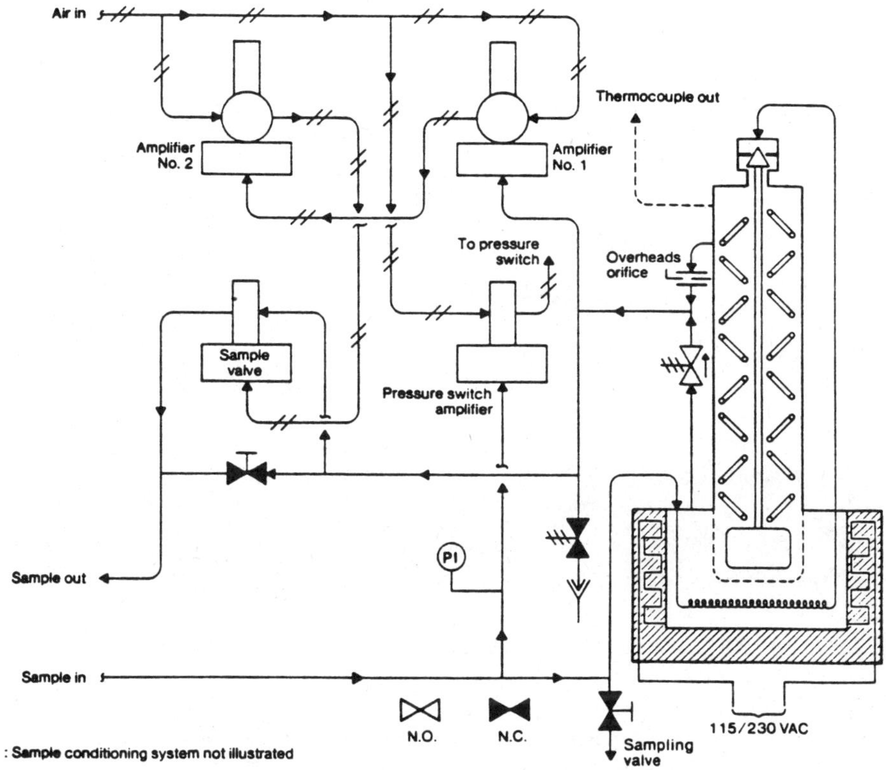

FIGURE 3.8 IBP Analyzer Model 41269: flow diagram. (Courtesy of GCA Corporation)

to maintain a preset overheads-to-bottoms ratio. The calibrated orifice is factory selected to provide the correct boiling point measurement in accordance with customer requirements (from IBP to approximately the 30% point). Vapor temperature at the top of the distillation column is measured by a thermocouple that provides a signal for remote recording.

A two-stage pneumatic amplifier system maintains constant column pressure (normally set at about 10 psig) by regulating the flowrate of the combined overheads and bottoms streams leaving the analyzer (operation of the sample discharge valve). This provides the necessary pressure drop across the orifice to maintain a constant ratio of overheads to bottoms at the desired setpoint value. Analyzer safety features include automatic power shut-off to the heater in the event of high boiling pot temperature, low instrument air pressure, or low sample pressure.

The analyzer shown in Fig. 3.9 is suitable for rack mounting in NEC Class 1, Group D, Division 1 hazardous locations and consists of two parts, the distillation column assembly and the pneumatic amplifier assembly, together with the necessary electrical equipment. Sample handling systems are available as an option.

FIGURE 3.9 IBP Analyzer Model 41269: equipment layout. (Courtesy of GCA Corporation)

(b) Final Boiling Point Analyzer, Model 46500

This instrument, shown in Fig. 3.10, is the updated version of the original Totco Model 6500. It is designed to measure any preselected point on the ASTM distillation curve from the 30% point to the FBP of a hydrocarbon liquid, covering a boiling point temperature range of 100–650°F. Repeatability is quoted as ±1% of sample boiling range and readings may be correlated with ASTM D86 distillation data. Operation is almost identical to that of the Model 41269, except for the following differences:

- Overhead vapor flows through an air-cooled condenser and the condensate is combined with the liquid bottoms from the boiling pot.
- The calibrated orifice is located in the bottoms outlet line before it combines with the overhead condensate line.
- The temperature of the boiling liquid bottoms is measured by a thermocouple.

The layout of the analyzer unit is the same as the Model 41269 and is suitable for installation in NEC Class 1, Group D, Division 1 hazardous locations.

Note that although these analyzers are quoted as being able to measure the IBP or FBP, in reality this is not possible. By virtue of their design, there must be a flow through the calibrated orifice; therefore, the ratio of overhead

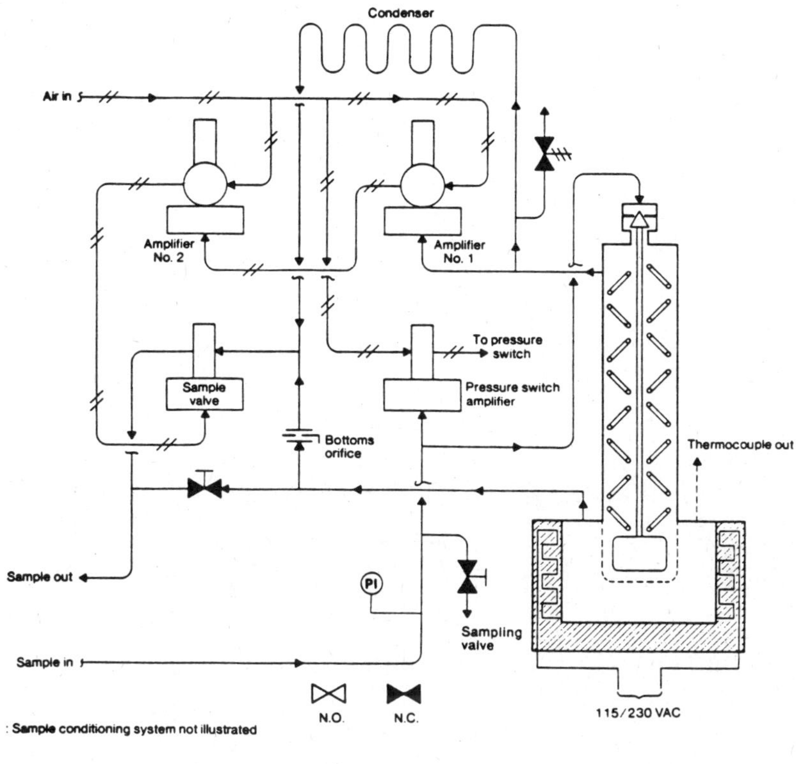

FIGURE 3.10 FBP Analyzer Model 46500: flow diagram. (Courtesy of GCA Corporation)

to bottoms flowrate allows only measurement of, say, the 5% point rather than IBP, and the 95–98% point rather than FBP.

(c) Vacuum Distillation Analyzer, Model 41468

This instrument, shown in Fig. 3.11, is designed to measure the specific boiling points of liquid hydrocarbons that would normally decompose if distilled at atmospheric pressure. The sample is split in a constant, preselected ratio of vapor overheads to liquid bottoms at any point between 5% and 95%. The bottoms temperature is measured and may be correlated with ASTM D1160 distillation data within the atmospheric boiling point temperature range 650–1000°F. Repeatability is quoted as ±1% of sample boiling range.

Sample flowrate through the analyzer is controlled by a float-actuated inlet valve that regulates flow into the distillation column at a rate equal to the combined discharge capacity of two gear pumps driven from the same motor. One unit pumps condensed vapor overheads, the other pumps liquid bottoms. The sample is fractionated in a packed distillation column with

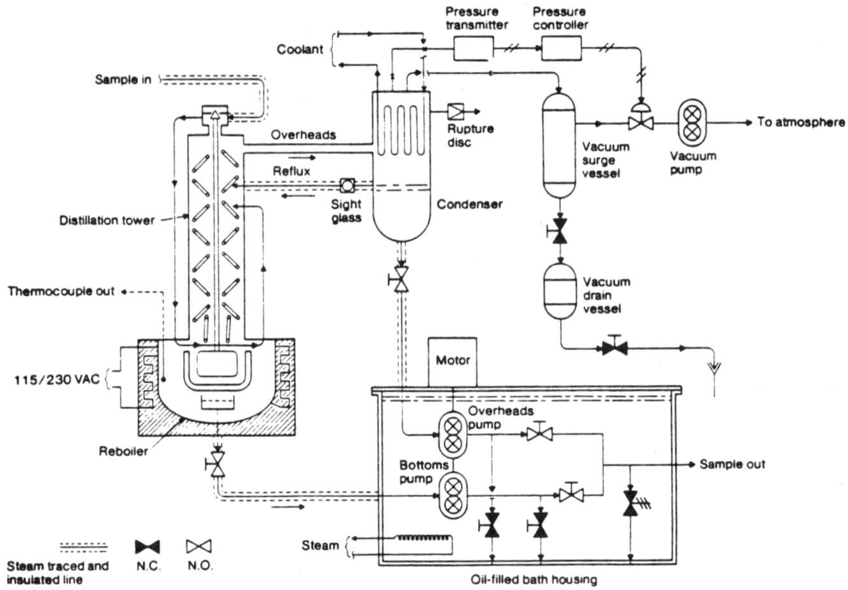

FIGURE 3.11 Vacuum Distillation Analyzer Model 41468. (Courtesy of GCA Corporation)

reboiler and electrical heater assembly. Vapor passing overhead is condensed in a liquid-cooled condenser. A vacuum pump and pressure control system maintain a reduced pressure within the analyzer at any preselected value that will permit fractionation without decomposition.

Overhead vapor is condensed in the overhead condenser and flows out of the analyzer through the overhead pump or back into the distillation column as reflux. The bottoms and overhead condensate pumps are preset at any desired rate (percent distilled). A thermocouple located in the reboiler produces an output proportional to temperature at the preselected fractionation ratio.

A rupture disc installed in the overhead condenser protects the analyzer from excessive pressures, and a pressure relief valve installed in the spent sample return line protects the pump against back pressures greater than 30 psig. This ensures accurate flow control and analyzer setpoint regulation.

The analyzer is mounted on a free-standing frame, as shown in Fig. 3.12, and is suitable for installation in NEC Class 1, Group D, Division 1 hazardous locations.

(d) Model 47500 BPA-II Boiling Point Analyzer

This analyzer is an updated version combining the basic principles of the Models 41269 and 46500 described earlier. The new instrument is designed to

FIGURE 3.12 Model 41468: equipment layout. (Courtesy of GCA Corporation)

monitor a preselected point on the distillation curve from 5% to 95% of the sample boiling range and within the temperature range 100–650°F. The analyzer consists of 3 modules, the sample conditioning, analysis, and control modules, mounted on a free-standing rack and suitable for installation in NEC Class 1, Group D, Division 1 hazardous locations.

Selected orifices in the sample bottoms and overheads lines establish the percent boiling point to be monitored. Boiling point temperature is measured by a thermocouple located in either the boiling tower (for initial BP ranges) or in the reboiler (for final BP ranges). Digital indication of boiling point temperature is provided locally, with a 4–20 mA dc output for remote recording purposes. A variable voltage control is provided for the heater to enable heater wattage to be set correctly in the field for a particular sample composition. A two-stage pneumatic amplifier system, as used in the earlier models, operates the sample discharge valve to maintain tower pressure at a setting of 10 psig. Safety facilities include operational status indicators, high temperature cut-off, system pressure relief, and automatic heater cut-off in the event of low sample pressure or low instrument air pressure. Sample requirements are 2–9 gph at 60–250 psig and 180°F maximum. Repeatability is quoted as ±1% of sample boiling range.

(e) Model 44520-11 Distillation Analyzer

This instrument is the latest addition to the GCA product line and utilizes microprocessor-based electronics. The analyzer uses a film evaporation technique to separate overheads and bottoms, and is capable of monitoring any point from 5% to 95% of sample boiling range within the temperature range 100–600°F.

Sample at a constant flowrate enters the evaporator column, and, as it flows down across the heater, light ends vaporize and condense on the evaporator walls and flow to an exhaust pump. Heavy ends (bottoms) pass over a thermocouple and out to a variable-speed bottoms pump.

Two operating modes are possible. In Mode I, % off at temperature, power to the heater is controlled to maintain the desired set point temperature. Bottoms flowrate is calculated according to the standpipe volume, subtracted from the total sample flowrate and the difference converted to % off and transmitted as the measured variable. In Mode II, temperature at % off, bottoms pump speed is controlled and heater voltage regulated to maintain constant standpipe volume. Thermocouple signals are corrected to 1 atmosphere and transmitted as the measured variable.

The microprocessor controls all unit operations, monitors all parameters, performs barometric correction of temperature measurement, and provides extensive system self-diagnostics. The local keyboard and LED display allow programming of measurement, control, and alarm parameters. The analyzer can operate under local or remote set point control and is computer compatible. Measurement and display can be in English or metric units and can be correlated with ASTM D86/IP123.

The analyzer consists of two modules, the analysis and electronics modules, mounted on a free-standing rack and suitable for installation in NEC Class 1, Group D, Division 1 hazardous locations. Sample requirements are 0.2 gph at 20–400 psig and 275°F maximum (but at least 40°F below IBP temperatures).

3.5.4 Hone-Addison Ltd. (Great Britain)

The Hone "Distillar" Distillation Analyzer is the latest of a range of distillation analyzers manufactured by Hone since the early 1960s. The operating principle is identical to the ASTM D86 Laboratory method. A schematic flow diagram of the analyzer is shown in Fig. 3.13.

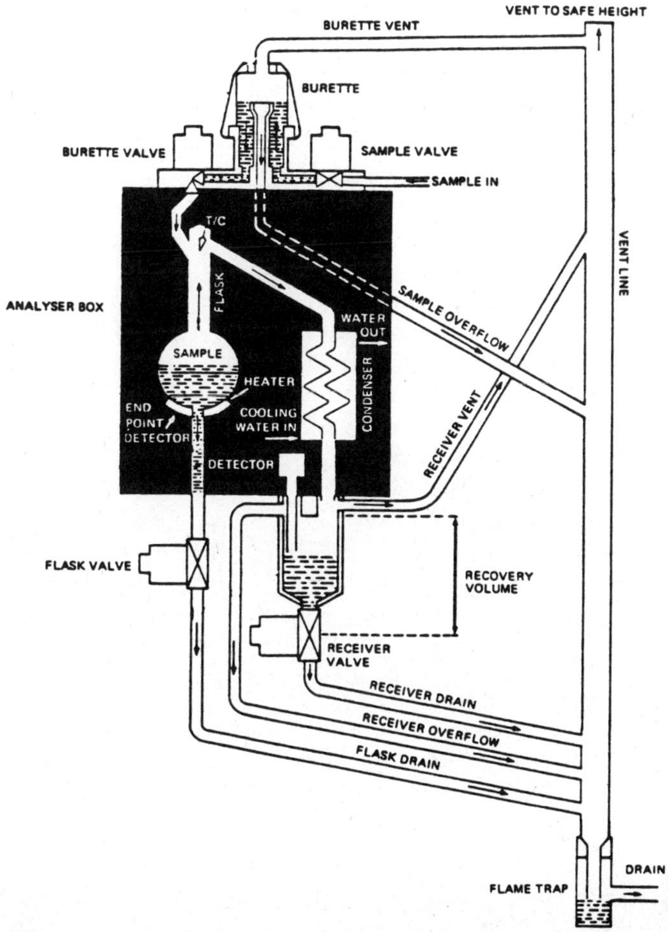

FIGURE 3.13 Hone "Distillar": flow diagram. (Courtesy of Hone-Addison Ltd.)

The equipment consists of a flask, heater, condenser, calibrated receiver, and temperature sensing element. In addition, there is a burette to charge 100 ml of sample to the flask and the electrical/pneumatic components required for automatic and safe operation of the analyzer. The burette, flask, and condenser are stainless steel. The flask is heated electrically and the detectors are fluidic devices (except for the end point detector). Filling, flushing, and draining are carried out by air-operated valves and the complete sequence is controlled electrically.

At the start of an analysis sequence, sample flow to the burette is stopped. The 100 ml of up-to-date sample in the burette is then transferred to the flask, which has already been flushed and cooled after the previous analysis. The flask heater is switched on and sample is distilled through the condenser into the calibrated receiver. Vapor temperature throughout the test is measured by a thermocouple located in the neck of the flask. The liquid level in the receiver is visible throughout the analysis.

When the liquid level in the receiver reaches the preset value (e.g., 90 ml = 90% point), the detector is operated and this initiates several operations:

- The heater is switched off.
- The programmer is started to carry out the next cycle of operations.
- The new boiling point temperature is recorded remotely via a peak-picker and mV/current transmitter. The programmer operates the pneumatic valves in the correct sequence to flush and drain the flask and receiver, ready for the next analysis. Each complete analysis takes from 5 to 30 minutes, depending upon the boiling point temperature and selected distillation point.

This description covers an analyzer designed to measure a single boiling point temperature. There are, however, a number of additional options available, depending on customer requirements.

(a) End-Point Model

Distillate collects in the receiver until no liquid remains in the flask. At this point, heater temperature increases rapidly and a resistance sensor on the heater trips, terminating the distillation and initiating a refill cycle.

(b) Two-Point Model

Distillate collects in the receiver until the level detector at the lower point is reached. This operates the first output transmitter and causes the level detector to travel up the receiver to the second preset point position. When distillate reaches this position, the second output transmitter is operated, a refill cycle is initiated, and the level detector returns to the lower position.

(c) Multi-Point Model

Distillate collects in the receiver and the distillate level is followed by the detector. As the detector passes the preset points, an output signal is given. When the last point is reached, a refill cycle is initiated and the detector returns to the base of the receiver.

(d) Percent Off Model

Distillate collects in the receiver until the preselected temperature is reached. When this occurs, the heater is switched off and the detector moves to the new level of distillate in the receiver. The amount of movement is related to the percent off at the selected temperature. A refill cycle is then initiated.

(e) Distillation Curve Model

Distillate collects in the receiver and its level is followed by the detector. Distillation temperature and recovered volumes are both transmitted to a suitable X-Y recorder. The distillation is terminated at any preselected temperature or volume and a refill cycle is initiated, during which time the detector returns to the base of the receiver.

The Distillar is mounted on a free-standing steel framework, as shown in Fig. 3.14, and is suitable for installation in NEC Class 1, Group D, Division 1 hazardous locations by means of air purging. Versions suitable for PTB and other classifications are also available. Purge failure results in automatic power disconnection and contacts are available for activation of external alarms.

The two large enclosures house the analysis unit and control unit respectively. The control unit contains the electrical power and control components, including a cam timer and solenoid valves, associated relays, detection, safety and heater controls, and the mV/current converter and peak-picker for temperature measurement. The signal termination box is located at the top left of the framework. Two-wire current systems (4–20, 0–

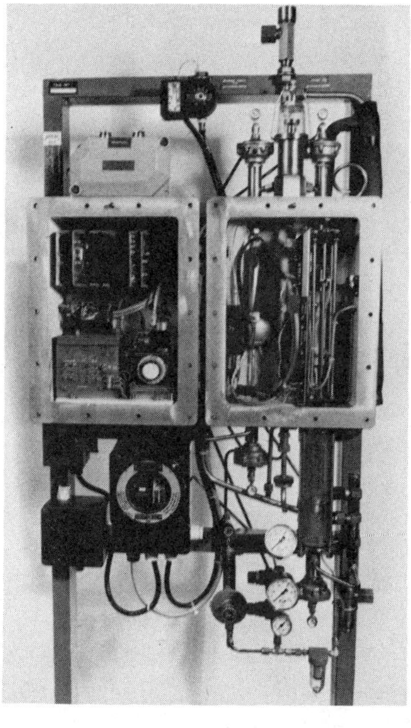

FIGURE 3.14 "Distillar": equipment layout. (Courtesy of Hone-Addison Ltd.)

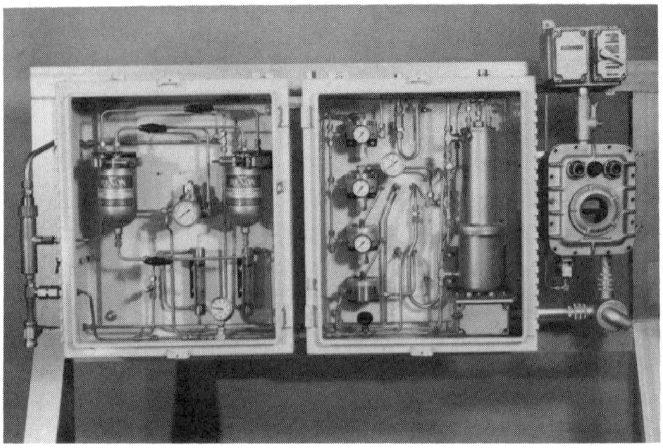

FIGURE 3.15 Model 52000 boiling point analyzer. (Courtesy of Rexon Corporation)

20, or 0–10 mA dc) can be provided, corresponding to a preselected temperature range to customer specification, minimum span being 100°C. Incoming process sample should be at least 30°C below the initial boiling point and at a pressure of 2–20 psig.

3.5.5 Rexon Corporation (United States)

The Rexon Model 52000 Boiling Point Analyzer is designed to measure any preselected distillation point of a hydrocarbon stream between the 5% and 95% points, over a temperature range of 100–650°F. Readings may be correlated with ASTM D86 distillation data. Repeatability is quoted as ±1% of sample boiling range.

This instrument, which is a relatively recent addition to the field of boiling point analyzers, is shown in Fig. 3.15. Its operating principle and design are similar to the GCA/PSG Models 41269 and 46500. The analyzer consists of three modules—the analyzer unit proper, the electronics unit, and an optional sample handling system. The complete system is packaged for good maintenance and servicing accessibility and is designed for mounting on a free-standing framework. It is suitable for installation in NEC Class 1, Group D, Division 1 hazardous locations.

The 52000 has a factory-calibrated variable heater control that will allow field adjustment in the event the analyzer application is changed. Several safety features are provided:

- A relief valve is provided to protect the distillation column against pressures in excess of 25 psig.
- The heater shuts off in the event of low instrument air pressure, low sample pressure, or boiling pot high temperature.

- A discharge valve will flush the reboiler, preventing coke buildup in the event of loss of instrument air pressure.

A duplex thermocouple is installed in the tower (low % point measurement) or in the reboiler (high % point measurement), with one pair of leads connected to a local digital temperature display, the other to a terminal block in the electrical enclosure for output to a remote recorder. Again, by virtue of its design, this instrument is not able to measure the IBP or FBP of a process sample.

3.5.6 Universal Oil Products Inc. (United States)

The UOP Monirex Series 404 Boiling Point Monitor (BPM) is a microprocessor-based, temperature programmed gas chromatograph system, designed to continuously determine the boiling point characteristics of a hydrocarbon process stream. The results are automatically correlated to provide a distillation curve comparable with that obtained by ASTM D86 procedure. Early versions of the BPM incorporating mechanical constants and analog electronics were described in a paper presented to the ASTM Committee D-2 in 1973. The current model was described in a paper presented to the ISA in 1977.

The instrument operates on the principle that if starting temperature and programmed heating rate of the chromatographic column are kept precisely constant for each analysis, then fixed times after sample injection will always represent the same boiling point temperatures. When a long column is used, a chromatogram peak is produced for each component; summing the areas under the peaks and recording the running total, produces a true boiling point curve, as shown in Fig. 3.16. Typical laboratory analyses of this type may take 20 to 45 minutes to complete. The Monirex BPM compresses the analysis, producing what is known as the 'hot column response' (shown in Fig. 3.17), resulting in a separation that takes much less time (about 4 min) and that may be correlated with ASTM D86 data.

Periodically, a microliter-size sample is injected into the helium carrier gas, where it flashes to vapor and is conveyed through the column. This separates components as a function of carbon chain length, which is related to boiling point. To generate a distillation curve, the output signal from the thermal conductivity detector is fed to an integrator and the component signals continuously summed and stored. The total integral is used to normalize the percent-off for each selected boiling point and the results are stored and displayed in indicating digital registers. A calibration sample, where the percent-off at each selected boiling point to be displayed is known, is used for analyzer calibration. A timer thumbwheel switch for each boiling point is adjusted until the selected known percent-off is indicated on the register readout.

The BPM is packaged in three separate units—the analyzer unit, which is

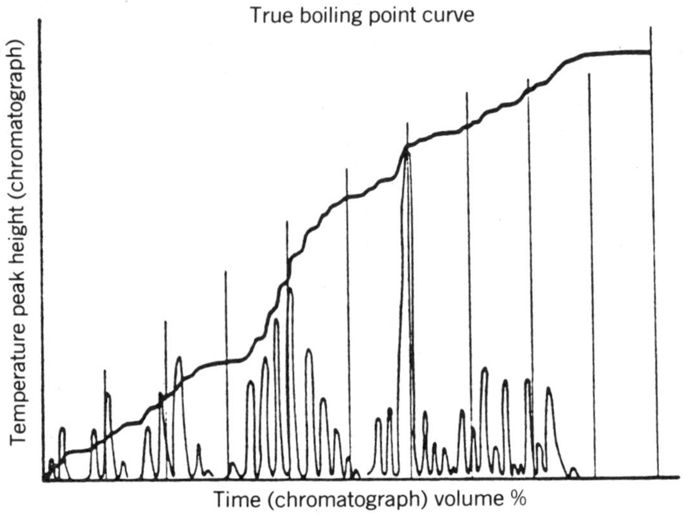

FIGURE 3.16 True boiling point curve. (Courtesy of Universal Oil Products Inc.)

field mounted and suitable for installation in NEC Class 1, Group D, Division 1 hazardous locations, and the programmer and the display unit, both of which are designed for remote mounting in non-hazardous locations. The analyzer unit interacts with the programmer over nine digital lines, six of which are used for analyzer status information, three for analyzer control. There is one analog signal loop for the 4–20 mA dc amplified signal from the GC detector.

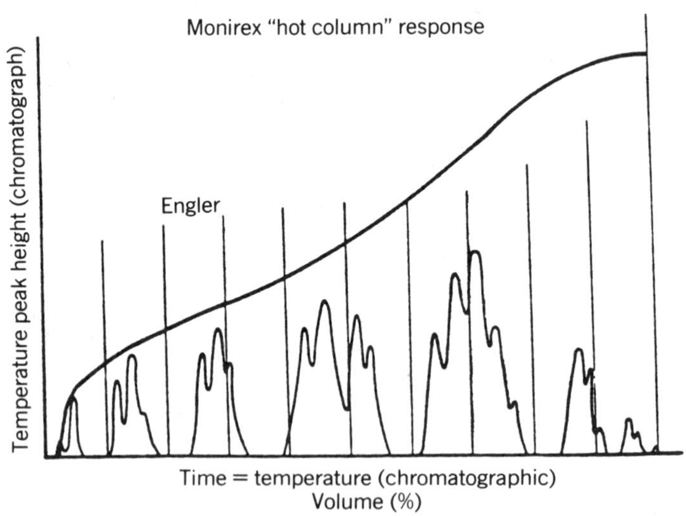

FIGURE 3.17 Monirex BPM hot column response. (Courtesy of Universal Oil Products Inc.)

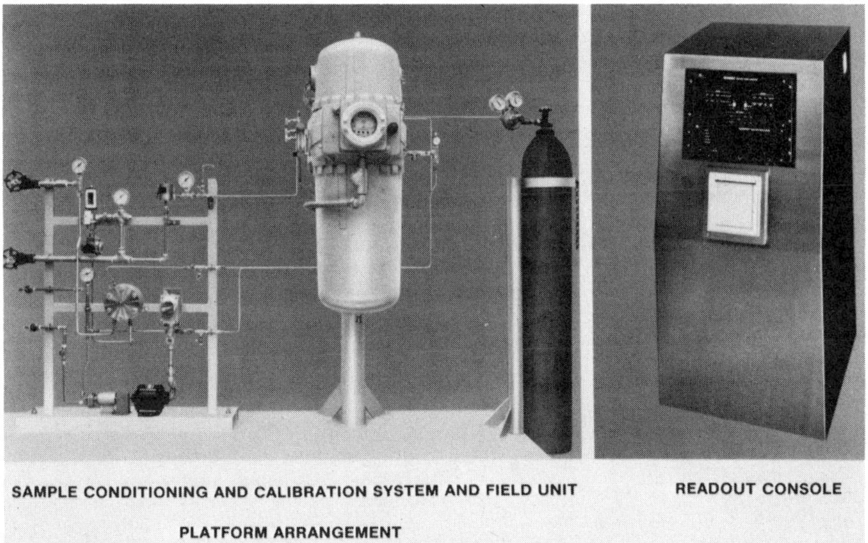

FIGURE 3.18 Monirex boiling point monitor. (Courtesy of Universal Oil Products Inc.)

The complete system is shown in Fig. 3.18. The BPM is capable of displaying digitally up to 18 boiling points, each with a reproducibility quoted as ±1%. The analysis range is normally hydrocarbons from C_3 to nominal nC_{22}, (although others are also available on request). Total instrument cycle time is 4 min, with a response time of less than 2 min after sample injection.

BIBLIOGRAPHY

Bajek, W. A., Sampson, R. W., and McLaughlin, J. H. An on-stream boiling point monitor employing a microcomputer. *8th IMECO Congr.*, Moscow, USSR (May 1979).

Bajek, W. A., Kuchar, P. J., and Remec, A. A. B. Increased production of valuable products made directly through the use of full distillation range monitors. *Instr. Conf.*, Shanghai, China (April 1983).

Clevett, K. J. Process stream analyzers for flash- and boiling-point measurement. *Instr. Rev.* (Dec 1966, Jan, Feb 1967).

4

Flash Point Measurement

4.1 INTRODUCTION

Flash point is an important characteristic of petroleum distillates; therefore, its application is mainly confined to the petroleum industry. The flash point of a sample may be defined as the lowest temperature of the sample, corrected to a pressure of 760 mm Hg, at which the application of an ignition source causes the vapor to ignite under the specified conditions of the test. At the flash point temperature, the sample liquid produces sufficient vapor to form an ignitable mixture with air.

Flash point measures the response of the sample to heat and flame under controlled laboratory conditions. It is only one of a number of properties that must be considered in assessing the overall flammability hazard of a particular material.

The presence of a highly volatile and flammable material in a relatively non-volatile or non-flammable material can be indicated by a decrease in flash point temperature. An example is the contamination of kerosene by gasoline. Determination of the flash point temperature of middle distillate products such as kerosene, gas oil, white spirit, and diesel fuel thus provides a criterion of product quality. Too high a flash point indicates that the product may not burn properly, too low a flash point indicates that the product may ignite prematurely.

4.2 LABORATORY METHODS OF MEASUREMENT

Flash point is a *man-made* parameter and the methods devised to measure it have been used in the petroleum industry for many years. The ASTM standard test methods for flash point determination we are concerned with here (D56, D92, and D93) are among the oldest and most widely used of all tests on petroleum products. Therefore, calibration of on-line flash point analyzers must be made with reference to these tests and the analyzer engineer should be familiar with the basic procedure, the repeatability, and reproducibility of the results.

4.2.1 Tag Closed Cup Method, ASTM D56-79

This method covers determination of the flash point of all mobile liquids with a flash point below 200°F and a viscosity of less than 5.5 cSt at 104°F or less than 9.5 cSt at 77°F. A typical manual Tag closed tester is shown in Fig. 4.1. The sample, at a temperature at least 20°F below the expected flash point, is placed in the test cup. With the lid closed, heat is supplied so that the sample temperature increases by 2°F/min ±6 sec for samples with flash points below 140°F. For flash points above 140°F, the rate of increase should be 5°F/min ±6 sec. A number of ASTM thermometers are available for temperature measurement, depending on the range to be covered.

When the sample temperature is about 10°F below the expected flash point, the lid of the test cup is removed and a small test flame inserted momentarily into the vapor space. Insertion and removal should take 1 sec. Application of the flame is repeated after each 1°F rise in temperature (2°F rise for samples with flash points above 140°F). The source of heat is removed and testing discontinued when a distinct flash is observed in the interior of the flash cup. The lowest temperature when such an ignition occurs is recorded as the sample flash point. The flash point temperature is corrected to a barometric pressure of 760 mm Hg. The repeatability of the test is 2°F for flash points below 140°F and 3°F for flash point between 140 and 199°F.

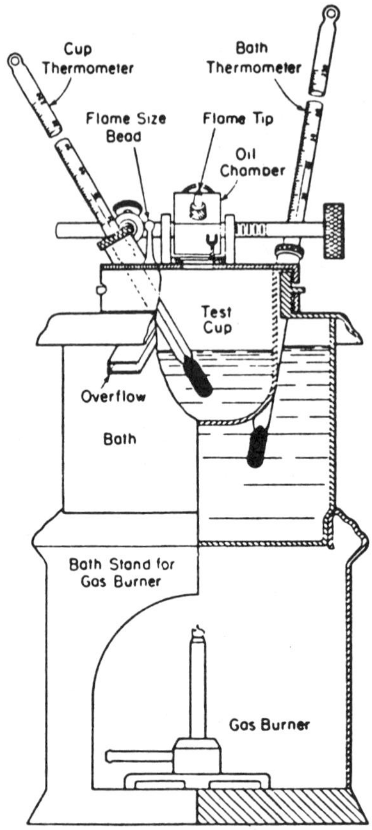

FIGURE 4.1 Manual Tag closed cup tester.

4.2.2 Pensky-Martens Closed Cup Method, ASTM D93-80

This method covers determination of the flash point of fuel oils, lubricating oils, and other viscous materials with flash points up to 700°F. The method is not applicable to drying oils, solvent-type liquid waxes, or cut-back bitumen. The test apparatus shown in Fig. 4.2 is very similar to the Tag apparatus, except that a stirrer assembly is included to ensure complete mixing of the sample and the sample test cup is heated directly rather than being immersed in a heated bath.

The sample, at a temperature at least 30°F below the expected flash point, is placed in the test cup and heat is supplied so that the sample temperature increases by 9–11°F/min, the sample being continuously stirred. There is a choice of two ASTM thermometers covering the ranges 20–230°F and 200–700°F, depending on sample flash point. At certain times during the heating cycle the heating is discontinued and a test flame inserted into the cup vapor space. The operation of insertion and removal should take 1.5 sec.

If the sample is known to have a flash point of 230°F or below, the test

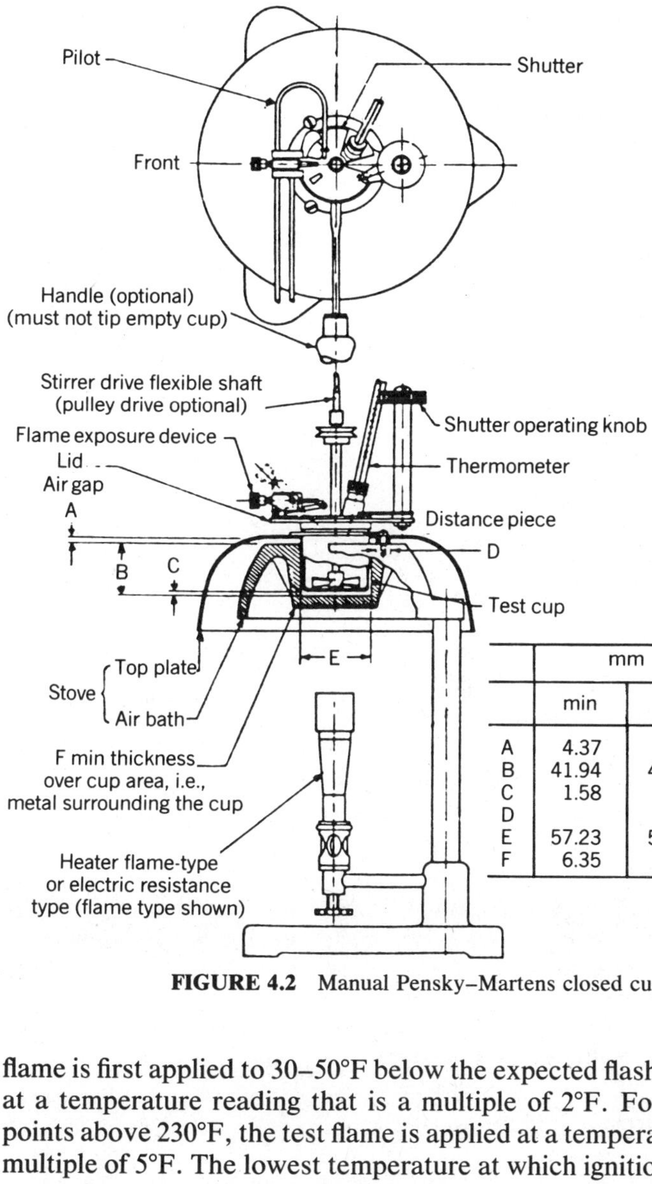

	mm		in.	
	min	max	min	max
A	4.37	5.16	0.172	0.203
B	41.94	42.06	1.651	1.656
C	1.58	3.18	0.062	0.125
D		9.52		0.375
E	57.23	57.86	2.253	2.278
F	6.35		0.25	

FIGURE 4.2 Manual Pensky–Martens closed cup tester.

flame is first applied to 30–50°F below the expected flash point and thereafter at a temperature reading that is a multiple of 2°F. For samples with flash points above 230°F, the test flame is applied at a temperature reading that is a multiple of 5°F. The lowest temperature at which ignition occurs is recorded as the flash point of the sample. The repeatability of this test is 4°F for flash points between 95 and 220°Fa and 10°F for flash points above 220°F.

4.2.3 Cleveland Open Cup Method, ASTM D92-78

This method covers the determination of the flash and fire points of all petroleum products, except fuel oils, with an open cup flash point below 175°F. The sample, at a temperature at least 100°F below the expected flash point, is placed in the open test cup and heat is applied so that the sample temperature rise is 25–30°F/min. When the temperature is 100°F below the expected flash point, the heat is decreased to give a sample temperature rise of 9–11°F/min. Starting at least 50°F below the expected flash point, a test flame is passed across the center of the cup and thereafter applied at a temperature reading that is a multiple of 5°F. The lowest temperature at which a flash appears at any point on the surface of the liquid is recorded as the open cup flash point. To determine the *fire point,* the sample is heated further and the test flame applied every 5°F until the liquid ignites and continues to burn for at least 5 sec. The temperature at which this occurs is recorded as the fire point of the sample.

4.2.4 Automatic Flash Point Testers

The three ASTM methods are described as they would be carried out manually in the laboratory by a technician. There are, however, commercially available automatic laboratory-type instruments for these tests. A number of factors indicate their use—manpower savings, shorter testing time, and using smaller samples. They are particularly justified in laboratories with a heavy flash point testing load.

The automatic ASTM D56 apparatus is designed to run accelerated tests. The operator is required to pour a quantity of sample into the sample entry point; thereafter, the test proceeds automatically. Sample and air are heated continuously at a controlled rate and the air-vapor mixture is exposed to an intermittent, high-voltage spark. A recorder plots the temperature rise of the sample. Ignition is detected by a sensitive differential mass thermocouple, at which time the spark is stopped, the sample drained, and the tester cooled, ready for the next test run.

The automatic ASTM D92 apparatus is similar in operation, except that ignition is detected by a capacitance probe and sample temperature is measured by a platinum resistance element.

4.3 PROCESS FLASH POINT ANALYZERS

4.3.1 GCA Precision Scientific Group (United States)

PSG manufactures the widest range of on-line flash point analyzers available. In 1958, Precision Scientific Development Company, as the company was then known, produced the first on-line flash point recorder under license from Standard Oil Company (Indiana). This analyzer rapidly became a leader in its field. Since that time, a wealth of operating experience has been gained, leading to considerable improvements in design over the years. There are now four different models to choose from, depending upon the application. Note that the GCA/PSG range of flash point analyzers are not truly continuous in operation, since there is a finite sample heating time prior to vapor ignition and flash point detection.

(a) General Flash Point Analyzer, Model 44650

This is designed for measurement of the flash point of hydrocarbon streams with viscosities up to 2000 cSt at 100°F and flash point temperatures from 140 to 600°F. The instrument supersedes the original High Range Flash Point Monitor, Model 44614, and is manufactured under license from Shell Development Company. The analyzer's principle applications are in heavy fuel oil blending and solvent refining. Readings may be correlated with ASTM D92 (Cleveland) and D93 (Pensky-Martens) test results.

The Model 44650, shown schematically in Fig. 4.3, consists of three main assemblies—the pump assembly, the measuring chassis, and the control

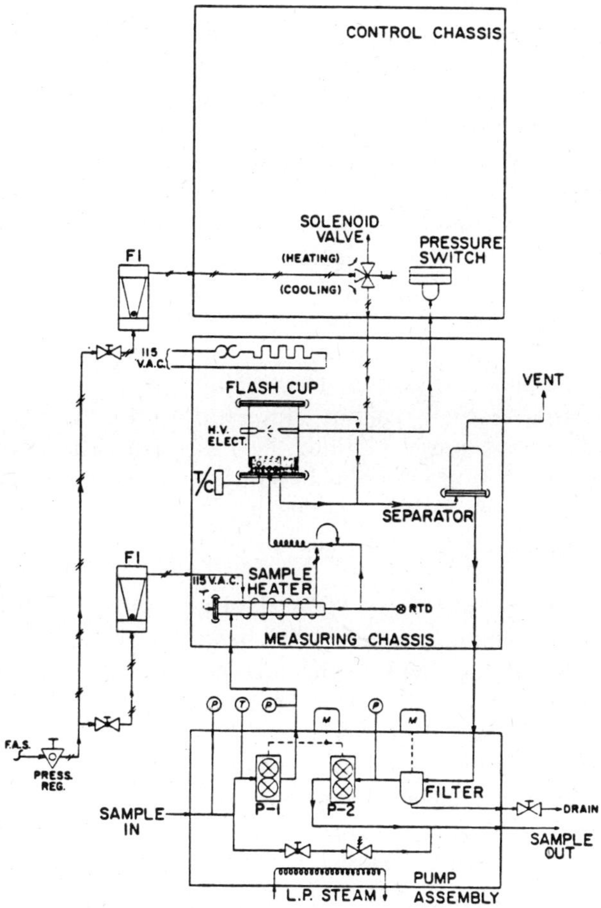

FIGURE 4.3 General flash point analyzer Model 44650. (Courtesy of GCA Corporation)

chassis, mounted together on a free-standing framework with panel. Process sample enters the pump assembly where one head of a duplex pump (P-1) delivers sample to the measurement assembly at a constant flowrate of 25 ml/min. The sample passes through a horizontal heater and is immediately mixed with a regulated flow of combustion air (1000 ml/min max.) preheated in a coil of tubing wrapped around the sample heater. Heated sample and air enter the flash cup, flooding the base of the cup, then pass over an annular orifice before overflowing to drain.

The orifice provides for uniform distribution of hot vapor throughout the flash chamber. The vapor rises past an electrode that produces a high-voltage (HV) spark each second during the heating cycle. Vapor then passes out of the flash cup into a separator together with the liquid flow. Liquid and any condensed vapor flow by gravity to the pump assembly, where the second head (P-2) of the duplex pump pulls spent sample through a motor-operated, self-cleaning filter (to remove coke particles and protect the pump) and returns it to the fast loop circulating system.

The control chassis contains the set point potentiometer rotated by a reversible motor. At the start of the heating cycle, the potentiometer begins to increase the set point to a solid-state temperature controller, which continuously monitors sample temperature as measured by a resistance bulb (RTD) at the outlet of the sample heater. The controller applies continuously increasing voltage to the heater, increasing sample temperature at a constant rate until the heated vapor ignites when exposed to the HV spark at the electrode. The force of ignition is sensed by a pressure switch, which then initiates the cooling cycle.

During the cooling cycle, normally less than 30 sec duration, three actions occur: the HV spark circuit is de-energized; the set-point motor reverses, reducing the set-point temperature to the controller; and a solenoid valve actuates, diverting a high volume flow (up to 8 1/min max) of unheated control chassis purge air into the flash cup. The purge air flows over the electrode (to retard coke build-up), flushes spent gas from the flash cup, and cools the walls of the flash cup so that the incoming vapor temperature is not affected by the heat of ignition that initiated the cooling cycle.

Since the cooling cycle is of short duration, there is little significant change in liquid sample temperature in the base of the flash cup. Thermocouples located in this liquid measure the flash point temperature. The control logic therefore tends to maintain the sample temperature at its flash point at all times.

The analyzer is suitable for operation in NEC Class 1, Group D, Division 1 hazardous locations, but should be installed where it is protected from direct exposure to ambient conditions. Analyzer output is from dual isolated thermocouples, both Type J (iron-constantan) and Type K (chromel alumel). Repeatability is quoted as $\pm 3°F$.

(b) Flashpointer, Model 44607

This is designed for the continuous, on-line measurement of the flash point, within the range 50–250°F, of hydrocarbon streams with viscosities up to 215 cSt at 100°F. Analyzer readings correlate with ASTM D-56 (Tag closed cup) test results. Correlation with the ASTM D-93 (Pensky-Martens) method can also be established by laboratory analysis within the measuring range of the analyzer. The analyzer's main applications are in crude distillation and blending operations.

The instrument, shown in Fig. 4.4, operates on the same basic principle as the Model 44650, except that the flash cup design is somewhat different and a differential thermocouple assembly is used to detect vapor ignition. Sample and combustion air enter the base of the flash cup assembly at a constant, controlled flowrate and at a temperature at least 30°F below the expected flash point. A variable transformer applies a continuously increasing voltage to the sample heater during the heating cycle. The sample/air mixture passes around a sheathed liquid thermocouple, then into the flash chamber. Vapor and air separate from the liquid, pass around a splash shield, then past a low-mass vapor thermocouple and HV electrode before escaping to vent. The temperature difference between the two thermocouples remains relatively

FIGURE 4.4 Flashpointer Model 44607. (Courtesy of GCA Corporation)

constant until the hot vapor-air mixture ignites in the presence of the HV spark.

At this point the temperature difference increases dramatically because of the heat of ignition and the fast response of the vapor-space thermocouple. The increase in differential temperature produces a low-level dc current to saturate a magnetic amplifier, which in turn energizes a sensitive relay to initiate the cooling cycle. During this period the heating is stopped, the ignition circuits are shut down, the variable transformer is driven in a down-scale direction, and a dc voltage is imposed across the liquid thermocouple.

The cooling cycle may be manually adjusted and is usually set to reduce the liquid thermocouple temperature by about 30°F. The dc current on this thermocouple immediately reduces its output following ignition, thereby preventing overshoot above the flash point temperature. The analyzer is suitable for installation in NEC Class 1, Group D, Division 1 hazardous locations. Cycle time is quoted as 2 to 7 min (depending upon sample inlet temperature) and repeatability as ±2°F. Required analyzer sample flowrate is 30–50 ml/min.

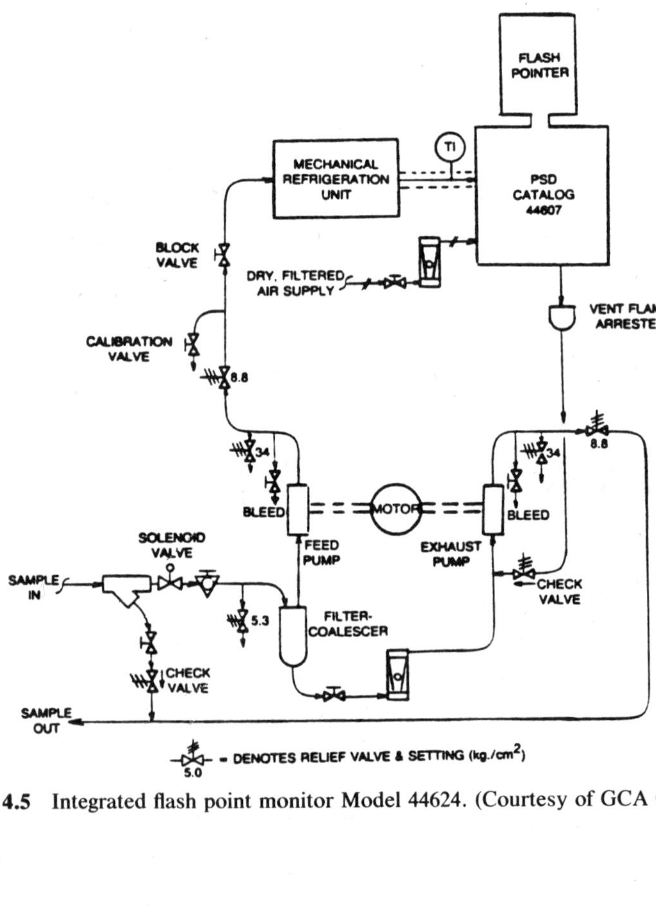

FIGURE 4.5 Integrated flash point monitor Model 44624. (Courtesy of GCA Corporation)

The standard analyzer output is a millivolt thermocouple signal (iron-constantan for Model 44607, chromel-alumel for Model 44610) for sample flash point temperature measurement. However, the analyzer can be adapted to provide a continuous 4–20 mA dc output proportional to flash point temperature by the use of a thermocouple-to-current signal converter and sample-hold module.

(c) *Integrated Flash Point Monitor, Model 44624*

This is shown in Fig. 4.5 and is essentially the Flashpointer described in Section 4.3.1(b). It is marketed as a complete analysis package, including sample handling system, which consists of fast loop filter, coalescer, mechanical refrigeration unit, and sample shut-off solenoid valve, connected as shown.

(d) *Pipeline Flash Point Monitor, Model 44620*

This is shown in Fig. 4.6 and is essentially the Integrated Flash Point Monitor just described, modified to permit operation of the system on intermittent service, with provision for local or remote start-up and for initial warm-up on a recirculating fuel sample. The modified system includes an air compressor, a fuel reservoir, and a recirculation system. The analyzer system was developed by PSG for the specific requirements of pipeline operations.

Actuation of the "standby" or recycle switch, energizes three solenoid valves. Two establish recirculation from the reservoir, the third closes off the coalescer bypass flow to prevent circulation of this stream back to the reservoir. The reservoir has a capacity of about 4.5 gal, which, at 35 ml/min sample flowrate, give a recycle time for the reservoir of about 8 hr.

4.3.2 Hone-Addison Ltd. (Great Britain)

Hone manufactures one process flash point analyzer, the Flashar, which is designed to measure the flash point of petroleum products up to a maximum of 480°F with an accuracy and repeatability equal or better than the ASTM D-93 test method. The operating principle of this instrument is considerably different from that of other process analyzers described in this chapter; therefore, the Flashar may be classed as a truly continuous analyzer.

It is a well-known fact that certain metals such as palladium and platinum catalyze the oxidation of hydrocarbons. Petroleum industry processes such as catalytic cracking and catalytic reforming are based on a similar principle. Very complex reactions take place on the surface of the catalyst and have been the subject of research for many years. The effect of these reactions is to generate heat, the quantity of which depends on the number and size of the hydrocarbon molecules adsorbed on the surface of the catalyst. Normally, for cracking or oxidation to take place, hydrocarbons must be heated

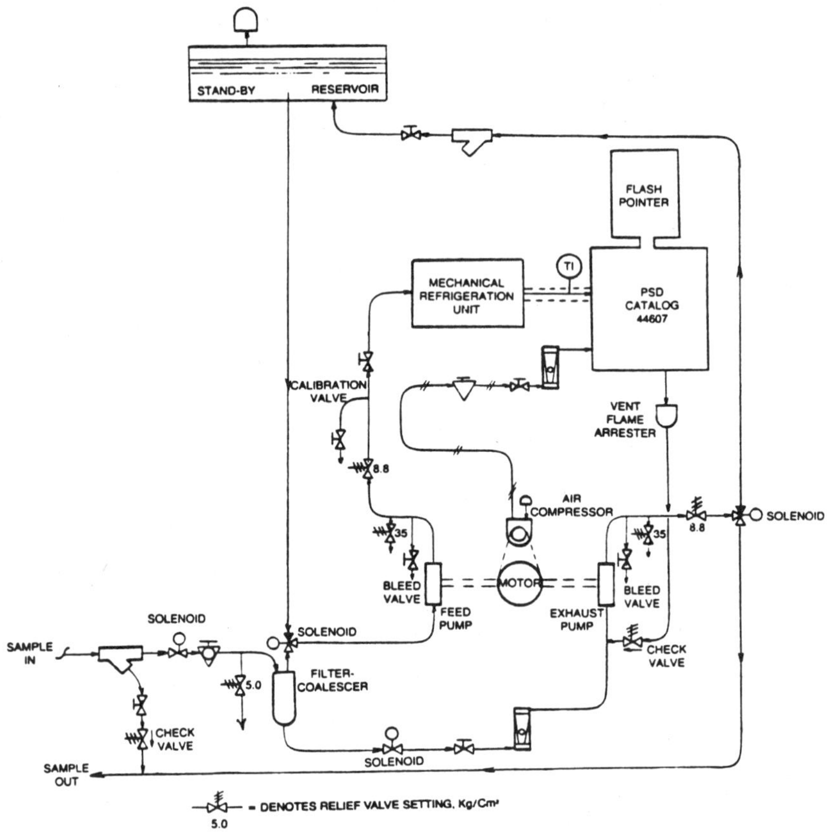

FIGURE 4.6 Pipeline flash point monitor Model 44620. (Courtesy of GCA Corporation)

to elevated temperatures (as in the thermal cracking and reforming processes). However, in the presence of a catalyst, these reactions occur at much lower temperatures; if sufficient heat is generated to offset the heat lost to the surroundings, the reaction then becomes self-propagating.

When hydrocarbon vapors are mixed with air, an vapor/air concentration at which the components will react together spontaneously when exposed to a flame or spark is eventually obtained. This is the principle of the laboratory tests for flash point, described earlier, where the hydrocarbon sample is heated to produce the critical vapor/air concentration required for the vapor to ignite. The *Flashar* flash point analyzer relies on the fact that the vapor concentration above a hydrocarbon liquid is essentially constant at the flash point and is independent of the actual flash point temperature.

The analyzer uses a catalytic detector to monitor the hydrocarbon vapor concentration that corresponds with the flash point. The vapor reacts with the catalyst to produce a reproducible catalyst temperature; the output of the detector is the input to a control system that seeks to maintain the sample temperature at a value that will sustain the required vapor concentration above the liquid. The sample temperature so maintained is the flash point temperature of the sample.

A schematic diagram of the Flashar is shown in Fig. 4.7. Process sample, which should be free of water and entrained solids, enters the sample heater from the metering pump, where its flowrate is set at approximately 2 l/hr. The sample then flows into the mixing chamber, where it is mixed with a constant flow of preheated air. Excess sample passes to the overflow chamber and then to drain via a flame trap. The vapor/air mixture produced passes through the detector housing (where the vapor concentration is measured) and then to atmospheric vent.

The catalytic detector, which is in the vapor/air stream, produces a variable voltage output depending upon its temperature (and hence the vapor concentration). This voltage signal is the input to the sample heater controller, where it is compared with the set point temperature value and an output produced that is proportional to the error. This output is used to control the electrical supply to the sample heater. The sample temperature is measured by a platinum resistance thermometer attached to the liquid section of the mixing chamber. The temperature of the detector housing is maintained within close limits during analyzer operation to eliminate the effects of ambient temperature variations.

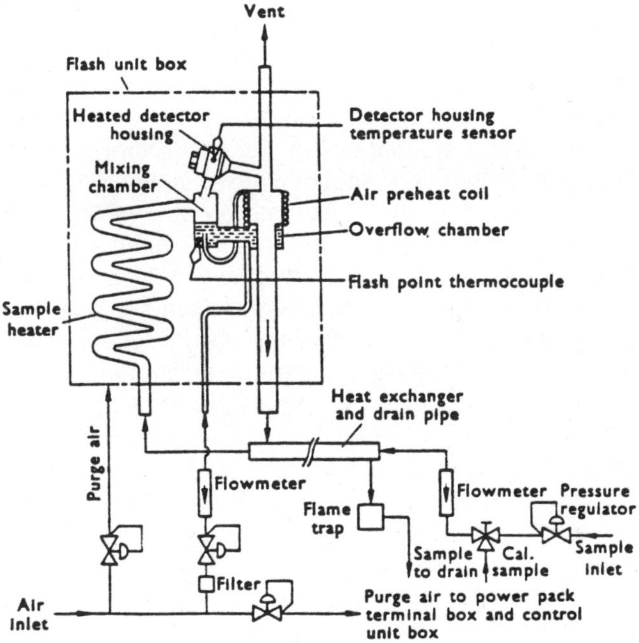

FIGURE 4.7 Flashar flash point analyzer (flow diagram). (Courtesy of Hone-Addison Ltd.)

FIGURE 4.8 Flashar flash point analyzer: equipment layout. (Courtesy of Hone-Addison Ltd.)

The analyzer consists of four flameproof enclosures mounted on a free-standing framework, as illustrated in Fig. 4.8. The right-hand enclosure contains the flash unit and the upper left-hand enclosure the control unit. The power supply unit is mounted below the control unit. At the base of the frame is a fourth enclosure containing the drive motor for the sample metering pump, the starting capacitor, fuse, and on/off switch. The control unit contains a transmitter consisting of a high-gain amplifier that translates the input from the platinum resistance thermometer into a 4–20 mA dc output. The transmitter is ranged to customer requirements.

A high temperature cut-out isolates the power supply in the event of a detector failure or loss of sample. In addition, an interlock is provided in the detector temperature control circuit, which prevents power being supplied to the main heater until the detector housing has reached its correct operating temperature. An additional safety feature is the standby facility. The standby switch isolates the temperature control circuit supplying power to the sample heater but maintains power to the detector housing preheater. This is useful if the analyzer is to be used in batch type operation, since it considerably reduces analyzer start-up time.

The analyzer is normally supplied with one of two alternative safety systems. Both systems operate on the air purge and pressurization principle and are suitable for installation in either Zone 1 or Zone 2 hazardous locations. One system uses an intrinsically safe timer relay and the analyzer has full PTB approval to VDE 0171. The second system uses a pneumatic timer relay designed to provide protection to the new CENELEC standard En 50-016. The manufacturer can also provide custom engineered sample handling systems for the Flashar flash point analyzer.

4.3.3 Carlo Erba, Scientific Instruments Division (Italy)

The Carlo Erba Flash-o-Matic Series 250 Flash Point Analyzer is available in two versions, the low temperature (LT) version, suitable for measurement of flash points within the range 20–120°C, and the high temperature (HT) version, covering the range 100–270°C. The two instruments operate on the same principle, the only difference between them being the design of the sample heater and associated electronics, which has a higher rating for the Model 250 HT. The analyzer consists of three separate units—the analyzer unit, which is field mounted, the control unit, and recorder, both of which are remotely located. A schematic diagram of the analyzer unit is shown in Fig. 4.9.

Process sample is drawn from the sampling system by a constant rate pump, cooled by a heat exchanger to reduce its temperature to 20–30°C below the minimum expected flash point, then passed to the test cup. There it is mixed with a constant flow of air. The mixture passes through an annular space in the test cup, where it is heated by an electrical heater fitted around the cup. The heated sample/air mixture then overflows into the main body of the test cup, where the liquid temperature is measured by a thermocouple

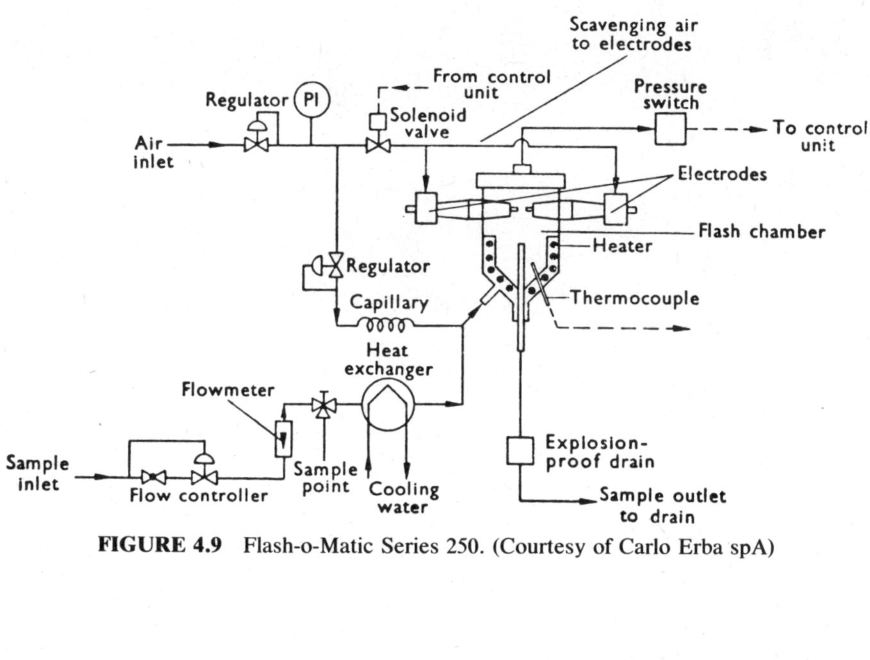

FIGURE 4.9 Flash-o-Matic Series 250. (Courtesy of Carlo Erba spA)

(iron-constantan or chromel-alumel) before it passes to drain, and the vapors are exposed to an HV spark. The negative (grounded) electrode of the spark system is a capillary tube through which scavenging air passes during the cooling cycle. The spark frequency is adjustable and is usually set at 40–50 sparks per minute.

The flash detector is a diaphragm-type pressure switch that senses the pressure wave produced by vapor ignition and actuates a relay, which in turn locks the recorder, starts the cooling cycle, and switches a solenoid valve to allow scavenging air to enter the test cup. This action reduces carbon build-up on the electrodes and vents hydrocarbon vapors from the test cup. A more detailed view of the test cup is shown in Fig. 4.10. The cooling cycle is adjustable by means of a timer.

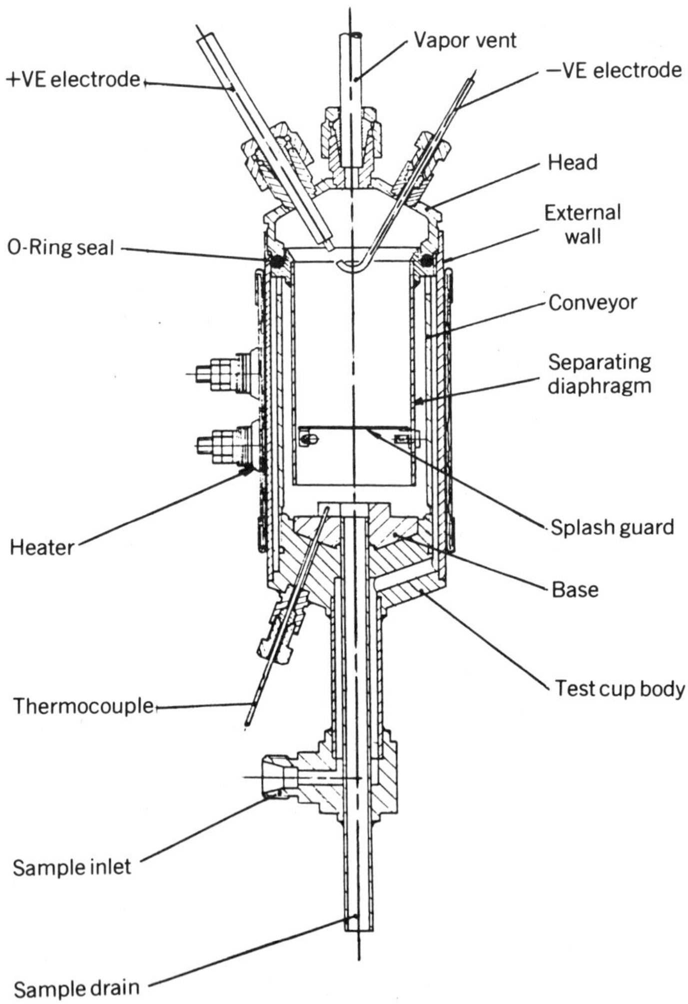

FIGURE 4.10 Flash-o-Matic Series 250 (test cup). (Courtesy of Carlo Erba spA)

The analyzer unit, which is suitable for installation in Zone 1 hazardous locations, consists of two enclosures, one housing the measurement assembly, the other housing all electrical control equipment such as power regulation and spark generation circuitry. The complete system is mounted on a free-standing metal framework, together with sampling accessories. The control unit, which is suitable for panel mounting in a non-hazardous location, contains the auxiliary electrical equipment for automation of the analyzer unit, including power switch, pump stop-start switch and lamp, the master timer for the cooling cycle, and pilot lamps for cycle indication. Also included is a trend/bar switch, which allows the recorder to provide either a bargraph or trend presentation. For trend, the recorder balancing motor is locked during the analytical cycle (heating and cooling) and released for a very short time only at the flash point.

5

Pour Point, Cloud Point, and Associated Measurements

5.1 INTRODUCTION

Pour point and *cloud point* are two important characteristics of petroleum products. The pour point temperature of a material is the lowest temperature at which the material will pour (or flow) under a prescribed set of conditions.

The cloud point temperature of a material is the temperature at which a waxy cloud or haze appears in the liquid when cooled under a prescribed set of conditions.

Pour point and cloud point are important specifications of fuels for jet engines, marine and autodiesel engines, and domestic heating equipment. They define the lowest temperature at which the product can be handled without causing pumping difficulties and wax plugging of filters. They are man-made specifications, and the standard laboratory tests described in this chapter are lengthy, tedious, of low accuracy, and subject to human error and judgment. In addition, the delay incurred between taking the sample and obtaining the test result is unacceptable in many cases, particularly when product is being blended in-line to a pour or cloud point specification.

Frequent and accurate measurement of these two parameters is required by petroleum refineries for a number of reasons. Continuous on-line measurement enables finished product quality to be controlled precisely and allows more economic use of available feedstocks, thereby aiding optimization of refinery operations. By operating closer to the pour point or cloud point specification of a particular product, on-grade production can be increased significantly, with consequent savings in valuable feedstocks and blendstocks. Applications are mainly confined to the petroleum industry and include monitoring of finished products from in-line blending processes, monitoring of feedstocks for jet fuels, kerosine, marine diesel fuels, and autodiesel fuels, and in lubricating oil production. A good example of the use of an on-line pour point analyzer is in the blending of heavy naphtha with catalytic gas oil to produce home heating oil or diesel fuel. The repeatability of the process analyzer permits fine tuning of the blending system to control the product to within 2°F of the specification. This saves a significant amount of heavy naphtha, which can then be used for jet fuel or kerosine production or for feedstock to a catalytic reformer to produce high-octane gasoline blendstock. All of these products carry a higher market value.

5.2 LABORATORY METHODS OF MEASUREMENT

ASTM test method D97-66 for pour point measurement is one of the most traditional tests and has been used as a standard in the petroleum industry since its adoption in 1914 as method GO11. The method was also adopted as a joint Institute of Petroleum standard in 1965 with the designation IP15/67. It has remained relatively unchanged to this day. The procedure for cloud point measurement originally formed part of method D97, but a separate ASTM method D2500-66 was adopted in 1967 and became a joint ASTM/IP standard in the same year, with the designation IP219/67. The method was reapproved in 1976.

As with most instruments that measure physical properties, process analyzers must finally be correlated to the standard laboratory test; therefore some knowledge of the procedure and repeatability is useful to the analyzer engineer.

5.2.1 ASTM Method D97-66

This method may be used to determine the pour point of any petroleum oil, with certain reservations regarding reproducibility and repeatability. In the case of waxy oils such as fuel oils, the low-temperature flow characteristics of the material depend to a large extent upon handling, storage, and thermal history. These characteristics may not, therefore, be truly indicated by pour point, and it may be difficult to obtain reproducible results.

For the test, a cylindrical, flat-bottomed test jar of clear glass is used, marked to indicate the level to which the vessel should be filled with the test sample. The test jar is mounted in a jacket with a gasket fitting snugly around it and loosely inside the jacket. The test jar is seated on a disc and fitted with a thermometer of specified range, conforming to the requirements prescribed in ASTM specification E-1. The complete assembly is immersed in a cooling bath, the size and shape of which are optional. For determination of pour points less than 50°F, two or more baths are required as laid down in the following procedure.

The sample is first heated to 115°F, then cooled to a temperature based on the expected pour point of the sample. The test jar is immersed in a cooling bath maintained at between 30 and 35°F. Testing commences at a temperature 15°F above the expected pour point for products with pour points above 100°F or, for other products, at a temperature 20°F above the expected pour point. At each temperature that is a multiple of 5°F, the test jar is removed from the jacket, tilted just enough to determine whether there is movement of the liquid, and replaced. The complete operation of removal, tilting, and replacement should take no longer than 3 sec. When wax crystals have formed in the liquid, great care must be taken not to disturb the mass of oil or allow the thermometer to shift in the sample; any such disturbance will lead to low and fictitious results.

If there is still movement of the sample when its temperature has reached 50°F, the test jar is transferred to a second cooling bath maintained at between 0 and 5°F. If the sample has not ceased to flow when its temperature has reached 20°F, the test jar is transferred to a third cooling bath maintained at between −30 and −25°F. For measurement of very low pour points, further cooling baths are required, each bath being maintained at a temperature 30°F below the preceding bath. In each case the test jar is transferred when the sample has cooled to a temperature 50°F above that of the new bath.

If the sample does not flow when the test jar is tilted, it is held in the horizontal position for 5 seconds and the temperature noted; 5°F is added to that temperature and the result recorded as the pour point of the sample. The repeatability of the test method is 5°F and its reproducibility (the same test conducted by different laboratories) is 10°F.

If the above procedure is followed for fuel oils and similar products, the result obtained is termed the upper pour point. The lower pour point is determined by heating the sample to 220°F while stirring, cooling it to 95°F, and carrying out the procedures described. Although the lower pour points show approximately the reproducibility quoted, the upper pour points are entirely dependent upon the thermal history of the sample.

5.2.2 ASTM Method D2500–66

This method covers only petroleum products that are transparent and have cloud points below 120°F. The apparatus used for the test is identical to that used for the pour point determination, except that the bulb of the thermometer is located at the base of the test jar rather than at the midpoint level of the liquid.

The sample is brought to a temperature at least 25°F above the expected cloud point and any moisture removed by a suitable method until the sample is perfectly clear. The test jar is then filled to the mark and immersed in a cooling bath maintained at between 30 and 35°F. At each temperature that is a multiple of 2°F, the test jar is removed, inspected for cloud, and replaced. If no cloud is observed when the sample has cooled to 50°F, the test jar is transferred to a second bath maintained at between 0 and 5°F.

Thereafter, the procedure is the same as for the pour point test, except that inspections for cloud are made at intervals of 2°F. When inspection reveals a distinct haze or cloud in the bottom of the test jar, the temperature reading is taken and recorded as the cloud point. For distillate oils, the repeatability of the test is 4°F and reproducibility is 8°F. For other oils, both repeatability and reproducibility are 10°F.

5.2.3 Correlation with Process Analyzer Readings

From these descriptions, it can be seen that the laboratory test methods for pour and cloud point measurement are traditional tests requiring meticulous attention to detail, they can be long and tedious and subject to human error. The process analyzers described later in this chapter are far more precise and not subject to human error. Therefore, it is often very difficult to obtain a precise correlation with the corresponding laboratory test and a compromise must be made.

5.2.4 Other Associated Laboratory Measurements

(a) Additional Tests for Residual Fuel Oils

As mentioned earlier, it is difficult to obtain reproducible results using the standard pour point test on residual fuel oils and heavy lubricating oils. In the past, pour point has been used as a guide for the pumpability of heavy oils, but it has been found that viscosity is a better criterion for determining the minimum temperature at which these products may be handled. The Institute of Petroleum carried out considerable work in this area with various laboratories and recommended that the viscosity of residual fuel oils should be used as a measure of *pumpability*.

The standard laboratory test for pumpability utilizes a rotational viscometer and is covered by ASTM method D3245-76 and by IP 230/74. The test is designed to give the storage and handling points for a given fuel oil, that is, the minimum temperatures required to store and handle the oil in normal oil-fired boiler installations. The ASTM method recommends the use of the Ferranti Model VH three-speed viscometer. The sample is placed in a bath set at a predetermined temperature and cooled at 1°F/min. A plot of viscosity versus temperature is made and the temperatures read off corresponding to viscosities of 6 P and 25 P respectively. The 6 P temperature is recorded as the *handling point* and the 25 P temperature as the *storage point*. The test can take up to an hour to complete, but it does give better data on product characteristics than the pour point test.

Another standard test for residual fuel oil is the measurement of *maximum fluidity temperature,* covered by ASTM method D1659–65. A sample of the fuel oil is heated to 220°F, divided into six parts, and cooled until solid. The aliquots are preheated to 100, 120, 140, 160, 180, and 200°F and each subdivided test sample poured into separate U-tubes. These samples are then cooled stepwise by immersion in successive baths. The six U-tubes containing the samples are then placed in the test bath and the samples tested for movement under prescribed pressure conditions at intervals of 5°F.

The approximate fluidity temperatures thus determined are confirmed using the additional U-tubes and the maximum fluidity temperature is reported. The fluidity temperature for a given preheating treatment is the lowest temperature (to 5°F) at which the fuel will flow 2 mm in 1 min in a 12.5 mm U-tube under a maximum pressure of 15.2 cm Hg. The fluidity test bath consists of a reservoir, stirrer, pump, and motor, a mercury manometer calibrated in 1-cm divisions, with a mark at 15.2 mm, and an automatic vacuum controller that gradually increases the vacuum applied to one end of the U-tube at the specified rate of 1 cm/4 sec. This is a long and involved test requiring meticulous attention to detail.

(b) Additional Tests for Distillate Fuels

Neither cloud point nor pour point tests really give a good indication of how well a distillate fuel will perform in vehicle service. The temperature at which the operability limit is reached lies somewhere between the two. The pour point is normally not sensitive enough, whereas the cloud point is too sensitive, particularly where flow improvers are used. These additives work by changing the growth habit of n-paraffin wax crystals formed when the distillate is cooled below its cloud point. The modified crystals are smaller and less cohesive and therefore flow more easily.

Considerable research, including field trials, was conducted in the mid 1960s in Europe and resulted in the introduction of the *cold filter plugging point* (*CFPP*) test method. This test has subsequently been shown to give an excellent indication of low temperature performance of diesel and heating oils, particularly when they are treated with flow improvement additives. The procedure for carrying out the CFPP test is covered in Standard AFNOR M70/042 and in IP.309. Other similar test methods are DIN 51428 and SIS 155122.

The apparatus for the IP.309 test is shown in Fig. 5.1. A sample of prescribed volume is placed in the test jar and cooled under specified condi-

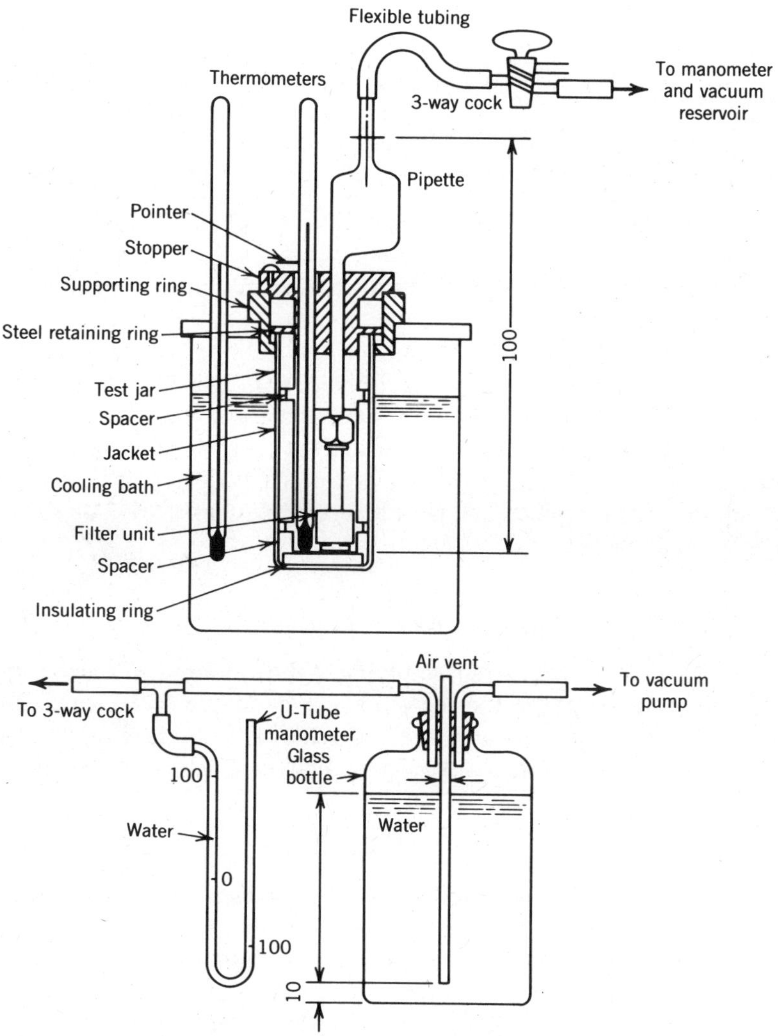

FIGURE 5.1 CFPP test apparatus per IP.309.

tions. At every 1°C fall in sample temperature, a suction of 200 mm water is applied to the pipette to aspirate the sample through a metal filter. The CFPP of the sample is the temperature in °C at the commencement of the aspiration, when it is no longer possible to draw 20 ml of sample through the filter in 60 sec. Three cooling baths are required to cover the full temperature range and the test apparatus must be moved from one bath to the next as the test proceeds. The specified bath temperatures are as follows:

CFPP	Bath Temp
Above −20°C	−34°C ±0.5
−20 to −35°C	−51°C ±0.5
Below −35°C	−67°C ±0.5

At present, the IP.309 method does not quote repeatability or reproducibility limits for CFPPs above 0°C.

5.3 PROCESS POUR POINT ANALYZERS

5.3.1 GCA Precision Scientific Group (United States)

The PSG General Pour Pointer, Model 44670 is designed to measure the pour point of hydrocarbon streams with viscosities up to 450 cSt at 100°F and pour points as low as −58°F. Readings may be correlated with ASTM D97/IP15 test results. A flow diagram of the analyzer is shown in Fig. 5.2. Operation of the analyzer is cyclic in nature and consists of five basic sequences. The analyzer is programmed so that it resets to sequence No. 1 on start up or upon any interruption.

(a) Forced Drain Sequence

Valves 4, 5, and 8 are energized, applying high-pressure (HP) air to the test cell. The electric current to the thermo-electric module (T/E) is reversed so that the test cell is warmed and force-drained of all residual sample. Valve 5 is de-energized to terminate this sequence; valves 4 and 8 remain energized.

(b) Fill Sequence

Valves 6, 7, and 9 are energized. The T/E module continues to heat the sample. Sample flows into the cell and out to drain via the overflow. Either HP or low-pressure (LP) air is applied to the test cell, depending upon sample viscosity. This air pressure prevents sample from entering the sensing lines of the analyzer. Valve 6 closes to terminate the sequence.

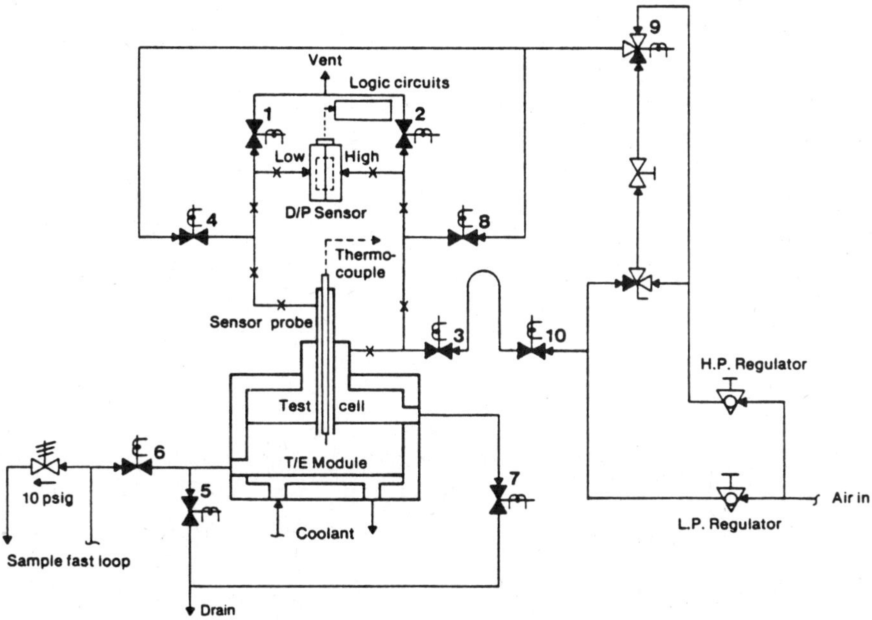

FIGURE 5.2 General pour pointer, Model 44670. (Courtesy of GCA Corporation)

(c) Level Sequence

Valves 4, 7, 8, and 9 remain energized, pressuring the surface of the sample and forcing surplus sample out to drain. Power to the T/E module is applied at normal polarity to initiate sample cooling. At the termination of this sequence, all valves de-energize and the sensor probe is submerged below the surface of the sample.

(d) Equalizing Sequence

Valves 1, 2, and 10 are energized. The test cell, DP sensor, and all connecting lines are vented to atmosphere through valves 1 and 2, establishing a condition of zero differential pressure in the sensing system. At the end of this sequence, all valves close; valve 10, in closing, traps LP air in the tubing between valves 3 and 10.

(e) Pulse Sequence

Valve 3 energizes. The LP air trapped between valves 3 and 10 expands into the test cell, producing a momentary pressure pulse against the liquid surface outside the sensor probe and against the HP diaphragm of the DP sensor, the output voltage of which rises immediately, reflecting this pressure pulse. The pressure pulse on the surface of the liquid also tends to push liquid up into the sensor probe, compressing the air inside the probe. As the sensor probe is connected to the LP side of the DP sensor, there is a reduction in DP sensor output voltage.

The control logic of the analyzer has a built-in fixed time delay, which gives the sample time to compress the air inside the sensor probe after each pressure pulse. The output of the DP sensor is then compared with an adjustable set-point voltage. If the output is below the set-point, another equalizing sequence and pulse sequence occur.

As the process sample cools, its resistance to flow increases and its ability to compress the air inside the sensor probe decreases, resulting in an increase in DP sensor output. Eventually, output exceeds set-point, tripping a peak holding relay and resetting the analyzer to the forced drain sequence to begin another cycle. A thermocouple installed inside the sensor probe measures sample temperature continuously. The pour point temperature is the lowest temperature attained by the sample in each measuring cycle.

The analyzer is mounted on a free-standing framework as shown in Fig. 5.3 and is suitable for installation in NEC Class 1, Group D, Division 1

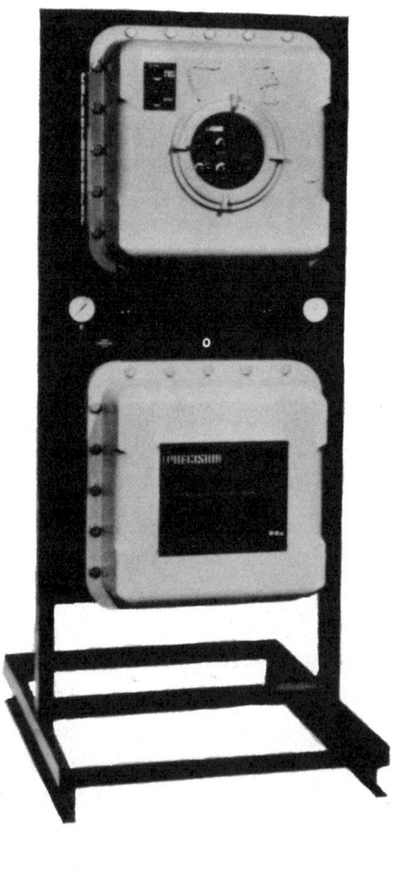

FIGURE 5.3 Model 44670: equipment layout. (Courtesy of GCA Corporation)

hazardous locations. It consists of two compartments housing the analyzer unit and the electronics unit respectively. Sample flowrate should be controlled at about 50 ml/min and pressure at 20 psig max. Spent sample must be discharged to atmospheric pressure. Sample temperature should be at least 20°F but not more than 122°F above the expected pour point temperature. Analyzer measuring range is −50 to +50°F and repeatability is quoted as ±5 deg F. Analyzer cycle time depends upon the pour point, but may be manually selected at 10, 30, or 60 min. By using a refrigerated cooling system, operating temperatures as low as −75°F may be attained.

5.3.2 Fluid Data Inc. (United States and Great Britain)

The Hallikainen Model 444B Pour Point Analyzer is designed to measure the pour point of hydrocarbon streams within the range −30°C to +38°C using 15°C cooling water (−35°C using 5°C coolant). A schematic diagram of the measurement and detection system is shown in Fig. 5.4.

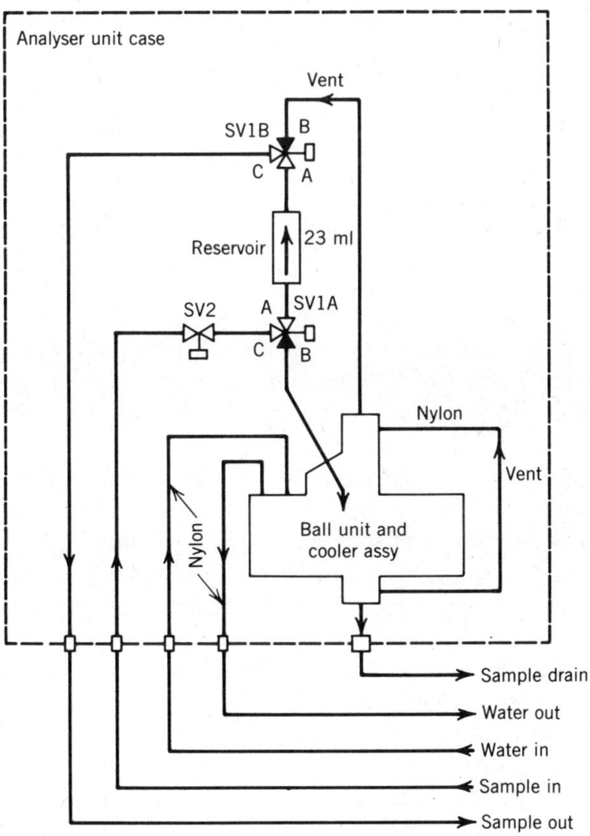

(*a*)

FIGURE 5.4 (*a*) Hallikainen Model 444B pour point analyzer. (Courtesy of Fluid Data Inc.)

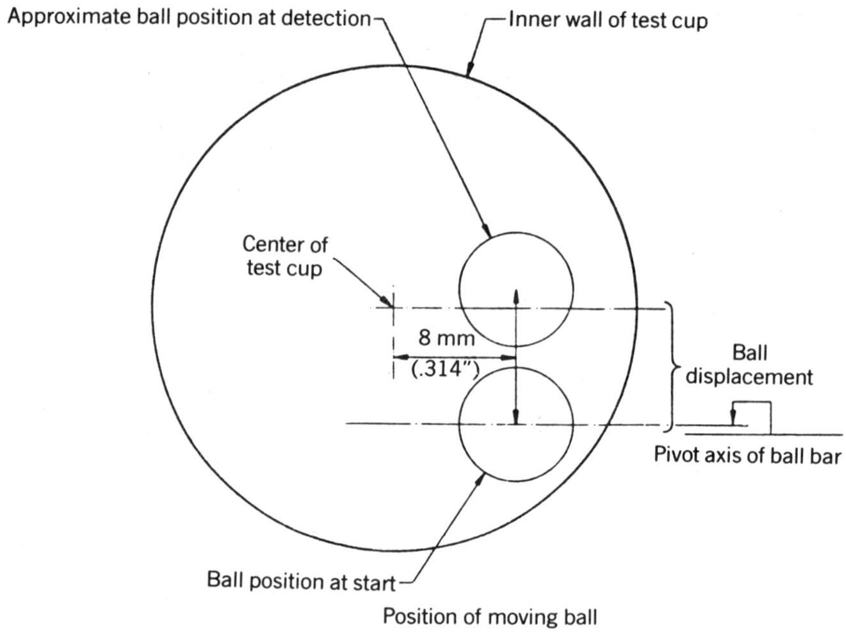

(b)

FIGURE 5.4 (*b*) Hallikainen Model 444B pour point analyzer. (Courtesy of Fluid Data Inc.)

Process sample flows continuously through a 23-ml reservoir by direction of three solenoid valves. SV-1 is normally energized (open) except under fault conditions. When SV-2 and SV-3 are energized, sample is cut off in the reservoir and feeds under gravity into the sample cup, while an equivalent volume of air passes up the vent line from the reservoir. The sample cup is positioned at the center of the cooling unit consisting of two dual-Peltier assemblies. The sample cup is rotated at $\frac{1}{6}$ rev/min by a motor mounted below the cooler unit. A second motor operates a drain valve fitted in the stem of the sample cup. This drain valve is held open for 1 min at the start of each analysis cycle.

The detection system consists of a tungsten carbide ball suspended from low-friction conical pivots by a thin steel tube and located as shown in Fig. 5.4. A thermocouple is brazed to the surface of the ball and measures sample temperature. Initially this is 46°C. As the sample cup rotates, sample temperature is reduced by the cooling unit and the force exerted on the ball by the rotating sample increases. This force acts on the ball against gravitational force and eventually causes sufficient displacement to operate an optical detection system, consisting of a lamp, shutter, and photocell assembly. When the final calibrated displacement is reached, the shutter breaks the light beam, the detection circuit is switched, and a peak picker operates. The control circuits are then switched to reverse the current through the Peltier cooler, which now acts as a sample heater, and a new measurement cycle is initiated.

The temperature at which this action occurs is the solid point. By definition (ASTM D97) the pour point is 3°C higher than the solid point measured in the laboratory test. Since the solid point is measured in 3°C steps, the solid point may occur at any temperature within the 3°C band between solid point and pour point. A mid-point correction would be 1.5°C. The measurement thermocouple signal is amplified to give two separate outputs:

- 2–10 V dc corresponding to a sample temperature of −40 to +60°C.
- A 4–20 mA dc isolated output corresponding to one of the following pour point scales:
  1. −40 to +60°C.
  2. 0 to +50°C.
  3. −40 to +10°C.

FIGURE 5.5 Model 444B: equipment layout. (Courtesy of Fluid Data Inc.)

The 2–10 V dc output of the thermocouple amplifier controls the analyzer program. The 4–20 mA dc output is derived from each solid point temperature of the sample that is detected. When the peak picker is operated, sample temperature plus 1.5°C is stored electronically and converted to a 4–20 mA dc signal, providing a trend output of the sample pour point. Visual readouts of the sample temperature and pour point temperature are provided on two front panel meters.

The analyzer is mounted on a free-standing framework as shown in Fig. 5.5 and consists of two compartments, the analyzer unit and the control unit, both of which may be provided with a suitable purge system to meet hazardous area installation requirements. Five safety devices are fitted as standard:

1. Cooling water temperature alarm.
2. Sample temperature alarm.
3. Sample cup high-level alarm.
4. Drain high-level alarm.
5. Empty cup or pour-point failure protection circuit.

Analysis time depends upon cooling temperature, pour point and sample wax content, but may generally be quoted as about 30 minutes for 0°C pour point samples (20-min cooling time and 10-min heating time). Sample flowrate should be about 10–100 ml/min at a maximum temperature of 75°C and a maximum viscosity of 250 cSt at 122°F. Repeatability is quoted as ±1 deg C. Readings may be correlated with ASTM D97/IP15 test results.

5.4 PROCESS CLOUD POINT ANALYZERS

5.4.1 GCA Precision Scientific Group (United States)

The PSG Cloud Point Analyzer Model 42668, the current version of the original Totco analyzer, is microprocessor controlled and designed to measure the cloud point of hydrocarbon streams within the range −22°F to +77°F (using 68°F cooling water). Readings may be correlated with ASTM D2500/IP219 test results. The analyzer detects the temperature at which wax crystals first appear as the hydrocarbon sample is cooled. A schematic diagram of the measurement system is shown in Fig. 5.6.

Like others, this instrument operates on a cyclic basis. A measuring cycle begins when a three-way solenoid valve energizes to introduce process sample into the test cell. The cell consists of a vertical glass tube in an aluminum block to which two Peltier cooling modules are attached. The solenoid valve de-energizes after the flushing period, trapping fresh sample in the test cell, and a cooling cycle is initiated.

The detection system consists of an incandescent lamp that directs light rays through a polarizing film, the sample cell, a second polarizing film, and

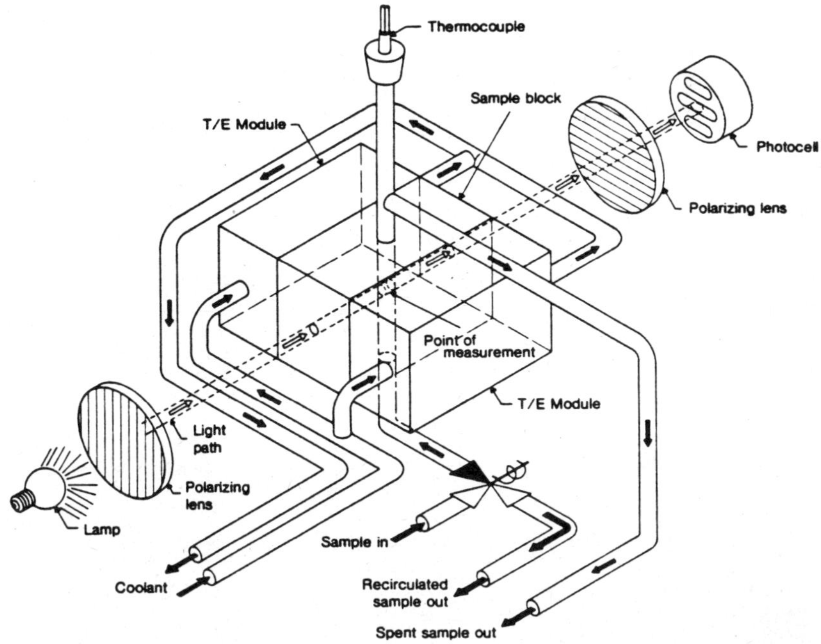

FIGURE 5.6 Cloud point analyzer, Model 42668. (Courtesy of GCA Corporation)

onto a photocell. The first polarizing film limits the vibration of the light rays to one plane. The second film lies perpendicular to the first, limiting the passage of non-refracted light rays onto the photocell. As the sample cools, it eventually reaches a temperature at which ice and/or wax crystals start to form in the liquid. These crystals reflect some light rays and refract others so that the plane of vibration of some light rays changes as they pass through the sample.

The second polarizing film is no longer capable of blocking all passage of light. The wax crystals have significantly greater angles of refraction and influence the passage of light rays to a much greater degree than the ice crystals. The refracted light rays pass through the second polarizing film at a level proportional to the degree of refraction. The electrical resistance of the photocell varies with the amount of light falling on it. By adjusting the sensitivity level of the detection circuits, the system can be made selective to refraction attributable to the presence of wax crystal growth only, thereby eliminating the effect of moisture in the sample. The photocell thus signals the first appearance of wax crystals in the sample. The sample temperature is measured by a thermocouple positioned in the sample immediately above the light path.

The microprocessor controls the operation of the analyzer in one of three modes. In Mode 1, the flushing cycle is set at 30 sec and power to the cooling modules is controlled to achieve maximum cooling rate; this quickly establishes the cloud point temperature. After each measurement cycle, the cooling power is adjusted so that the time interval between termination of the flush cycle and detection of the cloud point is about 90 sec. This feature allows the analyzer to react to changes in feedstock. Should a cloud point not be detected, an auto-dump sequence is initiated. If three cycles occur without cloud point detection, a 9-min flush cycle commences.

In Mode 2, the flushing cycle is set at 60 sec and the cooling rate is gradually increased to simulate laboratory test conditions. If a cloud point is not detected at full power and minimum temperature and 20 min elapses, the analyzer dumps the sample, flushes for 3 min and repeats the cycle.

Mode 3 is a combination of Modes 1 and 2, consisting of 120 cycles in Mode 1 followed by a single Mode 2 cycle. The power required to reach the cloud point in Mode 2 is sensed by the microprocessor and this setting is used for testing in Mode 1.

The analyzer is mounted on a free-standing framework, as shown in Fig. 5.7, and is suitable for installation in NEC Class 1, Group D, Division 1

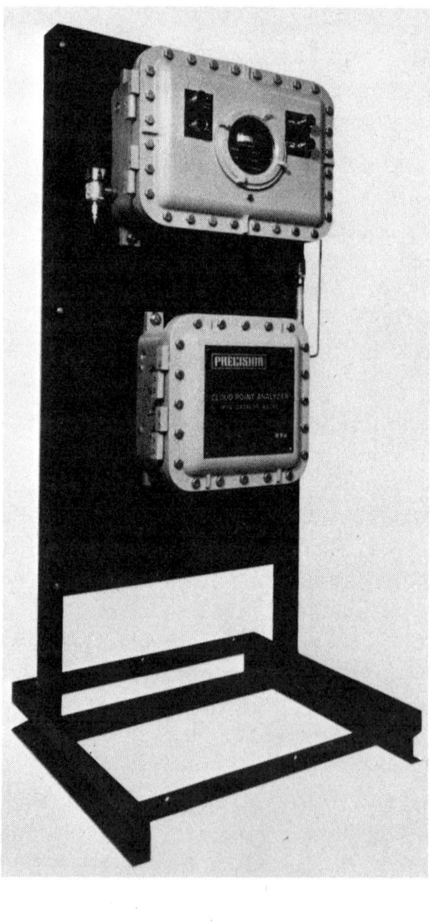

FIGURE 5.7 Model 42668: equipment layout. (Courtesy of GCA Corporation)

hazardous locations. Two explosion-proof enclosures house the analysis unit and the microprocessor control unit. Analyzer cycle time is nominally 120 sec in Mode 1 and 20 min in Mode 2. A sample flowrate of 200–300 ml/min is required at a pressure of 5–25 psig and a temperature at least 20°F above the expected cloud point temperature. A clean dry air supply is required for purging of the enclosures in order to cool electronic components and prevent moisture condensation on supercooled measuring components.

The influence of color and water on the measurement is zero below ASTM 8 color and 5000 ppm respectively. The analyzer output is a type J or type K thermocouple and therefore cyclic, reflecting the alternate heating and cooling cycles. A continuous 4–20 mA dc output can be provided by the use of signal converters and peak holding devices. Contacts for alarm actuation are also provided.

5.4.2 Hone-Addison Ltd. (Great Britain)

The first models of the Hone Cloud Point Analyzer, a photometric instrument, were produced in the late 1960s. Hone reported substantial success with the analyzer. After the benefit of over ten years experience, Hone introduced the first Cloudar Cloud Point Analyzer. The basic photometric principle remained unchanged but refinements were incorporated, such as solid-state technology, an improved cooling system to give faster cycle times and greater temperature differential, a continuous 4–20 mA dc output for process control, and an improved equipment layout to facilitate installation and maintenance. The latest, second generation, Cloudar has been developed under license from CFP (Total) after two and a half years of evaluation.

Whereas in the original Cloudar light is passed through the test cell and the detector responds to loss of light as the wax cloud forms, the latest Cloudar embodies a reflective principle. The detector is placed at right angles to the light beam so that there is no output initially. When the wax cloud forms at the cloud point, light is reflected from the cloud and received by the detector. A temperature sensor mounted on the test cell reads the cloud point temperature of the sample. Cleaning of the cell is now enhanced by instrument air injection, and flushing times can be adjusted to suit the particular sample. A new microprocessor-based electronics package has been incorporated to give improved reliability and to provide self-diagnostic features.

A schematic flow diagram of the analyzer is shown in Fig. 5.8. The operating sequence of the Cloudar has three steps:

1. Fill Cycle. The test cell is flushed with the process sample at normal temperature for 20–45 sec.
2. Cool cycle. The cell is cooled by Peltier type cooling modules, the heat being extracted by a cooling water system.
3. Read cycle. The cloud point is detected by the photocell system. This

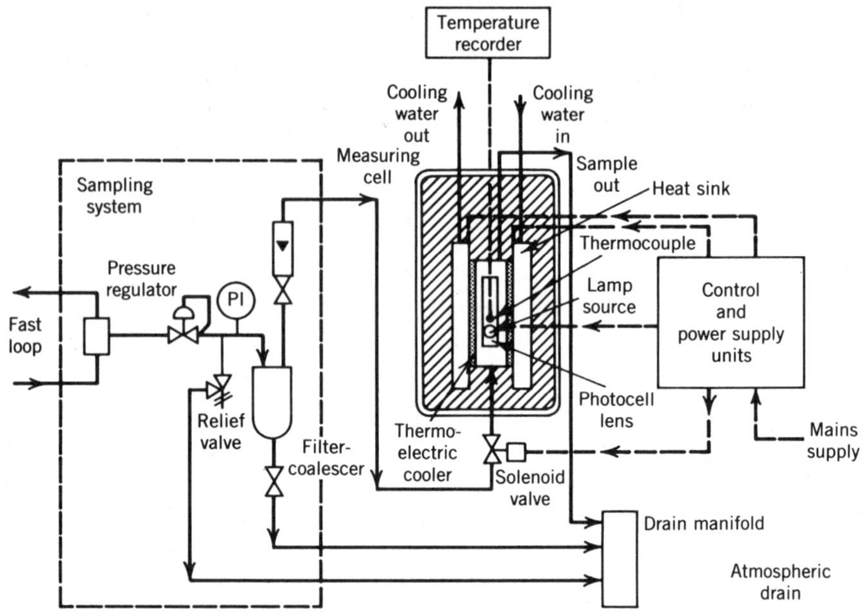

FIGURE 5.8 Hone "Cloudar" cloud point analyzer. (Courtesy of Hone-Addison Ltd.)

> initiates a 1-second "read" cycle during which the cloud point temperature is read and stored. The output signal is updated and held at this value until the next read cycle. If no cloud is detected in the sample within 25 min, an electronic timer automatically recycles the analyzer. The sequence of operation is then repeated continuously.

The minimum cloud point temperature for the analyzer is 140°F below the temperature of the cooling water used. Analyzer cycle time is between 1 and 4 min and repeatability is quoted as better than ± 1°F. Sample flowrate required is up to 5 gal/hr at pressures up to 14 psig and at a temperature at least 45°F above the expected cloud point. Cooling water consumption is about 5 gal/hr at an inlet temperature not exceeding 77°F. The analyzer flushing time may be adjusted for any period up to 255 sec.

The analyzer is mounted on a free-standing framework as shown in Fig. 5.9 and consists of three flameproof enclosures—the test cell (top right), the microprocessor control unit (top left), and the main power supply transformer (lower left). In addition there is an enclosure housing the safety equipment for hazardous area compliance. The analyzer is normally supplied with one of two alternative safety systems. Both operate on the air purge and pressurization principle and are suitable for Zone 1 and Zone 2 hazardous locations. One system uses an intrinsically safe timer relay and is fully PTB-approved to VDE 0171. The second uses a pneumatic timer relay

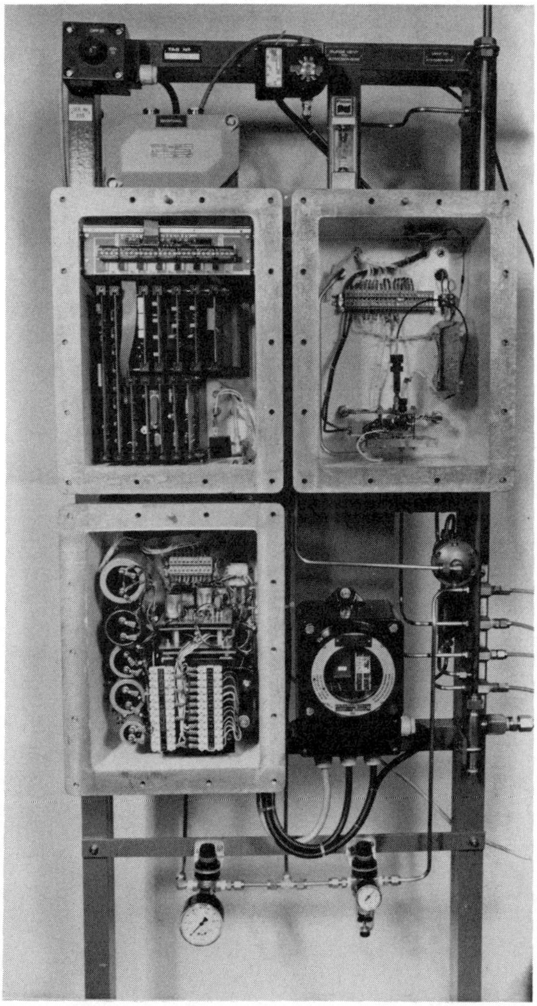

FIGURE 5.9 "Cloudar": equipment layout. (Courtesy of Hone-Addison Ltd.)

designed to comply with the new CENELEC standard En 50016. LED status indicators are provided on the control unit, visible through the front cover during normal operation, for the following: sample control valve operation, cell cooling, cloud point detection, alarm, light source failure, over temperature switch operation, and stand-by operation.

Tests conducted at Hone have shown that with samples of water content of up to 1000 ppm there is no deviation from the true cloud point. In addition, tests with samples containing flow-improvement additives have shown deviations of less than 1.0°F.

5.5 PROCESS ANALYZERS FOR ASSOCIATED MEASUREMENTS

5.5.1 Hone-Addison Ltd. (Great Britain)

The Hone *Filtar* is an automatic on-line analyzer designed to measure the CFPP of a petroleum product as defined in Section 5.2.4(b). The analyzer is manufactured under license from Total Oil Company and operates on the same principle as the laboratory test. It can be arranged to conform to any of

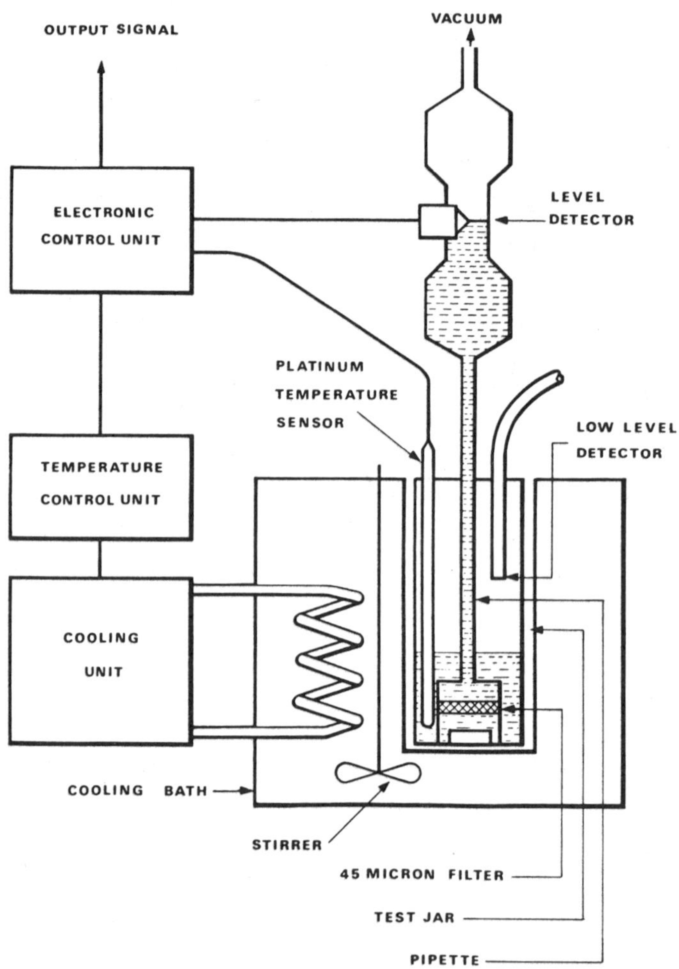

(a)

FIGURE 5.10 Hone "Filtar" CFPP analyzer. (*a*) Operating principle. (Courtesy of Hone-Addison Ltd.)

(*b*)

FIGURE 5.10 Hone "Filtar" CFPP analyzer. (*b*) equipment layout. (Courtesy of Hone-Addison Ltd.)

the following test methods: AFNOR M70/042, DIN 51428, SIS 155122, and IP309.

The analyzer is shown schematically in Fig. 5.10 and consists of cooling bath, test jar, pipette, and temperature sensor, as in the laboratory test equipment. A level detector replaces the human eye used in the lab test and there is a refrigerated bath and the control equipment necessary for fully automatic operation. The analyzer is available in two versions. With the single bath analyzer, designed for measurement of CFPP down to −20°C, the test cell is immersed in the cooling bath throughout the analysis. With the two bath version, the analysis starts in the −34°C bath and automatically transfers to the −51°C bath when the temperature reaches −20°C without the filter blocking. The test then proceeds as before until the CFPP is reached.

The analyzer operates on a repeating cycle in the following sequence:

1. Flushing. Test cell and filter are flushed with fresh, clean sample for a period of 8–10 min to remove old sample, clear any wax deposits, and warm up the test cell for the next analysis.

2. Filling. Flow of fresh sample is shut off and the correct level of sample in the test jar is established by the syphon action of the drain system.
3. Analysis. Sample temperature in the test jar falls under the action of the cooling bath. At every 1°C a vacuum is applied at the top of the cell unit and sample is aspirated through a 45 micron filter until its surface reaches the optical detector at the 20-ml level. The analyzer may be set up to start aspirating 10°C above the previously measured CFPP or at +10°C.

   If the 20-ml level is reached within 60 sec, a signal from the optical detector terminates the aspiration and sample is allowed to flow back through the filter in preparation for the next aspiration. An electrical signal from the detector resets the digital control system so that the next aspiration occurs when the sample temperature has fallen by a further 1°C.

   The above sequence is repeated at every 1°C fall in temperature until the wax formed on the filter obstructs the sample flow and the 20-ml level is not reached in 60 sec. When this occurs, the temperature of the sample is measured and converted to a 4–20 mA dc signal for remote recording. The test cell is then flushed and refilled in preparation for the next analysis.

   The analyzer is designed to measure CFPP temperatures down to −35°C and has a typical cycle time of 15–30 min. Sample flowrate required is up to 30 l/hr at a temperature at least 25°C above the expected CFPP but not exceeding 80°C and at an inlet pressure up to 7 psig. Two safety systems are available for the Hone Cloudar, suitable for Zone 1 and Zone 2 hazardous locations.

6

Vapor Pressure Measurement

6.1 INTRODUCTION

A gas will fill any space to which it has access. Liquids do not possess this property. When placed in a vessel, a liquid will take the shape of the vessel and a liquid surface is formed. If a volatile liquid is placed in a closed container, the liquid will evaporate and the pressure above the liquid will increase. After a time, the liquid level does not change and the pressure does not increase further.

Although the system appears to be static, the gas molecules are in constant motion, colliding with the surface of the liquid, losing kinetic energy, and sometimes returning to the liquid. At the same time, some molecules gain sufficient kinetic energy to leave the liquid. When the rate at which molecules leave the liquid is equal to the rate at which they return to the liquid, the system is said to be in *equilibrium*. It is a *dynamic* equilibrium because, although no overall change is observed, two exactly equal and opposite changes are taking place.

The pressure of the gas molecules above the liquid at equilibrium is known as the *vapor pressure* of the liquid at the temperature specified. Vapor at the vapor pressure is said to be *saturated.* If the available space above the liquid is too large for the vaporized liquid to fill it at the vapor pressure, the liquid will continue to evaporate until it disappears entirely into the vapor phase.

The vapor pressure of a liquid increases with increase in temperature until a point is reached where the pressure is equal to the external (atmospheric) pressure. This temperature is known as the boiling point of the liquid. The rate of increase of vapor pressure with temperature is given by the Clausius-Clapeyron equation:

$$\frac{d \log p}{dT} = \frac{\Delta H}{2.303 RT^2}$$

where p is the vapor pressure
ΔH is the latent heat of evaporation
T is the absolute temperature
R is the universal gas constant

Integrating and assuming that the latent heat of evaporation ΔH is constant:

$$\log p = -\frac{\Delta H}{2.303 RT} + C$$

where C is a constant.

In practice, ΔH is found to be constant within the accuracy of most experimental work, over temperature spans not exceeding about 50°C. If $\log p$ is plotted against $1/T$, a straight line is obtained over spans of 50°C, with a slope of $-\Delta H/2.303R$. For wider temperature spans, there is slight curvature of the plot due to the gradual increase of ΔH with rising temperature.

It follows that the latent heat of evaporation of a liquid may be determined from measurements of the vapor pressure at two or more temperatures. The accuracy of this method is equal to or better than the direct calorimetric method.

Although these theoretical considerations bear little relevance to on-line process vapor pressure analyzers, they show that vapor pressure measurement forms an important part of experimental laboratory analysis and can lead to the determination of other liquid properties.

In the petroleum and petrochemical industries, vapor pressure is an important parameter. The relative volatilities of the components in a hydrocarbon mixture forms the basis of all fractionation processes, as we have seen in Chapter 3. Since the vapor pressure of a product is an indication of its purity, vapor pressure is an important product specification and quality control parameter. The vapor pressure of a product will increase as the concentration of the low boiling components in the product increases. The concept of vapor pressure is also important in the design of sample handling systems, where a liquid must be vaporized prior to analysis. This subject is covered in detail in Chapter 20.

Another parameter indirectly related to vapor pressure is the *vapor-liquid* (*V-L*) *ratio*. This is the volume of vapor produced at a given temperature and an absolute pressure of 760 mm Hg from a given volume of liquid. The vapor-liquid ratio is sometimes used as a specification point for gasolines (some companies require a 20 : 1 ratio). Alternatively, the vapor temperature at a specified vapor-liquid ratio may be measured.

6.2 LABORATORY METHODS OF MEASUREMENT

Correlation and validation of process vapor pressure analyzers is carried out with reference to two laboratory test methods, the Reid method which is an empirical test, and the Micro method, which measures the absolute or true vapor pressure.

6.2.1 The Reid Method

The standard laboratory test method for vapor pressure of petroleum products by the Reid Method, ASTM D323-79, was first introduced in 1930; since then it has been the universally accepted standard for the petroleum industry. The method covers determination of the vapor pressure of volatile crude oils and volatile non-viscous petroleum products, except liquefied petroleum gases, which is covered by ASTM Method D1267-79.

The test apparatus consists of two vessels (usually two bombs)—the sample vessel and the air vessel. The ratio of the volumes of the air and sample vessels must be held to within the limits 3.8–4.2. For testing of aviation gasolines the ratio must be within closer limits (3.95–4.05). The sample vessel is filled with chilled sample and connected to the air vessel. The complete apparatus is immersed in a constant temperature bath, maintained at 100°F ±0.2°F and shaken periodically until equilibrium is attained. The reading of a pressure gauge attached to the apparatus is recorded as the Reid vapor pressure (RVP) of the sample.

Normally, the air vessel is immersed in the constant temperature bath prior to connection to the sample vessel so that it may attain the operating temperature of the test. If this is not done, additional stabilization time is required or a correction is applied to the gauge reading. Since the external (atmospheric) pressure is counteracted by the atmospheric pressure initially present in the air vessel, the Reid vapor pressure is an absolute pressure at 100°F in lb/in.^{2}

It differs from the true vapor pressure of the sample because of some small sample vaporization and the presence of water vapor and air in the confined space. The repeatability of the Reid method is defined by the statement that two tests by the same operator with the same apparatus should give results that differ by less than 1.8 lb/in.^{2} nineteen times out of twenty (95% probability).

From this test description, it would appear that the Reid method is fairly simple and straightforward. In fact, it is a method that requires meticulous attention to detail and preparation. The most important part of the test is not the test itself, but the collection, handling, and transfer of samples. The measurement of vapor pressure is extremely sensitive to losses by evaporation or to slight changes in sample composition. Therefore, the utmost care must be taken in the handling of samples. The author remembers well the considerable time expended in the summer of 1962 trying to correlate a new RVP analyzer to the laboratory tests, only to find out by accident that the low reading obtained by the lab test was as a result of incorrect sampling. The significant point was that the process had been operating for about ten years using incorrect lab data!

The precise procedure for sampling is laid down in ASTM D323, which makes reference to ASTM standard D270, "Sampling of Crude Petroleum and Petroleum Products." The procedure for sample transfer, introduction of the apparatus into the constant temperature bath, and measurement of the vapor pressure must be followed carefully otherwise gross errors will result and the test will be meaningless.

For products with an RVP greater than 26 lb/in.^{2} the standard Reid apparatus is modified slightly to overcome inaccuracies and hazards to personnel. This test is covered in ASTM Method D1267-79.

The bourdon-type pressure gauge used for RVP measurement is checked against a manometer before each test to confirm readability within 1% of scale range. For RVPs above 26 lb/in.^{2}, a dead weight tester is used to calibrate a test gauge. The gauge is then fitted to the apparatus and its reading compared with the measurement gauge to determine the gauge error, which is then used in the gauge error correction procedure.

6.2.2 The Micro Method

The Micro Method, ASTM D2551-80, was first introduced in 1966 and covers measurement of the total pressure exerted "in vacuo" by air-containing, volatile, non-viscous petroleum products at preselected vapor-to-liquid ratios and temperatures. The method is applicable within the range 0.5–17 lb/in.^{2} Vapor pressure equivalent to RVP is obtained by a correlation.

The apparatus for the test may be made of metal or glass (or both) and consists of a bulb of known volume and a valve block or manifold, connected to a measuring device (e.g., a manometer), and a vacuum system. The specifications for the test apparatus are laid down in an annex to the ASTM standard.

The procedure may be summarized as follows. A known volume of the sample is introduced into an evacuated, temperature-controlled bulb, also of known volume. The relative volumes of the bulb and sample pipette are chosen to give the required vapor-to-liquid ratio. The rise in pressure in the bulb after introduction of the sample is measured by a mercury manometer. The pressure is equivalent to the sum of the vapor pressure of the sample and the partial pressure of the dissolved air. As in the Reid method, great care must be taken in sample handling and procedure if accurate and meaningful results are to be obtained.

Commercial equipment for carrying out the ASTM D2551 test is available from Stanhope-Seta Ltd. in England and from Manufacturing Engineering and Equipment Corp. (MEECO) in the United States.

6.2.3 Vapor-Liquid Ratio

The standard laboratory test method for measurement of vapor-liquid ratio is covered in ASTM D2533-82, specifically for gasoline. Here is a summary of the procedure. A measured volume of the sample at 32–40°F is introduced through a rubber septum into a glycerin-filled burette. The charged burette is then placed in a temperature-controlled water bath. The volume of vapor in equilibrium with the liquid is measured at the desired temperature and at a specified pressure, usually 760 mm Hg. The vapor-liquid ratio is then calculated. If the temperature corresponding to a given V-L ratio is the required parameter, the V-L ratio is determined at several temperatures and the selected pressure. The results are then plotted and the temperature read at the given V-L ratio.

6.3 PROCESS VAPOR PRESSURE ANALYZERS

Process analyzers for vapor pressure measurement have been available commercially for about twenty years. The first analyzers were manufactured by Hallikainen in the United States and were marketed in Europe by GEC Elliott, London. The Hallikainen range of process analyzers is now marketed by Fluid Data Inc. in the United States and Great Britain and by GCA Precision Scientific Group in the United States. The selection of process vapor pressure analyzers is consequently rather limited.

6.3.1 GCA Precision Scientific Group (United States)

GCA/PSG currently manufactures two analyzers, the Reid Monitor Model 44770 and the Absolute Vapor Pressure Analyzer Model 41350. The latter is an updated version of the original Hallikainen Model 1350.

(a) Reid Monitor Model 44770

This instrument is designed to duplicate the ASTM D323 method requirements, namely, that measurement is made at a temperature of 100°F using an air-saturated sample of vapor-liquid ratio 4 : 1. The analyzer readout is made directly in RVP units.

A schematic diagram of the analyzer is shown in Fig. 6.1. Process sample enters the analyzer via the feed pump, (one head of a duplex pump assembly), which establishes a constant sample flowrate of 100 ml/min. From the feed pump, the sample enters a temperature-controlled oil bath, passes through a heat exchanger and the measurement chamber, and then to an insulated and jacketed exhaust pump, the second head of the duplex pump assembly. The exhaust pump is heated by a continuous circulation of bath oil

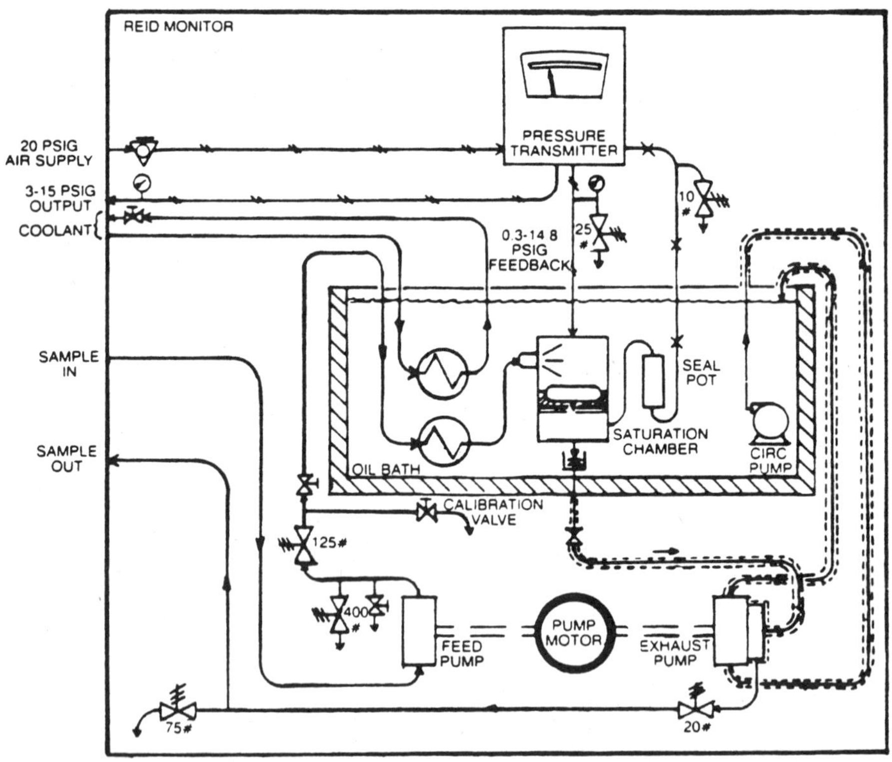

FIGURE 6.1 Reid monitor, Model 44770. (Courtesy of GCA Corporation)

to maintain a constant temperature of 100°F and is designed to maintain an extremely accurate flowrate of 500 ml/min of two-phase product. This establishes the 4 : 1 vapor-liquid ratio required by the ASTM D323 method.

As sample flows into the saturation chamber, it lifts the float and needle valve assembly away from the orifice between the top and bottom sections of the chamber. Since the inlet flowrate is 100 ml/min and the outlet flowrate is 500 ml/min, the exhaust pump creates a vacuum in the lower section of the chamber and causes the sample to flash across the orifice into a 4 : 1 vapor-liquid state. The pressure in the lower section is sensed by a pressure transmitter, the impulse lines of which are sealed with kerosine to isolate the two-phase sample from exposure to ambient temperature.

The transmitter consists of an absolute pressure bellows and two transmitters; one produces a 3-15 psig output signal, the other a feedback signal. The latter provides a saturation air supply, which is introduced into the sample as it enters the top section of the saturation chamber through a spray nozzle. Instrument measurement range is 2–19 RVP and repeatability is quoted as ±0.1 RVP. Response time is 60 seconds for a 63% response to a step change in sample RVP at the analyzer inlet.

(b) Absolute Vapor Pressure Analyzer (Hallikainen) Model 41350

This instrument is designed to measure the absolute vapor pressure of hydrocarbon streams under conditions of constant temperature and carefully controlled pressure drop across a Venturi-type ejector system. By reducing the ejector efficiency, the analyzer may also be calibrated to measure RVP, if desired for specific applications. The analyzer is normally calibrated to read in absolute pressure units within the range 2–200 psia. Standard range is 0–20 psia. Any preset temperature can be selected over the range 100–160°F, but is normally set at 100°F, in accordance with ASTM D323.

The principle of the instrument, illustrated in Fig. 6.2, is similar to the common laboratory ejector pump. The sample is forced through a jet orifice in which the increase in fluid velocity produces a pressure drop sufficient to induce incipient vaporization of the sample. The minimum pressure reached

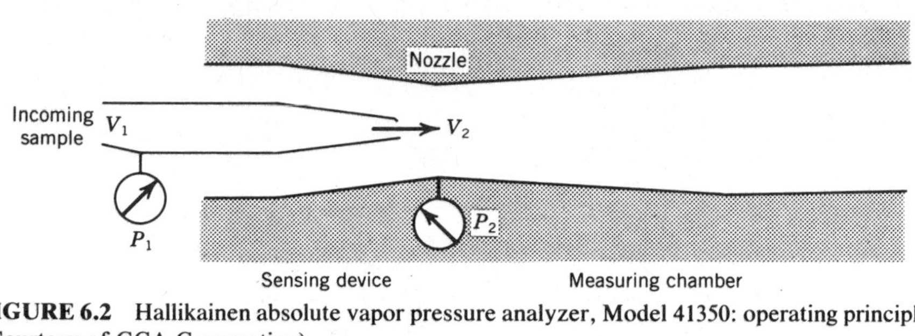

FIGURE 6.2 Hallikainen absolute vapor pressure analyzer, Model 41350: operating principle. (Courtesy of GCA Corporation)

occurs when the static pressure of the liquid at the nozzle exit reaches the effective vapor pressure of the liquid. The relationship between the fluid velocity and pressure is given by the flow energy equation:

$$\frac{V_1^2}{2g} + \frac{P_1}{\rho} + F_1 = \frac{V_2^2}{2g} + \frac{P_2}{\rho} + F_2$$

where V_1 and V_2 are the velocities at the inlet and outlet of the jet orifice respectively, P_1 and P_2 are the inlet and outlet pressures, g is the gravitational constant, ρ is the density of the liquid, and F_1 and F_2 are the fluid friction losses. If the outlet velocity V_2 can be made sufficiently large while the inlet velocity V_1 and P_1 are held constant, then, ignoring friction, the pressure of the fluid at the nozzle outlet can be reduced to the effective vapor pressure of the sample.

A schematic diagram of the analyzer is shown in Fig. 6.3. Conditioned process sample enters the constant temperature bath through a pressure regulator, then passes through a parallel-path heat exchanger and a nozzle assembly, and out of the analyzer through a second (back pressure) regulator for return to the process or recovery system. The bath is controlled to within ±0.01°F of the set point by a temperature control system using an RTD as the sensing element. A stainless steel cooling coil is provided for constant

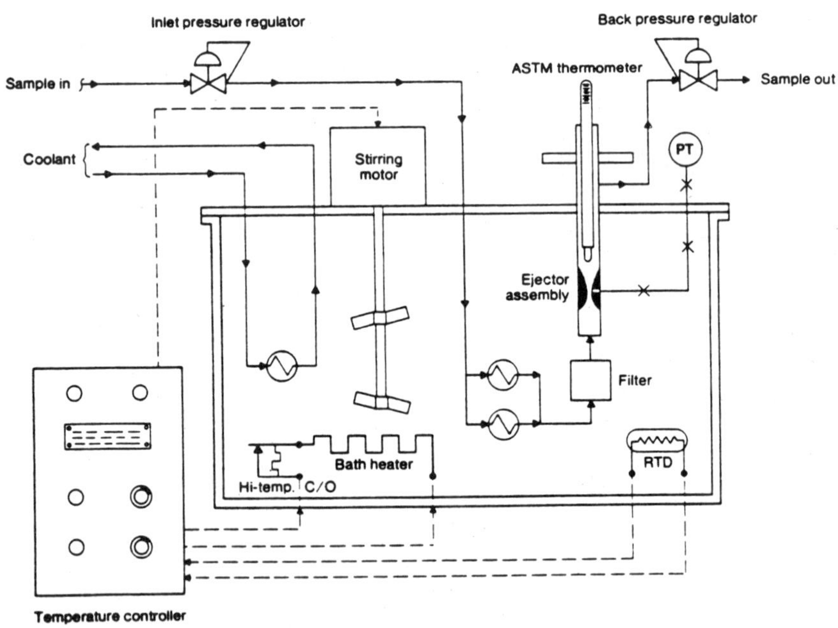

FIGURE 6.3 Model 41350: flow schematic. (Courtesy of GCA Corporation)

cooling water circulation. Two hollow-blade, motor-driven impellers provide complete mixing of bath oil to ensure even temperature distribution.

As the sample passes through the nozzle assembly, its velocity increases and the pressure downstream of the nozzle is the effective vapor pressure of the sample. The nozzle position is readily adjustable. Insertion of adjusting shims modifies the distance between the throat and the nozzle tip, thereby changing the measured vapor pressure and permitting output readings to be correlated against ASTM D323 or ASTM 2551 results as desired.

The vapor chamber of the ejector is connected to an absolute pressure transmitter, which can be supplied with either pneumatic or electronic transmission (Models 41354 and 41355 respectively). The analyzer can also be supplied calibrated but without a transmitter (Model 41356). Nozzle suction pressure is indicated on a small pressure gauge and sample temperature by a thermometer located in a thermowell downstream of the nozzle assembly.

Analyzer response time is quoted as 43 sec and repeatability as ±0.1 psi. Sample viscosity should be less than 50 cP at bath temperature. Sample flowrate is higher than most analyzers, namely 750–4000 ml/min, depending upon the back pressure. However, this may be an advantage in some applications where response time is critical, although it does not allow the instrument to be calibrated by injecting standard samples.

FIGURE 6.4 Model 1354 (Hallikainen) vapor pressure analyzer. (Courtesy of Fluid Data Inc.)

6.3.2 Fluid Data Inc. (United States and Great Britain)

Fluid Data currently manufactures the Hallikainen Models 1354 and 1354/10 Vapor Pressure Analyzers, which operate on exactly the same principle as the GCA instrument described in Section 6.3.1(b). The basic difference between these two models is that the 1354 has a check valve installed downstream of the nozzle assembly, whereas the model 1354/10 utilizes a back-pressure regulator as with the GCA instrument. Packaging of the analyzer systems is basically the same; a typical example is shown in Fig. 6.4.

6.4 VAPOR-LIQUID RATIO ANALYZERS

The importance of V/L for gasoline volatility control has been recognized for many years. However, this basic physical property of gasoline could not be monitored or controlled effectively during in-line blending operations until the commercialization of on-line analyzers for the measurement. Empirical correlations to obtain V/L values are based on RVP and distillation point data and refineries usually control to these values for day-to-day operations.

National surveys of service station gasolines have shown a wide variety of temperature-V/L and RVP characteristics for finished gasolines. Some show excess volatility, indicating the possibility of vapor locking problems; others indicate that additional light hydrocarbons could have been added to the blends.

The limits on RVP and distillation specifications are often set so that it is almost impossible to produce a gasoline in excess of 20 V/L. Often, only one of these specifications is being effectively used for blending control; commonly, the RVP specification becomes the controlling factor, and target values are conservatively set to ensure that the product will pass the final inspection.

Normal butane is probably the most widely used blendstock for front-end volatility control. In view of its cost relative to finished gasoline, it is economically attractive to maximize butane addition to gasoline blending. This increases gasoline production, resulting in increased profitability. On the other hand, there may be situations where butane is in short supply, or where fuels for refinery processes must be purchased. In these cases, the economics may require minimizing butane addition. In either situation, on-line monitoring of V/L will provide the capability for improved front-end volatility control.

6.4.1 Ethyl Corporation (United States)

Ethyl manufactures the *VOLRAC* On-line V/L Analyzer which is designed to measure the V/L ratio of in-line blended gasolines, providing excellent

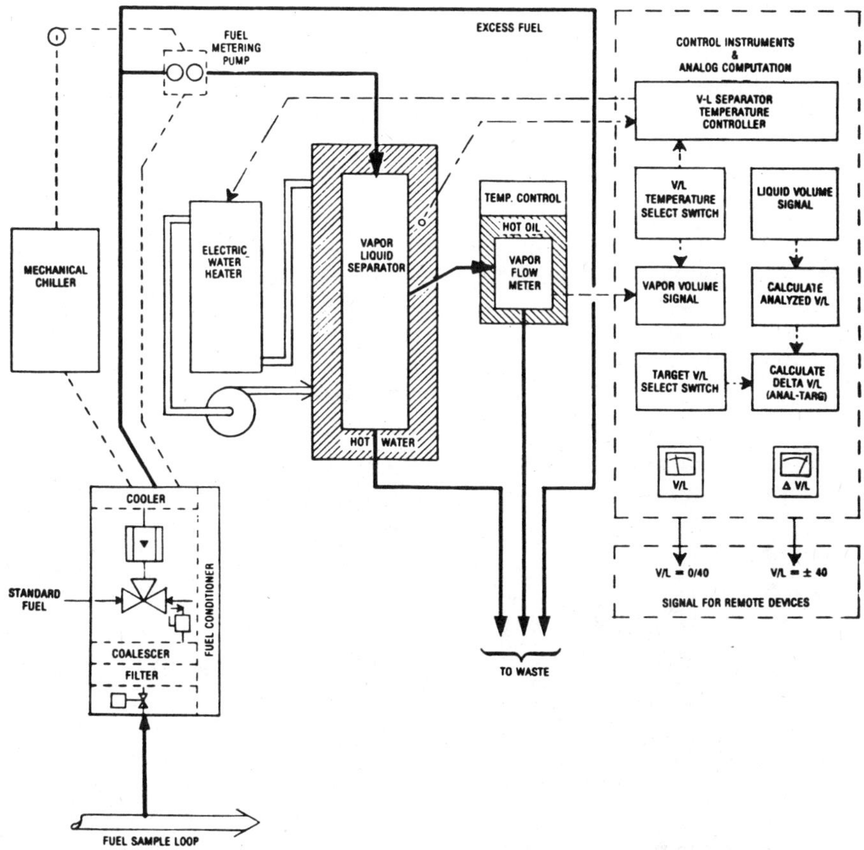

FIGURE 6.5 "Volrac" vapor-liquid ratio analyzer. (Courtesy of Ethyl Corporation)

correlation with ASTM Method D2533 (described earlier). To the author's knowledge, this is the only commercially available on-line V/L analyzer.

A schematic diagram of the instrument is shown in Fig. 6.5. Gasoline sample first enters the fuel conditioning system, where it is filtered, passed through a coalescer, and its pressure reduced to 3–5 psig. The sample then passes through a mechanical chiller to the top of the Volrac unit, where its pressure is reduced to atmospheric and excess passed to the waste outlet. The fuel metering pump delivers a constant flowrate of chilled liquid gasoline to an orifice located at the inlet to the vapor-liquid separator.

This device is the key component of the analyzer. The liquid gasoline is quickly heated to an extremely thin film dispersed over a relatively large surface area, resulting in an equilibrium vapor-liquid relationship. Liquid is removed to waste, and vapor is continuously removed from the side connection and passed through a highly accurate laminar flowmeter, using a differential pressure, reluctance-type transducer.

Since liquid gasoline is introduced at a constant flowrate, calculation of the V/L ratio is a direct function of the vapor flowrate. Two forms of compensation are provided. One corrects the vapor volume signal for temperature, the other corrects for variations in barometric pressure. The analyzer provides a delta V/L signal, where

$$V/L = \text{analyzed } V/L - \text{target } V/L$$

Target V/L is the desired quality for the fuel. Any one of five values may be selected. In addition, a temperature switch allows selection of any one of six preset V-L temperature values. Controls for these functions are mounted inside the analyzer enclosure.

The analyzer is mounted on a free-standing rack as shown in Fig. 6.6 and is suitable for installation in NEC Class 1, Groups C and D, Division 2 hazardous locations by purging of the electrical enclosures using instrument air. Sample requirements are 6–8 gal/hr at a pressure of 50–150 psig. Liquid

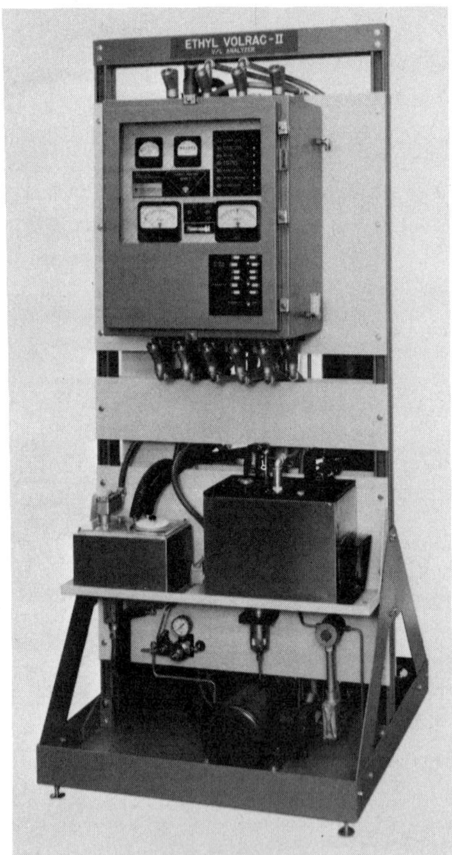

FIGURE 6.6 "Volrac": equipment layout. (Courtesy of Ethyl Corporation)

drainage from the analyzer is at atmospheric pressure, therefore some type of sample recovery system is required to return sample to the process.

V-L ratio range is 10 to 30 standard, and temperature values may be selected within the range 100–150°F standard. Two delta V/L signals are provided, covering the range −4 to +4 V/L, and may be 4–20 or 10–50 mA dc. A "valid data" feature is incorporated, whereby an alarm contact is actuated should any of several critical operating conditions be incorrect. At the same time, the two V/L signals are set to zero. Pilot lamps on the instrument indicate valid or non-valid data. Precision is quoted as ±0.5 V/L and the instrument has a fast response time of under 60 sec.

Currently the VOLRAC is in use on in-line gasoline blending facilities at 21 refineries in the United States. It is reported to be performing well and is capable of detecting V/L changes as small as 0.2 V-L. Calibration of the instrument must be carried out initially using a standard on which multiple determinations have been made by the ASTM D2533 procedure to provide a statistically sound average V/L.

7

Oxygen Measurement

7.1 INTRODUCTION

Oxygen, first discovered by Priestley in 1774 and named by Lavoisier three years later, is the most abundant element in the Earth's crust, including the oceans and the atmosphere. It is essential to most forms of life and is a prerequisite for combustion. Measurement of its concentration in other gases has assumed increasing importance with modern advances in technology and is now a routine requirement in many branches of industry and research. Oxygen is often used as a reactant in chemical processes and its concentration must be controlled to within very close limits. In some cases, the presence of oxygen must be avoided for reasons of safety, corrosion prevention, and the elimination of secondary reactions. The term "oxygen-free atmosphere" is rather vague, however, and can mean any concentration from 1% to 1 ppm or less. Consequently, industrial oxygen analyzers must cover a very wide range of measurement. Oxygen is not an easy element to detect, particularly when it is present in very low concentrations.

With the accent today on energy conservation and efficient, economic process operation, one of the main applications for oxygen measurement is in combustion efficiency monitoring and control of furnaces and boilers. This application covers a wide spectrum of industrial processes.

Any method used to determine oxygen concentration must, of course, be specific to oxygen alone. General analysis techniques such as thermal conductivity, gas chromatography, and heat of adsorption can be used, particularly for laboratory work. However, for industrial on-line analysis, more specificity is required (except, of course, in the case of GLC). Process oxygen analyzers are based on four fundamental methods of measurement:

- Measurement of the paramagnetic property of oxygen.
- Catalytic combination of oxygen and hydrogen.
- Oxidation of certain characteristic chemical compounds by oxygen.
- The effect of oxygen on the operation of a galvanic cell.

7.2 THE PARAMAGNETIC PROPERTY OF OXYGEN

The physical property that distinguishes oxygen from the majority of other gases is its paramagnetism. Faraday in 1845 discovered that substances other than iron, nickel, and cobalt can be magnetized. He found that some materials, which he called *paramagnetics,* tend to be attracted by a magnetic field, while others, which he called *diamagnetics,* are repelled by a magnetic field. Paramagnetism and diamagnetism are not confined to solids, but occur in fluids also. In one experiment, Faraday demonstrated that a hollow glass sphere at the end of a horizontal rod supported by silk fibers was attracted by a magnetic field when filled with oxygen.

The paramagnetism of oxygen arises from its electronic structure. The Pauli Exclusion Principle requires that two paired electrons in the same orbital shall have opposite spins; thus the magnetic moment of each electron is neutralized. The oxygen atom contains two unpaired electrons in its outer (valency) shell. Since the presence of an unpaired electron in the outer shell gives rise to paramagnetism—and oxygen has two such unpaired electrons—it is strongly paramagnetic. Three other gases (nitric oxide, nitrogen dioxide, and chlorine dioxide) exhibit paramagnetism, but to a lesser extent

TABLE 7.1 Relative Magnetic Susceptibilities of the Common Gases

Gas	Percent[a]	Gas	Percent[a]
Acetylene	−0.24	Hydrogen chloride	−0.30
Ammonia	−0.26	Hydrogen fluoride	+0.10
Argon	−0.22	Hydrogen sulfide	−0.39
Bromine	−1.30	Methane	−0.20
1,3-Butadiene	−0.49	Neon	+0.13
n-Butane	−1.30	Nitric oxide	+43.0
iso-Butene	−0.85	Nitrogen	0.0
Carbon dioxide	−0.27	Nitrogen dioxide	+28.0
Carbon monoxide	+0.01	Nitrous oxide	−0.20
Chlorine dioxide	+25.5	Oxygen	100
Ethane	−0.46	Propane	−0.84
Helium	+0.30	Water vapor	−0.02
n-Heptane	−2.10	Xenon	−0.95
Hydrogen	+0.24		

[a] Equivalent percent oxygen, normalized to the scale oxygen = 100, nitrogen = 0.

than oxygen. This may be explained by the fact that each contains oxygen and each has an odd number of electrons.

The relative magnetic susceptibilities of some of the more common gases are listed in Table 7.1, expressed as equivalent percent oxygen normalized to the scale oxygen = 100 and nitrogen = 0. It can be seen that the majority of gases listed are either slightly diamagnetic or paramagnetic (with the exception of the three given in the last paragraph). Therefore, methods exploiting the paramagnetic property of oxygen can be used effectively for the measurement of oxygen concentration in gases. Some methods also make use of the fact that oxygen tends to lose its paramagnetism when heated (thermomagnetic effect).

Three methods are widely used today for the measurement of oxygen concentration in gases, using the paramagnetic property of oxygen.

7.2.1 The Magnetic Wind Principle

The so-called "magnetic wind" principle makes use of the thermomagnetic effect. Two types of detector may be used, the annulus-type and the filament-type.

In the annulus-type detection cell shown in Fig. 7.1 the analysis chamber consists of a metal block housing an annulus with sample gas inlet at the bottom and outlet at the top. A thin-walled glass tube is fixed across the horizontal diameter of the annulus as shown and carries two heating coils connected to form two arms of a Wheatstone bridge network. The poles of a permanent magnet are positioned at one end of the cross-tube as indicated

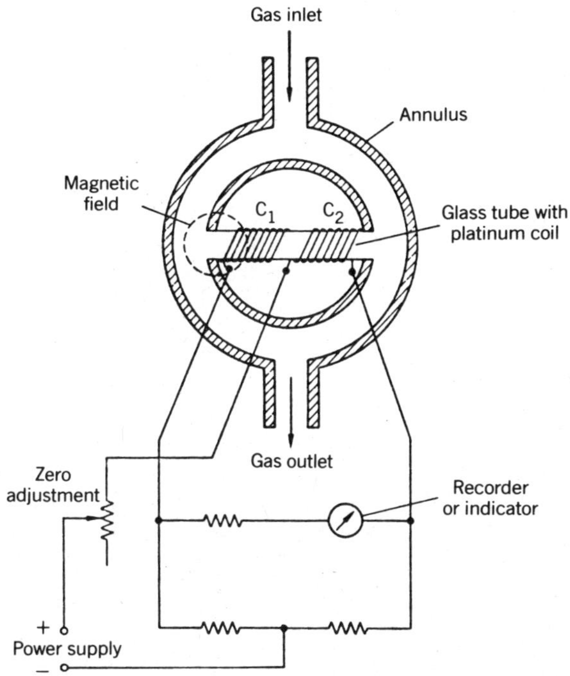

FIGURE 7.1 Annulus-type detection cell.

by the dotted line. When a current passes in the bridge network, the coils are heated and the temperature in the cross-tube increases over that in the annulus.

When a sample containing oxygen passes through the cell, the oxygen is attracted into the magnetic field, the attraction being more pronounced at the left-hand end of the field, which is outside the heated zone of the coils. Consequently, there is a flow of gas through the cross-tube, maintained by fresh, unheated oxygen being constantly attracted into the cooler part of the magnetic field and displacing the heated oxygen. The gas flow cools coil C_1 in relation to coil C_2 and the resultant bridge unbalance is directly proportional to the oxygen concentration in the sample gas.

In the filament-type detection cell shown in Fig. 7.2, the analysis chamber consists of a metal block with two identical cavities, each of which is open to the sample gas. A filament is mounted in each cavity and they are connected to form two arms of a Wheatstone bridge network, the filaments being heated by the bridge current. The poles of a permanent magnet are located across one cavity such that the magnetic field is partly within and partly outside the heated zone of the filament.

When sample gas containing oxygen passes through the detection cell, the oxygen is attracted into the magnetic field, the attraction being more pronounced at the cool edge of the magnetic field. A flow of gas is therefore

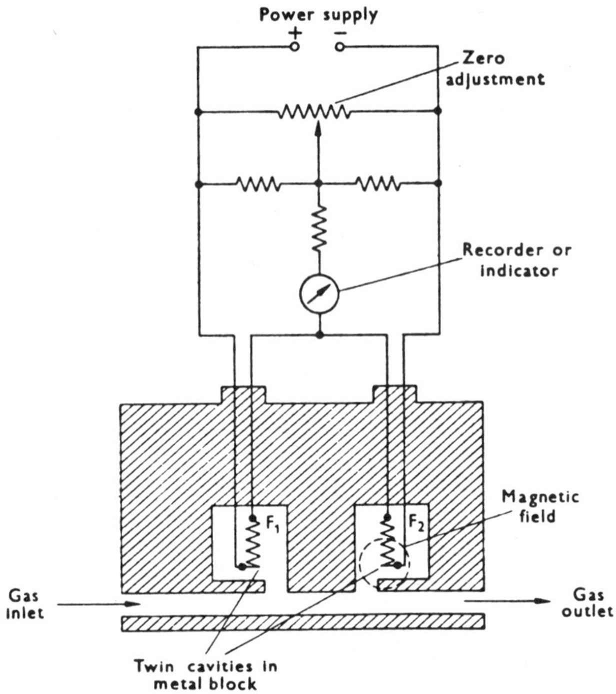

FIGURE 7.2 Filament-type detection cell.

induced in cavity B, cooling filament F_2 in relation to F_1. The resulting bridge unbalance is directly proportional to the oxygen concentration in the sample gas.

Measurement systems based on the magnetic wind principle rely on the paramagnetism of oxygen, but are also affected by other properties such as thermal conductivity and thermal convection of the sample gas. Generally, the sample gases encountered in process streams, except for the three gases mentioned earlier, do not affect the paramagnetic properties of the sample. However, they do affect the thermal conductive and convective properties and must be compensated for. These "background" effects result in some scale non-linearity and a loss in sensitivity. In filament-type detection systems, the loss in sensitivity may be restored by using four filaments, two as measuring and two as compensating filaments.

The magnetic susceptibility and density of oxygen are also influenced directly by pressure; therefore, many instruments of this type provide compensation for changes in atmospheric pressure (cell pressure). Hydrocarbons and other combustible gases present in the sample can cause measurement error since they may react on the heated filaments, causing changes in temperature and therefore resistance. Care must be taken during installation of annulus-type systems to ensure that the cross-tube is mounted precisely horizontally to avoid errors due to gravitational "chimney" effects.

7.2.2 The Quinke Principle

The operating principle of this type of system (often called the magneto-pneumatic principle) is illustrated in Fig. 7.3. A reference gas (e.g., nitrogen) flows past the filaments, which are connected into a Wheatstone bridge network. The reference gas flows are adjusted so that the bridge is balanced. The right-hand reference stream passes through a magnetic field, the left-hand stream does not. The two reference streams combine into a common outlet. Sample gas is introduced upstream of the common outlet and mixes with the reference gas as it emerges from the two arms. If oxygen is present in the sample gas, it is attracted into the magnetic field, generating a small back-pressure on the right-hand reference arm, which in turn distorts the flow pattern of reference gas around the detector filaments. The resulting bridge unbalance is proportional to the oxygen concentration in the sample gas.

This principle of measurement has the advantage that the sample gas does not come into contact with the heated filaments, thus preventing errors due to thermal conductive and convective effects or the presence of hydrocar-

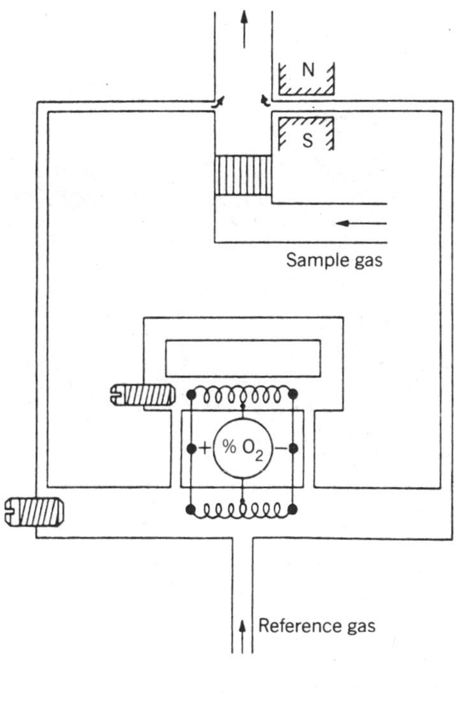

FIGURE 7.3 Quinke-type detection system.

bons. However, a continuous flow of reference gas is required and accurate alignment is necessary to ensure that there are no gravitational "chimney" effects.

7.2.3 Direct Measurement of Paramagnetism

The total magnetic susceptibility of a gas sample is the summation of the susceptibilities of individual gases present in the sample. As shown in Table 7.1, this total is in most cases entirely due to the oxygen present; therefore, the measurement is an accurate indication of oxygen concentration.

The total magnetic susceptibility of a sample may be measured by a method that is an extension of the original Faraday arrangement. The first "magneto-dynamic" measurement system, developed by Linus Pauling, is the Pauling Dumb-Bell System, shown in Fig. 7.4. The operating principle makes use of the fact that a diamagnetic body located in a non-uniform magnetic field will be repelled by that field, the force acting upon it being dependent upon the strength of the field and the magnetic susceptibility of the gas that surrounds it. The presence of a paramagnetic gas such as oxygen increases the susceptibility, and therefore the force, acting upon the diamagnetic body, in proportion to the oxygen partial pressure.

The Pauling cell consists of two diamagnetic spheres of glass filled with nitrogen and mounted at the ends of a bar to form a dumb-bell arrangement. The dumb-bell is mounted horizontally on a vertical quartz fiber torsion

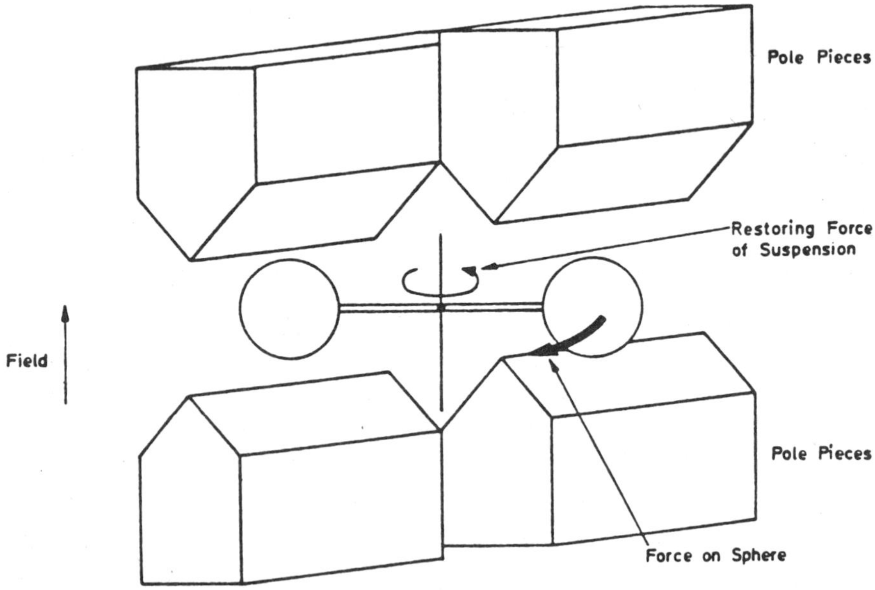

FIGURE 7.4 Pauling Dumb-Bell detection system.

suspension. The complete cell operates inside a strong non-uniform magnetic field. The spheres are repelled from the strongest part of the field and therefore rotate horizontally until the force produced by the twist of the suspension is balanced by the force acting on the spheres. The twisting force can be measured directly, using a lamp/mirror system, or by electromagnetic feedback applied to restore the dumb-bell to the zero position. The current required to maintain the dumb-bell in the zero position is proportional to the magnetic susceptibility of the gas present in the measuring cell and therefore to the oxygen partial pressure. This cell is the basis of the Taylor Servomex range of oxygen analyzers.

The magneto-dynamic measurement method has one significant advantage over other paramagnetic methods in that no physical property of the sample gas other than paramagnetism has any significant effect on the measurement. However, the method cannot be used for oxygen concentration measurement in samples containing other paramagnetic gases, unless facilities are provided for their removal before analysis.

7.3 CATALYTIC COMBUSTION METHODS OF MEASUREMENT

Oxygen and hydrogen react exothermically in the presence of a catalyst such as palladium, to form water:

$$O_2 + 2H_2 \rightarrow 2H_2O$$

The heat generated by the reaction is directly proportional to the oxygen concentration, provided that enough hydrogen is present to ensure that all the oxygen is used up in the reaction. The method may therefore be used to determine oxygen concentrations in gaseous samples. If the sample does not contain sufficient hydrogen for complete reaction of the oxygen present, hydrogen is mixed with the sample before combustion, supplied from a cylinder, or generated in an integral electrolytic cell. The temperature rise may be measured by a thermopile or similar device, with the cold junction located in the inlet gas stream and the hot junction in the outlet stream.

The reaction of hydrogen and oxygen to form water takes place stoichiometrically, in accordance with the exothermic equation, and is accompanied by a decrease in gas volume. This, again, is proportional to the oxygen concentration in the original sample, provided that an excess of hydrogen is present to ensure complete reaction. This method of measurement has also been used successfully in industry.

A typical example of a catalytic combustion-type oxygen detector is shown in Fig. 7.5. This device operates like the combustible gas detector. The sensor consists of a measuring cell and a reference cell, with a filament in each. The measuring filament has a catalytic (noble metal) surface to

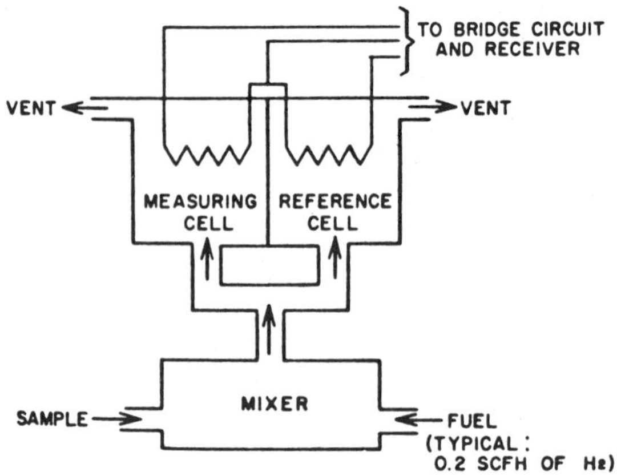

FIGURE 7.5 Catalytic combustion type oxygen detector.

oxidize the fuel (hydrogen), while the reference filament acts as a compensator for variations in sample temperature and thermal conductivity. The unit is usually mounted in a temperature-controlled enclosure. The filaments are connected as two arms of a Wheatstone bridge network. Sample and fuel are mixed and enter both cells. In the measurement cell, oxidation takes place on the filament and the resulting temperature change produces a bridge unbalance proportional to the oxygen concentration in the gas sample.

7.4 ELECTROCHEMICAL METHODS OF MEASUREMENT

Measurement of oxygen using electrochemical sensors falls into three main types:

- Polarographic cells.
- Galvanic or fuel cells.
- High-temperature, in-situ ceramic sensors.

All three methods are widely used in industrial applications for measurement in process gas streams and, in some cases, for measurement of dissolved oxygen in aqueous streams (e.g., boiler feedwater).

7.4.1 Polarographic Cells

A schematic diagram of a typical polarographic sensor is shown in Fig. 7.6. This is a membrane-covered type, comprising a gold cathode and a silver anode immersed in a common electrolyte gel (e.g., neutral potassium chloride). A PTFE membrane constitutes the base of the cell and is in close

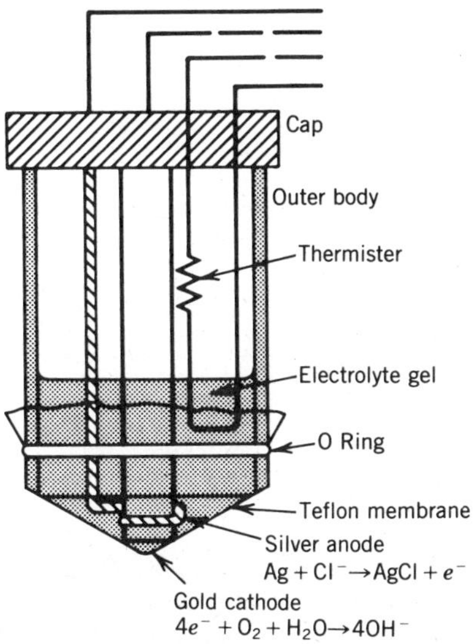

FIGURE 7.6 Polarographic sensor. (Courtesy of Beckman Industrial)

contact with the gold cathode. Oxygen diffuses through the membrane and reacts at the cathode to produce hydroxyl ions:

$$4e^- + O_2 + 2H_2O \rightarrow 4OH^-$$

The corresponding reaction at the anode is

$$Ag + Cl^- \rightarrow AgCl + e^-$$

An external voltage of approximately 0.7 V is applied across the electrodes to promote these reactions. The output current is proportional to the amount of oxygen reacting at the cathode, in accordance with Faraday's Law. This, in turn, is dependent upon the partial pressure of oxygen on the sample side of the PTFE membrane. Since the diffusion of oxygen across the membrane determines the rate at which oxygen reaches the cathode, the output of the cell is independent of sample flowrate. However, the rate of oxygen diffusion is temperature dependent, therefore requiring either close control of sample temperature or some form of temperature compensation.

There are two main disadvantages associated with the polarographic sensor:

1. The electrolyte composition changes with time as a result of the cell reaction with oxygen. Thus, regular replacement of the cell electrolyte is required.

2. Background current produced by the externally applied potential results in some electrolysis of water, which introduces significant errors (up to 0.05% oxygen equivalent) when measuring low oxygen concentrations.

7.4.2 Galvanic or Fuel Cells

Galvanic or fuel cells differ from the other electrochemical-type oxygen sensors in that they require no external source of power to drive them; they are power devices in their own right. There are basically three types, described next.

(a) The Micro-Fuel Cell

A typical example, manufactured by Teledyne Analytical Instruments (TAI) Inc., is shown in Fig. 7.7. The cell consists of a lead anode, configured to provide the maximum amount of metal available for reaction, and a convex, disc-shaped cathode of approximately 0.75 in. diameter with numerous perforations to facilitate continuous wetting of the surface with electrolyte and to ensure minimum internal resistance during the oxygen sensing reaction. The external surfaces of the cathode are plated with an inert or noble metal, normally gold. A PTFE membrane covers the cathode and both electrodes are immersed in the electrolyte (aqueous potassium hydroxide).

Diffusion of oxygen through the membrane allows the following reaction to occur at the cathode:

$$4e^- + O_2 + H_2O \rightarrow 4OH^-$$

The corresponding anode reaction takes place:

$$Pb + 2OH^- \rightarrow PbO + H_2O + 2e^-$$

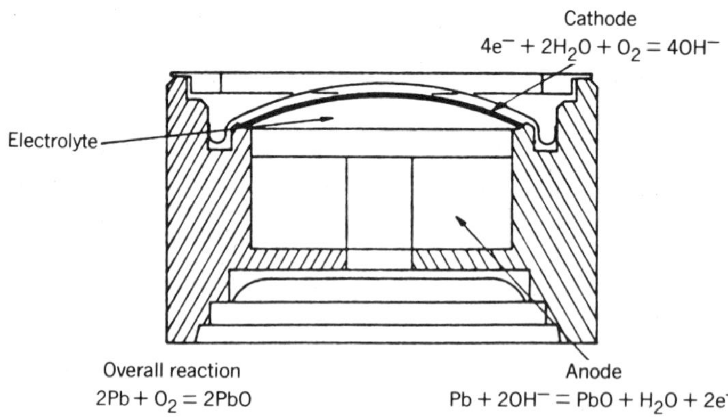

FIGURE 7.7 Micro-fuel cell. (Courtesy of Teledyne Analytical Insts. Inc.)

and the overall cell reaction is therefore

$$2Pb + O_2 \rightarrow 2PbO$$

Analogous with the polarographic cell described in Section 7.4.1, the output of the micro-fuel cell is related to the partial pressure of oxygen on the sample side of the PTFE membrane. However, in this case, since there is no applied potential and therefore no electrolysis of the electrolyte, the absence of oxygen in the sample gives zero cell output. This, combined with a linear cell response with respect to oxygen partial pressure, makes the micro-fuel cell suitable for oxygen measurement over a wide range. The major limitation at low oxygen partial pressures is the rate of oxygen diffusion across the PTFE membrane, which determines the speed of response. Again, cell response is independent of sample flowrate, but sensitive to temperature variations.

(b) The Open-Type Trace Level Fuel Cell

This is a variation on the micro-fuel cell, designed to provide better accuracy and response at low levels of oxygen (e.g., 1 ppm). The principle of operation is the same, but an increased cathode surface area is provided and there is no PTFE membrane. The main disadvantage with this cell is that some means of sample humidification is required if prolonged operation on dry sample is envisaged. Without humidification, the water-based electrolyte in the cell will eventually dry out. A typical example, manufactured by TAI, is shown in Fig. 7.8.

(c) The Hersch-Type Cell

The Hersch-type galvanic cell forms the basis of oxygen analyzers manufactured by Anacon Inc.; a typical example is shown in Fig. 7.9. It differs from

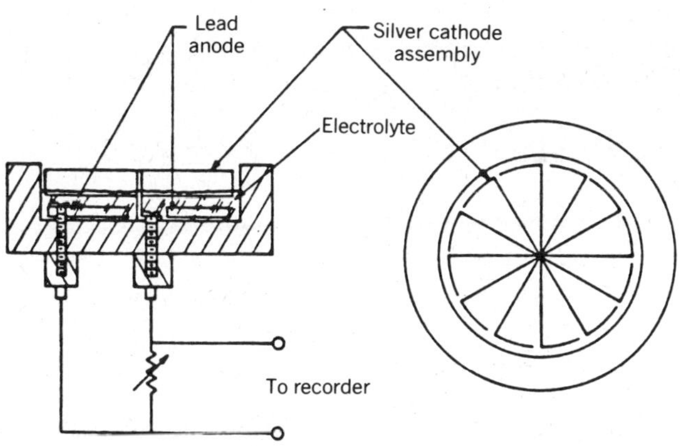

FIGURE 7.8 Open-type trace level fuel cell. (Courtesy of Teledyne Analytical Insts. Inc.)

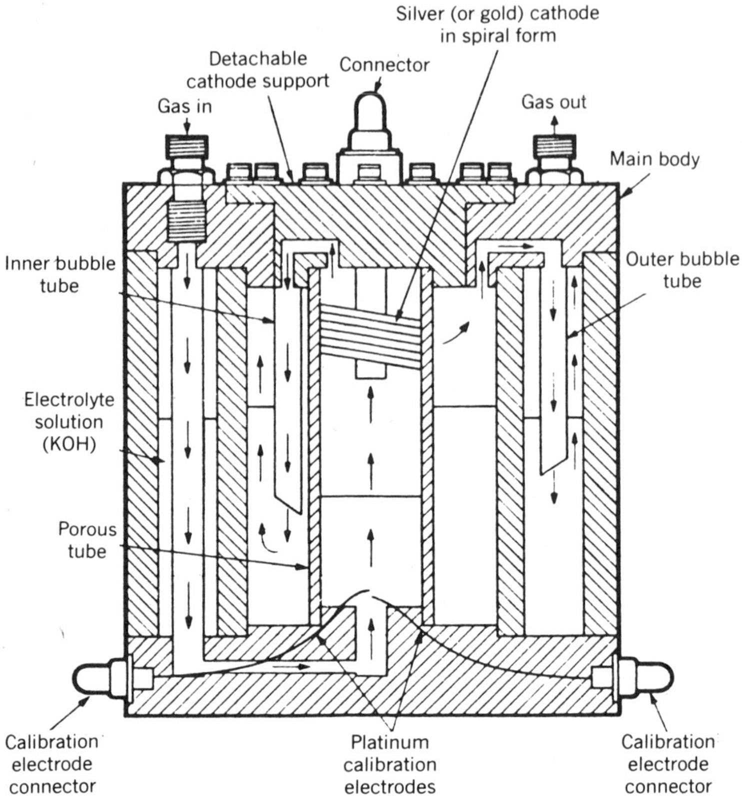

FIGURE 7.9 Hersch-type cell. (Courtesy of Anacon Inc.)

other fuel cells in that the cathode is made of silver and the anode of cadmium. While the cell reaction is basically similar to the micro-fuel cell, a third platinum electrode is added and a potential applied to provide anodic protection of the cadmium. Limiting the cathode current to a few microamps extends the life of the cadmium electrode. However, the cell output is nonlinear with respect to oxygen partial pressure, thus introducing the need for signal linearizing circuitry.

7.4.3 The High-Temperature Ceramic Sensor

In recent years we have seen a tremendous increase in the number of analyzers using the ceramic sensor, primarily in the combustion control area. This sensor is, in principle, an oxygen concentration cell consisting of two electrically conducting, chemically inert electrodes (normally platinum) attached to each side of a solid electrolyte, closed-ended tube. The tube is completely gas-tight and made of a ceramic material (normally calcium-stabilized zirconia), which, at the temperature of operation, conducts electricity by means of the passage of oxygen ions through it. The potential (E) across the cell is given by the Nernst equation:

$$E = 2.303 \frac{RT}{4F} \log \frac{P_1}{P_2}$$

where R is the universal gas constant
T is the absolute temperature
F is the Faraday constant
P_1 is the oxygen partial pressure in the sample
P_2 is the oxygen partial pressure in the reference gas

Provided that the oxygen partial pressure in the reference gas is known, the potential difference between the two electrodes is related to the partial pressure of oxygen in the sample gas. In practice, air is normally used as the reference gas and is allowed to come into contact with one electrode, while the other electrode is exposed to the sample gas. The response of the ceramic sensor, as given by the Nernst equation, holds over a very wide range

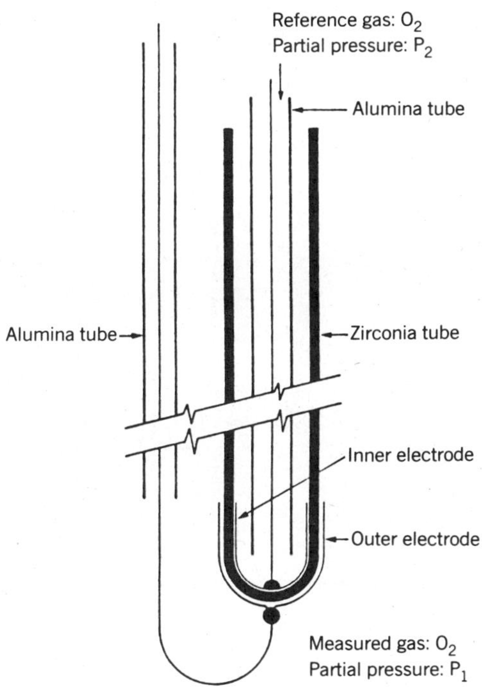

(a)

FIGURE 7.10 High-temperature ceramic sensor. (*a*) Operating principle. (Courtesy of Kent Industrial)

of oxygen partial pressures, allowing measurements to be made in the percentage oxygen range, down through ppm concentrations, to partial pressures as low as 10^{-25} atm. The output of the sensor increases with decreasing oxygen partial pressure, as shown by the response curve in Fig. 7.10*b*. A schematic diagram of a typical sensor, manufactured by Kent Process Control, is shown in Fig. 7.10*a*.

An important point about these sensors is that a temperature in excess of 600°C is required for their operation, but the development of new ceramics in the near future may lead to similar sensors capable of operating at lower temperatures. Operation at high temperature may be an advantage in some applications, notably flue gas oxygen measurement for furnace/boiler combustion control. Bear in mind, however, that at these temperatures, combustible gases present in the sample (e.g., methane, carbon monoxide, etc.) will react with oxygen under catalytic action of the platinum electrode, so the sensor output will therefore be a measure of *residual* oxygen.

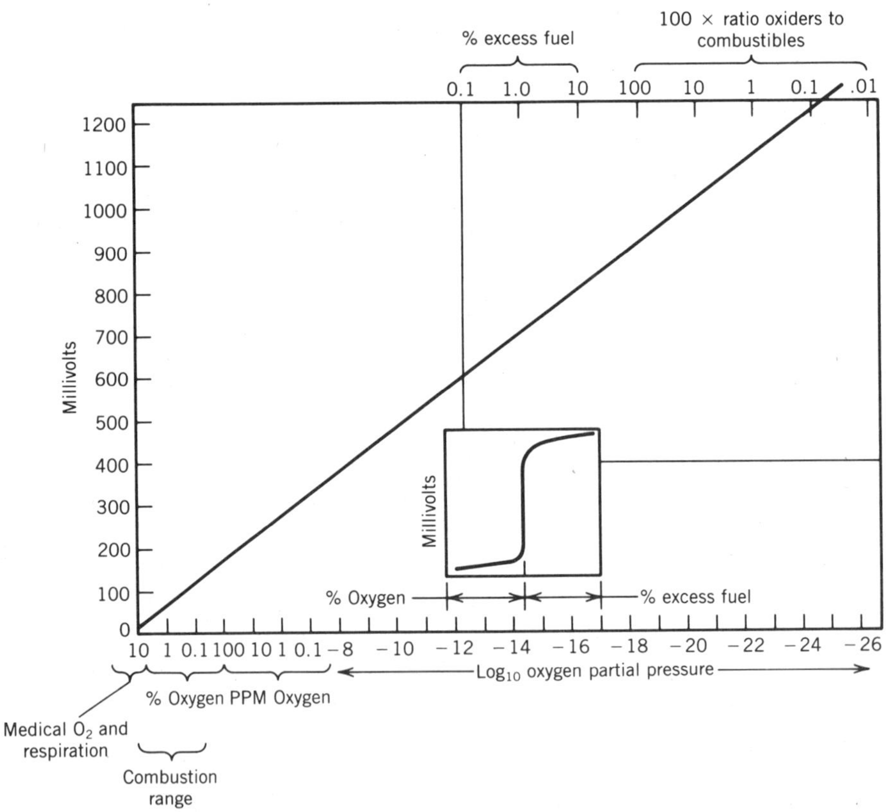

(*b*)

FIGURE 7.10 High-temperature ceramic sensor. (*b*) Cell output vs. oxygen concentration. (Courtesy of Kent Industrial)

7.4.4 Electrochemical Oxidation Methods

In the presence of oxygen, certain chemical compounds are oxidized, resulting in a solution in which both the unoxidized and oxidized forms of the compound are present. As the extent of the oxidation is increased, the electrochemical properties of the solution are changed. Measurement of the electrochemical potential is therefore a measure of the extent of oxidation which, in turn, is related to the amount of oxygen present.

An example is the sodium salt of 9,10-dihydroxy-anthracene-2-sulfonic acid, which is oxidized to form the corresponding quinone:
The concentration of the oxidized form relative to the reduced form is a measure of the oxygen that has passed through the solution.

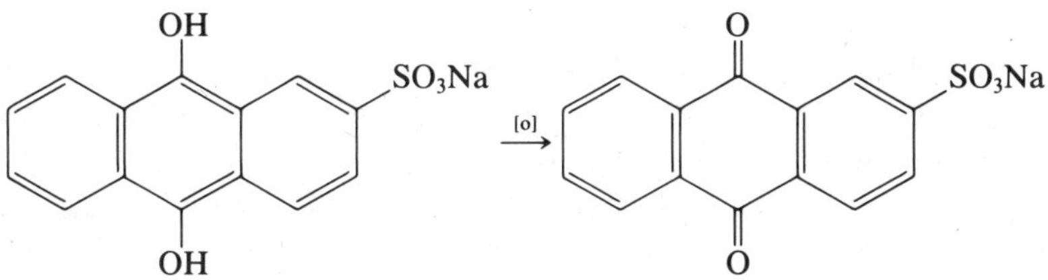

7.5 LABORATORY METHODS OF OXYGEN MEASUREMENT

Methods used in the laboratory to measure oxygen concentrations in gaseous and liquid samples are limited to the following main types:

1. Chromatographic analysis, based on the use of molecular sieves type columns (GSC). This is the standard method for analysis of permanent gases.
2. Titrimetric methods. This is the standard for dissolved oxygen analysis in aqueous samples.

TABLE 7.2 ASTM Standard Test Methods for Oxygen Measurement

ASTM No.	
D888-81	Dissolved Oxygen in Water
D1136-	Oxygen Content of Natural Gas by Volumetric Chemical Analysis
D1945-81	Analysis of Natural Gas by Gas Chromatography
D1946-82	Analysis of Reformed Gas by Gas Chromatography
D2223-66	Oxygen Impurity in Nitrogen (Electrochemical Method)
D2504-	Oxygen Content of C_3 and Lighter Hydrocarbons by Gas Chromatography
E379	Apparatus for Direct Microdetermination of Oxygen in Organic Compounds

3. The use of a galvanic cell. This is the standard method for oxygen concentration in nitrogen.
4. Gravimetric-volumetric analysis, a very tedious affair, especially at very low levels of oxygen.

The electrochemical methods use apparatus similar to the process analyzer. The commonly used ASTM standard methods for oxygen analysis are listed in Table 7.2.

7.6 PROCESS OXYGEN ANALYZERS FOR GASES

In view of its importance in water quality monitoring, the measurement of dissolved oxygen in aqueous samples (e.g., boiler feedwater) is discussed in detail in Chapter 13, together with other water and steam quality parameters. This section is limited to process analyzers for the measurement of oxygen in gaseous streams.

In view of the large number of oxygen analyzers commercially available, a few typical examples are presented, based on the different operating principles discussed earlier. A complete list of commercially available equipment is given in Table 7.3 on pages 245–249.

7.6.1 Development of the Zirconia Oxygen Sensor

As mentioned earlier, one of the major multi-industry applications for oxygen measurement is in combustion efficiency monitoring and control of industrial furnaces and boilers, which involves sampling the flue gas in the convection section or stack.

The first oxygen analyzers to appear in the early 1960s were either paramagnetic or electrochemical types, which required extractive sample systems, that is, the process stream was sampled by a probe mounted in the furnace duct, then transported to the analyzer and conditioned prior to analysis. The sample is very difficult to handle since it is at high temperature, water-laden, and contains acid gases such as sulfur dioxide, nitrogen oxides, and so on. The sampling system is therefore susceptible to corrosion and failure, making it a high maintenance item.

Nevertheless, oxygen analyzers in these applications are justified on the basis of the significant savings attributed to increased operating efficiency and energy conservation. Much effort was expended by users and manufacturers alike during the 1960s to improve sampling system design, but maintenance requirements still remained too great. Obviously, a radical breakthrough was needed to improve the performance of analyzers in these applications, particularly with the advent of process computer control.

The breakthrough occurred with the development of the zirconia sensor, described in Section 7.4.3, which came out of the basic research work on fuel cells for spacecraft and led to the introduction of the zirconia sensor for on-line oxygen measurement by Westinghouse in 1970. At about the same time, Kent in England and Ametek in the United States pioneered the work on their zirconia sensors.

The zirconia sensor, capable of direct installation and operation in or near the flue gas duct, removed the need for a complex sample handling system, which was the major problem with this type of application. The development was hailed as a major advance in design and resulted in a significant reduction in maintenance requirements. Several of the major oil companies undertook evaluation of the zirconia analyzer, and within a few years analyzers were in extensive use in the United States and in Europe.

Since that time, the popularity of this approach is demonstrated by the fact that the author knows of at least 19 companies that manufacture zirconia-type oxygen analyzers.

7.6.2 Paramagnetic-Type Analyzers

(a) Beckman Instruments Inc. (United States)

The Model 755 Process Oxygen Analyzer uses the magneto-dynamic dumbbell principle to measure the oxygen concentration in gaseous process streams. The test body is mounted on a platinum suspension in a nonuniform magnetic field. A diamond-shaped mirror mounted on the suspension reflects light from a prefocused source equally onto two photocells

FIGURE 7.11 Model 755 paramagnetic oxygen analyzer. (*a*) Paramagnetic sensor assembly. (Courtesy of Beckman Industrial Inc.)

FIGURE 7.11 Model 755 paramagnetic oxygen analyzer. (*b*) Control unit. (Courtesy of Beckman Industrial Inc.)

when the test body is in the neutral position. As the magnetic susceptibility of the gas surrounding the test body increases, a force is applied to the test body, tending to rotate it out of the field. The mirror also rotates, so that the photocells become unequally illuminated; through an amplifier, feedback current is applied to oppose the magnetic force and bring the test body back to the neutral position. The current is a linear function of the oxygen concentration in the gas sample and is converted to an output suitable for remote recording.

The detector assembly and control unit layout are shown in Fig. 7.11. The unit is available in a NEMA 4 enclosure (with optional ISA type-Z purge) or in an explosion-proof case suitable for NEC Class 1, Group B, C, D, Division 1 hazardous locations. Many range options are offered from 0–1 to 0–100%, including suppressed ranges such as 20–21%, 90–100%, and so on. Response time is quoted as 90% full scale in 7 sec at a sample flowrate of 200 ml/min. Voltage output is standard; optional current output is available by the use of plug-in circuit boards (4–20 and 10–50 mA dc).

(b) Hartmann and Braun (West Germany)

This company manufactures probably the widest range of process oxygen analyzers available today, using the paramagnetic property of oxygen and a variety of different detection systems.

The Magnos 2T and 2T-Ex operate on the magnetic wind principle, using an annulus-type detection cell as shown in Fig. 7.12*a*. The Magnos 2T is designed for installation in a non-hazardous location, whereas the 2T-Ex is

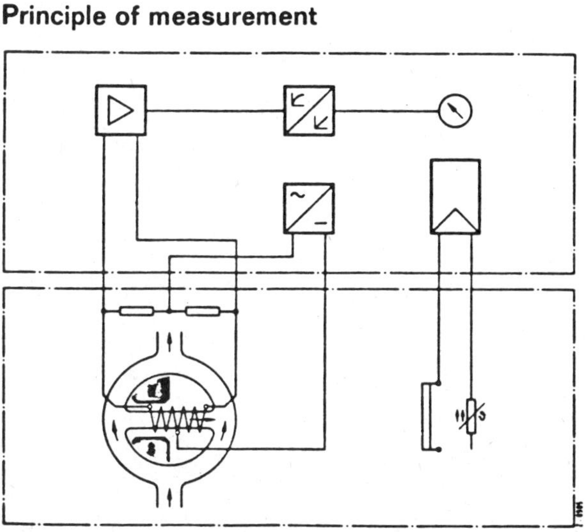

(a)

FIGURE 7.12 Hartmann & Braun paramagnetic oxygen analyzers. (*a*) Magnos 2T and 2T-Ex. (Courtesy of Hartmann & Braun)

designed as intrinsically safe with PTB approval for hazardous area installation. Several measuring ranges are available from 0–1 to 0–100%, with suppressed ranges having a minimum span of 2% (e.g., 19–21%). A current output (0–20 or 4–20 mA dc) is provided. Repeatability is quoted as ±0.2% of full scale).

The Magnos 3 and 3-Ex operate on the principle of direct paramagnetic measurement using the dumb-bell type detection system, as shown in Fig. 7.12*b*. A mirror and photocell detect the angular displacement of the dumb-bell, and the current required to restore the dumb-bell to its neutral position is proportional to the oxygen concentration in the sample gas. The Magnos 3 is designed for non-hazardous area installation; the Magnos 3-Ex is suitable for hazardous areas by inert gas purging to PTB standards. Five standard measurement ranges are offered, 0–1, 3, 10, 30, and 100% oxygen, switchable electrically. Suppressed ranges are not possible. A current output (0–20 or 4–20 mA dc) is provided and repeatability is quoted as ±0.5% of span.

The Magnos 4G is a magneto-pneumatic device, the operating principle of which is shown in Fig. 7.12*c*, and is similar to the Quinke principle described in Section 7.2.2. Sample and reference gas flow from opposite directions into a magnetic field generated by an intermittently operated electromagnet. When the magnetic field is switched on, each gas undergoes a change in pressure proportional to the oxygen concentration of the respective gas in the inhomogeneous part of the field. A pressure differential builds up between the gases and is measured by a diaphragm capacitor. The resulting

Principle of measurement

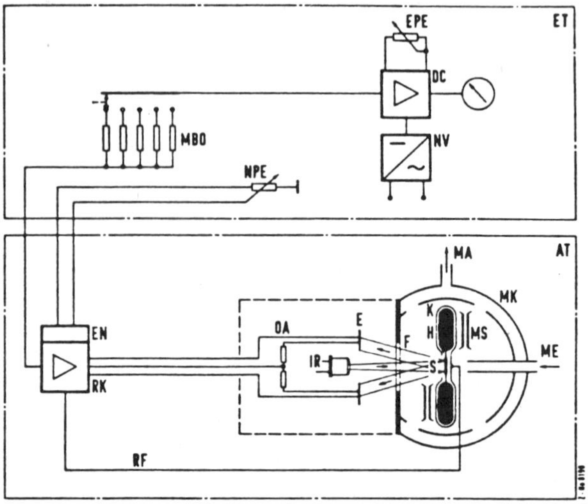

AT = Analyzer section	ET = Electronics section
ME = Sample gas inlet	MBO = Measuring range organization
MA = Sample gas outlet	NPE = Zero adjustment
MK = Measuring chamber	EPE = Span adjustment
H = Dumbbell	DC = DC amplifier
K = Compensation loop	NV = Power supply
S = Mirror	
MS = Magnet gap	
F = Window	
OA = Optical pickoff	
IR = Infrared diode	
E = Detector	
RK = Control circuit	
RF = Feedback	
EN = Electr. zero	

(b)

FIGURE 7.12 Hartmann & Braun paramagnetic oxygen analyzers. (*b*) Magnos 3 and 3-Ex. (Courtesy of Hartmann & Braun)

capacitance change is amplified, conditioned by a phase-sensitive rectifier, and displayed by a meter in terms of oxygen concentration. A gas distribution system provides gas flow stability and the effect of temperature variations is compensated for by an electronic circuit.

Four measurement ranges are offered, 0–2, 5, 10, and 25% oxygen, with suppressed ranges also available. Voltage (0–10 V dc) and current (0–20 or 4–20 mA dc) outputs can be provided and repeatability is quoted as ±0.5% of full scale.

The Magnos 5T operates on the magnetic wind principle, using a filament-type detection cell as shown in Fig. 7.12*d*. The temperature-sensitive ring resistors W_1 and W_2 are protected against corrosion and catalytic ignition by

Principle of measurement

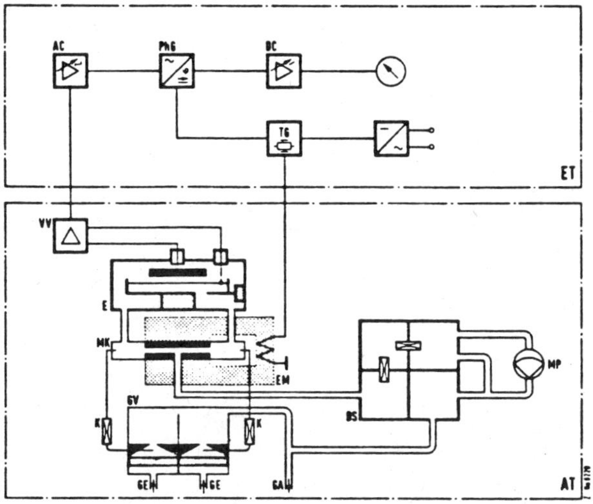

AT = Analyzer section
GE = Gas inlet
GA = Gas outlet
GV = Gas distributor
K = Capillaries
EM = Electromagnet
E = Detector with diaphragm capacitor
MK = Measuring chamber
DS = Attenuation network
MP = Diaphragm pump
VV = Preamplifier

ET = Electronics section
AC = AC amplifier
PhG = Phase-sensitive rectifier
DC = DC amplifier
TG = Clock generator

(c)

FIGURE 7.12 Hartmann & Braun paramagnetic oxygen analyzers. (*c*) Magnos 4G. (Courtesy of Hartmann & Braun)

means of thin-walled capillary tubes. Presence of oxygen in the sample gas results in an increased circulation flow in the measurement chamber and consequent unbalance of the Wheatstone bridge network. The bridge output voltage is amplified and linearized (if required) to produce a current proportional to the oxygen concentration in the sample gas.

The Magnos 5T is designed for surface or panel mounting in non-hazardous locations. For measurement of flammable gases, which can form explosive mixtures with air or oxygen, special precautions such as purging of the enclosure and the use of flame arrestors must be taken. Eight measurement

Principle of measurement

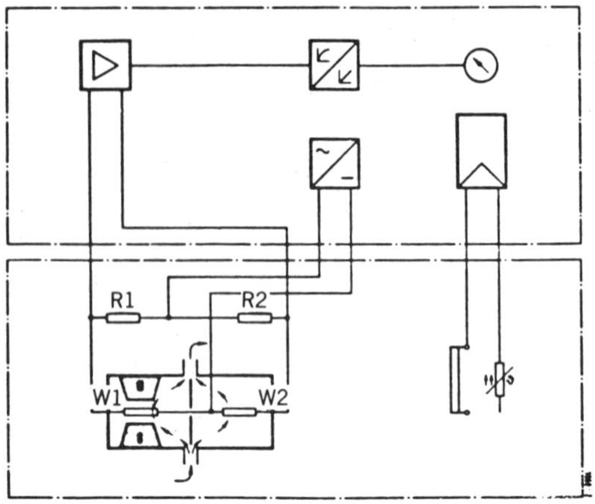

R1:R2 Fixed resistors
W1:W2 Temperature-dependent ring resistors

(d)

FIGURE 7.12 Hartmann & Braun paramagnetic oxygen analyzers. (*d*) Magnos 5T. (Courtesy of Hartmann & Braun)

ranges are offered, from a minimum of 0–3% to a maximum of 0–100% oxygen. Repeatability is quoted as less than 1% of full scale.

(c) Servomex Ltd. (Great Britain)

Servomex manufactures a range of process oxygen analyzers based on the magneto-dynamic, dumb-bell principle outlined in Section 7.2.3 and illustrated in Fig. 7.4.

The Model 540A shown in Fig. 7.13*a* is BASEEFA approved for installation in Zone 2 hazardous locations, except when used to measure flammable or oxygen-rich gases, when additional protection is required. The unit is designed to accept ISA type X, Y, and Z purging to suit specific application requirements. Measurement ranges offered are 0–1, 2.5, 5, 10, 25, 50, and 100% oxygen, switch selectable from the front panel. Alternatively, a fixed range may be specified. Sampling equipment, provided as standard with the analyzer, includes sample bypass rotameter, analyzer rotameter (both with needle valve), and micro-fiber filter (95% retention of 0.6-micron particles). Response time is quoted as 8 sec for 90% response and repeatability as ±1% of span.

Various accessory modules are available, covering altitude compensation, remote range selection, barometric pressure compensation, sample flow alarm, analyzer failure alarm, and range suppression. Voltage and current outputs are available, including 4–20 mA dc. The analyzer is suitable for surface, panel, or relay rack mounting.

The Model 541A, shown in Fig. 7.13*b*, is designed specifically for mea-

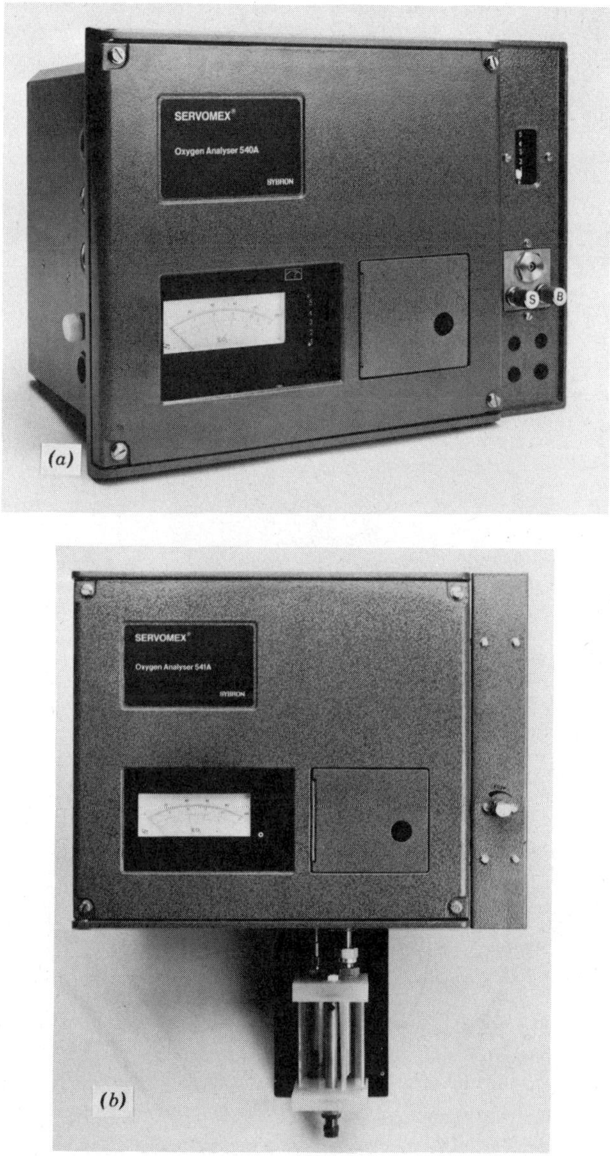

FIGURE 7.13 Servomex paramagnetic oxygen analyzers. (*a*) Model 540A; (*b*) Model 541A. (Courtesy of Servomex Ltd.)

surement of oxygen concentration in combustion gases and process streams with a high concentration of condensible vapors at ambient temperature and pressure. It has the same basic features as the Model 540A, except for special sample conditioning equipment.

The Model OA-292 shown in Fig. 7.13*c* consists of three parts—the sample conditioning, analysis and electronics enclosures—and is designed for

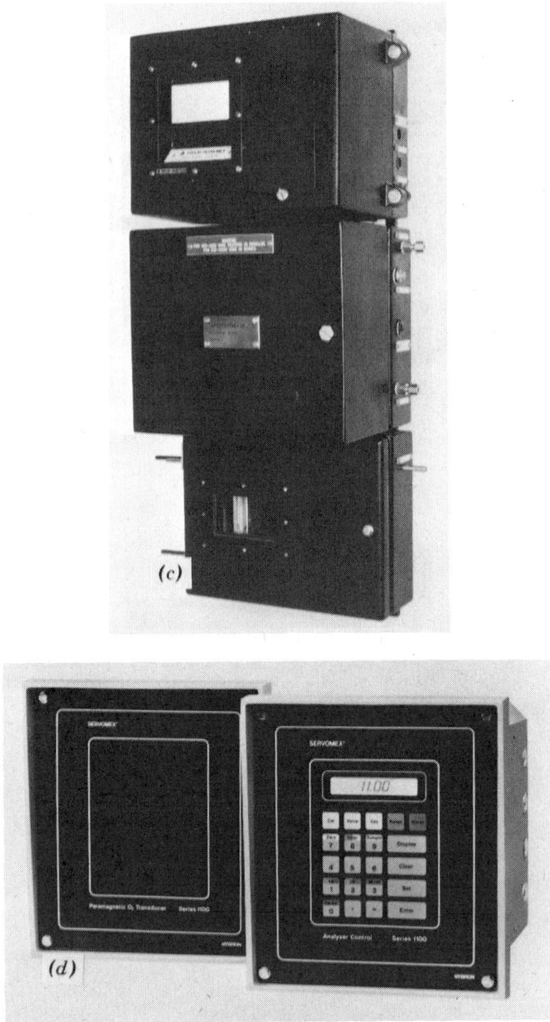

FIGURE 7.13 Servomex paramagnetic oxygen analyzers. (*c*) Model OA292; (*d*) Series 1100A microprocessor control unit. (Courtesy of Servomex Ltd.)

oxygen measurement of high-temperature gases containing condensible vapors with dew points between 122 and 221°F. Purge connections are provided on the upper and center enclosures for use when the analyzer is to be installed in a hazardous location. Local indication, alarm options, field-selectable ranges and output options are also included with the OA-292.

A recent introduction by Servomex is the Model 1100A Oxygen Analyzer, shown in Fig. 7.13*d*. This instrument utilizes the well-proven Servomex magneto-dynamic measurement technology, coupled with a microprocessor-based control unit. The analyzer consists of two units. The transducer unit houses the measuring cell, heaters, mechanical zero adjustment, and an optional pressure compensation transducer. The control unit digitizes and processes the signal from the transducer unit to provide a liquid crystal display of oxygen concentration. The microprocessor performs zero and span calibration corrections, generates an isolated 4–20 mA dc output for remote recording purposes, and provides internal self-diagnostics for the system. It may also be programmed to set alarms, carry out automatic calibration at preset intervals, and pressure compensation calculations. A data transmission board may be fitted as an option to provide a serial data link to a host computer (RS232C or 20 mA current loop).

Analyzer sample requirements are 50 mmWG–4 psig pressure and 14–122°F at a flowrate of 150 ml/min. Two standard sample handling systems are available for wet or dry samples respectively. Analyzer spans offered are 2.5, 5, 10, 25, 50, and 100%, with zero suppression in 1% steps up to a maximum of 99%.

The analyzer is designed for installation in NEC Class 1, 2, and 3, Groups A–G, Divisions 1 or 2. Certification has also been obtained from CENELEC and BASEEFA.

7.6.3 Electrochemical-Type Analyzers

(a) Anacon Inc. (United States)

Anacon Models 76 and 79 Oxygen Analyzers are based on the Hersch principle (Patent No. 707323), but incorporating certain improvements. The principle of cell operation is described in Section 7.4.2 and illustrated in Fig. 7.9.

The Model 76 contains the analysis section and electronics section in a single enclosure, designed for installation in a non-hazardous location. Standard measurement ranges are 0–10, 100, and 1000 ppm oxygen, switchable by illuminated pushbuttons. Other range options are also available. The Series A is fitted with a gold analysis cell for use with acetylene-containing gases and the Series G for carbon dioxide or gases containing carbon dioxide.

Calibration of the instrument is accomplished by passing the sample gas through a purifier to remove any oxygen present and then adding an accurately measured quantity of oxygen generated electrolytically. The purifier and oxygen generator are built into the instrument for this purpose. Calibration normally takes about 10 min.

Analyzer response time at 8 l/hr sample flowrate is quoted as 15 seconds for initial indication of change and 100 seconds for 90% response. Standard outputs are 0–5 mV dc, 0–10 V dc and 4–20 mA dc. Other options are also available.

The Model 79 has the same specifications as the Model 76, except that it is intended for use in hazardous locations. It consists of two units, the analysis unit and the control unit, which may be separated by distances up to 300 feet. The analysis unit has facilities for purging; the control unit is suitable for safe area installation only.

(b) Beckman Instruments Inc. (United States)

The Models 7003, 741 and 778 Process Oxygen Analyzers are all based on the polarographic sensor illustrated in Fig. 7.6 and differ basically in the packaging of the control unit.

Depending on the particular application, the sensor, shown in Fig. 7.14*a*, can be supplied in one of three configurations. The flow chamber is used to mount the sensor in variable pressure, gaseous, or liquid streams up to 50 psig. The submersion assembly is used to place the sensor at depth in a tank or for closed vessel monitoring. The fast response gas flow assembly is designed for use with gaseous samples at pressures slightly above atmospheric. The polarographic cell will operate continuously for up to 6 months or more, after which the electrolyte will require recharging.

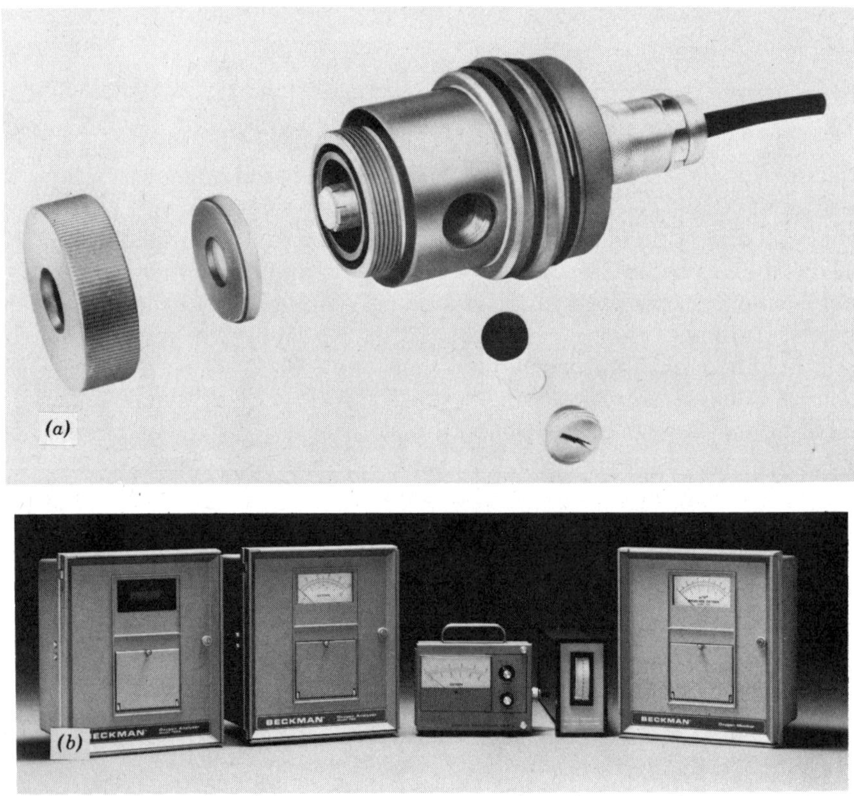

FIGURE 7.14 Beckman polarographic oxygen analyzers. (*a*) Polarographic sensor; (*b*) range of control units. (Courtesy of Beckman Industrial Inc.)

The Model 7003 is a solid-state control unit mounted in a NEMA 4 enclosure suitable for rack or panel mounting in a non-hazardous location. An air purge kit is available for the provision of ISA type Z purging. The unit may be located at distances up to 1000 feet from the sensor. Measurement ranges quoted are 0–1, 5, 10, and 25% oxygen with a repeatability of ±1.0% of full scale. Voltage output is standard (0–1, 0–10 V dc and 0–10, 0–100 mV dc). Current output (4–20 or 10–50 mA dc) is available from plug-in circuit boards. The unit can be equipped with dual alarm SPDT contacts, adjustable over the full range, with individual dead-band adjustment between 1% and 20% of full scale.

The Model 741 has the same features as the model 7003, except that it is a miniature 3 in. × 6 in. unit, suitable for panel mounting in a non-hazardous location and with integral indicating meter. Various control units are shown in Fig. 7.14*b*.

The Model 778 is suitable for use in gaseous or liquid applications and may be calibrated in percent oxygen, mm oxygen partial pressure, or ppm dissolved oxygen. The 6 in. × 6 in. unit is suitable for panel mounting in a non-hazardous location, at distances up to 125 ft from the sensor. If an extension cable is used, a connector box is required. A 0–50 mV dc output is available for remote recording.

(c) Teledyne Analytical Instruments (United States)

TAI probably manufactures the widest range of process oxygen analyzers in industry today, covering percent and ppm measurement in gaseous process streams and using the patented Micro-fuel cell described in Section 7.4.2 and illustrated in Fig. 7.7.

The Series 320 Analyzers are designed for measurement in the range 0–100% oxygen and the various models available differ only in the way they are packaged. A selection is illustrated in Fig. 7.15.

FIGURE 7.15 Series 310 and 320 oxygen analyzers. (Courtesy of Teledyne Analytical Insts. Inc.)

FIGURE 7.16 Model 322 multichannel system. (Courtesy of Teledyne Analytical Insts. Inc.)

The Model 322 is a multi-channel system of modular design, suitable for rack or panel mounting as shown in Fig. 7.16. Also shown is a typical 64-channel system mounted in a custom-built, free-standing rack unit. Each model in the 320 Series is available with optional single or dual alarm capability, current outputs of 1–5, 4–20 or 10–50 mA dc, and three measurement ranges as standard, either 0–5, 10, 25% or 0–5, 25, 100% oxygen.

The Series 310 Analyzers are designed for trace oxygen measurement in the range 0–10 to 0–10,000 ppm. Again, the various models differ only in packaging, also illustrated in Fig. 7.15. Each model is offered with four measurement ranges: 0–10, 100, 1000, and 10,000 ppm, except the model 315, a miniature general-purpose unit, which has two ranges. Also provided are alarm facilities and current outputs of 1–5, 4–20 or 10–50 mA dc.

7.6.4 Zirconia-Type Analyzers

As mentioned earlier, development of the zirconia sensor was a significant step forward in the application of oxygen analysis to combustion control of furnaces and boilers. Some manufacturers developed in-situ sensors designed for installation directly in the flue gas duct or stack; others designed a system for mounting directly outside the duct and provided a probe and aspiration facilities to take a small sample of the flue gas for measurement by the externally mounted zirconia sensor. However, the basic operating principle of the sensor remains the same as described in Section 7.4.3.

(a) Ametek, Thermox Instruments Division (United States)

Ametek manufactures a range of oxygen analyzers based on the externally mounted zirconia sensor. The WDG analyzer was first introduced in 1970, specifically for combustion oxygen measurement. The THERMOX oxygen analyzers are designed for process stream oxygen measurement using the same principle.

The Model WDG-III Combustion Oxygen Analyzer, a third-generation instrument first introduced in 1976, consists of two units, the sensor unit designed to be mounted directly to the flue gas duct or stack, and the control unit, which is usually remote mounted. The standard control unit is suitable for wall or rack mounting in a safe area, but is also available in explosion-proof design for NEC Class 1, Group D, Division 1, or in purge-type designs. Figure 7.17 shows a schematic diagram of the sensor unit.

The WDG-III is suitable for oxygen measurement in wet, dirty, and corrosive flue gases up to 3200°F, generally without sample filtering, cooling, or other conditioning. The sensor assembly consists of a short heated flow path, controlled temperature furnace, and the zirconia sensor. The main gas path is a short U-shaped tube through which the sample is drawn by an air-operated aspirator and discharged back to the process. Extending upward from the main flow tube is another loop containing the sensing element enclosed in the temperature-controlled furnace. A small but constant amount of sample flows past the sensor by thermal convection and back down to the main flow tube. All surfaces in contact with the sample gas are maintained at 400°F (or even higher in a special version). Heat is supplied to

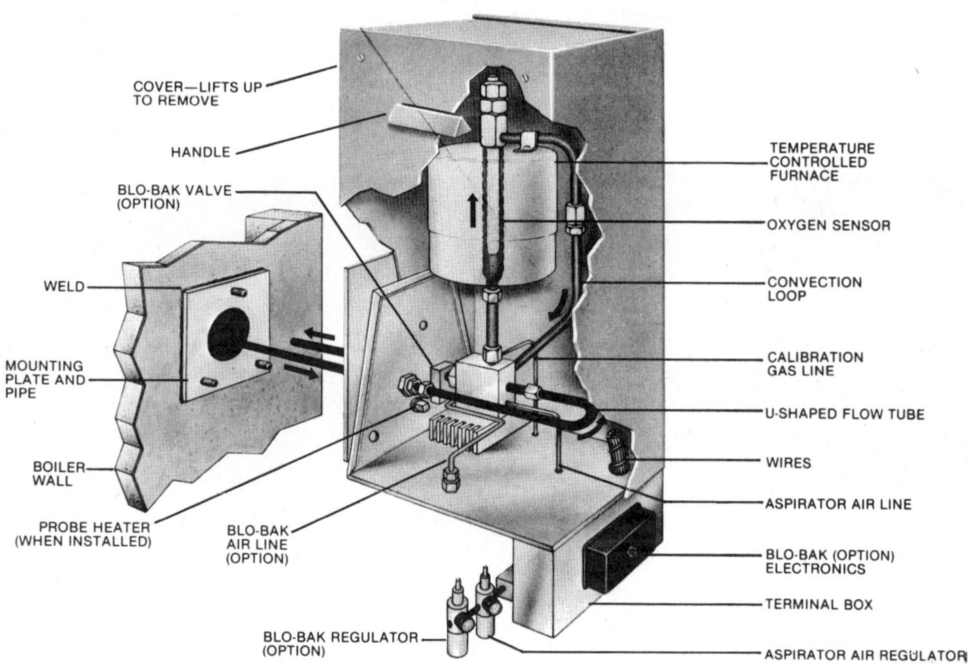

FIGURE 7.17 Thermox Model WDG-III oxygen analyzer. (Courtesy of Ametek Inc.)

the enclosure and aspirator air by cartridge heaters inserted into a finned aluminum block. Three types of probe are available, depending upon sample temperature:

- 316 stainless steel—up to 1300°F.
- Alonized 310 stainless steel—up to 1700°F.
- Ceramic—up to 3200°F.

The continuous output from the sensor is transmitted without amplification to the control unit and displayed on an integral meter. Instrument span can be any portion of the range from 100% down to 0.1% oxygen. Repeatability is quoted as ±0.2% of reading. The response of the system is 90% of full scale in less than 5 sec (not including sample transport time). Standard output is 0–10 mV dc logarithmic, but other voltage and current, log, and linear outputs are available as options.

The Thermox II Oxygen Analyzer operates on the same principle as the WDG-III but is designed for use with a sample handling system. Instrument span can be any portion of the range from 100% down to 0.1% oxygen. The sensor unit may be located at distances up to 1000 ft from the control unit.

The Thermox III is a multi-point oxygen analyzer, originally developed for installation in air separation plants. It is suitable for panel mounting and incorporates the sensor unit, flowmeters, and control unit as a single system. Three-way ball valves are provided for manual switching of the three streams. The system has the same options as the Thermox II.

(b) Westinghouse Electric Corp. (United States)

Westinghouse manufactures a range of oxygen analyzers based on both the in-situ and externally mounted zirconia sensor.

The Model 218 uses the Westinghouse-Hagen in-situ zirconia probe, designed for direct installation in the flue gas duct or stack. The sensing package shown in Fig. 7.18, consists of the filter-type probe, a control unit which may be mounted in a NEMA 12 enclosure or in an enclosure suitable for NEC Class 1, Group B, Division 1, and 20 feet of interconnecting cable.

The probe is suitable for use at process temperatures from 50 to 1400°F and is available in 18 in., and 3, 6, 9, and 12-ft lengths. A two-decade logarithmic range is standard, normally chosen as 0.25–25% or 0.1–10% oxygen. Sensing element operating temperature is 1550°F. The probe is fitted with an "in-place" calibration facility, consisting of a direct tubing connection between the outside of the probe head and the area inside the diffusion element surrounding the zirconia cell. Thus, calibration gas may be introduced directly onto the sensing element. In addition, an optional package is available for checking probe calibration automatically on a periodic basis. The system response time is quoted as 3 sec and an analog current output (4–20 mA dc) is provided.

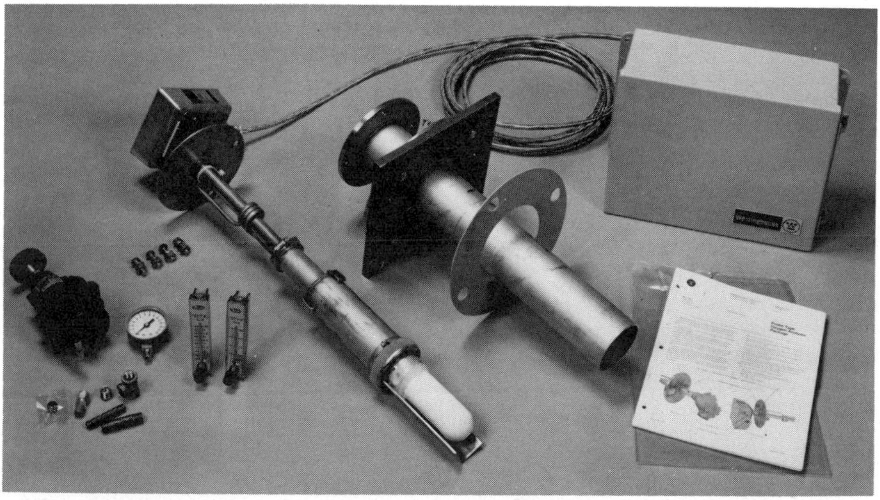

FIGURE 7.18 Model 218 oxygen analyzer. (Courtesy of Westinghouse Corporation)

The Model 225 has the same features as the Model 218, except that the sensor probe is suitable for process temperatures up to 1000°F only.

The Model 450 High-Temperature Combustion Analyzer is designed to measure the net excess oxygen or net excess combustibles in flue gases, using a single, in-situ zirconia probe, at process temperatures between 1472 and 2732°F (800–1500°C). Construction of the sensor allows it to respond to air-rich or air-deficient combustion conditions; by proper calibration, the analyzer output can be made proportional to excess or deficiency of air over a wide range of air-fuel ratios. The operating principle is illustrated in Fig. 7.19.

In the diffusion limited current mode, the zirconia cell is a cavity within which the oxygen partial pressure is reduced to near zero when a small excitation voltage is applied. As oxygen diffuses into the cavity and is "pumped" out by action of the excited cell, positive current that is proportional to the oxygen concentration in the sample gas flows in the excitation circuit. When the sample gas becomes reducing (air deficiency and presence of combustibles), the voltage produced by the cell exceeds the excitation voltage and the oxygen flow and current are reversed. Oxygen flows from the reference side through the cell into the cavity to combine with the combustibles present. This results in continuous bathing of the sample side electrode in oxygen, even when the sample may be strongly reducing, thereby providing electrode protection. Therefore the Model 450 analyzer is able to measure combustion conditions on either side of the stoichiometric point, providing an output current proportional to excess oxygen or excess combustibles. Zero cell current occurs at the chemically correct ratio.

The electronics unit of the Model 450 converts the positive and negative

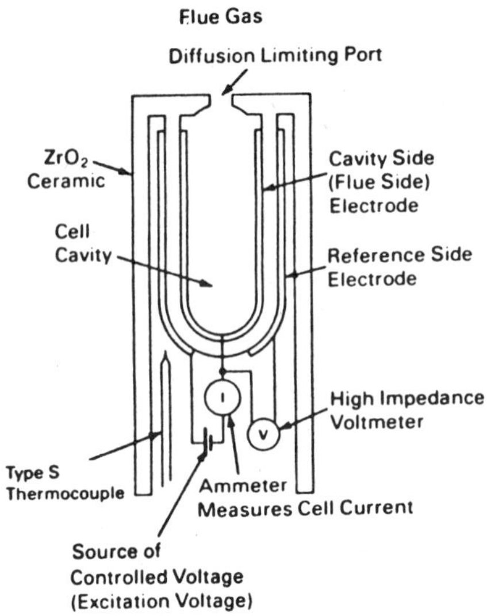

FIGURE 7.19 Model 450 high-temperature combustion analyzer. (Courtesy of Westinghouse Corporation)

currents to a standard 4–20 mA dc output signal proportional to the excess or deficiency of air. The electronics unit can be supplied as an indicating or non-indicating package mounted in a NEMA 12 enclosure or as an explosion-proof unit suitable for installation in NEC Class 1, Group B, Division 1 locations. Standard ranges are 0–10% oxygen and 0–10% combustibles. The control unit may be located at distances up to 300 feet from the sensor.

The Model 240 Excess Oxygen/Excess Combustibles Analyzer operates in the same manner as the Model 450, except that the sensing probe is limited to process operating temperatures up to 1400°F.

The Model 215 (Sentinel) Oxygen Analyzer is based on the sample extraction principle and designed for close coupling to the flue duct, stack, or process, using a zirconia sensor. The analyzer consists of a sample probe and necessary mounting hardware, a temperature-controlled furnace unit containing the sensor, a pump unit or air set for sample aspiration, an electronics unit, and 20 ft of interconnecting cable. The electronics unit is also available in an NEC Class 1, Group B, Division 1 enclosure, as for the other models. A variety of sample probes are offered to suit the particular application. Standard instrument ranges are 0–10% (linear), 0.1–10% (log), 0.25–25% (log), 0–5% (linear), and 0–25% (linear). An analog 4–20 mA dc output is standard.

The Model 132 Mini-Probe Oxygen Analyzer is designed for small packaged boilers and suitable for process temperatures up to 1000°F. It uses an in-situ zirconia sensor 12 in. long and is packaged in the same manner as the Model 218 and 225 analyzers. Standard measurement ranges are 0–5, 10 or

TABLE 7.3 Process Oxygen Analyzers (Gases)

Manufacturer	Country of Origin	Model Number	Sensor Type	Sampling Configuration	Typical Ranges	Remarks
1. Ametek-Thermox	USA	WDG-III	Zirconia	Ex situ	Various within overall range 0.1 ppm–100% (log and linear)	Max. process temp. 3200°F
		WDG-III-C	Zirconia	Ex situ		Also measures combustibles
		Thermox II	Zirconia	Extractive		
		Thermox III	Zirconia	Extractive		
2. Anacon Inc.	USA	76/79	Hersch (electrochemical)	Extractive	0–10, 100, 1000 ppm	
		78/8081	Zirconia	In situ	0.01 ppm–100%	Max. process temp. 2900°F
3. Anarad Inc.	USA	AR 200 SM	Fuel cell	Extractive		Measures CO also
4. Bailey Controls	USA	Type OL	Zirconia	Ex situ	0–10, 25, 50% linear	Measures CO also Max. process temp. 1500°F
5. Beckman Industrial	USA	755	Paramagnetic	Extractive	0–1, to 0–100%	
		7003/741	Polarographic	Extractive	0–1, 5, 10, 25%	
6. Combustion Eng., Process Analytics	USA	304	Zirconia	Extractive	0–10, 25%	
		305	Polarographic	Extractive	0–5, 10, 25%	
7. Comsip-Delphi Inc.	USA	A	Electrochemical	Extractive	ppm–% dual range (ratios 5 : 1 or 10 : 1)	
		G	Electrochemical	Extractive	0–5 to 0–100% (three ranges)	
		J	Paramagnetic	Extractive	0–5, 10, 25, 100%	

TABLE 7.3 (*Continued*)

Manufacturer	Country of Origin	Model Number	Sensor Type	Sampling Configuration	Typical Ranges	Remarks
8. Data Test Inc.	USA	301	Zirconia	In situ	0–10, 25% linear 0.25–25% log	Max. process temp. 2850°F
		303	Zirconia	Extractive	0–10, 25% linear 0.25–25% log	
9. Delta F Corp.	USA	Various (based on packaging)	Electrochemical	Extractive	Various (ppm and percent)	Nondepleting electrode design
10. Dynatron Inc.	USA	2100	Zirconia	In situ	0–25% log	Single or multipoint versions available Max. process temp. 1400°F
11. Econics Corp.	USA	7210	Zirconia	In situ	0–21%	Microprocessor based Also measures CO Hydrocarbons & opacity
12. Fuji Electric Co.	Japan	Zircomat-P	Zirconia	In situ	0–5, 10% log	Max. process temp. 1100°F
13. Hartman & Braun	West Germany	Magnos-3	Paramagnetic	Extractive	0–1, 3, 10, 30, 100%	PTB approved (pressurized)
		Magnos-3-EX	Paramagnetic	Extractive	0–1, 3, 10, 30, 100%	
		Magnos-2T	Thermomagnetic	Extractive	Min. 0–1%	Suppressed ranges available
		Magnos-2T-EX	Thermomagnetic	Extractive	Min. 0–1%	PTB approved (15)
		Magnos-5T	Thermomagnetic	Extractive	Min. 0–3%	Suppressed ranges available
		CGRGM 1	Zirconia	In situ	0.25–25% log or linear	Max. process temp. 1400°F
		Magnos-4G	Magnetopneumatic	Extractive	0–2, 5, 10, 25%	

14. Horiba Ltd.	Japan	PMA-31/31F	Paramagnetic	Extractive	0–10, 25%	
		MPA-21	Magnetopneumatic	Extractive	0–5, 10, 25%	
		PMA-200	Paramagnetic	Extractive	0–10, 20%	Dual range
		MPA-31/31F	Magnetopneumatic	Extractive	0–1, 5, 25%	
15. Kent Industrial Measurements Ltd.	Great Britain	Z-GP2/Z-GE	Zirconia	In situ	0.1–20.95% log	Max. process temp. 2200°F
					0–10, 20% linear	General purpose only
		Z-FG	Zirconia	In situ	0.25–25% log	Linear ranges available
					0.10–10% log	Max. process temp. 1100°F
		Z-LT	Zirconia	In situ	0.1–10, 0.2–20% log	Max. process temp. 1100°F
					0–10, 20% linear	
16. Lear Siegler	USA	CM-50	Zirconia	In situ	0–2.5, 10, 25% linear	Max. process temp. 1400°F
17. Leeds & Northrup	USA	7875	Zirconia	In situ	0–2, 5, 10, 25%	Max. process temp. 1300°F
		7860 Series	Thermomagnetic	Extractive	Min. 0–2%	Uses 2 cells (meas. & ref.)
18. H. Maihak AG	West Germany	Oxygor 6N	Paramagnetic	Extractive	0–0.5% min. up to 0–100%	Suppressed range available Dual range available
19. M.S.A. Company	USA	802	Paramagnetic	Extractive	0–1, 10, 100%	Switchable ranges
		803	Zirconia	Ex situ	0.1–21% log	
		4000	Fuel cell	Extractive	0–1 to 0–100%	
20. Milton Roy	USA	Oxyprobe	Zirconia	In situ	0.1–21% log	Max. process temp. 1200°F Also available as multi-probe system (up to 8 probes)
		Hi-Temp	Zirconia	In situ	0.1–21% log	Max. process temp. 3000°F
		Oxy-Mizer	Zirconia	In situ	0.1–21% log	Max. process temp. 1000°F For small package boilers

TABLE 7.3 (*Continued*)

Manufacturer	Country of Origin	Model Number	Sensor Type	Sampling Configuration	Typical Ranges	Remarks
21. Neotronics Ltd.	Great Britain	OTOX 90 Series	Electrochemical	Extractive	0–35%	Digital control unit
22. Schlumberger/Sereg	France	Semacox	Thermomagnetic	Extractive	0–5, 7, 10% Standard	
		Oscar	Paramagnetic	Extractive	0–1% to 0–100%	5 switchable ranges
23. Servomex Ltd.	Great Britain	500 Series	Paramagnetic	Extractive	0–1 to 0–100%	Model 292A for hot samples
		700 Series	Zirconia	Ex situ	0–10, 25% log or linear	
		1100	Paramagnetic	Extractive	Various spans from 1–100%	Microprocessor-based control unit
24. Siemens AG	West Germany	Oxymat 2	Magnetopneumatic	Extractive	0–2 to 0–100%	Suppressed ranges
		Oxymat 3	Magnetopneumatic	Extractive	0–0.5 to 0–100%	Suppressed and dual ranges

25. Teledyne Analytical Insts.	USA	310 Series	Micro-fuel cell	Extractive	0–10, 100, 1000, 10,000 ppm	
		320 Series	Micro-fuel cell	Extractive	0–5, 10, 25% or 0–5, 25, 100%	
		9400	Zirconia	Ex situ	0.1–10% log	Max. process temp. 1200°F Measures combustibles also
		9700	Micro-fuel cell	Extractive	0–5, 10, 25% O_2 0–5% combustibles	
26. Westinghouse	USA	218	Zirconia	In situ	0.25–25%, 0.1–10% (log) or 0–10% (linear)	Max. process temp. 1400°F
		225	Zirconia	In situ		Max. process temp. 1000°F
		450	Zirconia	In situ	0–10% O_2, 0–10% combustibles	Max. process temp. 2730°F
		240	Zirconia	In situ	0.25–25% or 0.1–10% O_2, 0–10% combustibles	Max. process temp. 1400°F
		215	Zirconia	Ex situ	0.1–10% or 0.25–25% (log)	
		Mini probe	Zirconia	In situ		Max. process temp. 1000°F
		209	Zirconia	Extractive	0.8–100%	
27. Yokagawa Electric Corp.	Japan	Z021	Zirconia	In situ	0–5, 10, 25%	Max process temp. 2550°F

25% (linear), and 0.1–10 or 0.25–25% (log). Standard output is 4–20 mA dc.

(c) *Servomex Ltd. (Great Britain)*

In addition to their well-known range of magneto-dynamic oxygen analyzers described earlier, the Series 700 range of oxygen analyzers are offered. They use the Servomex zirconia oxygen sensor, developed about five years ago, and are used mainly in flue gas oxygen measurement applications.

The Series 700 analyzer units are designed for mounting on the outside of the flue gas duct or stack, thus allowing simpler replacement of and avoiding thermal shock to the zirconia sensor. A positive sample flow is achieved by using an aspirator system which draws sample through the probe flowtube, around the sample loop, and back to the flue, thus providing a consistently fast response. All units are fitted with flame arrestors as standard to isolate the sensor from the process.

The Series 700 control units are provided with diagnostic indicators to allow rapid identification of equipment status; these include sample flow failure, zirconia sensor failure, probe enclosure temperature sensor failure, and high/low temperature alarms. All analyzer units and control units are purgeable to meet hazardous area installation requirements (Probe Enclosure: Division 1 or 2; Control Unit: Division 2). Analyzer ranges offered as standard are 0–25% logarithmic and 0–10 or 0–25% linear. Two-zone temperature control of the probe enclosure is provided to avoid acid dewpoint problems.

The Model 750 A has a repeatability of ±0.05% oxygen typical and a response time of less than 5 sec to 90% of full scale. The analyzer is suitable for sample temperatures up to 800°C.

The Model 760A is a low-cost analyzer and may be fitted with an integral pump to drive the aspirator system.

The Model 770A measures simultaneously both oxygen and combustibles (utilizing a catalytic type combustibles sensor). The analyzer is suitable for sample temperatures up to 500°C (as for the Model 760A).

The Model 780A is a high-temperature flue gas oxygen analyzer. Using a standard 310 stainless steel probe, it is suitable for sample temperature up to 1000°C. For temperatures up to 1800°C an aluminum oxide probe is used. The aspirator system is automatically shut off if the sample temperature drops below 200°C.

REFERENCES

1. Pauling, L., Wood, R. E., Sturdivant, J. H. *J. Amer. Chem. Soc.*, **68,** 795 (1946).
2. Quinke, G. *Ann. Phys.*, **24,** 347 (1885); **34,** 401 (1888).
3. Luft, K. *Z. Angew. Phys.*, **3,** 300 (1951).

BIBLIOGRAPHY

Guide to selecting combustion stack gas analyzers. *InTech* (April 1982).

Heisig, C. G. Process oxygen analysis: More than just picking an instrument. *InTech* (Oct 1984).

Langridge, S. Application of zirconia oxygen analyzers in the iron and steel industry. *Steel Times*, England (Oct 1982).

Langridge, S. Approach to oxygen analysis in glass melting furnaces. *Glass*, England (May 1983).

Slevin, M. J. Boiler oxygen trim: Why it usually doesn't work. *InTech* (Oct 1984).

Tyler, C. M., and Powers, J. Refinery seeks $1000/day fuel savings using bypass oxygen analyzers. *Chem. Proc.* (Oct 1984).

8

Density and Specific Gravity Measurement

8.1 INTRODUCTION AND DEFINITIONS

One often hears the expressions "as light as a feather" and "as heavy as lead." Equal volumes of different substances vary considerably in mass. A good example is the use of aluminum alloys in aircraft manufacture. These materials are as strong as steel but, volume for volume, weigh significantly less.

The *density* (ρ) of a substance is defined as the mass (M) per unit Volume (V) at a specified temperature:

$$\rho = \frac{M}{V}$$

and may be measured in a variety of different units, such as g/cm^3, lb/ft^3, lb/gal, and kg/m^3.

The *relative density* (*specific gravity*) of a liquid is defined as the ratio of the density of the liquid at a specified temperature to the density of water at the same temperature. Specific gravity (SG) is often referred to a base temperature of 60°F (15.5°C) or 39.2°F (4°C). The density of water at 4°C is at its maximum of 1.000 g/cm^3, so the specific gravity of a liquid at 4°C is numerically equal to its density at that temperature.

The *specific gravity* of a gas is defined as the ratio of the density of the gas at a specified temperature and pressure to the density of air at the same temperature and pressure. Usually, the conditions are specified as Standard Temperature and Pressure (STP): 0°C and 14.7 psia.

8.1.1 Temperature and Pressure Effects

Temperature, pressure, and the nature (composition) of a fluid all affect the density. Except for liquids such as liquefied petroleum gas (LPG), liquids can be considered as incompressible fluids, and pressure variations have an insignificant effect on the density. However, variations in temperature have a considerable effect on the density and it is generally necessary to provide temperature compensation if accurate measurement is required.

Specific gravity is a dimensionless quantity, but a number of units related to specific gravity are in common industrial use. Degrees API is used widely in the petroleum industry and degrees Baume in the chemical industry. These units are defined by the following equations:

$$°\text{API} = 141.5 \,/\, \text{SG } 60/60 - 131.5$$

$$°\text{Baume (for liquids less dense than water)} = 140 \,/\, \text{SG } 60/60 - 130$$

$$°\text{Baume (for liquids more dense than water)} = 145 \,/\, \text{SG } 60/60 - 145$$

Hydrometers used in the laboratory for specific gravity measurement are calibrated in these units.

The density of a gas is dependent on the temperature, pressure, and nature of the gas, and may be described in terms of the ideal gas laws and the deviations from them. An ideal gas may be defined as a gas in which the volume occupied by the gas molecules is negligible compared with the total system volume, and attractions between molecules are extremely small under all conditions. The ideal gas law is described by the equation,

$$PV = nRT$$

where P is the absolute pressure
V is the volume
n is the number of moles of gas
R is the universal gas constant
T is the absolute temperature

For real gases, the ideal gas law applies only at low pressures and at relatively high temperatures. One parameter commonly used as a measure of the deviation of a gas from ideality is the compressibility factor (Z), where

$$Z = \frac{PV}{nRT}$$

The density of a gas is given by:

$$\rho = \frac{Mn}{V}$$

where M is the molecular weight. Substitution in the non-ideal gas equation gives

$$\rho = \frac{MP}{ZRT}$$

Thus, the density of a gas may be calculated from this equation. One must, of course, ensure that all units used in the equation are compatible. Special attention should be paid to the value of R used. Typical values of R are listed below, with the units of measurement.

	R	*Units*
*	10.731	ft^3, $lb/in.^2$, $lb\text{-}mole^{-1}$, $°R^{-1}$
	1545.3	ft^3, lb/ft^2, $lb\text{-}mole^{-1}$, $°R^{-1}$
*	0.7302	ft^3, atm, $lb\text{-}mole^{-1}$, $°R^{-1}$
*	0.08205	liter, atm, $g\text{-}mole^{-1}$, $°K^{-1}$
	82.06	cm^3, atm, $g\text{-}mole^{-1}$, $°K^{-1}$
	62.36	liter, mm Hg, $g\text{-}mole^{-1}$, $°K^{-1}$
	62.363	cm^3, mm Hg, $g\text{-}mole^{-1}$ $°K^{-1}$
*	1.9872	g-cal, $g\text{-}mole^{-1}$, $°K^{-1}$
*	1.9872	BtU, $lb\text{-}mole^{-1}$, $°R^{-1}$
	8.314	Joules, $g\text{-}mole^{-1}$, $°K^{-1}$

* Most commonly used values of R.

8.2 LABORATORY METHODS OF MEASUREMENT

8.2.1 Density and Specific Gravity of Liquids

Traditional laboratory methods for the measurement of density and specific gravity of liquids use two different techniques:

- Methods involving the use of pycnometers (weighing principle).
- Methods involving the use of hydrometers (buoyancy principle).

These principles are used in the following ASTM Standard Laboratory Test Methods:

D287-67	API gravity of crude petroleum and petroleum products (hydrometer method)
D941-55	Density and relative density of liquids by Lipkin bicapillary pycnometer
D1217-81	Density and relative density by Bingham pycnometer
D1298-80	Density, relative density, or API gravity of crude petroleum and liquid petroleum products by hydrometer method
D1657-81	Specific gravity of light hydrocarbons by pressure hydrometer
D3142-78	Specific gravity or API gravity of liquid asphalts by hydrometer method

The *pycnometer,* often known in school jargon as the "density bottle," is a small borosilicate glass vessel of accurately defined volume with a mass not greater than about 30 g. The pycnometer is cleaned, dried, and weighed to the nearest 0.1 mg on a balance. It is then filled with the liquid sample under test and immersed in a constant temperature bath maintained at the test temperature (usually 20 or 25°C). When the sample has equilibrated to the test temperature, the pycnometer is removed from the bath, cleaned and dried outside, allowed to cool, and then weighed again to the nearest 0.1 mg. The density (g/ml) of the sample is then calculated from the expression,

$$\rho = \frac{(W_2 - W_1)}{V}$$

where W_1 = mass of pycnometer empty (g)
W_2 = Mass of pycnometer and sample (g)
V = Volume of pycnometer (ml)

The Bingham pycnometer, referred to in ASTM D1217-81, is shown in Fig. 8.1. During the temperature equilibration period the sample liquid will expand and excess sample must be removed to bring the sample level to the

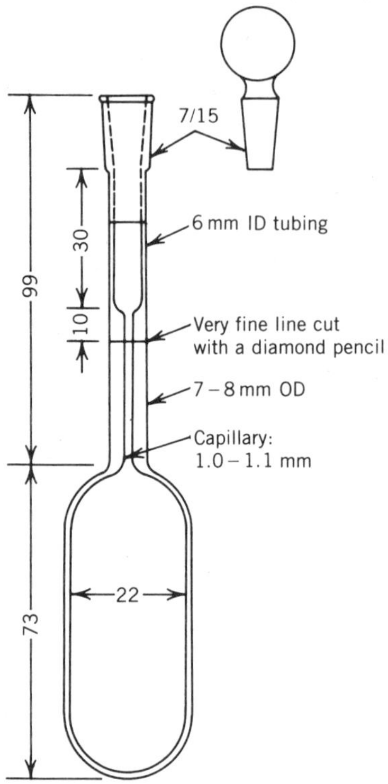

FIGURE 8.1 Bingham pycnometer.

calibration mark on the pycnometer. This method requires meticulous attention to obtain accurate results.

Based on Archimedes' principle, the *hydrometer* is usually made of glass and weighted so that it floats upright in the liquid. The upper stem incorporates a graduated scale. When immersed in the liquid under test, the hydrometer sinks to a depth dependent on its weight in the liquid and therefore on the density or specific gravity of the liquid. The scale may be calibrated in specific gravity or other units such as °API or °Baume. Many different types of hydrometer are available; ASTM Standard E100-72 covers the various types required for standard ASTM test methods. Hydrometers are also available with a built-in thermometer and with the scale in the hydrometer body or the stem. ASTM hydrometers are calibrated in specific gravity or °API.

8.2.2 Density and Specific Gravity of Gases

Methods for laboratory measurement of density/SG of gases are based on the use of commercially available instrumentation normally designed to measure the specific gravity of the sample gas. ASTM Method D1070-73 lists the methods and the recommended equipment. The majority of these instruments are similar in design to process analyzers. Many are direct-reading instruments, requiring no calculations subsequent to initial calibration.

Since SG is equal to the ratio of the density of the sample gas at the test temperature and pressure to the density of air at the same temperature and pressure, the density of the sample gas may be calculated by multiplying the gas SG, as measured by the laboratory instrument, by the density of air at the same temperature and pressure, as obtained from tables.

Instruments covered by ASTM D1070:

- Ac-Me gravity balance.
- Ac-Me recording gravitometer
- Arrco-Anubis recording gas gravitometer.
- Arcco-Anubis portable gas balance.
- Kimray gravitometer.
- Ranarex recording gravitometer.
- UGC gas gravitometer.

The design of the Arcco-Anubis gas gravitometer and the UGC Gas gravitometer is virtually the same as the process analyzers manufactured by these companies.

The principles of operation of these instruments may be classified as follows.

(a) Pressure Balance Principle

In this method, a beam carrying a bulb and counterweight is brought to balance successively in air and in the sample gas by adjusting the pressure within the balance case. The absolute pressures are determined by means of a barometer and mercury-filled manometer and the sample gas specific gravity computed from the ratio of the absolute pressures.

(b) Displacement Balance Principle

This method depends upon the principle of balancing the weight of a given volume of sample gas at atmospheric pressure by displacement of the center of gravity of a balance beam. The amount of deflection, subject to correction, is a measure of the gas specific gravity.

(c) Kinetic Energy Principle

This method measures the ratio of the change in kinetic energy between an impeller and an impulse wheel operating in the sample gas and another impeller and impulse wheel operating in a reference gas (air). Measurements are made in terms of the torques of the impulse wheels, which are proportional, respectively, to the densities of sample gas and air.

(d) Centrifugal Force Principle

This method measures the difference in centrifugal force between the sample gas and a reference gas (air) as both are accelerated by a specifically designed wheel.

The Arcco-Anubis gas gravitometer shown schematically in Fig. 8.2 is an example of the direct-weighing instrument and is designed to produce a temperature, barometric pressure, and humidity-compensated recording of gas specific gravity. The equipment consists of two inverted, thin-walled "bells" suspended on knife-edge links and with the lower edges of the bells submerged in oil to form a gas-tight seal. The bells are of equal size and weight and are placed at equal distances from the fulcrum of a balance beam that drives the recorder mechanism. One bell contains the sample gas, the other the dry reference air. The difference in weight produces an unbalance of the beam that is proportional to the sample gas specific gravity. In addition to its laboratory application, this instrument is used as an on-line specific gravity analyzer.

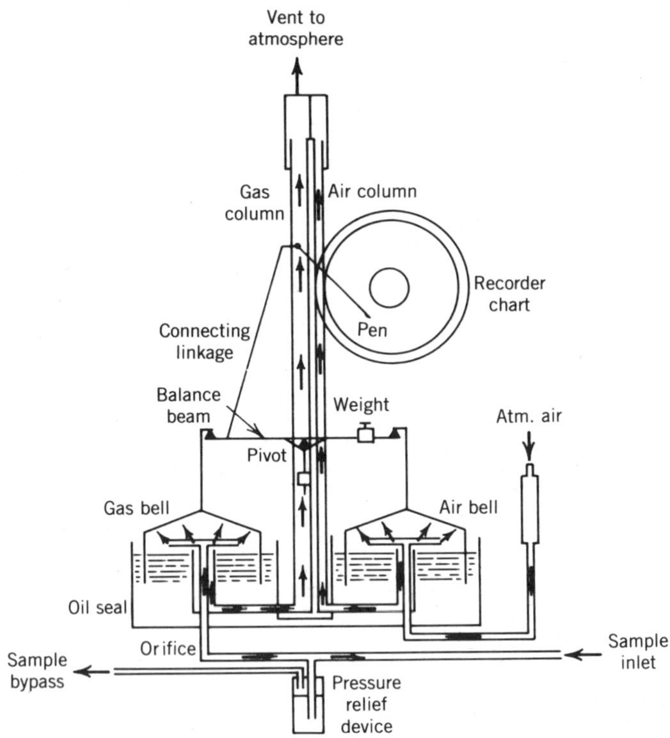

FIGURE 8.2 Arcco Anubis gas gravitometer. (Courtesy of Arcco Inst. Co.)

8.3 PROCESS ANALYZERS FOR DENSITY AND SPECIFIC GRAVITY MEASUREMENT

Process analyzers for density and specific gravity measurement of process fluids may be conveniently classified into the following categories according to operating principle:

- Buoyancy (Archimedes) principle.
- Vibration principle.
- Balance beam (weighing) principle.
- Radiation absorption principle.
- Kinetic energy principle.
- Aerostatic weighing principle.
- Turbojet principle.

Many instruments are suitable for measurement in both liquids and gases, whereas others are suitable for both liquids and solids. The specificity of analysis is covered next.

8.3.1 Buoyancy Principle

(a) Princo Instruments Inc. (United States)

Princo manufactures the Densitrol W-700 range of liquid density analyzers, the operating principle of which is shown in Fig. 8.3. The instrument consists of two units, a sensing unit and a remote-mounted integrator that combines the electrical density and temperature signals received from the sensor.

The sensing chamber contains a totally submerged plummet fastened by chains to a fixed reference point. The plummet, chains, and sensing chamber are available in a variety of materials to suit the particular application requirements (e.g., 316 stainless steel, Pyrex, PVC). The plummet is so weighted that at the middle of its calibration range it assumes an equilibrium position where the weight of the calibration chains is equally supported by the plummet and the reference point.

Any change in density will cause the plummet to move to a new equilibrium position. For example, on an increase in sample density from the midrange point, the plummet will rise due to the additional buoyancy. As the plummet rises, additional chain weight is transferred from reference point to plummet until a new equilibrium position is attained, where added chain weight equals additional buoyancy. Changes in the vertical position of the plummet and its core are sensed by a linear variable differential transformer, which transmits a low-voltage ac signal to the integrator. Temperature compensation is provided by a resistance thermometer located in the sensing chamber, and its output is also transmitted to the integrator.

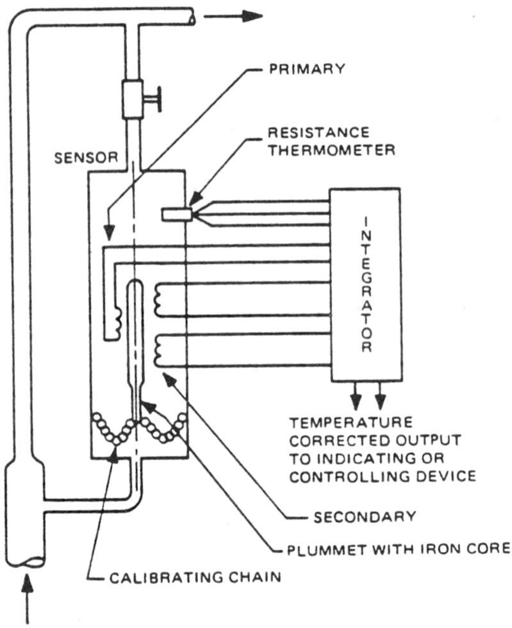

FIGURE 8.3 Princo W-700 Series densitrol. (Courtesy of Princo Insts. Inc.)

The integrator is a solid-state unit that may be located at distances up to 1000 ft from the sensing unit and connected to it by a nine-conductor cable. The integrator produces a temperature-corrected output signal proportional to sample density. Standard output is −5 to +5 mV but optional current outputs are also available (1–5, 4–20, or 10–50 mA dc). With current output, instrument span is adjustable typically from 50% to 150% of original span, and zero by as much as 100% of span. The integrator is available as a general-purpose, dust-tight unit (Model CRI); a weather-proof, NEMA-4 surface mounted unit (Model WPI); or an explosion-proof Class 1, Group D, Division 1 surface mounted unit (Model EPI).

(b) UGC Inc. (United States)

The UGC Electronic Densitometer Model 241 shown in Fig. 8.4 is designed to measure accurately the density of a flowing fluid under actual operating conditions. The instrument has been used successfully in both liquid and gaseous services.

The measurement system consists of a beam suspended at a pivot point. Attached to one end of the beam is a float. The buoyant force exerted by the sample on the float provides a moment around the pivot point that varies in direct proportion to the density of the sample surrounding the float. When the summation of the moments around the pivot is not zero, there is an angular displacement of the float beam.

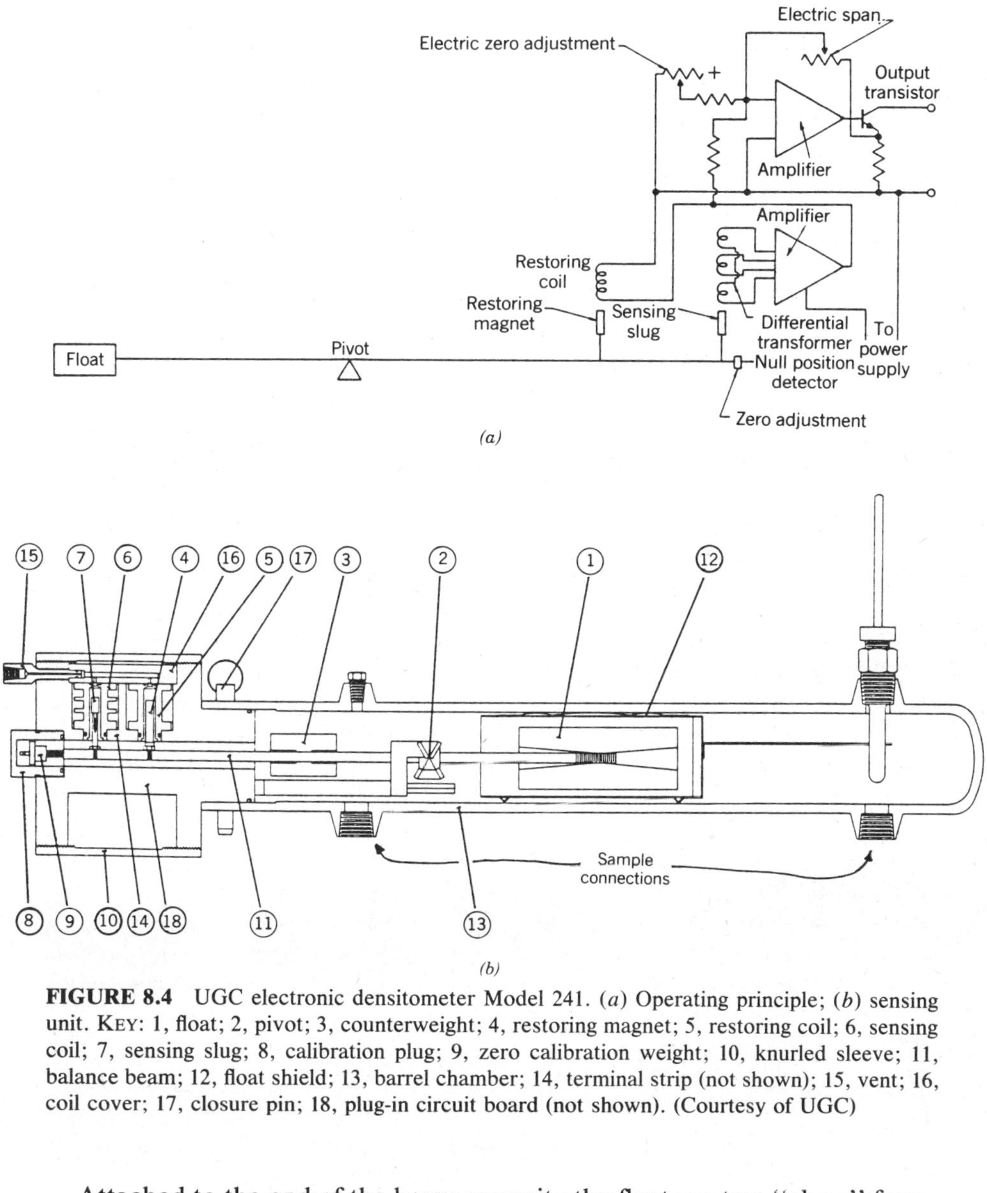

FIGURE 8.4 UGC electronic densitometer Model 241. (*a*) Operating principle; (*b*) sensing unit. KEY: 1, float; 2, pivot; 3, counterweight; 4, restoring magnet; 5, restoring coil; 6, sensing coil; 7, sensing slug; 8, calibration plug; 9, zero calibration weight; 10, knurled sleeve; 11, balance beam; 12, float shield; 13, barrel chamber; 14, terminal strip (not shown); 15, vent; 16, coil cover; 17, closure pin; 18, plug-in circuit board (not shown). (Courtesy of UGC)

Attached to the end of the beam opposite the float are two "slugs" forming part of an electrical force balance system. One senses the position of the deflected beam, the other is coupled to a restoring coil that drives the beam into the balanced, or null, position.

The electronic circuitry produces a high-frequency ac signal, which is applied to a differential transformer detector that senses the position of the beam. As the sensing slug moves in or out of the magnetic field of this detector, the amplitude of the signal is changed. The resultant voltage output is rectified and controls the current passing through the restoring coil. The current thus applied causes the magnetic restoring coil to move the restoring magnet position, forcing the beam into the null position.

The current necessary to achieve balance is highly dampened to give an instrument response of about 30 seconds. The instrument output current (1–5, 4–20, or 10–50 mA dc) is proportional to the density of the process fluid being measured. The analyzer is suitable for operating pressures up to 2500 psig and has a temperature range of 0–140°F. A high-temperature version is available for up to 350°F. The analyzer is suitable for installation in NEC Class 1, Group D, Division 1 hazardous locations. Minimum range is 0–1 lb/ft^3 and repeatability is quoted as $\pm 0.25\%$ of full scale.

8.3.2 Vibration Principle

(a) Sarasota Automation Inc. (Great Britain and United States)

This company, formerly known as Agar Instrumentation Ltd., manufactures a range of density analyzers based on the vibration principle and designed for measurement in gases, liquids, and cryogenic fluids.

In the 700 Series density meters, the sensing element is a thin-walled cylinder, or "spool," of a magnetic stainless alloy, maintained in circumferential oscillation by an electromagnetic circuit. The spool is totally surrounded by the process fluid and is mounted in the meter body. The spool is secured so that one end is fixed and the other free. The drive coil sets the spool into oscillation and the movement is detected by the pickup coil. The resulting signal is amplified and fed back to the drive coil, thus maintaining the spool in oscillation. Since the process fluid completely surrounds the spool, it too is in a state of oscillation. The resonant frequency is dependent on the total mass of the vibrating system; therefore any change in sample density will result in a change in the frequency of oscillation. This change is measured very accurately by the density converter, which provides an output proportional to the sample density. A schematic diagram of the instrument is shown in Fig. 8.5.

By using frequency modulating techniques, the analyzer overcomes problems of hysteresis, changes in electrical conductivity, and pressure and viscosity effects. Since the spool is suspended in the process fluid, it is not under stress and the oscillation amplitude is well below fatigue level, giving long-term stability.

The sensing unit transmits a pulse frequency signal in accordance with the following equation:

$$d = 2d_0 \frac{\Delta T}{T_0}\left[1 + \frac{K\Delta T}{2T_0}\right]$$

where d = measured density (g/l)

d_0 = meter constant (g/l)

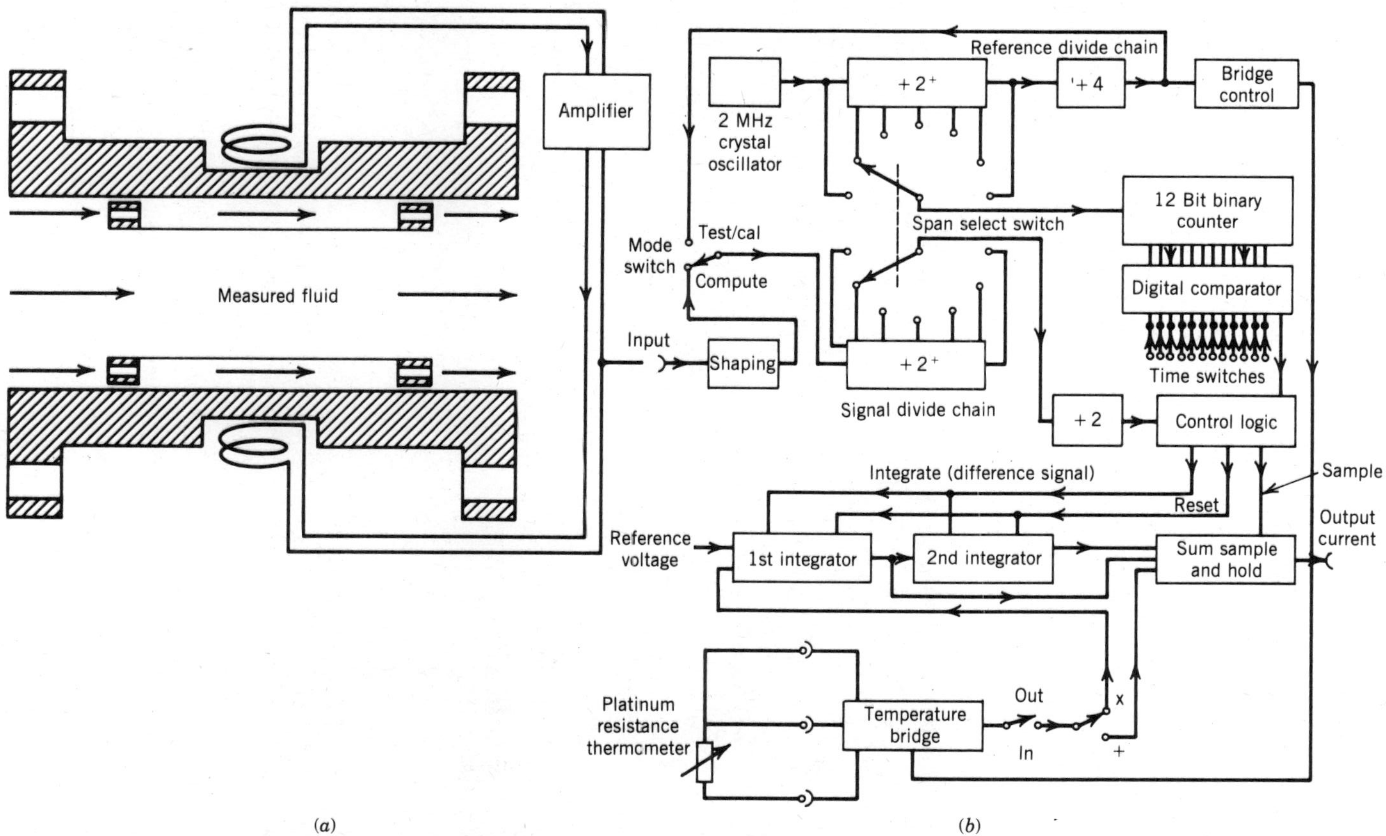

FIGURE 8.5 700 Series density meter. (*a*) Sensor; (*b*) density converter. (Courtesy of Sarasota Automation Inc.)

T = periodic time at density d (microsec)
T_0 = periodic time in vacuum (microsec)
$\Delta T = T - T_0$
K = calibration constant

The density converter compares the periodic time of the output signal with a highly stable periodic time reference signal derived from a quartz crystal oscillator. A temperature reference circuit is also included to allow the density signal to be referred to a particular temperature for specific gravity measurement. The analog output signal of the density converter represents measured density in accordance with the preceding equation. A variety of current outputs are available to meet customer requirements.

In addition to the Model FD700 series density meter shown in Fig. 8.6, which is designed for use in bypass (on-line) installations for gas and clean non-viscous liquids, the Model ID700 series meter is also available. This is designed for direct insertion in the process line and may be fitted with a

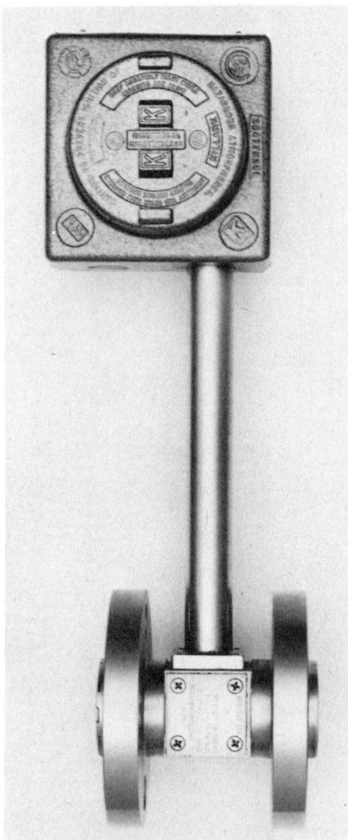

FIGURE 8.6 Series FD700 density meter. (Courtesy of Sarasota Automation Inc.)

retraction device to allow the meter to be removed from the line through a suitable valve without the need for line shut-down. This type of design has the advantage that the measuring element is exactly at the process temperature and pressure, particularly important for high-accuracy gas measurement.

The Series 700 density meters have a repeatability of 0.01% of range, a fast dynamic response, and an operating temperature range dependent upon the spool material selected, as follows:

A · Ni-Span C	−270 to +75°C
B · Ni-Span C	−20 to +75°C
C · FV520 B	−20 to +200°C

The meter body is suitable for operating pressures up to 2000 psig.

The Model FD800 Series density meter shown in Fig. 8.7 uses the vibrating tube technique in which a thin walled tube is set into transverse oscillation by an electromagnetic circuit. Since the process fluid only passes through the center of the tube, this instrument is largely immune to particulates. Also, the transverse mode of oscillation gives negligible viscosity effect, allowing measurement of viscous process streams. All wetted parts are made from corrosion-resistant stainless steel. The instrument has a fast dynamic response, a repeatability of ± 0.03 kg/m^3, and an operating temperature range of -30 to $+200$°C. The meter body is suitable for operating pressures as defined by Class 600 rating.

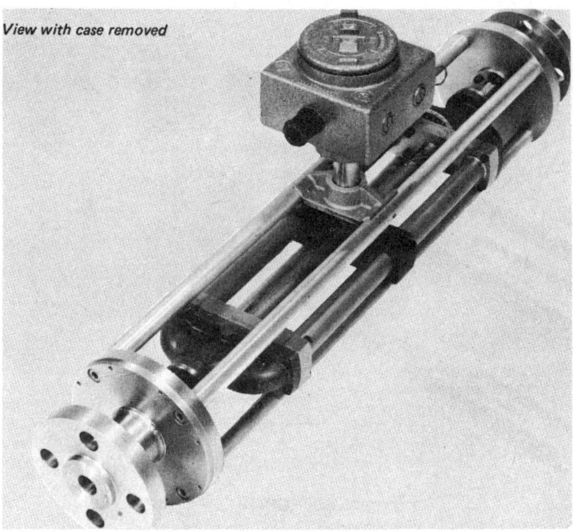

FIGURE 8.7 Series FD800 density meter. (Courtesy of Sarasota Automation Inc.)

(b) Automation Products Inc. (United States)

The Dynatrol Series CL-10TY systems manufactured by this company are designed for the measurement of density or specific gravity of process liquids or percent solids content of slurries. The operating principle is illustrated in Fig. 8.8. The process sample flows through a U-tube of approximately $\frac{1}{2}$-in. diameter welded to the meter body at the node points. A drive coil, electrically excited by a pulsating current, drives the U-tube into mechanical vibration, which is a function of the mass of the U-tube and its contents. As the sample density increases, the effective mass of the U-tube increases and vice-versa.

The vibration is sensed by the pick-up end of the sensor, which consists of an armature and coil arrangement similar to that at the driver end. The ac voltage induced in the pick-up coil is transmitted to a remote-mounted solid-state converter, where it is converted to a current output (1–5, 4–20, or 10–50 mA dc) proportional to sample density. Temperature compensation is provided by circuitry in the converter and a resistance temperature element is mounted as an integral part of the U-tube. Span and zero controls are located at the converter.

The sensing unit is suitable for installation in NEC Class 1, Group D, Division 1 hazardous locations. Pressure rating is 1000 psig at 100°F (standard) and temperature rating is 300°F (standard). The converter is available in either general purpose or Class 1, Group D, Division 2 design. A wide

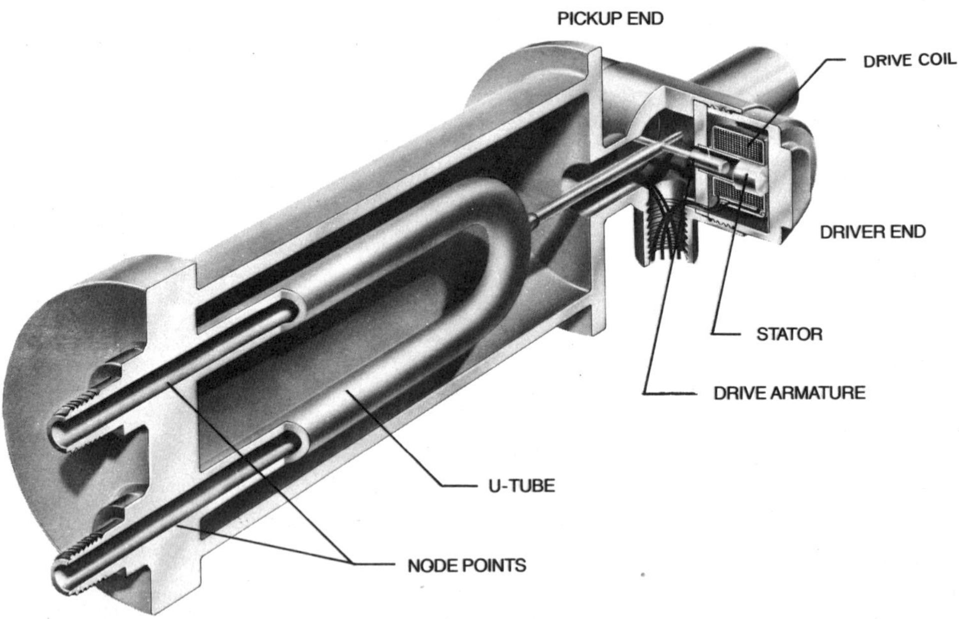

FIGURE 8.8 Dynatrol Series CL-10TY. (Courtesy of Automation Products Inc.)

variety of corrosion-resistant materials are available for sensing unit construction. A variety of sensing units cover the overall range of 0.5 to 2.0 SG units; each sensor has a base range of 0.5 SG units and a span of 0.05–0.5 SG units within the base range. Systems can be provided with or without temperature compensation, depending upon the application.

(c) *Yokagawa Corporation (Japan and United States)*

The Yokagawa vibration type liquid density measurement system is designed for oil, chemical, and food industry applications. An operational diagram of the equipment is shown in Fig. 8.9.

The density of the process sample is measured by detecting the lateral free vibration frequency of the tube containing the sample. The tube is set into vibration by a piezoelectric element and maintains mechanical oscillation at resonant frequency in a positive feedback loop, by which the ac voltage generated by the pickup electrode is amplified and fed back to the piezoelectric element. The difference between the frequency of a standard oscillator and that of the vibrator is converted by a frequency-to-voltage converter to a dc voltage signal proportional to sample density. A current output (4–20 mA dc) is also available. Temperature compensation for specific gravity measurement is accomplished by appropriate circuitry in the converter and a resistance temperature element is located in the sensing unit as shown.

Measurement range is 0.5–1.5 g/cm^3, with a span of 0.05–0.5 g/cm^3. Repeatability is quoted as 1% of span. The sensing unit is rated for pressures up to 20 kg/cm^2, temperatures within the range -10 to 100°C, and is designed for installation in NEC Class 1, Group D, Division 1 hazardous locations. The converter is designed for remote mounting in non-hazardous locations.

Yokagawa can also supply complete rack-mounted systems including sample handling equipment such as flowmeters, valves, filters, pressure, and temperature gauges, designed to meet customer requirements.

(d) *ITT Barton (United States)*

ITT Barton manufactures the Series 660 Range of densitometers, which are suitable for density measurement in gaseous or liquid process streams. The densitometer consists of an in-line sensing probe and a remote mounted electronics unit, as shown in Fig. 8.10.

The sensing probe consists of a cylinder that supports a symmetrically positioned vane that is set into vibration and moves with a simple harmonic motion about its principal axis at its resonant frequency. A small piezoelectric sensor imbedded in the support cylinder converts the mechanical motion into an electrical signal, which is amplified in the probe and sent to the transmitter, where it is conditioned into a digital or analog display of density. The transmitter performs two functions: it provides a signal proportional to

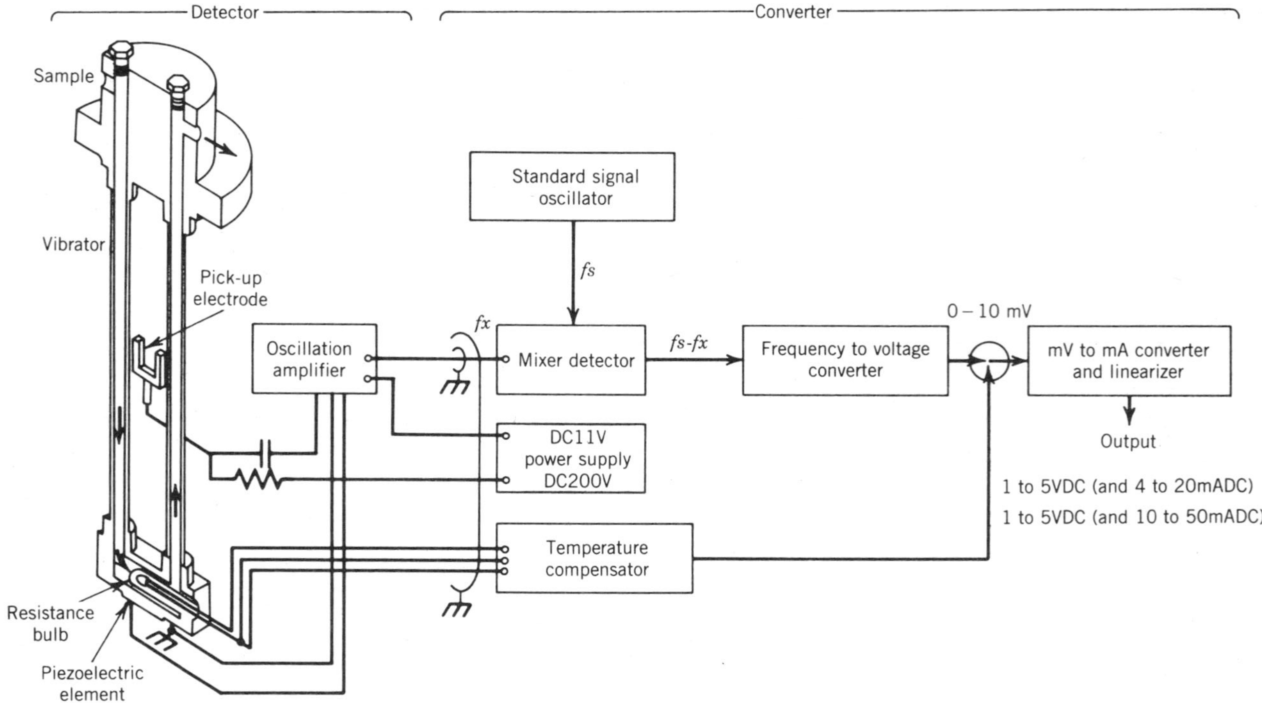

FIGURE 8.9 Model VD6D liquid density meter. (Courtesy of Yokagawa Corporation)

FIGURE 8.10 Series 660 densitometer. (Courtesy of ITT Barton Instruments Inc.)

sample density and provides a feedback signal to the probe to maintain the resonant condition. A change in sample density produces a change in the resonant frequency of the vane in accordance with the equation,

$$\rho = \frac{A}{f^2} + B$$

where ρ = fluid density, f = frequency, and A and B are calibration constants.

For the liquid density probes Models 664 and 666, the density range is 0.3–1.2 g/cm^3, with spans within the limits 0.1 and 0.8 g/cm^3. Repeatability is quoted as $\pm$0.025% of range. Flow velocity should not exceed 33 ft/sec and fluid viscosity is limited to 20 cP. For the gas density probes Models 663 and 665, the density range is 0.5–7.5 lb/ft^3, with spans within the limits 0.8 and 5.0 lb/ft^3. Repeatability is quoted as $\pm$0.025% of range. Flow velocity should not exceed 40 ft/sec. Various construction materials are available, depending upon the particular application. Probes are available for operating pressures up to 1440 psig (600 lb rating) or to 2160 psig (900 lb rating) and operating temperatures of -30 to 200°F. Fixed or retractable probe designs can be supplied.

The Model 653 Electronics Unit is mounted in a NEMA-12 enclosure suitable for rack mounting in non-hazardous locations. Analog (4–20 or 10–50 mA dc) or frequency outputs are available.

The Model 668 Fixed Integral Drive Liquid Densitometer is designed for continuous in-line density measurement over a wide range of process liquids. It draws a low current, so it can be powered by most available microprocessors. The output frequency signal can be interfaced with electronics to calculate density or with flow computers for mass flow calculation. Specifications are the same as for the previous liquid densitometers, except that the probe is available in 600, 900, and 1500 lb ANSI flanged versions. The probe is designed for use in NEC Class 1, Group C and D, Division 1 hazardous locations. Maximum power consumption is 100 mA (typically 50 mA for normal operation).

(e) Solartron Transducers (Great Britain and United States)

The Solartron Transducer Group manufacture analyzers for measurement of gas density, liquid density, and gas specific gravity. The change in resonant frequency with density variation, used as the operating principle, was developed by Solartron during the late 1950s.

The Models 7830 and 7840 Liquid Density Transducers are similar in design and consist of the sensing element, a straight, smooth-bore tube, and a microprocessor based electronics unit. The sensing unit shown in Fig. 8.11 is a pipeline quality, totally sealed unit of all-welded construction, which is vibrated at one of its natural frequencies, so that the frequency is dependent upon the density of the fluid within the tube. Process connections are 1-in. flanges (300 or 600 RF).

In the Model 7830 the vibrating tube is manufactured from Ni-span-C material, providing good stability and low temperature coefficient. In the

FIGURE 8.11 Models 7830/7840 liquid densitometers. (Courtesy of Solartron Transducers Inc.)

Model 7840 the tube is made of 316 stainless steel to give better corrosion resistance, but there is some loss in accuracy. A 100-ohm platinum resistance temperature sensor is included in both instruments for temperature correction of density. The frequency signal from the sensor is amplified and transmitted to the electronics unit for further processing. Overall density range is 0–3000 kg/m^3, with repeatability quoted as $\pm$0.02 kg/m^3 for the Model 7830 and $\pm$0.1 kg/m^3 for the Model 7840. IS Certification to BS standards is pending.

The Models 7810 and 7811 Gas Density Transducers shown in Fig. 8.12 use the well-proven vibrating cylinder technique and are suitable for measurement of dry gases at pressures up to 2000 psig and densities up to 400 kg/m^3. The sensing units are designed for insertion mounting into the gas flow path, either directly or in a thermowell pocket, but with the gas sample brought to the instrument via a sample loop. The transducer element consists of a thin-walled Ni-span-C cylinder set in resonance at its natural frequency and contained within a constant-volume 316 stainless steel chamber.

A solid-state amplifier, electromagnetically coupled to the sensing element, maintains the conditions of vibration and provides the output frequency signal, which is transmitted to the electronics unit for further processing. The amplifier and signal output circuits are encapsulated in epoxy resin. IS approval has been obtained from CENELEC, PTB, and CSA. Overall density range is 0–60 kg/m for the Model 7810 and 0–400 kg/m for the Model 7811, with repeatability quoted as 0.01% of reading. Operating temperature range is -20 to $+85$°C.

FIGURE 8.12 Models 7810/7811 gas density transducers. *(a)* Sensing unit. (Courtesy of Solartron Transducers Inc.)

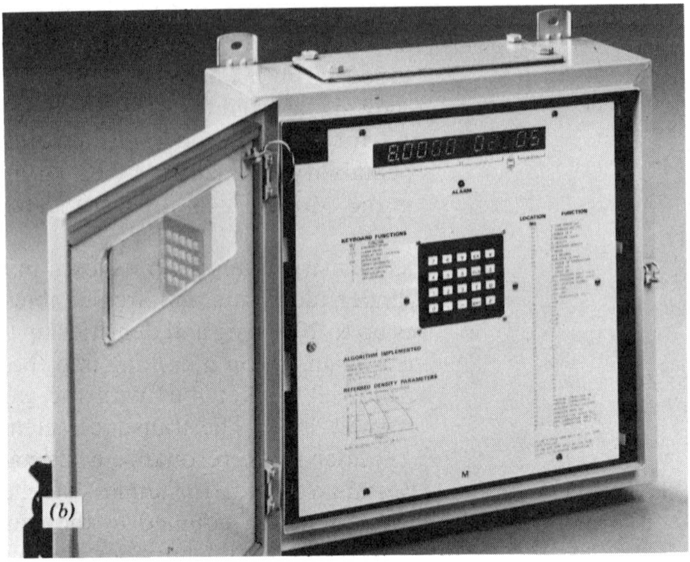

FIGURE 8.12 (*b*) Model 7940 signal converter. (Courtesy of Solartron Transducers Inc.)

The Model 3096 Gas Specific Gravity Transducer consists of a vibrating cylinder gas density transducer and a constant-volume reference chamber, providing automatic compensation for gas compressibility and variations in sample temperature and pressure and giving a frequency output signal proportional to gas specific gravity. IS approval has been obtained from CENELEC, PTB, and CSA.

The Model 7940 Signal Converter shown in Fig. 8.12b is a microprocessor-based unit designed specifically to process signals from the density transducers first described. The converter can receive up to three input signals—the frequency signal from a density transducer, temperature, and pressure. All calibration data, including input and output scaling factors and transducer calibration constants, are entered via a calculator-style keyboard mounted on the instrument front panel. A 4–20 mA dc fully floating output and relay contacts for high and low alarm conditions are provided and may be dedicated to any measured or calculated parameter. System malfunction is indicated by an alarm lamp and a separate relay contact is provided for external alarm. Intrinsic safety barriers are available to meet IS certification for the system.

8.3.3 Balance Beam (Weighing) Principle

(a) Arcco Instrument Company (United States)

In addition to the gas gravitometer, described earlier in connection with laboratory measurements, Arcco manufactures the Anubis range of liquid

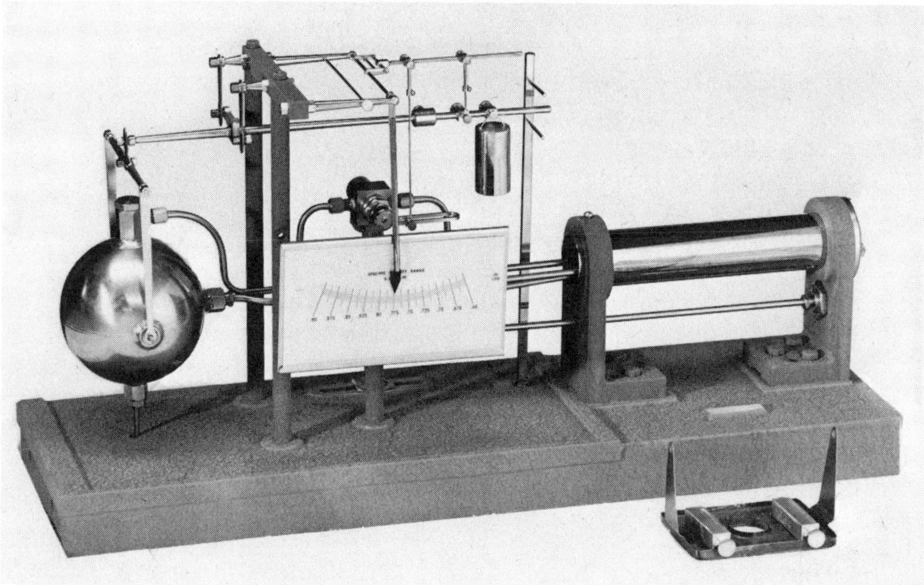

FIGURE 8.13 Arcco Anubis liquid gravitometers. (Courtesy of Arrco Instrument Company)

gravitometers, designed to measure the specific gravity of flowing process liquids with high accuracy. A typical example is shown in Fig. 8.13. The instrument can be considered to be a spring balance that measures the weight of a fixed volume of the process liquid in terms of its specific gravity.

The closed liquid system consists of a bulb of approximately 1000 ml volume, inlet tube, and outlet tube. These tubes also act as the spring elements of the weighing mechanism. The bulb is suspended by suitable links and fulcrums at the left end of the main beam. At the opposite end is a counterweight and a vernier weight (used for adjustment). The counterweight balances the weight of the bulb and liquid at the minimum range of the chart scale. As the specific gravity of the sample liquid increases, the bulb tends to move downward, the movement being resisted by the spring action of the inlet and outlet tubes. The motion of the bulb is thus directly proportional to the changes in specific gravity of the liquid passing through the bulb. This motion is transferred to the main beam and from there by suitable linkages and fulcrums to the recorder, indicator, or transmitter.

Temperature correction is accomplished by the use of a special bi-metallic coil enclosed in a thin tube. The coil is immersed in and surrounded by the flowing sample under pressure. The actual density of the sample in the bulb and its temperature are coordinated to give the correct integrated value, which is then recorded on the chart as specific gravity.

A wide range of gravitometers based on the balance beam operating principle are available for the following purposes, in general purpose or explosion-proof design.

- Local recording.
- Local indication.
- Pneumatic transmission.
- Electronic transmission.
- Pneumatic control.
- Monitoring specific gravity only.
- Monitoring specific gravity and temperature.
- Monitoring specific gravity and pressure.

The majority of Arcco-Anubis liquid gravitometers appear to be installed in oil industry pipeline applications and in the chemical and food industries.

(b) KDG Instruments (Great Britain)

The Gravitrol Density Meter shown in Fig. 8.14 is designed for continuous measurement of liquids or slurries. The process sample flows through a constant-volume tube-loop pivoted on flexures about a horizontal axis that passes through flexible connectors. The weight of the tube-loop and contents is transferred to a weigh beam and counterpoised by a balance weight, adjustable along the beam. A change in liquid density produces a directly

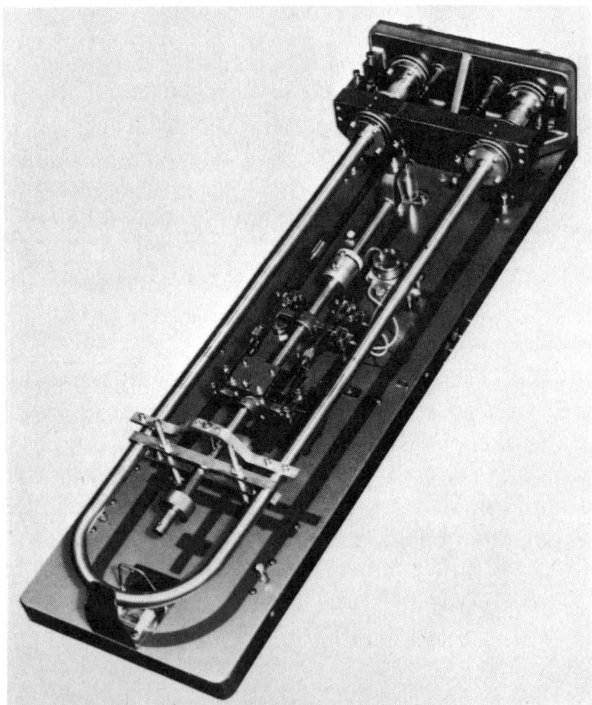

FIGURE 8.14 Gravitrol density meter. (Courtesy of KDG Instruments Ltd.)

proportional change of force on the weigh beam. This force is measured by a pneumatic or electronic force-balance system, providing the appropriate pneumatic (3–15 psig) or electronic (4–20 mA dc) output for remote recording purposes.

The density span of the instrument is determined by the position of a span adjustor, which acts on a force bar and is adjustable between 0.02 and 0.5 g/cm^3. The range for a given span depends upon the position of the balance weight. Range and span can be verified by using a scale pan to hang known weights on the pivot to simulate density changes.

The standard gravitrol has an overall range of 0.4–2.5 g/cm^3. Instruments are available in two nominal tube sizes, 25 mm and 40 mm. Recommended flowrates for clean liquids are 25–90 and 60–230 l/min respectively. For slurries, sample flowrate should be sufficient to avoid deposition of solid material. Temperature compensation is achieved by converting the signal of an external temperature transmitter, through a compensation unit, into a force acting on the weigh beam. Standard tube-loop material is 316 stainless steel. Borosilicate glass is available as an option, as well as other materials such as monel and tantalum. PTFE flexible connectors are usually used with borosilicate glass. The maximum working pressure of the analyzer, which depends upon the stability of the flexible connectors, is 150 psig for 25 mm and 70 psig for 40 mm nominal size. Repeatability is quoted as ±0.25% of span. An intrinsically safe version to BS1259 can be supplied complete with barrier unit.

Care should be taken when installing the unit to ensure that a solid, vibration-free foundation is provided; anti-vibration mountings may be required. In addition, all connecting pipework should be attached with flexible connectors. Some protection against ambient conditions is also recommended.

(c) SPRI Limited (Great Britain)

The Gravitymaster Density Meter shown in Fig. 8.15 is designed for continuous measurement of liquid process streams and slurries. The process sample is circulated through a 316 stainless steel U-tube supported at the open end by a cross-leaf suspension. The combined weight of the tube and contents (of chosen nominal density) is counterbalanced in such a way that density variations about the nominal are indicated. Deviations in sample density result in beam displacement, which is detected optically. The resultant electrical error signal is amplified and used to control the current passing through the coil of a force transducer to produce an equal and opposite correcting force. The current required is proportional to the same density deviation, which is indicated on the control unit indicator. The counterbalance weight is adjustable, enabling a wide range of densities to be covered.

The overall operating range of the instrument is 400–1400 g/l. Operating spans are the ±full-scale readings about the nominal operating range set-

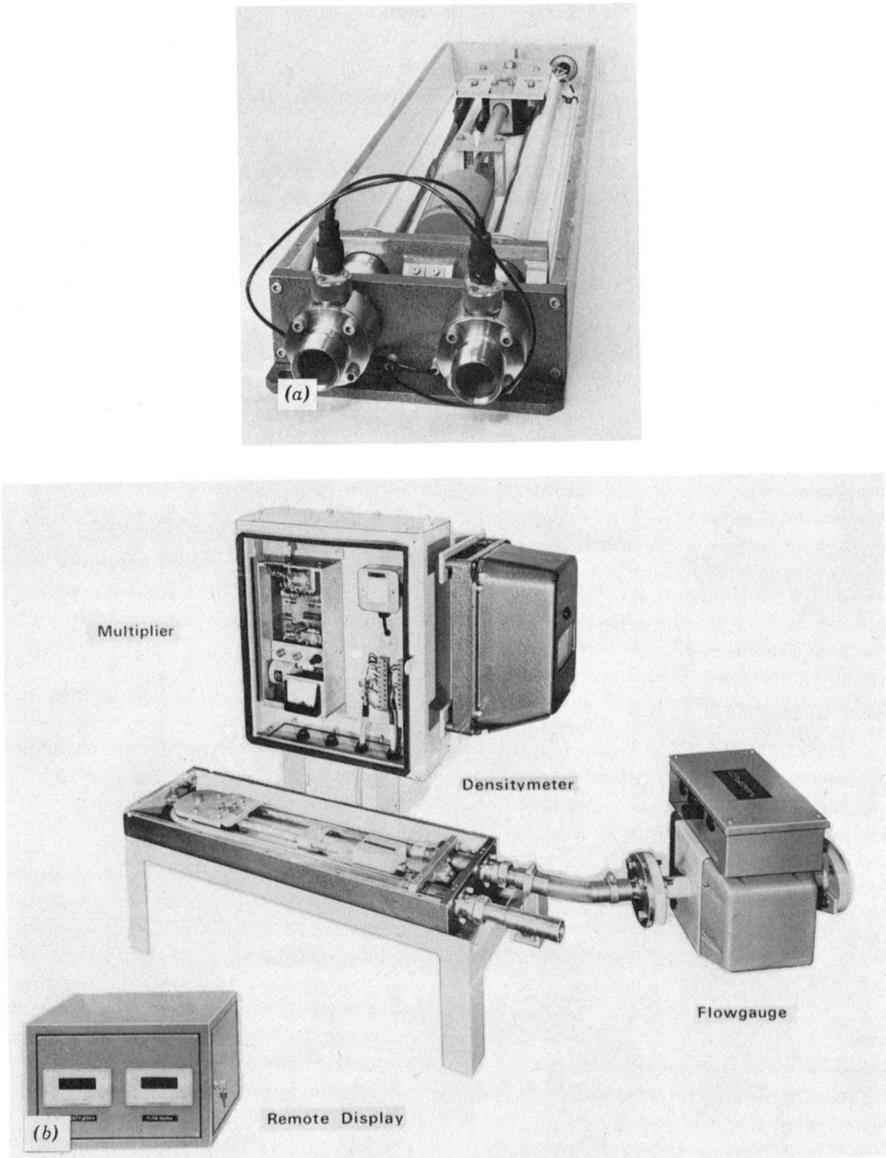

FIGURE 8.15 Gravitymaster density meter. (*a*) Sensing unit; (*b*) component parts of system. (Courtesy of SPRI Ltd.)

point. Two spans are available as standard, +−5 g/l and +−50 g/l; other spans can be provided, with corresponding decrease in meter reading accuracy. A special dial-a-density feature permits changes of nominal setpoint density to be made without adjustment of the counterbalance weight.

A high-resolution control provides for nominal density offsets of ±100 g/l or ±200 g/l, calibrated for direct density settings. This feature, combined with an external calibration facility, provides an operating range of 2000 g/l in five spans each of 0–400 g/l. Supplementary weights adjust the calibration of the beam unit for five nominal settings of 200, 600, 1000, 1400, and 1800 g/l, with the dial-a-density control providing adjustment of the nominal set-point by ±200 g/l. Recalibration takes about 2 min and the supplementary weight may be added without removing the beam unit cover plate.

Compensation is provided for sample temperature variations from ambient to 100°C to allow referencing to a standard temperature. In addition, compensation is provided as standard for mechanical errors caused by sample temperature variations. The error is less than ±0.01 g/l/°C over the range 5–100°C. As with all balance-beam-type analyzers, this instrument requires a firm, level surface for mounting and a vibration-free environment. The control unit is designed for wall mounting and may be located remote from the beam unit.

8.3.4 Radiation Absorption Principle

Density analyzers of this type operate on the principle that gamma energy will penetrate all materials, but is attenuated in proportion to the mass of the material through which it passes. Thus, any variation in the energy received by the detector is indicative of a proportional change in the density of the process stream.

These devices are non-contacting in nature, that is, the source and detector are mounted on the outside of the process pipe. They are therefore eminently suitable for applications involving process streams that are highly corrosive or difficult to handle, and are used in a wide variety of industrial processes in the food, petrochemical, pulp and paper, plastics, rubber, fertilizer, and petroleum refining industries.

(a) Kay-Ray Inc (United States)

The measurement principle of the Model 3600 Liquid Density Analyzer is shown in Fig. 8.16. A small source of gamma radiation is housed in a lead-shielded holder. When the shutter mechanism on the source holder is opened, a highly collimated beam of gamma energy is emitted directly through the pipe wall and the process fluid to a gamma radiation detector mounted directly opposite the holder. The detector generates an electrical signal, proportional to sample density, which is transmitted to a remotely located electronics unit for amplification, scaling, and calibration in density units.

Cesium 137 is used as the radioactive source because of its unique combination of good penetrating ability, its 30-year half-life, and its ease of handling and safety. The source material is doubly encapsulated in stainless steel and located in the center of a lead-filled welded steel holder. The specific source size is determined for each application by computer analysis based on pipe diameter, pipe wall material, density span, desired accuracy,

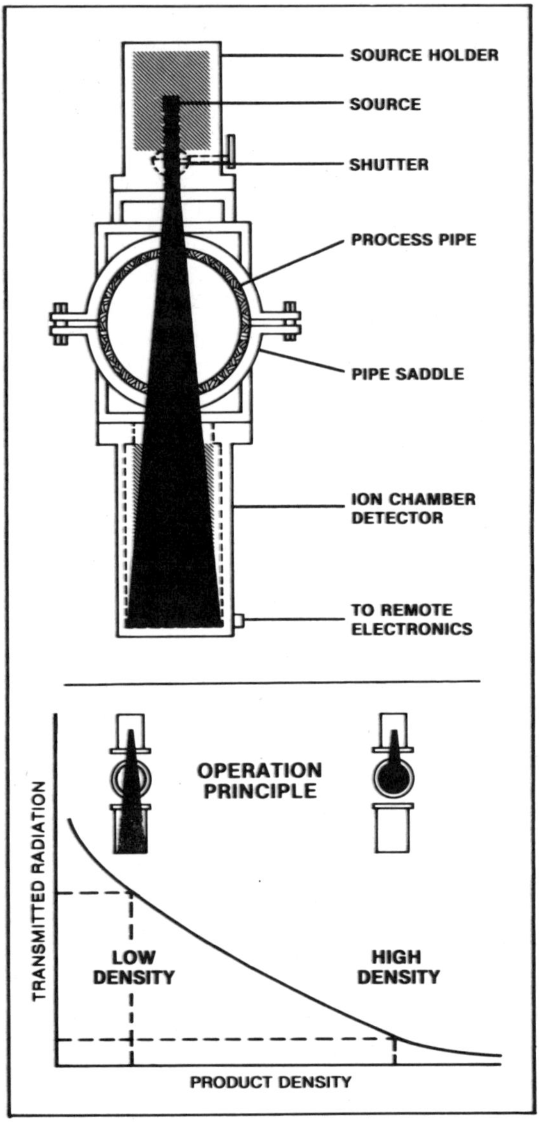

FIGURE 8.16 Model 3600 liquid density analyzer: operating principle. (Courtesy of Kay-Ray Inc.)

and response-time requirements. A lockable shutter mechanism operated by an external handle isolates the source in the closed position, making it completely safe during shipping, installation, and maintenance.

An ion chamber detector is used, mounted in a rugged enclosure and containing an inert gas at electrical potential. Incident radiation causes the gas to ionize, producing a continuous current output proportional to radiation strength. The detector is temperature-stabilized to minimize the effects of ambient temperature variations and is designed for installation in NEC Class 1, Group D, Division 1 hazardous locations. The detector output signal is transmitted directly to the electronics unit; no measurement electronics are housed in the detector.

The solid-state electronics unit, which is mounted in a NEMA 4 enclosure, amplifies the detector signal, scales it, and produces an output signal (0–10 V dc or 4–20 mA dc) proportional to sample density. Independent zero and span controls and an indicating meter are provided. Other optional features include isolated 4–20 mA dc output, log function linearizer, source decay compensation, single or dual alarm contacts, and temperature compensation. All electronic components are interchangeable with Kay-Ray belt scale and level systems.

The source holder and detector may be bolted to any existing pipe with a simple mounting saddle. The components of a typical system are shown in Fig. 8.17. Since the system is completely non-contacting, it is unaffected by variations in sample temperature, pressure, corrosivity, abrasiveness, viscosity, solids buildup, and so on. Because of its inherently simple design, installation does not require process downtime; maintenance is quoted as being minimal.

Accuracy levels are quoted as ±0.0005 to 0.001 SG units and response time is adjustable between 5 and 60 sec. All density systems are designed to minimize personnel hazards and meet the requirements of the Occupational

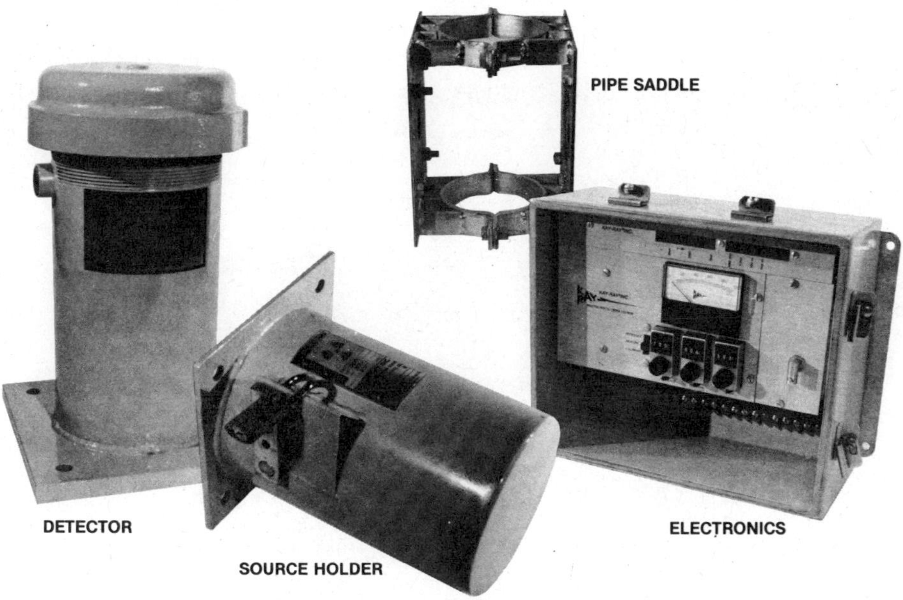

FIGURE 8.17 Model 3600 density analyzer: component parts. (Courtesy of Kay-Ray Inc.)

Safety and Health Act, the Nuclear Regulatory Commission (NRC), and international regulatory standards. The NRC permits sale and installation of Kay-Ray density systems under a general license.

The Model 3600X Digital Density System uses the same non-contacting radiation sensor described and a microprocessor-based control unit preprogrammed with a wide range of selectable features. The unit has an integral keyboard as the operator interface. The microprocessor continuously and automatically performs input signal linearization, scaling, compensation, and so on to provide a density output signal. Internal diagnostic routines are also carried out. In addition, the system has integral 3-mode and proportional controllers for single-loop automated control and can communicate with, or be controlled by, a host computer for data logging and report generation via a 1200-baud ASCII code serial data link. Calibration of the Model 3600X is accomplished by entering the appropriate calibration curve into the system via the keyboard. A digital display is provided for readout of process parameters.

(b) Ohmart Corporation (United States)

The Densart 3000 Series Density Gages, designed for density, specific gravity, or percent solids measurement in process liquids and slurries, are the latest of a line of radiation absorption-type instrumentation and an outgrowth of the well-proven level gages (Levelart series).

Three basic types are available. The Densart 3400 and 3300 series, the most common, are clamp-on types. The Densart 3100 series is a flanged axial gage with integral source, source holder, and detector. Each system consists of the sensing unit and an electronics unit, the latter suitable for local or remote, surface, or panel mounting. The sensing unit, a typical example of which is shown in Fig. 8.18, consists of a radioactive double-encapsulated

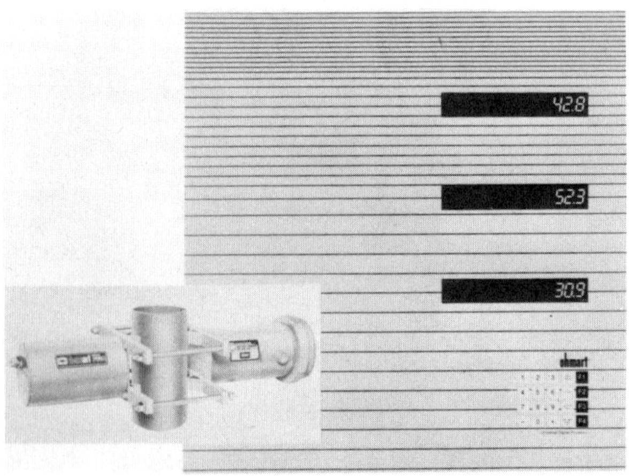

FIGURE 8.18 Densart 3000 series density analyzer. (Courtesy of Ohmart Corporation)

source contained in a lead-filled source holder, a low-voltage ionization chamber radiation detector with amplifier and power supply, and a mounting assembly. Three radioactive sources are available—cesium 137 (half-life 30 years), cobalt 60 (half-life 5.3 years), and radium 226 (half-life 1620 years).

The 3400 series has many options, including source decay compensation, process temperature compensation, and a special calibration feature, known as PreCal. This stores a predetermined calibration curve for the density analyzer (analyzer indicated density versus actual process density) and is used prior to shipment from the factory. For calibration verification in the field, all that is necessary is to take one process sample anywhere within the instrument range. If PreCal is offset, one potentiometer adjustment will move the entire PreCal computerized curve to line up with the process curve, reducing calibration time.

The 3300 series has a more rigid design, but does not have the compensation features of the 3400 series. The 3100 series uses the same electronics as the 3400 series and therefore has the same options. Standard outputs for remote recording are 0–50 V dc, 0–10 V dc, and 4–20 mA dc.

The Universal Digital Processor, a recent introduction by Ohmart, is capable of handling inputs from multiple analyzers, providing the user with one central processor for monitoring instruments for measurement of density, weight, mass flow, moisture, level, or any combination thereof. The processor forms the heart of a fully digitized system. The operator interface is a simple keypad and digital displays giving process measurements in engineering units.

Source decay compensation is a software driven function, as are calibration curves for many processes and products. The system can be linked to a host computer by an RS 232C serial data link and to a hard-copy printer. Analog output signals (0–10 V dc, 0–100 mV dc, or 4–20 mA dc) are available for remote recording. The processor is housed in a NEMA 4 enclosure suitable for surface mounting and is connected to each sensor by a twisted pair cable. Maximum transmission distance is 1 mile.

(c) Texas Nuclear (United States)

The E-ZCal Density System shown in Fig. 8.19 is a non-contacting type system designed for density measurement of process liquids and slurries. The system consists of a source head assembly, detector and digital density transmitter. The radioactive source is a ceramic cesium 137 bead, doubly encapsulated in stainless steel and mounted in a lead-filled steel encased enclosure. Various source sizes are available, ranging from 100 to 4000 mCi, depending upon application requirements. The detector can be an ion chamber or scintillation crystal and includes a preamplifier, heater with control circuitry, and regulated dc power supplies. Source and detector are mounted on the process pipe using a special pipe saddle.

The digital density transmitter is a microprocessor-based unit with many features, including alarm and relay functions, source decay compensation,

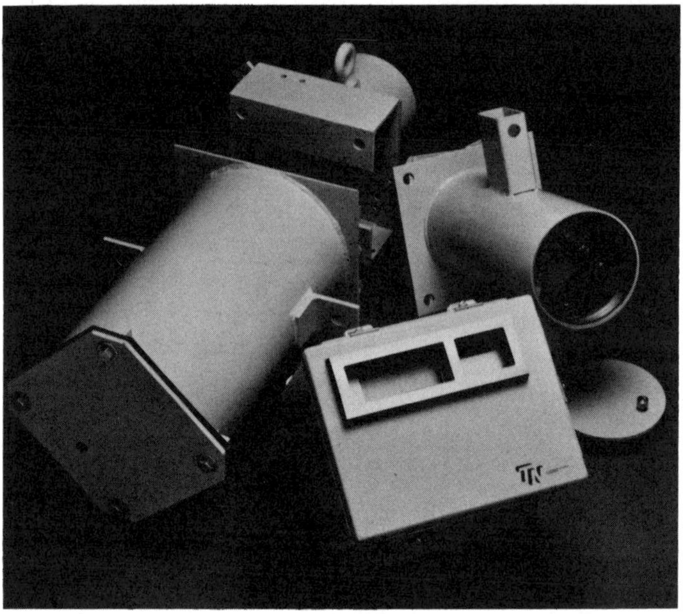

FIGURE 8.19 E-Zcal density system: component parts. (Courtesy of Ramsey Engineering, Texas Nuclear)

process temperature compensation, and so on. The system can operate in a dual time constant mode to provide the best precision and tracking of the process. Up to eight density calibration ranges can be stored in memory; the four analog outputs available (4–20 mA dc) can be assigned a primary and an alternate calibration range from these eight ranges. A maximum of three displays are available in a variety of engineering units, depending on the application. The operator interface is a simple keyboard.

8.3.5 Kinetic Energy Principle

(a) Permutit, Ranarex Division (United States)

The operating principle of the Ranarex Range of Gas Specific Gravity Analyzers is shown in Fig. 8.20. The instrument consists essentially of two cylindrical gas-tight measuring chambers, each with separate inlet and outlet sample connections. Each chamber contains an impeller and an impulse wheel, both with axial vanes. These wheels are mounted on separate shafts, facing each other but not touching, similar to the arrangement used in automotive fluid couplings. An electric motor and drive belt rotate the impellers at the same speed and in the same direction.

The impellers draw continuous flows of the process gas sample and dry reference air into their respective chambers and rotate them against the vanes of the companion impulse wheels. As the rotating gas and air impinge

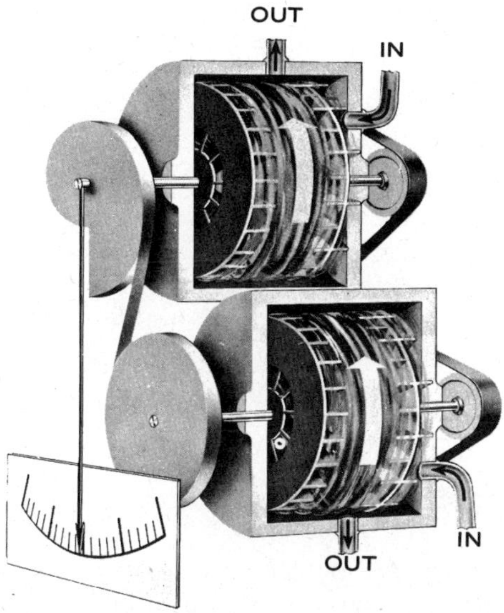

FIGURE 8.20 Ranarex gas-specific gravity analyzer. (Courtesy of Permutit Co.)

on the impulse wheel vanes, they create torques proportional to the density of the gas and air respectively. These torques are transmitted from the chamber by pivot shafts on which wheels are mounted as shown in Fig. 8.20. The wheels are restrained from continuous motion by a flexible tape that is wrapped over the wheel rims in the crossed direction. The torques, which are a measure of the densities, are thus opposed to each other.

Referring to Fig. 8.21, in the "zero" position (dry air flowing through both

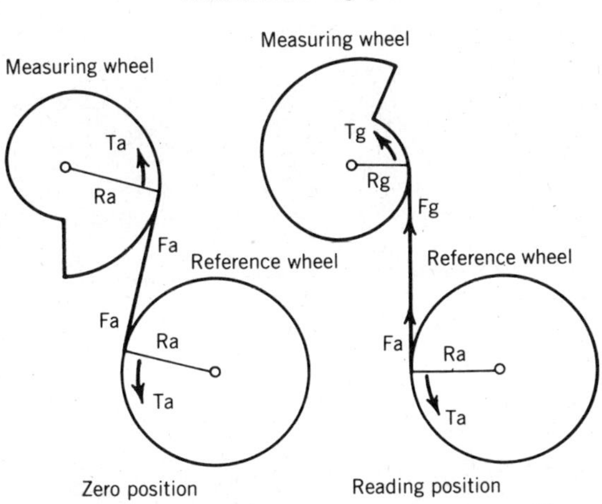

FIGURE 8.21 Ranarex: operating principle. (Courtesy of Permutit Co.)

chambers) the two torques T_a are equal, and the wheels will take up a position in which the forces F on the tape will equalize (see left view). To measure the specific gravity of gases below 1.000 the process sample is admitted to the upper chamber and reference air to the lower chamber. For specific gravities above 1.000 the situation is reversed. Assuming the first case, when the process sample is admitted the force F_g developed by the rim of the upper wheel is temporarily smaller than the force F_a developed by the rim of the lower wheel. The reference wheel and tape then rotate the measuring wheel clockwise and the measuring wheel seeks the position (see right view) where F_g and F_a are equalized and motion ceases.

The measuring system divides the torque of sample gas by the torque of dry air. Since

$$\frac{\text{torque of sample gas}}{\text{torque of dry air}} = \frac{\text{density of sample gas}}{\text{density of dry air}}$$

the instrument measures the specific gravity of the sample gas. The rim of the measuring wheel is contoured so that it rotates through equal angles for equal changes in specific gravity. The uniform motion of the gas measuring wheel is transmitted by a linkage to the readout element, which may be an integral indicator, recorder, pneumatic or electronic transmitter, or high-low alarm contacts, as selected by the customer. Accuracy of measurement is unaffected by the length of the measuring tape, since the circular reference air measuring wheel also functions as a take-up device, winding or unwinding the tape as required.

Ranarex analyzers are available as portable types or as instruments for permanent installation (wall or flush panel mounting).

8.3.6 Aerostatic Weighing Principle

As mentioned in Section 8.2, the Arcco-Anubis gas gravitometer is an example of the direct-weighing type of instrument and is also used for on-line, process measurements.

(a) UGC Inc. (United States)

The UGC Model GR6-01 Gravitometer is designed for continuous on-line measurement of gas specific gravity, with automatic compensation for changes in ambient temperature, humidity and barometric pressure. The operating principle is shown in Fig. 8.22.

Two identical tanks (sample and reference) are suspended on taut wires at equal distances from the pivot axis of a balance beam. As the specific gravity of the sample gas varies, the sample tank weight varies accordingly and is reflected as a rotational movement of the balance beam assembly, which is transferred to the pen arm. The effects of ambient temperature variations are eliminated by having the reference tank pressure regulate the sample tank

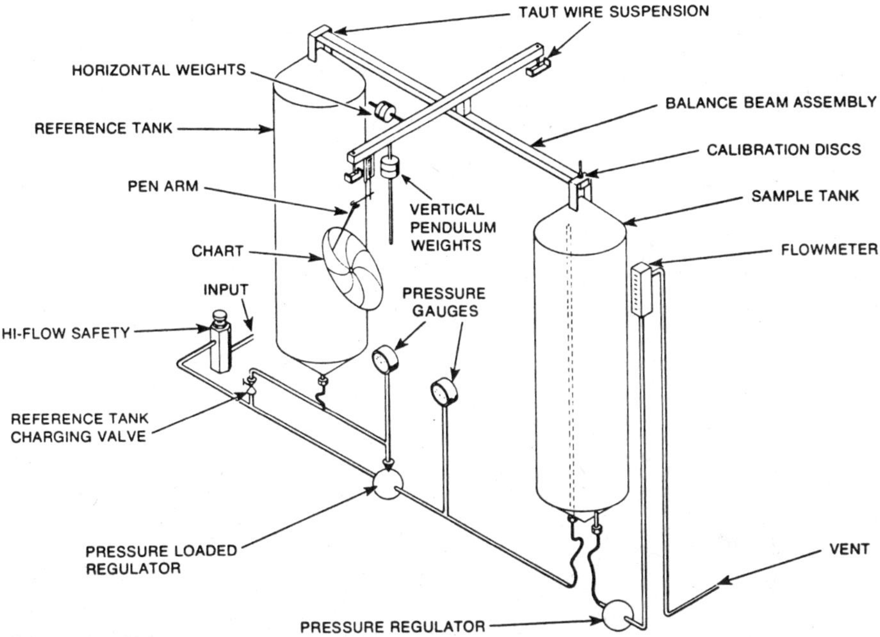

FIGURE 8.22 Model GR6-01 gas gravitometer. (Courtesy of UGC)

pressure with a special regulator. Sample flowrate is set at about 10–12 ft/hr at an inlet pressure of 10–25 psig. A high-flow safety valve, set at 22 ft^3/hr, is incorporated into the design to shut off sample gas in the event of regulator or tank hose failure.

Calibration discs are placed at the sample tank end of the balance beam assembly; their weights are determined by the specific range of the instrument and are used to calibrate for gravity span. The horizontal weights shown serve the dual purpose of compensating for any unbalance forces in the system (e.g., differences in tank weights) and for setting up the instrument to give the correct reading for a sample of known specific gravity.

The basic gravitometer is adaptable to any range in 0.5, 0.2, and 0.1 spans as standard. Repeatability is quoted as $\pm 0.25\%$ of span. Current output (4–20 mA dc) for remote recording is available as an option. Chart drives are available as electric motor (115 or 230 V ac), battery, or clockwork types.

8.3.7 Turbojet Principle

(a) Fluidynamic Devices Ltd. (Canada)

The Model 601D Turbojet Densitometer is a simple device designed for measurement of gas density; the operating principle is shown in Fig. 8.23. The working element is a turbulent jet of a suitable clean supply gas, such as instrument air or nitrogen. The primary sensor incorporates a single nozzle

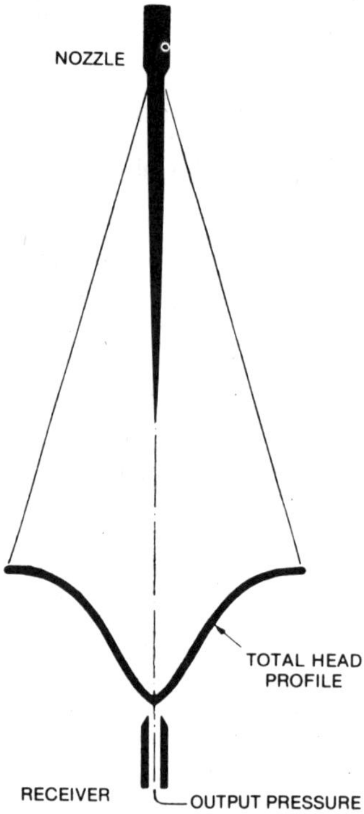

FIGURE 8.23 Model 601D turbojet densitometer: operating principle. (Courtesy of Fluidynamic Devices Ltd.)

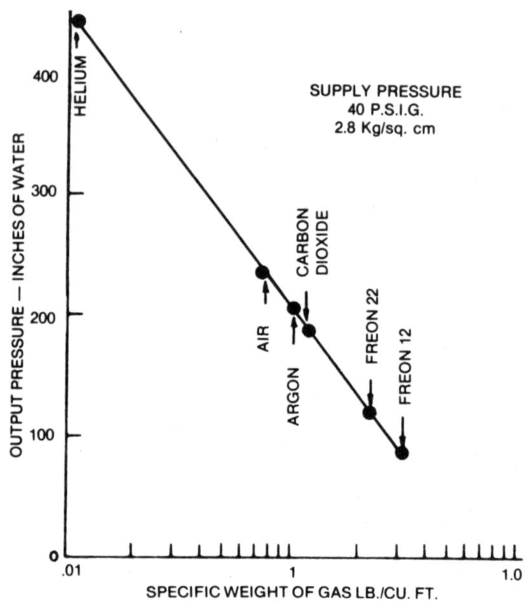

FIGURE 8.24 Turbojet densitometer-calibration curve. (Courtesy of Fluidynamic Devices Ltd.)

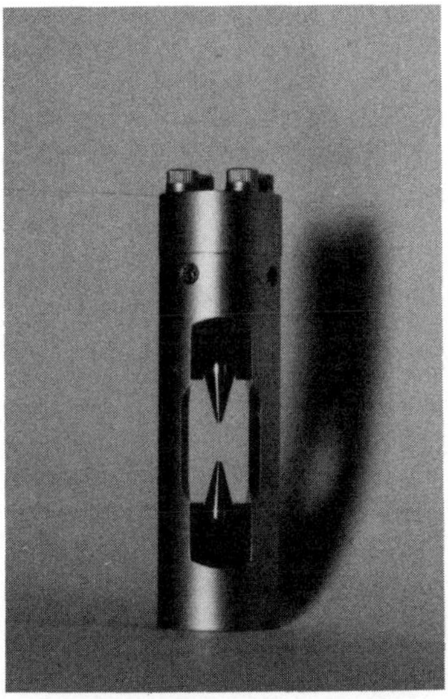

FIGURE 8.25 Turbojet densitometer-sensor/probe assembly. (Courtesy of Fluidynamic Devices Ltd.)

with a coaxial receiver. The turbulent jet issuing from the nozzle impinges on the receiver port, applying dynamic pressure that is inversely proportional to the density of the ambient gas stream. The greater the density of the gas entrained by the turbulent jet, the greater the spreading angle and the flatter the pressure profile of the jet at the receiver port. The device has high sensitivity and rangeability, as evidenced by the calibration curve for a specific gas pressure shown in Fig. 8.24.

The densitometer consists of the sensor/probe assembly and an electronics unit. The sensor/probe assembly (see Fig. 8.25) is of retractable design for direct insertion into a pipe or duct. Standard construction material is type 316 stainless steel. Maximum process temperature for the standard version is 450°F (1200°F is available as an option).

BIBLIOGRAPHY

Agar J. Vibrating spool densitometer. *Insts. & Contr. Syst.* (Jan 1970).

Bajek, W. A. Instrument system for density profile determination. *23rd Ann. ISA Conf.*, New York (Oct 1968).

Ferguson, B. C. Monitor boiler fuel density to control air/fuel ratio. *Hcbn. Proc.* (Feb 1974).

Ord, R. Jr. Digital solutions in densitometry, concentration analysis and mass flow computation. *Proc. ISA Symp.*, Boston, MA (May 1985).

Stansfeld, J. W. The application of on-line density meters for fiscal measurement on oil and gas. *Petr. Rev.*, England (Nov 1982).

Weeks, D. E. Density measurement of cryogenic fluids. *Proc. CryoTech Conf.* (1973).

Withey, P. The application of process control to North Sea oil. *Contr. & Instr.* (Dec 1981).

9

Hydrogen Sulfide and Total Sulfur Measurement

9.1 INTRODUCTION

The detection and measurement of trace quantities of hydrogen sulfide (H_2S) and other sulfur-containing compounds in high-purity process gases, natural gas, LPG, and gaseous fuels is important in many industries. Demand for high-purity products has increased the incentive to keep off-grade production to an absolute minimum, making the on-line trace H_2S or total sulfur analyzer an important tool in maintaining maximum process operating efficiency. Light hydrocarbon gases and propane/butane mixtures (LPG) are usually manufactured to a tight specification with respect to sulfur concentration, because in many cases the product is supplied directly to a customer. Under such conditions, continuous on-line analysis is an absolute necessity if customer satisfaction is to be maintained at all times.

If sulfur is present in more than trace quantities in the feedstock to certain processing plants, this can result in deactivation of catalysts used in the particular process. If the catalyst is badly contaminated, it will have to be replaced at considerable expense and with loss in production. Acetylene converters used in the production of high-purity ethylene are particularly prone to deactivation by sulfur compounds.

Measurement of trace hydrogen sulfide concentration may be used to monitor the efficiency of gas sweetening processes such as caustic treating. A feature of such processes is the sudden breakthrough that may occur as a result of overloading. It is therefore essential that a buildup of H_2S is detected promptly and the necessary action taken to correct the situation.

Sulfur, in the form of hydrogen sulfide, mercaptans (organic compounds of the general formula R-SH), sulfides, and other compounds, occur in crude oil at concentrations up to 5% by weight and persists at concentrations up to several percent in refined products. Since sulfur is usually an unacceptable contaminant, its concentration must be reduced to an acceptable level, depending on the intended use of the product. In view of the strict legislation set up in many countries to prohibit the burning of high-sulfur fuels, there is an increased incentive for rapid and accurate analysis of sulfur in petroleum products in order to meet customer and statutory specifications. Accurate chemical determination in the laboratory is a time-consuming operation, and short-term changes in process stream sulfur level may not be detected. Therefore continuous on-line sulfur measurement is justified.

Two important applications for trace hydrogen sulfide measurement are in environmental monitoring and personnel protection. Both of these applications come under the category of ambient air monitoring (see Chapter 12).

The removal of sulfur from petroleum and natural gas is a necessary part of many refining and processing operations. Production of marketable elemental sulfur is usually carried out in a Claus sulfur recovery unit, where the sour gas is partially burnt with oxygen to form sulfur dioxide (SO_2). The H_2S and SO_2 are then reacted over a catalyst to form elemental sulfur and water:

$$2H_2S + SO_2 \rightarrow 2H_2O + 3S$$

For optimum efficiency, a ratio of 2:1 must be maintained, as shown in the equation. One of the major on-line analyzer applications for this process is the measurement of the H_2S/SO_2 ratio. A number of analyzer manufacturers can provide analyzer systems for this application, using photometric analysis as described in Section 9.4.3.

9.2 SULFUR COMPOUNDS IN PETROLEUM PRODUCTS

The impurities present in practically all crude oils are sulfur and sulfur compounds, lesser amounts of oxygen, nitrogen and chlorine compounds, and traces of metallic contamination. Sulfur compounds are the major impurity and processes for their removal play an important part in refinery/ petrochemical plant operations.

Considerable work has been done by the American Petroleum Institute (API) to identify and study the various types of sulfur compounds present in petroleum products. Here are some of the major forms of sulfur present in crude petroleum and its distillates:

Sulfur Form	General Formula
Elemental sulfur	S
Hydrogen sulfide	H_2S
Alkyl mercaptans	$C_nH_{2n+1} \cdot SH$
Cyclic mercaptans	$C_nH_{2n-1} \cdot SH$
Aromatic mercaptans	$C_nH_{2n-7} \cdot SH$
Alkyl sulfides	$(C_nH_{2n+1})_2 \cdot S$
Alkyl disulfides	$(C_nH_{2n+1})_2 \cdot S_2$
Cyclic sulfides	$(CH_2)_n \cdot S$
Thiophene	$C_4H_4 \cdot S$

Here are some of the sulfur forms found in products from cracking, coking, and alkylation processes:

Sulfur dioxide	SO_2
Sulfuric acid	H_2SO_4
Alkyl sulfates	$(C_nH_{2n+1})_2 \cdot SO_4$
Sulfonic acids	$C_nH_{2n+1} \cdot SO_3H$
Carbon disulfide	CS_2
Carbonyl sulfide	COS

In lower boiling distillates, the molecular weight of the sulfur compounds present is relatively low and the chemical structures not particularly complex, consisting principally of mercaptans, sulfides, and disulfides. The quantities, relative proportions, and exact composition of these compounds occurring in petroleum distillates vary considerably, depending on the crude oil source and the nature of previous operations carried out on the distillate.

Extensive work has also been done on evaluating the structure of products obtained from the various sulfur forms, when the crude oil is distilled or cracked at elevated temperatures. General reactions of the various types of compounds can be summarized as follows:

1. Elemental sulfur reacts with hydrocarbons to form hydrogen sulfide at temperatures above 325°F.
2. Mercaptans decompose to form hydrogen sulfide and hydrocarbons at higher temperatures.
3. Organic sulfides form hydrogen sulfide, mercaptans, thiophene derivatives, and hydrocarbons.
4. Disulfides decompose to mercaptans when heated in the crude distillation process (pipestill). The mercaptans can continue to decompose as in 2.
5. Thiophene is very stable and not normally changed in any thermal process, although substituted thiophenes can break up into smaller molecules.

The major problems with H_2S contamination are the contribution to corrosion and the hazard caused by its extreme toxicity. Mercaptans lead to odor problems when present in finished products. Disulfides, sulfides, and especially polysulfides in naphthas act in varying degrees to inhibit octane improvement when using tetra-ethyl lead (TEL). Thiophenes contribute to the overall sulfur content of a particular fraction and to the amount of corrosive sulfur oxides formed during combustion.

9.3 LABORATORY METHODS OF MEASUREMENT

Traditional methods used in the laboratory for measurement of hydrogen sulfide and total sulfur at percentage or trace levels in petroleum products generally involve complicated time-consuming procedures, based on combustion or pyrolysis of the sample, followed by gravimetric or titrimetric analysis. Various ASTM test methods are listed in Table 9.1; the following is a brief summary of some of the methods.

ASTM D129-64 covers the total sulfur (S) determination in petroleum products of low volatility with a total S content greater than 0.1%. The sample is oxidized in a bomb containing oxygen under pressure. The sulfur present is oxidized to sulfate, which is determined gravimetrically as barium sulfate ($BaSO_4$).

ASTM D1072-80 covers total S determination in fuel gas. The sample is burned in a closed system and the sulfur oxides thus formed are absorbed in sodium carbonate solution. The sodium sulfate is then titrated with barium chloride solution using tetrahydroquinone as indicator.

ASTM D1266-80 (*lamp method*) cover total S in petroleum products in the range 0.01–0.4 %wt. The sample is burned in a 70% CO_2/30% O_2 atmo-

TABLE 9.1 ASTM Standard Test Methods for H_2S and Total Sulfur Measurement

ASTM	
D129-64	Sulfur in Petroleum Products by General Bomb Method
D1072-80	Total Sulfur in Fuel Gases
D1266-80	Sulfur in Petroleum Products by the Lamp Method
D1552-79	Sulfur in Petroleum Products by the High Temperature Method
D2385-81	H_2S Mercaptans in Natural Gas by Titration
D2420-81	H_2S in LPG by the Lead Acetate Method
D2622-82	Sulfur in Petroleum Products by X-Ray Spectrometry
D2725-70	H_2S in Natural Gas by the Methylene Blue Method
D2747-81	Trace Total Sulfur in Volatile Organic Liquids by the Combustion Method
D2784-80	Sulfur in LPG by Oxyhydrogen Burner or Lamp
D2785-80	Trace Total Sulfur by Wickbold/Beckman Combustion Apparatus
D3031-81	Total Sulfur in Natural Gas by Hydrogenation Method
D3120-77	Trace Sulfur in Light Liquid Hydrocarbons by Oxidative Microcoulometry
D3177-75	Total Sulfur in Coal and Coke
D3227-73	Mercaptan Sulfur in Gasoline and Distillate Fuels by Potentiometric Titration
D3246-81	Sulfur in Petroleum Gases by Oxidative Microcoulometry
D3961-80	Trace Sulfur in Aromatic Hydrocarbons by Oxidative Microcoulometry
D4045-81	Sulfur in Petroleum Products by Hydrogenolysis and Ratiometric Coulometry
D4084-82	H_2S in Gaseous Fuels by Lead Acetate and Reaction Rate
E257-67	Sulfur in Organic Compounds by Sodium Peroxide Bomb Ignition
E443-80	Sulfur in Organic Compounds by Oxygen Flask Combustion
E840-81	Sulfur by the Use of Flame Photometric Detectors and Gas Chromatographic Analysis

sphere and the sulfur oxides thus formed are absorbed in hydrogen peroxide solution where they are converted to sulfuric acid. This is followed by titration with standard sodium hydroxide solution. Variations on this method include gravimetric determination of sulfate as barium sulfate after burning in air, and, for trace S, measurement of the turbidity of the solution using a photometer set for 450 nm absorbance.

ASTM D1552-79 is a high-temperature method for petroleum products with boiling points greater than 350°F and containing more than 0.06% S. The sample is burned in oxygen at a temperature high enough to convert 97% or more of the sulfur to sulfur dioxide. The combustion product is absorbed in potassium iodide solution, followed by iodometric titration using starch indicator.

ASTM D2420–81 covers measurement of H_2S in LPG. The vaporized sample, at a flowrate of 2.3 ± 0.2 l/min, is passed over moist lead acetate paper for exactly 2 min and the stain intensity measured. This test, as written, is qualitative only. However, laboratory instruments operating on the lead acetate staining principle are now used for quantitative determination of H_2S and total S (see ASTM D4045 and 4084 following).

ASTM D2622-82 covers measurement of total S in petroleum products by an X-ray spectrographic method. The sample is irradiated by X-rays from a low-power source and the intensity of the $K\alpha$ line at 5.373 Å is measured. The intensity of a selected background at 5.190 Å is subtracted from the measured intensity at 5.373 Å. The instrument is calibrated with a certified di-n-butyl sulfide standard. This method is also used as a basis for on-line sulfur analyzer design.

ASTM D3227-73 covers the measurement of mercaptan sulfur in gasoline, kerosine, aviation fuel, and other distillate fuels within the range 0.0003–0.10 wt%. Hydrogen sulfide is removed first, since it interferes with the test. The sample is then dissolved in alcoholic sodium acetate solution, followed by potentiometric titration with silver nitrate. The end-point of the reaction is a large change in cell potential.

ASTM D4045-81 covers the measurement of total S in petroleum products within the range of 0.02–10 mg/kg. The sample is pyrolyzed with hydrogen at 1300°C. Sulfur compounds are converted to hydrogen sulfide. Readout is by radiometric determination of the colorimetric reaction of H_2S with lead acetate-impregnated paper tape.

ASTM D4084-82 covers the measurement of H_2S in gaseous fuels within the range 0.1–16 ppm using the lead acetate reaction rate method. The gaseous sample is humidified and passed over lead acetate-impregnated paper tape automatically driven by a motor. The rate of reaction and resulting rate of color change is proportional to the H_2S concentration in the sample. The method is designed specifically for the use of instruments such as the Houston-Atlas Model 825, which operates on the same principle as its process analyzer counterpart, described later in this chapter.

From these brief descriptions, it can be seen that for the most part the laboratory test methods laid down by the ASTM are traditional chemical analyses, requiring considerable laboratory equipment, time, and strict adherence to the procedures laid down.

The X-ray spectrographic method for sulfur in liquid hydrocarbons is probably the most popular, since it is an instrument method with relatively high precision. As mentioned earlier, this principle is used also in some process analyzers for sulfur measurement in liquids.

Another commonly used method involves chromatographic analysis utilizing the flame photometric detector, which is highly specific to sulfur compounds. This technique can be used to measure trace concentrations of *individual* sulfur compounds previously separated on the chromatographic column.

9.4 PROCESS ANALYZERS FOR H_2S/TOTAL SULFUR

There are many commercial instruments for on-line measurement of H_2S and total sulfur in process gases and liquids. These instruments use a variety of different measurement principles, as follows:

- Gas chromatography with flame photometric detector.
- Lead acetate paper tape staining.
- Photometric.
- Titrimetric.
- Absorption and conductivity measurement.
- Solid state sensor.
- Photoionization.
- X-ray absorbance.
- Neutron-excited gamma-ray absorption.

9.4.1 Gas Chromatography with FPD

Most of the major manufacturers of process gas chromatographs, particularly Amscor, Applied Automation, CE Process Analytics, Foxboro Analytical, and Beckman Instruments in the United States, have been involved in designing systems for environmental monitoring of H_2S at the ppm level using the sulfur-specific flame photometric detector. See Chapter 1 for the basic principles of chromatography and a more detailed description of the flame photometric detector.

9.4.2 Lead Acetate Tape Staining Type

This type of analyzer has been used extensively for both H_2S and total sulfur measurement in process gases and for ambient air monitoring, both at the ppm level.

(a) Tracor-Atlas Inc. (United States)

Tracor-Atlas specializes in the manufacture of H_2S and total S analyzers for laboratory use, for on-line process applications (including capability for sample handling system design), and for environmental monitoring.

The Model 722-AEX is designed for trace H_2S measurement in process gases and for environmental monitoring. All Tracor-Atlas analyzers operate on the same photometric principle, as illustrated in Fig. 9.1a. Sample enters the instrument, and its flowrate is precisely controlled prior to entering a humidification chamber, where it bubbles through a 5% acetic acid solution. The sample then passes over the top surface of the lead acetate-impregnated sensing tape. Tape movement is accomplished by a precision synchronous motor. A soft-seating pressure bar provides positive tape seal against the reaction window. The 100-ft tape reel provides approximately 2 weeks of continuous operation.

The optical detection system consists of a light source (14 V tungsten lamp), the light from which is focused onto the reaction window area, with

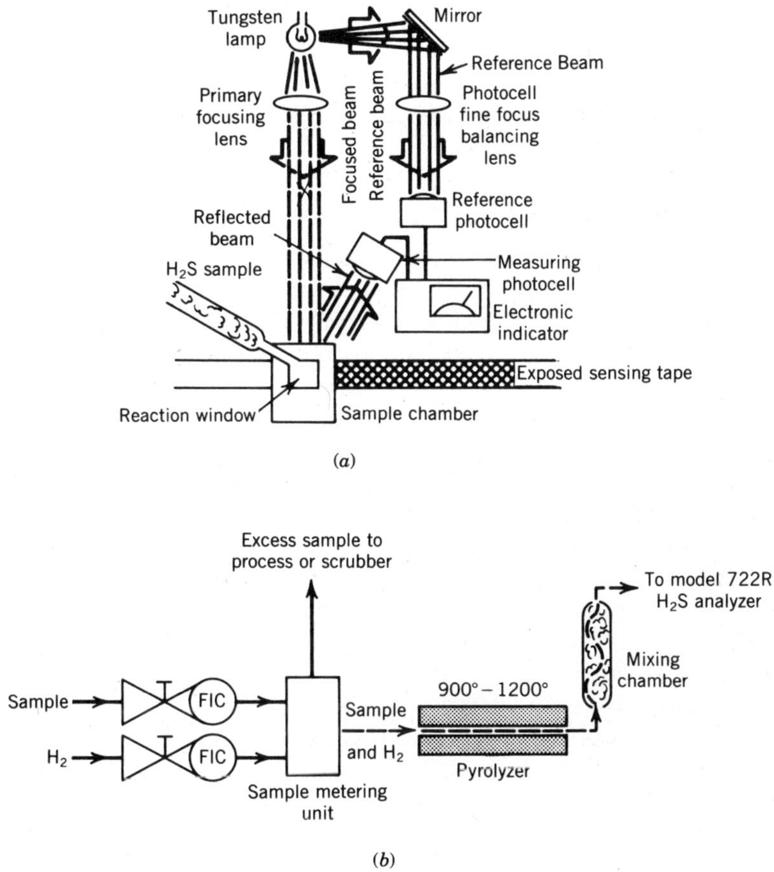

FIGURE 9.1 Model 722-AEX (*a*) and hydrogenator (*b*). (Courtesy of Tracor-Atlas Inc.)

the incident beam striking the sensing tape and reflecting to the measurement photocell. Simultaneously, light from the lamp is reflected from a mirror and focused on a reference photocell. The reference beam provides compensation for variations in light intensity. The photocells are connected to form a resistance bridge network, the unbalance output of which is amplified by a solid-state electronic unit to provide an output proportional to the H_2S concentration in the sample gas.

Standard measurement ranges fall within the limits 0–2 and 0–24 ppm full scale. Higher ranges are also available. Speed of response at the specified flowrate is quoted as 63% of full scale in 3 min and repeatability as ±3% of full scale. Analyzer output is 0–100 mV dc standard, with 4–20 mA dc as an option. The analyzer is designed for NEC Class 1, Groups B and D, Division 1 or 2 hazardous locations.

For measurement of total sulfur, the Model 756 Hydrogenator shown in Fig. 9.1b is used in conjunction with the Model 722-AEX. The gas sample is precisely metered into a continuously flowing stream of hydrogen. The combined sample then enters the pyrolysis chamber, operated at 900–1200°C, where sulfur compounds undergo hydrogenation to H_2S. The effluent from the hydrogenator then passes to the Model 722-AEX where the H_2S is analyzed as described before.

(b) H. Maihak AG (West Germany)

The Monocolor MC1N is designed for on-line measurement of H_2S in air or process gases within the range 0–1.6 to 0–64 ppm. The standard measurement range is 0–16 ppm with a sample gas flowrate of 70 ml/min and an exposure time of 1 min. The instrument and its operating principle are illustrated in Fig. 9.2.

The lead acetate-impregnated tape passes at a constant feed rate, first through a reference chamber (V) and then a measurement chamber (M). The reference chamber is filled with air, while a constant volume of sample gas flows through the measurement chamber. The tape is exposed in both chambers to the light from a single source, and reflected light from the paper falls onto two photocells (L_1 and L_2). The resulting differential current is proportional to the stain intensity on the tape in the measurement chamber and therefore to the H_2S concentration in the sample gas.

The tape is advanced in steps of 1 mm/min and the peak measured value of staining is electronically stored. The hold amplifier provides a continuous 0–20 mA dc output signal as standard; 4–20 mA dc is available as an option. The MC1N is designed for installation in non-hazardous locations only, but

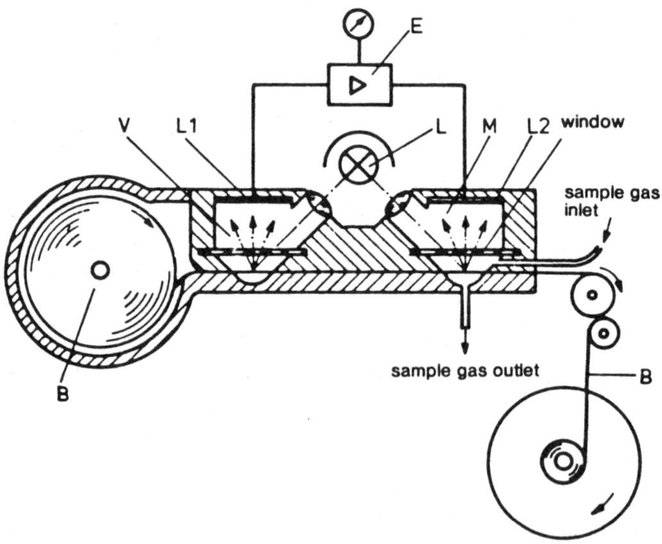

FIGURE 9.2 Monocolor MC-1001. (Courtesy of Maihak)

can be provided with an automatic purge system or with intrinsically safe measurement and control circuit.

The Model MC-1001, identical in operation, is designed for use in hazardous locations.

These analyzers may also be used to measure other gases such as phosgene, hydrogen chloride, and hydrogen cyanide by using different tapes and adapted optics.

9.4.3 Photometric Type

The measurement of H_2S using photometric analyzers is based on the characteristic absorption of this compound in the UV-visible portion of the electromagnetic spectrum. It is only one of a wide range of different compounds that may be measured using the photometric technique. The principles of spectrometry and its application to on-line process analysis in the infrared, visible, ultraviolet, and other portions of the electromagnetic spectrum, are discussed in Chapter 10. The following are two typical examples of process instruments used specifically for sulfur measurement.

(a) DuPont Analytical Instruments Division (United States)

The Model 400 Photometric Analyzer operates in the 214–1000 nm wavelength range and measures the H_2S concentration in gaseous process streams within the range 0–1000 ppm. The instrument is shown in Fig. 9.3.

The Model 400 is a split-beam analyzer consisting of the analyzer unit, which may be of general-purpose design or designed for installation in NEC Class 1, Group D, Division 1 hazardous locations, and the control unit, which is designed for remote mounting in non-hazardous locations at distances up to 1000 ft from the analyzer unit.

The analyzer unit consists of three individual compartments—the lamp,

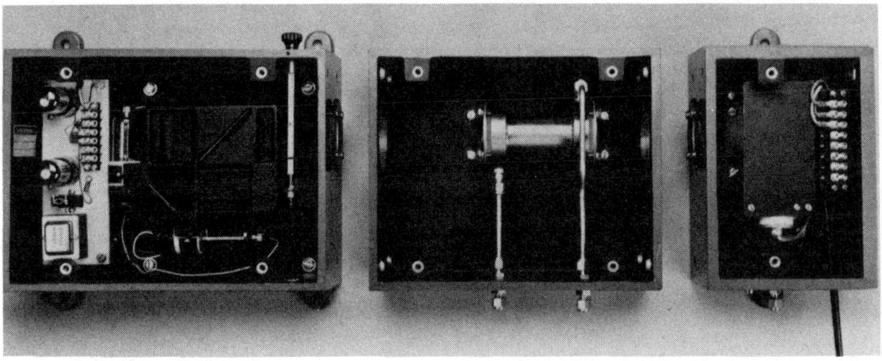

FIGURE 9.3 Model 400 photometric analyzer. (Courtesy of DuPont Company)

the sample, and the photometer housings. Light from the single light source passes through the sample cell and into the photometer, where it is split into two beams by a semi-transparent mirror. One beam is directed onto the measuring phototube through the measuring wavelength filter, while the other beam is directed onto the reference phototube through the reference wavelength filter. The two filters are selected so that the component of interest absorbs strongly at the measurement wavelength and only weakly at the reference wavelength.

The modular design of the analyzer unit, which separates the sample cell from the optics, allows operation at higher sample temperatures and pressures (up to 1000°F and 1000 psig). It also allows accommodation of long sample cells for low concentration measurement. Standard outputs are 0–10 mV dc, 4–20 mA dc, and 10–50 mA dc. Repeatability is quoted as ±1% of full scale.

DuPont also manufactures the Model 462 analyzer system for the measurement of the H_2S/SO_2 ratio in the tail gas from a Claus sulfur recovery plant (see Section 9.1). Analysis is based on the UV absorption of each component of interest using two split-beam photometric analyzers as shown in Fig. 9.4. Ranges are 0–2% H_2S, 0–1% SO_2, and 0–4 ratio. Analyzer units

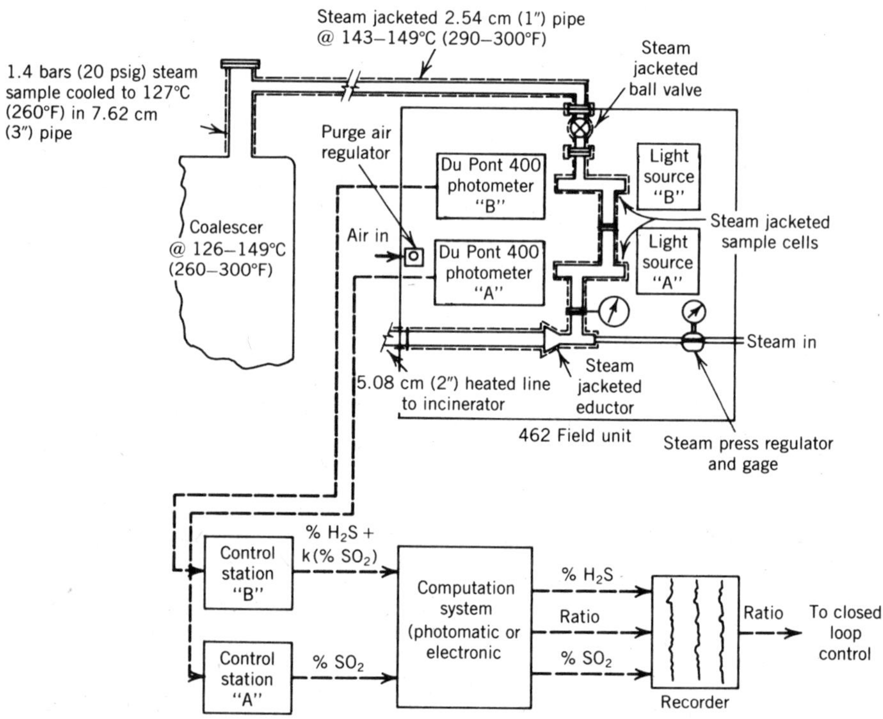

FIGURE 9.4 Model 462 H_2S/SO_2 ratio analyzer. (Courtesy of DuPont Company)

may be designed for NEC Class 1, Group D, Division 1 or 2 hazardous locations. The system includes sample handling and conditioning facilities.

(b) Teledyne Analytical Instruments (United States)

The Model 600 UV/Visible Photometric Analyzer operates over the 210–1000 nm wavelength range and may be used for the measurement of H_2S in gaseous streams within the range 0–300 ppm. The analyzer is available in two basic configurations—the Model 610, suitable for NEC Class 1, Group D, Division 1 hazardous locations, and the Model 611, suitable for non-hazardous locations only. The control unit of both models may be integrally or remotely mounted.

The operating principle is illustrated in Fig. 9.5. The optical system consists of a lamp source, sample cell, and single detector. In front of the detector is a motor-driven filter wheel containing two optical filters (measurement and reference) mounted 180° apart, which alternately rotate into and out of the optical path. The measurement filter is selected at an H_2S

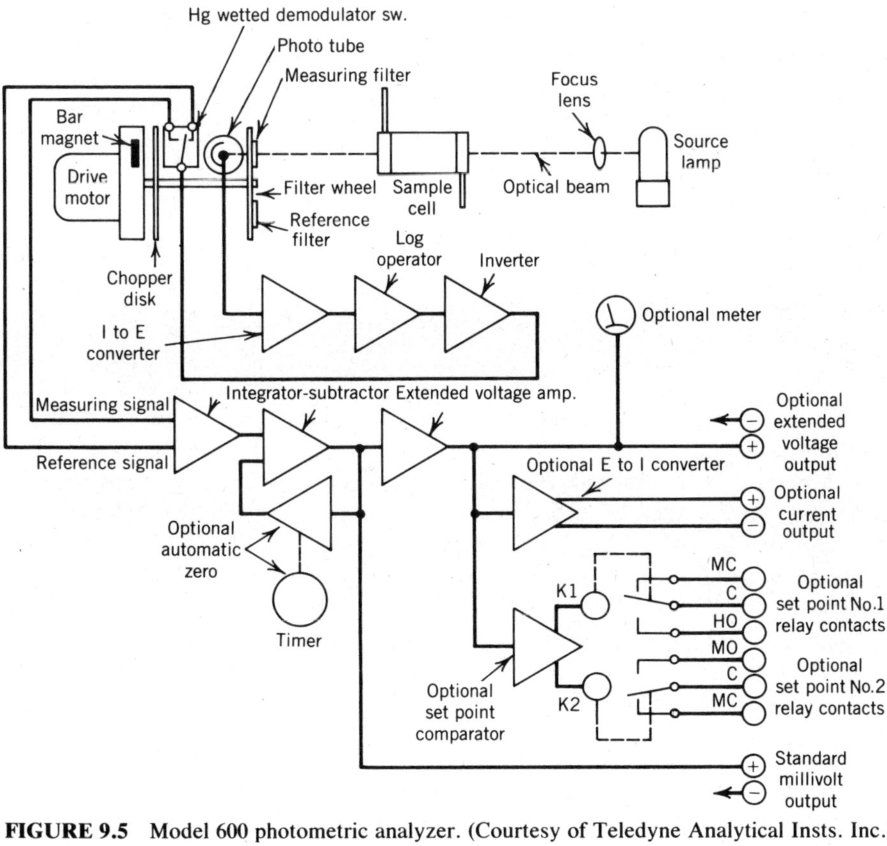

FIGURE 9.5 Model 600 photometric analyzer. (Courtesy of Teledyne Analytical Insts. Inc.)

absorbance peak and the reference filter at a point of no absorbance. This design is inherently stable and compensates for lamp aging, window deposits, and so on. All models are solid-state with analog output and alarm facilities. Repeatability is quoted as ±1% of full scale.

9.4.4 Titrimetric Type

(a) ITT Barton (United States)

ITT Barton manufactures a range of on-line process analyzers for trace sulfur measurement in gaseous process streams, using the titrimetric principle.

The electrolytic generation of bromine as a titrating reagent for the measurement of sulfur compounds was introduced about 20 years ago. The development of solid-state electronic circuits and a practical, coulometric bromine-sensing electrode system led to the development and design of a wide-range electrolytic titrator for continuous on-line sulfur monitoring. The Barton sulfur analyzers operate on the principle of generating a reactive agent, either bromine or iodine, by passing an electrical current through a suitable electrolytic solution. A schematic diagram of the titration cell and basic titrator circuit is shown in Fig. 9.6.

To measure sulfur concentration, the sample gas is passed through the titration cell at a fixed flowrate. The inner compartment of the cell is filled to a fixed level with hydrobromic acid solution, through which the sample gas bubbles. Three platinum electrodes are immersed in the solution. Direct current from a solid-state control circuit passes through the solution between two of these electrodes, generating hydrogen at the cathode and bromine at the anode, according to the following equations:

$$2H^+ + 2e^- \rightarrow H_2 \text{ (generating cathode)}$$

$$2Br^- - 2e^- \rightarrow Br_2 \text{ (generating anode)}$$

Hydrogen escapes with the sample gas flowing from the cell, while the bromine is dissolved and reacts with the sulfur compounds in the sample.

When electrical current is increased until the rate of bromine generation is balanced by the rate of entry of sulfur compound with the sample, any increase of current will produce an excess of bromine. In the titrator, a second low-voltage source of direct current is connected across the bromine generator electrode and the third electrode (sensing cathode). In the absence of bromine, no current flows in the sensing circuit, but when bromine is present, the current flow is proportional to the electrode surface and the voltage. The reactions in the sensing circuit are,

$$Br_2 + 2e^- \rightarrow 2Br^-$$

$$2Br^- - 2e^- \rightarrow Br_2$$

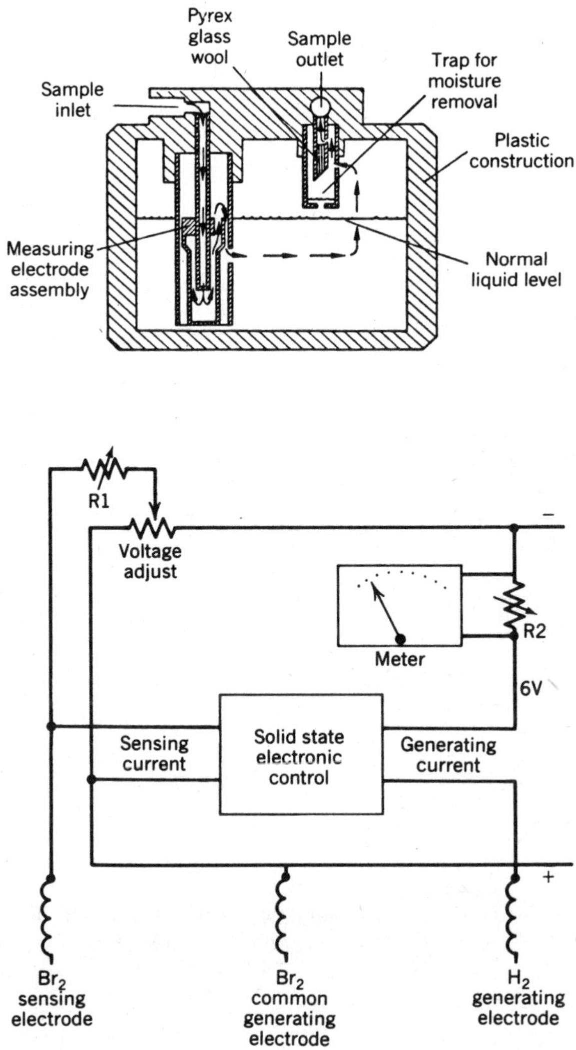

FIGURE 9.6 ITT Barton sulfur analyzers: operating principle. (Courtesy of ITT Barton Instruments Inc.)

These reactions are equal and opposite and produce no change in the system. The current flowing operates the control circuit to switch the generating current supply on and off as required, to maintain an indicating concentration of bromine in the cell as set by the zero control of the instrument. The difference between the instrument "blank" current and the current during titration of sulfur compounds is directly proportional to the sulfur concentration in the sample gas.

Typical reactions that take place with various sulfur compounds are shown by the following equations:

$$H_2S + 4H_2O + 4Br_2 \rightarrow H_2SO_4 + 8HBr$$

$$RSH + 2H_2O + 2Br_2 \rightarrow RSO_2H + 4HBr$$

$$RSR + H_2O + Br_2 \rightarrow RSOR + 2HBr$$

$$RSSR + 4H_2O + 3Br_2 = 2RSO_2 + 6HBr$$

Reactions at the ppm level are not all complete as indicated in these equations. Therefore it is necessary to calibrate the titration cell for known mixtures. Also, the amount of bromine required varies with the type of sulfur compound; therefore, sample composition must be known in order to provide a general cell factor for total sulfur calibration.

Selective analysis of sulfur compounds is made by inserting filter/absorbers upstream of the titration cell, which selectively remove the various sulfur compounds. The drop in titration level as compounds are selectively removed is proportional to the concentration of the absorbed compound. Here are some typical filter/absorber solutions used with success:

Filter/Absorber Solution	Compounds Absorbed
2% Boric acid 1% Cadmium sulfate	H_2S
1% Sodium hydroxide	H_2S, RSH
0.5% Silver nitrate	H_2S, RSH, RSR

Carbonyl sulfide (COS) is unreactive with bromine or other commonly used electrolytes. To measure COS, it is catalytically reduced to H_2S by hydrogen in the presence of platinum at 600–700°C. Under these conditions, however, saturated light hydrocarbons are converted to olefins, which are weakly titrated by bromine and interfere with the measurement. Therefore an iodine-based electrolyte is used.

A selection of some of the Barton range of process sulfur analyzers is shown in Fig. 9.7.

The Model 286 is a portable instrument consisting of three modules—the titration cell, electronics unit, and a strip recorder, all mounted in a rigid PVC enclosure. The cell, electrode assembly, and piping are made of polyethylene, polypropylene, teflon, or PVC. Seven measurement ranges are provided over the range 0–1000 ppm sulfur, with a minimum range of 0–1.0 ppm and a repeatability of ±2% of range. Sample flowrate is specified as 250 ml/min.

The Model 342 consists of an explosion-proof titration module, suitable for installation in NEC Class 1, Group D, Division 1 hazardous locations,

FIGURE 9.7 ITT Barton range of sulfur analyzers. (*a*) Model 407 (weather proof); (*b*) Model 408 (ex-proof; IS design). (Courtesy of ITT Barton Instruments Inc.)

and a general-purpose control module for remote mounting at distances up to 500 ft from the titration module. Construction, materials, and operating specifications are the same as for the Model 286. An analog output of 0–100 μA, 0–10, or 0–100 mV dc is available for remote recording.

The Model 407 is a weatherproof unit packaged for rack or wall mounting and consists of titration module, electronics module, and strip chart recorder mounted in one enclosure, similar to the Model 286.

The Model 408 consists of two units, an explosion-proof titration module of intrinsically safe design and a general-purpose control unit.

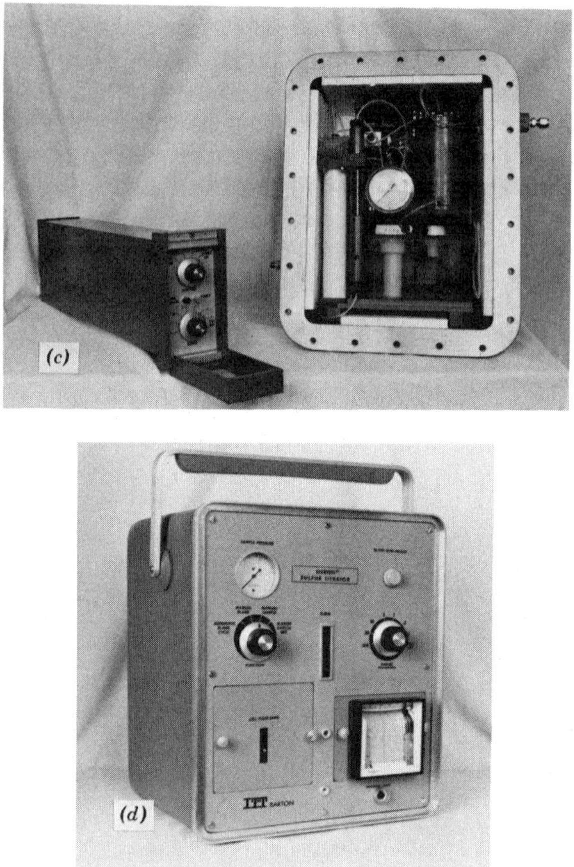

FIGURE 9.7 ITT Barton range of sulfur analyzers. (*c*) Model 342 (ex-proof); (*d*) Model 286 (portable). (Courtesy of ITT Barton Instruments Inc.)

The Model S053 is a multi-channel sulfur analyzer designed for continuous measurement of each of four sulfur compounds in natural gas or process streams. Direct reading, recording, and alarm features are available for hydrogen sulfide, mercaptans, sulfides, disulfides, and total sulfur. Four separate detection cells are used.

Several configurations can be offered for various application requirements, based on the analyzer's modular design concept. Selective scrubbers are employed to correspond to the sulfur species of interest. The system is mounted in a free-standing cabinet, as shown in Fig. 9.7, and consists of two sections, the analysis section and the electronics section. The analysis section contains the gas selection system (sample, cal gas, sulfur-free gas), flow controllers, scrubbers, titration cells, and gas venting/return system. The electronics consists of three sections—the flow control module, processing

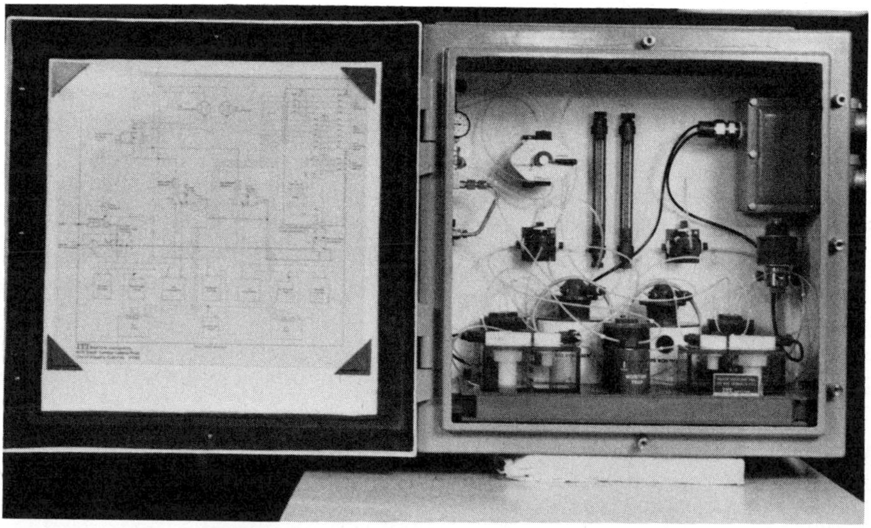

FIGURE 9.8 Model 406 H_2S/SO_2 ratio analyzer. (Courtesy of ITT Barton Instruments Inc.)

module, and the recording section. Microprocessor-based electronics is used for computation, control, and data manipulation.

The Model 327 Automatic Filter Unit, used with Barton sulfur analyzers, is designed to automatically pass the sample gas stream through filters that selectively absorb sulfur compounds by classes. This device is installed upstream of the titration cell.

The Model 406 Analyzer System shown in Fig. 9.8 is designed to measure the H_2S/SO_2 ratio in tail gases from a Claus sulfur recovery plant. The system consists of three modules—a dilution sampling control probe, a dual channel bromine titration analysis module, and an electronics module for control and recording. The probe and analysis modules are designed for NEC Class 1, Groups C and D, Division 1 hazardous locations; the control unit is designed for remote mounting in a non-hazardous location. Two titration cells are used, one for H_2S, one for SO_2. The gas sample is split into two streams; each passes through a scrubber, one to remove H_2S and the other to remove SO_2. The control unit controls the titrations and computes the original constituent concentrations from the data generated. Various outputs are available, a 4–20 mA dc ratio signal, a 4–20 mA dc deviation output scaled from -1 to $+1\%$ ($1H_2S-2SO_2$), and a 4–20 mA dc or 1–5 V dc output for each component scaled 0–2%.

9.4.5 Conductimetric Type

Instruments categorized as "conductimetric" are based on a chemico-physical method in which the component of interest in the sample gas is absorbed by and reacts with a suitable reaction solution and the resulting change in conductivity of the solution is measured. Hydrogen sulfide is only one of a variety of different compounds that can be measured using this technique.

(a) Hartmann and Braun AG (West Germany)

The Picoflux 4 Trace Gas Analyzer is designed for continuous measurement of trace components such as hydrogen sulfide, sulfur dioxide, ammonia and chlorine, in a wide variety of carrier gases, including atmospheric air. Selectivity for a particular component and measurement sensitivity are obtained by suitable choice of the reaction solution.

The operating principle of the analyzer is shown in Fig. 9.9. Constant flowrates of sample gas and reaction solution are mixed and react together in the reaction tube. The analysis unit design maintains the flows constant by means of pressure and level balancing systems. The conductivity of the reaction solution is measured in the compensation cell before reaction and in

Principle of measurement

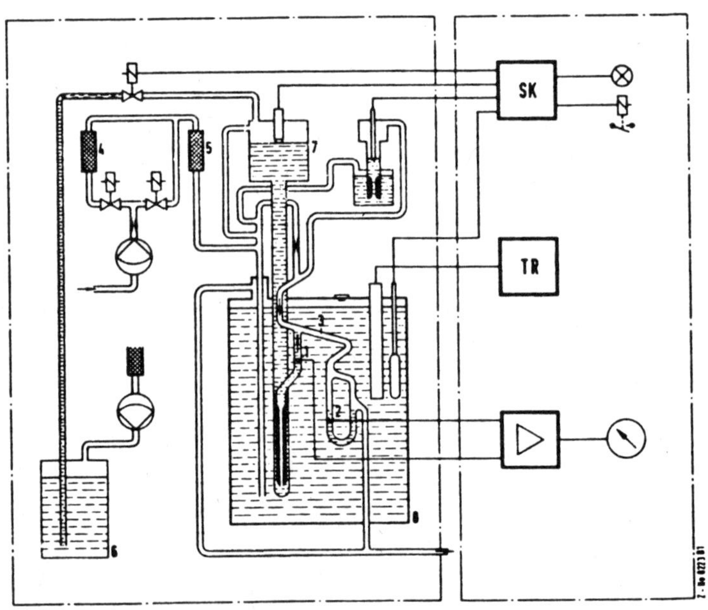

1 Compensation cell
2 Measuring cell
3 Reaction tube
4 Zero gas absorber
5 Selective filter
6 Reagent supply bottle
7 Level regulator
8 Liquid thermostat
SK Status signals
TR Temperature controller

FIGURE 9.9 Picoflux 4 trace gas analyzer. (Courtesy of Hartmann & Braun)

the measuring cell after reaction. An alternating voltage is applied to both cells, and the electrical current flowing through the two cells is a measure of the concentration of the component of interest in the sample gas. The resulting signal is amplified and converted to a 0–20 or 4–20 mA dc output.

The analysis unit is mounted in a temperature controlled enclosure to eliminate the effects of temperature variations on conductivity measurement, reaction rate, viscosity, and density of the reagent. Typical ranges for H_2S measurement fall within the limits 0–2 ppm to 0–750 ppm. The analyzer response time is quoted as 200 sec for a 90% response to a step change and repeatability as ±3% of reading. Sample gas flowrate is approximately 300 ml/min and reagent consumption about 10 liters/month.

The analyzer is designed for installation in non-hazardous locations and consists of two units, the analyzer unit and the control unit, which is designed for rack mounting.

(b) H. Wosthoff K.G. (West Germany)

The Mikrogas Trace Gas Analyzer is designed for continuous measurement of a wide variety of trace components in process gases and in the atmosphere. Typical examples include hydrogen sulfide, carbon monoxide, carbon dioxide, sulfur dioxide, hydrogen chloride, ammonia, and chlorine. A schematic diagram of the measurement system and electronic circuit is shown in Fig. 9.10.

Flowrates of sample gas and reaction solution are kept constant by means of motor-driven proportioning pumps constructed as a multi-plunger system, each plunger performing a specific pumping operation. The sample gas is proportioned by one or several plungers, whereas the reagent is propor-

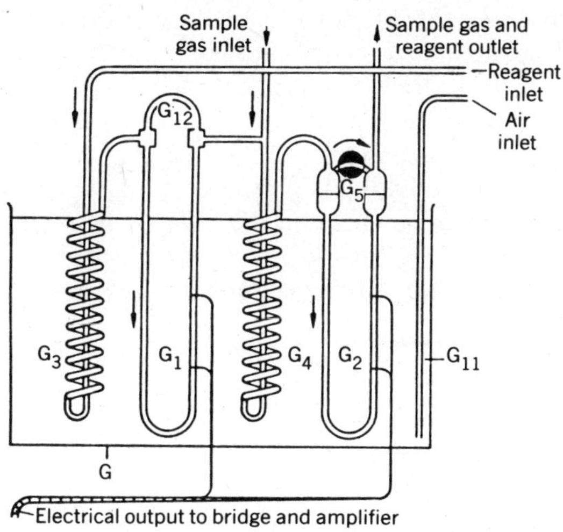

FIGURE 9.10 Mikrogas trace gas analyzer. (Courtesy of Wosthoff)

tioned by a diaphragm pump, which is part of the same system. The measurement assembly proper consists of two glass spirals, two electrode measuring cells, and facilities for gas-liquid separation. The complete assembly, including the proportioning pumps, is immersed in a temperature-controlled oil bath to minimize the effects of temperature variations.

Reagent is pumped by the diaphragm pump at a constant rate from a storage vessel into the first glass spiral, which acts as a heat exchanger. Reagent then passes through the first electrode measuring cell, where its conductivity is measured prior to reaction. Reagent and sample gas mix at the inlet to the second glass spiral, which acts as a mixing and absorption chamber. The gas-liquid mixture then passes to the gas-liquid separator, where reagent flows down through the second electrode measuring cell; here, its conductivity is measured after reaction. Liquid then flows to drain via a liquid trap and sample gas is discharged to vent via one plunger of the pump assembly. Another plunger bubbles air through the oil bath to ensure even temperature distribution.

The two electrode systems are connected into an ac bridge network as shown in Fig. 9.10. The resulting bridge unbalance produced by the reaction of H_2S with the reagent is amplified, and a 0–20 mA dc output proportional to the H_2S concentration in the sample gas is produced.

The analyzer is mounted in a weatherproof enclosure, consisting of two sections, the analysis section and the electronics section, suitable for installation in non-hazardous locations. Facilities for purging can be provided to meet requirements for hazardous area installation.

9.4.6 Specific Ion Type

(a) Delphian Corporation (United States)

Delphian manufactures the Model 130 Hydrogen Sulfide Analyzer, which operates on the specific ion principle and is designed for continuous on-line measurement of the dissolved H_2S concentration in process liquids. Its major application so far has been in monitoring and control of amine-type gas sweetening processes. To the author's knowledge, this system is the only one of its kind commercially available today.

The operating principle is based on the measurements of sulfide ion concentration using an electrode specific to the species. In soluble form, sulfur exists mainly in three states in equilibrium with one another:

1. Dissolve H_2S.
2. Hydrosulfide ion (SH^-).
3. Sulfide ion (S^{2-}).

The total soluble sulfur is the sum of these species, and the relative distribution of these states is a function of the solution pH in accordance with the following equilibria:

$$K_1 = \frac{[H^+][S^{2-}]}{[HS^-]} \cong 10^{-13}$$

$$K_2 = \frac{[H^+][HS^-]}{[H_2S \text{ dissolved}]} \cong 10^{-7}$$

$$K_3 = \frac{[H_2S \text{ (ppm in atmosphere)}]}{[H_2S \text{ dissolved}]} \cong 10^{7}$$

Specific ion electrodes are available for measurement of sulfide ion (S^{2-}) and H^+ (standard pH glass electrode). It follows from these equilibria, that if these two parameters are measured, they can be used for determination of hydrosulfide ion (SH^-) and dissolved H_2S.

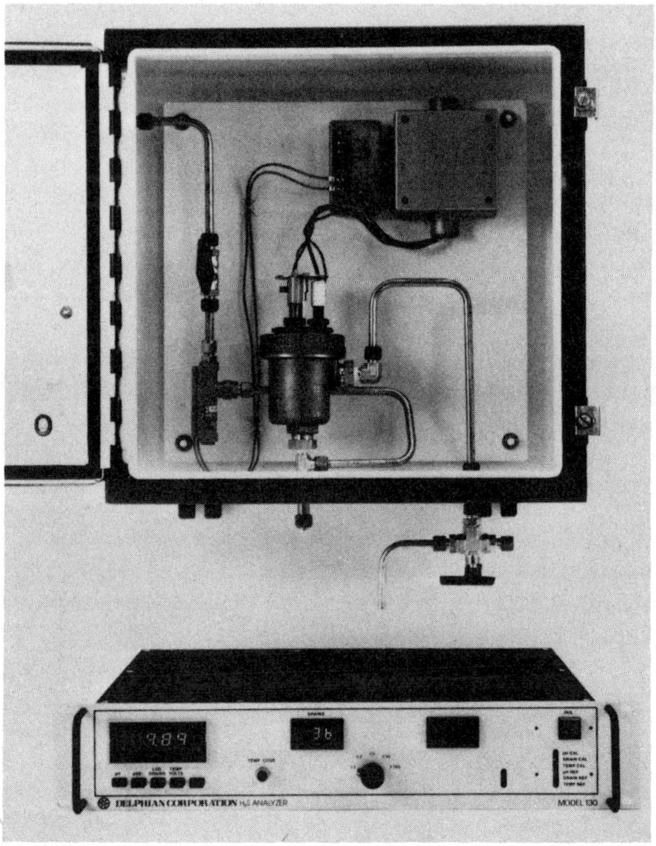

FIGURE 9.11 Model 130 H_2S analyzer, showing sensing unit and control unit. (Courtesy of Delphian Corporation)

The Model 130, using appropriate electronic circuitry, combines the signals from a sulfide ion (S^{2-}) electrode, a pH (H^+) electrode, and a reference electrode to provide simultaneous readings for the three soluble sulfur species and for pH. Also included is a temperature sensor (RTD). The signals from the sensing probes are temperature-compensated electronically and converted to the appropriate engineering units (ppm, grains H_2S/US gal, etc.) which are displayed digitally.

The system consists of two units, the analyzer unit and the control unit, as shown in Fig. 9.11. The analyzer unit contains the electronic signal transmitter, the chemical sensors, and necessary sample piping (with flow switch), all installed in a NEMA 12 enclosure. The transmitter is FM-approved intrinsically safe for NEC Class 1, Group D, Division 1 hazardous locations when used with safety barriers (MTL) between the transmitter and control unit.

The transmitted signals are converted by the control unit to a 4–20 mA dc output proportional to H_2S concentration in the sample liquid. The control unit also displays sample pH, pHS^- (negative logarithm of the hydrosulfide ion concentration), log of H_2S concentration reading, and a voltage reading from the temperature probe. These displays are used for diagnostic tests of the equipment or for trend recording. System response time is quoted as less than 50 sec to 90% of final reading, and repeatability as ±5% of reading (H_2S). The analyzer unit may be located up to 2000 ft from the control unit. Sample stream requirements are 2–10 l/min at a temperature and pressure within the limits 100–131°F and 0–25 psig respectively.

9.4.7 Solid-State Sensor Type

The advent of solid-state semiconductor technology gave a new perspective to atmospheric monitoring of trace gases such as sulfur compounds, ammonia, chlorine, hydrocarbons in general and, more specifically, hydrogen sulfide. It was found that the metallic and non-metallic oxides of elements in Groups III and IV of the periodic table, particularly silicon, change their resistive/conductive properties when exposed to trace quantities of the gases. By suitable treatment, the semiconductor material can be made specific to certain compounds, and the change in electrical properties measured by suitable electronic circuitry to provide a display and output proportional to the concentration of the measured component.

Many of the manufacturers of combustible gas detection equipment are now using this technology for the measurement of H_2S in ambient air at the ppm level. The solid-state H_2S sensor may be used in conjunction with the modular electronics developed for combustible gas monitoring. The subject of combustible gas detection, primarily associated with personnel and equipment safety, is covered in detail in Chapter 12.

The major advantage of the solid-state sensor approach is that there is no sampling system involved. The sensor itself is the only field-mounted unit,

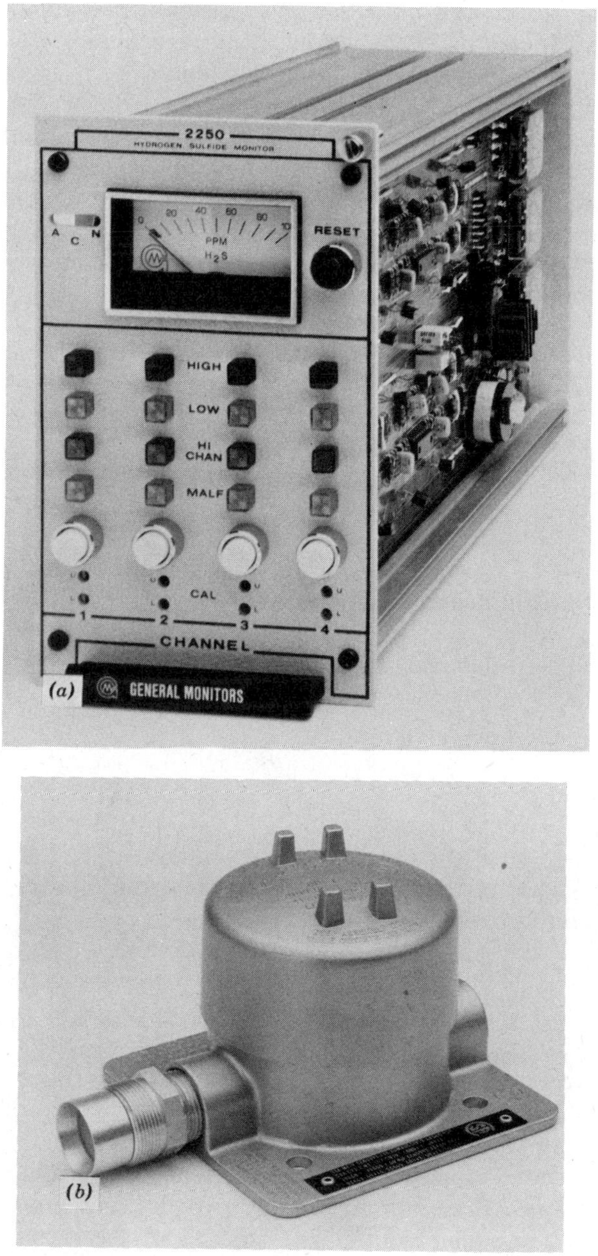

FIGURE 9.12 Model 2250 four-channel H_2S monitor. (*a*) Control module; (*b*) sensing head. (Courtesy of General Monitors Inc.)

and it is connected electrically to the control module, mounted remotely in a non-hazardous location. Shown in Fig. 9.12 are examples of the types of modular control units available for H_2S monitoring, as manufactured by General Monitors Inc.

The actual sensor, shown in Fig. 9.12*b*, is normally mounted within a small flameproof enclosure. Connection between sensor and control unit is by a four-conductor cable, the maximum length of which is determined by a 20-ohm loop resistance (for AWG 18 wire size, this is about 1350 ft).

The solid-state sensor approach raises the question of how selective one can make the device to the particular component of interest, what components constitute poisons to or interfere with the measurement, etc. Improvements will certainly be made as more information becomes available on this relatively new technology. This will result in more confidence on the part of the user.

9.4.8 Photoionization Type

(a) HNU Systems Inc. (United States)

The Model 201 Continuous Trace Gas Analyzer manufactured by HNU uses the "photoionization" principle, so called since it is based on the absorption of UV light by a molecule, leading to ionization in accordance with the following equation:

$$R + h\nu \rightarrow R^+ + e^-$$

where R^+ is the ionized species and $h\nu$ represents a photon with energy greater than or equal to the ionization potential of the species. Hydrogen sulfide is only one of a variety of compounds that may be detected and measured quantitatively using this technique.

A schematic diagram of the measurement system is shown in Fig. 9.13. The sensor consists of a sealed UV light source that emits photons that are energetic enough to ionize many trace species (particularly organics) but do not ionize the major components of air (e.g., O_2, N_2, CO, CO_2, and H_2O). A chamber adjacent to the UV source contains two electrodes. When a positive potential is applied to one electrode, the field created drives any ions formed by the absorption of UV light to the collector electrode, where the current measured is proportional to the concentration of the particular species in the sample. Measurement range is 0.1–1500 ppm and repeatability is quoted as $\pm 1\%$ of full scale. The sensitivity of this technique makes it particularly applicable to trace gas monitoring in ambient air, stack gases, and process streams.

The standard Model 201 analyzer is suitable for panel, bench, or wall mounting in non-hazardous locations. The wall-mounted version (Model 201W) is also available with air purge facilities, flame arrestors, and pres-

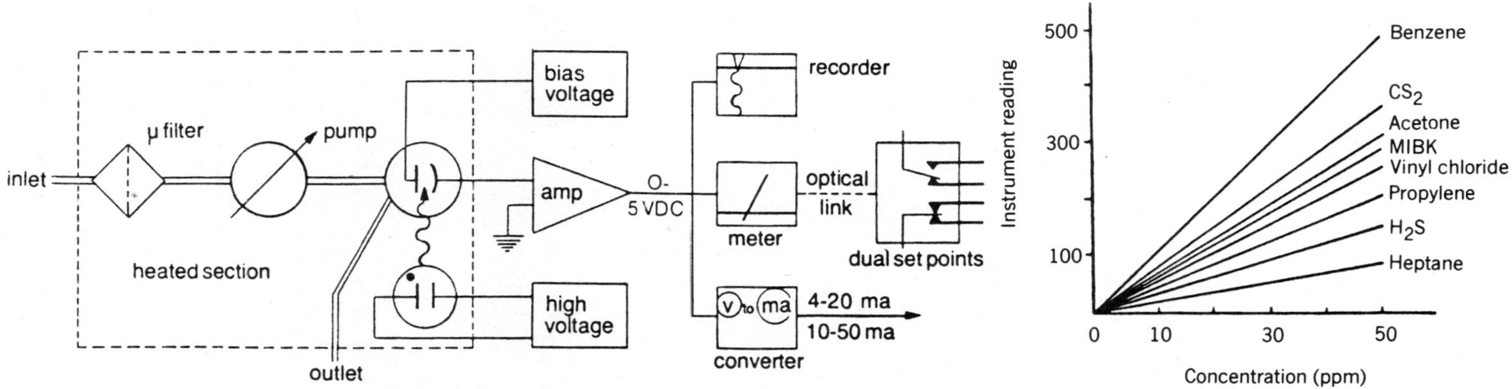

FIGURE 9.13 Model 201 trace gas analyzer. (Courtesy of HNU Systems Inc.)

sure-interlocked power switch to conform to requirements for installation in NEC Class 1, Group D, Division 1 hazardous locations. Multi-point sampling systems are available that convert the basic single-point monitor into a 2–20 multi-channel system. Outputs of 4–20 mA dc or 10–50 mA dc are available.

9.4.9 X-Ray Absorption Type

The technique of X-Ray absorption has been used for many years in the laboratory and on-line to measure percent sulfur in liquid petroleum products. As we saw earlier, laboratory methods other than ASTM D2622 involve accurate measurement of weight and volume and skilled estimation of titrimetric end-point and are time-consuming and subject to operator error and judgement. X-ray absorption has the advantage of being an instrument method with consequently higher precision.

The intensity of X-rays from a source of constant energy and intensity, detected after transmission through a liquid hydrocarbon sample, is dependent upon the density and elemental composition of the sample. For X-ray energies in the region 10–20 KeV, sulfur has a much greater mass absorption coefficient than carbon and hydrogen. At an energy of about 20 KeV, the mass absorption coefficients of carbon and hydrogen are approximately equal, so that changes in X-ray absorption due to changes in the C : H ratio (to the extremes found in petroleum products) are negligible and related to changes in density and sulfur content only. Commercially available sulfur analyzers for laboratory and process applications, operate on the above principle.

(a) Fluid Data Inc. (Great Britain)

Fluid Data recently assumed responsibility for ISM-1 On-Line Sulfur Monitor originally manufactured by Nuclear Enterprises Ltd. Shown in Fig. 9.14, this analyzer uses a secondary emission source combining americium-241 with a silver target to produce a compact, robust source of 22 KeV X-rays with a half-life of 458 years. The radiation energy of the primary source (60 KeV) is sufficiently low to allow the use of simple shielding for personnel protection, and risk of source dispersal in the event of a fire is minimized by providing the radioactive material in the form of a ceramic enamel of very high melting point.

The analyzer is supplied as a single unit with the source holder and flow cell mounted on the outside of the electronics enclosure, which is suitable for purging to meet requirements for installation in hazardous locations (BASEEFA certified). The flow cell is fitted with beryllium windows 1.6 mm thick, capable of withstanding sample pressures up to 300 psig and temperatures up to 100°C. Standard flange connections are provided and the flow cell

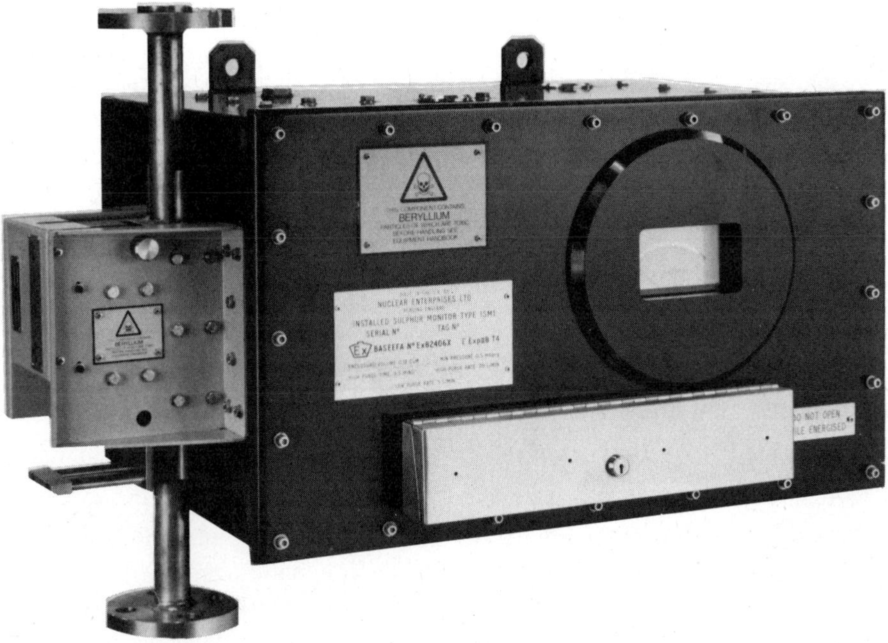

FIGURE 9.14 Type SM-1 on-line sulfur monitor. (Courtesy of Nuclear Enterprises Ltd. and Fluid Data Inc.)

may be removed without disturbing the source. Steam tracing or electrical heating may also be provided. For density measurement, the Solartron density meter is standard, but other alternatives are available.

The detector is a high-resolution proportional counter Type RSG61. All electronic circuits are modular on printed-circuit boards. The count rate is processed in a simple analog computer to give a direct reading of sulfur content on an integral meter, and outputs of 0–10 V dc or 4–20 mA dc are available for remote recording. Integrating time may be preset between 50 and 400 sec, depending on response time and precision required. All controls and test points are readily accessible from the front of the enclosure.

Measurement range is 0.5–6.0 wt% sulfur with a precision of ±1% of full scale. Variations of C : H ratio from 5 : 1 to 10 : 1 do not affect precision. Density range is 0.6–1.1 g/cm^3 measured with a precision of ±0.1% and sulfur reading corrected automatically. Calibration may be verified using a specially designed four-position manually operated shutter installed in the measuring head. In the two "calibration" positions, aluminum absorbers, corresponding to a sample of specified sulfur content, are inserted into the X-ray beam. This permits periodic calibration verification without the need for special samples of known sulfur content.

(b) Yokagawa Corporation (United States)

Yokagawa manufactures the System 8820 Sulfur-in-Oil Analyzer, which is based on the X-ray absorption principle, using the Density Analyzer Model VD6DF for density compensation. The analyzer system and its component parts are shown in Fig. 9.15.

The sulfur detector consists of the liquid cell section, radiation source section, and ionization chamber, mounted in a cast-iron enclosure designed for installation in NEC Class 1, Groups C and D, Division 1 hazardous locations, as shown in Fig. 9.15*a*. Fig. 9.15*b* shows the radiation source capsule and assembly. The X-rays emitted from an americium-241 source bombard a target, producing secondary X-rays at 20 KeV. These penetrate a beryllium back-plate 5 mm thick and a thin Teflon plate, pass through the sample cell, and enter the ionization chamber. In addition to its explosion-proof safety features, the sulfur detector is designed to meet all requirements for personnel protection against radiation sources.

The current from the ionization chamber detector is the input to the sulfur converter, where it is compared with the zero set signal from a standard current source. The difference between these signal currents is the input to the pico-diode electrometer, where the differential current (10^{-12} to 10^{-11} A) is amplified to provide an output signal to the S-Set unit.

There, the signal is linearized and subjected to density compensation at the zero point by the signal from the density converter, and converted to a voltage signal proportional to the sulfur concentration and density. After span adjustment, the signal is transmitted to the divider unit. This unit divides the sulfur concentration signal from the S-Set unit (function of sulfur and density) by the density signal also from the S-Set unit (calibrated at the

FIGURE 9.15 System 8820 sulfur-in-oil analyzer. (*a*) Sensing cell. (Courtesy of Yokagawa Corporation)

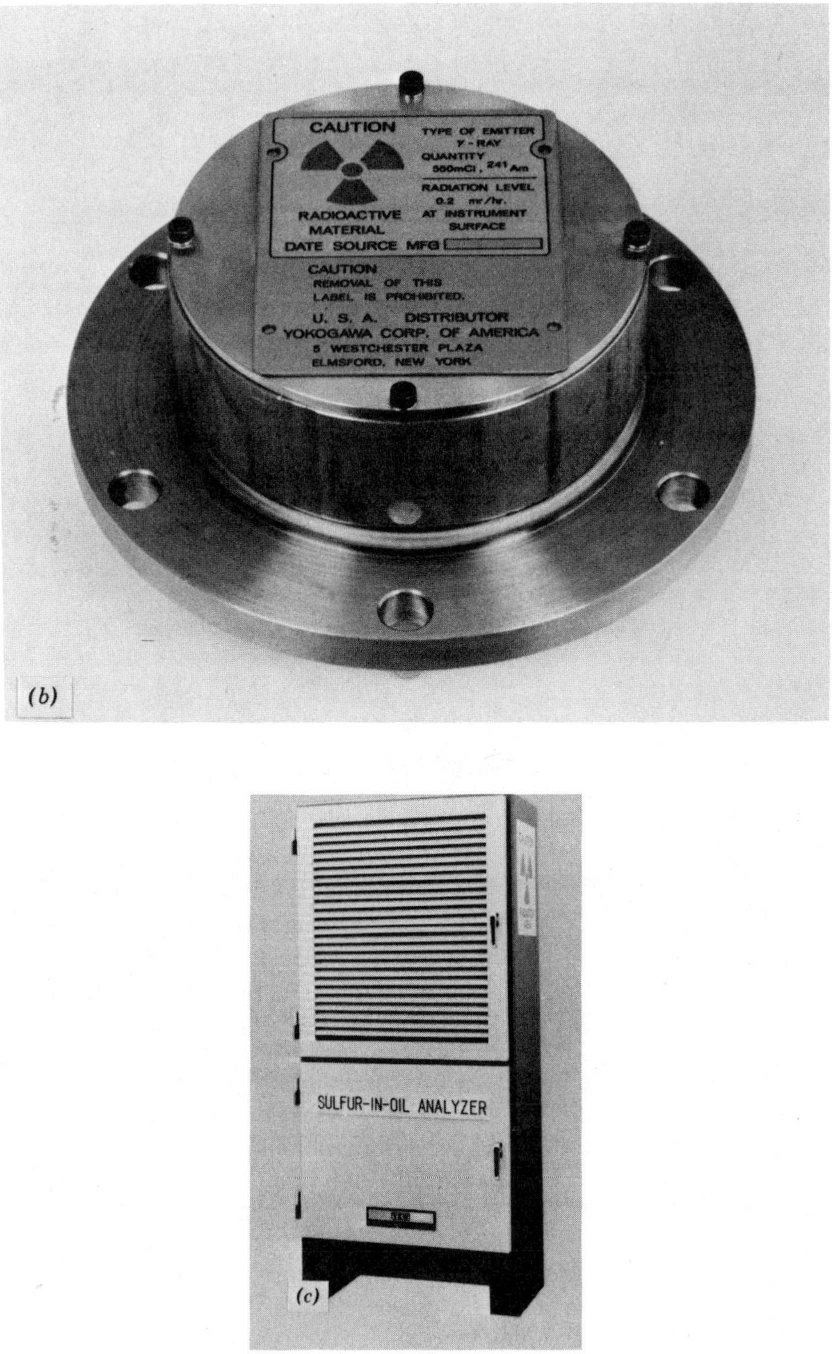

FIGURE 9.15 System 8820 sulfur-in-oil analyzer. (*b*) Radioactive source assembly; (*c*) cabinet mounted installation. (Courtesy of Yokagawa Corporation)

(*d*)

FIGURE 9.15 System 8820 sulfur-in-oil analyzer. (*d*) Explosion proof S-set unit. (Courtesy of Yokagawa Corporation)

sample temperature of the sulfur detector). The divider unit produces a 4–20 or 10–50 mA dc or 1–5 V dc output signal proportional to sulfur concentration.

The Model 8824 Processor is a rack mounted unit containing the sulfur converter, S-Set unit, density converter, and divider, designed for installation in non-hazardous locations. If the transmission distance from analyzer unit to control unit is greater than 200 m, the sulfur converter is mounted in an explosion-proof case (Fig. 9.15*d*) installed adjacent to the analyzer unit in the field. When a two-range system is required, an optional multi-unit is provided.

The most sensitive range of the system 8820 is 0–0.5 wt% sulfur, with a repeatability of 0.01 wt%. The density measuring system VD6DF has a sensitivity of 0.0005 g/cm^3.

(c) *Laboratorium Prof. Dr. Berthold (West Germany)*

Berthold has manufactured process analyzers for measurement of sulfur in oils in the low percent range for many years, using the absorption of low energy gamma rays. The latest instrument is the Model LB375 Continuous Sulfur Analyzer.

A ^{244}Cm source is sealed in a capsule with ISO classification and installed in front of a sample flow chamber, which has two windows made of 1.2 and 2.5 mm thick beryllium panes. The chamber has an explosion-proof design, suitable for installation in Zone 1 hazardous locations. The source emits low-energy gamma rays with energies between 12 and 21.4 keV, with a mean energy of 18 keV. The absorption coefficient for sulfur is about 17 times higher than that of oil; 1% sulfur weakens the passing radiation by about 20–30% depending upon which sample flow cell is used. Radiation passing through the sample cell is detected by a scintillation counter. The source, sample cell, and scintillation counter are mounted together as a single unit.

The radiation absorption is influenced also by variations in sample density and carbon/hydrogen ratio. Changes in C/H ratio can be caused by variations in sample composition and temperature, but may be considered negligi-

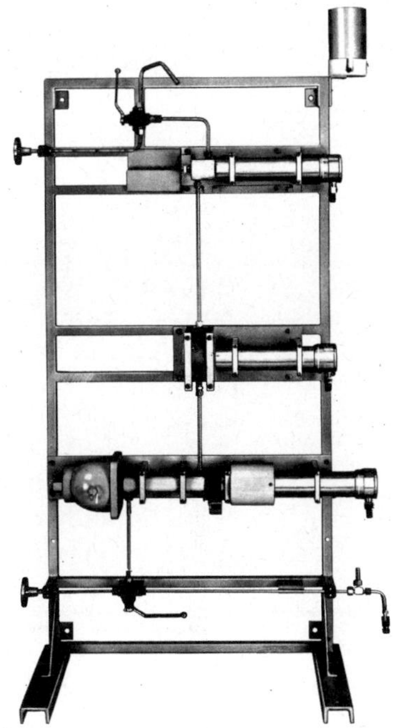

FIGURE 9.16 Berthold sulfur analyzer: rack mounted sensing system. (Courtesy of Laboratorium Prof. Dr. Berthold)

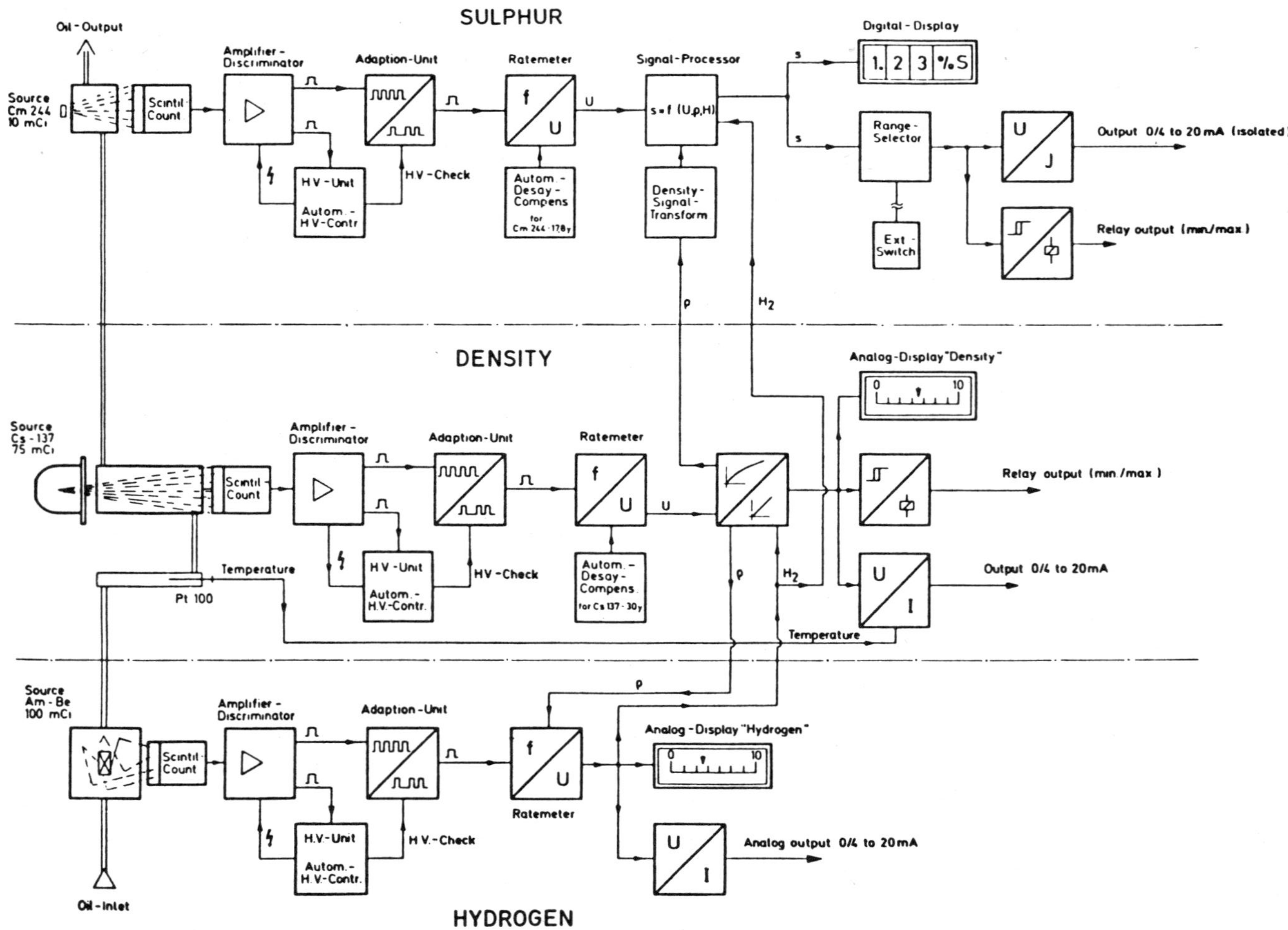

FIGURE 9.17 Berthold sulfur analyzer: block diagram of electronics system. (Courtesy of Laboratorium Prof. Dr. Berthold)

ble compared with the influence of density. A density change of 1% weakens the radiation by the same amount as a sulfur content change of 0.06%; therefore, density compensation is necessary.

The sample density is measured by an independent, non-contacting sensor. The source in this case is cesium 137. The radiation detector is a scintillation counter. The output of the density sensor is an input to the analog portion of the sulfur measuring system.

If compensation for C/H ratio is required, a third sensing system is installed, consisting of a fast-neutron source, sample cell, and scintillation counter, all mounted as a single unit.

An example of an analysis system consisting of the three units described is shown in Fig. 9.16, mounted on a free-standing frame. Sample enters the system at the base and flows upward through the density cell, C/H ratio cell, then the sulfur cell, exiting at the top of the system.

The analyzer control unit (measurement system) is suitable for rack or panel mounting in non-hazardous locations only and may be installed at distances up to 1500 m from the analysis unit. It is modular in design and consists of three separate parts—the density measurement system, the hydrogen content measurement system, and the sulfur measurement system. A block diagram of the measurement electronics is shown in Fig. 9.17. Source decay compensation is provided for the sulfur and density sources. Sulfur content is displayed on a digital meter with a resolution of 0.1% sulfur. Analog displays are provided for density and hydrogen content. Current outputs (0–20 or 4–20 mA dc) are available for remote recording of all three variables.

BIBLIOGRAPHY

Beamish, M. C. J. Controlling the Claus process. *Western Research and Development,* Calgary, Canada (Sept 1982).

Cuff, B. Plant optimization in practice. *Oil Week,* **28,** 40–46 (1978).

H_2S measurement in lean amine solutions. *Delphian Corporation Tech. Note No. 23.*

Kaminski, C. L., and Poli, A. Electrochemical or solid state H_2S sensors: Which one is right for you? *InTech* (June 1985).

Palm, J. W., and Carruthers, N. M. Guidelines aid control of SRU tail gas composition. *Oil & Gas J.* (Nov 20, 1978).

Paskall, H. G. Capability of the modified Claus process. *Western Research and Development,* Calgary, Canada (1979).

Saltzman, R. S., and Dell, C. G. Photometric analyzer systems for monitoring low levels of hydrogen sulfide. *Proc. ISA Symp.,* Des Plaines, ILL (May 1984).

10

Spectrometric Measurements

10.1 INTRODUCTION

Spectroscopy is one of the oldest of the specialized scientific techniques, but in the past forty years it has probably seen the greatest change in sophistication and improvements in technology.

The first basic optical requirements for spectroscopic observation were presented by Newton [1] in the early 18th century. A few years later, Bouguer [2] observed the change in intensity of a beam of light as it passed through various thicknesses of absorbing material. In the 19th century, Bunsen developed a simple spectroscope and established that the emission and absorption spectra of an element are characteristic of that element and may be used for identification purposes.

During the latter part of the 19th century, the discoveries of eminent spectroscopists using emission and absorption techniques laid the foundations for the early 20th century theories of atomic and molecular structure [7]. Investigation of the ultraviolet region, discovered by Ritter in 1801, as well as the visible region, was made possible by the substitution of silica for glass in spectroscopes. The infrared region, first recognized by Herschel in 1800, was studied extensively by Coblentz [3], resulting in the recognition that functional groups in organic molecules have characteristic absorption frequencies.

The period from 1905 to the mid 1930s saw an explosion of scientific knowledge that was applied vigorously to spectroscopic studies in all regions of the electromagnetic spectrum. Microwave spectroscopy grew out of the radar techniques used during World War II. The so-called resonance techniques such as nuclear magnetic resonance (NMR) and electron spin resonance (ESR) were introduced shortly thereafter. The introduction of the laser in the early 1950s had a profound effect on the development of Raman spectroscopy.

During this period, these scientific developments were aided by advances in technology that made possible the production of more sophisticated instrumentation. The advent of the computer provided a powerful tool for rapid data acquisition, an essential feature of some spectroscopic measurements.

Spectroscopy, as we know it today, is a vast subject with many facets. Each individual analytical technique is a specialized field in its own right and it is beyond the scope of this book to deal with them in great detail. The reader is therefore referred to the very extensive and excellent literature on each subject.

This chapter provides the reader with the historical background and basic theoretical principles of the various techniques and the major laboratory applications, along with an understanding of how these principles are used for on-line process analysis. Some of the techniques described are used only in laboratory research. However, because of the rapid development of the field in recent years, it is expected that there will be further development directed toward the use of these techniques for on-line process analysis.

10.2 THEORETICAL CONSIDERATIONS

This section covers the fundamental theory associated with the study of spectroscopy. Additional theory relevant to the specific types of spectroscopy is discussed in Section 10.3.

10.2.1 The Quantum Theory

During most of the 19th century, scientists assumed that energy was infinitely subdivisible and could be gained or lost by matter in indefinitely small amounts, with no lower limit to the quantity of energy that could pass to or from the matter. Classical mechanical theory can be used to provide an accurate and consistent description of the behavior of physical systems that are large compared with atoms and molecules. However, this approach did not provide a theoretical explanation for the phenomena associated with cavity radiation and the manner in which energy is distributed among the various wavelengths in the radiation emitted by a hot body. A number of capable physicists advanced theories based on classical physics, but only with limited success.

The quantum theory, evolved by Max Planck [4] in 1900, made the assumption that energy can only be absorbed or emitted by matter in integrals of an elementary, indivisible quantum of energy. This energy quantum, the smallest unit in which energy can be exchanged between light and a vibrating molecule, is exactly proportional to the frequency of the radiation and is given by,

$$E = nh\nu \tag{10.1}$$

where E is the quantum of energy in Joules, ν is the vibration frequency in Hz, h is a constant (now called Planck's constant, the currently accepted value of which is 6.6262×10^{-34} Joules-sec), and n is a number that can take on only integral values (now called a quantum number).

This radical and novel theory was accepted with considerable reserve initially; indeed, Planck himself resisted wholehearted acceptance for many years. Nevertheless, the theory was applied to explain various other phenomena associated with energy emission. For example, light falling on alkali metals causes electrons to be emitted. In 1902, Lenard [5] discovered the vitally important fact that the velocities of the electrons emitted depend upon the frequency of the light and not upon its intensity.

Planck's ideas soon received reinforcement from Einstein [6], who, in

1905, applied the concepts of energy quantization to the *photoelectric effect*. Monochromatic light falling on a metal plate liberates photoelectrons that can be detected and measured as a current if they are attracted to a metal cup by means of a potential difference applied between the plate and the cup. If the potential difference is made large enough, the photoelectric current reaches a limiting value at which all photoelectrons emitted from the metal plate are collected by the cup. This phenomenon could not be explained in terms of the wave theory of light. Einstein succeeded in explaining the photoelectric effect by making a remarkable assumption for that time—that energy in a light beam travels through space in concentrated "bundles" called *photons*.

The energy of a single photon is given by,

$$E = h\nu \tag{10.2}$$

Each photon is capable of giving the whole of its energy to an electron. Since the number of electrons emitted is proportional to the light intensity, the intensity must be proportional to the number of photons arriving per second on the metal.

Planck believed that light, although emitted from its source discontinuously, travels through space as an electromagnetic wave. Einstein's photon theory suggested that light traveling through space behaves not like a wave at all but like a stream of particles. Experimental work by Millikan verified Einstein's ideas in every detail.

Light is unquestionably propagated as if it were wave motion; this is proved by the phenomena of interference and diffraction. However, when light interacts with matter, it behaves far more like a stream of particles carrying definite amounts of energy. In large-scale material phenomena there is nothing resembling this combination of wave propagation and arrival at a given point as a concentrated package of energy. It is important to understand that all electromagnetic radiation, including X-rays, ultraviolet (UV), visible and infrared (IR) light, has this dual wave-particle nature.

10.2.2 The Electromagnetic Spectrum

As we have seen, the absorption and emission of energy in the electromagnetic spectrum occurs in discrete packets called photons. The intensity of a beam of radiation is characterized by its radiant power, which is proportional to the number of photons per second propagated in the beam. A beam of radiation of one discrete wavelength is said to be *monochromatic;* a beam of several wavelengths is said to be *polychromatic*.

For spectroscopic measurements, it is more convenient to consider radiation as a continuous wave motion in which the wavelength represents the interval between nodes in the wave pattern. The relationship between wavelength and frequency is given by:

$$\nu = \frac{c}{\lambda} \tag{10.3}$$

where ν is the frequency in Hz, λ is the wavelength in cm, and c is the velocity of light in a vacuum, the currently accepted value of which is 2.9979×10^{10} cm/sec.

In spectroscopy, wavelength is expressed in a variety of different units because individual areas of the field were developed independently:

angstrom (Å) $= 10^{-10}$ m $= 10^{-8}$ cm
micrometer (μm) $= 10^{-6}$ m $= 10^{-4}$ cm $=$ 10,000 Å
nanometer (nm) $= 10^{-9}$ m $= 10^{-7}$ cm $=$ 10 Å
millimicron (mμ) $= 10^{-9}$ m

Another commonly used unit is the wavenumber (ν'), which is the reciprocal of the wavelength in cm (i.e., cm^{-1}). Since $\nu = c/\lambda$ and $\nu' = 1/\lambda$, then $\nu = c\nu'$ and the Planck equation becomes,

$$E = hc\nu' \tag{10.4}$$

Equations 10.1 and 10.4 show that there is a definite amount of energy corresponding to any wavelength.

The energy units most frequently used in spectroscopic measurements are electron volts per atom or per molecule, kilocalories per molecule, and kilojoules per molecule. The conversion factors are,

$$1 \text{ eV/mol} = 23.06 \text{ kcal/mol} = 96.4 \text{ kJ/mol}$$

The relationship between wavelengths, wavenumbers, and energy is shown in Table 10.1.

TABLE 10.1 Relationship Between Wavelength, Wave Number, and Energy

Wavelength (nm)	Wave Number (cm^{-1})	Energy (kcal/mole)
0.1	10^8	3×10^5
100	10^5	300
200	5×10^4	150
400	2.5×10^4	70
800	1.25×10^4	35
2000	5×10^3	15
7000	1.4×10^3	4.1
25,000	400	1.1

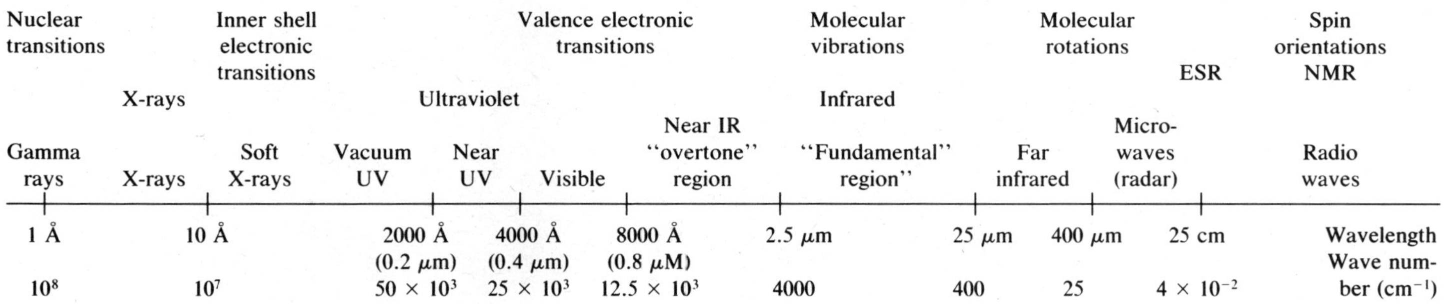

FIGURE 10.1 The electromagnetic spectrum.

The various regions of the electromagnetic spectrum are shown in Fig. 10.1, together with the nature of the energy changes involved. Visible light represents only a small part of the electromagnetic spectrum generally considered to extend from 380 to 780 nm. The ultraviolet region extends from 185 nm to the visible region. Shorter wavelengths are considered to be the far (or vacuum) UV region. The extreme UV overlaps the soft X-ray portion of the spectrum. The infrared region of the spectrum consists of three portions—the near IR (0.8–2.5 μm), the mid IR (2.5–25 μm) and the far IR (25–450 μm). Regions of longer wavelength include the microwave region (400 μm–25 cm) and the radio region (above 25 cm).

10.2.3 The Laws of Photometry

Photometry is based on two fundamental laws. Bouguer [2] (1729) and Lambert (1760) found that the energy absorbed by a thin layer of material is dependent upon the type of substance and the frequency of the incident radiation and is proportional to the thickness of the absorbing material.

When a beam of radiant energy impinges upon a substance, several things may happen to it:

- It may pass through the substance with little absorption taking place and therefore little energy loss.
- The direction of propagation of the beam may be altered by reflection, refraction, or diffraction.
- The radiant energy may be absorbed by the substance entirely or in part.

Referring to Fig. 10.2, the rate of decrease in radiant power of a beam of parallel, monochromatic radiation with a path length *b* through the absorbing substance is proportional to the radiant power (intensity) of the beam. The intensity will be diminished by geometrical (or exponential) progression. That is, if a certain thickness absorbs 50% of the radiation, a thickness that follows the first and is equal to it will absorb another 50% of the radiation.

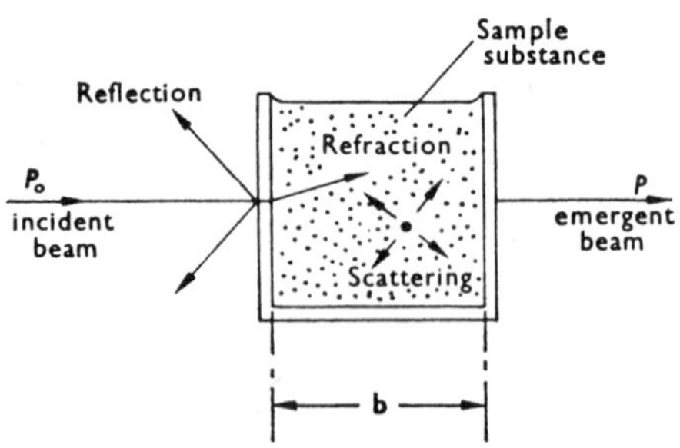

FIGURE 10.2 The laws of photometry.

Thus, the intensity will be reduced to 25% of its original value. Therefore,

$$\frac{\delta P}{P} = k\delta b \qquad (10.5)$$

Integrating and putting $P = P_0$ when $b = 0$ gives,

$$\ln\left(\frac{P_0}{P}\right) = kb$$

where k is a constant, or

$$P = Pe^{-kb} \qquad (10.6)$$

Changing to base 10 logarithms,

$$2.303 \log_{10}\left(\frac{P_0}{P}\right) = kb$$

or

$$P = P_0 e^{-0.434kb} \qquad (10.7)$$

Bernard (1852) and Beer (1852) found that the intensity of a beam of parallel, monochromatic radiation decreases as the concentration of the absorbing constituent in the substance increases. Therefore,

$$2.303 \log_{10}\left(\frac{P_0}{P}\right) = k'C \qquad (10.8)$$

where k' is a constant and C is the concentration of the absorbing constituent.

Equations 10.7 and 10.8 may be combined and written with a single constant:

$$\log_{10}\left(\frac{P_0}{P}\right) = abC \qquad (10.9)$$

where a is the absorptivity, a constant depending upon the wavelength of the radiation and the nature of the absorbing constituent, and C is the concentration of the absorbing constituent in g/l. This combined law is known as the Beer-Lambert Law.

$$\text{Absorbance } (A) = \log_{10}\left(\frac{P_0}{P}\right) = abC \qquad (10.10)$$

$$\text{Transmittance } (T) = \frac{P}{P_0} \tag{10.11}$$

The units of *a,* which must be specified, will depend upon the units of *b* and *C* (it is common practice to use a 1-cm path length and express C in mole/l). In order to obtain some consistency in a field where much discrepancy existed, the above terms were endorsed by the advisory board of Analytical Chemistry in 1962.

10.3 TYPES OF SPECTROSCOPY

10.3.1 Infrared Absorption Spectroscopy

(a) Theoretical Basis

The region of the electromagnetic spectrum extending from 0.8 μm to about 400 μM is referred to as the infrared. Spectra in the 0.8–50 μm region originate primarily from the vibrational stretching and bending modes within molecules. Most organic and inorganic materials show absorption; in all but a few cases, this absorption includes several characteristic wavelengths.

The modes of vibration of atoms in polyatomic molecules can best be visualized using a mechanical model of the molecule, such as that for acetaldehyde shown in Fig. 10.3. The atoms are represented by spheres, their weight being proportional to the corresponding atomic weight and arranged in accordance with the space geometry of the molecule. Bonds between atoms are represented by springs that are proportional to the bonding forces within the molecule. In Fig. 10.3, Hookes Law may be applied to the calculation of the stretching vibration frequency of the two bonded atoms.

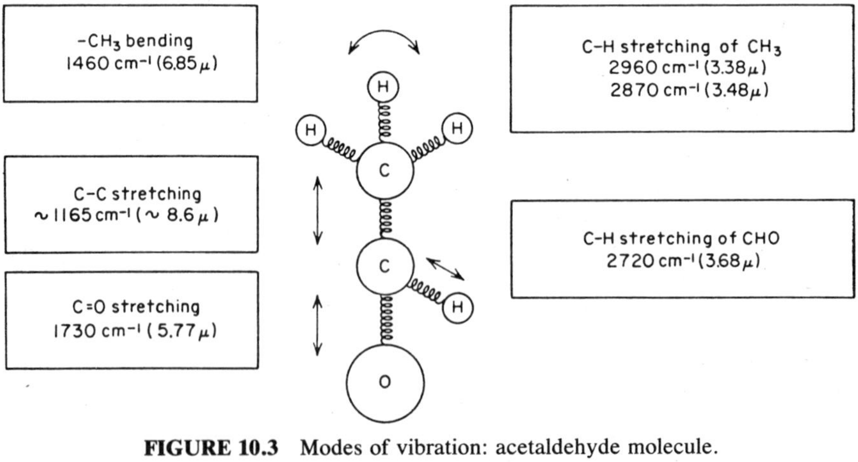

FIGURE 10.3 Modes of vibration: acetaldehyde molecule.

$$\nu = k\sqrt{\frac{f}{M_0}} \quad \text{or} \quad \lambda = k'\sqrt{\frac{M_0}{f}} \tag{10.12}$$

where

$$M_0 = \frac{M_1 M_2}{M_1 + M_2}$$

and f is the force constant.

The vibration frequency is thus directly related to the force constant of the bond, which is a measure of its resistance to vibration or, roughly, to the bond energy.

Three coordinates in space are specified for each nucleus to describe the normal modes of vibration. If there are n nuclei, the system as a whole is said to have 3n degrees of freedom of motion. Three degrees must be deducted for translation of the molecule as a rigid unit and a further three for rotation of the molecule about each principal axis (two if the molecule is linear). Therefore, a non-linear molecule has $3n - 6$ degrees of freedom of vibrational motion ($3n - 5$ for linear molecules). However, this number of fundamental absorption bands may not necessarily exist in the IR spectrum of the molecule. Some of these vibrations may be IR-inactive; in addition, the symmetry of the molecule may be such that certain pairs (or triads) of fundamental frequencies may be exactly equal. This gives rise to what are known as *doubly* (*or triply*) *degenerate bands,* which are observed as only one band.

The fundamental stretching and bending vibrations of the H—C—H group in hydrocarbon molecules are shown in Fig. 10.4. These usually pro-

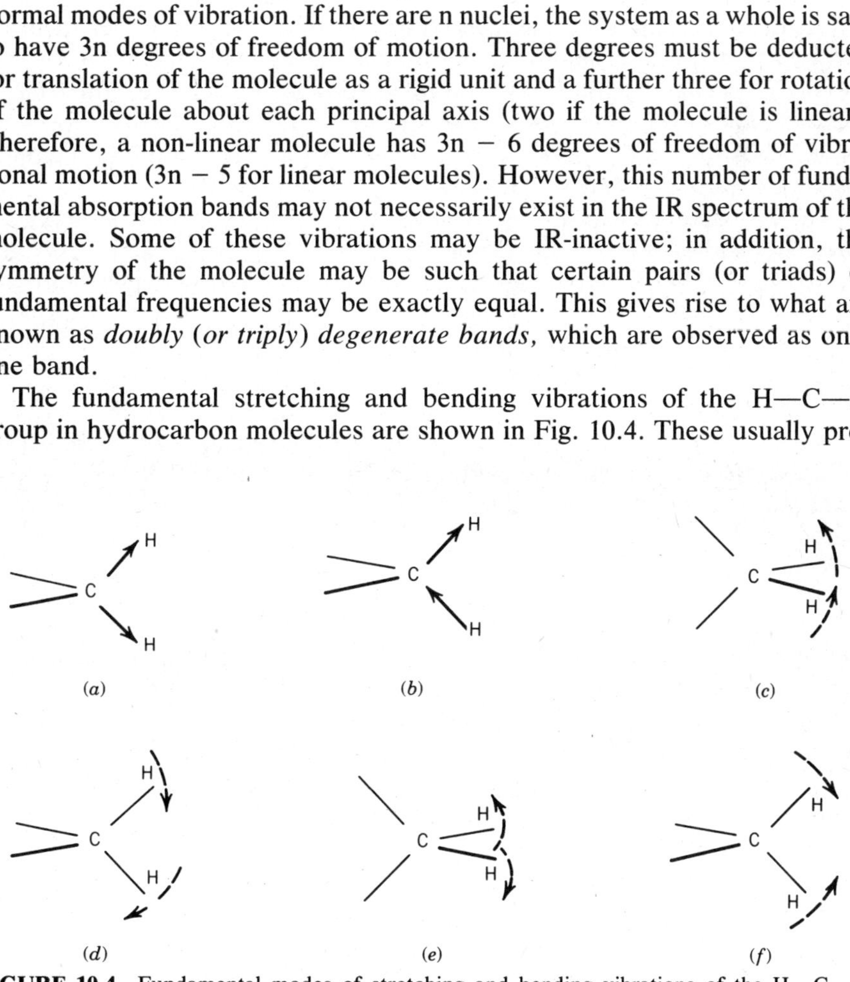

FIGURE 10.4 Fundamental modes of stretching and bending vibrations of the H—C—H bond in hydrocarbons.

duce two absorption bands because of symmetrical and asymmetrical movements of the same vibration type, as follows:

(a) Symmetrical stretching, where the H atoms move in unison.
(b) Asymmetrical stretching, where the H atoms move in opposite directions.
(c) Out-of-plane bending or wagging.
(d) Asymmetrical in-plane bending or rocking.
(e) Out-of-plane deformation or twisting.
(f) In-plane deformation or scissoring.

It can therefore be seen that for complex molecules many different vibrational modes are possible, each giving rise to an absorption band in the infrared spectrum. In addition, the number of bands is increased by harmonic (overtone) and combination vibrations of much lower intensity. A combination band frequency is the sum of or the difference between the frequencies of two fundamental or harmonic bands. It is from these combination bands that the uniqueness of the IR spectra arises.

The fundamental feature of IR spectra, of great value to the analytical chemist, is that most covalent bonds have characteristic and essentially invariant absorption wavelengths, no matter what their molecular environment is. This means that the presence of a band in the IR spectrum (particularly below 6.5 μm) indicates the presence of a particular bond in the molecule, and vice versa. Therefore, group frequencies are a great help in identifying molecules from the IR spectra. IR analysis is thus most straightforward when the component molecules of the sample have significantly different atomic groups. The portion of the IR spectrum offering the best discrimination between molecules is the region between 7 and 15 μm, often called the *fingerprint region*.

(b) Laboratory Instruments

Grating Spectrometers. The standard laboratory IR instrument is a double-beam optical null spectrometer, shown schematically in Fig. 10.5. The IR beam is separated into its component wavelengths by a diffraction grating. A heated "black-body" source emits radiation at all IR wavelengths and the detector responds roughly equally at all wavelengths. The complete spectrum is scanned by rotating the grating and measuring the radiation intensity passing through its exit slit. Sample cells are placed in the "sample" beam and, as the spectrum is scanned, the instrument drives an attenuator into the "reference" beam until the detector sees equal energy from both beams. With the introduction of computerized data handling, these instruments are able to carry out both qualitative and quantitative analyses.

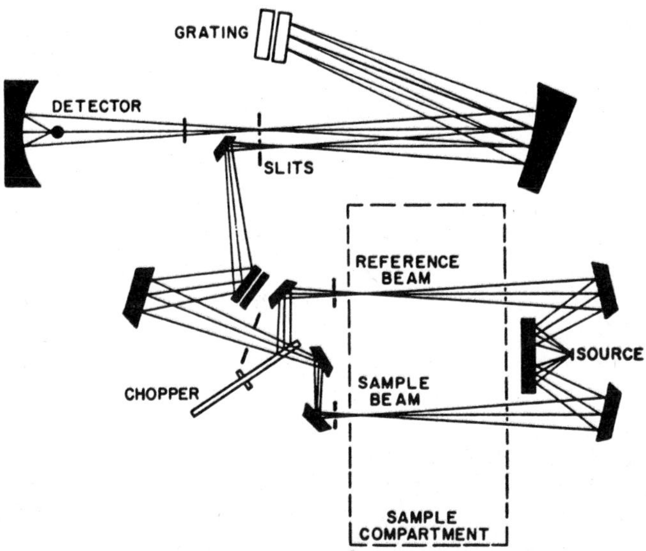

FIGURE 10.5 Double-beam optical null spectrometer.

Filter Spectrometers. These single-beam instruments use a circular variable filter to select the measurement wavelength; with the addition of microcomputer-based control, they are capable of providing quantitative analysis of mixtures containing up to 10 components. Measurements are made at each wavelength in sequence. Although spectral resolution is lower than that obtained with a grating spectrometer, the signal-to-noise ratio is higher. This operating principle is also used in on-line IR analyzers, whereas the grating-type instruments have not been used on-line until quite recently.

10.3.2 Fourier Transform Infrared (FTIR) Spectroscopy

The subject of FTIR spectroscopy is treated separately here, in view of its importance as a high-speed, highly sensitive analytical technique. The Fourier transform is named for the French mathematician J. B. J. Fourier, who first presented its principles in 1822. Basically, any property varying periodically with time can be analyzed by splitting the property into its harmonic components. Sound waves are a classical example. The complex wave emitted from a specific instrument can be imitated by mixing together all harmonic oscillations related to the note in question, but varying in amplitude and phase (this is, in effect, Fourier synthesis). The human ear and brain perform something akin to a Fourier analysis on the sound of an orchestra when the individual tries to identify individual instruments playing. The Fourier transform technique is of course used extensively in the computer control and instrumentation fields for the development of control system theory.

As mentioned earlier, the traditional method of obtaining IR spectra is to use a dispersive spectrophotometer (monochromator), the heart of which is a diffraction grating that spatially disperses the radiation from the source into its component frequencies. The second method, widely accepted today, is to use an instrument known as an interferometer. The origin of the interferometer can be traced back to the turn of the century, when Michelson first described the instrument that bears his name. Although both techniques produce IR spectra, their methods of measurement are quite different. Where the monochromator is designed to measure a spectrum resolution element by resolution element, an interferometer is designed to measure optical interference.

The Michelson design of interferometer is shown schematically in Fig. 10.6. This design has two elements in common with the dispersive spectrophotometer—the source and the detector. In place of the monochromator is the interferometer module. This includes the source, detector, beam splitter (B), and two mirrors (M_1 and M_2), which form the two arms of the interferometer.

The beam splitter is essentially a single crystal of potassium bromide coated with germanium, designed to transmit 50% of the radiation to the mirror in one arm of the interferometer and reflect 50% of the radiation to the other mirror. The fixed mirror reflects the radiation back to the beam splitter along a fixed and constant path. The moving mirror, on the other hand, is designed to create path lengths different from the fixed path arm of the interferometer. The effect of moving the mirror is to introduce a path differ-

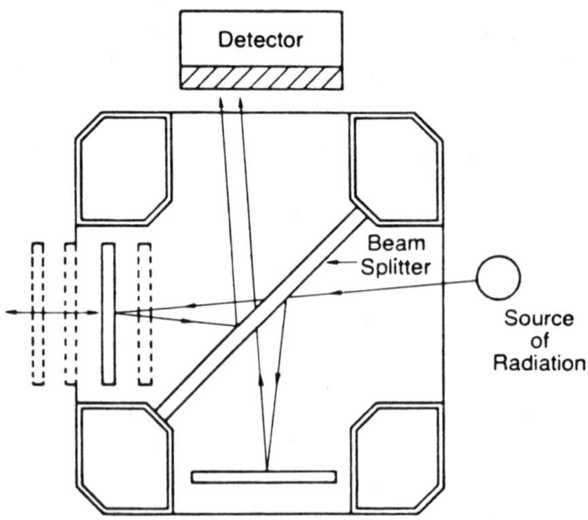

FIGURE 10.6 Michelson interferometer.

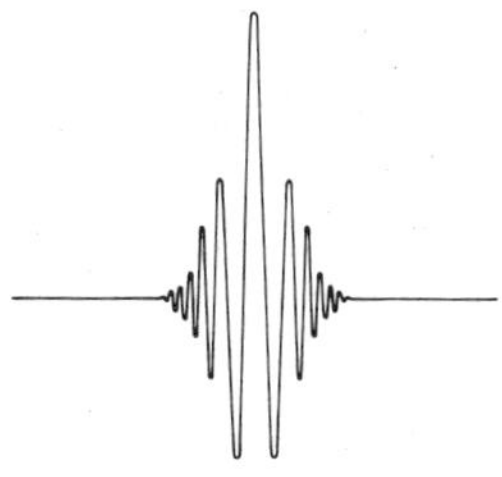

FIGURE 10.7 Typical interferogram.

ence, and hence a phase difference, between the two beams on recombination at the splitter, so that interference between the beams occurs.

For a monochromatic source and a mirror moving at constant velocity, the signal measured at the detector (known as an interferogram) is sinusoidal. For a polychromatic source, the interferogram is the resultant of the sinusoidal signals of the individual wavelengths emitted by the source. Only when the fixed and moving mirrors are equidistant are the signals emitted by the source in phase. At this point, the interferogram has a very large amplitude; this region is often known as the *centerburst*. A typical interferogram is shown in Fig. 10.7.

The interferogram results because all of the incident wavelengths are modulated differently due to velocity and mirror position. To avoid distortion in the interferogram, the moving mirror must be maintained plane and parallel to the fixed mirror; alignment errors will affect the intensity at particular points in the interferogram. Since the detector is sensitive to the modulation frequency (again affecting intensity), constant mirror velocity must also be maintained.

If the radiation from a monochromatic source (line source) is passed through a suitable interferometer, the detector output will be a cosine wave of constant frequency and intensity for the distance the mirror is moved. This process is, in essence, equivalent mathematically to the Fourier transform. The only difference is that the Fourier transform extends from $-\infty$ to $+\infty$, where as the interferogram is limited by the distance travelled by the mirror. Therefore, a Fourier transform on the equation of the line sources could be performed to generate the equation of the cosine waves; conversely, performing a Fourier transform on the cosine waves will generate the line sources. In view of the practical limitation mentioned, the Fourier transform of the interferogram is terminated at a finite point, often termed the *boxcar truncation*.

In practice, an interferogram is measured and Fourier-transformed to produce the spectrum. Since the Fourier transform is performed by a computer, the interferometer can measure IR spectra of the same quality as spectra measured by a grating spectrophotometer, but in about one-thousandth of the time. This is the major advantage of the technique, made possible by the use of computer technology.

Essentially, all interferometers are single-beam instruments, that is, they measure background and sample sequentially, not simultaneously. First, an interferogram of the background (i.e., scan without sample) is obtained, followed by Fourier transformation. Then a sample is inserted and another interferogram generated, followed by Fourier transformation. Now the two single-beam transmission spectra are ratioed to produce the desired result (the transmittance of the sample). The negative logarithm of this spectrum is often computed to yield the absorbance of the sample.

The resolving power of an interferometer depends upon several factors:

1. The maximum retardation (optical path difference).
2. The apodization function used in computing the spectrum. This function is used to minimize unwanted side lobes produced by the boxcar truncation.
3. The divergence of the beam passing through the interferometer,
4. The stability of the mirror drive system.

There are several advantages in measuring IR spectra using an interferometer. First is the multiplex (or Felgett) advantage, which is derived from the fact that radiation at all wavelengths reaches the detector simultaneously throughout the measurement. This allows spectra of the same signal-to-noise ratio to be measured much faster on an FTIR spectrometer than on a monochromator.

Second is the optical throughput (or Jacquinot) advantage, which arises from the fact that much more energy is directed onto the detector for a given resolution.

Third is the FTIR spectrometer's extremely high wavenumber precision.

Fourth is the ease with which spectra can be manipulated using computer data handling technology.

10.3.3 Ultraviolet and Visible Absorption Spectroscopy

The UV region of the electromagnetic spectrum is considered to extend from 100 to 400 nm and is divided into two portions, the quartz UV (200–400 nm) and the vacuum UV (100–200 nm). The visible region extends from 380 to 780 nm.

When molecules interact with radiant energy in the UV and visible regions, the absorption displaces an outer electron in the molecule. If the radiation is of a frequency corresponding to one of the natural frequencies of the molecule, the radiant energy may be absorbed to increase momentarily the energy content by an amount equal to that of the quanta absorbed. If the electronic energy is increased sufficiently, a transition to an upper electronic state may take place.

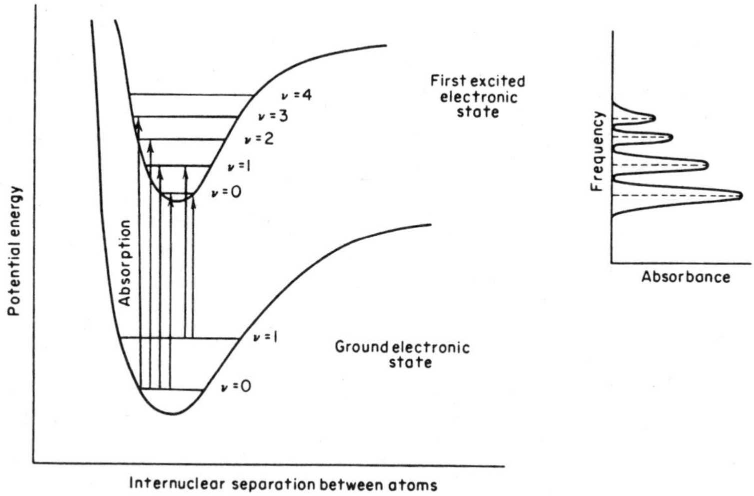

FIGURE 10.8 Energy diagram of a diatomic molecule.

The absorption is a highly specific property of the molecule, and the frequency range within which energy can be absorbed is specifically dependent upon the molecular structure of the absorbing material. The more mobile or loosely bound the electrons, the smaller the energy difference between the ground state and the electronically excited state, and the longer the wavelength of the absorption band (lower frequency).

A schematic energy diagram of a diatomic molecule is shown in Fig. 10.8. The lower curve represents the potential energy of the molecule in its ground (unexcited) state as a function of the internuclear separation between the two atoms. Permissible energies associated with the ground state are represented by a set of vibrational levels ($\nu = 0, 1, 2, 3$, etc.). An electronically excited state of the same molecule is represented by the upper curve, with a different value of internuclear distance at the point of minimum potential energy and different slopes on either side of the minimum. Absorption transitions from the lowest vibrational level, the most populated in the ground state, will produce the most intense peak.

Electronic motion is so much faster than nuclear motion (Franck-Condon principle) that no appreciable change in internuclear distance can take place during an electronic transition. Energy absorption is accomplished in about 10^{-15} sec, or within the period of the frequency of the radiation. The electronic transition takes the form of a group of lines essentially vertical to the vibrational sublevels reached in the excited state.

For structural analysis of molecules containing more than a few atoms, the number of vibrational sublevels becomes so large and their spacing so close that discrete peaks tend to coalesce into broad absorption bands or merely a general envelope. This tendency is enhanced in the liquid state and in solution, where the vibrations are further damped by physical interaction between neighboring molecules.

UV and visible absorption is particularly sensitive for measurement of unsaturated organic compounds, especially those containing aromatic or carbonyl groups. Many organic molecules show strong fluorescence when irradiated with UV light, thereby allowing identification of these substances in trace concentrations. Many substances that do not absorb in a convenient region of the spectrum may be determined quantitatively by forming suitable colored derivatives, followed by colorimetric (photometric) analysis. The wet chemical analyzers described in Chapter 13 operate on this principle, as do the corresponding laboratory tests. The limitations of these colorimetric procedures usually lie in the chemical reactions on which the procedures are based, and include:

1. Specificity of the color-forming reaction.
2. System and reagent stability with time.
3. The effects of pH, temperature, and so on.
4. Conformity to the Beer-Lambert law.

The color of a chemical compound is one of its more obvious characteristics, and the relation between color and chemical constitution has been the subject of scientific investigation for many years. As a result of these investigations and the wealth of information generated, it is usually possible to predict from the structural formula of a chemical compound whether or not it will be colored and the region of the UV/visible spectrum in which it will absorb.

One approach to the prediction of the region of electronic absorption is the chromophore-auxochrome theory of Witt (1876), who observed that certain chemical groups appeared to be associated with the appearance of color in molecules; he called these groups *chromophores*. The color depends upon the relative position in the molecule of the chromophoric group, and on certain other groups, which he called *auxochromes,* which do not of themselves produce color, but tend to shift the absorption associated with the chromophoric group.

Some of the common chromophoric groups are given in Table 10.2. The majority of these are unsaturated groups; however, there are some saturated groups with lone-pair electrons in the valence shell of one of the atoms. Molecules containing structurally similar groups often have spectra resembling each other (e.g., both alcohols and ethers absorb at the edge of the accessible UV region, the latter showing the stronger absorption).

If two or more chromophores are present in the same molecule, the spectrum will show the characteristic absorption of each chromophore if they are insulated by more than one carbon atom or carbon-carbon single bond. If the

TABLE 10.2 Common Chromophoric Groups

Chromophore	Group	Example
Ether	—O—	C_2H_5—O—C_2H_5
Thioether	—S—	CH_3—S—CH_3
Amine	—NH_2	CH_3—NH_2
Mercaptan	—SH	C_2H_5—SH
Alkyne	—C≡C—	CH≡C—CH_2—CH_3
Aldehyde	—CHO	CH_3—CHO
Alkene	>C=C<	CH_3—C=CH_2
Alcohol	—OH	C_2H_5—OH
Ketone	>C=O	$(CH_3)_2C$=O
Nitro-compound	—NO_2	CH_3—NO_2
Azo-compound	—N=N—	CH_3—N=N—CH_3
Benzene		
Naphthalene	$(—C=C—)_n$	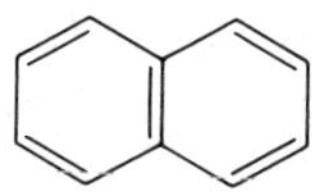
Anthracene		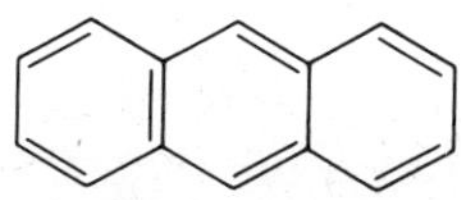

groups are conjugated (e.g., 1 : 3 butadiene), the spectrum may differ considerably from that characteristic of the individual chromophores. Conjugation usually results in an absorption shift to lower frequencies (auxochromic effect).

Qualitative identification of chemical compounds by means of visible absorption spectroscopy involves establishing the wavelength of maximum absorption, followed by a search of tables (similar to but more comprehensive than Table 10.2) to identify the chromophoric group. Further information may then be deduced from the shape, intensity, and precise location of the absorption bands. Finally, the spectrum is compared with spectra in standard compilations of UV spectra.

Quantitative methods are based on a comparison of the extent of absorbance (or transmittance) of radiation at a particular wavelength by a solution of the test sample with the absorbance of a series of standard solutions.

Filter spectrophotometers are suitable for routine work involving non-complex spectra, but more precise work is done using spectrophotometers employing narrow band widths and designed to handle the UV region.

It is interesting to note that among inorganic compounds, the chromophoric group may involve several atoms (e.g., permanganate and dichromate groups) or only single atoms with incomplete outer d-electron shells, where closely spaced, unoccupied energy levels are available (e.g., rare earth and transition elements in coordination compounds). The absorption spectra of these compounds are believed to result from a charge transfer from one part of the ion to another. The spectra are altered considerably by changes in the complexing group.

Color Measurement. The measurement of color utilizes the part of the electromagnetic spectrum that is visible or sensed by the human eye; this region extends from 380 to 780 nm. In the visible spectrum there is a definite relationship between wavelength and color, as illustrated in Table 10.3 showing transmitted colors and their complementary hues. If a material has a characteristic color, its complementary hue may indicate the proper wavelength region. Unfortunately, the human eye is not always a very good judge; color properties often change with concentration and optical path length.

Instrument designs for measurement of color are similar to those used for UV measurements. Interference filters are selected for desired wavelengths determined from optical absorbance curves; detectors are chosen to give spectral response in the visible region. Instruments capable of measurement between 150 and 1000 nm (UV colorimeters) are often called *photometers*.

Industrial Color Standards. There are three main one-dimensional color standards used in industry—the American Public Health Association (APHA) standards, the Saybolt Color Scale, and the ASTM Color Scale.

TABLE 10.3 Relationship Between Absorption of Light and Color

Wavelength (nm)	Transmitted Color	Complementary Hue
<380	Ultraviolet	
380–435	Violet	Yellowish-green
435–480	Blue	Yellow
480–490	Greenish-blue	Orange
490–500	Bluish-green	Red
500–560	Green	Purple
560–580	Yellowish-green	Violet
580–595	Yellow	Blue
595–650	Orange	Greenish-blue
650–780	Red	Bluish-green
>780	Near infrared	

APHA standards are based on a series of known concentrations of platinum-cobalt solutions. These are defined by the weight concentration of metallic platinum per liter of solution. The APHA color unit is 1 mg/l (a standardized combination of potassium chloroplatinate and cobalt-2 chloride). The laboratory color test involves comparing the sample solution with the standard solutions in Nessler tubes.

The Saybolt color scale, used in the petroleum industry, is covered by ASTM test method D156 and gives a color number ranging from −16 to +30, −16 being the darkest. The test uses a set of standard glass discs that serve as color comparison standards. The sample under test (usually a hydrocarbon liquid) is added to a measuring flask or tube until the colors match, a light-colored oil requiring greater depth than a darker-colored oil to give the comparison color. The color number is then obtained from the table provided in the test method. A special apparatus, called the Saybolt chromometer, is specified for this test.

The ASTM color scale is based on ASTM Test Method D1500, which is used for measurement of the color of hydrocarbon liquids with color darker than Saybolt −16, and is therefore the complement of ASTM D156. The color standards consist of a range of 16 colored discs ranging in value from 0.5 to 8.0 in intervals of 0.5. Using a standard light source, the liquid sample is placed in a test container and compared with each of the colored discs. If

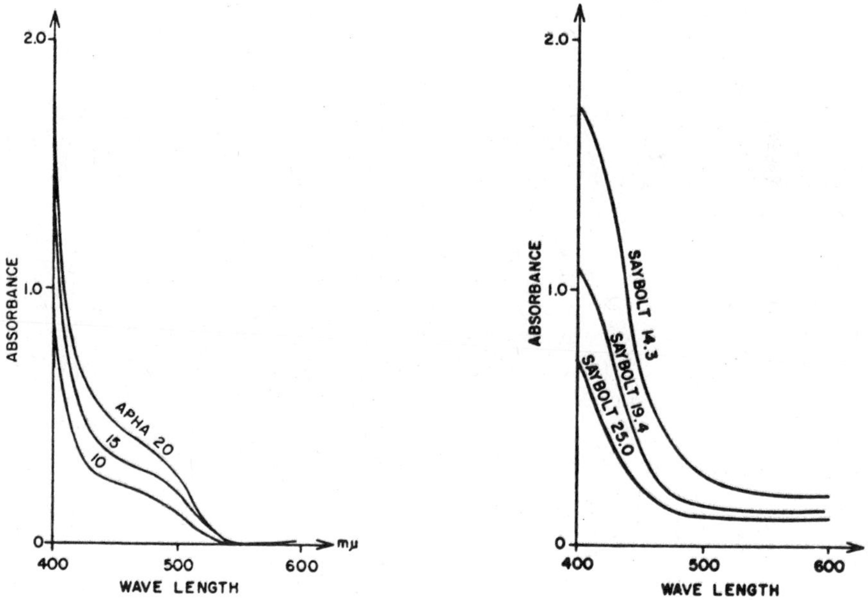

FIGURE 10.9 Spectral response of color standards. *(a)* APHA color standards; *(b)* Saybolt color standards.

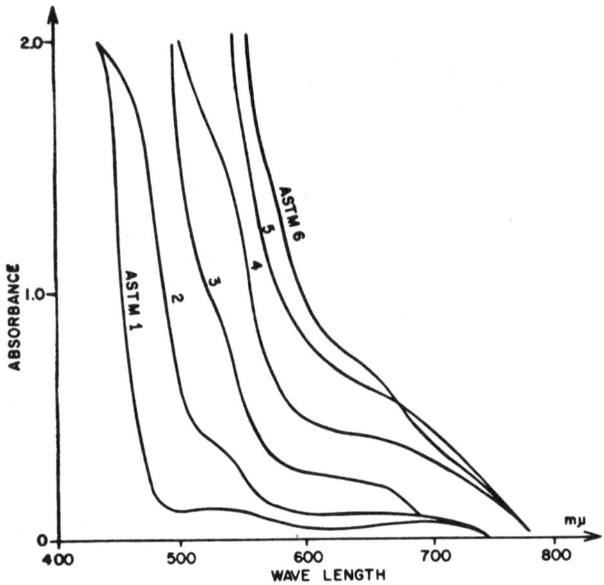

(c)

FIGURE 10.9 Spectral response of color standards. (c) ASTM color standards.

an exact match cannot be found and the sample color falls between two standard discs, the higher (darker) color is reported. If the color of the sample is darker than ASTM 8, it is diluted 15 : 85 with kerosine. The colorimeter used for the test permits observation of the sample and any one of the color standards simultaneously, either by direct viewing or by the use of an optical eyepiece.

The approximate spectral responses of these three color standards are shown in Fig. 10.9.

10.3.4 Raman Spectroscopy

The Raman effect was discovered by Sir C. V. Raman in 1928; for this work he received the Nobel Prize in 1930. The Raman effect arises when a beam of monochromatic exciting radiation passes through a transparent medium containing molecules capable of undergoing a change in *polarizability*. Polarizability is a measure of the elastic deformation of the vibrational configuration of a molecule and, in the Raman effect, measures the dipole moment induced in the molecule by the exciting radiation. Some of the incident photons collide with molecules in the sample, most of the collisions being elastic, that is, the frequency of the scattered radiation is the same as that of the incident radiation (Rayleigh scattering). However, some of the collisions are nonelastic and involve an exchange in energy between the molecule and the incident photon.

The incident photon elevates the scattering molecule into a quasi-excited state with a potential energy above the initial state equal to the energy of the exciting radiation. Upon return to the ground electronic state, a vibrational quantum of energy may remain with the scatterer, producing a *decrease* in the frequency of the scattered radiation.

On the other hand, if the scattering molecule is already in an excited vibrational state, a vibrational quantum of energy may be taken from the scatterer, thus *increasing* the frequency of the scattered radiation. The normal process (a decrease in energy of the incident photon) results in scattered radiation known as Stokes lines; the reverse process gives anti-Stokes lines. A schematic diagram of the energy interchange involved is shown in Fig. 10.10. The scattered lines in the Raman spectrum are characteristic of the vibrational modes of the molecule irradiated and therefore act as a "fingerprint" of that molecule.

To observe the Raman effect, monochromatic radiation is concentrated to give an intense illumination in the sample under test, and the scattered radiation, usually at right angles to the incident radiation, is studied spectroscopically. The Raman spectrum consists of a group of lines on both sides of the very intense exciting line, the Raman lines on the long wavelength side usually being of greater intensity. The spacings between the Raman lines are closely related to the vibrational-rotational spacings in the IR spectrum of the sample.

Raman spectra provide a means of studying vibrational-rotational frequencies of molecules in the visible and UV regions where photography can be used. In addition, they have further importance because a number of frequencies, "forbidden" by wave mechanics for vibrational-rotational spectra in the IR (and therefore not in these spectra), are found in Raman spectra. For instance, diatomic molecules such as H_2, O_2 and N_2, which do

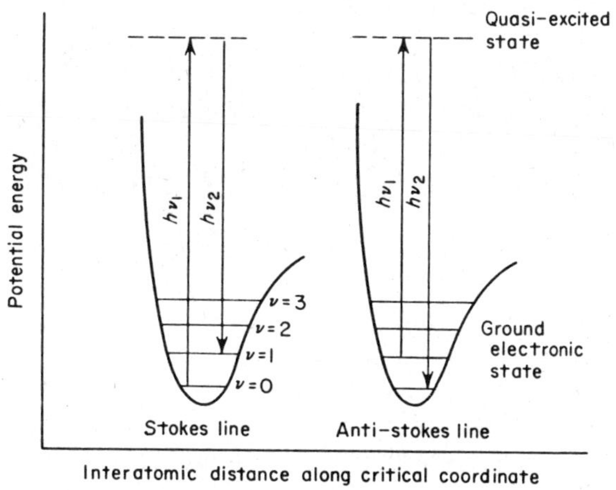

FIGURE 10.10 Energy interchange: Raman spectroscopy.

not give IR spectra, do have Raman spectra. Often it is easier to discover the vibrational frequencies of complex polyatomic molecules from Raman spectra than from IR spectra.

Since the first appearance of Raman spectrometers in 1946, the Raman effect has become an important research method of elucidating molecular structure, for locating various functional groups or bonds in molecules, and for the quantitative analysis of complex mixtures, particularly for major components. Although Raman spectra are related to infrared absorption spectra, they are generated in a quite different manner and thus provide complementary information.

(a) Laser Raman Spectroscopy

The invention of the laser (an acronym of light amplification by the stimulated emission of radiation) and the maser (for the microwave region) in the mid 1950s had a profound effect on Raman spectroscopy. Until then, the traditional source was the mercury arc, which provided useful lines at 2537, 3650, 4047, 4358, and 5461 Å.

The line at 4358 Å is the most commonly used, but its efficiency is very low. For instance, of 2.5 kW generated by the source, about 50 W is radiated in the 4358 Å line and about 1 W is responsible for excitation of Raman spectra.

The major advantages of the laser are:

1. It is highly directional.
2. It is highly monochromatic.
3. It is extremely bright.
4. It is spatially and temporally coherent.

The last of these is very important; conventional sources are incoherent, that is, electromagnetic waves associated with any two photons are out of phase. In addition, the laser's very high intensity and monochromaticity allow much weaker Raman scattering to be observed and provide much improved resolution. Use of the laser in Raman spectroscopy was therefore one of the immediately obvious applications of this device, and most of the development was directed toward the laser's use in this area of spectroscopy.

The ammonia maser was first introduced in 1954, providing a source of 10^{-10} W, with a line at 23.870 MHz. The ruby laser was first demonstrated in 1960 by Maiman. This is a pulsed, solid-state laser operating in the visible region. The lasing constituent is the Cr^{3+} ions, which are in so low a concentration as to act as free ions. Lines are produced at 510–600 nm or 360–450 nm with a laser line width of about 300 GHz. The helium-neon laser, first introduced in 1963, is a gas laser, the laser action occurring between excited energy levels of the neon atom, upper energy levels being populated partly through collisions with helium atoms in metastable excited states. The energy output is about 100 mW and the exciting line is at 6328 Å. In dye lasers, the active medium is a liquid; these devices have the advantage that they are continuously tunable over quite a large wavelength range. Many dyes are available, each of which can be used over a 20–30 nm range. Together, they will cover the range from about 365 nm in the UV to 930 nm in the near IR.

A discussion of the theory of the laser is beyond the scope of this book and the reader is therefore referred to the great wealth of literature on this subject and its relevance to Raman spectroscopy. At present, this area of spectroscopy is limited to laboratory research; no attempt has yet been made to utilize the technique for on-line process analysis, the main problems being packaging for hostile environments and the current extent of commercialization of available lasers for this type of work. Nevertheless, with the rapid developments that are taking place in this field, commercial on-line Raman spectrometers may well appear in the next few years.

10.3.5 Mass Spectrometry

(a) Historical Background

All mass spectrometers have the ability to separate ions of a particular atomic or molecular species by their mass/charge ratios (m/e).

The first work in this interesting branch of spectroscopy dates back to the beginning of the century, with the extensive investigations by Sir J. J. Thomson into the nature and properties of the so-called *positive rays* conducted at the Cavendish Laboratory, Cambridge, England. He used what was known as the "parabola method" of analysis, by which the rays were allowed to pass through a very narrow tube and the fine beam so produced analyzed by electric and magnetic fields.

The first mass spectrograph, shown schematically in Fig. 10.11, was built by F. W. Aston at Cambridge in 1919. The mass spectra of many atoms and

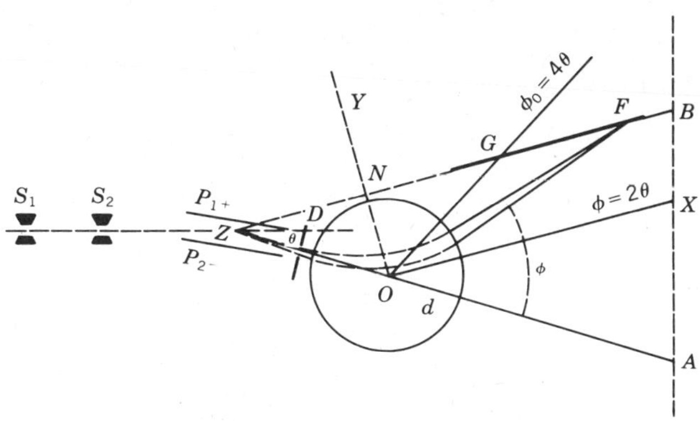

FIGURE 10.11 F. W. Aston's first mass spectrograph.

compounds were studied using this instrument and laid the foundations for future work in this field. Considerable development took place in mass spectrography during the period 1920 to 1937. The discovery of artificial disintegration in 1932, followed by that of neutron transmutation, led to a period of the greatest activity in nuclear physics, with a demand for greater and greater accuracy in the mass measurement of nuclei. This in turn led to the introduction of more precise and powerful mass spectrographs by Aston (1925 and 1932) in England, Dempster and Jordan in the United States, and Costa in France. All of these instruments used the camera and photographic plate for detection and recording of mass spectra, hence the name mass spectro*graph*.

The first mass spectrometer was built by Nier in the United States in 1932 and served as the forerunner of modern instrumentation. By means of a valve amplifier, Nier was able to detect concentrations as low as 10 ppm and provide reliable data on over twenty elements, including the discovery of several rare isotopes.

(b) Principle of Operation

The mass spectrometer produces charged ions, consisting of the parent ion and ionic fragments of the original molecule, and sorts these ions according to their mass/charge ratios. The mass spectrum is thus a record of the number of different kinds of ions present, the relative numbers of each being characteristic for every compound, including isomers. Only very small quantities of sample are required, provided that the material is able to exist in the gaseous state at the temperature and pressure existing in the ion source chamber and analysis portion of the instrument.

From the mass spectra, a wealth of information can be obtained regarding the composition of mixtures of organic compounds and the elemental analysis of samples. With the more sophisticated spectrometers, molecular weight can be determined directly to fractions of a mass unit. A detailed interpretation of the mass spectrum frequently makes it possible to place functional groups into certain areas of the molecule and to see how they are connected together.

Several distinct methods of obtaining mass spectra have been developed; however, most instruments have four components in common—the sample inlet system, the ion source, the electrostatic accelerating system, and the ion collection, amplification, and readout system.

A high vacuum is maintained throughout the instrument, from the entrance to the ion source to the detector (or ion collector). The sample gas, which is essentially at atmospheric pressure, diffuses through a "molecular leak" into the ion source chamber. For analytical studies, molecular flows are usually employed whereby collisions between molecules and the walls of the chamber are much more frequent than collisions between molecules.

The ion source chamber is maintained at a very low pressure (10^{-7} torr) and at a temperature of about 200°C. An electron gun is located perpendicular to the incoming gas stream. Electrons emitted by a glowing tungsten filament pass through a pair of positively charged slits into the body of the chamber and are accelerated by the electric field maintained between the slits. Collisions of these electrons with the gas molecules produce ionization and fragmentation. The electric field can usually be varied from 6 to 100 V. A range of 6–14 V is generally employed for molecular weight determinations, whereas 50–70 V is used to obtain mass spectra. After passage through the ion source chamber, the electron beam is collected in an anode trap; the current to this trap is used to control the electron beam intensity. Although electronic bombardment is the most popular ionization method, others such as spark generation and optical and chemical ionization methods are also used.

The positive ions formed in the ionization chamber are accelerated by a series of electrostatically charged slits and emerge from the final slit as a collimated beam of ions. The current carried by the ion beam is very small (10^{-14} A) and therefore must be amplified. The ion beam then passes to the resolving section of the instrument. The major difference between the various designs of mass spectrometer lies in the method used for separating the ions according to their m/e as described next. More details of instrument design are given in Section 10.4.3, where it is related specifically to on-line mass spectrometers.

(c) Mass Spectrometer Design

Mass spectrometers may be classified into the following types, according to operating principle.

Magnetic-Deflection Mass Spectrometer. In this instrument, sometimes called the magnetic sector mass spectrometer, the ions are accelerated by application of an electrostatic field and are deflected by a magnetic field through an angle of 60, 90, 120, or 180 degrees. This is a popular design and is used in some on-line mass spectrometers (see Section 10.4.3).

The velocity attained by an ion of mass m and charge e on passage through the electrostatic field can be expressed in terms of the accelerating voltage V and the kinetic energy of the ion on leaving the field:

$$\frac{1}{2} mv^2 = eV \tag{10.13}$$

or

$$v = \left(\frac{2eV}{m}\right)^{1/2} \tag{10.14}$$

The ion next enters a uniform magnetic field of constant intensity H, imposed at 90 degrees to the direction of ion motion, which diverts the ion into a circular orbit. Equating centripetal and centrifugal forces:

$$\frac{mv^2}{r} = HeV \tag{10.15}$$

where r is the radius of curvature of the trajectory and is given by $r = mv/eH$. Therefore

$$r = \left(\frac{2V}{H} \cdot \frac{m}{e}\right)^{1/2} \tag{10.16}$$

Only those ions that follow the path that coincides with the arc of the analyzer tube in the magnetic field will be focused on the slit of the detector. Ions of different mass/charge ratio will strike the analyzer tube, be neutralized, and pumped out of the system. The magnetic field thus classifies and segregates the ions into beams, each of a different m/e. To obtain a mass spectrum, either the accelerating field voltage or the magnetic field strength is varied (usually the former) to bring each beam of specific m/e successively to focus on the detector.

Double Focusing Mass Spectrometer. In this instrument, focusing is provided for energy (direction) and velocity, using electrostatic and magnetic deflecting fields. The improved focusing allows entrance and exit slits to be narrowed, with corresponding increase in resolving power.

Cycloidal Focusing Mass Spectrometer. In this instrument, a uniform electrostatic field is applied perpendicular to the magnetic field and parallel to the direction of ion acceleration. On entering the cross field region, all ions are made to travel cycloidal rather than circular paths. As either of the two cross fields is varied, ions of identical m/e focus on the fixed collector, even though they are not equivalent in energy.

Quadrupole Mass Spectrometer. In this instrument, the ions are separated according to their masses by the quadrupole mass filter, which consists of four high-precision cylindrical rods located precisely in an orthogonal array. Diametrically opposed rod pairs are electronically connected to form positive and negative pairs of rods. The positive pair has an RF voltage superimposed on the positive dc voltage; the negative pair has a negative dc and RF voltage that is 180 degrees out of phase with the positive pair.

As an ion moves at uniform speed along the filter axis, it undergoes complex oscillatory motion transverse to the filter axis. For a given set of voltages, all ions below a given mass number are neutralized on the positive set of rods, while larger ions are passed. The negative set of rods passes low-mass ions and neutralizes ions above a given mass. By adjusting the RF-to-dc ratio, the passbands can be made to overlap so that only ions of a very specific mass can get through; all others are neutralized by one set of rods or the other.

A unique feature of the quadrupole mass filter is that the mass of the ion passed is directly proportional to the voltage applied to the rods. Thus by changing the voltages, masses of interest are allowed to pass through the filter to the ion collector (detector) where they are measured. This design is also used in on-line mass spectrometers.

Time-of-Flight Mass Spectrometer. In this instrument, ions of different m/e are separated by the difference in time required for them to travel an identical path from ion source to detector, which are situated at opposite ends of an evacuated tube. The ion source is energized for 0.25 μsec every 100 μsec and the pulses of ions generated by electron bombardment are accelerated to attain a velocity that is a function of their m/e. Their drift time down the evacuated tube is measured. With the use of an oscilloscope, the time base of which is synchronized with and triggered by the production of the original ion pulses, a complete mass spectrum can be scanned in microseconds. Thus, speed of anlaysis is the major advantage of this type of mass spectrometer. The design is also used in on-line mass spectrometers.

10.3.6 Nuclear Magnetic Resonance Spectroscopy

Nuclear magnetic resonance (NMR), sometimes called nuclear spin resonance, is a form of spectroscopy in which, as a result of the magnetic properties of nuclei arising from their axial spin, radio-frequency radiation is absorbed in a magnetic field.

The existence of resonant absorption in bulk matter was demonstrated by Bloch and Purcell in 1946; for this work they were awarded the Nobel Prize in 1952. However, the first work in this area goes back to the early 1920s. With the production of mass spectrographs of higher and higher resolving power (see Section 10.3.5), Pauli observed the so-called hyperfine splitting in mass spectra. Stern and Gerlach, in their molecular experiments, demonstrated the existence of quantization of magnetic moments. It was suggested that certain nuclei had angular momentum and thus, as a spinning electric charge, a magnetic moment. The existence of NMR was also predicted by Gorter in 1936.

(a) Theoretical Basis

Classically, a rotating charge will behave as a magnetic dipole, the dipole moment being given by,

$$\mu = iA \qquad (10.17)$$

where i is the current and A the loop area. Also,

$$i = \frac{q}{t_r} \tag{10.18}$$

where q is the charge and t_r is the time for one rotation. For a nucleus of mass m_n, radius r, and rotational speed ω (rad/sec), the angular momentum is given by,

$$P = \omega m_n r^2 = \frac{2\pi r^2 m_n}{t_r} \tag{10.19}$$

Since rotation is in a circle, $A = \pi r^2$. Therefore,

$$A = \pi r^2 = Pt_r/2m_n \tag{10.20}$$

The magnetic moment (μ) is given by,

$$\mu = \frac{Aq}{t_r}$$

$$= -\frac{Pt_r}{2m_n} \times \frac{q}{t_r} = -\frac{qP}{2m_n} \tag{10.21}$$

The magnitude of the nuclear magnetic moment is often expressed in terms of a constant called the *gyromagnetic ratio* (γ), where,

$$\gamma = \frac{\mu}{P} \tag{10.22}$$

When the nucleus is placed in a magnetic field, an interaction will take place and it will acquire energy:

$$E = -\mu H_0 \tag{10.23}$$

where H_0 is the magnetic field strength.

However, the nucleus is not merely a miniature bar magnet; it also has angular momentum. It aligns with the magnetic field, but also *precesses* about it at an angle θ as shown in Fig. 10.12. For angular motion with velocity ω_0,

$$\frac{dP}{dT} = P\omega_0 = \mu H_0 \tag{10.24}$$

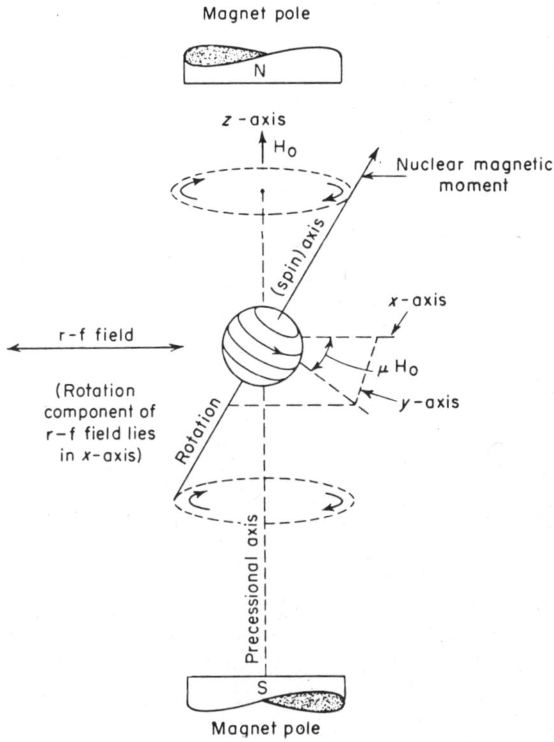

FIGURE 10.12 Basis of NMR spectroscopy: the spinning nucleus in a magnetic field.

Therefore,

$$\omega_0 = \gamma H_0 \qquad \text{or} \qquad \nu_0 = \gamma' H_0 \tag{10.25}$$

This is called the *Larmor equation*.

Two everyday examples of precession are the child's spinning top and the gyroscope. Nuclear precession occurs as a result of interaction between the torque generated by rotational motion of the nucleus and the effect of the magnetic field on nuclear magnetic moment. The Larmor equation describes the basic phenomenon of NMR, that is, the magnetic field is proportional to the frequency, and the proportional constant is the gyromagnetic ratio. Also system energy does not depend upon μ, but only the projection of μ onto the Z axis (see Fig. 10.12).

The angular momentum of the charge created by the spinning electrons can be described in terms of *spin numbers* (I), which have integral or half-integral values (in units of $h/2\pi$, where h is the Planck constant) from $\frac{1}{2}$ to at least $\frac{9}{2}$ for different nuclei; $I = 0$ denotes no spin. The spin number is obtained by the addition of individual proton and neutron spin numbers of $\frac{1}{2}$

each, with the restriction that neutrons can cancel out only neutrons and protons can cancel out only protons. Nuclei can therefore be classified into three types:

1. Those with zero spin, in which both the number of neutrons and the number of protons are even (e.g., C^{12}, O^{16}, and S^{32}). Nuclei in this category produce no NMR signal.
2. Those with half-integral spin, in which either the number of protons or the number of neutrons is odd (e.g., H^1, B^{11}, F^{19}, P^{31}, Cl^{35}, and Br^{79}).
3. Those with integral spin, in which both the number of neutrons and the number of protons are odd (e.g., H^2, N^{14}).

Most of the research effort in NMR spectroscopy has been directed to the study of the second category of nuclei, particularly H^1.

Excitation is applied in the form of an alternating radio-frequency magnetic field (H_1), rotating about the basic field H_0, which is conventionally along the Z axis (see Fig. 10.12). When the frequency of rotation of H_1 is equal to the frequency of a nuclear energy level (ν_0) and in the same sense, resonance will occur, energy will be absorbed by the nucleus from the exciting field, and the nucleus will precess about H_0 at a larger value of γ. The Larmor condition may be detected by measuring the absorption energy from the exciting field.

So far we have considered the behavior of a single, isolated nuclear spin; the NMR spectrum, however, records the mean behavior of an ensemble consisting of a large number of similar nuclei. In optical spectroscopy, the excited molecule returns to the ground state almost instantaneously (10^{-13} sec). In NMR spectroscopy, return to the ground state occurs in the order of seconds. There are two important consequences of this:

1. NMR spectra are very sharp
2. Saturation (i.e., equalization of the population in the ground and excited states) can occur, because the relaxation from the excited state is slow and, with a strong exciting field, an equilibrium can be set up.

(b) NMR Spectrometers

A typical example of an NMR spectrometer is shown in Fig. 10.13. The NMR spectrum can be scanned either by changing the oscillator frequency or by making a small change in the applied magnetic field. Common practice is the use of a fixed oscillator and a variable magnetic field.

The sample is contained in a glass tube of small cross section placed in a sample holder, which is spun at several hundred rpm to increase homogeneity of the magnetic field throughout the sample volume. The sample is placed between the poles of an electromagnet. An oscillating RF field supplied by a

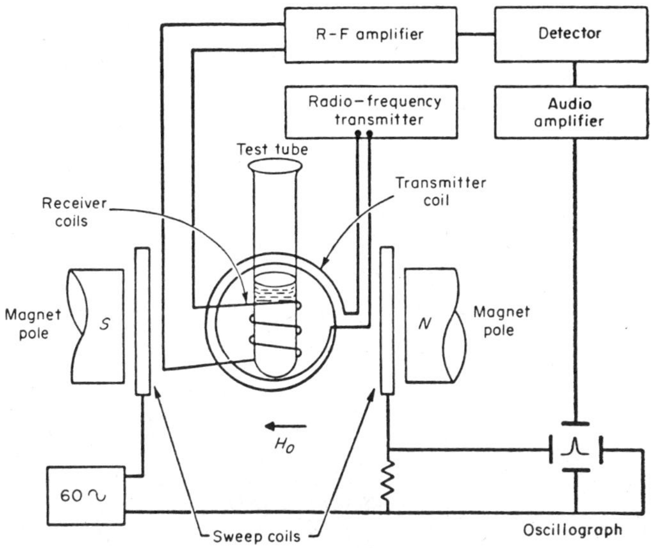

FIGURE 10.13 Typical NMR spectrometer.

crystal-controlled oscillator is imposed at right angles to the magnetic field (i.e., perpendicular to the plane of the paper in Fig. 10.13). Sweep coils wound around the poles of the magnet or located within the pole gap allow a sweep to be made through the applied magnetic field. This produces resonance in the range of anticipated precession frequencies and the rate of sweep, which is about 0.5–2.0 Hz for most modern spectrometers. The range of sweep for protons is about 1000 Hz.

The receiver coil, which consists of a few turns of wire wound tightly around the sample tube, picks up the resonant signal from the sample. Absorption of energy, produced by nuclear transitions, causes the RF voltage across the receiver coil to drop. This voltage change is amplified and diode detector-tuned to the same frequency as the RF transmitter. The resultant dc voltage is used to produce a graph of intensity versus frequency on an oscilloscope.

(c) NMR Spectra

Application of NMR to chemical investigations depends upon the fact that the exact field experienced by a nucleus is affected by the electronic environment of that nucleus (i.e., diamagnetic shielding by electrons in the same molecule). Resonance will occur in a different part of the spectrum for each chemically distinct group.

The condition for resonance is expressed in terms of a *chemical shift,* the difference between the field necessary for resonance in the sample and that for a chosen reference standard. Typical reference standards for proton

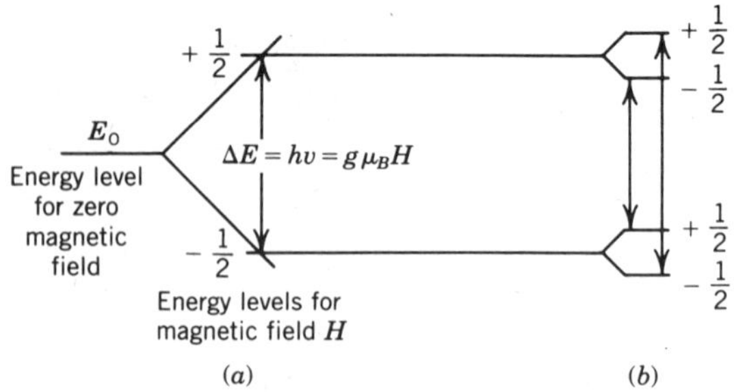

FIGURE 10.14 NMR spectra: spin-spin coupling. Splitting of electron energy levels by a magnetic field. (*a*) Effect of electron spin only; (*b*) effect of adding a nucleus with spin $I = \frac{1}{2}$. Allowed transitions are indicated by arrows.

spectra, each having a single resonance line, are tetramethylsilane (TMSi), water, chloroform, cyclohexane, and benzene. The chemical shift (δ) is expressed as a function of increasing field strength necessary to achieve resonance:

$$\delta = \frac{H_S - H_R}{\mathrm{H}_1} \times 10^6 \tag{10.26}$$

where H_S and H_R are the positions of the absorption peaks for sample and reference in Hz and H_1 is the RF signal used. The term δ is dimensionless and is usually expressed in ppm. Chemical shift is affected substantially by factors such as the nature and temperature of the solvent, solution concentration, and of course the structure of the molecule.

Another feature of NMR spectra is what is known as *spin-spin coupling*. The spin state of one nucleus affects the resonance of another, and vice versa; this causes mutual splitting of their absorption bands into multiplet structures. The separation between adjacent multiplet peaks is called the *coupling constant* (J). For example, the NMR spectrum of a substance whose molecules contain two nuclei with spin-spin coupling consists of four lines of equal intensity, as shown in Fig. 10.14. The spin-spin coupling constants of protons in various organic compounds has been extensively researched and covered in the literature.

The techniques of NMR and IR spectroscopy are used in the same way to elucidate the chemical structure of organic compounds and therefore complement each other. When used together, they provide a powerful method for structure investigation, electronic distribution, and reactivity of molecules. Whereas IR spectroscopy is used widely in on-line analysis, NMR spectroscopy is used mainly as a research tool at present.

10.3.7 Electron Spin Resonance Spectroscopy

Electron spin resonance (ESR), sometimes called electron paramagnetic resonance (EPR), is based on the fact that atoms, ions, or molecular fragments with an odd number of electrons exhibit characteristic magnetic properties. These properties arise from either the orbiting or spinning action (or both) of unpaired electrons about the nucleus. Unpaired electrons, a relatively unusual occurrence, are present in free radicals, triplet electronic states, and transition element ions. Since an electron possesses a spin, associated with the spin is a magnetic moment. When a strong magnetic field is applied to the unpaired spins of an electron, the torque acts to make the electron dipoles line up parallel or antiparallel to the direction of the magnetic field (the only two orientations allowed), and precess about the magnetic field axis at a frequency proportional to the applied magnetic field and the electron magnetic moment. This is similar to the behavior of protons in NMR spectroscopy (see Section 10.3.6).

ESR spectroscopy uses microwave/waveguide technology and was developed from experience with radar techniques used in World War II. A typical ESR spectrometer is shown schematically in Fig. 10.15, with the reflection microwave impedance bridge drawn to a larger scale for clarity.

The sensing element is a rectangular microwave cavity constructed to maximize the applied magnetic field along the sample dimension. The microwave field is generated at 9500 MHz by a klystron oscillator. The field is applied to the resonant cavity, which is connected to one arm of the microwave bridge. The resonant cavity is located between the poles of an electromagnet, which provides a homogeneous magnetic field that can be varied from near zero to about 5000 gauss, with a stability of 1 part in 10^6. The magnetic field is swept over a small range and through the resonant condition by varying the current in a pair of sweep coils provided around the resonant cavity and driven by a 100-kHz sweep generator. Standing waves are set up, the cavity serving as a very long path-length cell in which the waves are reflected to and fro thousands of times.

The *hybrid tee* shown in Fig. 10.15 is a device that does not allow microwave energy to pass from one arm to the arm opposite. Thus, energy emitted from Arm 1 divides equally between Arms 2 and 3; Arm 3 usually contains a

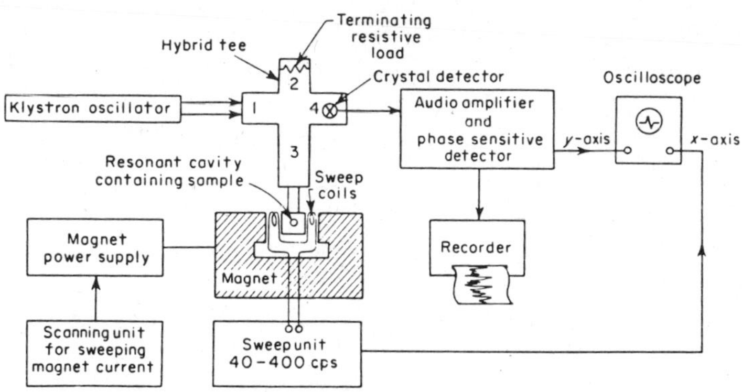

FIGURE 10.15 Typical ESR spectrometer.

balancing load. When all energy is absorbed and none reflected, no energy enters Arm 4; therefore, the detector (usually a silicon-tungsten crystal rectifier) receives no energy. At resonance, the sample absorbs energy from the microwave field, resulting in an imbalance in the impedance of the cavity, and a signal is reflected to the detector.

If the dc magnetic field is varied slowly through the value required for resonance, an absorption peak as shown in Fig. 10.16*a* is obtained. However, the 100-kHz sweep generator provides modulation of the dc magnetic field at the sample position. Then, if the magnetic field is swept slowly through resonance over a period of several minutes, a dynamically recurring imbalance in the hybrid tee is detected and amplified to produce the derivative of the absorption spectrum, as shown in Fig. 10.16*b*.

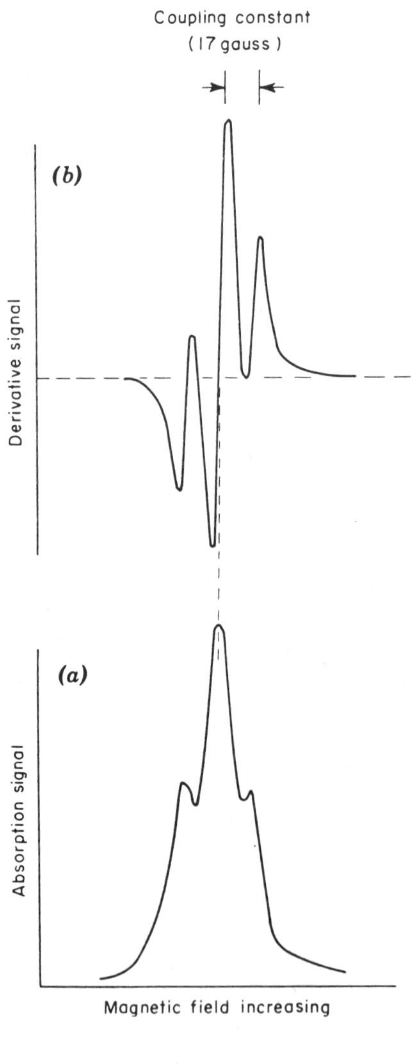

FIGURE 10.16 ESR spectra. (*a*) Absorption peak spectrum; (*b*) derivative spectrum.

For analysis by ESR spectroscopy, the sample may be in the form of a single crystal, solid powder, liquid, or solution. Although the peak height has little significance, the total area enclosed by either the absorption signal or the derivative signal is proportional to the number of unpaired electrons in the sample.

Comparison is made with a standard containing a known number of unpaired electrons with the same line shape as the unknown. The usual standard is 1,1-diphenyl-2′-picryl-hydrazyl (DPPH) which contains 1.53×10^{21} unpaired electrons per gram. No counterpart to the "chemical shift" in NMR spectroscopy exists in ESR; therefore, the spectra must be analyzed from hyperfine splittings and from the number and relative intensities of multiplet structures.

ESR spectroscopy is basically a research tool, and no attempts have been made so far to use the technique for on-line analysis.

10.3.8 X-Ray Fluorescence Spectroscopy

X-Ray spectroscopy is divided into three separate and distinct techniques—X-ray absorption, X-ray diffraction, and X-ray fluorescence. Although the theory of X-rays is common to all three techniques, the last is discussed in more detail here since it has been used successfully in on-line analysis.

(a) Historical Background

In 1895, Röntgen found that when a beam of high-speed electrons falls on a target (or "anticathode") in a vacuum tube, very penetrating rays are produced. He called them X-rays and suggested that they were a very short-wave radiation, similar to light except that their frequency was much higher. They were not deflected by electric or magnetic fields and therefore were not charged particles. The final proof of their wave nature was provided in 1912 when Laue suggested (and Friedrich and Knipping showed experimentally) that beautiful but complex diffraction patterns can be obtained when a beam of X-rays is passed through a small thickness of a crystal.

In 1913, Moseley measured the wavelength of "characteristic" X-rays from a large number of anticathodes made of different elements. He showed that the square root of the frequency is nearly a linear function of what he called the atomic number (N) of the element, as given by the following expression:

$$\frac{c}{\lambda} = R(N - \sigma)^2 \tag{10.27}$$

where R is a proportionality constant and σ is a constant whose value depends on the particular series of emission lines in the X-ray spectrum of the element. Moseley's discovery was an important one for chemistry, since it led to the replacement of the chemical atomic weight by the atomic number as the fundamental quantity that determines chemical behavior.

The atomic number (now Z rather than N) is identified with the number of positive units of electricity contained in the atomic nucleus. In the older Periodic Table, with the elements arranged in order of increasing atomic weights, there were three anomalies—potassium came before argon, nickel before cobalt, and iodine before tellurium, but chemical properties would require a reversal of this order. Using Moseley's atomic numbers, the order agreed perfectly with the chemical properties for every element known at that time.

In his second paper on the subject, Moseley made the following statement in his summary: "Every element from aluminum to gold is characterized by an integer N, which determines its X-ray spectrum. Known elements correspond with all the numbers between 13 and 79, except three. Thus there are here three possible elements still undiscovered." This prediction proved to be correct.

The diffraction of X-rays by crystals was studied extensively by W. H. and W. L. Bragg in 1913, and they described an X-ray spectrometer, by means of which the angle at which a beam of monochromatic X-rays is diffracted from planes of atoms parallel to a crystal face could be measured. X-ray diffraction thus became an important technique for the elucidation of crystal structures.

(b) The X-Ray Spectrum

The radiation emitted by an anticathode depends on the material used and consists of two types of different "hardness" or penetrating power—the very penetrating K-radiation and the "softer" or less penetrating L-radiation. Kossel explained X-rays as the result of expulsion of an electron from a completed *inner* shell by collision with the bombarding electrons, followed by an electron falling into the vacant space from a shell further out.

The relative simplicity of the characteristic X-ray spectra is explained by the limited number of energy levels in the atom, together with certain "prohibited" transitions, and by the association with the *inner* electrons of the atom, rather than the outer valence electrons. The features of the spectra depend only on the atomic number, not on any chemical properties of the elements or the compound in which they may be present or on the physical state of the sample.

The ionized atom returns to its normal state in a series of steps, in each of which an X-ray photon of definite energy is emitted (fluorescent radiation). These transitions give rise to the line spectrum characteristic of the material in the anode or of a specimen pasted onto the target. The so-called K-series of lines is obtained when electrons in the innermost K shell are dislodged; they arise from electrons dropping down from the L and M shells into the vacancies in the K shell. Corresponding vacancies in the L shells are filled by electron transitions from outer shells, giving the L-series of lines, and so on. The energy required to eject an electron completely from an atom is related to the wavelength of the "absorption edge" for a particular transition. A spectral series will be excited by continuous radiation only when the wavelength limit is shorter than the absorption edge of the series.

(c) X-Ray Fluorescence Analysis

The characteristic X-ray spectrum of an element can also be excited by irradiation of a sample with a beam of X-rays, provided that the primary X-radiation is sufficiently energetic to ionize an inner shell. The intensities of the resultant fluorescent X-rays are smaller by a factor of about 1000 than the X-ray beam obtained by direct excitation with a beam of electrons. Consequently, X-ray fluorescence spectroscopy is only feasible with the availability of high-intensity X-ray tubes, very sensitive detectors, and suitable X-ray optics.

A typical laboratory X-ray fluorescence spectrometer is shown in Fig. 10.17. The sample, contained in a sample holder and often rotated to improve uniformity of exposure, is irradiated with a beam of primary X-rays, causing the elements in the sample to emit their characteristic fluorescence lines. The sample is supported by a thin film of Mylar and an aluminum disc is used to restrict the area of sample irradiated. A portion of the scattered fluorescence beam passes through the entrance slit and is directed onto the plane surface of an analyzing crystal. The line radiations pass through the exit slit to the detector, where X-ray quanta are converted to electrical impulses, or counts. The primary slit, analyzer crystal, and exit slit are placed on the focal circle so that Bragg's law will always be satisfied as the

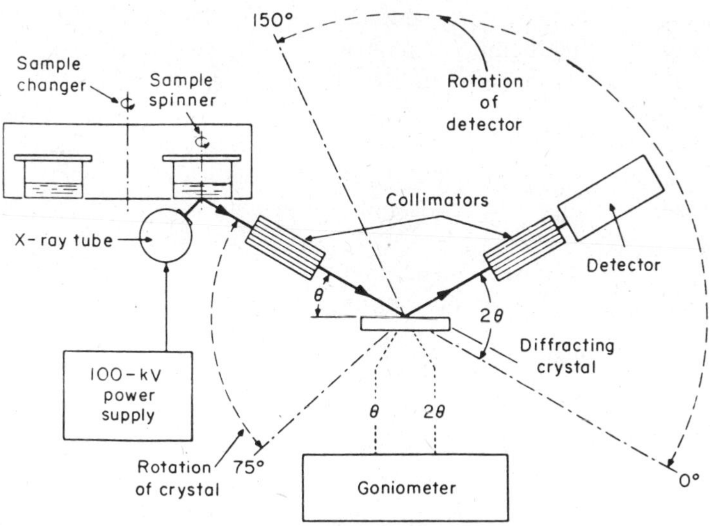

FIGURE 10.17 Typical X-ray fluorescence spectrometer.

goniometer is rotated (that is, the detector is rotated to *twice* the angular change in the crystal setting).

For quantitative analysis, the intensity of a characteristic line of the element under analysis is measured. Counts are collected for fixed periods of time, or the time is measured for the period required to collect a specified number of counts. The goniometer is then set at a nearby portion of the spectrum where only the background contributes. For major elements, 200,000 counts can be accumulated in 1–2 min; however, background counts can take considerably longer. The net intensity (peak minus background) is then related to the concentration of the element via a calibration curve.

Fluorescent X-ray analysis is inherently very precise and rivals the accuracy of wet chemical methods for major constituents, but concentrations below about 1% are difficult to detect.

10.4 ON-LINE PROCESS ANALYZERS

10.4.1 Introduction

On-line process analyzers using spectrometric techniques are used in industry for a wide variety of applications, involving mostly the selective measurement of the concentration of one or more components of interest. Measurement is made primarily in the gaseous phase, but instruments are also available and have been used successfully for measurement in the liquid and solid phases.

Many of these applications and some of the commercially available equipment are described elsewhere in this book.

In Chapter 9, the technique of UV/visible absorption is used for the measurement of hydrogen sulfide in gaseous process streams. Instruments by DuPont and Teledyne are described.

Measurement of moisture in solids, using IR absorption techniques, is covered in Chapter 11. Analyzers by Ohmart, Moisture Register Company, and Moisture Systems Corporation are described.

Chapter 12, which deals with air quality monitoring, includes descriptions of a variety of spectrometric-type analyzers, used for the measurement of sulfur dioxide, carbon monoxide, carbon dioxide, ozone, and opacity, and for automotive emissions monitoring.

Chapter 13, which covers water quality monitoring, abounds with descriptions of analyzers utilizing UV/visible and IR absorption, photometric, and UV fluorescence techniques. Wet chemical analyzers with photometric detectors and TOC analyzers using IR detectors for measurement of CO concentration are also included.

Chapter 18 covers the measurement of CO in stack flue gases using in-situ and external sampling-type IR analyzers. The emphasis is on the in-situ type, in view of its increased popularity at present. However, almost any one of the wide variety of commercially available IR analyzers may be used in this application; the primary problem here is sample handling.

In view of the extensive coverage in other chapters, this chapter covers commercially available on-line spectrometric analyzers from the viewpoint of the various different instrument designs, with emphasis on some of the more recent additions to the marketplace. Mass spectrometers and X-ray fluorescence analyzers, which are specific to this chapter, are covered in more detail.

10.4.2 Process Infrared Analyzers

IR analyzers have been used for industrial on-line process analysis for over 25 years, so there is a wealth of experience with them. The most commonly used, the non-dispersive infrared (NDIR) analyzer, really has no analog in laboratory instrumentation and is used almost exclusively for the measurement of a single component in gaseous process streams.

As with many other areas of process analyzer technology, the IR field has seen many new developments over the past 20 years, such as improvements in source and detector technology, optical filter design, packaging, electronics, and so on. Consequently, at present a wide variety of process IR analyzers is available commercially for industrial applications, using various design configurations. It is not feasible to cover all of them in detail in the space available. The following information therefore illustrates the basic designs in use.

(a) Non-Dispersive Infrared (NDIR) Analyzers

These instruments are usually dual-beam; use the gas selective, Luft-type detector, which is filled with the component to be analyzed; and usually operate in the positive filtering mode. IR radiation from a single or dual source is chopped and passed in phase through the sample and reference cells to the detector. Absorption of IR radiation by sample molecules results in a detector imbalance, which is sensed as a change in capacitance and amplified to provide an output corresponding to component concentration in the sample.

The selectivity of the analyzer may be improved by using narrow band-pass filters that select only wavelengths of interest from the source radiation. Filter cells may also be added, filled with interfering sample components, which remove undesirable wavelengths from both sample and reference beams. Spectral resolution of these instruments is set by the width of the selected absorption lines and for many applications is much better than the typical laboratory grating spectrophotometer. The Luft-type detector is highly sensitive, but is unfortunately susceptible to temperature variations and external vibration.

FIGURE 10.18 LIRA Model 3000 NDIR analyzer. (Courtesy of Mine Safety Appliances Company)

Mine Safety Appliances Company (United States). The LIRA Model 3000 shown schematically in Fig. 10.18 is designed to measure a single component in a complex gas mixture. Two nichrome IR sources, which emit a broad range of IR frequencies, are used. The radiation is channeled in parallel paths through a sealed reference cell and a flowing sample cell. A chopper is located between the source block and the cell block, providing an alternating optical signal of 2 Hz at a supply frequency of 50 Hz (2.2 Hz at 60 Hz).

The detector block contains a condenser microphone. The microphone membrane divides the detector into two sections, one of which has windows that accept energy from the two IR beams. As the chopper alternately blocks each beam, energy is received by the detector first from one path, then the other. When IR energy at the desired frequency enters the detector, it is absorbed by the sensitizing gas (usually the component of interest), resulting in a temperature/pressure increase.

When a zero gas such as nitrogen is passed through the sample cell, the IR energies in the two beams are equal. However, when the component of interest is present in the sample, it absorbs some of the IR energy at the desired frequency and the detector therefore sees less energy via the sample path. A pressure difference occurs inside the detector, causing the membrane to move at the chopper frequency. The membrane movement varies the capacitance of the condenser microphone, generating an electrical signal proportional to the difference in energy received from the two paths. The detector output is amplified, demodulated, and further amplified to drive the local indicating meter and provide an output for remote recording.

The standard Model 3000 consists of an optical bench, amplifier circuitry, analog or digital master, power supply, and zero/span controls, all housed in a general purpose enclosure suitable for purging.

The LIRA Model 202 Series of NDIR Analyzers are designed on the same principle as the Model 3000, except that analyzer and control units are also available as separate units, including explosion-proof versions for installation in hazardous locations.

Siemens AG (West Germany). The Ultramat 3 gas Analyzer is a dual-beam NDIR instrument using a single IR source and a Luft-type detector, as shown in Fig. 10.19, and covering the spectral range 2–8 μm. The detector features a microflow sensor consisting of two nickel grid resistors forming part of a Wheatstone bridge network. The detector cell consists of four single chambers filled with the component of interest, with the microflow sensor located at the pneumatic gravity center of these chambers as shown.

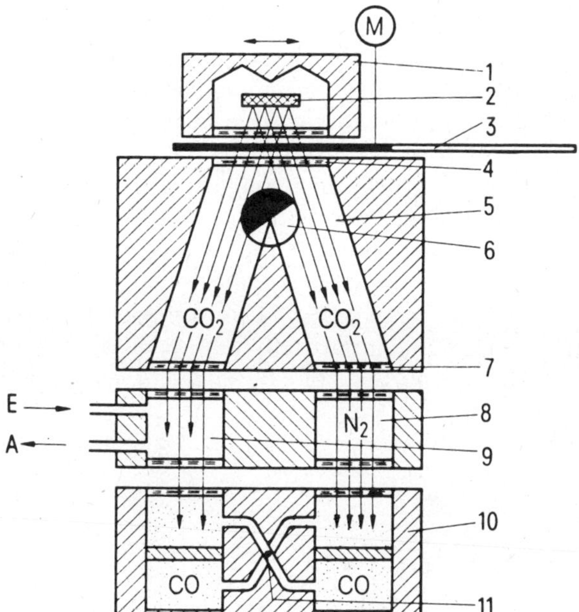

1 Deflector } adjustable
2 Radiator } adjustable
3 Diaphragm wheel
4 Holder for optical filter
5 Filter chamber
6 Reflection diaphragm
7 Window
8 Reference chamber
9 Measuring cell
10 Receiver chamber
11 Microflow sensor
E Gas input
A Gas output

FIGURE 10.19 Ultramat 3 NDIR analyzer. (Courtesy of Siemens AG)

When absorption of IR energy occurs in the sample cell due to presence of the measured component, a balancing flow is generated in the connecting channel, which pulses at chopper frequency (22 Hz) and produces a temperature difference in the grid resistors of the microflow sensor. This upsets the balance of the Wheatstone bridge, resulting in a flow-proportional ac signal.

The analyzer is housed in a two-compartment enclosure suitable for surface mounting, one compartment containing the analysis unit, the other the control electronics. The enclosure is suitable for purging to meet hazardous area requirements. Measurement ranges from 0–100% down to 0–5 ppm are possible, depending upon the particular application. Two switch-selectable ranges are provided, with span ratios up to 1 : 10. Repeatability is quoted as ±0.05% of full scale.

Horiba Instruments (United States and Japan). Horiba manufactures a wide range of process NDIR analyzers, examples of which are the EIA and TIA Series. The EIA Series Analyzers cover full-scale ranges between 500 ppm and 100% for various gases. The TIA Series Analyzers are long path-length models of high sensitivity, providing full-scale ranges down to 50 ppm. Both series use dual IR sources, dual beams, and Luft-type detectors as shown in Fig. 10.20. The chopper is located between the sample/reference cells and the detector. Solid filters can be installed as shown to eliminate interferences.

A dual detector technique is used in applications where there is significant interference from a component in the sample or from external disturbances such as vibration. The main detector receives the signals of the measured sample component and the interfering component; the compensation detector, mounted behind the main detector, receives only the signal from the interfering component. The output from both detectors is amplified and the gains adjusted so that the portions of each signal caused by interferences are equal. A differential amplifier electronically subtracts the output of the compensation detector from the main detector to give an interference-free signal.

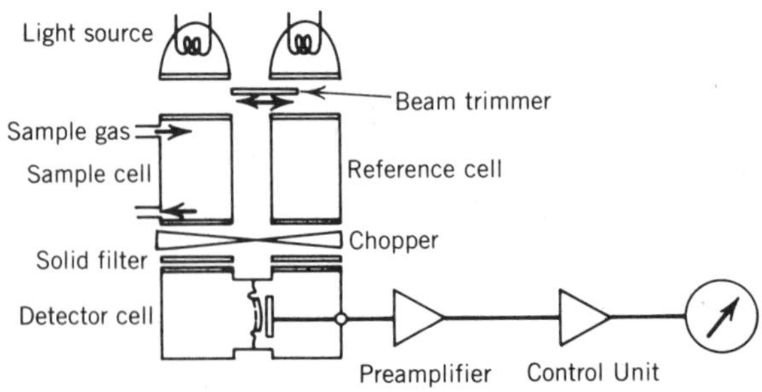

FIGURE 10.20 EIA & TIA Series NDIR analyzers. (Courtesy of Horiba Inst.)

Hartmann and Braun (West Germany). H & B manufactures a range of dual-beam NDIR analyzers using a single IR source and Luft-type detector, operating over the spectral range 2.5–12 μm.

The URAS 2T-Ex shown schematically in Fig. 10.21 is a single-component analyzer suitable for use in hazardous locations, providing a current output (0–20 or 4–20 mA dc) for remote recording. The analysis and control units are installed together as one system suitable for surface mounting.

The URAS 3 is available in two versions, one for surface mounting (3G), the other for 19-in. rack mounting (3E). The Luft-type detector in these analyzers is a four-chamber detector consisting of two sets of chambers mounted one behind the other, the front and rear chambers being separated by partly transparent and partly reflecting layers. Up to four measurement ranges can be provided with a maximum ratio of 1 : 10. These analyzers are suitable for installation in general purpose areas only.

The URAS 7N is available in surface mounting and 19-in. rack mounting configurations and is suitable for general purpose areas only. Design is similar to the URAS 2T-Ex.

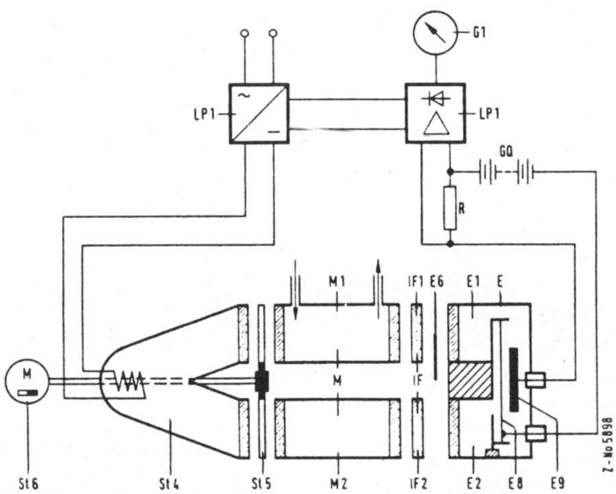

E	Detector	LP1	Main amplifier and power supplies
E1	Measuring chamber	M	Sample cell
E2	Reference chamber	M1	Analysis chamber
E6	Detector shutter	M2	Reference chamber
E8	Metal diaphragm	R	High-value resistor
E9	Counterelectrode	St4	Beam splitter
G1	Meter	St5	Chopper wheel
GQ	Direct voltage source	St6	Chopper wheel motor
IF	Interference filter		

IF1 } Optical filter with identical transmission characteristics
IF2 } in the measuring and reference radiation paths

FIGURE 10.21 URAS 2T-EX NDIR analyzer. (Courtesy of Hartmann & Braun)

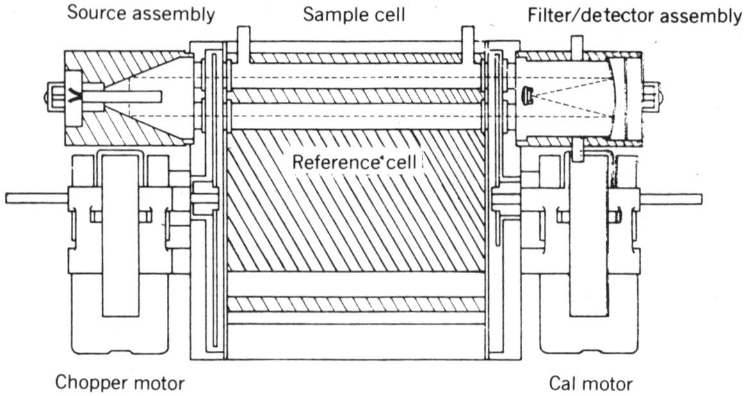

FIGURE 10.22 Anarad NDIR measurement system. (Courtesy of Anarad Inc.)

Anarad Inc. (United States). Anarad manufactures a wide range of NDIR analyzers based on the analysis system shown schematically in Fig. 10.22. This system uses state-of-the-art optical filter technology and a solid-state photodetector and was developed from the inexpensive exhaust emissions analyzers provided for the automotive industry. The single IR source is a nichrome wire wound and housed in a soft steel sheath. The source cavity itself is designed to collimate the IR energy through the sample and reference cells, eliminating the need for separate optical systems or surfaces. The analyzer has a variable path length up to 36 in., selected in accordance with the absorption characteristics and concentration of the component of interest.

The output end of the instrument, although requiring some optics, is not a focused system but rather a tuning cavity containing a solid-state filter and lead selenide photodetector. This detector is sensitized to a specific component by means of an optical interference filter with a band-pass centered around the absorption peak of the component of interest. Sapphire is used as the window substrate since it covers the range 0.18–6 μm and has excellent physical properties and chemical resistance. The analyzer has no exposure of the optical path to ambient air, allowing measurement of gas concentrations below those in the environment about the instrument. Internal calibration is provided by an optical wedge, which provides a full-scale signal with zero gas in the sample cell.

The AR-400 Series Analyzers are portable instruments with digital display, linearized output, and internal pump. The AR-500 Series Analyzers are designed for single component measurement and are provided in various packaging configurations to suit customer requirements. The AR-600 Series Analyzers are dual component instruments, again available in various packaging configurations.

The Model AR-701 is designed for measurement in liquid process streams. One filter/detector is used for the sample and a second filter/detector is used for generating the reference signal, allowing compensation for window deposits and sample turbidity. Analysis is carried out within the spectral range 2–5 μm.

Anarad also manufactures the AR-800 Series Multi-Stream Systems with automatic sequencing, which can be provided in various configurations to suit customer requirements. The AR-8000 Analyzer is a microprocessor-based system capable of handling up to 24 process streams, with automatic calibration, sequencing, and so on, and has a CRT/keyboard as the operator interface.

Stack CO monitoring systems are available, consisting of extractive sample handling system and analyzer and control units mounted in a single enclosure. The standard AR-500 series analyzers or the microprocessor-based AR-1500 Series analyzers may be used for this application.

(b) Filter Spectrometers

These instruments use narrow band-pass filters or non-specific thermal or solid-state detectors and are available in various configurations:

- Single-beam, dual-wavelength.
- Dual-beam, single-wavelength.
- Dual-beam, dual-wavelength.

The purpose of the dual-beam design is to compensate for source output variations and detector/electronics gain changes. However, it does not account for changes in sample cell transmission due to dirt accumulation on optics or sample turbidity.

The purpose of dual-wavelength design is to use a reference wavelength not absorbed by the sample component(s) but affected by the above sources of error to the same extent as the sample (analytical) wavelength.

The single-beam design is the simplest from an optical and mechanical viewpoint. This design is easily expandable to multi-component (multi-wavelength) analysis; filters may be mounted on a rotating filter wheel or inserted sequentially into the beam by a cam mechanism.

Combustion Engineering Process Analytics (United States). The Model 501 IR Analyzer shown in Fig. 10.23 is a single-beam, dual-wavelength instrument designed for measurements in process gases or liquids, operating over the spectral range 0.75–14.5 μm and using time-separated signals and ratio measurement. The analyzer consists of an analysis module and a control module, which may be separated by up to 500 ft. Installation of an optional communications board will allow extension beyond 500 ft.

FIGURE 10.23 Model 501 IR analyzer. (Courtesy of Combustion Eng., Process Analytics)

The analysis module is suitable for installation in NEC Class 1, Groups B, C, and D, Division 2 locations, with the capability for Type-Y purge to meet Division 1 requirements. Sample cells are available in lengths varying from 0.1 mm up to 512 mm, with rating to 150 psig operating pressure. The single detector is a pyroelectric device.

The control module contains all necessary circuitry to process the analyzer signal, generate analysis information, monitor the total system performance, and diagnose system failures. Standard output is 0–10 V dc; current output (4–20 mA dc) is available as an option. Precision is quoted as $\pm 1\%$ of full scale.

Anatek Instruments Ltd. (Great Britain). The Model PSA402 IR Analyzer shown in Fig. 10.24 is a single-beam, dual-wavelength analyzer consisting of a sealed analysis unit and an electronics unit, which may be separated by up to 1500 ft and connected by a standard 12-core cable, screened overall. The analyzer is suitable for measurement in process gases or liquids and covers the spectral range 1–10 μm. The IR source is a hot glazed platinum filament mounted inside the source body. IR radiation is collected by the primary lens and condensed through the filter wheel, which is spun at 2400 rpm.

Four filters mounted on the filter wheel (two measurement, two reference) are passed through the beam alternately such that a ref-meas-ref-meas beam is generated. Radiation from the filters is focused down the sample cell and condensed onto a solid-state detector. The ratio of the reference signal to the measurement signal is measured and amplified to give an output proportional to component concentration. A timing pulse from the filter wheel is used to synchronize the digital sampling pulses to the detector signal.

The analysis unit consists of the source and detector compartments, separated by the sample cell. The system is sealed to prevent ambient ingress to the optical path. The sample cell can be heated and temperature-controlled. Temperatures up to 70°C are standard, but a 150°C option is offered. Other options include isolated outputs, four-level alarm, automatic zeroing and span correction, dual range, and sample pressure/temperature compensation. The control unit is suitable for surface or panel mounting and contains

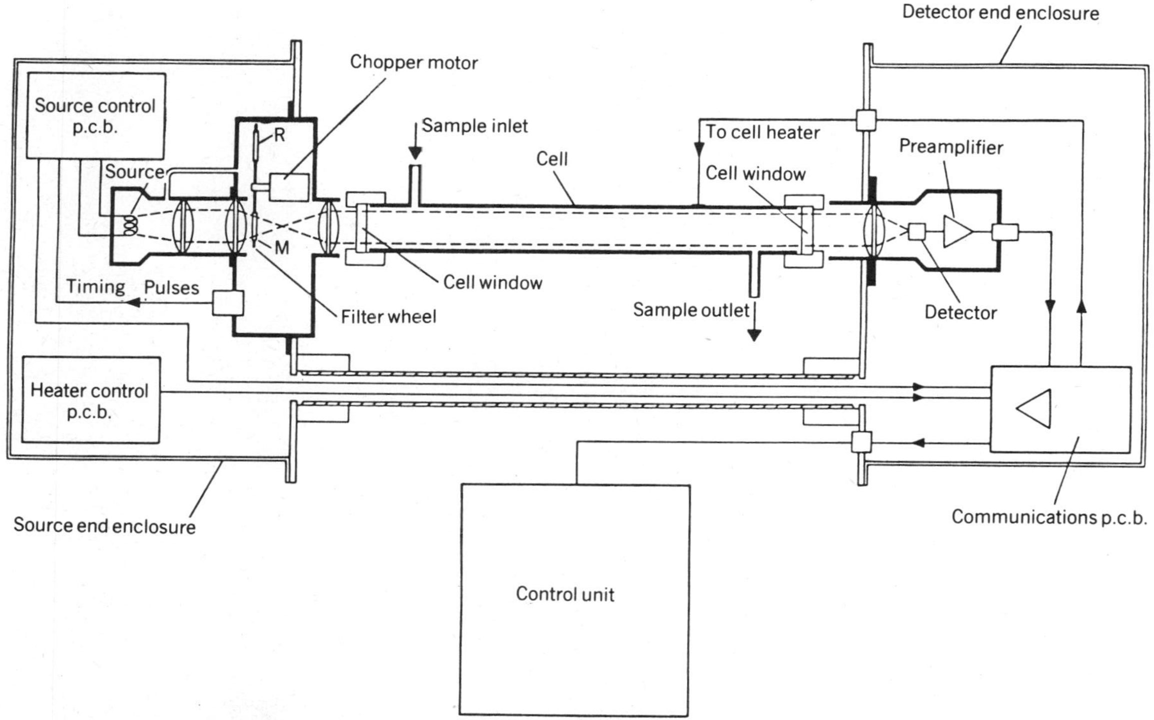

FIGURE 10.24 Model PSA402 IR analyzer. (Courtesy of Anatek Inst. Ltd.)

the display and plug-in circuit cards, which process the analyzer signal, generate analysis data, and monitor system performance. Repeatability is quoted as ±0.5% of full scale and response time as 90% of full scale in 5 sec.

Foxboro Analytical (United States). The Model 973 MIRAN IR Analyzer shown in Fig. 10.25 is a single-beam, dual-wavelength instrument operating over the spectral range 2.5–14.5 μm. IR radiation from the nichrome source is collected by a concave focusing mirror and the beam directed to the chopper wheel, which is an eight-segment disk. The beam is focused onto a slit after passing through one of two narrow band-pass filters alternately positioned in the beam. The beam then passes through the sample cell and is focused onto the detector, which is of the pyroelectric type, composed of a single crystal of lithium tantalate.

With every rotation of the chopper, a cam positions each filter segment in the IR beam. The filter flag and chopper wheel interrupt reference LEDs and the pulsating signals so produced are converted to electrical signals for use in the detector signal demodulation process.

The analyzer is modular in design and consists of optical bench and sample cell, transmitter, and control unit. The transmitter mates with either of two optical benches, which together accept the entire Foxboro range of IR sample cells, including a multiple internal reflection (MIR) cell (MIR is discussed in the next section). The transmitter provides balancing, demodulation, and filtering of the sample and reference signals. The output signals are then transmitted to the control unit via two 0–20 mA current loops. The control unit provides zero and span adjustments, digital indication, and internal system diagnostics, and contains the electronics for signal processing and for the various available options, such as dual range, autozero, linearizing, dual alarm, and so on. Analyzer and control units may be separated by up to 1000 ft.

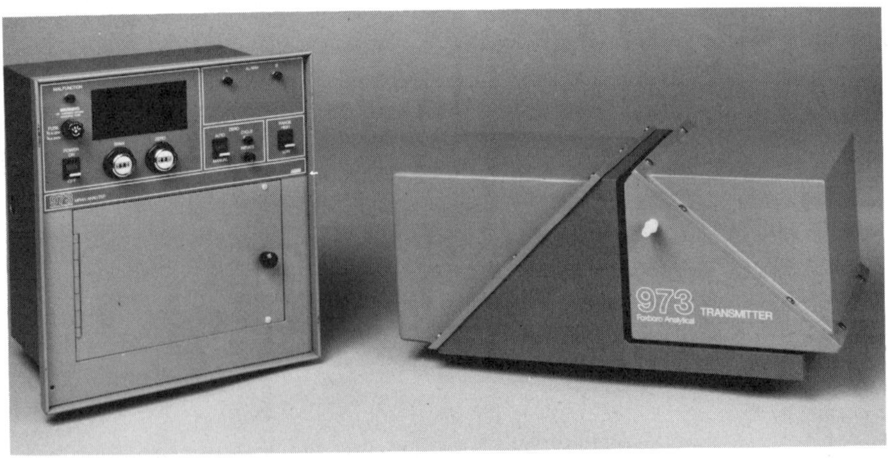

FIGURE 10.25 Model 973 MIRAN IR analyzer. (Courtesy of The Foxboro Company)

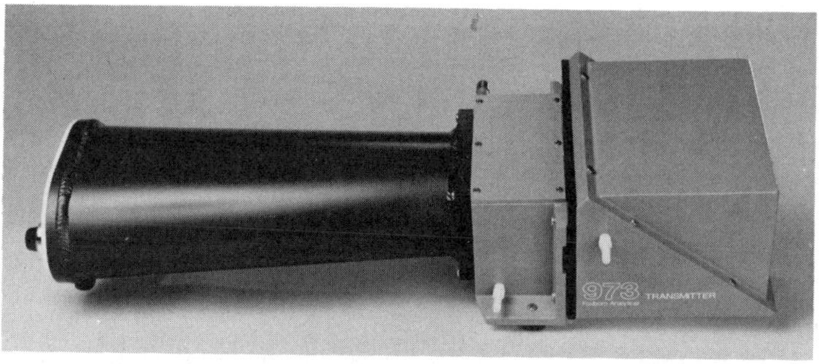

FIGURE 10.26 Foxboro Model 432 IR sample cell shown fitted to Model 973 analyzer. (Courtesy of The Foxboro Company)

The Model 420 Sample Cells range from 0.1 mm up to 3.0 mm pathlengths; pathlength is determined by spacer thickness and is changeable in the field. The Models 421, 422, 423 and 424 Sample Cells are 5 mm, 1 cm, 3 cm, and 5 cm respectively, not field-changeable. The Model 431 Cell has a 50-cm pathlength, while the Model 432 Sample Cell shown in Fig. 10.26 has an adjustable pathlength from 0.75–20.25 m in 1.5 m increments. This long pathlength is provided in a relatively small package by using mirrors to reflect radiation through multiple paths within the cell.

(c) Multiple Internal Reflection (MIR) Analyzers

The technique of MIR, also called attenuated total reflection, is used primarily for the analysis of liquid samples since it avoids the severe plugging problems associated with the very thin transmission-type cells necessary for these applications. The IR beam in this case makes multiple internal reflections at the surface of a high-refractive index crystal wetted by the sample liquid.

The MIR technique makes use of the fact that when a beam of radiation moving within a transparent medium is reflected internally from a surface a portion of the energy in the beam projects slightly beyond the reflecting surface. When air is in contact with the surface there is negligible attenuation of the beam energy. However, if an absorbing liquid is brought into contact with the surface, energy will be absorbed from the beam at the wavelengths at which the liquid normally absorbs (as in a transmission cell). The energy of the beam projects only a few micrometers into the sample; therefore, any sample beyond that point has no effect on the measurement.

Since one reflection does not normally produce enough absorption to be measured easily, the optical system must provide multiple reflections (usually about 10). The MIR crystal can be made from materials such as zinc sulfide, sapphire, germanium, or silicon; the crystal selection is based on the type of sample under analysis and the inertness of the crystal material to this sample.

General Analysis Corporation (United States). The LAN-1 Liquid Stream Monitor operates on the MIR principle and is shown schematically in Fig. 10.27. The heart of the system is a polished internal reflection cylinder with cone-shaped ends made, for example, from a zinc sulfide crystal. Surrounding this crystal is a glass or stainless steel sample chamber sealed to the cylinder by teflon O-rings. A pair of short focal length, high-aperture mirrors collect radiation from the nichrome source and focus it at the correct angle on the entrance cone of the cylinder so that the radiation reflects internally along the cylinder. The emerging radiation at the exit cone of the cylinder is collected by a similar pair of mirrors and focused onto two detectors. A two-sided flat mirror directs half of the radiation to each detector. Narrow band-pass filters, which are the detector windows, admit only the selected analytical and reference wavelengths. A chopper driven by a clock motor chops both beams simultaneously. The system has a very high energy throughput and provides high signal-to-noise ratio with very little drift.

The analyzer consists of two modules, the analysis and control modules, mounted together to form one complete system. The analysis module is sealed and temperature controlled at 43°C. The control unit has a digital indicator, power switch, zero/span controls, and diagnostic malfunction indicators. Internal controls are accessible by removing the front cover plate. Voltage or current outputs are available for remote recording.

A recent addition by GAC is the LAN-IV Liquid Stream Monitor, shown in Fig. 10.28, which uses the same basic operating principle as the LAN-I but is capable of providing absorbance data continuously at four different wavelengths. Wavelength selection is made by interference filters mounted as an integral part of each detector container. A microprocessor or computer may be used to compute the concentration of up to three components in a process stream using these absorption data.

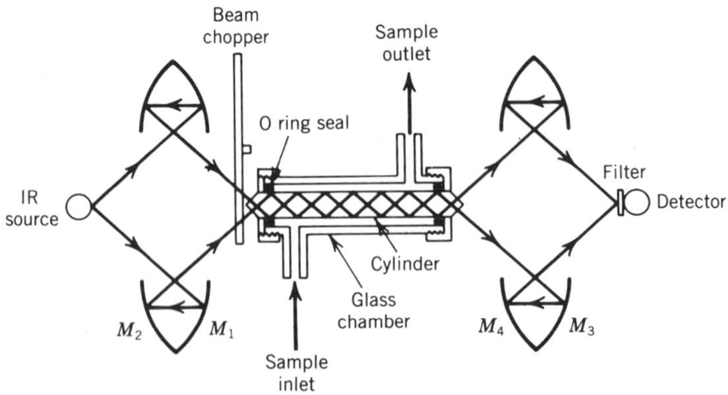

FIGURE 10.27 LAN-1 liquid stream monitor. (Courtesy of General Analysis Corporation)

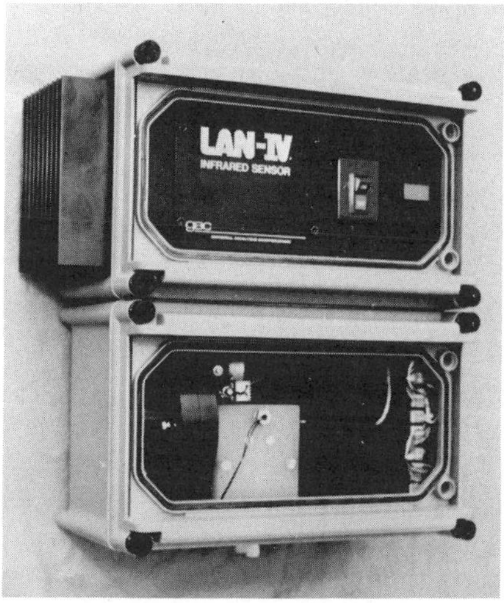

FIGURE 10.28 LAN-IV: equipment layout. (Courtesy of General Analysis Corporation)

(d) Infrared Gages for Paper Coating and Moisture Measurement

One of the major limitations of coating technology is the difficulty in measuring the amount of coating applied to paper. Paper is a notoriously variable material, showing fluctuations in basis weight and the ability to absorb solvents. The traditional instrument for coating or moisture measurement is the beta absorption gage, generally using two measuring heads in a differential configuration (for before and after coating). In the past decade, much work has been done on the use of IR gages (often called Infragages) in this and other applications involving measurement in solids. Some devices for the measurement of moisture in solids are described in Chapter 11. The Infragage is much simpler, since only one measuring head is required downstream of the coating head.

Infrared Engineering (Great Britain). Infrared Engineering has been involved in the measurement of moisture in paper, tobacco, and other solid products using IR techniques since about 1970. Apart from the UK, instruments are now in operation in Europe, the Far East and the United States. The latest range of analyzers includes the Model MM4, a low cost gage; the Model SM4, a more sophisticated instrument; and microprocessor-based analysis and control systems.

The Model MM4 is shown schematically in Fig. 10.29. IR radiation from the source is focused by lenses onto the conveyed product (e.g., tobacco) after passing through a filter wheel carrying two optical filters (analytical and reference). The filter wheel is rotated at a speed between 1000 and 4000 rpm. Scattered radiation is collected by a concave mirror and focused onto a solid-state detector. The detector signal is split into components representing scattered radiation intensity at the reference and analytical wavelengths;

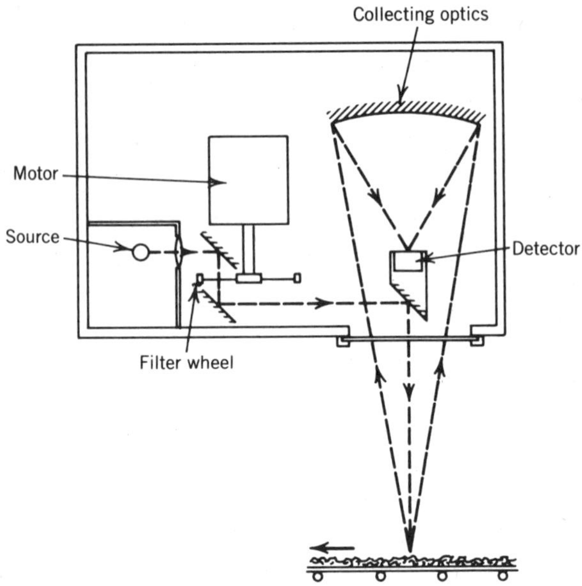

FIGURE 10.29 Model MM4 infragage. (Courtesy of Infrared Eng. Ltd.)

the ratio of the components is taken and processed electronically to generate an output proportional to moisture content.

The Model SM4 incorporates a number of special features. An optical modulator causes the detector signal to be superimposed on a carrier frequency, providing better noise and interference rejection. Four optical filters are used, two of which provide reference wavelengths, to minimize blend sensitivity and improve precision and long-term stability. Both models incorporate temperature compensation. Control units are modular in construction, incorporating a digital display and built-in diagnostics.

The Model DS4 is a microprocessor-based data reduction and display system with CRT/keyboard operator interface. It will accept a wide range of analog and digital inputs from plant and ancillary equipment. The Model DP4 is a microprocessor-based control unit designed to interface with the Series 4 Infragages.

10.4.3 Process Ultraviolet/Visible Analyzers

Process UV/visible analyzers generally operate on the same basic principles as process IR analyzers, except that the source and detector are designed specifically for the UV/visible region. By suitable interchange of the source and the detector, it is possible to extend the spectral range of some instruments to include the near IR.

Two typical examples of UV analyzers (often called photometric analyzers), the DuPont Model 400 and the Teledyne Model 600, are described in detail in Chapter 9 (Section 9.4.3) and shown in Figs. 9.5 and 9.6 respectively. As with process IR analyzers, these instruments are applied to measurement of the concentration of specific components in a gas stream. Visible absorption analyzers (also called photometers) are used to monitor the opacity (turbidity) or color of process liquids. Instruments by Sigrist, Monitek, Great Lakes, and Hach are described in Chapter 13.

Pier Electronic GmbH (West Germany). The Alternating Light Photometer operates on the chopped light principle and is designed to compensate for source and detector aging and sample cell contamination. Measurements can be made within the following three ranges:

Ultraviolet	250–400 nm
Visible	400–700 nm
Near infrared	700–2000 nm

One of the principal applications of this analyzer is the measurement of water in a variety of different substances (e.g., alcohols, acids, ether, acetone, benzene, etc.), using the strong absorption line for water at 1940 nm, with water concentrations measurable to about 1000 ppm.

The analyzer unit is shown schematically in Fig. 10.30. The reference and measuring beams are generated by two optical filters mounted on a filter wheel that rotates at 3000 rpm. The two beams are directed alternately along a path containing the sample cuvette and onto the photodetector. The detector output signals are ratioed to give an output proportional to the measured component concentration.

The measuring head contains the light source, filter wheel and synchronous motor, sample cuvette compartment, detector, and preamplifier,

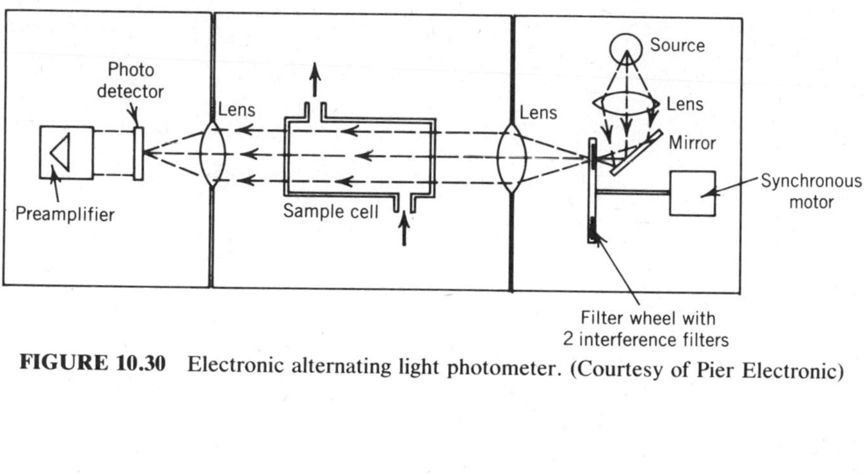

FIGURE 10.30 Electronic alternating light photometer. (Courtesy of Pier Electronic)

housed in a light metal enclosure. Both static and flow-through cuvettes are available, the latter made of glass or stainless steel. The measuring head and control unit may be separated by distances up to 100 m.

For the UV region, the standard source is a high-pressure mercury vapor lamp; measurement is restricted to the particular lines emitted—254, 302, 313, 366, and 436 nm. Special sources can be supplied for other wavelengths. The physical arrangement of the UV instrument is slightly different from that shown in Fig. 10.30 because of the heat generated by the source.

Fluid Data Inc (Great Britain and United States). The Hallikainen Model 7612 Opacity Alarm is shown schematically in Fig. 10.31*a*. Light from a tungsten lamp illuminates the input to a fiber optic "Y" piece. One arm leads to the sample cell, the other to a reference light attenuator. The two emerging beams are transmitted via fiber optics to a photocell, where a chopper selects the beams alternately. The detector signals are stored and compared;

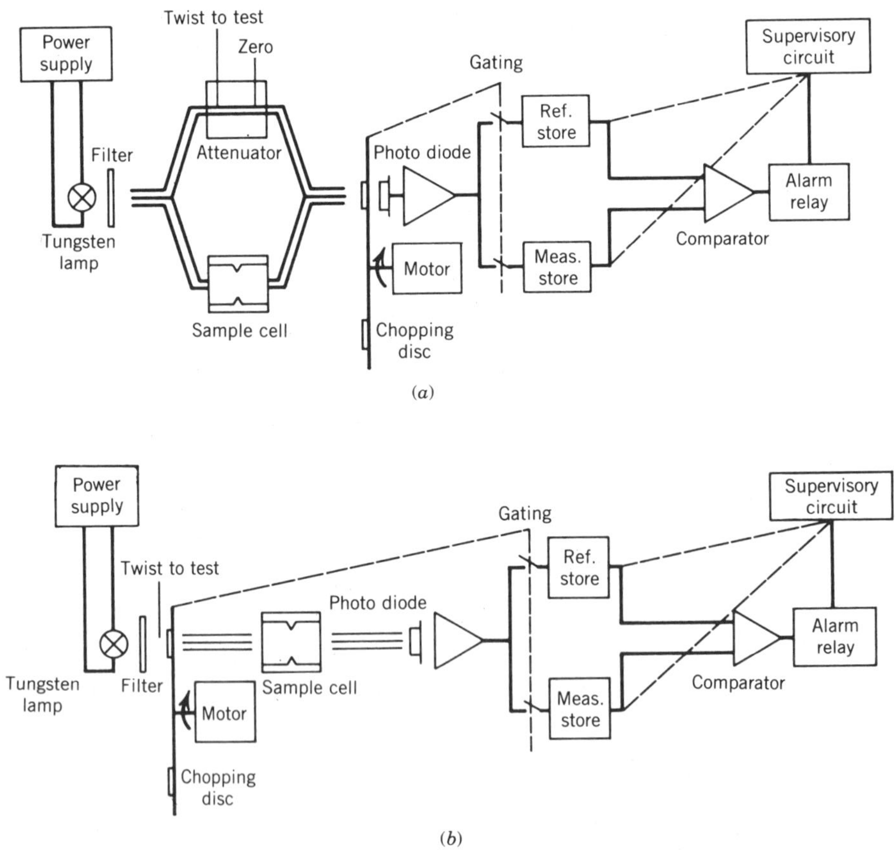

FIGURE 10.31 Hallikainen alarms. (*a*) Model 7612 opacity alarm; (*b*) Model 7613 color alarm. (Courtesy of Fluid Data Inc.)

if they are different by more than a preset amount, an alarm is sounded. The system is designed for the measurement of turbidity (or opacity) in process liquids.

The Hallikainen Model 7613 Color Alarm shown in Fig. 10.31*b* differs from the Model 7612 in that the chopper wheel is located after the light source and has two filters chosen to provide a suitable reference and measurement wavelength for the particular application. The beams alternately pass through the sample cell and onto a photocell. The comparator and alarm circuitry is similar to the Model 7612. The analyzer is designed to measure the color of process liquids in accordance with the Saybolt or ASTM color scales and provide an alarm if the color exceeds a preset limit (light or dark). Analyzer ranges are:

Saybolt Color +30 to −16	(using the 40 mm cell)
ASTM Color No. 0–3	(using the 40 mm cell)
ASTM Color No. 0–7	(using the 6 mm cell)

A current (4–20 mA dc) output is also available as an option.

10.4.4 On-Line Mass Spectrometers

(a) CVC Products Inc (United States)

CVC has been involved with the engineering and manufacture of time-of-flight and quadrupole mass spectrometers since their origins and has over 75 years experience in the design of high-vacuum components and systems.

The SuperSpec 600 Process Gas Analyzer is a time-of-flight mass spectrometer designed specifically for industrial on-line process monitoring and control, the development of which began in 1977. The analyzer has a wide mass range (1–600 amu) with resolution of better than 1 amu over the range 1–200 amu and mass scale stability of ±0.1 amu over 8 hr. Detectability limit is quoted as 5 ppm for most gases. The analyzer allows very fast response in monitoring up to 16 components per stream and capability to handle up to 16 streams.

A schematic diagram of the analyzer is shown in Fig. 10.32. A two-stage inlet system is used, consisting of a capillary restriction and a gold foil molecular leak. The ion source uses a tungsten filament and an open "stacking" configuration of its elements, allowing easy and rapid cleaning and replacement.

The analysis portion of the instrument is an empty stainless steel tube, housing no components requiring alignment and insensitive to contamination. The detector is a patented magnetic electron multiplier with high gain and stability and very low noise level. Ion source, analyzer, and detector are located within a high-conductance, horizontal tubular vacuum housing suitable for continuous operation at 200°C. A choice of five pumping systems is

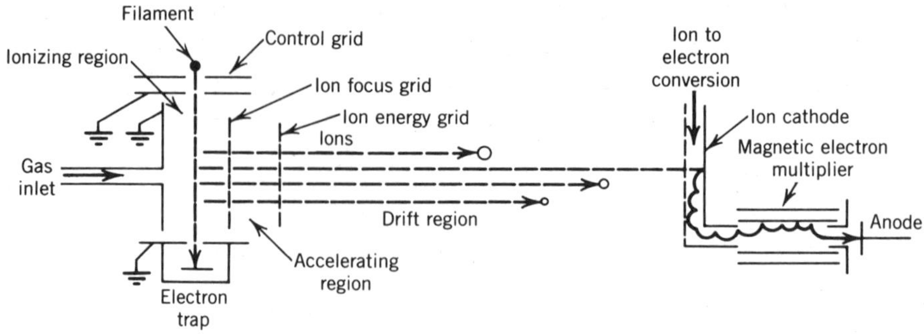

FIGURE 10.32 SuperSpec 600 mass spectrometer. (Courtesy of CVC Inc.)

offered to maintain the high vacuum required in the instrument. The standard is an oversized turbo-molecular pumping system (120 l/sec). Full electronics protection is provided against high vacuum failure.

The system can provide a complete mass spectrum in real time on an oscilloscope. This display is available at all times, even during slow scanning and peak sequencing. An analog scanner is provided for recording the complete analog mass spectrum where needed. Minimum analog scan time is 3 sec per complete mass range. A multiple component monitor (MCM) is provided to allow the operator to determine up to six mass-peaks to be monitored sequentially, with variable dwell time of 0.1–10 sec per peak. The basic SuperSpec 600 can be provided with an unlimited number of MCMs.

The microprocessor package is designed to provide complete control of the mass spectrometer operation, including mass spectra data acquisition, processing, and readout, allowing the user to exercise maximum control of the mass selection in any run, while permitting great flexibility in acquisition and reduction of data. The system is capable of analyzing up to 16 components in a 16 × 16 matrix algebra program that provides concentration readings in a few seconds.

Five critical outputs are provided in tabular mode and continuously updated on a 19-in. color CRT:

1. Process stream monitored.
2. Concentration of each of 16 components.
3. Stream number.
4. Time of day.
5. Run number.

Printer interfaces are provided in Serial ASCII, EIA, and 20 mA current loop format, along with an RS232C interface for a host computer. The operator conversational interface is a standard keyboard. Other output modes include bargraph, reconstructed analog spectrum, and trend display of each component for each stream.

The system includes extensive internal diagnostic capability for mass spectrometer electronics, vacuum system, and host/CVC factory computer diagnostics (via a remote communications modem) for ease of maintenance. An auto recalibration scheme is also incorporated in the computer program to ensure accurate, reliable data without the need to shut down and manually calibrate, maximizing uptime. In addition, the microprocessor package controls automatic stream switching/programming.

The SuperSpec 600 is housed in a NEMA enclosure meeting NEC Class 1, Group D, Division 2 requirements. The CRT/Terminal is designed for general purpose locations only.

(b) ExtraNuclear Laboratories Inc. (United States)

The Model EL-150 Quadrupole Mass Spectrometer is designed specifically for on-line process analysis and operates on the quadrupole mass filter principle discussed earlier. A schematic flow diagram of the analyzer is shown in Fig. 10.33. Instrument range is 1–200 amu, with a lower concentration detection limit of 10 ppm. The standard EL-150 can monitor up to 12 components per stream, with a minimum response time of 100 msec per component, and has multistream capability.

The analyzer data and control system is based on the DEC PDP-11 series of computers, with 1 Mbyte dual floppy disk storage and a VT-110 terminal with keyboard and CRT display. Software and hardware are designed specifically for process use, require minimal operational training, but provide the flexibility required for a variety of applications. The system enables the user to design analysis, calibration, and sample sequencing methods to optimize conditions for a particular application. These methods, once designed, can be stored on floppy disk, allowing the user to design different procedures for each stream to be analyzed. A sample sequencing method can be designed for automatic sequencing of up to 16 valves.

The EL-150 has three modes of operation. In the sweep mode, via CRT/keyboard command, it can be made to sweep the quadrupole over a specified range, with data displayed on an oscilloscope. This mode is useful for setup, tuning, and determining the masses present in a particular sample stream.

In the manual mode, real-time concentrations are displayed on the CRT. The operator can change all analyzer parameters for fine tuning purposes. Data can be presented as intensity (raw signal voltages), percent concentrations, or as a ratio to a specified component. Data can also be printed or stored on disk.

In the automatic mode, analysis, calibration, and stream sequencing are carried out automatically by the EL-150 in accordance with the methods stored on disk.

The analyzer is housed in a NEMA 12 industrial cabinet fitted with casters for mobility and suitable for installation in NEC Class 1, Group D, Division 2 hazardous locations. The VT-100 Terminal is designed for General Purpose Locations only.

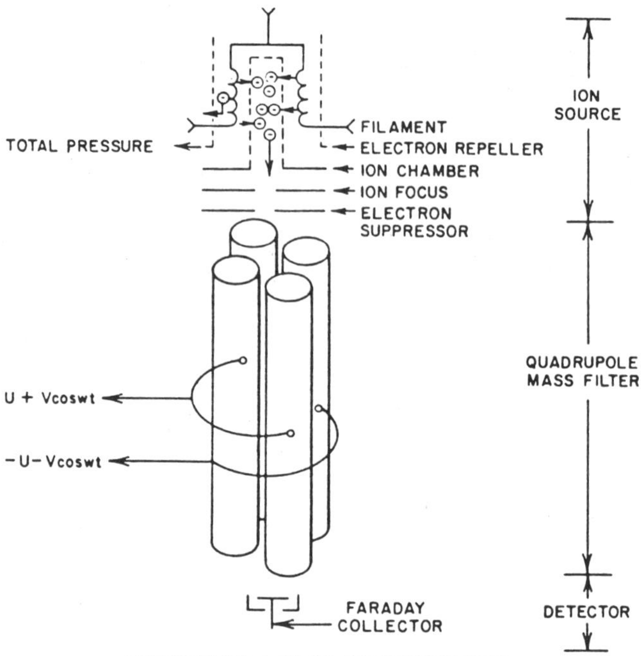

QUADRUPOLE MASS SPECTROMETER

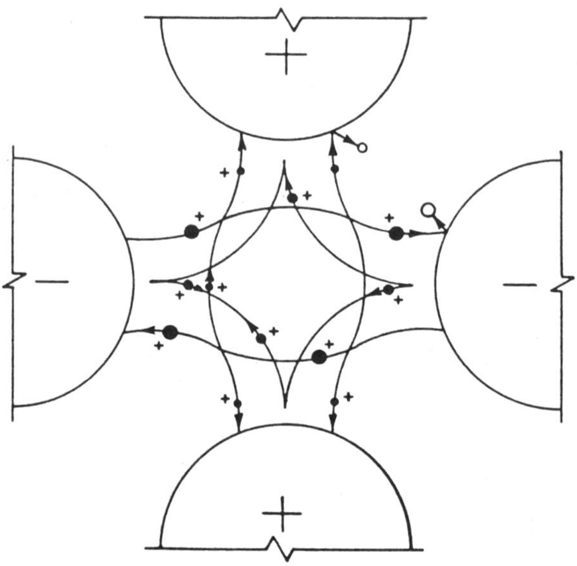

○, ◯ NEUTRAL GAS SPECIES
• SMALL MASSES NEUTRALIZE ON POSITIVE RODS
● TUNED MASSES PASS THROUGH FILTER
⬤ LARGE MASSES NEUTRALIZE ON NEGATIVE RODS

QUADRUPOLE MASS FILTER

FIGURE 10.33 EL-150 mass spectrometer. (Courtesy of ExtraNuclear Labs.)

(c) *Perkin Elmer Corporation (United States)*

The multiple gas analyzer (MGA) is a magnetic deflection type on-line mass spectrometer, originally developed in the late 1960s by Perkin Elmer's Aerospace Division. The Model MGA 1200 was introduced in the early 1970s, initially in the medical field, and was extensively evaluated by Dow Chemical U.S.A. in 1974. In 1980, over 400 of these instruments were reported to be in operation. A new instrument, the Model MGA 1400, was introduced in 1983 and exhibited at the ISA Show in Houston, Texas, in October of that year.

The Basic Model MGA 1200 is designed specifically for on-line monitoring of up to eight components in a gaseous process stream and covers the mass range 2–135 amu. Capability is also provided for monitoring up to four process streams. A schematic diagram of the analyzer is shown in Fig. 10.34. Sample enters the ion chamber from a flowby inlet system and molecular leak (a sintered metal disc). A capillary inlet system is also available, driven by the MGA rough vacuum pump. The ion source design shown in Fig. 10.35 has two filaments to provide backup in the event of filament burnout. Only one electron gun is used in actual operation; should that filament fail, the instrument can be manually switched to the second filament. System vacuum is maintained by a 20 l/sec ion pump, with a typical operating life of 10,000–15,000 hr.

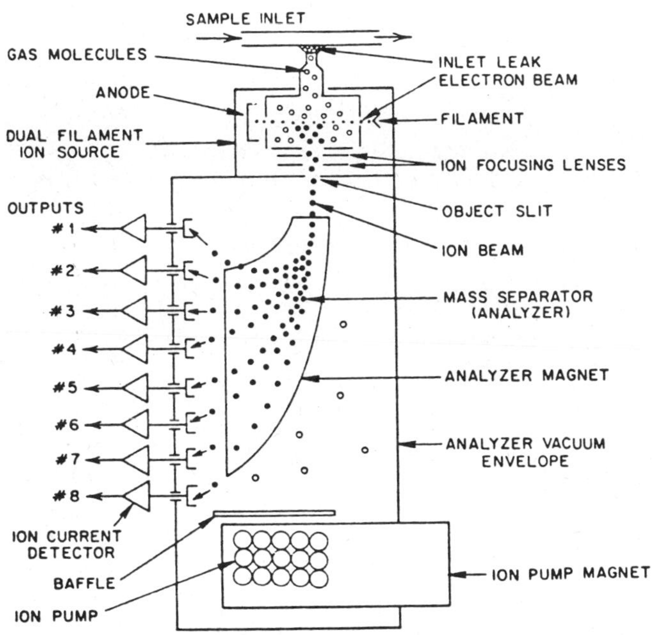

FIGURE 10.34 MGA-1200 mass spectrometer. (Courtesy of Perkin Elmer Corporation)

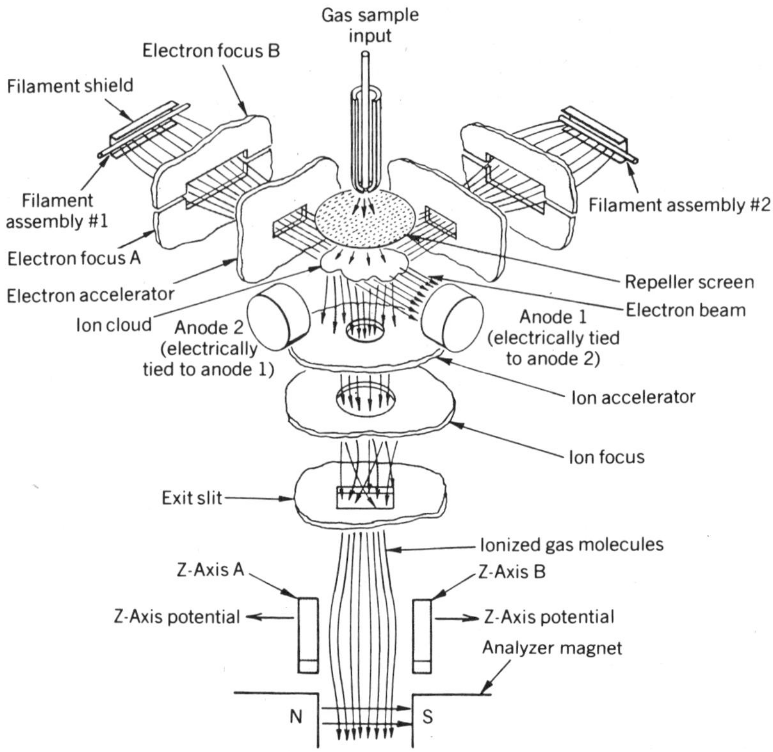

FIGURE 10.35 MGA-1200: ion source design. (Courtesy of Perkin Elmer Corporation)

The analyzer is designed with multiple Faraday collectors, rather than one collector with a scanning system. For each specific application, collectors are located at predetermined points in the focal plane of the instrument. A special patented span calibration technique is used to compensate for common-mode, drift-type errors inherent in multiple-collector mass spectrometers.

An analog output (0–10 V dc or 4–20 mA dc) is available for each measured component for remote recording purposes. The front panel digital display is used for readout of individual components (in percent composition) and during calibration; an interrogation switch allows the operator to select any component. On the display panel there are up to four status indicators for the inlet system, the vacuum system, and the interrogation switch position. The user may manually select any inlet for analysis with the front panel selector switch; in addition, automatic selection is available on a time-multiplexed or random-selection basis via an external computer.

Time-multiplexed selection is accomplished by a separate printed circuit board in the MGA. Each of four selection channels may be programmed for a dwell time of 1–9 sec in 1-sec steps, 10–90 sec in 10-sec steps, or 100–900 sec in 100-sec steps.

The analyzer is housed in a wheel-mounted cabinet, suitable for installation in general purpose locations, or may be suitably purged to meet hazardous area requirements.

The Model MGA 1400 is a 90° magnetic deflection mass spectrometer with four fixed Faraday collectors and is microprocessor controlled. The analyzer is capable of monitoring up to 16 components per process stream and up to eight process streams, with a response time of 10-sec per process stream (based on 12 components). Two standard external stream switching manifolds are available, one providing for four-stream sampling, the other for eight-stream sampling. These manifolds are housed in a NEMA 12 enclosure, electrically heated to 75°C, with a Z-type purge. Two standard calibration check gas manifolds are available, for eight and 11 binary calibration gas mixes plus a check gas.

The microprocessor and its associated CRT display and keyboard provides the operator interface, and controls accelerating voltage, monitors and displays key operating voltages, and provides automatic stream switching and system calibration. All changes to operating parameters are made via the keyboard. Interfaces are provided for printer and host computer communications (RS232 or RS422).

The analyzer is housed in a general purpose or X-purged, wheel-mounted cabinet. The CRT display and keyboard are designed for remote location in a general purpose area.

10.4.5 X-Ray Fluorescence Analyzers

X-ray fluorescence (XRF) is used extensively in the mining and metallurgical industries for rapid on-line elemental analysis. It is especially popular as an analytical technique where bulk material handling is involved (e.g., alloy separation and verification).

Until a few years ago, the technology was virtually unknown outside of these industries. Today, XRF is being used in the petroleum industry in applications such as the measurement of sulfur in fuel oil, lead in gasoline, and sulfur in petroleum coke. Among its industrial applications are the analysis of coal slurries in synthetic fuels production, silicone coating weight on paper and silver on photographic film. It has widespread use as a laboratory technique for elemental analysis on a variety of solid and liquid process samples.

As with any other on-line analyzer, sampling can be a major problem and is inherently more difficult with slurries than with liquids, gases, or finely pulverized, homogeneous solids. One other problem in applying XRF to on-line process analysis, particularly in the petroleum and petrochemical industries, is the low pressure (virtually atmospheric) required at the analyzer sample cell. This can create major difficulties when the process stream operates at high pressure (e.g., 1500 psig) as in many catalytic processes. Therefore, to date, the use of XRF analyzers for on-line process monitoring and control in petroleum and petrochemicals processes has been somewhat lim-

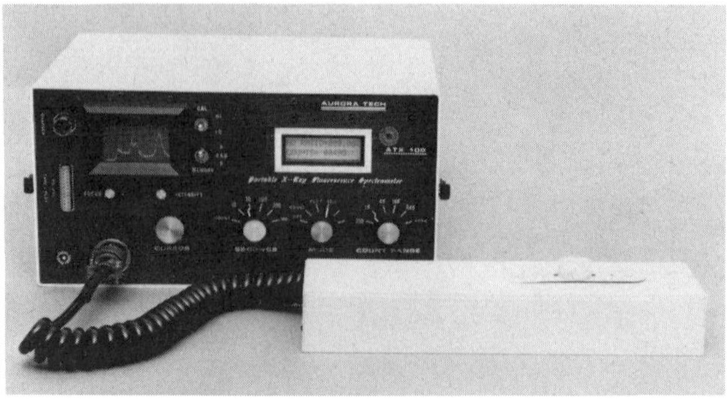

FIGURE 10.36 ATX-100 portable XRF spectrometer. (Courtesy of Aurora Tech Inc.)

ited. However, as technology advances, we may well see a change in this situation during the next few years.

(a) Aurora Technology Inc. (United States)

The Model ATX-100 Portable X-Ray Fluorescence Spectrometer shown in Fig. 10.36 is a miniaturized, self-contained instrument that can detect and analyze over 60 elements. The ATX-100 was originally designed by Martin Marrietta, NASA, and the Bureau of Mines for the Viking Lander's mission to Mars and performed successfully in this very harsh environment for over two years. The instrument is suitable for elemental analysis in the chemical and mineral industries, in metal recovery and pollution monitoring, and in the medical field.

The system consists of a sensor head and a portable control unit. The sensor head, using isotopic emitters (standard sources ^{109}Cd and ^{55}Fe), excites X-ray fluorescence in the sample, which is measured by a high-resolution proportional detector. The control unit has a small CRT on which the element spectrum is displayed, and a 32-character liquid crystal alphanumeric strip that displays elements, count, and ratio simultaneously. The analyzer records the element energy level and percentage concentration in the sample, performing the analysis in about 30 sec. The unit may be mains or battery-powered and weighs about 16 lb.

(b) Kevex Corporation (United States)

The Analyst 6600 is used extensively in the metallurgical industry, primarily in materials identification (e.g., alloy sorting and scrap metals).

An X-ray source contained in a sample-handling subsystem (front end) excites the sample, producing characteristic X-ray emissions for each element. A sensitive detector digitizes the signals and transmits them to a computer-based control unit. Analysis results are computed and displayed in 5 sec or less.

A variety of sample chambers and accessories are offered for testing materials in almost any form, including powders and liquids. The X-Site 9900 is a complete hand-held excitation-detection system, which can be separated from the control unit by distances up to 75 ft. A built-in eight-character display provides on-the-spot readings when the unit trigger is pulled. The Model 9010 is designed for handling standard size samples and is an 8 × 17 × 23-in. chamber fitted integral with the control unit. X-ray tube voltage and current can be adjusted for various types of samples. The Model 9025 is designed for oversized samples and is ideal for many automatic testing applications. The Model 9250B is designed for analysis of light elements such as silicon and phosphorus and provides an inert gas or vacuum environment within a sample drawer.

The 6600 console contains all necessary electronics for data processing, system control, and internal self-diagnostics. A CRT and keyboard provide the operator interface. Computational software is available in two versions. One analysis program accurately identifies iron, nickel, copper, cobalt, and titanium alloys. A more sophisticated program computes composition in weight percent for up to 18 elements (from a listing of 26) and, by using preprogrammed standards, is capable of identifying up to 470 different alloys.

The Analyst 6700 shown in Fig. 10.37 was introduced at the 1982 Pittsburgh Conference and is designed to extend the range of applications of XRF

FIGURE 10.37 Analyst 6700 XRF analyzer: control unit. (Courtesy of Kevex Corporation)

spectroscopy already covered by the Analyst 6600. Improved software and hardware emphasize compactness, versatility, maintainability, and analytical power, together with extensive internal self-diagnostics. The same range of sampling subsystems is available as for the 6600, except that the model numbers are slightly different. A 5-in. video screen with improved resolution and a three-position tilt base form the operator interface. The front panel also contains a thermal printer and a membrane keyboard with special function keys. System options include a redesigned tape drive and a modem, which allows communication between 6700s at different locations or directly to Kevex Systems for troubleshooting, and so on. Analysis programs provide improved accuracy, flexibility, and advanced data correction routines.

(c) Princeton GammaTech Inc. (United States)

The Model PGT-100 X-Ray Fluorescence Analyzer shown in Fig. 10.38 is a bench-top, push-button XRF spectrometer, designed to measure up to three elements in a liquid, solid, or powder sample. By virtue of its modular design, this analyzer is adaptable for use as an on-line process analyzer and has been used successfully in this way.

The probe head, which is covered by a shielded enclosure on the top of the analyzer, houses the excitation source and X-ray detector. The analog electronics are factory tuned to select those detector pulses representing the elements of interest; the digital circuitry converts this information into the form required for display. When the analysis is completed (usually from 10 to 200 sec, depending on the accuracy required), the result is shown on a digital LED display in the customer-specified engineering units.

Two basic count modes may be selected via a thumbwheel switch—preset time or preset count. Automatic or manual modes are also switch-selectable.

FIGURE 10.38 Model PGT-100 XRF analyzer: field unit installation. (Courtesy of Princeton Gamma Tech Inc.)

In the automatic mode, used primarily in on-line analysis, the analyzer repeats sample measurements without further instructions and displays the result after each measurement. Safety facilities include a safety interlock, which prevents the source shutter from opening if the cover shield is open.

The Model PGT-800 is an XRF analyzer designed specifically for oil analysis. Typical applications include elemental analysis of engine oils, cutting oils, gear oils, hydraulic oils, and auto transmission fluids. The system is housed in a compact, free-standing cabinet. The X-ray source is a low-power air-cooled X-ray tube with a silver target; the detector is a high-resolution lithium-drifted silicon device.

(d) Texas Nuclear (United States)

The INSCAN System is a multi-probe XRF analysis system designed primarily for mineral processing applications involving measurement of the concentrations of several elements in process liquids or slurries. The system is capable of supporting up to eight analysis probes. These are in most cases placed directly in the process stream, although optional facilities are available to allow slip-stream configurations. The system has a central controller, based on the DEC PDP 11/23 computer, for data acquisition, total system control, and internal self-diagnostics.

The central controller is housed in a two-door, NEMA 12 enclosure, which is temperature controlled to protect the electronics. All analysis probe support electronics is located within the controller enclosure. The data analysis system consists of the PDP 11/23 computer, dual floppy disk system, video graphics terminal, and a printer for analysis report printing.

The Model 9800 Analysis Probe Assembly shown in Fig. 10.39 contains the isotopic X-ray source (typically Pu is used), a high-resolution lithium-drifted silicon detector, and the dual-mylar window, source holder/shutter assembly. Moisture sensors are located between the windows and behind the inside window. A renewable dessicant chamber is mounted inside the probe shroud to absorb condensate. A 10-l liquid nitrogen dewar and associated controls are provided for cooling the detector.

The INSCAN operating program, written in Fortran IV, is dedicated to setup, testing and operation of the system and features the following standard modules:

- Automatic analysis module. Used for automatic data collection and analysis from previously calibrated analysis probes, reporting analysis results, and report/alarm generation.
- Calibration module. Used to acquire raw data for calibration of each probe.
- Setup module. Used for initial setup procedures
- Diagnostics module. A stand-alone program on a separate disk, which is used for system test and troubleshooting.

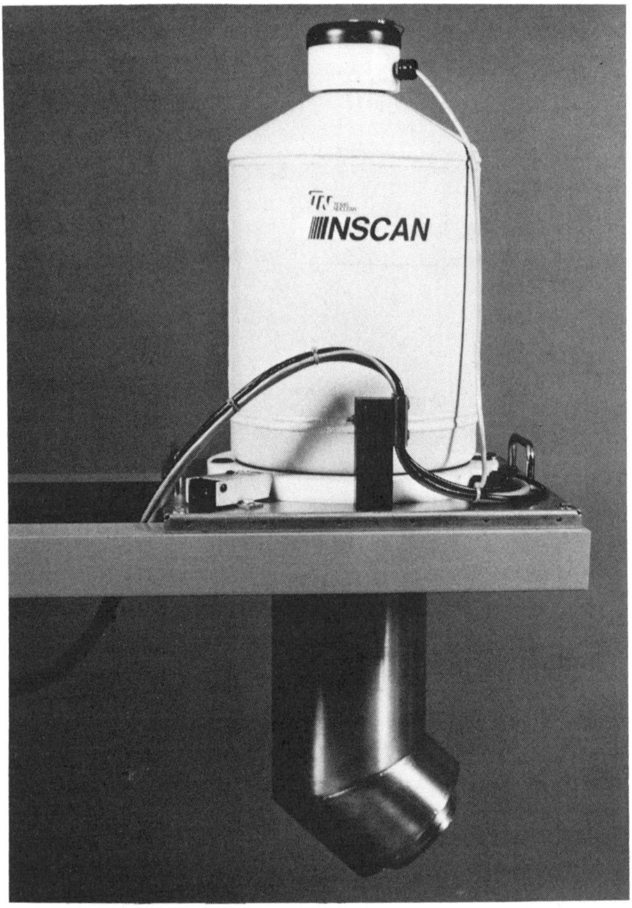

FIGURE 10.39 Model 9800 analysis probe. (Courtesy of Ramsey Engineering, Texas Nuclear)

10.5 RECENT DEVELOPMENTS

As mentioned earlier, the field of spectroscopy has probably seen more development in the last decade, as a result of advances in technology, than any other analytical technique. Research is currently proceeding at a fast and furious pace in many areas of spectroscopy and will almost certainly continue into the future. As a result, it is rather difficult to keep up with the situation. In relation to the development of commercially available on-line process analyzers, there are two areas of recent development that are worthy of note here.

10.5.1 On-Line FTIR Spectrometers

As discussed in Section 10.3.2, Fourier Transform Infrared Spectroscopy (FTIR) has given the spectroscopist the ability to solve all classical IR problems, and yields results in seconds or fractions of a second. The introduction of the microcomputer in the mid 1970s paved the way for low-cost, high-speed FTIR data systems. FTIR has therefore rapidly become an accepted analytical technique in the laboratory, but until very recently appeared to be limited to that environment. This is because of the instrument's susceptibility to vibration and temperature effects and the need for highly accurate alignment of the moving mirror in the Michelson interferometer design. The introduction of two on-line FTIR spectrometers at the Pittsburgh Conference in 1984 changed that picture completely.

(a) Analect Instruments Inc. (United States)

Analect has been involved in FTIR spectroscopy for some time and has been responsible for several significant developments in the areas of optical design, computer architecture, and display technology. The Models FX-6160 and FX-6260 FTIR spectrometers represent Analect's state-of-the-art laboratory line, which incorporates high-speed data processing capability. The Model FX-6260 utilizes a patented interferometer design called TRANSEPT, which does not have a moving mirror; instead, scanning is accomplished by means of a reciprocating wedge of refractive material that intersects one of the interferometer arms. Recent additions to the Analect line include the Model EVM-60 Environmental Vapor Monitor, which is a laboratory-type instrument, and the Model PCM-80 Process Control Monitor, which is an on-line instrument, capable of carrying out FTIR monitoring of gaseous, liquid and solid process streams.

In the Model PCM-80 Process Control Monitor, operation is based on the TRANSEPT interferometer principle, shown in Fig. 10.40, where path length scanning takes place by movement of one of the two matched wedges of transparent material, to effectively change the thickness of a uniform layer of this material in one arm of the interferometer. The optical path length change is given by:

$$\Delta p = \Delta y(n - 1)\sin \theta$$

where Δy is an increment of wedge translation, n is the refractive index of the wedge material, and θ is the apex angle of the wedge. The beam splitter, whose substrate is identical to that of the scanning wedge, is symmetrically placed in the second interferometer arm. The TRANSEPT design is reputed to eliminate the drawbacks of conventional Michelson-design interferometers, namely, angular alignment and thus vibration and temperature sensitivity.

FIGURE 10.40 Model PCM-80 process control monitor showing field unit and control cabinet. (Courtesy of Analect Instruments Inc.)

The MP-40 Data Processing Package is configured to provide a standard system for rapid production of FTIR spectra, thus providing the basis for real-time composition monitoring. The FTIR spectral array can also be communicated to a second data processing system, based on the DEC PDP 11/23 computer system, which contains chemometric statistical analysis software.

A number of optical sensing heads are available to allow the PCM-80 to cover a variety of industrial process applications. Diffuse reflectance and specular reflectance heads are used for monitoring thickness of plastic sheets, analysis of thin organic films or inorganic oxide layers on a reflective substrate, and analysis of conveyed materials. For liquid applications, a cylindrical MIR cell or the conventional transmission cell may be used. For gaseous applications, short-path and long-path cells are available.

The operation of the PCM-80 is based on empirical determination of the best monitoring method, making use of 1500 spectral data points. Once the process stream is modelled in this way, the instrument can make accurate measurements of complex chemical systems. During on-line operation, the analyzer automatically carries out the measurement procedure established by the modeling process, including spectra acquisition, data reprocessing, data point selection and scaling, composition readout, and so on. This process requires only a fraction of a second.

The spectral region of operation is 2.3–22 μm (4400 to 450 cm^{-1}), with a resolution of 4 cm^{-1}. The frequency reference system is a HeNe laser. Separation between optical system and data processing system can be up to 50 ft.

(b) Janos Technology Inc. (United States)

The Janos 1000 Series of FTIR Spectrometers, now under commercial development, are portable, battery-operated instruments with a compact 4 × 4 × 6-in. optical measuring head. The refractive scanning principle, illustrated in Fig. 10.41, is used rather than the reflective mode of the conventional Michelson interferometers. Radiation from the nichrome source is collimated by lenses onto an amplitude beam splitter, which consists of a partially reflective coating on a crystalline substrate. A compensator plate is mounted immediately adjacent to and a millimeter or so from the beam splitter and slightly out of parallel with it, to allow for escape of spurious reflections and provide for refractive correction.

Incident energy, separated into two beams by the beam splitter, is directed to mirrors 8 and 5 and then to the rotating refractive window, positioned to intercept both beams. The beams are received at opposite faces and are transmitted in parallel toward mirrors 6 and 7. Rotation of the refractive plate causes optical path differences between the two beams. Both beams retrace their paths to the beam splitter; the portion of the beam used is that directed toward the focusing lens and detector. The compartment containing the sample under investigation is located between the beam splitter and the focusing lens. When using potassium bromide optics, the spectral range of the instrument is 5000–400 cm^{-1}, with a resolution of better than

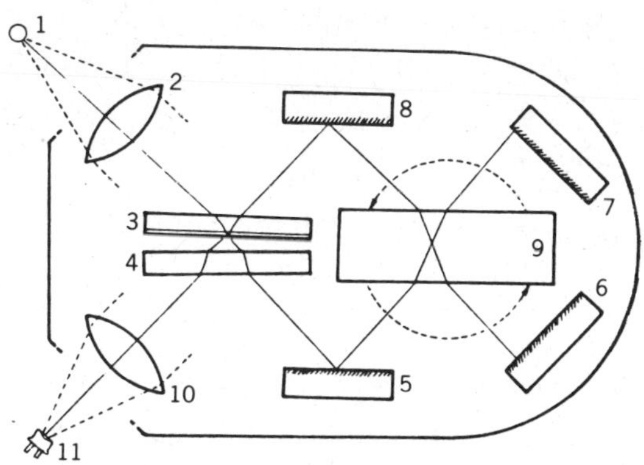

FIGURE 10.41 Series 1000 FTIR spectrometer. (Courtesy of Janos Inst. Tech.)

5 cm^{-1}. Four interferograms are determined during each 30-sec rotation of the refractive plate. The instrument is microprocessor-controlled, providing fast Fourier transform (FFT) of 1 to 15 sec, depending upon the desired resolution. Data are plotted using a typical X-Y recorder.

For field use, a battery-operated motor is used to drive the refractive plate. Motor speed is controlled via a lock to a laser frequency. For application studies, an instrument with calcium fluoride optics has also been used, appears to give better than 5 cm^{-1} resolution, and is completely insensitive to water and many other chemical reagents. Sampling accessories are available to allow measurement in gases, liquids and solids, including MIR cells, transmission cells, and heads for scanning moving film sheets.

10.5.2 Fiber Optic Spectrometers

In recent years there has been a high level of activity in the important new technology of fiber optics. Most of this activity up to now has been in the communications industry, for the high-speed transmission of information such as telephone calls and computer data. Telephone companies on a worldwide basis are installing optical fibers to replace conventional wiring.

Fiber optics technology is also beginning to enter many other fields, including the development of new types of physical and chemical sensors for on-line analysis (commonly called *optrodes*). These devices can be installed directly in the process fluid to be monitored, with the advantages of low cost, rugged construction, corrosion resistance, and noise immunity. Systems combining laser-induced fluorescence spectroscopy techniques with fiber optics (remote fiber fluorimetry, RRF) are already being developed for industrial use, and research is proceeding rapidly. The limitations at present are in the need to develop optrodes suitable for specific measurements. One such instrument, which combines spectroscopy and fiber optic technology and which is now commercially available for industrial applications, is described next.

(a) Guided Wave Inc. (United States)

The Optical Waveguide Spectrum Analyzer shown in Fig. 10.42 is a general-purpose spectrophotometer using large-core, single-strand optical waveguides to provide optical paths between the sensor and the instrument. One waveguide transmits the reference radiation to the sample; a second waveguide returns sample-modified radiation from the sensor to the instrument.

The analyzer consists basically of five functional assemblies.

1. The monochromator contains a direct-drive servo that positions a concave holographic grating serving as the spectral scanning device.
2. The radiation source is a tungsten-halogen lamp, focused to the source port adaptor.

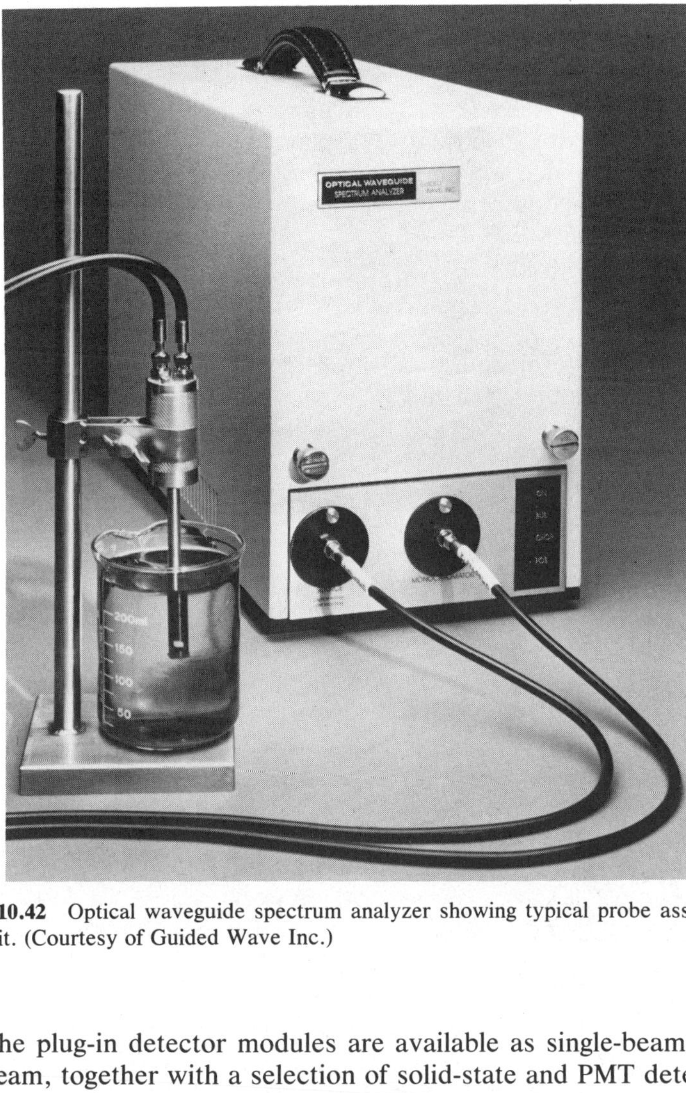

FIGURE 10.42 Optical waveguide spectrum analyzer showing typical probe assembly and control unit. (Courtesy of Guided Wave Inc.)

3. The plug-in detector modules are available as single-beam or dual-beam, together with a selection of solid-state and PMT detectors.
4. The control logic and analog circuitry provides an interface to the computer for timing, grating position commands, detector conversions, and status logic.
5. The power supply unit provides power for the lamp source, logic and servo functions, switches, and fuses.

The spectral range of the analyzer is 190–2000 nm, with an accuracy of 2 nm or less, and provides ultra-fast scanning at up to 9600 nm/min in 8-nm increments. Gratings are interchangeable, allowing the instrument to cover and optimize different spectral ranges. A number of probe (sensor) options are available, including general purpose probes for hand-held or portable clamp-on applications and special purpose versions for permanent or semi-permanent tank or in-line applications.

The analyzer is designed for compatibility with the IBM Personal Computer as standard, for spectral data display and printout. Complete software programs are included for display, scaling, and post-analysis of data. The software is designed to provide the functions of analyzer control, data collection, data storage, and data display, and is packaged in three levels.

- Level I covers initial setup and provides a user-friendly, menu-driven display.
- Level II governs data acquisition, display, display scaling, and data storage (printout or disk file).
- Level III is concerned with post-processing of spectral data and is application specific.

Three optional programs are available for Level III—spectrophotometry, spectra/radiometry, and liquid absorbance.

REFERENCES

1. Newton, I. *Optiks,* 4th Ed. London (1730), Dover, New York (1952).
2. Bouguer, P. *Essai D'optique sur la graduation de la Lumière* (1729).
3. Coblentz, W. W. Investigations of infrared spectra. *Carnegie Inst. Tech. Bull.,* **35** (1905); **65** (1906); **97** (1908); *Nat. Bur. Stds. Bull.,* **11,** 131 (1914).
4. Planck, M. *Ann. Physik,* **1,** 69 (1906); **4,** 553 (1905).
5. Lenard, P. *Ann. Physik,* **8,** 149 (1902).
6. Einstein, A. *Ann. Physik,* **17,** 132 (1905); **20,** 199 (1906).
7. Bohr, N. *Phil. Mag.,* **26,** 1, 476, 867 (1913); **27,** 506 (1913); **29,** 332 (1915).

BIBLIOGRAPHY

A Village Scene in the Infrared. *Process. Eng.* (Oct 1982).

Adams, V. H. The mass spectrometer on-line analyzer: Where does it stand? Where is it going? *Proc. ISA Conf.,* Houston, TX (Oct 1980).

Ahlstrom, R. C., and Shaver, F. A. The development and application of an on-line process mass spectrometer. *31st Ann. Instr. Symp. for the Proc. Ind.,* Texas A & M Univ. (Jan 1976).

Bellamy, L. J. *The Infrared Spectra of Complex Molecules,* 2nd Ed. Wiley, New York (1958).

Bourg, W. Performance of the Perkin Elmer MGA1200 Spectrometer in a 1000 T/D Kellog ammonia plant. *Proc. 28th Ann. ISA-AID Symp.,* Baton Rouge, LA (April 1982).

Brant, W. C., and Kiviat, F. E. Rapid determination of total aromatics by IR spectroscopy. *Proc. ISA Symp.,* Boston, MA (May 1985).

Cardis, T. M. Photometric analysis of aqueous process streams. *Proc. ISA Symp.,* Boston, MA (May 1985).

Cardis, T. M., and Lastrapes, T. M. Continuous monitoring of sodium hydroxide by IR analysis. *Proc. ISA Symp.,* Des Plaines, ILL (May 1984).

Cook, B. W., and Jones, K. *A Programmed Introduction to Infrared Spectroscopy*. Heyden & Sons, London (1972).

Cooper, H. R. On-stream X-ray analysis. *A. M. Gaudin Memorial,* Vol. 2, pp. 865–894. Amer. Inst. of Mining, Metall. & Petr. Eng., New York (1976).

Cooper, H. R. Using on-stream X-ray fluorescence for slurry composition analysis. *InTech* (July 1981).

Edgar, R. F. On-line measurement of aqueous- and solvent-based coatings by infrared absorption gauge. *Paper Tech. & Ind.* (Nov, Dec 1977).

Edgar, R. F., and Hindle, P. H. The on-line use of infrared measuring instruments for paper moisture and grammage. *Paper Tech. & Ind.* (April 1975).

Emsley, J. W., Feeney, J., and Sutcliffe, L. H. *High-Resolution Nuclear Magnetic Resonance Spectroscopy*. Pergamon, New York (1966).

Ewing, G. W. *Instrumental Methods of Chemical Analysis*. McGraw-Hill, New York (1960).

Fisher, J. F. Reduction of the water interference in the near IR measurement of alcohols. *Proc. ISA Symp.,* Des Plaines, ILL (May 1984).

Griffiths, P. R. *Chemical Fourier Transform Infrared Spectroscopy*. Wiley-Interscience, New York (1975).

Griffiths, P. R., and Fuller, M. P. *Advances in Infrared and Raman Spectroscopy,* Vol. 9, pp. 63–129. Heyden & Sons, London (1981).

Gross, M. L. (Ed.). *High-Performance Mass Spectrometry: Chemical Applications*. ACS, Washington DC (1978).

Gunderson, L. C., and Keck, D. B. Optical fibers: Where light outperforms electrons. *Tech. Rev.* (May/June 1983).

Harrick, N. J. *Internal Reflection Spectroscopy*. Wiley, New York (1967).

Harrick, N. G. *Internal Reflection Spectroscopy*. Harrick Scientific Corp., Ossining, NY (1979).

Hirayama, K. *Handbook of UV and Visible Absorption Spectra of Organic Compounds*. Plenum, New York (1967).

Hirschfeld, T., Hauger, G. R., Johnson, D. C., and Hrubesh, L. W. Remote techniques based on long-distance fiber optics. *Proc. 179th Nat. Mtg. ACS,* Washington DC (1980).

Jackman, L. M. *Applications of NMR Spectroscopy in Organic Chemistry,* 2nd Ed. Pergamon, New York (1969).

Jutila, J. M. Multi-component on-stream analysis for process monitoring and control. *InTech* (July 1979).

Kiurn, E. S. A new instrument for accurate on-stream XRF measurement. *Promecon '81, Proc., Meas. & Cat. Conf.,* ISA (1981).

Lang, M. C. Application of process mass spectrometer in urea manufacture. *Proc. ISA Conf.* (1983).

LeFebre, D. A., and Thomas, G. P. An in-line scanning UV-VIS-NIR process monitor. *Proc. ISA Symp.,* Boston, MA (May 1985).

Meneghelli, B. IR as a tool for occupational exposure monitoring. *Proc, ISA Symp.,* Boston, MA (May 1985).

Milanovich, F. P., and Hirschfeld, T. Process, product and waste stream monitoring with fiber optics. *Adv. Instr.,* **38,** ISA (1983).

Nafie, L. A., and Vidrine, D. W. *Fourier Transform IR spectroscopy,* Vol. 3, pp. 83–123. Academic Press, New York (1982).

Nakanishi, K. *Infrared Absorption Spectroscopy*. Holden-Day, San Francisco (1962).

Robertson, A. J. B. *Mass Spectrometry*. Wiley, New York (1954).

Robots will sniff chemicals. *Chem. Week* (March 1984).

Saltzman, R. S. Process stream analyzers based on absorption changes with pH. *16th ISA-AID Symp.,* Pittsburgh (May 1970).

Sonner, R. H., and Dailey, W. V. An open path hydrocarbon detector. *Proc. ISA Symp.,* Des Plaines, ILL (May 1984).

Taillez, B. Y., and Hume, S. H. The use of time-of-flight mass spectrometry (TOFMS) in process monitoring and control. In *Dynamic Mass Spectrometry,* Vol. 6. Heyden & Sons, London (1982).

Van Der Maas, J. H. *Basic Infrared Spectroscopy,* 2nd Ed. Heyden & Sons, London (1979).

Willis, H. A. *Advances in Infrared and Raman Spectroscopy,* Vol. 2, Chap. 3. Heyden & Sons, London (1976).

Wilks, P. Sampling method makes on-stream IR analysis work. *Ind. Res. & Dev.* (Sept 1982).

Wilks, P., and Rein, A. Cylindrical internal reflection cell. *Amer. Lab.* (Oct 1982).

11

Moisture Measurement

11.1 INTRODUCTION

After nitrogen and oxygen, water vapor is the most plentiful gas in the earth's atmosphere and is essential to our comfort and even the existence of life on earth. We can demonstrate this simply by running our house heating equipment in the winter without a humidifier and noting the resulting discomfort in breathing. Water vapor is present to some extent in any gas prepared industrially, unless extreme precautions are taken to remove it or the gas reacts with it. Water vapor modifies the properties of many materials and is one of the most difficult substances to remove completely.

In contrast to the relatively constant concentrations of oxygen and nitrogen in the atmosphere, the concentration of water vapor varies appreciably, depending on the temperature and pressure. Therefore, if the presence and precise concentration of water vapor is of importance in an industrial process, further quantitative information is usually required.

Water vapor, being a chemical substance, may be determined quantitatively using chemical methods, but these are usually tedious. The earliest recorded methods for determination of water vapor (moisture) concentrations involved physical measurements of the effect of moisture on some material or system. In 1500, Leonardo da Vinci described a device by which the quantity of moisture in the atmosphere was related to the change in weight of a ball of wool or hair. In 1783, De Saussurie measured the change in length of a human hair with variations in atmospheric moisture concentration. The hair hygrometer is still widely used today and the basic concept has remained essentially unchanged. In fact, the Foxboro Company still manufactures the Model 12 Relative Humidity Recorder, which uses a human hair element as the sensing device.

The measurement of moisture, then, is of importance in many industrial processes and, of course, in meteorological work. Water may be a reaction by-product, a catalyst poison, or contribute to equipment corrosion. In some processes, the presence of moisture may be required to catalyze certain reactions or provide desired product properties (as in paper production). In some cases, the presence of moisture results in undesired side effects in a reaction process and must therefore be rigidly excluded from the system.

With the accent now on improved product quality and efficient, economic operation of process plants, there has been a greatly increased demand for on-line process analyzers to measure moisture content of products and intermediates. Note that there are a number of techniques that may be used to measure moisture content, other than those described in this chapter. These include infra-red absorption (IR), gas-liquid chromatography (GLC), and mass spectrometry (MS). These techniques are covered more fully in other chapters.

11.2 BASIC THEORY AND DEFINITIONS

The science of *psychrometry* is concerned with the determination of the properties of gas-vapor mixtures, of which the air-water system is by far the most commonly encountered. Before proceeding to a discussion of methods used to measure moisture content, it is pertinent to establish exactly what is meant by "moisture content" and to define the terminology and relationships used in psychrometry.

It is a well-known fact that the atmosphere exerts a pressure, called atmospheric pressure, on any object with which it is in contact. Since there is always water vapor present in the air, part of the atmospheric pressure is exerted by the water vapor. This is known as the *partial pressure* of the water vapor and is a measure of the volume of water in the air. For example, if the atmosphere contains 1% by volume of water vapor, the partial pressure exerted by the water vapor will be 0.01 atm. Assuming atmospheric pressure is 14.7 psia, the partial pressure of water vapor will be 0.147 psia.

11.2.1 Definitions

(a) Absolute Humidity (H)

This is the mass of water vapor carried by unit mass of dry gas (often measured in pounds of water vapor per pound of dry gas). Assuming ideal gas behavior:

$$H = \frac{M_w p}{M_a(P - p)}$$

where M_w = mol. wt. of water
M_a = mol. wt. of gas
P = total pressure (atm)
p = partial pressure of water vapor

(b) Saturation Humidity

This is the humidity of the gas when saturated with water vapor at a given temperature. When the partial pressure (p) of water vapor in the gas at a given temperature is equal to the vapor pressure of water (p_s) at the same temperature, the gas is saturated and the absolute humidity is designated the saturation humidity.

(c) Percentage Absolute Humidity

Also known as percentage saturation, this is defined as the ratio of the absolute humidity (H) to the saturation humidity (H_s), that is:

$$\%H = \frac{100H}{H_s}$$

or

$$\%H = \frac{100p(P - p_s)}{p_s(P - p)}$$

(d) Humid Volume

This is the volume occupied by unit mass of dry gas and the water vapor it contains.

(e) Saturated Volume

This is the humid volume of saturated gas.

(f) Percentage Relative Humidity (RH)

This is defined as the ratio of partial pressure of water vapor in the gas to the vapor pressure of water at the given temperature, that is:

$$\%RH = \frac{100p}{p_s}$$

(g) Dew Point (or Saturation Temperature)

This is the temperature at which a given mixture of gas and water vapor is saturated, that is, the temperature at which water exerts a vapor pressure equal to the partial pressure of water vapor in the given mixture. It is the temperature at which condensation will first occur when the gas is cooled.

From the definitions of percentage absolute humidity (c) and percentage relative humidity (f), it can be seen that

$$\%H = \frac{(P - p_s)}{(P - p)} \times \%RH$$

Therefore the percentage absolute humidity and the percentage relative humidity are the same *only* when $P - p_s/P - p$ is equal to unity, a condition approached when the gas is nearly saturated ($p = p_s$) or when the partial pressure of water vapor is small compared with the total pressure.

11.2.2 Humidity Measurement

There are two methods commonly used for the manual measurement of humidity, namely, the dew point method and the wet-bulb method.

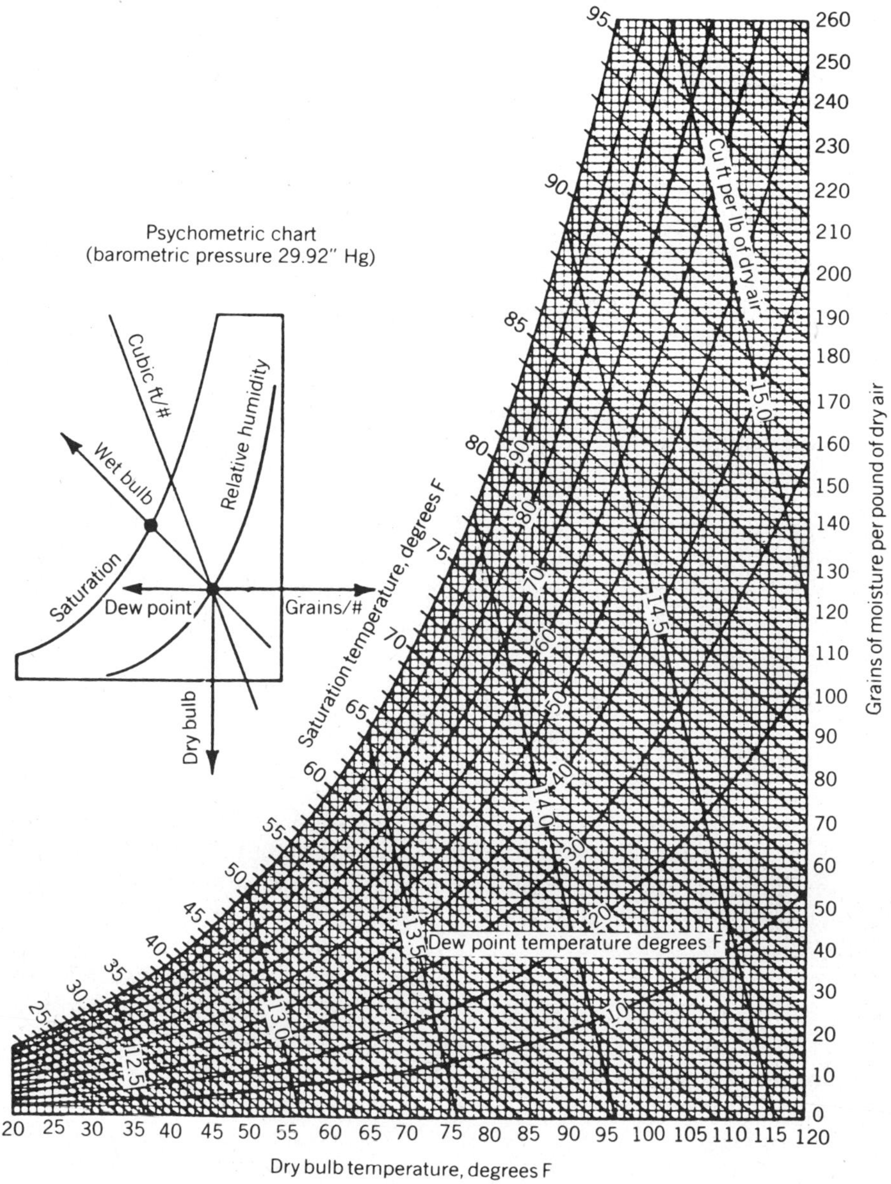

FIGURE 11.1 Typical psychometric chart.

(a) The Dew Point Method

Using this method, the dew point of wet air is measured directly by observing the temperature at which moisture begins to condense on an artificially-cooled, polished surface. The polished surface is usually cooled by evaporation of a low-boiling solvent such as ether or acetone, or by evaporation of a condensed permanent gas such as carbon dioxide.

Although this method may be considered a fundamental technique for measuring humidity, there are sources of error in its use:

1. It is not always possible to measure precisely the temperature of the polished surface or to eliminate gradients across the surface.
2. It is difficult to detect the appearance or disappearance of fog (condensed moisture). The usual practice is to take the average of the temperatures when condensation first appears on cooling and disappears on heating.

One portable instrument based on this principle is the Alnor Dewpointer, commonly used for spot dewpoint determinations in instrument air systems. There are also commercially available on-line process analyzers whose operation is based on "chilled mirror technology." These are discussed later in the chapter.

(b) The Wet-Bulb Method

This is probably the most commonly used method for determining the humidity of a gas stream and involves the measurement of wet- and dry-bulb temperatures. Anyone involved in meteorological work or who has a greenhouse is familiar with the technique. The wet-bulb temperature is measured by a thermometer that has a bulb covered by a wick saturated with water. The dry-bulb temperature is measured by a conventional mercury-in-glass thermometer. When the wet- and dry-bulb temperatures are known, the humidity is readily obtained from psychrometric tables or charts, such as the example shown in Fig. 11.1.

Care must be taken to ensure that the wet bulb remains wet and that radiation to the bulb is minimized. The latter is achieved by making the relative velocity between the wick and the gas stream high (usually about 15 ft/sec is adequate). For measurement of atmospheric humidities, the *sling psychrometer* is widely used. This device consists of both wet- and dry-bulb thermometers mounted in a sling, which is whirled manually to give the desired gas velocity across the wet bulb.

11.3 MOISTURE MEASUREMENT IN THE LABORATORY

Water vapor is one of the most common constituents of industrial process streams; therefore, periodic measurement of its concentration in various

TABLE 11.1 ASTM Standard Test Methods for Moisture Measurement

ASTM	
D95-70	Water in Petroleum Products by Distillation
D96-73	Water and Sediment in Crude Oils
D1142-63	Water Vapor Content of Gaseous Fuels by Dew Point Measurement
D1364-78	Water in Volatile Solvents by Karl Fischer Titration
D1631-80	Water in Phenol by Iodine Reagent Method
D1744-64	Water in Liquid Petroleum Products by Karl Fischer Titration
D2709-68	Water and Sediment in Distillate Fuels by Centrifuge Method
D2713-81	Dryness of Propane by Valve Freeze Method
D2867-70	Total Moisture in Coal
D3173-73	Moisture in Coal and Coke
D3240-75	Undissolved Water in Aviation Turbine Fuels
D3302-82	Total Moisture in Coal
D4006-81	Water in Crude Oil by Distillation
D4023-82	Humidity Measurement—Terms Related to
E123-78	Apparatus for Water by Distillation
E203-75	Karl Fischer Test Method
E337-62	Relative Humidity by Wet/Dry Bulb Psychrometer
E700-79	Water in Gases by Karl Fischer Test Method

process samples is a routine laboratory requirement for many industrial processes. Examples include petroleum refining and petrochemicals; coal and coke refining; iron and steel production; pulp and paper production; in the manufacture of chemicals such as antifreeze, plastics, electrical insulating materials, liquid and gaseous petroleum and petrochemical products, solvents, metals, building construction materials, industrial gases such as nitrogen, helium, oxygen, argon, and so on; and in atmospheric analysis studies.

With so many applications, laboratory methods of moisture measurement must be adapted to the type of sample under analysis and to the normal level of moisture expected in the sample. Of course, the above criteria apply equally to on-line process analyzers. One needs to bear in mind that validation (calibration) of process analyzers must be carried out by correlation with the routine laboratory tests summarized in this section. There are two major problems here; first, the difficulty of sampling, and second, the precision of the standard test method compared with the precision of the process analyzer. The subject of process analyzer calibration is covered in detail in Chapter 22. ASTM methods for moisture measurement are listed in Table 11.1.

11.3.1 Oven Drying Method

This is a "gravimetric" method commonly used to estimate the moisture content of solid samples in the percent range. It involves the precise weighing of the sample before and after removal of moisture in a drying oven. This method is used in the paper industry for moisture content of paper and paperboard (ASTM method D644-55), in the lumber industry for moisture content of wood (ASTM method D2016-74), and in the coal industry for moisture content of coal (ASTM methods D2961-79 and D3302-82).

11.3.2 Distillation Method

This is a classical technique, developed by Dean and Stark in England and covered under British Standard BS-756. The procedure is covered in ASTM method D95-70 and specifications for the test apparatus in ASTM E123-78. The apparatus consists of a glass or metal still with heater, a reflux condenser, and a graduated glass trap. A known quantity of the sample is heated under reflux with a water-immiscible solvent that co-distils with the water in the sample. Condensed solvent and water are continuously separated in the trap, the water settling in the graduated section of the trap and the solvent returning to the still. The solvent used is usually either industrial grade xylene or a mixture of 20% toluene and 80% xylene (both industrial grade).

11.3.3 Karl Fischer Titration

This is probably the most widely used laboratory method for the determination of low-level concentrations of moisture in the liquid or the gaseous phase and is detailed in the ASTM standards for many industries. The technique has been the subject of extensive research. The standard test for moisture in liquid petroleum products is covered in ASTM method D1744-64 and for gases in ASTM method E700-79. The technique itself is summarized in ASTM E203-75.

The sample is dissolved or dispersed in a suitable liquid (sample solvent) and titrated with Karl Fischer reagent (KF reagent), a mixture of iodine, sulfur dioxide, pyridine, and methanol (or glycol ether). As long as any water is present, the iodine is reduced to colorless hydrogen iodide. The end-point is the first appearance of free iodine, which may be determined either colorimetrically or electrometrically (more accurate). The fundamental reactions that take place are:

(1) $C_5H_5N \cdot I_2 + C_5H_5N \cdot SO_2 + C_5H_5N + H_2O \rightarrow 2C_5H_5N \cdot HI + C_5H_5N \cdot SO_3$

(2) $C_5H_5N \cdot SO_3 + ROH \rightarrow C_5H_5N \cdot HSO_4R$

The KF reagent may be prepared in the laboratory or obtained commercially. It is recommended that the reagent be standardized at least daily in the following manner. The reagent is titrated with 50 ml of the sample solvent until the end-point is reached. Then, an accurately-measured weight of distilled water is added to the pre-titrated solvent in the flask and further reagent added until an end-point is again reached. The "water equivalence"

(F) of the KF reagent is given by:

$$F = \frac{W}{V}$$

where W is the weight of water added (mg)
V is the volume of KF reagent in ml. required for titration of the added water.

To determine the water content of a liquid sample, the standardized KF reagent is first titrated to an end-point with 50 ml of the sample solvent. Then, an accurately measured weight of sample is added to the titration flask and titrated with KF reagent until an end-point is again reached. The water content (M) of the sample is then given by:

$$M(\text{ppm}) = \frac{1000CF}{W}$$

where C is the volume of reagent
F is the water equivalence of the KF reagent
W is the weight of sample (g)
1000 is a factor for conversion to ppm

By suitable choice of sample size, KF reagent concentration, and apparatus, this method is suitable for measurement of moisture over a very wide concentration range, from parts per million to pure water. As covered in ASTM method E700-79, the method is suitable for the measurement of moisture in gases in the concentration range 1 to 50 ppm.

The test itself is a tedious one, calling for precise measurements by the operator and frequent standardization of the KF reagent. It is consequently prone to operator error and judgement. Extreme care must be exercised in its application, especially when measuring moisture levels in the low ppm range, when ingress of moisture from outside the apparatus will have a disastrous effect on the result of the test. The method also suffers from the fact that a number of substances and classes of compounds interfere in the determination of water by this titrimetric technique. However, the subject of interference has been extensively studied and documented.

11.3.4 Other Methods

Several manufacturers of on-line process moisture analyzers—for example, Beckman, Panametrics, Alnor, Shaw, DuPont, Ondyne, General Eastern, and EG&G—also offer instruments designed specifically for use in the laboratory. These instruments operate in the same way as their process counterparts, described later in this chapter. The introduction of these specially designed laboratory monitors has led to problems of calibration, basically because the instrument precision is invariably far better than any laboratory test for moisture concentration (e.g., Karl Fischer). The manufacturers listed also offer moisture calibrators, the majority of which operate on the permeation tube principle. The use of these devices for primary calibration of moisture analyzers is really a question of one's confidence in the device performing in a reproducible manner.

11.4 PROCESS ANALYZERS FOR MOISTURE IN GASES AND LIQUIDS

Process analyzers for moisture measurement in gaseous and liquid process streams may be categorized as follows:

Electrolytic	Automated Karl Fischer
Quartz crystal	Lithium chloride
Thin-film aluminum oxide	Optical condensation type
Capacitance	

Some of these techniques have been used for many years; others are recent developments.

11.4.1 Electrolytic Type

The operating principle of these instruments is based on Faraday's law of electrolysis. The quantitative relationships between electricity and chemical change were first described by Michael Faraday in 1832. He showed that the *weight* of a chemical substance liberated by electrolysis is directly proportional to the amount of current passing through the electrolysis cell. If Avogadro's number of electrons (6.02×10^{23}) are consumed at an electrode, 1 gram equivalent of the substance is produced. The corresponding quantity of electricity is called the faraday (F) and is equal to 96,500 coulombs. Current, in amps, is defined as the number of coulombs of electricity passing per second.

For the electrolysis of *water*, Faraday's Law can be expressed as

$$I = 1.072 \times 10^4 \frac{W}{t}$$

where I is the current flowing (A)
W is the weight of water (g)
t is the time the current is passing (sec)

The following example illustrates how the electrolysis current may be calculated. Assume that a sample containing 1 ppm of moisture is passed through the electrolysis cell at a flowrate of 100 ml/min, temperature of 25°C, and pressure of 1 atm (14.7 psia). The flowrate of moisture (as a gas) is equal to:

$$100 \times 10^{-6} = 10^{-4} \text{ ml/min}$$

Since 1 g · mol of any ideal gas occupies 22.4 at STP and 1 g · mol of water is equal to 18.0 g, the weight of moisture (as a gas) is given by:

$$W(\text{g/min}) = 10^{-4} \times 18/22.4 \times 10^{3}$$

Therefore, the current required to electrolyze this quantity of water is given by:

$$I = \frac{1.072 \times 10^{4} \times 10^{-4} \times 18}{22.4 \times 10^{3} \times 60}$$

or $I = 14.36\ \mu\text{A}$ at STP. Since the operating temperature is 25°C, the current required at this temperature will be:

$$14.36 \times \frac{273}{298} = 13.16\ \mu\text{A}$$

This is the current required to electrolyze 1 ppm of moisture in the sample gas at 25°C, atmospheric pressure, assuming a gas flowrate of 100 ml/min. Note that *all three* of these parameters affect the measurement of moisture and the accuracy of the measurement can only be as good as the accuracy to which the sample flowrate is controlled. Note also that (in theory) using this technique, moisture content is measured in *mass* units.

One of the main problems with this method of measurement is a phenomenon known as the "recombination effect." The standard electrolytic cell uses platinum electrodes and the products of the electrolysis are hydrogen and oxygen. If the sample stream is rich in either of these components, the platinum acts as a catalyst for the recombination of hydrogen with the free oxygen of electrolysis to form more water, which is electrolyzed again. This results in inaccurate, high apparent moisture measurements. Another problem is that electrolytic cells have a drastically shortened life when used in applications where the sample moisture levels are relatively high. Manufacturers have extended cell operating life and minimized the recombination effect by using rhodium in place of platinum.

(a) Fluid Data Inc. (United States and England)

Fluid Data manufactures two moisture monitors, the Model 0133 designed for laboratory use with non-flammable gases, and the Model 153, which is an on-line process analyzer. Both operate on the electrolytic principle described above.

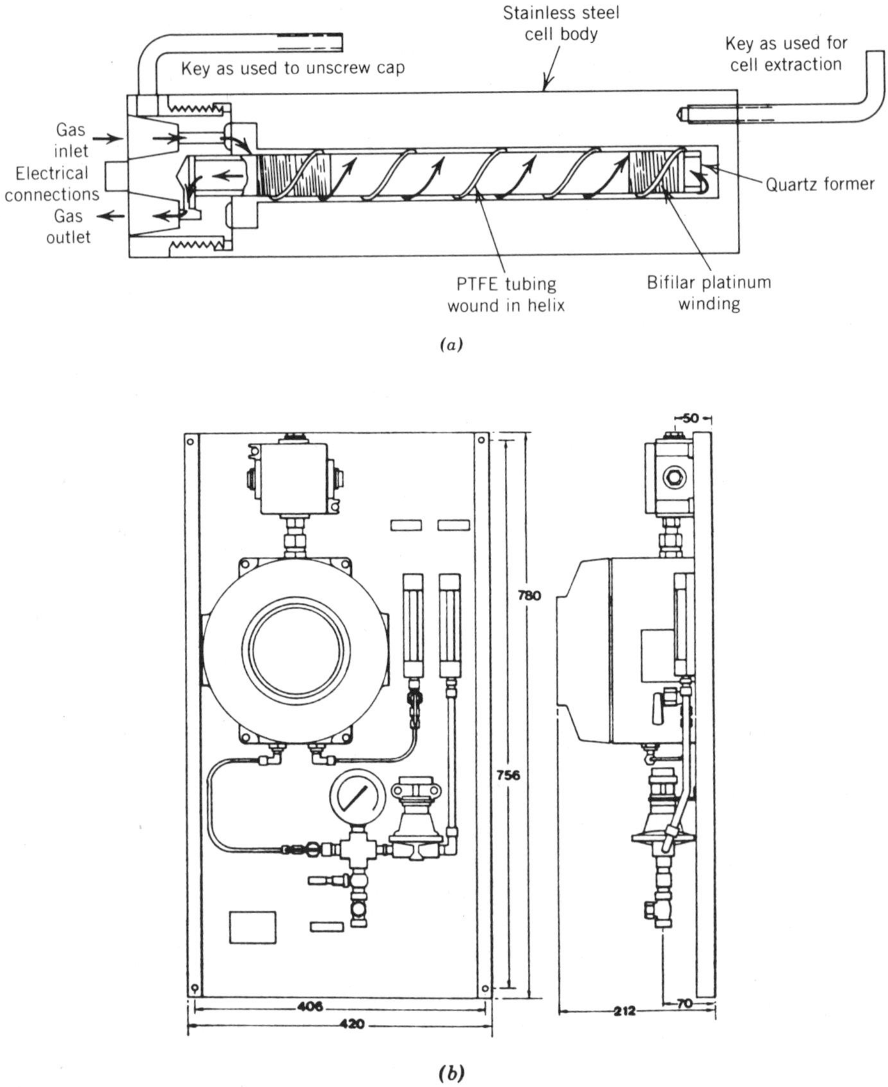

FIGURE 11.2 Model 153 moisture analyzer. (*a*) Probe design; (*b*) layout of sensing unit. (Courtesy of Fluid Data Inc.)

The sensing cell shown in Fig. 11.2*a* consists of two fine platinum (or rhodium) wires wound in bifilar form on a tubular quartz former. The wires and the spaces between them are coated with a thin film of phosphorus pentoxide, which is a good insulator when dry and readily absorbs moisture, that is, it is *hygroscopic*. Sample passes through the sensing cell at a controlled flowrate of 100 ml/min and the moisture is absorbed by the phosphorus pentoxide. This moisture is electrolyzed by a current passed between the two wires. The current required is directly proportional to the moisture concentration in the sample.

The analyzer unit is shown in Fig. 11.2*b*. The sample system includes a fast loop (flowrate about 1 l/min) and the analyzer sample flowrate is controlled at 100 ml/min using a differential pressure regulator on the outlet of the measuring cell. Two ranges are quoted, 0–1000 ppm and 0–10,000 ppm. Repeatability is ±5% of full scale and response time is better than 30 sec for a 63% response to a step change in either direction between 50 and 1000 ppm. An analog output (4–20 mA dc) is available for remote recording.

(b) Beckman Instruments Inc. (United States)

Beckman manufactures a range of trace moisture analyzers based on the electrolytic principle. The Models 193005 and 193006 are the portable variety, either ac mains or battery powered. These instruments are designed for general purpose locations and are therefore more suitable for laboratory use.

The Model 193000 is designed for panel mounting, with the analyzer unit and control unit located together in a general purpose enclosure. Air purge facilities can be provided as an option.

The Model 193004 consists of two parts, the analyzer unit which is suitable for NEC Class 1, Group D, Division 1 hazardous areas, and the control unit, which is designed for panel mounting in a safe area.

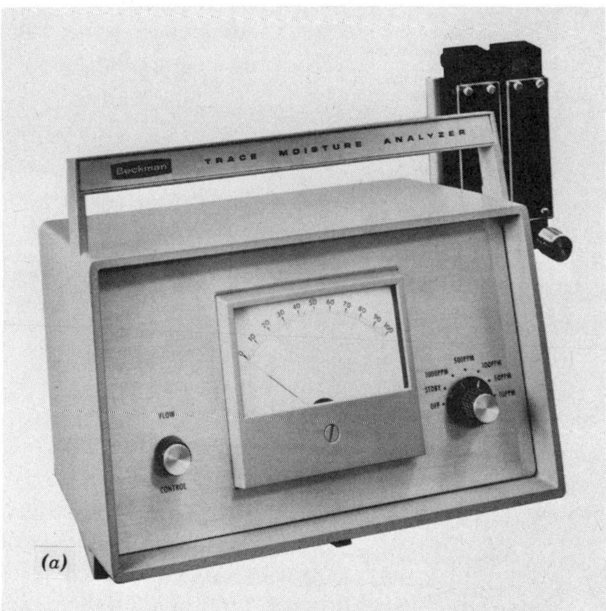

FIGURE 11.3 Beckman electrolytic hygrometers. (*a*) Portable bench-mount unit. (Courtesy of Beckman Industrial Inc.)

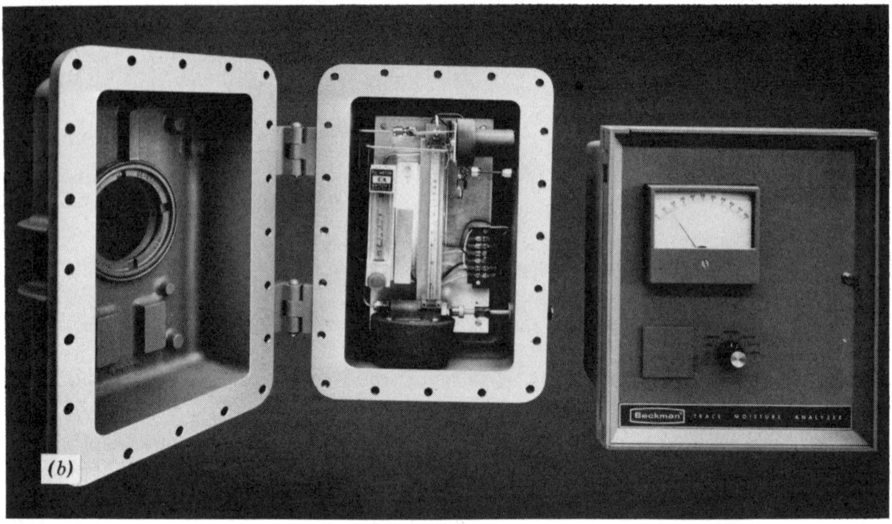

FIGURE 11.3 Beckman electrolytic hygrometers. (*b*) explosion proof sensing unit and remote mounted control unit. (Courtesy of Beckman Industrial Inc.)

The various packages available are shown in Fig. 11.3. Ranges offered are 0–10, 50, 100, 500, and 1000 ppm, and instrument repeatability is quoted as ±5% of full scale. Response time for a step change is given as 90% in 1 min (exclusive of sample system response). An analog 4–20 mA dc output is available for remote recording. Alarm facilities and sample system accessories are also available.

11.4.2 Quartz Crystal Type

The *DuPont Company* Model 560 Moisture analyzer operates on this principle. The moisture concentration in a gas stream is measured by monitoring the vibrational frequency change of a hygroscopically sensitized quartz crystal that is exposed alternately to wet and dry gas. The system is shown in Fig. 11.4.

The sample gas is split into two streams, "sample" and "reference," which are passed alternately across the measuring crystal. The standard cycle time is 30 sec for each. The reference gas, before reaching the crystal, is passed through a molecular sieves drier, which removes virtually all the moisture. As sample gas is passed over the crystal, moisture is absorbed by the hygroscopic coating, thereby causing a vibrational frequency change.

At the end of the 30-sec period, the microprocessor-based control unit reads and stores the frequency difference ΔF_1 between the measuring crystal and a sealed local oscillator crystal. The system is then switched to allow the dry reference gas to pass over the measurement crystal for the next 30 sec,

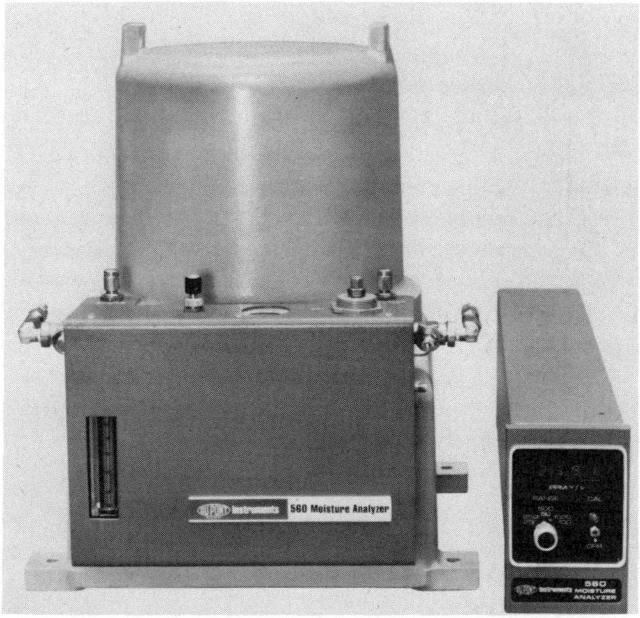

FIGURE 11.4 Model 560 moisture analyzer showing field unit and control unit. (Courtesy of DuPont Company)

at the end of which the frequency difference ΔF_2 is read and stored. These frequency differences are then subtracted to obtain ΔF_3, which is proportional to the moisture concentration in the sample gas. The moisture level is related to ΔF_3 through a polynomial expression, the constants of which are determined during factory calibration. Both measurement and local oscillator crystals are temperature controlled at 60°C for optimum repeatability and accuracy.

The analyzer unit is suitable for NEC Class 1, Group D, Division 1 hazardous locations. The microprocessor-based control unit is designed for remote mounting in a safe area at a maximum of 2000 ft from the analyzer unit. The instrument is calibrated 0–1000 ppm (vol), with selectable range segments of 0–25, 50, 100, 250, and 1000 ppm for the optional 4–20 mA dc analog output signal. Calibration in lb/MMSCF is also available.

Several options may be purchased:

1. An internal moisture calibrator may be fitted. The feed to this unit is reference gas, which is conditioned to a pre-determined moisture level by a single-point permeation-type moisture generator. This consists of a temperature-controlled water reservoir and a permeation tube of Teflon fluorocarbon resin, designed to provide either 20 ppm or 3 ppm moisture level, depending upon the application. Moisture

concentration in the outlet gas is determined by temperature, gas flowrate, and the dimensions and permeability of the tubing.

2. The Model 5600 is available for enhanced capability. This is a keyboard-operated unit, which greatly increases the accuracy, sensitivity, and range of the system.
3. Sample handling system components can be supplied either as a complete package designed for the application, or as individual components, such as filters, flowmeters, regulators, vaporizers, and driers.

11.4.3 Thin-Film Aluminum Oxide Type

(a) Panametrics Inc. (United States)

The Panametrics aluminum oxide humidity sensor was originally developed for meteorological work and for the aerospace industry. In view of its excellent performance under the extreme conditions experienced in these applications, it was adapted for use in on-line process stream applications for the measurement of water vapor concentrations.

The sensor consists of an aluminum strip, anodized to provide a porous oxide layer. A very thin gold coating is evaporated over this structure. The aluminum base and gold layer form the two electrodes of what is essentially an aluminum oxide capacitor (see Fig. 11.5). Water vapor is rapidly transported through the gold layer and equilibrates on the pore walls in a manner functionally related to the vapor pressure of water in the atmosphere surrounding the sensor. The number of water molecules adsorbed on the oxide structure determines the conductivity of the pore wall, thereby providing a distinct value of electrical impedance that is a direct measure of the water vapor pressure.

The sensor is also suitable for measurement of water vapor in organic liquids. The pore wall openings are small compared with the size of organic molecules; therefore admission into the pore cavity is limited to small molecules such as water. Parts per million by weight (or weight percent) of water

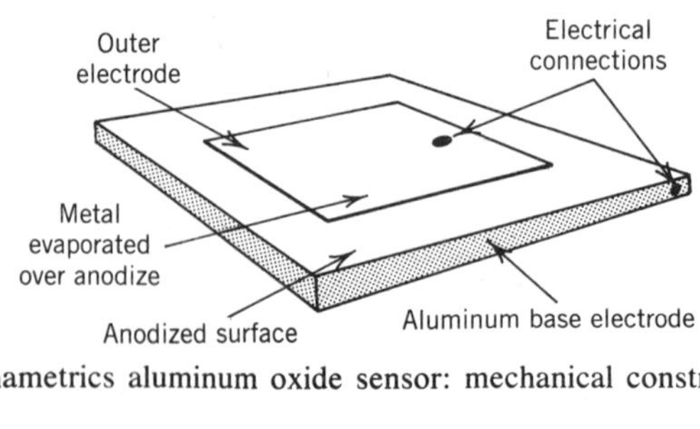

FIGURE 11.5 Panametrics aluminum oxide sensor: mechanical construction. (Courtesy of Panametrics Inc.)

in organic liquids is given by Henry's law, which states that the mass of gas dissolved in a given volume of solvent at constant temperature is proportional to the pressure of gas with which it is in equilibrium. That is:

$$\text{ppm water} = \text{partial pressure of water vapor} \times K$$

where K is the Henry's law constant. This constant must be computed separately for each liquid and is given by:

$$K = \frac{\text{saturation weight ppm of water for that solvent}}{\text{saturation vapor pressure of water}}$$

both at the temperature of measurement. Once the constant is known, the water concentration in the organic liquid is computed by multiplying this constant by the vapor pressure of water as measured by the sensor probe.

Probe Construction. Panametrics probes are designed for in-situ installation and are claimed not to be damaged by severe shock or vibration conditions. The probes offered differ only in fitting construction, required to meet various pressure limitations as follows:

Model M1R	5 microns Hg (75 psig)
Model M2R	5 microns Hg (5000 psig)

Probes may be located remotely at distances up to 2000 ft from the control unit. The speed of response of the probe is quoted as 63% in a few seconds over the entire dynamic range for gas applications. The sensor does not appear to be flow-sensitive, the only limitation being avoidance of mechanical damage.

Moisture Measurement Systems. A range of systems is offered by Panametrics. Systems I and II were developed from the Model 4000 computing moisture analyzer introduced by the company a few years ago.

The *System I* (see Fig. 11.6) provides for up to six parallel inputs from moisture, temperature, and pressure sensors. The maximum sensor-control unit distance for this system is quoted as 1000 ft, (although longer distances are available by consultation with the factory). There are three forms of output available from the control unit:

1. Digital front-of-panel indication.
2. Analog 4–20 mA dc or voltage output (one per channel).
3. A single digital serial output (two-wire), available either as RS232 standard interface or a 20 mA standard current loop. Information is transmitted as ASCII characters.

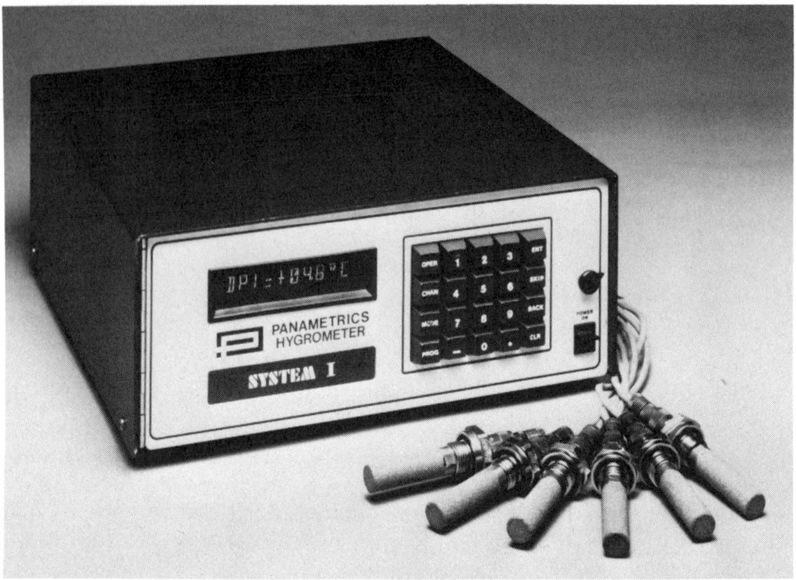

FIGURE 11.6 System I microprocessor-based moisture measurement system. (Courtesy of Panametrics Inc.)

Individual, isolated alarm relay contacts (SPDT) are available for high and low limits on each channel. The control unit can be designed for rack, bench, or panel mounting.

Output results can be calculated from input data to yield any of the following parameters on any channel:

ppm (volume)	ppm (weight)
relative humidity	dew/frost point (°C or °F)
lb water/MMSCF	pressure (bar, psig)
temperature (°C or °F)	

Programming of the system is by means of a front-of-panel keyboard. The main program is stored in EPROM and a battery back-up power supply is available as an option, suitable for 30 min of full system operation in the event of a power failure.

The System II offers the same facilities as System I, except that it is designed for *one* set of moisture, temperature, and pressure inputs only.

The Model 600 is suitable for NEC Class 1, Group D, Division 1 hazardous locations and for continuous single probe monitoring. Calibration is in dew point/frost point, covering the linear range from −110 to +60°C, with direct reading from −80 to +20°C.

The Model 700 is a general purpose unit, suitable for single probe continuous monitoring. Calibration is the same as the Model 600.

The Model 2100 is also a general purpose unit, which will accept inputs from up to six probes, calibrated as for the Models 600 and 700. However, this model also has options to allow installation in Division 1 and Division 2 locations.

Panametrics also manufactures the Models MG100 and MG101 moisture generators. These are portable, explosion-proof units that may be used to verify calibration of any moisture analyzer with a sample inlet pressure of 10 psig or less. They are also suitable for general use as precision moisture generators (e.g., in the laboratory).

(b) Ondyne Inc. (United States)

Ondyne manufactures two series of moisture analyzers, based on the aluminum oxide sensor technology, for use in gaseous and hydrocarbon liquid applications; a typical sensor probe is shown in Fig. 11.7, including a sectional view of the type DY20C probe.

The Model 1400 Series control units are solid-state types, suitable for field mounting (1450) and for rack or panel mounting (1440). Sensor probes may be located remotely at up to 3000 ft from the control unit. A single sensor input is standard, but options are available for multi-sensor inputs. The Model 1460 is a portable unit. The instruments are normally calibrated in dew point/frost point over the range −110 to +60°C, with a direct digital readout for −80 to +20°C as standard. Single or dual adjustable alarm facilities are available as an option. Output can be voltage (0–100 mV, 0–10 V) or current (4–20 mA dc).

The Model 1500 Series control units are capable of accepting up to six continuously active input channels. The units have UL and CSA approval for intrinsic safety with respect to the probe and cable only. Outputs (current

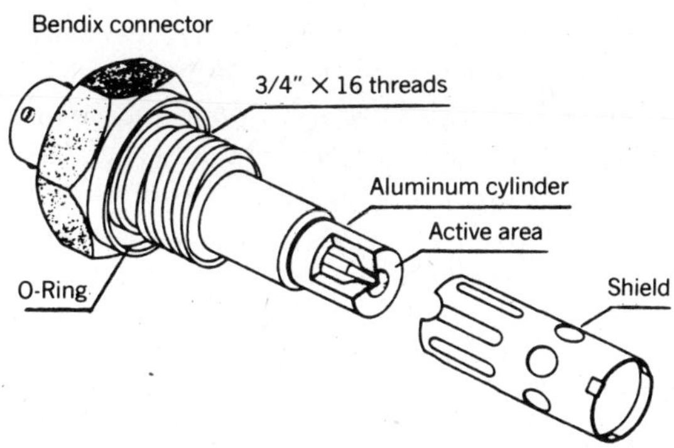

FIGURE 11.7 Typical aluminum oxide type moisture sensor. (Courtesy of Ondyne Inc.)

or voltage) are provided for all active inputs. Available options include ambient temperature display, multiple alarm facilities, and direct-reading dew point display with linear output signals (as for the Model 1400 series).

Ondyne also offers the Models 12 and 14 hygrometer calibrators, based on the permeation tube principle. The Model 12 employs a single-stage dilution system to produce mixtures in the range 10–100 ppm (vol) moisture, with a single permeation tube and an accuracy of ±5%. The Model 14 uses a two-stage dilution system for generation of mixtures containing less than 1 ppm (vol) moisture. Both units are available in either general purpose or explosion-proof designs. The outlet flowrate of the calibrator is 100–3000 ml/min at 0–15 psig.

11.4.4 Capacitance Type

Shaw Moisture Meters Ltd. (UK) manufactures a wide range of instruments operating on this principle. The heart of the system is the probe, which consists of a metal core, coated with a hygroscopic dielectric material and finally covered with a porous gold film, the probe acting as a variable capacitor. As the dielectric layer is only a few microns in thickness, it rapidly achieves equilibrium with the surrounding water vapor pressure, and the corresponding capacitance value is computed by the control unit for display in dew point temperature. All standard sensors have a linear response in dew point temperature and a logarithmic response in ppm moisture concentration. Four gas sensors are available, covering the following ranges:

	Dew point (°F)	Moisture (ppm vol)
Sensitive (red spot)	−112 to −4	0–1000
Medium (yellow spot)	−76 to +32	10–6000
Gold spot	−58 to +68	38.8–23,072
Green spot	−22 to +68	376–23,072

All units have intrinsic safety approval from F.M. and CENELEC when used with a zener barrier between the sensor and the control unit (mounted in a safe area). Control units are available in bench, rack, and panel mounted designs, as illustrated in Fig. 11.8. Analog output of 0–100 mV is standard; however, 4–20 mA dc output can be provided by use of a signal converter.

Model SHA is available in 4, 6, and 8 channel versions, with an individual sensitivity control for each channel and push-button switches for channel selection. Alarm facilities are also provided.

Model SAHA provides fully-automatic scanning of 4, 6, or 8 sensors, with a manual override, in addition to the facilities provided by the Model SHA. Dual-range instruments are also available (Model SHA-TR)

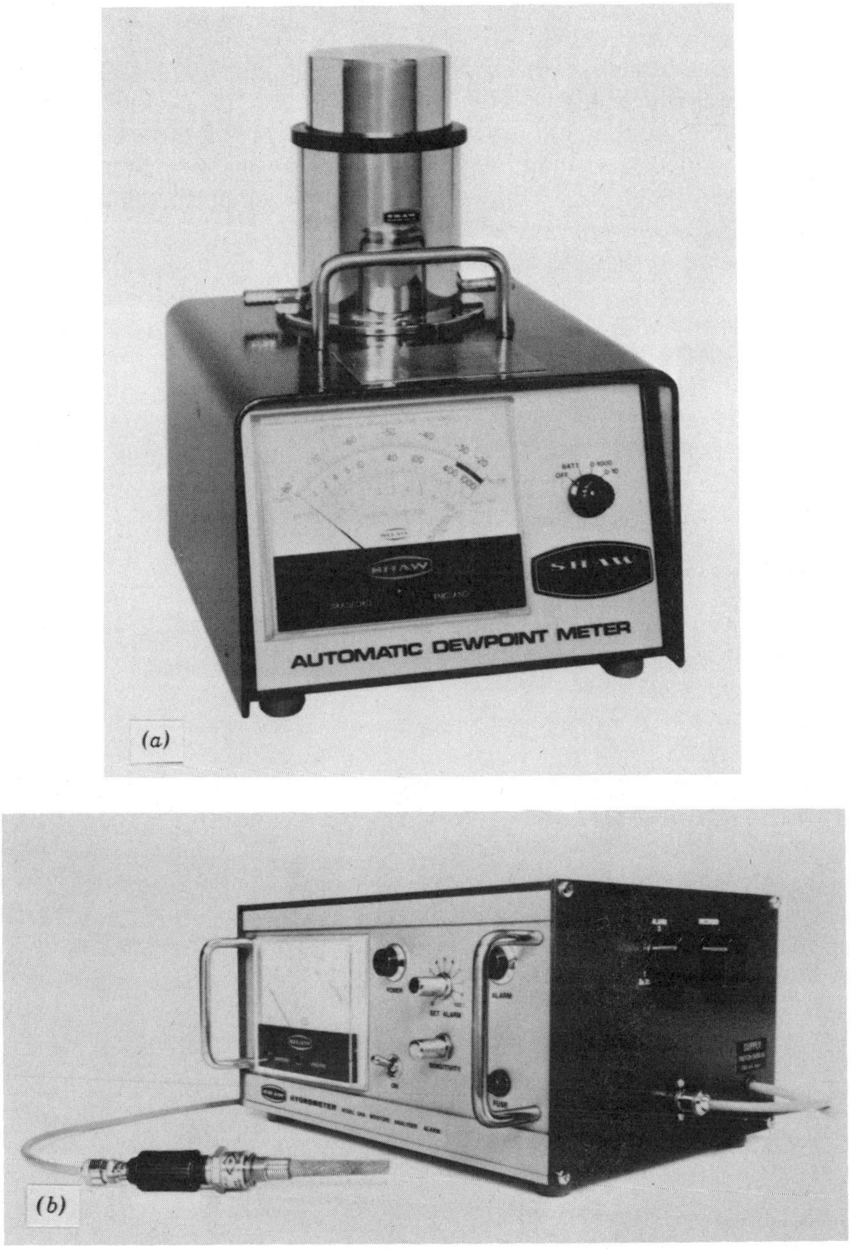

FIGURE 11.8 Shaw moisture analyzers. (*a*) Model SADP; (*b*) Model SHA. (Courtesy of COSA Corporation)

11.4.5 Automated Karl Fischer Type

U.G.C. Industries Inc. (United States) manufactures the Model 263 Moisture Titrator, which is based on the Karl Fischer test method outlined in Section 11.3.3. A schematic diagram of the instrument is shown in Fig. 11.9, outlining the theory of operation, together with details of the physical equipment.

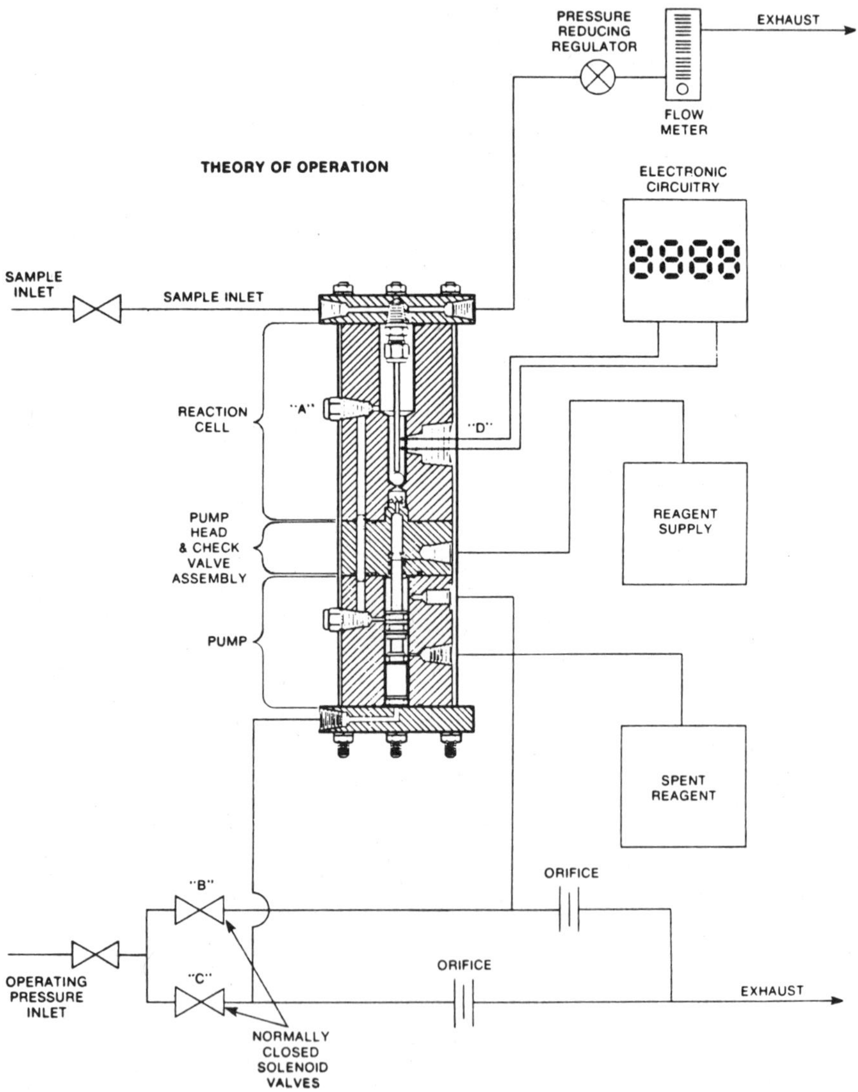

FIGURE 11.9 Model 263 moisture titrator. (Courtesy of UGC)

Sample enters the titrator at line pressure and passes to the reaction cell (A). Operating pressure of 300–600 psig enters at the operating pressure inlet and passes to solenoid valves B and C. These are normally closed, with the piston in the "down" position. The sample gas bubbles up through the KF reagent and leaves the reaction vessel through a reagent trap, then via a pressure reducing regulator and flowmeter. As water in the sample is absorbed by the titrant, platinum electrodes (D) sense a change in conductivity of the solution. When a slight excess of water exists, this causes a sharp decrease in current. When the current drops to a preset value, a relay operates to supply solenoid power to the reagent injection system. The electronic circuitry measures the time between end-points, computes the moisture content of the gas, and displays the value on a liquid-crystal display in the customer's selected units (e.g., ppm, lb/MMSCF, kg/MMm^3).

During the time the display is updated, solenoid valve C opens, forcing the piston up, collecting the overflow of spent reagent. Valve C now closes and B opens momentarily, forcing the piston down. In the full down position, fresh reagent is drawn into the pump head and spent reagent is exhausted, 0.5 ml of KF reagent is injected into the reaction vessel with each titration .The entire end-point sequence is completed in approximately 2 sec. Current to the electrodes returns to its former level and another titration begins.

The instrument is designed for NEC Class 1, Group D, Division 1 hazardous locations and has a standard range of 1–100 lb/MMSCF moisture concentration. Repeatability is quoted as ±3% of reading. A 4–20 mA dc output is available as standard. The moisture titrator is used primarily in natural gas pipeline applications, where the 300–600 psig pipeline pressure can be used as the pump operating medium.

11.4.6 Lithium Chloride Type

The Foxboro Company (United States) manufactures a dew point measurement system based on this operating principle. The standard humidity sensor is the Dewcell element, which consists of a varnish-insulated metal tube for the temperature-measuring element, a wick of woven glass tape wound around the tube, and two parallel gold overlay wires wound around the wicking. Anchorage for the windings is provided by bakelite end-caps. The wicking is impregnated with lithium chloride, a hygroscopic salt, and the wires are connected to a source of ac power. The Dewcell element operation depends upon the fact that the dry salt is a non-conductor of electricity, whereas the salt solution is a good conductor. If the amount of water vapor in the atmosphere surrounding the sensor increases, the salt absorbs moisture and heat must be applied to bring the solution to saturation at a new (higher) equilibrium temperature, and vice-versa.

The two electrodes, in contact with the wicking containing the salt, will

FIGURE 11.10 Foxboro Dewcell element. (Courtesy of The Foxboro Company)

therefore pass current only if the temperature of the salt is at or below the equilibrium temperature and the salt is moist. The passage of current through the thin salt film heats it quickly, raising the temperature to the equilibrium point. The change from moist to dry salt is almost instantaneous (a few seconds) so that there is very little or no conductivity above the equilibrium point. The amount of electrical energy supplied is a measure of the absolute humidity of the atmosphere surrounding the sensor. Power consumption is about 8 watts. The instrument is designed to read directly in dew point over the operating range −50 to +142°F.

The Dewcell element itself is shown on Fig. 11.10; graphs are available showing the precise working range of the device as limited by the Dewcell characteristic line and the saturation line. The temperature sensor in the Dewcell element may be a thermocouple (iron-constantan, chromel-alumel, or copper-constantan) or an RTD (platinum or nickel). The Model 2740 power unit provides a 25 V ac supply to drive a single Dewcell element. Other power units are available to drive up to six Dewcell elements.

11.4.7 Optical/Condensation Type

Two companies manufacture instruments operating on this principle (also known as "chilled mirror" technology).

(a) E.G & G, Inc. (USA)

In the Model 660 Dewpoint Hygrometer, dew point temperature measurement is made by a direct-measuring sensor using a Peltier element-cooled, gold-surfaced mirror, which is automatically held at the dew point temperature by a condensate-detecting optical system incorporating an LED light source and phototransistor. When condensate forms on the mirror as it is cooled, the optical sensing bridge detects the change in light intensity and controls the power to the Peltier element so that the mirror continuously tracks the dew point and will detect changes of 0.1°F or less. The dew point temperature is measured directly by a precision platinum resistance element

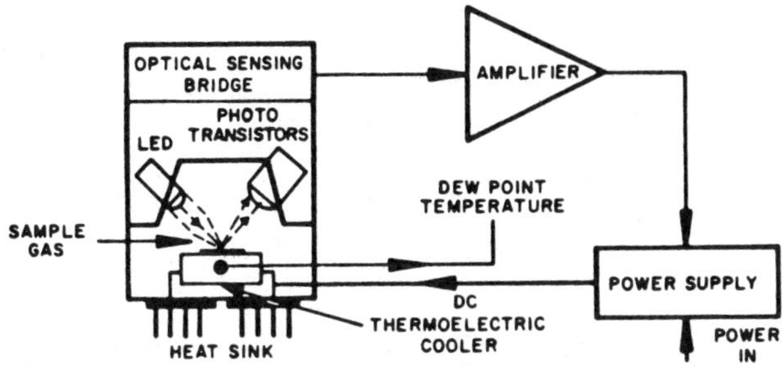

FIGURE 11.11 Model 660 hygrometer. (Courtesy of EG & G)

embedded in the mirror. A schematic diagram of the sensor assembly is shown in Fig. 11.11.

The Model 660 Control Unit is equipped with a timed electronic automatic self-standardization circuit, a track-and-hold output circuit, mode indicator, and alarm set feature. Standard outputs are −5 to +10 V dc and PRT (100 ohms nominal); optional outputs include parallel BCD and 4–20 mA dc via a plug-in module. The self-standardization circuit allows for correction of changing "dry mirror" reflectance as a function of contamination. The mirror is periodically heated to a temperature above the dew point and the optical bridge is electronically balanced for the present reflectance. The balance cycle may be initiated manually or pre-programmed at 6, 12, or 24 hr increments. The track-and-hold circuitry holds the data output(s) at the previous dew point reading during the balance cycle.

The sensor assembly is usually mounted at the rear of the control unit, but can be located remotely at distances up to 500 ft from the control unit. A remote mounting kit is available, consisting of a sensor mounting assembly and up to 500 ft of extension cable for the sensor/control unit interconnection.

The Model 300 Humidity Analyzer is a microprocessor-based system, designed to provide continuous, accurate dew point measurement over the range −100 to +140°F, or equivalent ranges of other parameters such as RH, ppm moisture, and so on. The control unit is suitable for either rack or panel mounting and has function and status switches in a keyboard array on the front panel.

The dew point sensor is different from the Model 660 and has been developed to give improved compensation for contamination build-up on the mirror surface. A schematic diagram showing the principle of operation is shown in Fig. 11.12, together with details of the sensor and control unit. Two reflective mirrors are placed in physical and thermal proximity to each other so that each is exposed to the same flow pattern of the gas sample; contaminant build-up on the "wet" and "dry" mirrors is therefore essentially equal. The dry mirror is maintained at a slightly higher temperature than the wet mirror and tracks the temperature of the wet mirror.

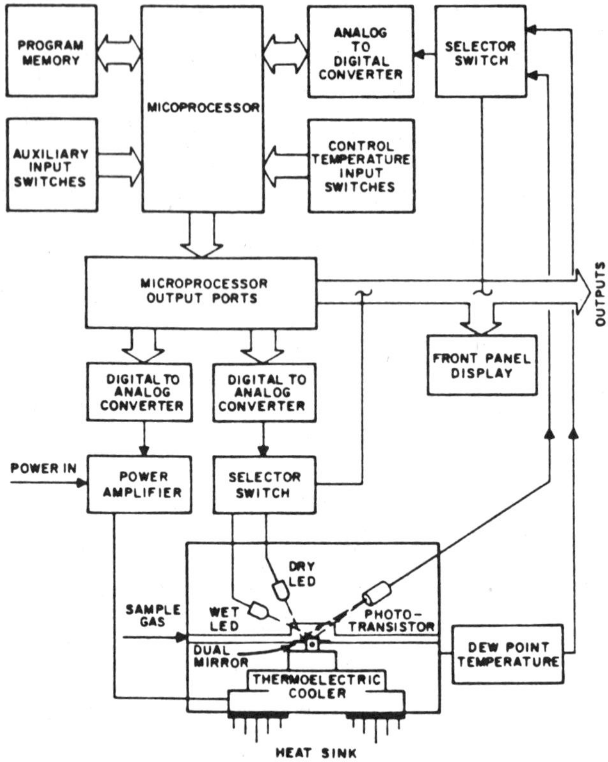

FIGURE 11.12 Model 300 humidity analyzer. (Courtesy of EG & G)

Two LED light sources are employed, with switching to allow one phototransistor to be used without mixing the two light signals. The continuous balance control system verifies sensor performance and corrects for contaminant build-up without the loss of real time data. On start up or manual activation, both mirror surfaces are heated to ensure that no condensate is present. During this period, the "wet';' and "dry" LEDs are energized alternately and the output of the phototransistor is compared with a common baseline output. The light intensity of each LED is adjusted independently by the microprocessor until signals from the two mirrors are the same. After a fixed time, or by manual activation, the control system will switch to the operate mode, where both wet and dry channels are sampled. The dry channel thus functions as a reference or calibrate channel. When the sensor approaches contaminant levels above which it can no longer compensate, an alarm signal is generated, indicating that mirror cleaning is now required.

(b) General Eastern Instruments Corp. (United States)

General Eastern manufactures a wide range of humidity analyzers based on the electrolytic, aluminum oxide, lithium chloride, and optical/condensation principles. Only those instruments operating on the optical/condensation principle will be described here.

The 1200 "S-Series" of Dew Point Hygrometers are designed for a wide variety of applications, covering dew point/frost point measurement over the range −100 to +200°F. They use a range of sensors and control units that are fundamentally identical and vary only in packaging and readout facilities. A schematic diagram of the measurement system is shown in Fig. 11.13*a*. The various control unit options are illustrated in Fig. 11.14.

All 1200 series instruments are equipped with a special circuit (termed the PACER auto-standardization circuit), which permits the instrument to check its own performance on a periodic basis (2, 6, 12, or 24 hr). PACER is designed to reduce measurement errors due to contaminant build-up on the mirror, thereby reducing the frequency of mirror cleaning; the time between mirror cleanings is reputedly extended by a factor of 10–100).

Operation of the PACER circuit is illustrated in Fig. 11.13*b*. When activated, the system first cools the mirror well below the prevailing dew point, causing an excess of water to deposit on the mirror surface and a "puddle" to form. The system then forces the mirror to quickly heat, evaporating the water. During heating, the large puddles break up into small puddles, which shrink in size, holding higher and higher concentrations of solute in solution, leaving bare, uncontaminated surface behind. Eventually, each puddle becomes saturated and solute begins to precipitate out in polycrystalline clusters. Most of the mirror surface (80–85%), however, is clean and thus available for undisturbed dew formation. PACER automatically adjusts the control loop to maintain a dew deposit that covers 40% of the mirror surface when the system reverts back to normal operation. During the time interval of the PACER action, the output data and display are sampled and held at the last prevailing dew point.

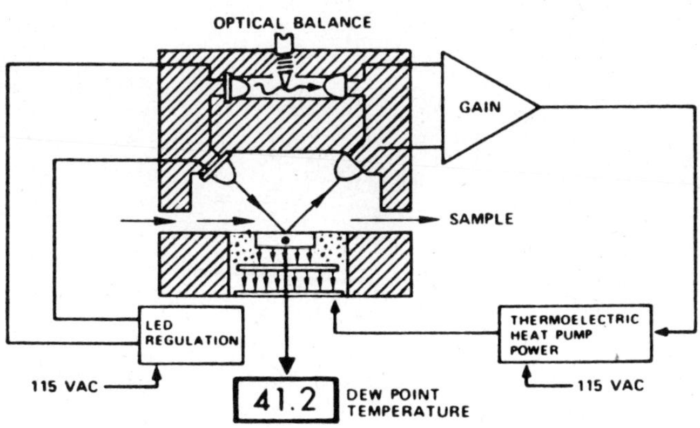

(*a*)

FIGURE 11.13 1200 S-Series dew point sensors. (*a*) Operating principle. (Courtesy of General Eastern Instruments Inc.)

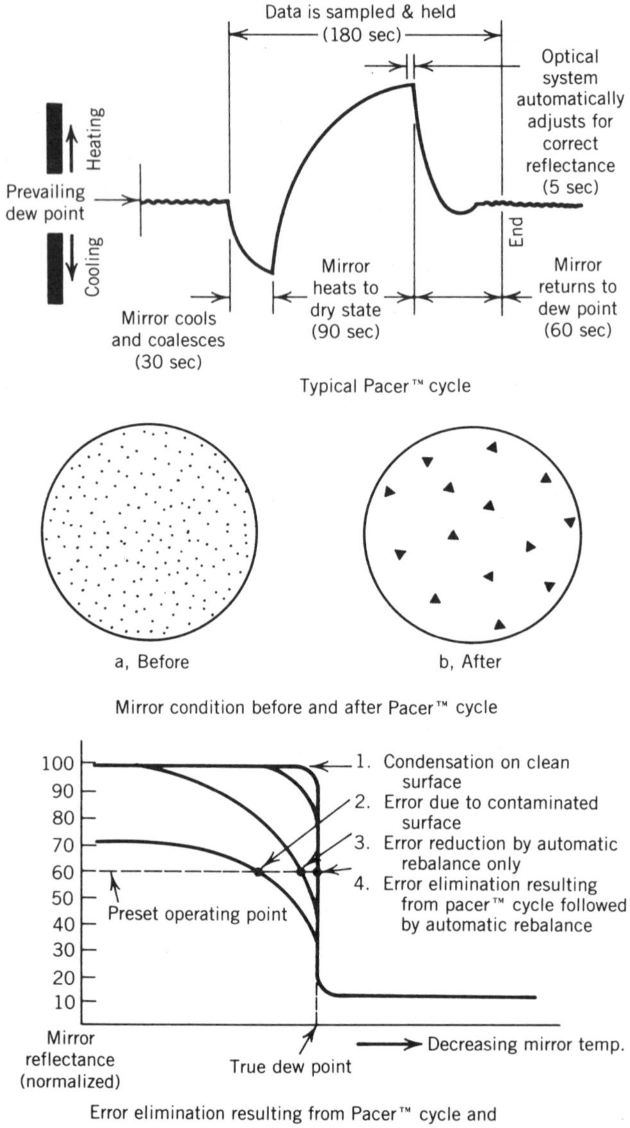

(*b*)

FIGURE 11.13 1200 S-Series dew point sensors. (*b*) Operation of the PACER system. (Courtesy of General Eastern Instruments Inc.)

The System 1500 Hygrocomputer is a microprocessor-based system designed for use with the Models 111D, 1211D, or 1311D sensors. The instrument provides current, voltage, or BCD outputs and alarms for temperature and humidity parameters such as dew point, %RH, ppm concentration, or grains per pound. Display can be in English or metric units, switch-selectable.

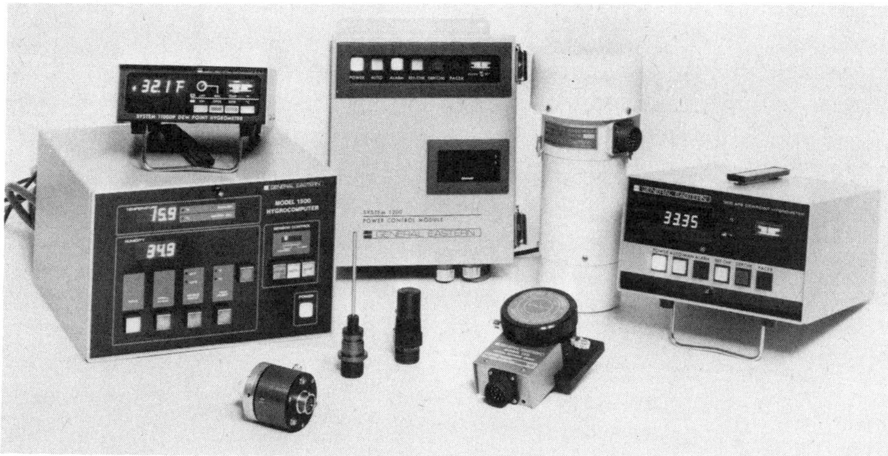

FIGURE 11.14 The range of General Eastern control units. (Courtesy of General Eastern Instruments Corp.)

11.5 PROCESS ANALYZERS FOR MOISTURE IN SOLIDS

Instruments for continuous measurement of moisture in solid materials are specialized, not because of a unique operating principle, but because of the difficulty in handling solid samples. Generally, instruments that infer moisture concentration from such properties as dielectric constant and conductivity are not as reliable as those that measure moisture directly, since these properties are also affected by other factors such as particle size and packing density, which are difficult to predict and to compensate for.

Commercially available process analyzers for moisture in solids accomplish the measurement using one of the following principles:

1. Neutron moderation.
2. Microwave absorption.
3. Infra-red absorption.

11.5.1 Neutron Moderation

Instruments operating on this principle use the moderation of fast neutrons by hydrogen atoms as the basis of moisture measurement. The *neutron* is a subatomic particle emitted by certain radioactive materials during the process of atomic decay. The neutron is electrically neutral, has nearly the same mass as the hydrogen atom, and is therefore only deflected or reflected on impact with the atomic nucleus. Such an impact involves a change in momentum of the neutron and the impacted nucleus.

Since energy and momentum are conserved (Newton's law), the energy of the neutron is decreased by an amount equal to the energy transferred to the impacted nucleus. The neutron energy after impact is a function of the mass of the impacted atom. Assume that m_1 and m_2 are the masses of the neutron and the impacted atom, V_1 and V_3 are the velocities of the neutron before and after impact, and V_2 is the velocity of the impacted nucleus (which is initially at rest). Then, by conservation of momentum:

$$m_1 V_1 = m_2 V_2 + m_1 V_3 \tag{1}$$

and, by conservation of energy:

$$\tfrac{1}{2} m_1 V_1^2 = \tfrac{1}{2} m_2 V_2^2 + \tfrac{1}{2} m_1 V_3^2 \tag{2}$$

Therefore, from (1):

$$V_2 = \frac{m_1}{m_2}(V_1 - V_3)$$

Substituting for V_2 in (2) and simplifying, we get,

$$V_3 = -V_1 \frac{(m_2 - m_1)}{(m_1 + m_2)}$$

Therefore, if the mass of the impacted nucleus is equal to the mass of the neutron, all of the neutron's kinetic energy is transferred to the impacted nucleus and V_3 is equal to zero.

Since the hydrogen atom is almost equal in mass to the neutron, it is the most efficient absorber (or moderator) of neutrons. The number of reflected or moderated neutrons (often called "thermal" neutrons) is thus a measure of the number of hydrogen atoms present in the process material. If all the hydrogen is contained in the moisture, or if there is a constant amount of hydrogen contained in other forms, the detector can be calibrated in terms of the weight of moisture per unit volume of process material. Unfortunately, density also affects the amount of neutrons reflected and therefore some means of density measurement is also required as part of the moisture measurement, in order to eliminate the effects of density variations.

(a) Texas Nuclear (United States)

The Texas Nuclear MNB Series MOLA (Moisture On-Line Analyzer) is designed to perform the non-contacting measurement of moisture in materials contained in bins and hoppers. The basic instrument consists of a moisture head and an amplifier-indicator, which provides power and signal processing circuitry. The moisture head contains a radio-isotope fast neutron source with appropriate shielding and two or more TEXLIUM detectors

FIGURE 11.15 MNB Series MOLA moisture analyzer. (Courtesy of Ramsey Engineering, Texas Nuclear)

(also manufactured by Texas Nuclear). The equipment is shown in Fig. 11.15. The detectors produce a pulse output for virtually every thermal neutron striking them, but are insensitive to fast neutrons, making the output pulse rate proportional to the hydrogen density of the process material. If the conditions outlined are met, the detector output may be calibrated in terms of weight moisture per unit volume of process material.

If the bulk density of the material varies, density compensation is necessary. The SGF Series density gauge is used, consisting of three components—a gamma-ray source, a gamma-ray detector head, and an amplifier-indicator. A special module computes the moisture/density ratio from the two detector signals and produces an analog current output (4–20 mA dc is an option) proportional to the percent by weight of moisture in the process material. The half-life of the moisture measurement source (^{241}Am-Be) is 433 years and so no source decay compensation is required. This is, however, provided as standard for the density measurement source (^{137}Cs).

The MOLA is also available with a microprocessor-based control unit that provides efficient control of the system, has internal diagnostics, and allows interfacing with peripheral devices such as a CRT terminal, line printer, and teletype.

Typical applications for the MOLA include the measurement of moisture in coke (1–20%), moisture in iron ore concentrates (1–7%), and water of hydration in alumina (0.5–3.0%).

(b) Ohmart Corp. (United States)

The Moistart Micro 5000 non-contact moisture analyzer is designed for measurement of the moisture content of solid materials moving on a belt conveyor (one prime application is the measurement of moisture in wood chips).

The instrument consists of a C-shaped measurement frame designed to fit over the conveyor, as shown in Fig. 11.16, and is connected to a microprocessor-based electronics unit. The measurement frame includes a fast neutron source and a gamma-ray source, enabling the system to compensate for

FIGURE 11.16 Moistart Micro 5000 moisture analyzer. (Courtesy of Ohmart Corporation)

variations in density of the process material. The instrument measures completely through and across the material on the conveyor, averaging automatically. It is not a surface measurement and therefore allows averaging of a large volume of material.

The Micro 5000 can measure moisture over the range 0–90%, dry basis, the minimum range being about 0–1%, depending upon the process material. Signal outputs available are a serial RS232 data link, 0–10 V dc, 0–100 mV, and 4–20 mA dc. Repeatability is quoted as ±0.5% of range.

11.5.2 Microwave (RF) Absorption

It is a well known fact that microwave energy is strongly absorbed by water. Therefore the amount of water present in a process material is indicated by a power reduction of the microwave signal. However, this power reduction is also affected by other factors such as bulk density, so it is necessary to measure the density in order to compute the moisture content.

(a) Kay-Ray Inc. (United States)

The Kay-Ray Model 8100 Moisture Analyzer uses the microwave absorption principle to measure the moisture content of solid materials. A special CHUTE sensor, illustrated in Fig. 11.17, is installed in the process line to facilitate sample flow. The chute is lined with a microwave absorber to eliminate stray reflections and houses the microwave transmitter and receiver for moisture measurement, together with a gamma-ray source and detector for density measurement. The sensor enclosures are heavy-gauge steel, constructed to NEMA 4 requirements, and are dust, water, and oil tight.

A microprocessor-based control unit accepts the sensor inputs and computes the percent moisture in the process material. Signal conditioning cir-

FIGURE 11.17 Model 8100 moisture analyzer. (Courtesy of Kay-Ray Inc.)

cuits provide indication of abnormal operating conditions, at which the output is frozen at the last valid reading. Calibration is automatically checked and re-referenced (if required) by the microprocessor every time the chute is empty. High and low alarm contacts are provided for moisture and density. Analog voltage and current outputs are available (0–5 V and 4–20 mA dc). Measurement may be made within the range 2–30% moisture and 10–50 lb/ft^3 density.

11.5.3 Infra-Red Absorption

The principle of measurement is the selective absorption of IR energy by water, based on the selection of suitable measurement and reference frequencies. However, in the case of moisture measurement in solids using this technique, the measurement and reference beams are reflected off the sample rather than transmitted through it. Since the IR energy does not penetrate very far beneath the surface of the solid, the measurement is only related to surface moisture. This may not always be representative of the true moisture level in the sample, especially when it is exposed to the atmosphere.

(a) Moisture Register Company (United States)

The Model MR is a non-contacting continuous moisture analyzer based on the IR absorption principle and consists of an IR sensor and an electronic control unit, as shown in Fig. 11.18. The IR source (quartz halogen) is focused on the sample and reflected light is filtered at two different near-IR wavelengths, one specific for water, the other a reference. The two signals are electronically ratioed and the resulting signal displayed and transmitted

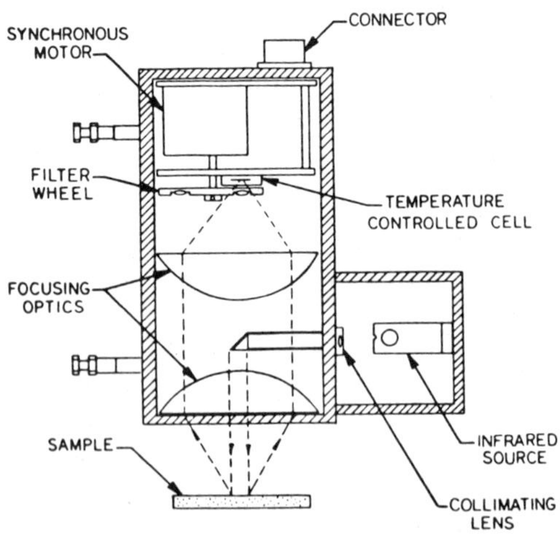

FIGURE 11.18 Model MR analyzer. (Courtesy of Moisture Register Co.)

as sample moisture content. The instrument is suitable for moisture measurement within the range 0–80%. Standard output is 0–10 V or 0–1 mA dc, with 4–20 mA dc available as an option.

(b) Ohmart Corp. (United States)

The Moistart Series 5200 is a non-contact, IR absorption-based instrument designed for moisture measurement in solids over the range 0–90%, with a minimum range of 0–0.2% and a quoted repeatability of ±0.1% of range. The analyzer consists of a measuring head and electronics unit, both mounted in water- and dust-tight NEMA 4 enclosures. Two IR beams are used (measurement and reference). The analyzer (see Fig. 11.19) covers a large area of illumination and thus allows for more accurate averaging of any surface moisture differences caused by discrete particle size variations. Increased illumination also combats the error effect of variable material loading and ambient light interference.

The control unit (5210) is capable of multi-product switching; it is able to store up to six sets of different materials data. Built-in diagnostics allows monitoring of optical-electrical parameters. High and low moisture alarm

FIGURE 11.19 Moistart Series 5200 moisture analyzer: measuring head. (Courtesy of Ohmart Corporation)

contacts are available. Analog output is either voltage (0–10 V dc or 0–100 mV) or current (4–20 mA dc or 10–50 mA dc).

(c) Moisture Systems Corp. (United States)

The MSC Quadra-Beam Model 475 is designed for the measurement of moisture in solid materials, using the IR absorption principle, covering the range 0–90% with a minimum range of 0–0.2%.

A *four-beam* feature is used to eliminate instrument drift and improve stability. A schematic diagram of the system is shown in Fig. 11.20. The instrument uses precision IR filters mounted on a rotating wheel, which allows pulses of "reference" and "measurement" light to pass alternately through the wheel. The optically chopped light beam is then focused on the sample and the reflected energy of these two beams measured by the detector. The two additional beams are of the same reference and measurement wavelengths, but these are totally internal and are not projected onto the sample. The *Prime* beams (as they are called) come from the same source, pass through the same filters, and impinge on the same detector as the main beams.

The signal strengths of the four beams, as measured by the detector, are inputs to a computation circuit that produces an output proportional to the water concentration of the process material, according to the following expression:

$$\text{Water concentration} = \frac{R}{M} \times \frac{M'}{R'}$$

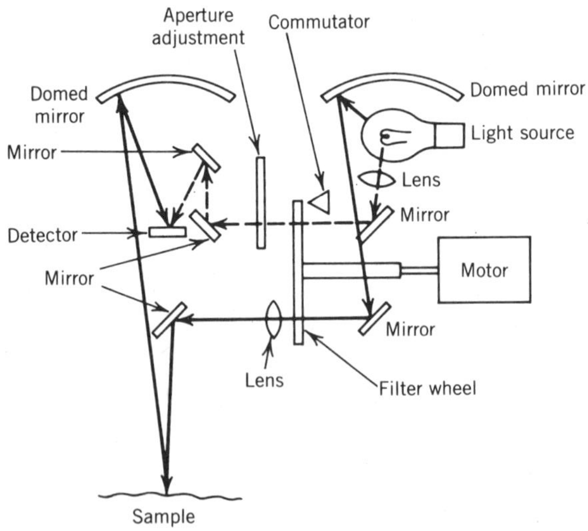

FIGURE 11.20 Quadra-Beam Model 475. Solid arrowed lines: main channel; dotted arrowed lines: prime channel. (Courtesy of Moisture Systems Corporation).

where R = main reference signal
M = main measurement signal
R' = prime reference signal
M' = prime measurement signal

Should a drift-causing condition occur (for instance, a change in light intensity at the measurement wavelength), both M and M' values will be affected proportionately and the effect is cancelled out. If the water concentration in the process material changes, then only the value of M is affected.

The instrument consists of two components, the measuring head and a microprocessor-based control unit. The measuring head, the various designs of which are suitable for measurement of conveyor-borne material, moving webs, or static samples, contains the IR optics, detector, and signal amplifier. The control unit provides signal averaging, integration, and true ratio computation. Moisture content is displayed digitally and alarm contacts are available for high and low moisture levels. Analog voltage (0–100 mV, 0–10 V) and current (4–20, 10–50, and 1–5 mA dc) outputs are available as standard. The control unit can also be interfaced with peripheral devices such as a printer.

Both sensor and control unit are available in explosion-proof enclosure design for NEC Class 1, Groups C/D, Division 1 and Class 2, Groups E/F/G, Division 1. Repeatability is quoted as ±0.5% of full scale.

BIBLIOGRAPHY

Blakemore, C. B. Piezoelectric moisture analyzer for semiconductor gases. *Proc. ISA Symp.*, Boston, MA (May 1985).

Botsco, R. Microwave moisture measurement. *Inst. & Contr. Syst.* (May 1970).

Bryant, R., and Scelzo, M. The impact of microprocessors on hygrometry. *Proc. Intl. Moisture & Humidity Symp.*, Washington DC (April 1985).

Cucchiara, O., and Mittelman, M. The measurement of dissolved water in organic liquids using an aluminum oxide hygrometer. *Proc. Intl. Moisture & Humidity Symp.*, Washington DC (April 1985).

Gardner, R. C. Moisture/basis weight infrared gage for paper. *InTech* (Jan 1968).

Hammond, R. H. Aluminum oxide hygrometry and water enhancement factors. *Proc. Intl. Moisture & Humidity Symp.*, Washington DC (April 1985).

Harding, J. C. Precision humidity measurement. *Adv. Instr., ISA Intl. Conf.*, Philadelphia, PA (Oct 1982).

Harding, J. C. Overcoming limitations inherent to aluminum oxide sensors. *Proc. Intl. Moisture & Humidity Symp.*, Washington DC (April 1985).

Harding, J. C. A chilled mirror dewpoint sensor/psychro-metric transmitter for energy monitoring and control systems. *Proc. ISA Symp.*, Boston, MA (May 1985).

Krigman, A. Humidity measurement and control: A hank of hair, a wet sock on a string. *Intech* (March 1985).

Mayeaux, D. P. Moisture analysis of contaminated gases. *Proc. Intl. Moisture & Humidity Symp.*, Washington DC (April 1985).

On-line moisture analyzers: Non-contact sensing comes of age. *InTech* (May 1984).

Ronchinsky, S. An electrochemical sensor for trace moisture in gases. *Proc. Intl. Moisture & Humidity Symp.,* Washington DC (April 1985).

Shinskey, F. G. Humidity, dew point and wet bulb temperature. *Inst. & Contr. Syst.* (July, Aug 1985).

Weiderhold, P. R. Humidity measurement. *Instr. Tech.* (Aug 1975).

Weiderhold, P. R. Which humidity sensor? *Inst. & Contr. Syst.* (June 1978).

12

Air Quality Monitoring

12.1 BACKGROUND AND DEFINITIONS

The subject of air quality monitoring is very complex and has intimately concerned scientists in universities, industries, and governments on a world-wide basis for many years. The aim of their investigations is to obtain a better understanding of how atmospheric pollution affects the environment in which we live and how the problems can be dealt with in an effective manner. Before describing how continuous air quality monitoring fits into this overall picture, some historical background on our global environment is appropriate.

The earth's atmosphere is made up of four layers, the troposphere, stratosphere, mesosphere, and thermosphere, as shown in Fig. 12.1. We are mainly concerned with the troposphere, the layer extending from the earth's surface to an altitude of 10–16 km. This layer contains the air we breathe, from it falls the water we drink, and through it passes the life-producing energy of the sun.

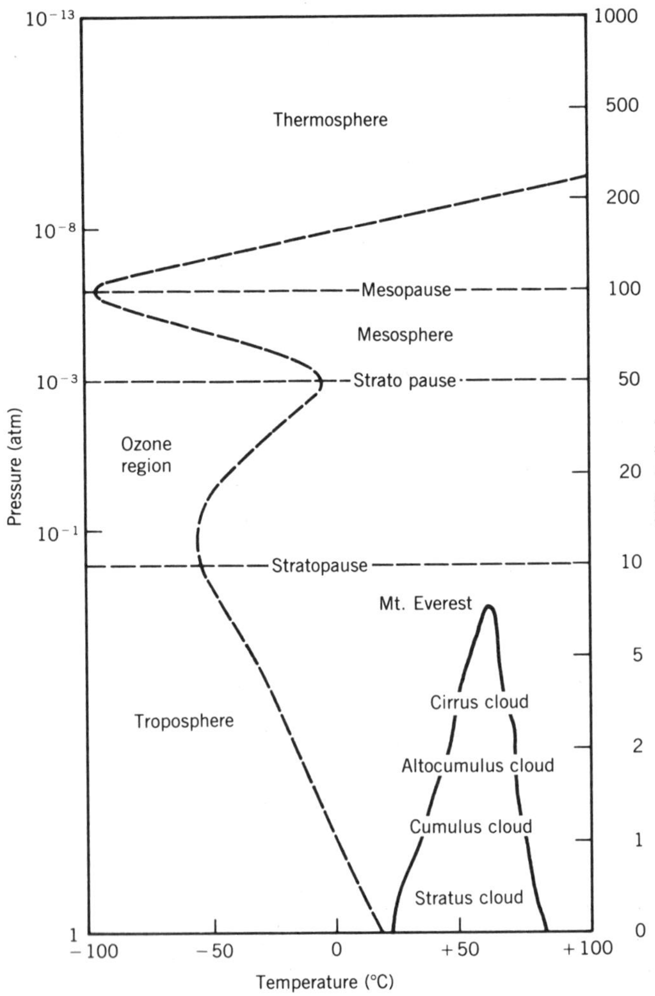

FIGURE 12.1 The earth's atmosphere.

Since the days of Greek civilization, philosophers, scholars, and scientists have puzzled over the mysteries of the earth's atmosphere. However, the first significant scientific advances in atmospheric chemistry were not made until the 18th century, when Priestley, Lavoisier, and Cavendish were the first to establish that the major constituents of the earth's atmosphere are nitrogen (79% by volume) and oxygen (20% by volume), with lesser amounts of water and carbon dioxide.

During the 19th and early 20th centuries, many prominent chemists and physicists further advanced our knowledge in this area. Ramsey documented the presence of the noble gases helium, neon, argon, krypton, and xenon in the atmosphere. Other investigators found evidence for small quantities of methane, hydrogen, ozone, carbon monoxide, and sulfur dioxide in the air and sulfate, nitrate, ammonium, and chloride ions dissolved in rain water. By the early 1950s, therefore, a considerable amount of qualitative data on the chemical composition of the atmosphere had been collected.

Since the 1950s, rapid advancement in analytical chemical techniques and the advent of high-speed computer technology has resulted in an explosion of information concerning atmospheric composition. The atmosphere has been found to be a veritable reservoir for a myriad of trace gaseous and aerosol species with concentrations below 1 ppm (vol) which, in spite of their relatively low levels, can have a major impact on the environment. Some gases, for instance, because of their toxicity, can affect plant and animal life; others can affect climate via the "greenhouse effect". Aerosols, on the other hand, can have a significant effect on cloud formation and precipitation patterns.

Simple calculations show that most trace species are present in the atmosphere at much higher levels than would be predicted solely on the basis of thermodynamic equilibrium with the main constituents nitrogen, oxygen, carbon dioxide and water. Clearly, then, kinetic mechanisms, involving chemical, physical, and biological processes, are controlling the trace gas composition of the atmosphere. It is the identity of these mechanisms and their relationship to natural life cycles that have been of fundamental interest to atmospheric chemists in recent years.

Global tropospheric chemistry is intimately concerned with atmospheric pollution, since pollution can move by the action of prevailing wind patterns from urban sources to remote locations. Chemists must therefore consider their effect on the natural global system; in some instances, this can be quite significant. On the other hand, tropospheric chemistry cannot be studied from a purely chemical viewpoint; a multi-disciplinary approach is often necessary. The trace gas and aerosol composition of the atmosphere is influenced strongly by a variety of physical, geological, and biological processes, examples of which are solar radiation, wind patterns, and earth-atmosphere surface interactions.

One significant example of the relationship between atmospheric pollution and its effect on our ecological environment is acid rain, which has been the subject of international political debate for a number of years and is continuing to grow in intensity. The small international community of acid rain researchers has been telling us for many years that we have a potentially devastating problem on our hands. Studies in Europe and North America have demonstrated that, for all the complications, the principle culprit is sulfur. When the amount of sulfur falling on an acid-sensitive ecosystem per year is greater than about 18 lb/acre, sooner or later damage will occur. Below that threshold, all but the most sensitive areas will be able to cope and will not be harmed. From this evidence, it is clear that in order to protect our environment the deposition of sulfur in sensitive areas must be reduced to below that critical level, which, in turn, means a reduction in sulfur dioxide emission to the atmosphere from industrial plant. The alternative is to play economic Russian roulette with the lakes, streams, forests, and their animal life, all of which are sensitive to acidification.

Environmental monitoring covers both atmospheric air quality measurement and water quality measurement (lakes and streams). These two are intimately connected since pollution of the atmosphere has a significant effect on the quality of our natural waters. This chapter is concerned with air quality monitoring only; water quality monitoring is covered in detail in Chapter 13. Nevertheless, one should bear in mind the interrelationship between the two.

12.2 THE ROLE OF CONTINUOUS AIR QUALITY MONITORING

As indicated, the monitoring of our environment has become a matter of governmental and industrial concern on an international basis. In the United States the three main governmental bodies concerned with the legal and procedural aspects of environmental monitoring are the Environmental Protection Agency (EPA), the National Institute for Occupational Safety and Health (NIOSH), and the Occupational Safety and Health Administration (OSHA). These organizations work together with a host of research laboratories, both academic and industrial, in formulating test methods, standards, and procedures for air quality monitoring. Two of particular interest are the ASTM Committee D-22 on Methods of Sampling and Analysis of Atmospheres and the National Center for Atmospheric Research (NCAR).

The EPA, ASTM, and NCAR sponsored two symposia on "Instrumentation for Monitoring Air Quality." The first was held in Philadelphia in October 1972, the second at the University of Colorado at Boulder in August 1973. The proceedings are summarized in ASTM Special Technical Publication 555, issued in June 1974, with a second printing in 1979. Both symposia were called at the request of EPA scientists, who felt the urgent need for public enlightenment on this important subject.

NIOSH and OSHA have responsibility within the federal government for ensuring that everyone is provided, as far as possible, with a safe and healthful working environment. The Standards Completion program (SCP), involving the cooperative efforts of several contractors and personnel from various divisions of NIOSH and OSHA, developed a series of occupational health standards for those substances with existing permissible exposure levels.

The output from the program after two and a half years work included about 400 substance-specific draft technical standards and supporting documentation, providing an extensive data base of information useful to the entire occupational health community and the worker. Listed in Table 12.1 are some of the more important air-borne contaminants, giving the permissible exposure limit (PEL) and the level at which the contaminant is immediately dangerous to life or health (IDLH), as established by the standards completion program of NIOSH/OSHA.

For convenience, the subject of air quality monitoring may be broken down into vehicle exhaust emission monitoring, workplace environment monitoring, and point source/fugitive emission monitoring. All three contribute to overall atmospheric pollution.

TABLE 12.1 NIOSH/OSHA Specifications for Airborne Contaminant Levels

Contaminant	PEL (ppm)	IDLH (ppm)
Ammonia	50	500
Benzene	1 (5 ceiling)	2000
Bromine	1	10
Carbon monoxide	35	1500
Carbon disulfide	10 (ceiling)	500
Chlorine	1	25
Chlorine dioxide	0.1	10
Hydrogen chloride	5 (ceiling)	100
Hydrogen fluoride	3	20
Hydrogen sulfide	20 (ceiling)	300
Methyl mercaptan	10	400
Nitric oxide	25	100
Nitric acid	2	100
Nitrogen dioxide	5	50
Ozone	0.1	10
Phosgene	0.1	2
Phthalic anhydride	2	10,000
Sulfur dioxide	5	100
Toluene	100	2000

PEL: Permissible exposure limit, based on time-weighted average level.

IDLH: Immediately dangerous to life or health.

12.2.1 Vehicle Exhaust Emission Monitoring

Nitrogen oxides and hydrocarbons are mostly a by-product of the internal combustion engine and natural gas burning furnaces. Nitrogen oxides are most evident when they react with other oxidants in the presence of sunlight to form photochemical smog, resulting in reduced visibility, eye irritation, and plant damage. This is a major problem in large urban areas such as Los Angeles. Federal standards have been set forth detailing test procedures for monitoring motor vehicle emissions and the specifications for continuous monitoring instrumentation. Annual inspection requirements of vehicles in the United States now stipulate maximum allowable levels of carbon monoxide and hydrocarbons in exhaust gases. Emission control equipment is now installed on all new motor vehicles.

12.2.2 Workplace Environment Monitoring

The threshold limit values set by OSHA for air-borne contaminants provide ample justification for continuous monitoring of the workplace environment. In this context, from an industrial viewpoint, we are primarily concerned with protection of personnel and equipment. Release of toxic gases such as hydrogen sulfide, chlorine, ammonia, benzene, vinyl chloride, and a host of other chemicals can severely endanger life. Release of hydrocarbons and other combustible gases can create a hazard to operating personnel and equipment.

12.2.3 Point Source and Fugitive Emission Monitoring

The proliferation of state and federal laws to control ambient pollution levels and stack NO_x and SO_2 emissions, again provide ample justification for continuous monitoring. As mentioned earlier, sulfur has been identified as the main contributor to acid rain. Thus, industry can expect even more stringent regulations to be imposed on NO_x and SO_2 emissions in the future, once legislation has been approved.

From this summary of the situation, it can be seen that continuous air quality monitoring instrumentation has in recent years played an important part in day-to-day industrial operations and is likely to be even more important in the future.

12.3 LABORATORY METHODS OF MEASUREMENT

The ASTM standard laboratory test methods for atmospheric analysis, some of which are EPA reference methods, are mostly traditional chemical test methods that are time-consuming and require meticulous attention to detail. The ASTM recognizes the use of continuous monitors by its recommended

TABLE 12.2 ASTM Standard Test Methods for Atmospheric Analysis

ASTM	
D1607-76[a]	NO_2 by Griess-Saltzman Reaction
D1609-60	Ozone Content of the Atmosphere
D1704-76	Particulate Matter in the Atmosphere
D2009-65	Particulate Content of the Atmosphere
D2010-65	Total Sulfation in the Atmosphere by Lead Peroxide Candle
D2011-65	Continuous Analysis and recording of the Oxidant Content of the Atmosphere
D2650-68	Gas Mixtures by Mass Spectrometry
D2913-75	Mercaptan Content of the Atmosphere
D2914-78[a]	Sulfur Dioxide Content of the Atmosphere by the West-Gaeke Method
D3162-78	Carbon Monoxide Content of the Atmosphere by NDIR Spectrometry
D3226-	Sulfur Oxides Content of the Atmosphere
D3249-79	Ambient Air Analyzers; Recommended Practice
D3416-78[a]	Total Hydrocarbons, Methane and Carbon Monoxide by Gas Chromatography
D3608-77[a]	Total Nitrogen Oxides (NO_x) Content of the Atmosphere by Griess-Saltzman Reaction
D3669-78	Carbon Monoxide Content of the Atmosphere by the Manual Turbidometric Method

[a] Indicates an EPA reference method.

practice D3249. Some of the more important test methods are summarized in Table 12.2 and a few of them are described here.

ASTM D1607-76 covers the determination of nitrogen dioxide within the range 0.002–5.0 ppm by the Griess-Saltzman reaction. The air is sampled for a period of 60 min at a flow rate of 0.4 l/min, using a fritted bubbler with a maximum pore size of 60 μm. A pump is used to draw the air sample through the bubbler and flowrate is measured by a dry test meter. The NO_2 is absorbed in an azo-dye-forming reagent. A red-violet color is produced within 15 min and its intensity is measured at 550 nm by a spectrophotometer.

ASTM D2914-78 covers the determination of sulfur dioxide within the range 0.01–4.0 ppm by the West-Gaeke method, which is based on the generation of the intensely colored pararosaniline methyl sulfonic acid. The intensity of the color developed is measured spectrophotometrically at 548 nm (Method A) or 575 nm (Method B). The two alternative methods are tedious and time-consuming. However, they have been automated by the use of instruments such as the Technicon AutoAnalyzer. Procedures are laid down for 1-hr and 24-hr sampling. Sample is collected at the rate of 0.5 l/min or 0.2 l/min respectively for the two methods.

ASTM D3608-77 covers the determination of nitrogen oxides (combined) by the same method as ASTM D1607. In this case, nitric oxide is converted to NO_2 quantitatively by a chromic acid oxidation prior to absorption by the Griess-Saltzman reagent.

ASTM D3416-78 covers the determination of total hydrocarbons, carbon monoxide, and methane by gas chromatography, using a flame ionization detector (FID). Total hydrocarbons are measured directly. Carbon monoxide and methane are separated on the GC column, the CO converted to methane using a methanator, and the two components measured individually by the FID after elution from the column.

12.4 CONTINUOUS ANALYZERS FOR ATMOSPHERIC MONITORING

From the foregoing discussion, it can be seen that atmospheric monitoring covers a very wide variety of different substances. In addition, many different analysis techniques are used. Several companies have specialized in air quality monitoring instrumentation and have received reference or equivalent method designation from the EPA for their equipment, based on three preferred analytical techniques—namely, infra-red absorption, fluorescence, and chemiluminescence.

It would be almost impossible to cover every continuous air quality monitor commercially available in the time and space available. Therefore, the more important atmospheric contaminants are discussed under the following headings:

- Combustible gases.
- Total hydrocarbons.
- Specific components by gas chromatography.
- Nitrogen oxides.
- Sulfur dioxide.
- Carbon monoxide.
- Ozone and oxidants.
- Ammonia.
- Chlorine and hydrogen chloride.
- Particulates/opacity.
- Vehicle exhaust emissions monitoring equipment.

Typical examples of commercially available analyzers are covered in each category and summarized more completely in tabular form.

12.4.1 Combustible Gas Detection

Combustible gas detection equipment is used in industry to monitor flammable gas release in process plants (fugitive emission) and for the purpose of protecting operating personnel and process equipment. Most instruments available today operate on the "single-point sensor" principle, whereby a single sensor is installed in a strategic location in the field (i.e., in the vicinity of suspect process equipment) and connected to a remote-mounted control unit. One obvious disadvantage of such a system is that in order to effectively monitor a process plant a large number of sensors are required. In addition, there is the question of exactly where on the unit these individual sensors should be located. As a consequence, several manufacturers have concentrated on modular, multi-channel design, some of the latest of which are computer-based.

The majority of sensors used operate on a catalytic principle. The sensor consists of two detector elements; one element is made sensitive, while the other is made insensitive to flammable (combustible) gases by chemical treatment. The sensor elements are connected to the control unit and form the arms of a resistance bridge network. The elements are heated to their operating temperature by an electrical current supplied from the control unit. In the presence of flammable gas, oxidation takes place at the sensitive element, raising its temperature and hence its electrical resistance. The resulting bridge unbalance generates a voltage proportional to the flammable gas concentration at the sensor. The voltage is used by the control unit to indicate gas concentration and to initiate alarm/control functions.

One manufacturer, International Sensor Technology, uses a solid-state semi-conductor type of sensor, in production since about 1972. Detection of combustible gases is accomplished by embedding two electrodes in the solid-state sensor cell and applying a voltage between the electrodes. Gas molecules are dissociated into ions and change the cell resistance, giving a change in electrical output proportional to the combustible gas concentration. One advantage with this type of sensor is that in the presence of combustible gas there is no increase in sensor temperature, thereby increasing operating life. Sensors can be made sensitive to one gas or group of gases and insensitive to others. Hence they are used by many manufacturers for toxic gas detection (e.g., hydrogen sulfide, chlorine, and ammonia). More detailed information on this application of the solid-state sensor is covered in Chapter 9.

Commercially available combustible gas detection equipment is listed in Table 12.3 on page 482, giving model numbers, number of channels, and sensor type. The list is fairly extensive and covers a wide variety of systems. Typical examples, described in more detail next, include the latest multi-channel, computerized systems.

(a) Bacharach Instruments (United States)

The Bacharach range of combustible gas detectors is based on the catalytic-type sensor. Single, dual, and four-channel systems are available in general-purpose or explosion-proof design. Multi-channel systems can be supplied, based on the modular, four-channel control unit shown in Fig. 12.2. The four-channel control module CD850/SD550 contains relay circuit outputs for switching external devices such as horns and buzzers. Alarm, warning, and failure indication are provided for each channel, with acknowledge and reset

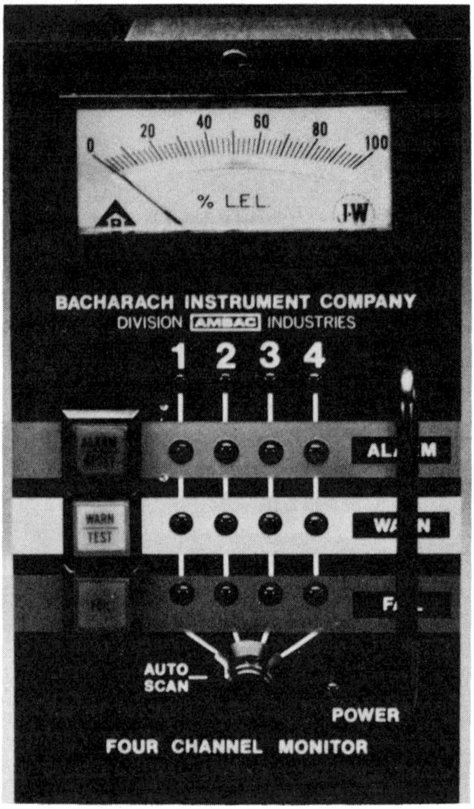

FIGURE 12.2 Model CD850/SD550 four-channel combustible gas detection module. (Courtesy of United Technologies, Bacharach)

buttons. The integral indicating meter is normally scaled 0–100% LEL. A new digital indicating version, Model CD860, is also available.

(b) International Sensor Technology (United States)

IST manufactures a wide range of combustible gas detection equipment based on the solid-state sensor. As mentioned earlier, these instruments may also be used for toxic gas detection. A wide variety of different gases are quoted; however, IST should be consulted regarding possible measurement interference. The 1, 2, 3, and 4-channel models are available in NEMA-12 wall mounted enclosures. The Model AG2200 is a single-channel explosion-proof unit suitable for wall mounting in NEC Class 1, Group D, Division 1 hazardous locations. The Model AG3102 shown in Fig. 12.3 is a general-purpose dual-channel modular control unit suitable for rack mounting and for multi-channel installations; a single-channel version Model AG3100 is also available.

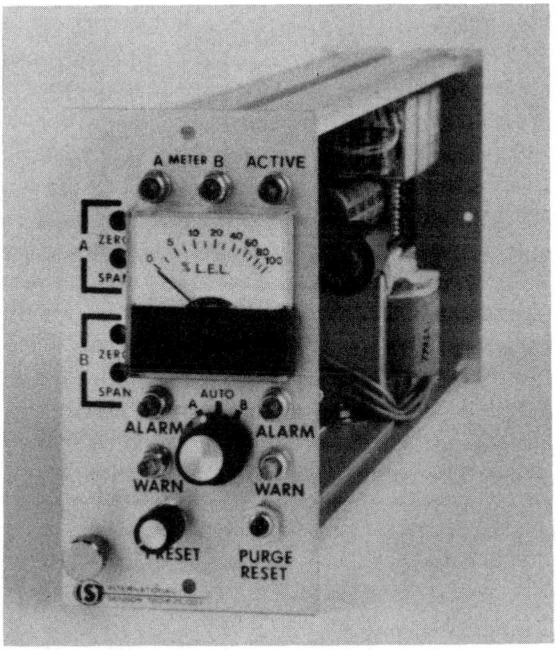

FIGURE 12.3 Model AG3102 combustible gas detectors. (Courtesy of International Sensor Technology)

The Model AG512 shown in Fig. 12.4 is a microprocessor-based, computerized system capable of handling up to 31 sensors (single system) or up to 255 sensors (duplex system) and consists of an assembly of the following component modules:

- Sensor modules (one per sensor, field mounted).
- Processor module.
- Scanner modules (1 for each 32 sensors).
- Status panel modules (1 for each 32 sensors).
- Sensor power supply modules (1 for each 64 sensors).

All are mounted in a free-standing cabinet. The system operating program is stored in ROM. Calibration data, warn and alarm threshold settings, and operating status of each sensor is contained in RAM. A back-up memory with battery power supply stores critical calibration data and set point values in the event of a power failure. A security key switch is provided to prevent against unauthorized damage to the calibration data. Operator data entry is by a front-of-panel keyboard. The system will interface to peripheral devices such as a printer for alarm and status report generation.

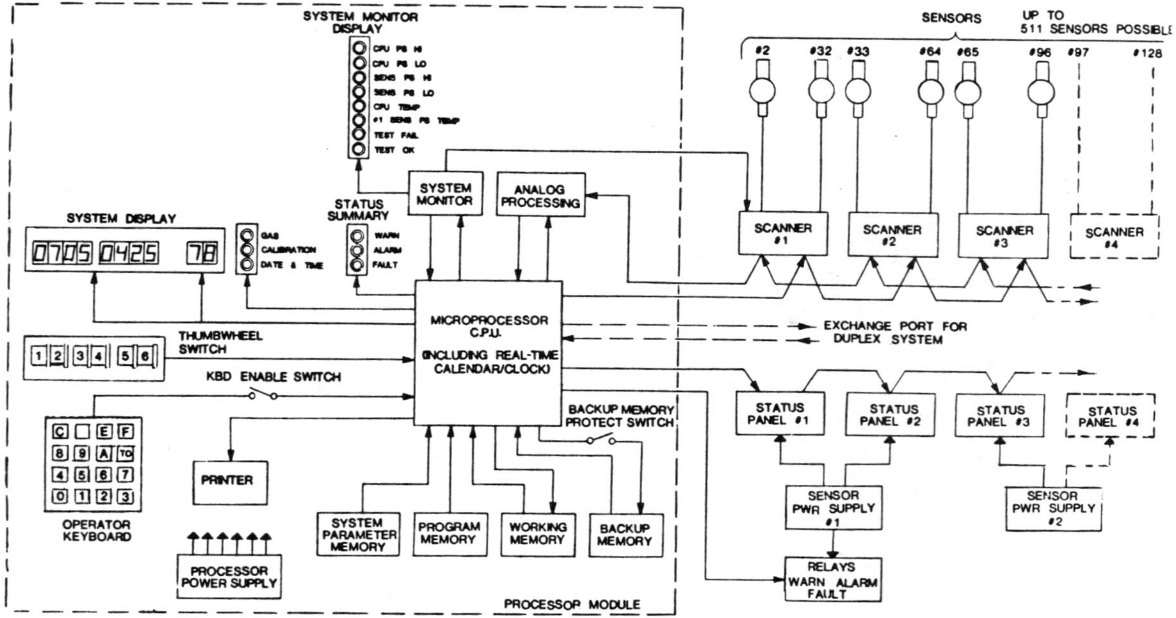

SIMPLIFIED BLOCK DIAGRAM OF THE AG-512 GAS MONITOR SYSTEM

(*a*)

FIGURE 12.4 IST Model AG512 multi-channel system. (*a*) System block diagram. (Courtesy of International Sensor Technology)

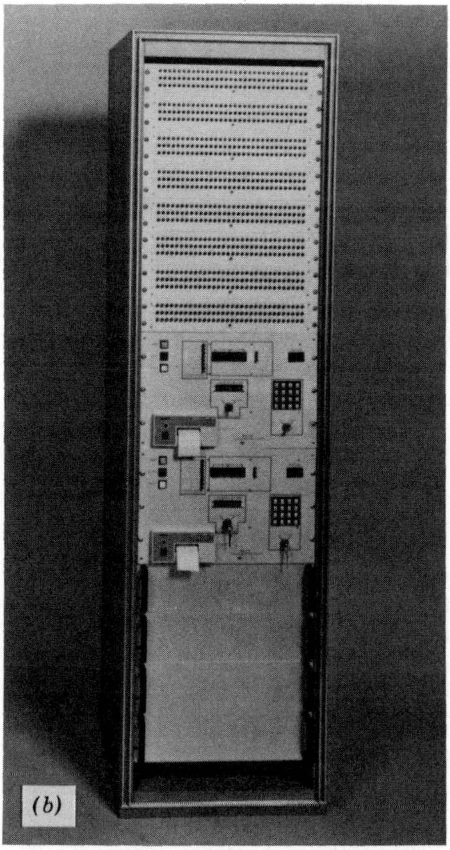

FIGURE 12.4 IST Model AG512 multichannel system. (*b*) Computer system cabinet. (Courtesy of International Sensor Technology)

(c) General Monitors Inc. (United States)

General Monitors manufactures a range of combustible gas detection equipment based on the catalytic-type sensor. The single and four-channel modular control units, shown in Chapter 9 (Fig. 9.12), are used for H_2S measurement. The combustible gas control units are identical in design and normally calibrated 0–100% LEL.

(d) Sieger Gasalarm (England)

Sieger has manufactured gas detection equipment since 1959. A variety of catalytic-type sensors is available and control units are of modular design. The Model FS1 is a single-channel module, allowing up to 10 modules per 19-in. tray for rack mounting, with automatic fault monitoring, double pole changeover alarm relays, alarm reset and test facilities, and an integral indicating meter calibrated 0–100% LEL. The Model FS4 is a four-channel module, allowing up to 32 channels per 19-in. tray for rack mounting.

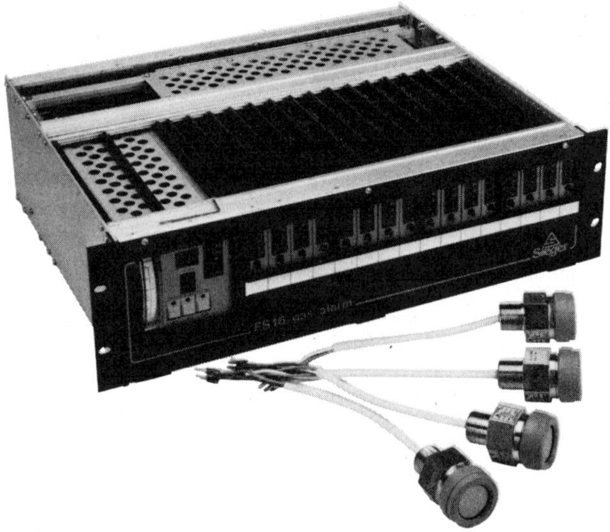

FIGURE 12.5 Model FS16 multi-channel system. (Courtesy of J & S Sieger Ltd.)

The Model FS16 is a multi-channel unit capable of taking up to 16 sensor inputs. Each sensor channel is serviced by a separate printed circuit module and a common meter module provides indication, reset switching, and functional checking. The system is mounted in a custom-built cabinet suitable for rack or wall mounting. A typical system is shown in Fig. 12.5.

(e) Delphian Corporation (United States)

Delphian manufactures two combustible gas detection systems—the Micro 550, which is a third generation modular system, and the computerized Sage System 512. Both use sensors of the catalytic type, but with a patented constant temperature circuit (Sensor-Saver) built into the sensor assembly, which is designed to overcome the problems caused by sensor temperature variations and consequent reduction in operating life. This circuit maintains reference and active sensor elements at a constant resistance (and therefore constant temperature) by varying the electrical power to each element. The difference in power provides the measurement signal to the control unit and gives a linear method of measuring combustible gas concentration. Sensor life has been shown to be considerably extended by the use of this measurement system.

The Micro 550 control module is shown in Fig. 12.6. Multi-channel systems can be built, taking up to 16 modules in a 19-in. rack assembly. Each module has an integral meter calibrated 0–100% LEL, span and zero controls, and LED indicators for high/low alarm, fail, and power.

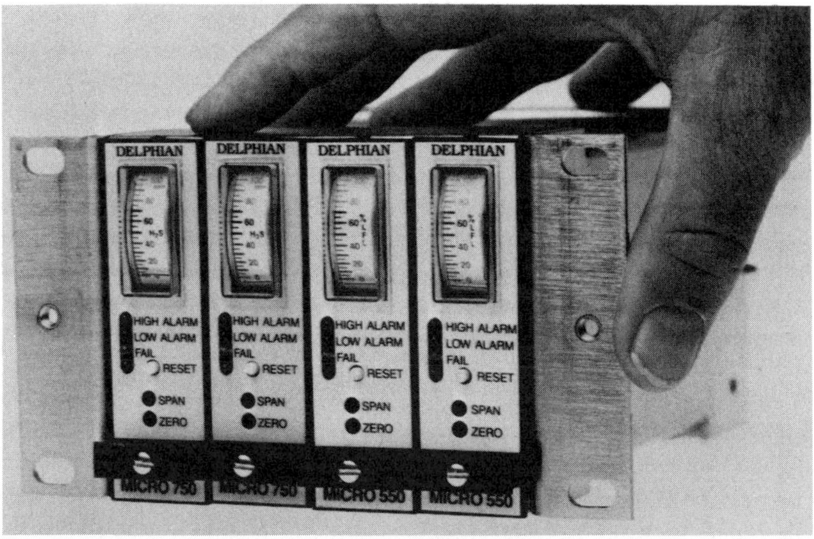

FIGURE 12.6 Micro 550 multi-channel system. (Courtesy of Delphian Corporation)

FIGURE 12.7 Sage System 512, showing computer console, CRT, and printer. (Courtesy of Delphian Corporation)

The Sage System 512 is a computerized monitoring and reporting system with alarm capability, event logging, automatic calibration, and built-in self-diagnostic features. The system consists of three functional parts—the central control console with CRT, ASCII keyboard, and printer, shown in Fig. 12.7; the remote satellite stations, each capable of accomodating 16 sensors; and the sensor transmitters themselves. The central control unit is capable of handling up to 32 remote satellites, making a total system capacity of 512 sensors. The remote satellites are connected to the central control console by a single twisted-pair cable in daisy chain configuration, at distances up to 1 mile.

(f) Scott Aviation (United States)

Scott manufactures two combustible gas detection systems, the Series 9000 modular system and the Quadraflex digital multi-channel system, both using catalytic-type sensors.

The Series 9000 control unit is a single-channel module available for individual panel, rack or wall mounting, or as multi-channel, modular systems built to customer requirements. Up to seven modules can be accomodated in a 19-in. rack. Each module has a vertical indicator calibrated 0–100% LEL and alarm, warn, fail/test status lights, and pushbutton and reset facilities.

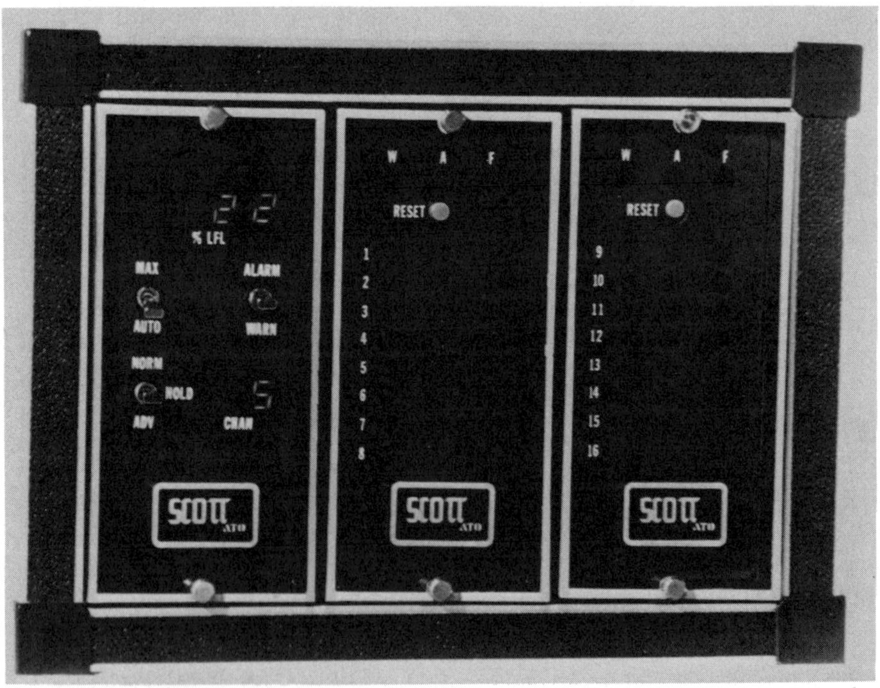

FIGURE 12.8 Quadraflex system control modules. (Courtesy of Scott Aviation)

The Quadraflex System shown in Fig. 12.8 uses the Series 9000 detectors in conjunction with quadraflex transmitters, which provide a 4–20 mA dc input to the control module. Each control module consists of three sections covering a total of 16 channels. The left-hand section contains all operator controls, digital displays, and plug-in digital printed-circuit cards. The center and right-hand sections contain LED alarm display status indications for each channel, together with flashing LED summary indications of any alarm condition on all channels and reset pushbuttons. The rear portion of the module housing contains the power supply PCBs and relays.

12.4.2 Total Hydrocarbon Measurement

Unlike most gaseous air pollutants, hydrocarbons comprise a class containing a large number of individual compounds. The most abundant hydrocarbon, methane, is essentially non-reactive in photochemical reactions and not a hazardous air pollutant. Other hydrocarbons vary widely in their reactivity. Thus, the type of hydrocarbon analysis to be used will largely depend upon the purpose for which the data are required. To obtain highly detailed analysis for numerous individual hydrocarbons, sophisticated and expensive techniques are required, such as mass spectrometry and gas chromatography. A simpler and less expensive approach is to measure total hydrocarbons. There are also methods of intermediate complexity that determine methane concentration separately from other hydrocarbons.

For the measurement of total hydrocarbons at the ppm levels found in polluted air, the flame ionization detector (FID) is the most popular device and is used in most total hydrocarbon analyzers available today. A description of the FID and its operating principle can be found in Chapter 1, Section 1.6.5 and a typical FID is shown in Fig. 1.17.

(a) Beckman Industrial Inc. (United States)

Beckman manufactures three total hydrocarbon analyzers, the Models 400, 402, and 404, all using the FID. These instruments are designed primarily for automotive emissions testing.

The Model 400, designed for panel, rack, or bench mounting in non-hazardous locations, has a dynamic range of 0–4 ppm to 0–10% full scale as methane and a precision of $\pm 1\%$ of full scale. Response time at a bypass flowrate of 3 l/min is quoted as 0.5 sec for 90% of full scale. Case temperature is controlled at 120°F to eliminate ambient temperature effects and enhance stability. The standard instrument has four selectable ranges of $\times 1$, $\times 10$, $\times 100$, and $\times 1000$. A seven-range kit is also available. In addition to the front panel indicating meters, voltage and current outputs are provided for remote recording.

The Model 402, designed for panel or bench mounting in non-hazardous locations, has a dynamic range of 0–5 ppm to 0–1% full scale as methane, with eight selectable ranges and a precision of ±1% of full scale. The electronics enclosure is attached separately to the analyzer and may be detached for remote mounting if necessary. Oven temperature may be adjusted within the range 200–400°F to prevent loss of high molecular weight hydrocarbons. Response time is quoted as less than 1 sec for 90% of full scale at the specified sample flowrate of 2.8 l/min. Sample flowrate is controlled at this

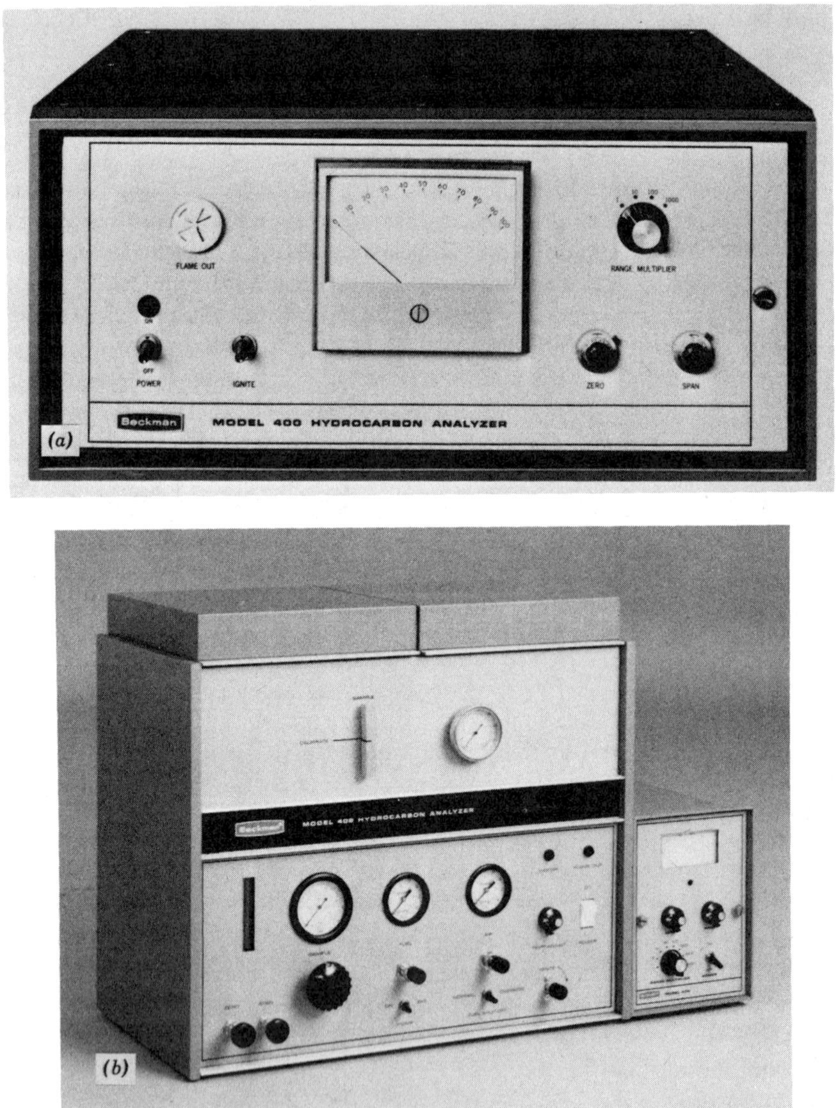

FIGURE 12.9 Beckman total hydrocarbon analyzers. (*a*) Model 400; (*b*) Model 402. (Courtesy of Beckman Industrial Inc.)

value by a positive displacement pump. Voltage outputs (0–10 mV, 100 mV, or 1 V dc) are provided for remote recording.

The Model 404 is a microprocessor-based unit with a dynamic range of 0–10 ppm to 0–1% as methane and a precision of ±1% of full scale. The analyzer consists of three modules—the detection, control, and sample modules, each designed for panel or console mounting in non-hazardous locations. Also available is an integrator module, which provides supervisory control of light and heavy diesel engine test sequencing (the system is designed specifically to meet the EPA requirements for diesel exhaust sampling and testing). Both analog and digital outputs are available for data acquisition and recording. The standard instrument has seven selectable ranges and response time is quoted as less than 1.5 sec for 95% of full scale.

The Beckman range of total hydrocarbon analyzers is illustrated in Fig. 12.9.

(b) Combustion Engineering Process Analytics (United States)

The Model 8401 Total Hydrocarbon Analyzer shown in Fig. 12.10 is designed primarily for ambient air monitoring and utilizes an FID. The analyzer is suitable for rack or bench mounting in non-hazardous locations. The FID cell is temperature controlled at 80 ±0.1°C to enhance stability. Minimum range is 0–1 ppm full scale as methane and a seven-range attenuation switch is provided. Response time is quoted as 6 sec for 90% of full scale with a 2-sec lag time. Precision is quoted as ±2% of full scale.

The Model 8201 Reactive Hydrocarbon Analyzer utilizes gas chromatographic separation together with an FID to measure the concentration of reactive (non-methane) hydrocarbons in ambient air. A flow diagram of the analyzer is shown in Fig. 12.11. The operational sequence of the analyzer is as follows. A metered quantity of sample is passed through the GC column, where the methane is separated and measured by the FID. Then a metered quantity of sample is passed directly to the FID for total hydrocarbon (THC) measurement. The two signals are electronically subtracted to give the measured value of reactive hydrocarbons. The cycle time is 200 sec. The three signals are stored in permanent memories and updated at the end of each cycle. Outputs (voltage or current) are provided for recording of the three variables. All equipment electrical controls are on the instrument front panel, including range attenuation, span and zero, function switch, and indicating meter. The analyzer is suitable for bench or rack mounting in non-hazardous locations.

(c) Teledyne Analytical Instruments (United States)

The Model 402 Hydrocarbon Analyzer uses an FID to measure total hydrocarbons in ambient air. Measurement is linear over the range 0–10,000 ppm with a minimum span of 0–2 ppm. The analyzer has three ranges—normally 0–10, 100, and 1000 ppm, switch-selectable from the front panel. The FID

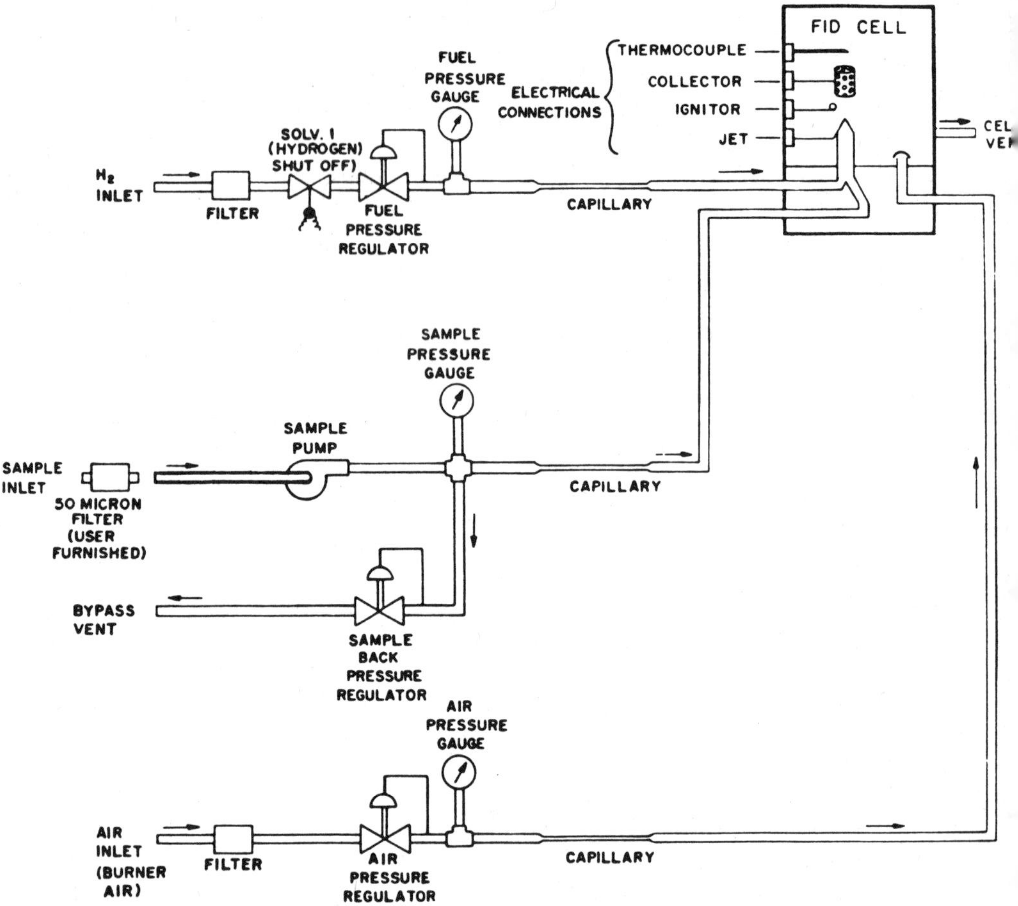

FIGURE 12.10 Model 8401 total hydrocarbon analyzer. (Courtesy of CE, Process Analytics)

fuel used is a mixture of 40% hydrogen and 60% nitrogen. Gas flows are regulated by maintaining a constant pressure drop across sintered stainless steel restrictors. A three-way valve is provided on the front panel for span, zero, and sample gas introduction. Response time is 3 sec for 90% of full scale. The instrument is designed for rack or panel mounting in non-hazardous locations. Standard output is 0–5 mV dc; other analog outputs are available as an option. The analyzer is shown in Fig. 12.12.

(d) Siemens A.G. (West Germany)

Siemens manufactures three FIDAMAT Series Total Hydrocarbon Analyzers based on the FID. The FIDAMAT 1 is designed for environmental moni-

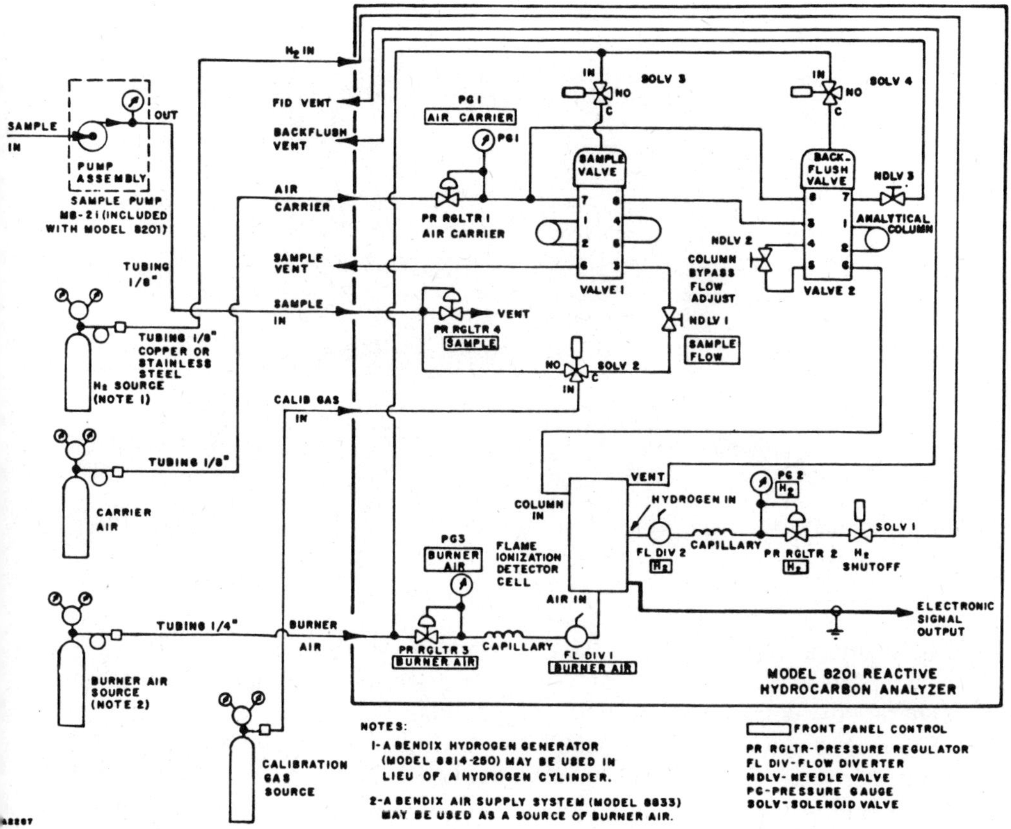

FIGURE 12.11 Model 8201 reactive hydrocarbon analyzer. (Courtesy of CE, Process Analytics)

toring and has a minimum range of 0–0.2 ppm with a linear output up to 10,000 ppm. Twelve standard ranges are manually selectable over the range 0–0.2 to 0–2000 ppm by a switch on the front panel. Repeatability is quoted as ±1% of full scale and instrument response time as 1.5 sec for 90% of full scale. The analysis chamber temperature control is adjustable to 110, 140, 170, or 200°C depending on the particular application. The analyzer is designed for rack mounting in non-hazardous locations.

12.4.3 Specific Contaminants by Gas Chromatography

For industrial plants to achieve compliance with OSHA regulations for the emission of toxic airborne contaminants, monitoring of total hydrocarbons is

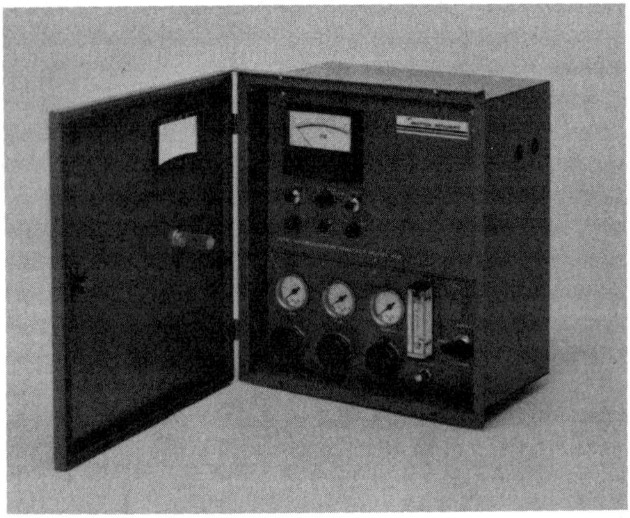

FIGURE 12.12 Model 402 hydrocarbon analyzer. (Courtesy of Teledyne Analytical Insts. Inc.)

not always acceptable, since the individual compounds endanger living organisms to varying degrees. It is therefore necessary to monitor the atmospheric concentration of specific compounds.

Gas chromatography is probably the most powerful technique for continuous environmental monitoring of individual compounds at the ppm level. Process chromatograph manufacturers such as Applied Automation, Beckman, Combustion Engineering, and Foxboro in the United States and Siemens in Europe, have been active in this area for several years. Specific applications include measurement of ethylene, benzene, ethyl benzene, toluene, vinyl chloride and other halogenated hydrocarbons, cumene, and a host of other toxic compounds. The power and simplicity of the GC technique is based on the ability of a particular column system to separate the component (or components) of interest, which may then be measured individually at high sensitivity using an FID (see Section 12.4.2). As described in Chapter 9, sulfur compounds can also be measured using the FPD.

The use of computer-based chromatograph systems and multi-point sampling techniques has resulted in systems providing data acquisition and report writing capability and facilitated computation of point averages and time-weighted averages, in compliance with OSHA requirements.

To measure individual contaminants in the atmosphere at the ppm level using the GC technique, the specified column system must separate the component(s) of interest from other components present in the sample. This sounds very straightforward, but in fact can pose a major problem in some cases, especially those where the GC is monitoring the atmosphere in a complex multi-product plant environment. The number and type of contami-

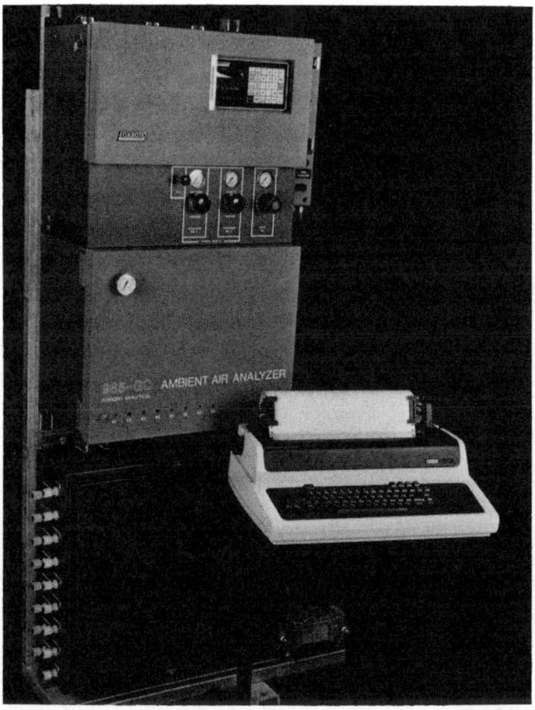

FIGURE 12.13 Model 985-GC ambient air monitoring system. (Courtesy of The Foxboro Company)

nants present will depend upon the type of process plant, wind strength and direction, normal or upset operating conditions, and so on, and will affect the composition of the sample taken for analysis. Consequently it is of the utmost importance that comprehensive data be obtained on atmospheric sample composition so that the information can be used as the basis for GC column design by the manufacturer. If this is not done, one could be faced with major column interference problems and consequent measurement inaccuracy.

Shown in Fig. 12.13 is a typical ambient air monitoring system supplied by Foxboro Analytical. This multi-point system has been used to monitor ppm levels of vinyl chloride in air, using GCs operating on 1-min cycle time. It can report analyses for each sample point every 15 min and can compute and record averages, time-weighted averages, and alarm levels as required.

The CE Microcomputer-based gas chromatograph air monitoring system shown in Fig. 12.14 has been used to monitor ppm levels of benzene, vinyl chloride, and acrylonitrile. Various methods of data reduction and analysis are possible with this system to meet particular application requirements. Magnetic tape or other mass storage facilities can be incorporated into the system for long-term data storage.

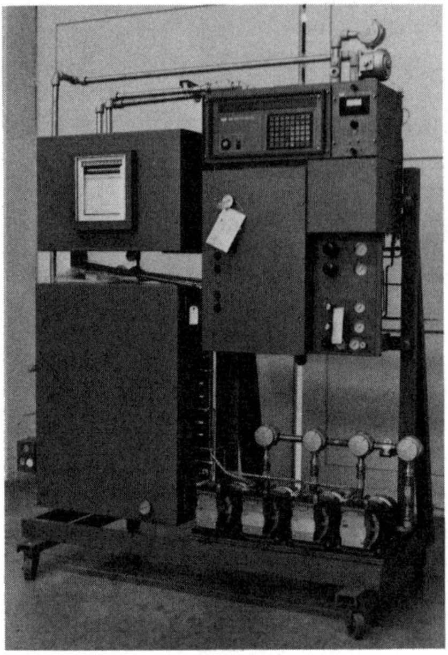

FIGURE 12.14 Typical ambient air monitoring system showing field-mounted rack unit. (Courtesy of Combustion Eng., Process Analytics)

12.4.4 Nitrogen Oxides Measurement

Most process analyzers for measuring nitrogen oxides in ambient air or stack gases operate on the chemiluminescence detection technique. This is based on the principle that nitric oxide (NO) reacts with ozone (O_3) to produce nitrogen dioxide (NO_2), 10% electronically excited nitrogen dioxide (NO_2^*), and oxygen. Following the reaction, the excited NO_2^* molecules revert to the unexcited state with the emission of light, the intensity of which is directly proportional to the NO concentration in the sample. For the measurement of NO_x (total NO + NO_2), the sample is passed through a special NO_2-to-NO converter prior to reaction with ozone, and the NO_2 concentration is calculated by difference.

(a) Beckman Industrial Inc. (United States)

Beckman manufactures three process analyzers for nitrogen oxides measurement, all operating on the chemiluminescence principle, with minor differences in packaging.

The Model 952A NO/NO_x/NO_2 Analyzer shown in Fig. 12.15, is designed specifically for ambient air monitoring and is EPA-approved for this service (EPA designation RFNA-0179-034). The analyzer is a compact, self-contained unit consisting of an analysis section and a solid-state electronics section. Reactant ozone is generated internally by exposing ambient air to

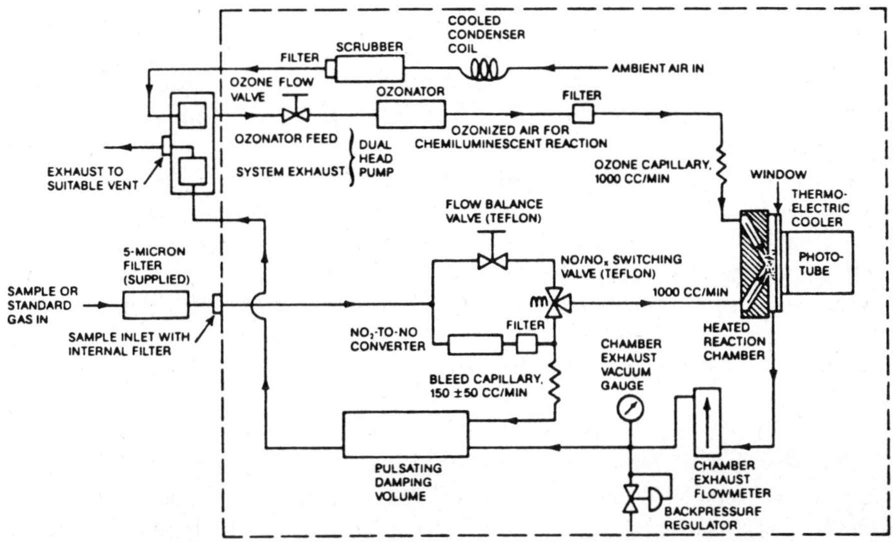

FIGURE 12.15 Model 952A NO_2 /NO_x /NO analyzer. (Courtesy of Beckman Industrial Inc.)

UV radiation from a source lamp. The reaction chamber is heated and temperature-controlled for stability. Sample and ozone flowrates are controlled at 1000 ml/min by the use of capillaries. For NO_x measurement, the sample is passed through a heated temperature-controlled carbon-base converter. A three-way solenoid valve located downstream of the converter automatically switches to allow either NO analysis (sample bypasses converter) or NO_x analysis (sample passes through converter).

The chemiluminescent reaction produces a signal current of 3×10^{-10} A/ppm concentration from the thermoelectrically cooled photomultiplier detector. A two-stage, high-impedance amplifier circuit provides signal gain, and memory amplifiers provide voltage outputs for each parameter (0–10 and 100 mV, or 0–1 and 5 V dc), switch selectable. Seven instrument ranges are available via a front panel selector switch (0–0.25, 0.5, 1.0, 2.5, 5.0, 10.0, and 25.0 ppm). The EPA designated range is 0–0.5 ppm.

The Model 951A NO/NO_x Analyzer is designed for exhaust emissions analysis of light-duty gasoline and diesel-fueled engines as well as for ambient air and process applications. The Model 955 NO/NO_x Analyzer is designed specifically for heavy-duty engine, turbine, and stationary (stack) source monitoring applications. All models are suitable for bench or surface mounting in non-hazardous locations.

(b) Combustion Engineering-Process Analytics (United States)

CE manufactures two nitrogen oxide analyzers, operating on the chemiluminescence principle.

The Model 8101-C NO/NO_x/NO_2 Analyzer is an EPA-approved device (EPA designation RFNA-0777-022). Standard measurement ranges are 0–0.2, 0.5, 1.0, and 2.0 ppm for each parameter, switch selectable from the front panel. Ozone is generated from an integral dry air supply. The carbon converter used for NO_2-NO reaction achieves measurement with almost no ammonia interference and is temperature controlled at 285°C. Air sample and ozone flowrates are controlled at about 200 ml/min by the use of capillaries. A carbon scrubber is installed in the analyzer exhaust to remove high levels of ozone exhausting from the reaction chamber.

The photomultiplier detector is thermostatically cooled for zero stability and minimum "dark current." In the solid-state electronics section, separate memory-type outputs are provided for each component: 0–10 mV or 0–1 V dc are standard; other outputs are available as an option. Instrument cycle time is 23 sec for full measurement.

The Model 8102 NO/NO_x Analyzer is similar in design except that standard ranges are 0–100, 200, 500, 1000, 2000, and 5000 ppm, and pure oxygen (99.5% at 150 ml/min) is required for ozone generation. Both models are suitable for surface or bench mounting in non-hazardous locations.

12.4.5 Sulfur Dioxide Measurement

The two major applications for SO_2 measurement in the process industries are ambient air monitoring and stack emissions monitoring. A variety of different techniques are used, namely, fluorescence, IR absorption, UV absorption, conductimetric, and polarographic. For reference, the four measuring principles most commonly used by federal, state, and local air pollution agencies are colorimetry, conductivity, coulometry, and flame photometry. The colorimetric method is discussed in Section 12.3 under ASTM D2914-78 (West-Gaeke method). The flame photometric detector, which is specific to sulfur compounds, is discussed in detail in Chapter 9. This method is acceptable for SO_2 measurement in areas where there are no known sources of sulfur compounds other than SO_2. Otherwise, these interfering compounds must be removed by selective absorption prior to analysis.

Conductimetric-type analyzers are discussed in Chapter 9 specifically in relation to H_2S measurement. However, as already stated, analyzers such as the Hartmann and Braun Picoflux 4 and the Wosthoff Mikrogas can also be made specific to other toxic gases, including SO_2. Another instrument of this type in common use in Europe is the Duktolyt Trace Gas Analyzer, shown in Fig. 12.16. This analyzer is marketed in Western Europe by Telsec Process Analyzers Ltd. (TPA) in Great Britain.

As with other analyzers of this type, reagent liquid is selected to have a specific reaction with SO_2, and its conductivity is compared before and after reaction with the test sample. The difference in conductivity is a measure of the SO_2 concentration in the sample gas. Measurement range limits are 0–2 ppm to 0–2000 ppm. Response time is quoted as 110 sec for 90% of full scale,

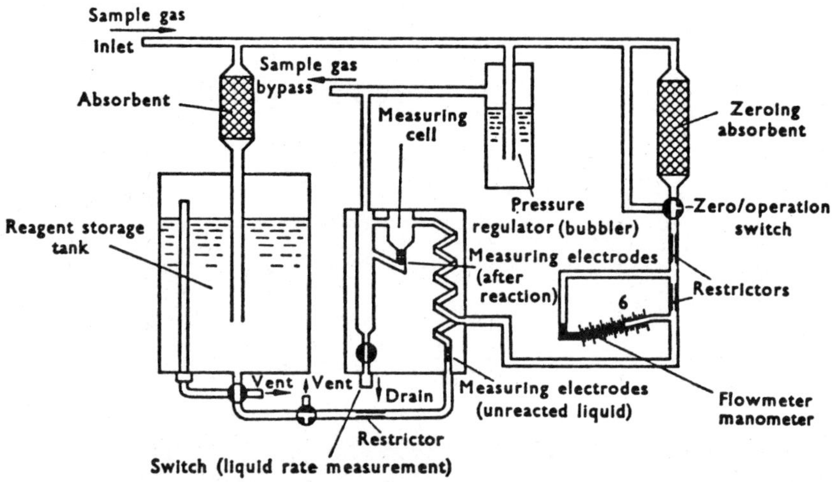

FIGURE 12.16 Duktolyt trace gas analyzer. (Courtesy of Telsec Process Analyzers Ltd.)

with a dead time of 25 sec. The instrument consists of two component parts, the analyzer unit, designed for panel or wall mounting, and the electronics unit designed for wall mounting. Standard design is suitable for non-hazardous locations only, although air purging facilities can be provided.

(a) Beckman Industrial Inc. (United States)

The Model 953 Ambient SO_2 Monitor is EPA-approved (EPA designation EQSA-0678-029) and operates on the fluorescence principle, based on the fact that SO_2 molecules fluoresce when irradiated by UV light in the 1900–3900 Å wave band. The model 953 transmits a broad UV light band via a quartz-deuterium lamp, and a narrow UV light band via a light-collimator assembly. The narrow UV light band passes through the sample reaction chamber, where a blue-sensitive photomultiplier detector measures the resulting SO_2 fluorescence.

The analysis section is temperature controlled at 41°C. An added feature is the interferent reactor that eliminates interference in the measurement from polynuclear aromatics (PNAs) that are typically found in samples taken from a dense auto traffic environment. Instrument ranges are 0–0.25, 0.5, 1.0, and 2.0 ppm SO_2. The EPA designated ranges are 0–0.5 and 0–1.0 ppm. Voltage outputs (0–10 mV, 100 mV, 1 V, and 5 V dc) are provided for remote recording. The analyzer is a single unit of split case design, suitable for bench or panel mounting in non-hazardous locations.

(b) Teledyne Analytical Instruments (United States)

The Model 690/691 Sulfur Dioxide Analyzer System is designed for stack emission monitoring and operates on the UV absorption principle. The system is shown in Fig. 12.17.

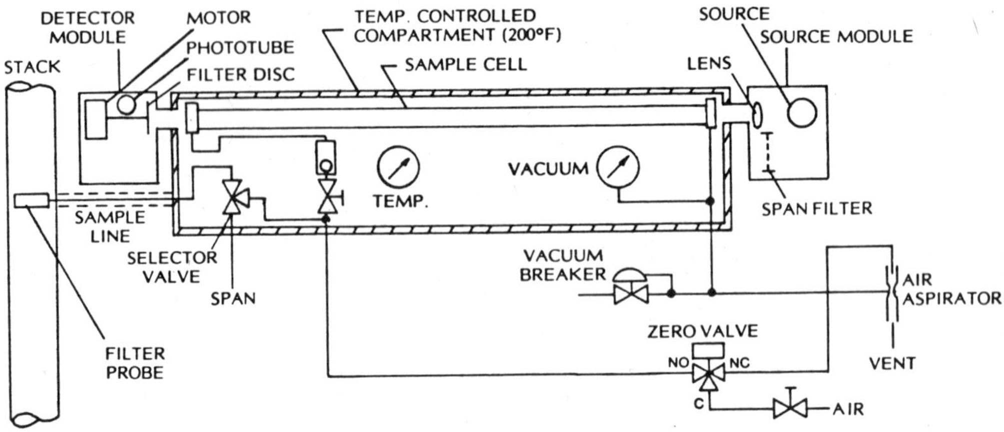

FIGURE 12.17 Model 690/691 sulfur dioxide analyzer. (Courtesy of Teledyne Anal. Insts. Inc.)

The analyzer consists of three interconnecting modules. The source module contains a deuterium arc lamp which emits radiation in the range 200–400 nm. A quartz lens focuses lamp radiation through a temperature controlled module containing the sample cell. After passing through the cell, lamp energy is chopped by a motor-driven rotating filter disc in the detector module and containing the measurement and reference filters. The phototube detector receives the alternate pulses at a rate of 60 pps (two pulses per revolution of the filter disc). The detector modulated output is processed by the electronic circuitry to produce an output signal that is linearly proportional to the SO_2 concentration in the sample. Since the analyzer output is the ratio of the absorbances at the measurement and reference wavelengths, interferences such as turbidity are ratioed out and do not affect the output signal.

The UV spectrometer and sample handling/conditioning system are mounted in a free-standing rack. The system is available in either general-purpose (Model 691) or explosion-proof (Model 690) design. Analyzer measurement range is 0–300 ppm to 0–100% full scale and repeatability is quoted as ±1% of full scale. Current or voltage output signals are available for remote recording.

Commercially available sulfur dioxide analyzers are summarized in Table 12.4 on page 485–487.

12.4.6 Carbon Monoxide Measurement

The asphyxiating properties of carbon monoxide are well known and OSHA has placed a maximum exposure limit of 50 ppm (8-hr time-weighted average) on CO concentration in the workplace atmosphere. Therefore continuous monitoring of CO concentrations in the atmosphere is an important application from a regulatory viewpoint.

Another important application is in furnace/boiler combustion efficiency monitoring, where savings in energy and increased operating efficiency are attributed to a reduction in excess oxygen down to a level based on the appearance of CO in the flue gas. The CO constraint is usually in the region of about 300 ppm, but is very largely dependent upon the furnace design and type of fuel being burnt. Consequently, we have seen a considerable increase in commercially-available in-situ CO analyzers (designed for direct measurement in the furnace duct or stack). These are described in Chapter 18, which covers combustion efficiency monitoring in greater detail.

Most process analyzers for measurement of CO in ambient air or stack gases use non-dispersive IR (NDIR) technology. Commercially available CO analyzers are summarized in Chapter 18, Table 18.1. Typical examples of the type of designs available are described next.

(a) Beckman Industrial Inc. (United States)

The Model 866 Ambient CO Monitoring System shown in Fig. 12.18 is an EPA-approved instrument (EPA designation RFCA-0876-012) operating on the IR absorption principle.

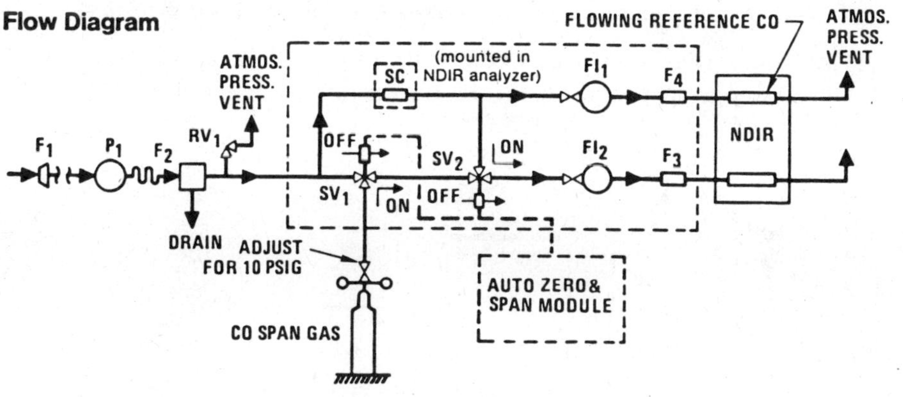

P_1—Sample Pump
F_1-F_4—Filters
RV_1—Relief Valve
SV_1, SV_2—3 Way Solenoid Valves
SC—CO Converter
FI_1—Reference Flowmeter
FI_2—Sample Flowmeter

NOTES:
1. Adjusts FI_1 & FI_2 to Indicate 1 CFH (Approx. 470 cc/min)
2.

VALVE	SAMPLE	ZERO	SPAN
SV_1	OFF	OFF	ON
SV_2	OFF	ON	OFF

⚠2 Converter to be Mounted Vertically Inside Analyzer Case

FIGURE 12.18 Model 866 ambient CO analyzer. (Courtesy of Beckman Industrial Inc.)

The system consists of the Model 865-17 NDIR analyzer, an automatic zero/span module, flowing reference panel, and sample handling module, combined into a single system. The NDIR analyzer is a two-beam unit with sample and reference flow cells. The sample cell continuously receives an ambient air sample containing CO, while the reference cell receives ambient air that has passed through a CO-CO_2 converter. A differential measurement of IR energy absorption between the two cells produces an accurate measurement of CO.

The automatic zero/span module provides periodic system standardization, zero sequence every 6 hr and span sequence every 24 hr, via a cam-operated timer. The recorder readout is corrected to the appropriate zero/span value.

The sample handling module, consisting of pump, filters, condenser coil, and pressure relief valve, supplies the ambient air sample to the automatic flowing reference panel. The unit is suitable for bench, panel, or surface mounting in non-hazardous locations. Voltage (0–10 mV, 100 mV, 1 V, or 5 V dc) and current (4–20 mA dc) outputs are available from the zero/span module for remote recording. Analyzer range is 0–5 ppm, with a precision of 0.2 ppm.

(b) Land Combustion Inc. (Great Britain and United States)

The Land CO Monitor Model 8000, shown in Fig. 12.19 is an example of an in-situ analyzer operating on the IR absorption principle and is designed

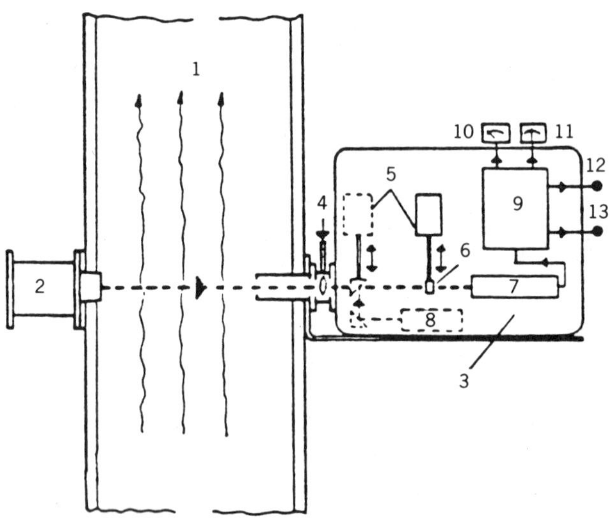

FIGURE 12.19 Model B in situ CO monitor. KEY: 1, flue gases; 2, infrared source; 3, receiver unit; 4, air purge; 5, solenoids; 6, gas filter; 7, photodetector; 8, self-checking system; 9, electronic signal processing; 10, CO ppm; 11, signal strength; 12, 0.20 mA output; 13, 4.20 mA output. (Courtesy of Land Combustion)

specifically for monitoring the CO content of flue gases from furnaces and boilers.

The instrument consists of two units, the IR source unit and the receiver/control unit, installed directly on and at opposite sides of the furnace flue duct or stack. The source unit emits IR radiation, which is absorbed over a certain band of wavelengths by the flue gas it passes through. The receiver unit employs negative filtration of the beam to obtain a reference and a measurement wavelength. The output is processed electronically to produce a 0–20 or 4–20 mA dc signal proportional to the CO concentration in the flue gas. To ensure high reliability, the analyzer incorporates an automatic self-checking system, which operates on an alternating basis with the measurement system by comparing the two channel outputs with an internal reference IR source. Standard measurement range is 0–1000 ppm CO, with options of 0–400 and 0–2000 pm. Repeatability is quoted as ±1%. Path length (source unit to receiver unit) is selectable within the limits 1.6 to 32 ft. Flue gas temperature range is 100–200°C.

(c) Other Methods of CO Measurement

In addition to the two specific types of CO analyzer described, the following techniques may also be used for CO measurement.

Process IR Analyzers. By suitable selection of measurement and reference filters, the very wide range of commercially available process NDIR analyzers can be made specific to CO. Normally the most sensitive range is about 0–1000 ppm.

Manufacturers include Siemens (Ultramat range), Hartmann & Braun (Uras range), Kent (Infragas range), Horiba, Anarad, Infrared Industries, and MSA (Lira range).

The subject of IR analyzers is covered in detail in Chapter 10, together with other analysis techniques of the spectrometric type.

Conductimetric Type Analyzers. By suitable choice of reagent, analyzers such as the Hartmann & Braun Picoflux 4, Wosthoff Mikrogas, and Telsec Duktolyt can be made specific to CO. Measurement ranges of 0–500 ppm are standard. For more details, see Section 12.4.5 and Chapter 9.

Combustible Gas Detectors. Since CO may be classed as a combustible gas, the standard catalytic-type combustible gas detection systems described in Section 12.4.1 may be used to measure CO. Sensitivity is limited to the %LEL range, however. In addition, the presence of other combustible gases will obviously affect the measurement, so the technique is not often used specifically for CO measurement, except in special circumstances.

12.4.7 Ozone Measurement

Three analytical techniques are in common use for the measurement of ozone in ambient air at the ppm level:

- UV absorption at 254 nm.
- Chemiluminescence.
- Coulometric.

(a) Beckman Industrial Inc. (United States)

The Model 950A Ozone Analyzer is EPA-approved (EPA designation RFOA-0577-020) and operates on the chemiluminescence principle, which is based on the fact that ozone, when mixed with ethylene, results in a chemiluminescent reaction that produces a light emission, the intensity of which is directly proportional to the ozone concentration in the sample. A flow diagram of the analyzer is shown in Fig. 12.20. The ambient air sample passes to the reaction chamber via a pump and flow control system at a constant flowrate of 400 ml/min, where it mixes with the reactant mixture of 90% CO_2/10% ethylene, which is controlled at 100 ml/min flowrate. The intensity of the resultant light emission is measured by a photomultiplier and associated electronics.

A standard for zero check is obtained by passing the air sample through a chemical scrubber to remove ozone. Span checks can be made using an optional ozone generator. The analyzer is suitable for panel or rack mounting in non-hazardous locations. A range switch provides manually selectable ranges of 0–0.025, 0.05, 0.1, 0.25, 0.5, 1.0, and 2.0 ppm ozone. Outputs of 0–10 mV, 100 mV, 1 V or 5 V dc are available for recording purposes. Precision is quoted as 0.003 ppm at 20% of span and 0.004 ppm at 80% of span.

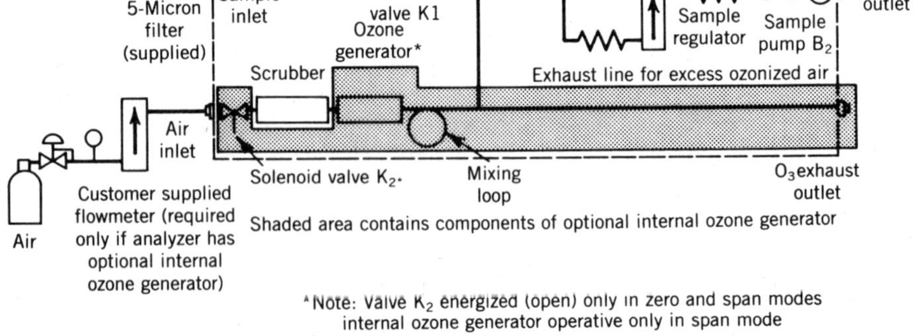

FIGURE 12.20 Model 950A ozone analyzer. (Courtesy of Beckman Industrial Inc.)

(b) Combustion Engineering-Process Analytics (United States)

The Model 8002 Ozone Analyzer is EPA-approved (EPA designation RFOA-0176-007), operates on the chemiluminescence principle, and is very similar in design to the Beckman instrument. Ranges are 0–0.01, 0.02, 0.05, 0.1, 0.2, 0.5, and 1.0 ppm, switch selectable. Outputs of 0–10 mV and 0–1 V dc are available for recording. Precision is quoted as ±2% from mean value on a 0–0.1 ppm range.

(c) Mast Development Co. (United States)

The Series 727 Ozone Monitors operate on the principle of specific UV absorption. The instrument operates on an 80-sec measurement cycle. In the zero mode, sample is diverted through a catalytic converter, where any ozone present is changed to oxygen and passes through the absorption chamber. The mercury source lamp intensity is automatically adjusted to set detector output equal to a reference voltage (zero ozone). Then the gas valve is actuated to allow sample containing ozone to pass through the absorption chamber. Absorption by ozone at a wavelength of 254 nm is measured by the detector, the output of which is electronically processed to provide a digital display and output calibrated in ppm ozone.

The Series 1724 Oxidant Monitors, an example of which is shown in Fig. 12.21, operate on a coulometric principle based on the oxidation/reduction

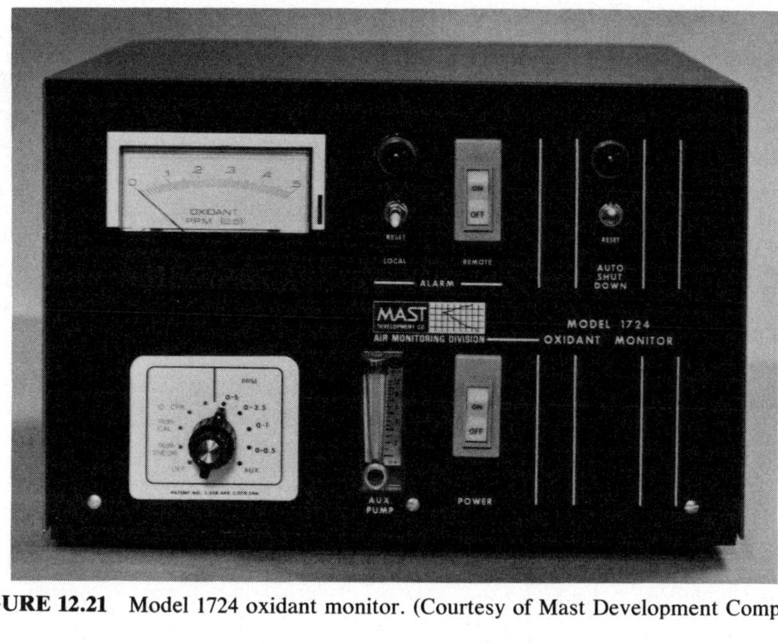

FIGURE 12.21 Model 1724 oxidant monitor. (Courtesy of Mast Development Company)

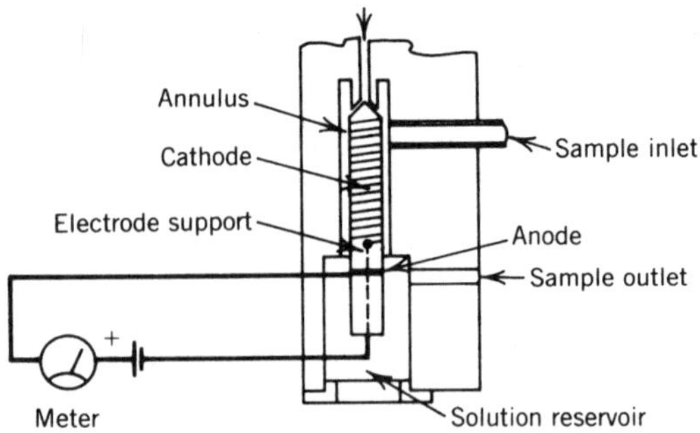

FIGURE 12.22 Microcoulomb oxidant sensor. (Courtesy of Mast Development Company)

of potassium iodide (KI) solution. The instrument is primarily used for the measurement of ozone and halogens (fluorine, chlorine, bromine), which liberate iodine by reaction with KI solution. These compounds give strong reactions, whereas nitrogen dioxide and hydrogen peroxide, for example, give much weaker reactions. If the iodide solution is replaced by an aqueous mixture of iodate and iodide, the same detector cell may be used to measure gases that produce protons (H^+) in aqueous solution. Examples are hydrochloric acid, sulfur dioxide, formic acid, and nitric acid.

The heart of the system, the Mast Microcoulomb Oxidant Sensor, is shown in Fig. 12.22. The chemical reagent solution is metered into the sensor and flows in a fine film down the electrode support, upon which are wound many turns of a fine wire cathode and a single turn of a wire anode. The ambient air sample is pumped through the sensor by the action of a precision vacuum pump and contacts the reagent on the electrode support, where the reaction with ozone occurs.

A thin layer of hydrogen gas is produced by a polarization current. When a voltage is applied to the electrodes, the hydrogen layer builds up to its maximum and the polarization current ceases to flow. Free iodine produced by the reaction of ozone with the reagent solution immediately reacts with the H_2 to form HI, causing a repolarization current to flow in the external circuit, re-establishing equilibrium. The rate of electron flow (current) is directly proportional to the mass per unit time of ozone entering the sensor. The basic reaction also occurs with the halogens, as well as ozone. The basic sensor output of 0–20 mA produces a range of 0–1 ppm ozone at 140 ml/min sample flowrate. Higher ranges are obtained by reducing sample flow.

12.4.8 Ammonia Measurement

Four techniques are commonly used for measurement of ammonia in ambient air:

- Use of solid-state sensors.
- Conductimetric.
- IR absorption.
- Electrochemical.

(a) *Solid-State Sensors*

Solid-state sensors of the type used for combustible gas and H_2S detection (see Sections 12.4.1 and 9.4.7) can by appropriate treatment be made specifically sensitive to other toxic gases, including ammonia. These sensors are used in conjunction with the electronic control modules described previously, except that they are calibrated for ammonia.

Companies manufacturing such systems include International Sensor Technology (IST) and Sieger Gasalarm. Ranges from 0–75 ppm up to %LEL are available. The catalytic-type combustible gas detector may also be used for ammonia detection, since ammonia may also be classified as a "combustible gas." However, sensitivity is limited to the %LEL range only and is subject to interference from other combustible gases.

In summary, solid-state sensors can be used with some reservations regarding interference by other gases; combustible gas detectors of the catalytic type may be used provided that other combustible gases are *not* present in the sample. In other words, specificity is the main problem with both of these types of detector.

(b) *Conductimetric Type Analyzers*

By suitable choice of reagent, the H & B, Wosthoff and Telsec trace gas analyzers can be made specific to ammonia (see Section 12.4.6). Measurement ranges of 0–10 ppm are possible.

(c) *Process Infra-Red Analyzers*

By suitable choice of measurement and reference filters, the wide variety of process IR analyzers can be designed for specific measurement of ammonia. Typical ranges are 0–1000 ppm.

(d) *Electrochemical Detectors*

Sensidyne Inc. manufactures an ammonia detection system based on an electrochemical, diffusion-type sensor. The sensor consists of a sensing electrode, a counter electrode and a suitable liquid electrolyte, enclosed together behind a membrane. Ammonia gas molecules permeate the membrane and react with the electrolyte, causing a change in electrode potential that is electronically processed to provide indication of ammonia concentration.

FIGURE 12.23 Ammonia detection system. (Courtesy of Sensidyne Inc.)

The control unit is modular in design, suitable for panel or rack mounting and is available in single, 6-channel, and 12-channel designs. The single-channel system consists of an alarm module and an indicating module. One alarm module will serve up to 12 indicating modules. Standard range is 0–75 ppm (logarithmic), with a precision of ±5% of full scale. The various options are shown in Fig. 12.23. Similar systems are also available for the detection and measurement of chlorine (Cl_2) and hydrogen cyanide (HCN).

12.4.9 Chlorine and Hydrogen Chloride Measurement

(a) Conductimetric Type Analyzers

Both Cl_2 and HCl may be measured in the ppm range by suitable choice of reagent and measuring cell (e.g., Hartman & Braun, Wosthoff, and Telsec). However, interference by other oxidants must be considered.

(b) Electrochemical Sensors

Draeger Safety manufactures a range of chlorine monitors based on the electrochemical principle. The *Chloralarm* is a fixed-head chlorine monitor, originally developed by Imperial Chemical Industries (ICI) in England. The sensor shown in Fig. 12.24 contains platinum electrodes and a calcium bromide electrolyte. The change in electrode potential produced by reaction of

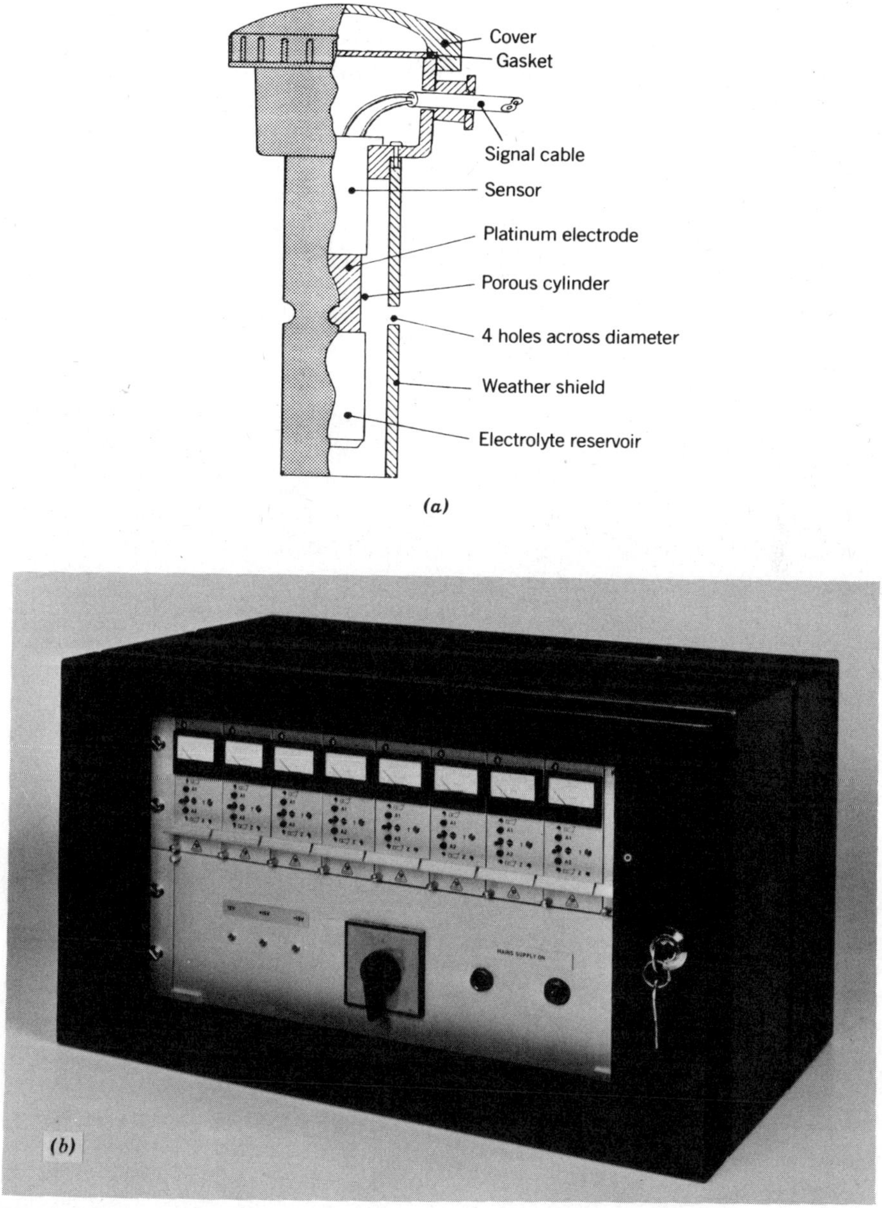

FIGURE 12.24 Draeger safety "Chloralarm" chlorine monitor. (*a*) Sensing cell; (*b*) cabinet mounted multichannel control unit. (Courtesy of National Draeger Inc.)

chlorine with the electrolyte, is electronically processed to give chlorine concentration indication, including alarm facilities. Control modules (see Fig. 12.24*b*) are available as one or two-channel, or these can be built into a modular wall-mounted case taking up to 24 modules. Standard measurement ranges are 0–5, 10, and 50 ppm Cl_2 and repeatability is quoted as ±5% of full scale. Sensor life is quoted as 1 year based on normal operation and treatment, after which the electrolyte must be replenished.

The Delta Scientific Model 5324 Chlorine Detection System consists of two units, an electrochemical sensor with integral cable and a transmitter/control unit. The sensor comprises a platinum cathode and silver anode immersed in a suitable electrolyte. The sensor is coated with a membrane through which chlorine molecules permeate to react with the electrolyte and generate a flow of current proportional to the chlorine concentration in the sample. Outputs of 0–1 V dc or 4–20 mA dc are available for recording purposes. Instrument range is 0–2 and 0–20 ppm, switch selectable, with a repeatability of ±2% of span.

The Anacon Trace Chlorine Monitoring System is designed to measure chlorine levels down to 0–2 ppm using an electrochemical sensor said to be specific to chlorine and unaffected by the presence of other chlorinated or halogenated compounds. The probe was originally developed for ICI in England and is now made under license by Anacon.

The sensor is essentially a polarographic cell consisting of two electrodes separated by a cellulose layer impregnated with a suitable electrolyte. The two electrodes are an inner silver winding and an outer platinum winding.

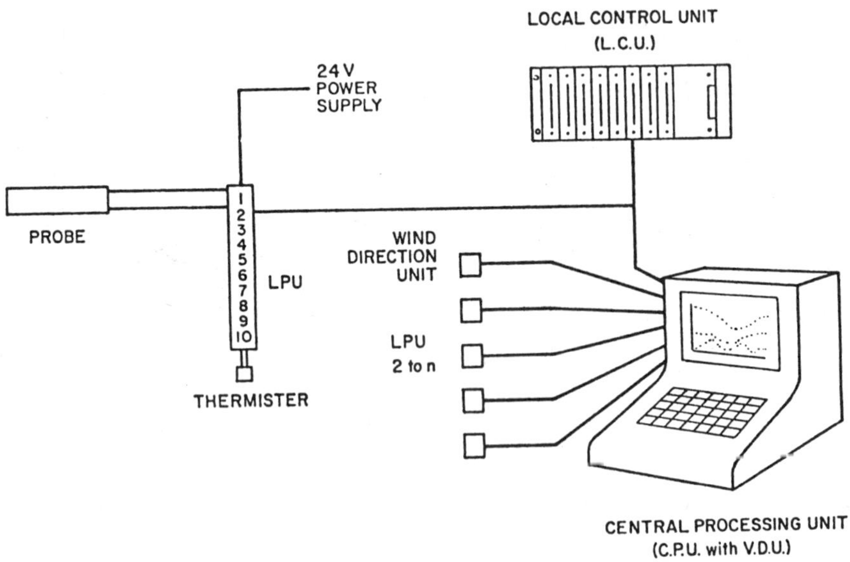

FIGURE 12.25 Multipoint trace chlorine monitor. (Courtesy of Anacon Inc.)

Chlorine gas is reduced by the electrolyte and the chlorine ions formed combine with the silver, which is thus slowly consumed. The current generated is amplified and transmitted by the probe as a voltage signal. Probes for measurement of 0–2, 5, 10, 25, 50, and 100 ppm Cl_2 are available.

In the multipoint system shown in Fig. 12.25, up to 10 probes are connected to an individual local multiplexer (LPU). Data from the microprocessor-based LPU is transmitted via an RS232C ASCII data link to a central processor (with CRT and printer), where time-weighted averages can be computed. Maximum probe/LPU distance is 3000 ft. Single-point units are also available using the same sensor. In this case, the control unit may be located up to 1000 ft from the probe.

(c) *Solid-State Sensors*

Capital Controls manufactures the Series 1030 Chlorine Gas Detector that uses a solid-state sensor specific to chlorine. Minimum detectable concentration is quoted as less than 1 ppm. The control unit is suitable for panel or rack mounting at up to 3000 ft from the sensor. Standard range is 0–10 ppm and response time is quoted as 30 sec. Separate alarm and warning circuits provide redundant monitoring and backup alert in the event of either slow gas accumulation or a major chlorine leak. Contacts are also provided for actuation of external safety devices.

(d) *Wet Chemical Analyzers*

Geomct Technologies Inc. (United States) manufactures the Model 401I Hydrogen Chloride Monitor, which is basically a wet-chemical analyzer operating on a chemiluminescence principle. The measurement of HCl in the low ppm range is based on the chemiluminescent reaction in which light is emitted during oxidation in an alkaline solution of 5-amino, 3-dihydro-1, 4 phthalazinedione (luminol). The instrument contains a single reaction cell, the light within which is monitored by a photomultiplier tube. The sample is passed through a sodium bromate/sodium bromide mixture to produce bromine, which initiates luminol oxidation, generating visible light at a wavelength of 430 nm.

A flow diagram of the instrument is shown in Fig. 12.26. Sample gas at a flowrate of 1800 ml/min is mixed in a 1 : 1 ratio with clean dilution air by a proportioning system. The mixture then passes through the bromide/bromate column and into the optical (reaction) cell. The luminol-in-potassium carbonate solution is pumped through the optical cell at a flowrate of 0.55 ml/min by a precision pump, via a pulse dampener. Liquid and gas pass to the waste container, where liquid is separated and gas passes to the exhaust line. The liquid reagent storage gives about 10 days full operation at the flowrate specified. The photomultiplier signal is processed by the electronics unit to provide an output proportional to the HCl concentration in the sample.

Instrument range is adjustable in four automatically controlled steps, nor-

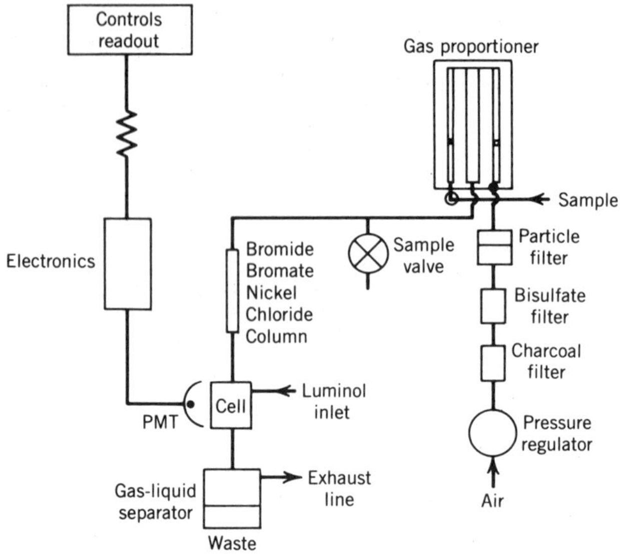

FIGURE 12.26 Model 401I hydrogen chloride monitor. (Courtesy of Geomet Technologies Inc.)

mally set at 0–0.1, 1, 10, and 100 ppm. Reproducibility is quoted as ±1% of span and response time as 5–10 min to 90% of full scale. The sensor unit consists of two NEMA-4 enclosures housing the analysis unit and the reagents respectively. The control unit may be mounted at up to 50 ft from the sensor unit.

12.4.10 Particulates/Opacity Measurement

Particulate matter is one of the most important classes of air pollution. A common effect of particulate air pollution is, of course, a reduction in visibility caused by light scattering out of the line of vision. The amount of light scattered in any direction depends on the size, quantity, and composition of the particles.

One method of measuring particulate concentration is by the use of an opacity monitor. Opacity is defined as the percentage of visible light attenuated due to the absorption and scattering by particulate matter in an optical medium (e.g., the flue gas from a furnace or boiler):

$$\%\ \text{opacity} = 100\% - \%\ \text{transmittance}$$

Opacity monitors therefore operate by detecting variations in light transmittance.

As a result of the ability of opacity monitors to demonstrate compliance and provide data regarding equipment operation, it is now mandatory on a national basis for certain categories of emission sources to install and maintain approved opacity monitoring systems. The latest federal (EPA) requirements for stationary emission sources are outlined in the Federal Register of October 10, 1979. In addition, many state and local authorities have implemented similar legislation on even broader categories of emission sources. Compliance regulations vary from State to State; however, some of the categories include emission sources also of relevance to the petroleum and petrochemical industries:

Bag houses	Residual oil-fired boilers
Coal-fired steam generators	Sulfuric acid plants
Fluid catalytic cracking units	Nitric acid plants
Industrial boilers	Turbine engine test cells

Opacity monitoring is an effective way to measure and control visible emissions (e.g., smoke). A key constituent of visible emissions are submicron particles that are a major cause of adverse health effects of air pollution, such as respiratory ailments and smog and gas phase reactions with other air pollutants. The opacity monitor can accurately duplicate the trained observer, with the advantages of 24-hr continuous operation and of being unaffected by night time, bad weather, and other adverse viewing conditions. Commercially available opacity monitors are summarized in Table 12.5 on page 488, two typical examples of which are described in more detail below.

(a) Lear Siegler Inc. (United States)

The Model RM-41 Visible Emission Monitoring System is designed to measure the opacity and amount of particulate matter in effluent from large industrial stacks for EPA compliance monitoring purposes.

The sensing system shown in Fig. 12.27 consists of two units, an optical transceiver mounted on one side of the stack or duct, and a reflector mounted on the opposite side. The transceiver contains a tungsten source, the light from which is modulated and alternatively gated into measurement and reference beams by rotating discs. The measurement beam is projected across the smoke channel to the retroreflector, which sends the beam back across the stack to a photodetector (double-pass measurement). The reference beam is projected within the transceiver onto the same photocell. The ratio of the intensities of measurement and reference beams is then used to establish the final measurement values.

Compensation is automatically made for fluctuations in lamp intensity, line voltage, detector sensitivity, and so on. Spectral response of the optical

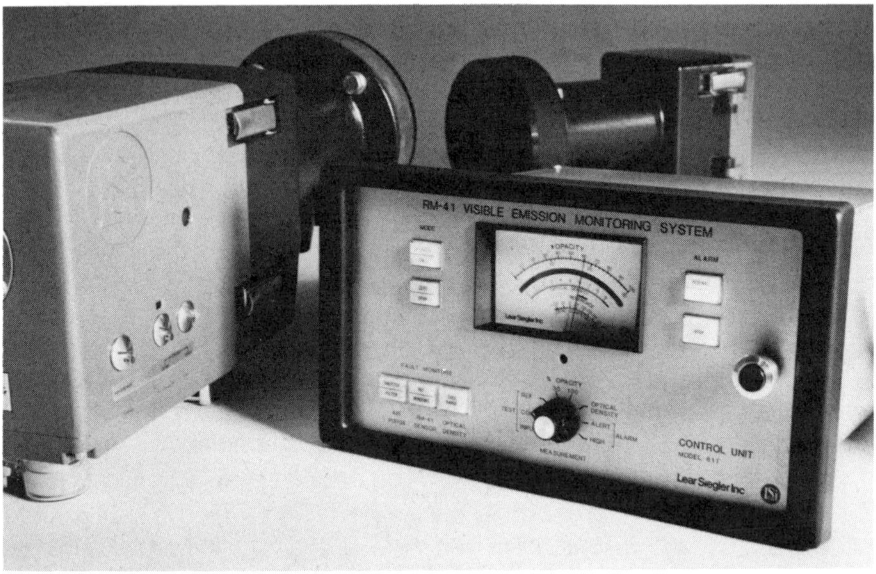

FIGURE 12.27 Model RM-41 emission monitor showing sensing units and control unit. (Courtesy of Lear Siegler Inc.)

system is greater than 90% within the 400–700 nm region comprising the visible light spectrum. The sensing system incorporates a forced-air purge system providing triple-filtered air to keep the optical windows free of dust and dirt deposits. Protective shutters are provided to protect the lenses in the event of an air purge failure.

The control unit, also shown in Fig. 12.27, is designed for control room panel or rack mounting and converts the transceiver output to linear opacity and optical density measurements, including correction of opacity measurements according to the ratio of stack exit diameter to depth of effluent at the measurement point (as required for EPA compliance for stack exit opacity value). Ten measurement ranges are provided (five for optical density, five for opacity value), independently selectable. Automatic on-stack calibration can be carried out from the control unit at 1, 2, 4, 8, or 24 hr intervals, switch selectable. Dual alarm and status/fault monitoring facilities are also provided. Outputs of 4–20 mA dc are available for remote recording.

(b) Dynatron Inc. (United States)

Dynatron manufactures two opacity monitoring systems, the Model 301 (single-pass) and the Model 1100 (double-pass). They are both designed for EPA compliance on intermittent and continuous emission sources respectively.

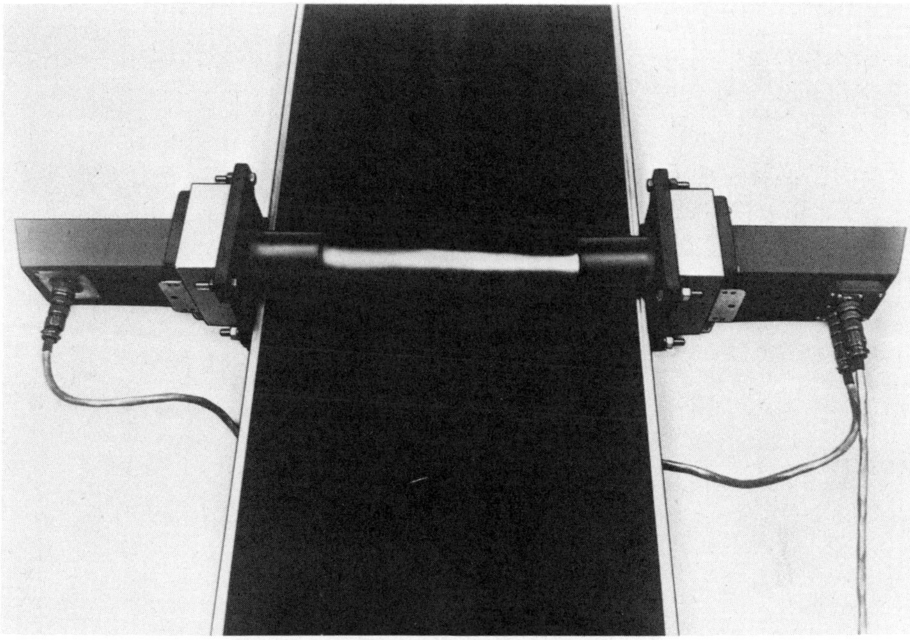

DYNATRON SINGLE PASS TRANSMISSOMETER STACK MOUNTING OPACITY SENSORS.

FIGURE 12.28 Model 301 opacity monitor. (Courtesy of Dynatron Inc.)

The Model 301 shown in Fig. 12.28 consists of a sensing unit and a remote mounted control unit. The sensing unit consists of a lamp source and an optical system, mounted on one side of the stack or duct, and the detector system mounted on the opposite side. Connections are provided on each unit for purge air entry from a regenerative blower air purge system, also available from Dynatron. The blower systems can be mounted separately or combined with the sensor system in a single enclosure. The Model 301 provides analog or digital opacity display, with alarm facilities and optional recorder. A 4–20 mA dc output is also available for remote recording.

The Model 1100 shown in Fig. 12.29 is a double-pass instrument consisting of a two-unit sensing system (transceiver and reflector units) and a remote mounted control unit. The control unit provides digital or analog display of opacity and optical density (switch selectable), automatic stack exit correlation, automatic on-stack calibration (1, 2, 4, 8, or 24 hr, manually selectable), alarm facilities, and a 6-min averaging circuit for EPA compliance reporting. Voltage (0–1 or 1–5 V dc) and current (0–20 or 4–20 mA dc) outputs are provided for remote recording.

Both instruments operate in the 515–585 nm spectral region, with greater than 90% of the peak response within 400–700 nm.

FIGURE 12.29 Dynatron Model 1100 opacity monitor. (*a*) Sensing unit configuration; (*b*) microprocessor control unit. (Courtesy of Dynatron Inc.)

12.4.11 Vehicle Exhaust Emissions Monitoring

With the federal government's current legislation regarding vehicle emissions has come the requirement for continuous monitoring of vehicle exhaust gases for a host of different compounds such as CO, CO_2, hydrocarbons, NO, NO_2, NO_x, SO_2, methane, oxygen, and air/fuel ratio. Several companies specialize in analytical instrumentation built specifically for this purpose, including Beckman in the United States and Horiba in Japan and the United States.

(a) Horiba Instruments Inc.

In addition to a range of process analyzers, atmospheric monitors and water quality monitors, Horiba manufactures probably the widest range of automotive emission analysis systems available today.

The MEXA—8020 Series Automotive Emission Analysis Systems are capable of carrying out continuous and simultaneous analysis for CO, CO_2, hydrocarbons, NO, NO_x, total hydrocarbons, and oxygen, using the standard methods of measurement such as IR, chemiluminescence, flame ionization, and so on. These systems are available in four versions, which differ in the method of sampling and measurement and are suitable for meeting federal specifications for gasoline and diesel-powered engines for modal mass and catalyst efficiency measurement.

The system can accommodate up to 14 analyzers, has a comprehensive set of alarm displays and interlock circuits, and is modular in design. There are two methods of sampling and measuring exhaust gases, namely, raw exhaust measurement and CVS (constant volume sampler) bag sample measurement. In the first method, the gas is measured direct from the automobile tail pipe; in the second method, the gas sample is diluted with ambient air.

The Model 1120/1220/1320 Series Analysis Systems are designed specifically for automotive emission measurement at end-of-assembly lines and engine tuning facilities at automobile manufacturing plants. The systems consist of sample handling, analytical equipment, and data processing equipment, and are modular in design, including castors for portability. Again, typical analyses include CO, CO_2, NO/NO_x, methane, non-methane HC, total HC, O_2, and particulates. For alcohol/gasoline engines, measurement of methanol and aldehydes can be incorporated.

A typical arrangement for an automotive emission test system, as designed by Horiba Instruments, is shown in Fig. 12.30. The system consists of the following component parts:

1. A chassis or engine dynamometer, which allows the test vehicle to reproduce a specified driving condition in the laboratory.
2. A continuous sampler, which collects exhaust sample from the test vehicle at a specified condition.
3. The emission analysis system, which accurately determines the concentrations of major components present in the exhaust sample collected.
4. A data processing system, which arithmetically processes the results of the measurements.
5. A facility that maintains the test environment in a specified condition.

With the addition of some accessory equipment, the system is capable of operating as a complete laboratory testing facility, in accordance with cus-

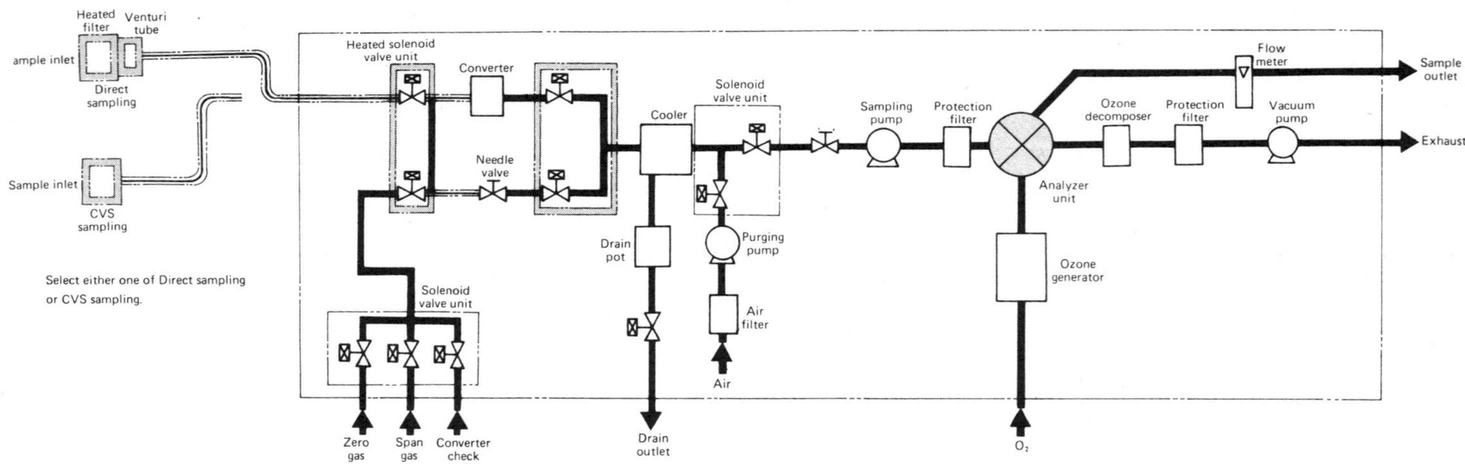

(a)

FIGURE 12.30 Horiba automotive emission test system. (*a*) Typical sampling system configuration. (Courtesy of Horiba Instruments Inc.)

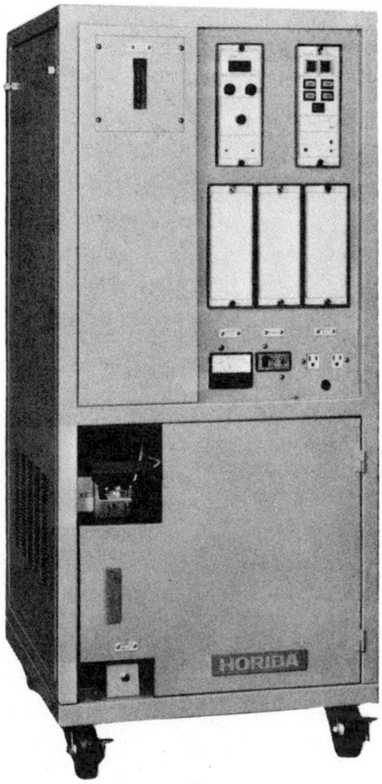

(b)

FIGURE 12.30 Horiba automotive emission test system. (*b*) Typical mobile cabinet installation. (Courtesy of Horiba Instruments Inc.)

tomer requirements. Horiba has experience in construction of major portions of key emission testing facilities in Japan, the United States, and Europe. The company also has its own emission test laboratory in Kyoto, Japan, and conducts auto emission tests there on an assignment basis. In addition to the systems described above, Horiba manufactures a range of portable automotive emission analyzers operating on the IR absorption principle.

(b) Beckman Instruments Inc.

Beckman has been manufacturing automotive emission testing equipment for over 25 years, from sub-assemblies that measure one component in a single test procedure to complete consoles capable of measuring several components for different test procedures. The well-proven Beckman IR, FID, and O_2 analyzers described in other chapters form the basis of the measurement technology. The use of modular electronics and building block packaging allows the user to select any number of components in any combination to meet specific requirements.

As an example, the Model 590 Infrared Exhaust Analyzer is a dual-channel NDIR instrument designed to measure CO and hydrocarbons in vehicle

TABLE 12.3 Combustible Gas Detectors

Manufacturer	Country of Origin	Response Time	Sensor Type	Model Number	Control Unit Configuration	Control Unit Mounting options
1. Bacharach Inst. Co. (United Technologies)	USA	<30 sec to 90% LEL	Catalytic	CD800W/XD500W	Single channel	Wall mounted
				CD802W/XD502W	Dual channel	Wall mounted
				CD830/XD530	Single channel	Modular (rack or panel)
				CD832/XD532	Dual channel	Modular (rack or panel)
				CD100W	Single channel	Wall mounted
				CD860	4 channels (digital)	Modular (rack or panel)
				CD850/XD550	4 channels	Modular (rack or panel)
2. Control Instrument Corp.	USA	<12 sec to 90% FS	Catalytic	CCS	4 channel	Modular (rack or panel)
				DCR	Multichannels	Max. 64 channels per per 19 in. rack
3. Crowcon Instrument Ltd.	Great Britain	10 sec to 25% LEL	Catalytic	Gaswarden 72GW	Single or dual channel	Wall mounted
				Gaswarden 73GW	1–8 channels	Modular (rack or panel)
				Gaswarden 74GW	1–8 channels	Modular (rack or panel)
				Gaswarden 75GW	Multichannel	Max. 8 channels per 19 in. rack
				Gaswarden 76GW	Multichannel	Max. 8 channels per 19 in. rack
				Gaswarden 77GW	Multichannel	Max. 8 channels per 19 in. rack
				System 2000	Multichannel (199 max.)	Modular (rack or panel)
4. Delphian Corp.	USA	20 sec to 90% of reading	Catalytic	Micro 550	Multichannel (miniature)	Modular: 16 channels per 19 in. rack
				Sage 512	Multichannel (512 max.) Computer-based	Console unit

5. General Monitors Inc.	USA	6 sec typical	Catalytic	160	Multichannel	Max. 100 channels per 19 in. rack
				180/180A	Single channel	Modular (rack or panel)
				480	Single channel	Modular (rack or panel)
				520	Dual channel	Rack or panel
				580	Dual channel	Modular (rack or panel)
				610	4 channel	Modular (rack or panel)
				4800	Multichannel	Max. 16 channels per 19 in. rack
6. International Sensor Technology (IST)	USA		Solid state	AG80/AG80-R	8 channel max.	19 in. rack or panel
			Electrolytic	AG 2000	Single channel	Wall mount
				AG 2000/3/4	2, 3, 4 channel	Wall mount
				AG 2200	Single channel (ex)	Wall mount
				AG 4000	Single channel	Bench mount
				AG 3100	Single channel	Modular (rack or panel)
				AG 3102	Dual channel	Modular (rack or panel)
				AG 512	Computer-Based Multichannel (max. 511)	Cabinet mount
7. M.S.A. Company	USA	<5 sec to 40% LEL	Catalytic	510 DH	Dual channel	Modular (rack or panel)
				510 Series	Single channel	Modular (rack or panel)
				511	Single channel	Wall mount
				512	Single channel (EX-P)	Wall mount
				511 DH	Dual channel	Wall mount
					Four channel	
8. Neotronics Inc.	Great Britain		Catalytic	AGM Series	Multichannel	Cabinet (rack or panel)
				CA-HL	Single channel (IS)	Modular
				C/3007	Single channel	Modular

TABLE 12.3 (*Continued*)

Manufacturer	Country of Origin	Response Time	Sensor Type	Model Number	Control Unit Configuration	Control Unit Mounting options
9. Rexnord/Dictaphone	USA	<20 sec to 90% FS	Catalytic	810	Single channel	Modular
				820	Dual channel	Modular
				880	8 channel	Modular
				910 (two wire)	Single channel	Modular
				920 (two wire)	Dual channel	Modular
10. Scott Aviation	USA		Catalytic	9000	Single channel	Modular (Wall, panel, rack, or cabinet) (7 channels per 19 in. rack)
			Catalytic (4–20 mA output)	Quadraflex	Digital Multichannel (max. 16 per unit) Also dual channel module available	Modular (rack or panel)
11. Sensidyne Inc.	USA		Catalytic	Combustion-Alert	Single channel	Modular (rack or panel)
12. J&S Sieger Ltd.	Great Britain	10 sec to 90% FS (typical)	Catalytic (various models)	FS-1	Single channel (10 per 19 in. rack)	Modular (rack or panel)
				FS-4	Four channel (32 per 19 in. rack)	Modular (rack or panel)
				FS-16	16 channel	Rack or panel
				606 Type II	6 channel	Wall mount
				2501	Single channel	Wall mount
13. U.S. Industrial Products Inc.	USA	<15 sec to 90% FS	Catalytic	8120	Single channel	Rack or panel
				8122	Dual channel	Modular (rack or panel)
				8124	4 channel	Modular (rack or panel)
				8100-M	Multichannel	Modular (rack or panel)

TABLE 12.4 On-Line Sulfur Dioxide Analyzers

Manufacturer	Country of Origin	Model Number	Measurement Principle	Sampling Configuration	Typical Measurement Ranges	Wavelength Range	Typical Applications
1. Anacon Inc.	USA	207	UV-vis absorption	Extractive		210–1000 nm	Stack flue gases
2. Beckman Industrial Inc.	USA	864/865	NDIR (Luft)	Extractive	0–500, 2000 ppm 0–10, 30%	2–15 μm	SO_2 in stack gases (sulfuric acid plant)
3. Comb. Eng. Process Analytics	USA	501	NDIR	Extractive	0–2000 ppm 0–15%	0.8–15 μm	SO_2 in stack gases & process gases
4. Contraves-Goertz	USA	100	NDIR	Cross-stack	ppm to % ranges		Stack flue gas emissions; also measures NO, CO, CO_2
5. DuPont Process Instruments	USA	460	UV absorption	Extractive	0–300, 3000 ppm (0–100 ppm min)	214–975 nm	SO_2 in stack gases (sulfur recovery plant)
6. Hartmann & Braun AG	West Germany	Uras 2T/2T-Ex	NDIR (Luft)	All are extractive type	0–200 ppm (min)	2.5–12 μm	SO_2 in process gases and for ambient air monitoring
		Uras 3G	NDIR		0–300 ppm (min)	2.5–12 μm	
		Uras 3E	NDIR		0–300 ppm (min)	2.5–12 μm	
		Uras 7N	NDIR		0–400 ppm (min)	2.5–12 μm	Stack flue gas monitoring
		Radas 1G	UV-vis absorption		0–200 ppm (min)		Exhaust gas monitoring
		Radas 1E	UV-vis absorption		0–200 ppm (min)		
		Picoflux 4	Conductimetric		0–0.2 to 0–200 ppm	N/A	Ambient air monitoring
7. Horiba Instruments Ltd.	Japan	PIR-2000	NDIR (Luft)	Extractive	0–500 ppm (min)	2–15 μm	Process gases
		Series 100/80	NDIR (Luft)	Extractive	0–500 ppm, 0–200 ppm		Stack flue gas monitoring
		EIA-31 & TIA-31	NDIR (Luft)	Extractive	0–500 ppm (min)		Process gases
8. Inficon, Leybold-Heraeus	USA	Binos range	NDIR, vis or UV absorption	Extractive	ppm to % ranges	200–1000 nm (UV-vis) 2–15 μm (IR)	SO_2 in stack gases & process gases
9. Infrared Industries Inc.	USA	IR-700 series	NDIR	Extractive	0–2% min		SO_2 in stack gases & process gases

TABLE 12.4 (*Continued*)

Manufacturer	Country of Origin	Model Number	Measurement Principle	Sampling Configuration	Typical Measurement Ranges	Wavelength Range	Typical Applications
10. Interscan Corp.	USA	1240 series	Electrochemical sensor	Integral sample pump	0–1, 5, 10, 25 % 50 ppm	N/A	SO_2 in process gases and ambient air monitoring
11. Kent Industrial Measurements Ltd.	Great Britain	Infragas 400 Infragas 500 6608	NDIR (Luft) NDIR Thermal conductivity	Extractive Extractive Extractive	0–300 ppm (min) Up to 5 components 0–5, 10, 15, 20%	2–15 μm 2–15 μm N/A	SO_2 in process gases SO_2 in process gases Designed for SO_2 monitoring in sulfuric acid plant burner gases
12. Lear Siegler Inc.	USA	SM810	UV absorption	In situ	0–250, 500, 750 ppm 0–500, 1000, 1500 ppm	SO_2: 218.5 nm NO: 226.5 nm	Stack flue gas monitoring; measures SO_2 & NO simultaneously
13. H. Maihak AG	West Germany	Unor 4N Unor 6N	NDIR (Luft) NDIR	Extractive Extractive	0–300 ppm (min) 0–300 ppm (min)	2–15 μm	Ambient air monitoring, process gases, stack gases, and exhaust gases
14. MDA Scientific	USA	PSM-8	Tape staining	Extractive	0–20 ppm	N/A	Ambient air monitoring
15. Measurex Inc.	USA	2001	UV absorption	Cross-stack	0–1500 ppm	200–1000 nm	Stack flue gas monitoring; also measures NO, CO, CO_2, & opacity
16. Mine Safety Appliances Co.	USA	LIRA 202 LIRA 204	NDIR (Luft) NDIR	Extractive Extractive	0–500 ppm 0–500 ppm	1–15 μm 1–15 μm	Process gases, stack gases, and auto exhaust monitoring
17. Sensidyne Inc.	USA		Electrochemical sensor	—	0–10 ppm	N/A	Ambient air monitoring

18. Siemens AG	West Germany	Ultramat 1	NDIR (Luft)	Extractive	0–0.5 to 0–100%	2–8 μm	SO_2 in stack flue gases, process gases, and ambient air monitoring
		Ultramat 2	NDIR (Luft)	Extractive	0–300 ppm to 0–100%	2–8 μm	
		Ultramat 3	NDIR (Luft)	Extractive	0–50 ppm (min)	2–8 μm	
		Ultramat 31	NDIR (Luft)	Extractive	0–0.1 to 0–0.5%	2–8 μm	
		Ultramat 32	NDIR (Luft)	Extractive	0–20 to 0–2000 ppm	2–8 μm	
19. Teledyne Anal. Insts.	USA	600	UV absorption	Extractive	0–300 ppm	210–1000 nm	Process gases
		690	UV absorption	Extractive	0–3000 ppm	210–1000 nm	Stack flue gases
		700	NDIR	Extractive	0–15%	2.7–10 μm	Sulfuric acid plant monitoring
20. Telsec Process Analyzers Ltd.	Great Britain	TPA330	NDIR	Cross-stack	0–250 ppm	2–15 μm	Stack gas monitoring
		TPA110 series	UV-vis-NIR photometer	Extractive	0–0.2 to 0–100%		Process gases
		TPA300 series	NDIR (Luft)	Extractive	0–100 ppm (min)	2–15 μm	Process gases & ambient air monitoring
		TPA460 (Duktolyt)	Conductimetric	Extractive	0–2 to 0–15 ppm		
21. Thermo-Electron Corp.	USA	40	Pulsed fluorescence	Extractive	0–50 to 0–5000 ppm	N/A	Process gases
		43	Pulsed fluorescence	Extractive	0–0.5, 1 & 5 ppm	N/A	Ambient air monitoring
		D12000	UV absorption	Cross-stack	0–2000 ppm		Stack flue gas monitoring
22. Western Research	Canada	720/721	UV absorption	Extractive	ppm & % ranges	200–1000 nm	Process gases, stack flue gases, & ambient air monitoring
23. Westinghouse	USA	260	Electrochemical (solid electrolyte) sensor	In situ	10–1000 ppm	N/A	Stack flue gas monitoring
24. H. Wosthoff	West Germany	Mikrogas-MSK	Conductimetric	Extractive	0–500, 800 ppm	N/A	Process gas & stack flue gas
		Ultragas	Conductimetric	Extractive	0–0.4, 0–4 ppm	N/A	Ambient air monitoring
25. Xertex Corp.	USA	OP-3300	NDIR	Cross-stack	0–250 ppm (min)	2–15 μm	Stack flue gas monitoring

TABLE 12.5 On-Line Stack Flue Gas Opacity Monitors

Manufacturer	Country of Origin	Model Number	Configuration	Path Length	Source and Rated Life	Peak Spectral Response	Measurements	Remarks
1. Contraves-Goertz Inc.	USA	400 (500 control unit)	Double-pass			515–585 nm	Opacity (0–100%) Optical density[a] Time average values	
2. DataTest Inc.	USA	900A	Single-pass	1–50 ft	Tungsten 3 years	515–585 nm	Opacity (0–100%) Optical density[a] Time average values Stack exit correction[b]	Microprocessor-based
		900RM	Double-pass	Up to 185 ft	Tungsten 3 years	515–585 nm	Opacity (0–100%) Optical density[a] Time average values Stack exit correction[b]	Microprocessor-based
3. Dynatron Inc.	USA	301 1100M	Single-pass Double-pass	1–25 ft 1–70 ft		515–585 nm 515–585 nm	Opacity (0–100%) Transmission Opacity Optical density[a] Time average values Stack exit correction[b]	Microprocessor-based
4. Econics Corp.	USA	7210 7220		Not cross-stack			Opacity	Part of a multi-parameter system (*see* Chapter 18) Microprocessor-based

5. Kent Industrial Insts. Ltd.	Great Britain	2010	Single-pass	Not cross-stack			Opacity (0–100%)	
6. Lear Siegler Inc.	USA	RM41	Double-pass		Tungsten 20,000 hr	500–600 nm	Opacity Optical density[a] Time average values Stack exit correction[b]	
7. Measurex Corp.	USA	2001	Double-pass				Opacity (0–100%)	Microprocessor-based Part of a multi-parameter system (*see* Chapter 18)
8. Siemens AG	West Germany	M52081 M52083		1–10 ft 1–11.4 ft			Opacity (0–25, 100%) Dust concentration (0–0.1, 0.2, 0.4, 0.8, 1.0 extinction)	

[a] *Optical density:* The optical density is defined by the expression

$$\text{Optical density} = \log_{10}\left(\frac{1}{1 - \text{opacity}}\right)$$

[b] *Stack exit correction:* The stack exit correction is given by

$$O_x = 1 - (1 - O_t)^{L_x/L_t}$$

where O_x = stack exit opacity, O_t = measured opacity, L_x = stack exit diameter, and L_t = path length of opacity monitor.

exhaust gases, specifically for use by the professional automotive mechanic in engine tune-up and diagnostic facilities. The analyzer is trolley mounted for mobility and includes a flexible sample probe for easy insertion into curved tail pipes.

A sample hose 25 ft long connects the probe to the analysis unit. Dual range switching is available for each component, 0–10%/0–2% for CO and 0–2000 ppm/0–400 ppm for hydrocarbons. A large 8-in. meter is provided for each measured component, with the same dual scales. Single-component and three-component analysis systems are also available. Analysis systems can also be custom built into complete consoles to meet customer requirements.

The Model 595 Engine Scope Tester is designed for complete engine diagnostic evaluation by the professional automotive mechanic. The instrument features an 11-in. CRT for information display in the many operating modes, with a four-times expansion giving the equivalent of a 30-in. wide screen. An eight-times expansion is also available. Two indicating meters are provided, one for voltage and spark advance, one for rpm, dwell degrees, and point resistance. Switches are provided for 4, 6, 8, and rotary engine selection. Cylinder selection switches are used for power balance testing or for selecting a cylinder of interest from the "parade" mode of the CRT. A power balance/kill button is used to disable selected cylinders for power balance testing or to kill all cylinders for cranking. The unit is trolley mounted for mobility.

BIBLIOGRAPHY

Affens, W. A., Carhart, H. W., and McLaren, G. W. Determination of flammable index of hydrocarbon fuels by means of a hydrogen flame ionization detector. *ACS Conf.*, Los Angeles, CA (March 1974).

Baum, R. M. Stratospheric science undergoing change. *Chem. & Eng. News* (Sept 13, 1982).

Bonady, F. M., and Kramer, K. L. Guide to air quality analyzers. *Chem. Proc.* (Sept 1981).

Callaghan, J. Performance standards for combustible gas detectors. *Inst. Tech.* (Dec 1981).

Chameides, W. L., and Davis, D. D. Chemistry in the troposphere. *Chem. & Eng. News* (Oct 4, 1982).

Clevett, K. J. Continuous on-line monitoring of trace gases in process streams. *Instr. Rev.* (April, May 1966).

Conner, W. D., Knapp, K. T., and Nader, J. S. Applicability of transmissiometers to opacity measurement of emissions from oil-fired power and Portland cement plants. *U.S. EPA-600/2-79-188* (Sept 1979).

Dailey, W. V. A novel NDIR analyzer for NO, SO_2 and CO analysis. *Anal. Instr.*, **12,** ISA (1977).

Debbrecht, F. J. Comparative technique for the quantitation of fugitive emissions. *75th Ann. APCA Mtg.*, New Orleans, LA (1982).

Duncan, J. Combustible gas detection instruments: History of their development and technical highlights. *CSA Std. C22.2, No. 152* (1976).

Ember, L. R. Government study admits man-made pollution: Acid rain tie. *Chem. & Eng. News* (June 20, 1983).

Gas detectors determine health hazards in sewage treatment. *Water & Sewage Works* (March 1977).

Gistello, J. C., and Walther, J. E. Evaluation of an on-stack transmissiometer as a continuous particulate monitor. *APCA 67th Ann. Mtg.*, Denver, CO (1974).

Herget, W. F., and Connor, W. D. Instrumental sensing of stationary source emissions. *Env. Sci. & Tech., **II**, No. 10* (Oct 1977).

Hollingsworth, G. A., and Chace, R. N. More effective electrolytic sensor monitors toxic, combustible gases. *Oil & Gas J.* (Feb 7, 1977).

Johanson, K. A. Gas detectors by the acre. *InTech* (Aug 1974).

Krigman, A. Monitoring air quality: More than buying an analyzer. *InTech* (Aug 1983)

Magison, E. Make sure your system is safe. *Instr. & Contr. Sys.* (Dec 1981).

Molloy, R. C. Smoke opacity monitoring systems: Pollution control and energy conservation. *ASHRA J.* (Sept 1976).

Morrow, N. L., Brief, R. S., and Bertrand, R. R. Sampling and analyzing air pollution sources. *Chem. Eng.* (Jan 24, 1972).

Nader, J. S., Jay, F., and Conner, W. D. Performance specifications for stationary source monitoring systems for gases and visible emission. *U.S. EPA-650/2-74-013* (Jan 1974).

Pacanowski, R. F., Bajek, W. A., and Browne, G. S. Monirex gas quality monitor: A unique device for gas analysis. *Intl. Conf. on Metering of Nat. Gas & LPG,* London (Feb 1984).

Precoat metals open new energy-saving coil line. *Indust. Finishing* (Feb 1982).

Schaeffer, M. J. The use of combustible detectors in protecting facilities from flammable hazards. *ISA Trans.*, **20,** No. 2 (1981).

Schranz, B. A. Guide to ambient air monitoring systems. *Insts. & Contr. Sys.* (Nov 1979).

Scott, W. C. Monitoring toxic and hazardous gases in the semiconductor industry. *Proc, ISA Symp.*, Boston, MA (May 1985).

Serdahl, E. C., and Sliger, E. J. Digital monitoring of combustible gases. *Computer Design* (Aug 1982).

Toxic gas monitoring: Simple instruments, complex measurements. *InTech* (May 1984).

Verdin, A. *Gas Analysis Instrumentation.* Macmillan, London (1973).

Wallace, M. J. Controlling fugitive emissions. *Chem. & Eng. News* (Aug 27, 1979).

Yaws, C. L. Sulfur oxides. *Chem. Eng.* (July 8, 1974).

13

Water Quality Monitoring

13.1 INTRODUCTION AND BACKGROUND

Water is all around us and forms an intimate part of our everyday lives. Today, we use water to conduct heat, transport nutrients, carry off waste, wash, irrigate, and so on. The average human being contains about 12 gal of water (about 100 lb). This amount is important, since the loss of just 15% of our body's water will almost certainly be fatal.

Most of the earth's surface is water; the oceans cover about 70 percent of the globe, not including underground aquifers. By one estimate, the earth's total water supply is more than 3.69×10^{20} gal. Thanks to rain and evaporation, our water recycles again and again. However, we are learning that even so enormous a supply does not guarantee that all waters will stay unspoiled or inexpensive to use. In fact, in some places, we are discovering that water can be poisonous even when it comes from the kitchen faucet.

As mentioned in Chapter 12, the subjects of air and water quality monitoring are very intimately connected; there is a close relationship between atmospheric pollution and its effect on the quality of our natural waters. In spite of the complexity of surface emissions resulting from biological and geological processes, the vast majority of trace species emitted into the atmosphere are in reduced oxidation states (e.g., hydrogen sulfide, ammonia, and methane). By contrast, the materials returned to the earth's surface from the atmosphere, usually by dissolution in rain or by dry deposition, are highly oxidized species (e.g., sulfuric acid, nitric acid, and carbon dioxide).

The "acid rain" issue has emerged as one of the top environmental issues of the 1980s and the terms "right as rain" and "pure as the driven snow" have become outmoded expressions in much of the industrialized world today. The reaction to acid rain is often one of scientific uncertainty. However, the phenomenon has been quite well understood for over 100 years, when the term was first coined. It was found then that acid rain is caused by emissions of sulfur dioxide and nitrogen oxides, which are chemically transformed in the atmosphere to sulfuric acid and nitric acid respectively (see Chapter 12).

As a consequence, water quality, like air quality, has become a matter of governmental and industrial concern on an international basis. Increasingly, we are learning that what society does in other spheres affects our water, often to its detriment; the press is continuously reporting on such incidents.

In recent years, the discovery of toxic chemicals in thousands of makeshift dumps has emerged as one of society's most serious problems. The catastrophe at Love Canal in the United States, where heavy rains caused the chemicals abandoned beneath the neighborhood to rise to the surface, is a good example. Another is the use of water in irrigation, where tons of pesticides, herbicides, fertilizers, and animal wastes leach into the soil with each application, flowing back into a below-ground aquifer and eventually feeding a river, estuary, or the homeowner's backyard well. In addition, powerful irrigation schemes often withdraw water from underground aquifers faster than they can be recharged by rainfall. This leaching effect also allows the entry of lead, aluminum, mercury, and other toxic chemicals into drinking water supplies.

Until recently, governments did little to restrain the dumping of poisonous chemicals into the water we drink. However, thanks to public outcry and action on the part of some responsible citizens, this situation is changing. The picture is not nearly so bleak today as it was a few years ago. Some bodies of water seem to be recovering, and frequent press reports indicate that fish are again returning to rivers and lakes that once were almost given up for lost. However, as a United Nations document points out, many of the fish that have returned to their traditional waters, are not fit to eat because they contain high levels of toxic chemicals. Thus, there still remains much work to remove the damage done by toxics to our natural waters, work that will take many years to complete.

13.2 THE ROLE OF CONTINUOUS WATER QUALITY MONITORING

In view of the increasing international governmental involvement in water quality, there is ample justification for continuous water quality monitoring, purely from an environmental standpoint. However, as governmental controls have become stricter, industrial management has become more involved in monitoring effluent from its process plants to ensure that concentrations of chemicals released do not exceed the governmental allowances. There have been many instances on a worldwide basis where companies have incurred large financial penalties for exceeding the limit. The general consensus is that the rules are going to get stricter and the permissible concentrations lower. Therefore, as with air quality monitoring, we can expect water quality monitoring to become even more important in the future.

Water quality monitoring is probably a subject of even more complexity than air quality monitoring. As a consequence, several companies have specialized in the manufacture of complete on-line water quality measurement systems capable of monitoring and reporting on a wide variety of different parameters, according to customer requirements, be this for industrial or environmental applications.

What is dissolved or suspended in water is very important to many people—water users, managers and operators of water treatment plants, waste water treatment plants, and steam generation plants (boilers), environmentalists, and a host of other interested parties, including the government agencies (e.g., EPA) and the general public.

A host of different chemical species and ions are measured and reported on a continuous basis. These measurements may be conveniently classified under three arbitrary headings—elemental analyses, specific ion analyses, and any other analyses that do not fall under the first two headings. Typical examples within these classifications are given here.

Elemental Analyses

Aluminum	Iron	Potassium
Antimony	Lead	Selenium
Arsenic	Lithium	Silver
Beryllium	Manganese	Sodium
Cadmium	Mercury	Strontium
Chromium	Molybdenum	Thorium
Cobalt	Nickel	Uranium
Copper	Phosphorus	Vanadium
		Zinc

Specific Ion Analyses

Bicarbonate	Fluoride	Phosphate
Bromide	Hydroxide	Sulfate
Carbonate	Iodide	Sulfite
Chloride	Nitrate	
Cyanide	Nitrite	

Other Analyses

Acidity	Conductivity (resistivity)
Alkalinity	Dissolved hydrogen
Ammoniacal N	Dissolved oxygen
Amines	Chemical oxygen demand (COD)
Hardness	Hydrazine
Morpholine	Halogenated hydrocarbons
pH and Redox	Phenolic compounds
Specific gravity	Total oxygen demand (TOD)
Residual chlorine	Total organic carbon (TOC)
PCBs	Oil, grease, and hydrocarbons
Particulates	Total Kjeldahl N

From an industrial viewpoint, the major applications for on-line water quality monitoring are found in raw water and process water treatment, cooling water systems, boiler feedwater treatment, internal boiler water and blowdown control, steam condensate and potable water production, and waste water treatment.

13.2.1 Raw Water Treatment

The impurities in raw water vary with the source of supply. Surface water is usually high in dissolved oxygen and sediment and low in total dissolved solids (TDS) and hardness, whereas ground water is usually high in TDS and hardness and low in suspended solids. Suspended solids are usually removed by a coagulation process, followed by clarification and filtration. Chlorination is sometimes required. The raw water pH must be controlled for efficiency control and the turbidity monitored to ensure correct coagulant dosage and effective filtration.

13.2.2 Process Water Treatment

Industries such as brewing, paper, food and beverages, metal plating, metal working, and other precision manufacturing activities have varying requirements for water quality. However, in most cases the water should be non-corrosive and free of turbidity, color, and metals such as iron and manganese. In many cases, taste and odor are important. Almost all process water is softened and chemically treated to adjust pH, control corrosion, and inhibit microbiological growth.

13.2.3 Cooling Water Treatment

In industrial plants, the cooling system usually represents the largest user of water. It is used for condensing steam and in exchangers for process cooling. Cooling water systems are generally of two basic types, the once-through type and the recirculation type, each of which has its own analysis requirements. For once-through systems, chlorination is usually all that is required to prevent biological fouling of heat exchangers.

For the recirculation type system, treatment is far more critical because of solids buildup, scaling, corrosion, and so on. Chelants or phosphates may be required. Water is often de-aerated to reduce electrochemical attack, or corrosion inhibitors such as chromates, polyphosphates, molybdate, and zinc compounds are added. Systems can usually be protected against chemical attack by controlling pH and setting proper levels of alkalinity and residual chlorine.

13.2.4 Boiler Feedwater Treatment

The main objectives of boiler feedwater treatment are to prevent introduction of contaminants into the boiler and to control the addition of conditioning chemicals. Without careful control, costly problems such as scaling, corrosion, carryover, and embrittlement can occur. In general, the higher the boiler pressure, the higher the feedwater quality required. Sources of boiler feedwater can vary from all makeup to all condensate, requiring various degrees of treatment between these extremes.

For low and medium-pressure boilers, Zeolite softeners are frequently used to lower calcium and magnesium hardness to acceptable levels. Considerable savings and increased operating efficiency can be achieved by continuous monitoring of the effluent to detect an increase in total hardness. Regeneration of the softeners only takes place when necessary, even though inlet water conditions change.

For high-pressure boilers requiring stringent water quality, demineralizers are generally used to remove unwanted impurities by ion exchange. The first impurity to break through when the demineralizer approaches exhaustion is usually silica. However, silica cannot be detected by typical dissolved solids (conductivity) analyzers because of its very low ionic strength. Therefore, an automatic silica analyzer is required, to detect increases in silica level prior to exhaustion and initiate a regeneration cycle. This will avoid costly scaling problems in the boiler.

13.2.5 Internal Boiler Water

Analysis and control of internal boiler water is necessary to eliminate scale, corrosion, and metal embrittlement, since even the best water treatment systems will allow some impurities to pass into the boiler. Scaling elements such as calcium and magnesium are generally controlled by phosphate addition/pH control, chelating agent addition, or a combination. Continuous monitoring of these residuals is important from a control standpoint. To fight corrosion and embrittlement, measures taken include close pH control, deaeration and the use of oxygen scavengers such as sodium sulfite and hydrazine. Often, ammonia and neutralizing amines are applied to neutralize carbon dioxide. Filming-type amines are sometimes added to steam to form a protective coating on steam system metal surfaces.

In high-pressure boilers, it is essential to hold silica to very low levels (less than 1 ppm) to prevent carryover into the steam and deposition on steam turbine blades that will result in loss in efficiency and costly shutdown for repairs.

13.2.6 Boiler Blowdown

Blowdown is used to control the accumulation of solids in a boiler and is basically the draining of some of the contaminated water and its replacement with purified feedwater. However, blowdown also results in a loss of heated water and costly treatment chemicals. Therefore, economical operation requires careful control of blowdown to maintain safe solids level and minimize heat and chemical losses.

Blowdown systems can be continuous or intermittent. In either case, continuous analysis is required to maintain this critical balance. Control parameters to be monitored include silica, hydrazine, phosphate, pH, and conductivity.

13.2.7 Steam Condensate

Steam condensate is very high purity water that is commonly returned to the boiler instead of expensive, treated make-up water. The condensate return system is a very good place to monitor overall water treatment performance, since the nature and levels of contaminants will help identify problems such as carryover and corrosion, thus initiating timely corrective action.

Steam condensate contamination can come from a variety of sources:

- Leakage of cooling water from heat exchangers.
- Leakage of dissolved gases and lube oil from pumps.
- Suspended metal particles, the results of corrosion.

Condensate polishers are frequently used to prevent these contaminants from entering the boiler. Continuous monitoring of these key parameters can aid in identifying the source of contamination and control of the water treatment processes. Such parameters include hardness, conductivity, turbidity, iron, copper, dissolved oxygen, sodium, and silica. Silica and turbidity analyzers can be used to indicate efficiency and regeneration points of polishing systems.

13.2.8 Potable Water

In the production of high quality drinking water, clarity, taste, odor, hardness, and bacterial activity are important parameters that must be closely controlled to ensure potability and safety. Other constituents, such as fluoride, iron, and manganese, must be limited to preserve esthetic quality. The turbidity of raw water may vary from day to day, requiring changes in treatment. Continuous turbidity monitoring throughout the treatment processes (flocculation, clarification, filtration) can aid the control of each process. Chlorination is used throughout the treatment process to prevent microbial growth and protect public health. Correct chlorine levels can best be maintained by the use of continuous residual chlorine monitors.

13.2.9 Wastewater Treatment

Important parameters in the biological treatment of waste are the pH, conductivity, and alkalinity of the incoming wastewater. The pH is normally measured continuously. Plants subject to septic wastes may also require chlorination of the influent. Turbidity is an important parameter throughout the waste treatment plant (e.g., density or opacity of sludge blankets). Where the final effluent is discharged to natural waters (rivers and estuaries), there are usually tight governmental specifications on the quality of the effluent involving continuous monitoring of parameters such as pH, turbidity, residual chlorine, dissolved oxygen, and total organic carbon.

Very often, this also involves proof that the effluent has not exceeded the time-weighted averages for certain parameters. Hence, there is a more than adequate justification for continuous monitoring, especially as there are usually heavy financial penalties involved in non-compliance with regulations.

13.3 LABORATORY METHODS OF MEASUREMENT

Measurement of water quality in the laboratory is a subject of very wide scope. A variety of different measurement techniques are used, ranging from traditional wet chemical analyses to fairly sophisticated instrumental methods and various combinations of the two. The importance and complexity of water quality monitoring is evidenced by the large number of ASTM standard test methods that have been devised for measurement of key parameters. See Table 13.1.

TABLE 13.1 ASTM Standard Test Methods for Water Quality Measurement

ASTM	
D511-82	Calcium and Magnesium
D512-81	Chloride Ion
D513-82	Carbon Dioxide, Bicarbonate, and Carbonate
D514-80	Hydroxide Ion
D515-82	Phosphorus
D516-82	Sulfate Ion
D857-79	Aluminum
D858-82	Manganese
D859-80	Silica
D992-71	Nitrate Ion
D1067-82	Acidity and Alkalinity
D1068-82	Iron
D1125-82	Conductivity and Resistivity
D1126-80	Hardness
D1129-78	Standard Definitions and Terminology
D1179-80	Fluoride Ion
D1192-70	Equipment for Sampling Water and Steam
D1246-82	Iodide and Bromide Ions
D1252-78	Chemical Oxygen Demand (COD)
D1253-76	Residual Chlorine
D1293-78	Hydrogen Ion (pH)
D1339-78	Sulfite Ion
D1385-78	Hydrazine
D1426-79	Ammoniacal Nitrogen
D1428-82	Sodium and Potassium
D1429-76	Specific Gravity
D1498-76	Oxidation-Reduction Potential
D1588-60	Dissolved and Gaseous Hydrogen
D1687-80	Chromium
D1688-82	Copper
D1691-77	Zinc
D1783-80	Phenolic Compounds
D1886-82	Nickel
D1888-78	Particulates and Dissolved Matter
D1942-80	Morpholine
D2036-82	Cyanides
D2327-80	Mono- and Dioctadecylamines
D2333-80	Thorium
D2579-78	Total Carbon (TC) and Organic Carbon (TOC)
D2580-80	Phenols by Gas Chromatography
D2777-77	Precision and Bias of Methods
D2791-77	Sodium
D2907-75	Uranium (trace)
D2909-80	Cyclohexylamine
D2972-78	Arsenic
D3113-80	EDTA (Tetrasodium Salt)
D3223-80	Total Mercury
D3250-77	Total Oxygen Demand (TOD)
D3352-74	Strontium
D3370-82	Sampling Practices

TABLE 13.1 (*Continued*)

ASTM	
D3371-79	Nitriles by Gas Chromatography
D3372-80	Molybdenum
D3373-80	Vanadium
D3534-80	Polychlorinated Biphenyls (PCBs)
D3557-78	Cadmium
D3558-82	Cobalt
D3559-78	Lead
D3560-83	Oxygen Demand
D3561-77	Lithium, Sodium, and Potassium
D3589-83	Total Nitrogen (Coulometric)
D3590-77	Total Kjeldahl Nitrogen
D3645-78	Beryllium
D3697-78	Antimony
D3859-79	Selenium
D3866-82	Silver
D3867-79	Nitrate and Nitrite Ions
D3868-79	Fluoride Ion in Brackish Water
D3869-79	Iodide and Bromide Ions in Brackish Water
D3875-	Alkalinity of Brackish Water
D3919-80	Trace Elements
D3920-80	Strontium
D3921-80	Oil, Grease and Hydrocarbons
D3973-80	Low Mol. Wt. Halogenated Hydrocarbons
D3977-80	Suspended Sediment

A knowledge of the chemical basis for some of the more important test methods will be useful to the analyzer engineer in appreciating the basis for development of some process analyzers. For instance, many of the wet chemical analyzers described later in this chapter are basically automated versions of the standard laboratory test method. Some of the more important test methods from Table 13.1 are described briefly, using the three classifications used previously.

13.3.1 Elemental Analyses

The most popular method of elemental analysis today is atomic absorption spectrophotometry. Many elements can be detected at low concentrations using this technique. Other methods involve the formation of intensely colored complexes, followed by photometric measurement of the color intensity.

ASTM D511-82: Calcium and Magnesium

(a) Gravimetric Method · Calcium is precipitated as calcium oxalate, then ignited and weighed as calcium oxide. Magnesium from the calcium filtrate is precipitated as magnesium ammonium phosphate, then ignited and weighed as magnesium pyrophosphate.

(b) Complexometric Titration · Calcium is titrated with EDTA at pH 10, then magnesium is titrated with EDTA at pH 12–13.

ASTM D515-82: Phosphorus

Four methods covering the determination of phosphorus in water and wastewater are described. All are based on reactions that are specific for the orthophosphate ion.

Method A. Ammonium molybdate and antimony potassium tartrate react with orthophosphate to form an antimony-phosphate-molybdate complex. The complex is reduced with ascorbic acid to form a deep blue-colored molybdenum complex. The range of measurement is 0.01–5 ppm of phosphorus. Photometric measurement of the color intensity is made at 880 nm.

Method B. Orthophosphate reacts with ammonium molybdate in acid solution to form a molybdophosphate that is in turn reduced to a molybdenum blue complex with amino-naphthol-sulfonic acid. When a bismuth salt is added to the sulfuric acid reagent, the intensity of the blue color developed in the test is increased about fourfold. Measurement range is 0.1–3 ppm phosphorus.

Method C. This method is best suited to boiler feedwater and boiler water containing 1–10 ppm of phosphorus. It is based on the photometric measurement of the yellow color of molybdovanado phosphoric acid formed in the test. Photometric measurement is made at 400–420 nm.

Method D. This method is a semiautomated procedure applicable to the measurement of total phosphorus in drinking water, surface water, and in domestic and industrial waste waters.

The sample is digested in the presence of sulfuric acid, potassium sulfate, and mercuric sulfate, where organic phosphorus is converted to orthophosphate. The orthophosphate is reacted with molybdate-antimony reagent, subsequently reduced with ascorbic acid, and the intensity of the blue phosphomolybdenum complex formed is measured photometrically at 650–670 nm.

ASTM D1068-82: Iron

(a) Photometric Method · Iron is put into solution with acid, followed by formation of an orange-red complex with o-phenanthrone, the intensity of which is measured with a photometer.

(b) Atomic Absorption Spectrophotometer.

ASTM D1428-82: Sodium and Potassium

(a) Flame Photometric Method.

(b) Specific Ion Electrode.

ASTM D1687-80: Chromium

(a) Photometric Method · Chromium is oxidized and a complex formed by reaction with diphenylcarbazide. The color intensity is then measured with a photometer. Three alternative methods of oxidation are listed, using permanganate, hypobromite or alkaline persulfate.

(b) Atomic Absorption Spectrophotometer.

ASTM D1688-77: Copper

(a) Photometric Method · A complex is formed with neocuproine, the intensity of which is measured with a photometer at 457 nm. Alternatively, cuprethol may be used, with photometric detection at 435 nm.

(b) Atomic Absorption Spectrophotometer.

ASTM D3223-80: Mercury

Wet chemical oxidation of mercury is followed by cold vapor atomic absorption (absorption of UV radiation at 253.7 nm by mercury vapor).

ASTM D3558-77: Cobalt

Cobalt is measured by atomic absorption spectrophotometry, either direct or by the chelation/extraction method.

ASTM D3559-78: Lead

As for cobalt.

13.3.2 Specific Ion Measurement

Methods for measurement of specific ion species often involve the formation of an intensely colored complex or dye. The intensity of the color, which is proportional to the concentration of the specific ion of interest, is measured using a spectrophotometer. Another popular method is the use of an ion selective electrode, which is identical to the on-line specific ion electrode.

ASTM D512-81: Chloride Ion

(a) Complexometric titration using diphenylcarbazone/bromophenol blue as indicator.

(b) Titration with silver nitrate at pH 8.3 using potassium chromate as indicator. Bromide, iodide, phosphate and sulfite ions interfere in this titration.

(c) Formation of ferric thiocyanate complex, followed by colorimetric determination.

(d) Determination using an ion selective electrode.

ASTM D516-82: Sulfate Ion

(a) Gravimetric determination as barium sulfate.

(b) Turbidimetric determination as barium sulfate.

(c) Volumetric titration with barium chloride using Thorin as indicator.

ASTM D1179-80: Fluoride Ion

(a) Formation of hydrofluosilicic acid and measurement of its bleaching effect on SPADNS dye.

(b) Determination using an ion selective electrode.

ASTM D1246-77: Bromide and Iodide

Method A. This method covers the determination of iodide ion in water in concentrations up to 100 ppb. It is based on the catalytic effect of traces of iodide on the rate of oxidation of arsenious ion by ceric ion in acid solution (ceric ions are yellow and cerous ions colorless in solution).

Method B. This method covers the determination of bromide ion in water in concentrations up to 100 ppb. It is based on the catalytic effect of traces of bromide on the rate of oxidation of iodine to iodate by potassium permanganate in sulfuric acid solution.

Method C. This method is applicable to essentially all waters, including brines and sea water. It is particularly recommended for waters containing iodide or bromide (or both) in concentrations in excess of 50 ppm.

Any iodide in the sample is oxidized with bromine to iodate in a buffered solution, the excess bromine being subsequently destroyed. Iodine equivalent to the iodate is liberated from potassium iodide and titrated with sodium thiosulfate.

Iodide and bromide occurring together are oxidized to iodate and bromate, respectively, with hypochlorite. Iodide equivalent to the combined reaction products is liberated and measured after destroying excess hypochlorite.

Bromide is calculated by difference between iodide and combined iodide-bromide determinations.

ASTM D1339-78: Sulfite Ion

(a) Reaction with iodine chloride, followed by titration with sodium thiosulfate.

(b) Reaction of potassium iodate to produce iodine, which in turn reacts with the sulfite ions present. The end point is detection of excess iodine using starch as indicator.

ASTM D3867-79: Nitrate and Nitrite Ions

Nitrate is reduced to nitrite. The mixture is diazotized to form an intensely colored dye, the intensity of which is measured photometrically.

13.3.3 Other Measurements

Measurement of pH, ORP (or REDOX), and conductivity fall under this classification. However, these tests are not described here since they are standard instrumental methods using the same type of apparatus as the on-line process analyzers. The reader is referred to Sections 13.4 and 13.5 for details.

ASTM D859-80: Silica

(a) Gravimetric Method. Measurement of the loss in weight when silicon is volatilized as silicon tetrafluoride.

(b) Colorimetric Method. Formation of a complex with molybdate ion, followed by reduction to a blue complex, the intensity of which is measured photometrically. This is the standard laboratory test, which has been automated for use as an on-line technique.

ASTM D1067-82: Acidity and Alkalinity

An electrometric titration is used to determine both of these parameters. For acidity, the sample is titrated with alkali; for alkalinity the sample is titrated with acid.

ASTM D1126-80: Hardness

The sample is titrated with EDTA (sodium salt) using Chrome Black T as indicator, which changes from red to blue at the end point.

ASTM D1252-78: Chemical Oxygen Demand (COD)

Oxidation by potassium dichromate, followed by titration with ferrous ammonium sulfate solution.

ASTM D1253-76: Residual Chlorine

(a) Amperometric titration using phenylarsine oxide.

(b) Iodometric titration using starch/iodide. Iodine liberated at the proper pH is proportional to the total chlorine present.

ASTM D1385-78: Hydrazine

Formation of a yellow complex with para-dimethylamino benzaldehyde, followed by photometric determination of color intensity.

ASTM D2579-78: Total Organic Carbon

(a) Oxidation to carbon dioxide, followed by IR measurement.

(b) Reduction, followed by determination using a flame ionization detector.

ASTM D3250-77: Total Oxygen Demand (TOD)

The small sample is mixed with a carrier containing a known amount of oxygen and pyrolized in a reactor at 900°C. The reduction in oxygen concentration following pyrolysis is measured. This method is also applicable to on-line analyzers for measurement of TOD.

ASTM D3560–83: Oxygen Demand

A small sample is mixed with carbon dioxide carrier and pyrolized in a reactor at 900°C. Carbon dioxide is reduced to carbon monoxide, which is measured using an IR detector.

13.4 pH AND REDOX MEASUREMENT

The measurement of pH has a wide variety of industrial applications, examples of which may be found in sugar refining, pulp and paper, food processing, metal plating, fertilizer manufacture, fermentation processes and mining, in addition to those in the water treatment industry covered in Section 13.2. Apart from its industrial applications, pH is also an important parameter in everyday life. Living organisms require a particular pH environment to survive and grow. When this environment is upset, growth and survival are threatened. Many plants, shrubs, and trees (e.g., azaleas and rhododendrons) require an acid soil for best performance; corn, wheat, and vegetables require certain soil conditions for maximum yield. The pH of human blood is highly critical; should the pH change by one unit, death is inevitable.

Although the on-line process pH analyzer is a relatively simple instrument, the theory behind its operation is somewhat complex. Nevertheless, measurement of pH is one of the most traditional of process measurements and has been used industrially for many years. Many process reactions are dependent on the pH, which must be closely controlled for optimum process operation. In other cases, certain pH values are detrimental, resulting in corrosion of process equipment and other problems.

There is a wealth of literature on the theory and application of pH, to which the reader is referred for more detailed information. The basic theory covered in the following section should be sufficient to provide an adequate understanding of the term pH and its significance in industrial measurement and control.

13.4.1 Theoretical Basis

Stable chemical compounds are electrically neutral, but when in aqueous solution, dissociate to form positive and negative ions. When a voltage is applied across electrodes immersed in the solution, the positive ions (cations) migrate toward the negative electrode (cathode) and the negative ions (anions) migrate toward the positive electrode (anode). The freedom of ions to migrate is measured as the *conductivity* of the solution (see Section 13.5). Strong electrolytes, such as strong acids and bases, are completely dissociated in solution, whereas weak electrolytes are only partially dissociated.

In developing his theory of ionization, Arrhenius established that the acidity of a solution is due to the concentration of hydrogen ions (H^+) produced by ionization of the acid, and not to the total acid present. Similarly, the alkalinity of a solution is due to the concentration of hydroxyl ions (OH^-). When in solution, a strong acid is ionized to produce hydrogen ions (protons) and anions. In the case of hydrochloric acid, protons and chloride ions are produced:

$$HCl = H^+ + Cl^-$$

In the case of a strong base, such as sodium hydroxide, sodium ions and hydroxyl ions are produced:

$$NaOH = Na^+ + OH^-$$

In the case of weak acids or bases, only partial ionization takes place, but the same rules apply. An example is acetic acid, a weak acid:

$$CH_3COOH \rightleftharpoons H^+ + CH_3COO^-$$

The double arrows in the equation indicate that the reaction may proceed in either direction and that an equilibrium exists between the ions and the unionized acid. The acid strength depends upon the number of hydrogen ions available (termed the *hydrogen ion activity*); the alkaline strength depends upon the number of hydroxyl ions available (termed the *hydroxyl ion activity*).

Water is a poor conductor of electricity and ionizes only to a small extent to give hydrogen and hydroxyl ions:

$$H_2O \rightleftharpoons H^+ + OH^-$$

The ionization constant (K) of water is determined by the ratio of the product of the ion concentrations to the concentration of the undissociated water:

$$K = \frac{[H^+][OH^-]}{[H_2O]}$$

The value of K for water at 25°C is 1.8×10^{-16} mol/l. Since water is dissociated to such a small extent, the concentration of undissociated water may be treated as a constant and only the numerator taken into account. Thus we may write:

$$K_w = [H^+][OH^-] = 1 \times 10^{-14} \text{ mol/l at } 25°C$$

where K_w is called the *ionic product of water*. Since there are equal quantities of the two ions present:

$$[H^+] = [OH^-] = 1 \times 10^{-7} \text{ mol/l at } 25°C.$$

More precisely, ion *activities* should be used rather than ion concentrations:

$$a_{H^+} = a_{OH^-} = 1 \times 10^{-7} \text{ mol/l at } 25°C$$

The product of the ion activities is always equal to 10^{-14} mol/l at 25°C. If the ion activities are equal, the solution is neutral and both ion activities are equal to 10^{-7} mol/l at 25°C. If an acid is added, hydrogen ion concentration will increase and hydroxyl ion concentration decrease; if an alkali is added, the reverse will occur. In either case, the ionic product will equal 10^{-14} mol/l at 25°C.

To avoid having to deal with such cumbersome numbers, Sorensen developed the concept of the pH scale during his investigations into methods of measuring hydrogen ion concentration. The term "pH" is a measure of the activity of hydrogen ions in a solution and is defined as the negative logarithm to the base 10 of the hydrogen ion concentration (or activity):

$$pH = -\log_{10}[H^+]$$

This gives us the well-known pH scale:

pH	0	1	2	3	4	5	6	7
$[H^+]$	1.0	0.1	0.01	10^{-3}	10^{-4}	10^{-5}	10^{-6}	10^{-7}
pH	8	9	10	11	12	13	14	
$[H^+]$	10^{-8}	10^{-9}	10^{-10}	10^{-11}	10^{-12}	10^{-13}	10^{-14}	

It can therefore be seen that the pH of pure water is 7, that acid solutions have a pH less than 7, and alkaline solutions a pH greater than 7. Some of the more common solutions, together with their approximate pH values, are given in Table 13.2.

The normal pH range is 0–14, but this does not mean that pH values outside these limits cannot exist. For instance, a 10N solution of a strong acid (i.e., fully dissociated) will have a pH of −1. Similarly the pH of strong alkaline solutions can be less than 10^{-14}. A 1N solution of an acid contains 1 g-equivalent of the acid per liter of solution, the equivalent being that amount of acid that contains 1 mol (1 Avogadro's number) of ionizable hydrogen atoms.

TABLE 13.2 pH Values of the More Common Solutions

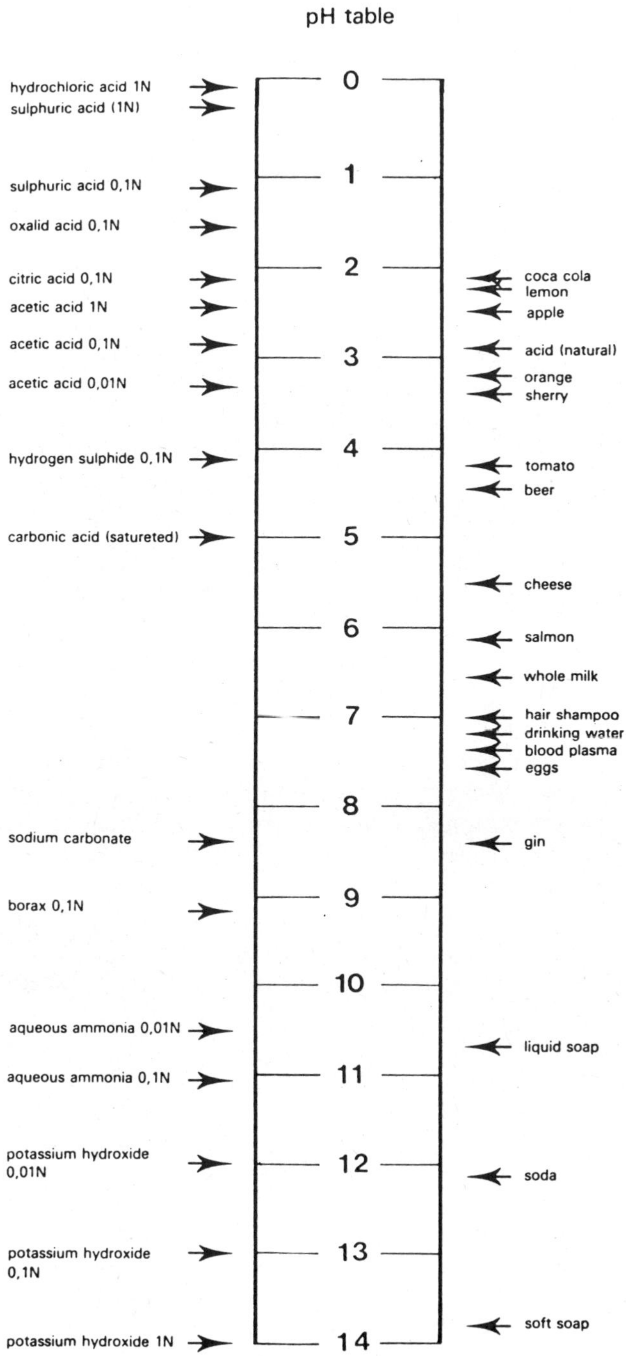

The term pH is therefore a mathematical symbol of convenience, but nevertheless one that is universally accepted by academia and industry as a means of expressing the acidity or alkalinity of aqueous solutions.

13.4.2 Buffer Solutions

As we have seen, pH is a measure of the dissociated hydrogen ions in a solution and not of the total acid or alkali present. A strong acid or base may be considered completely ionized in solution; therefore, equivalent quantities (in terms of the chemical reaction) of a strong acid and a strong base react in solution to give a neutral salt, and the solution has a pH of 7.

However, if a strong acid and a weak base, or a weak acid and a strong base, react together to form a salt, the resulting solution is not neutral. By mixing an acid and the salt of a weak acid, or a base and the salt of a weak base, the resultant solution, by virtue of its chemical constitution, will maintain a nearly constant pH, despite addition of relatively large quantities of acid or base. Such a solution is known as a *buffer solution*.

For instance, consider an aqueous solution containing ammonium hydroxide and ammonium nitrate. Ammonium ions (NH_4^+) are produced from both the salt and the base:

$$NH_4OH \rightleftharpoons NH_4^+ + OH^-$$

$$NH_4NO_3 \rightleftharpoons NH_4^+ + NO_3^-$$

This results in a shift in the ammonium hydroxide equilibrium to the left (termed the *common ion effect*), removing hydroxyl ions from the solution, and producing more of the reserve base, ammonium hydroxide. Addition of a little acid to the solution results in the removal of some hydroxyl ions by combination with hydrogen ions produced by dissociation of the acid. The reserve base equilibrium is shifted to the right, producing more hydroxyl ions to make up for the loss incurred by addition of the acid. The solution is therefore resistant to change in hydroxyl ion concentration. The more concentrated the buffer solution, the greater is its capacity for absorbing acid or base without pronounced change in pH value.

Buffer solutions thus provide standard reference solutions of accurately known pH that may be used for calibration and standardization of pH measurement equipment, both laboratory and on-line process types. The National Bureau of Standards (NBS) provides instructions for the preparation of buffer solutions reliable to ± 0.005 pH or better, which are usually retained as primary standards. Working standards are then prepared and compared with the NBS primary standards. Working standards usually have a reliability of ± 0.01 pH over a temperature range of 0–60°C. There are many buffer solutions available covering the whole pH range.

13.4.3 Methods of Measurement

Methods used for measurement of pH generally fall into two categories, colorimetric and potentiometric methods:

(a) Colorimetric Methods

Litmus Paper. When immersed in an acid solution the paper shows red; it changes to blue in an alkaline solution.

pH Paper (Universal Indicator Paper). This is paper impregnated with a suitable dye. After immersion in the test solution, the color of the paper is compared with a color disc, which shows the relevant pH value for varying shades of color.

Natural Indicators. Typical examples are red cabbage and mushrooms. Red cabbage is red in an acid medium and blue/violet in a natural medium; the color changes to green in a strongly alkaline medium. Mushrooms whiten when treated with vinegar (acid); in an alkaline medium they will turn brown.

(b) Potentiometric Measurement

This is the most accurate method for measurement of pH and is the universally accepted method for both laboratory and on-line process application.

The principle of potentiometric pH measurement can be explained by reference to Nernst's law. Nernst found that a potential difference occurs between a metal (electrode) and a solution containing ions of the same metal when the metal is immersed in the solution. The potential difference (E) caused by exchange of metal ions between the metal and the liquid was defined by Nernst according to the following equation:

$$E = E_0 + \frac{RT}{nF} \cdot \ln[M_n]$$

where R is the universal gas constant (8.314 J/mol · °K)
F is the Faraday constant (96,493 coulombs/mol)
n is the valency of the metal
Mn is the metal ion concentration
T is the absolute temperature (°K)
E_0 is the standard electrode potential

The standard electrode potential is the potential difference between the metal and a solution containing 1 mol of M per liter.

Hydrogen gas behaves in some ways like a metal; therefore, Nernst's law can also be applied to a "hydrogen electrode" immersed in a solution containing hydrogen ions. The ''normal hydrogen electrode'' can be made by coating a layer of platinum black on a platinum electrode, immersing it in a solution of unit hydrogen ion activity and passing a flow of hydrogen gas (99.8% purity) at one atmosphere partial pressure through the solution and over the electrode. The presence of platinum black results in the hydrogen gas being adsorbed on the electrode. For the hydrogen electrode, the Nernst equation may be rewritten as follows:

$$E = E_0 + \frac{RT}{F} \cdot \ln[H^+]$$

Using common logarithms, inserting numerical values for the constants, and assuming a temperature of 25°C, the Nernst equation for the hydrogen electrode becomes:

$$E = E_0 + 0.0591 \log[H^+]$$

and using the Sorensen expression for pH:

$$E = E_0 - 0.0591 \cdot \text{pH at } 25^\circ\text{C}.$$

Therefore, for any electrode system following the Nernst equation, one pH unit is equivalent to 59.1 mV at 25°C.

Since the early work of Arrhenius, Sorensen, Nernst, and others, developments in this field have led to the potential of an electrode being expressed in terms of ''activity'' rather than concentration. The theory of activity developed by Debye and Huckel is of particular importance in this context. However, this concept is rather complex mathematically and outside the scope of this book. For a detailed treatment of the subject, the reader is referred to any standard work on electrochemistry. The use of activity rather than concentration will result in a change in standard potential equivalent to approximately 0.04 pH units. Thus, activity may be considered as an ''effective'' concentration. Provided that the electrode system and the pH measuring device are both standardized with buffer solutions calibrated on the activity scale, the measured values will be activity values. The pH values measured in this way are often called ''operational'' pH values.

(c) *The Glass Electrode*

The glass electrode measurement system is now accepted as the only device accurate, durable, and universal enough for industrial measurement of pH. It dates back to the early 1900s, when it was found that certain types of glass gave a potential difference, the magnitude of which depended upon the acid value of the solution in which the glass was immersed. Later, it was established that this potential difference, within a fixed pH range, followed Nernst's law in the same manner as the "hydrogen electrode."

The main component of pH glass is silicon dioxide (SiO_2) with the addition of sodium oxide (Na_2O) and calcium oxide (CaO), commonly called soda-lime glass. The glass membrane consists of a matrix of silicon and oxygen atoms, where each silicon atom is surrounded by four oxygen atoms and each oxygen atom by two silicon atoms. The spaces within this matrix are taken up by cations, thus balancing the charges in the membrane. The strength of the cation bonding varies with the matrix and the type of glass. At high pH levels and high sodium ion concentrations, electrodes constructed of soda-lime glass exhibit the so-called "sodium ion error," with evidence of an attack on the silicon-oxygen lattice at the glass surface and a resulting decrease in the active life of the electrode. By replacing sodium by lithium and calcium by barium or strontium, pH glasses with considerably smaller sodium ion error can be produced.

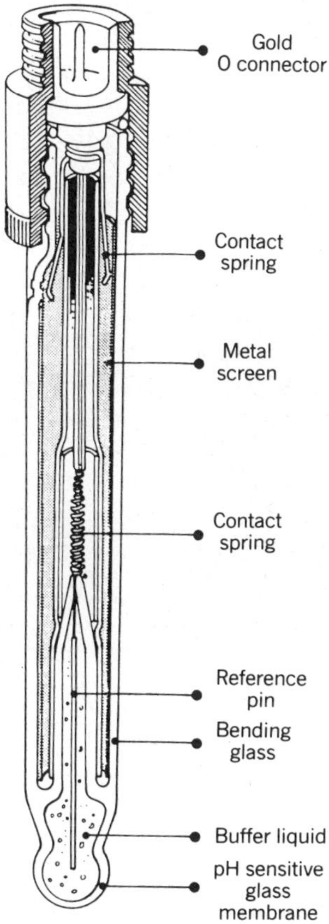

FIGURE 13.1 The glass electrode. (Courtesy of Yokagawa-Electrofact)

One common characteristic of all types of pH-sensitive glass is that water is adsorbed at the surface, giving rise to the following reaction:

$$Si—O—M + H_2O \rightarrow SiOH + MOH$$

where M represents a metal. This reaction results in the formation of a so-called "gel-layer" on the surface of the glass membrane; formation and maintenance of the gel-layer are essential to the functioning of the glass electrode. An equilibrium condition is established within one or two days where the resulting gel-layer has a thickness of about 10–40 nm, the actual value depending on the thickness and composition of the glass and the temperature of the solution in which the electrode is immersed. Generally, it is recommended that all new glass electrodes are "conditioned" in this way for at least 24 hr prior to use.

The glass electrode, a typical example of which is shown in Fig. 13.1, consists of a bulb-shaped membrane of pH-responsive glass "welded" to the stem of a non-pH-responsive, high-resistance glass tube. Variations in electrode response due to depth of immersion are eliminated, since the pH response is limited to the small area of the special glass membrane. Both surfaces (the inside and the outside) of the glass membrane are responsive to pH variations; therefore, the bulb is filled with a buffer solution of accurately known and stable composition. This forms an internal reference electrode, which is used to maintain a constant potential with respect to the inner surface of the glass membrane. To measure the potential difference between the outer surface of the glass membrane and the process liquid (which varies with the pH of the process liquid), it is necessary to use another electrode system, called the "reference electrode," to complete the circuit.

(d) The Reference Electrode

Measurement of the potential in the process liquid must be independent of changes in solution composition. If platinum or carbon electrodes are used, they will act as redox measuring electrodes and will be responsive to oxidizing or reducing agents in the process liquid and therefore will not give a potential relative only to pH. The reference electrode does this.

Nernst found that the combination of a metal, an insoluble salt of that metal, and a salt solution has a constant potential at a given temperature. By using such a reference electrode system, it is possible to measure the pH-dependent potential of the glass electrode very accurately.

An example is the silver/silver chloride electrode shown in Fig. 13.2. The metal element is silver coated with silver chloride and is immersed in a solution of potassium chloride. The potassium chloride is saturated with silver chloride to prevent the dissolution of the silver chloride coating. The saturated potassium chloride solution, often called the "salt bridge," is in contact with the process liquid via a liquid junction.

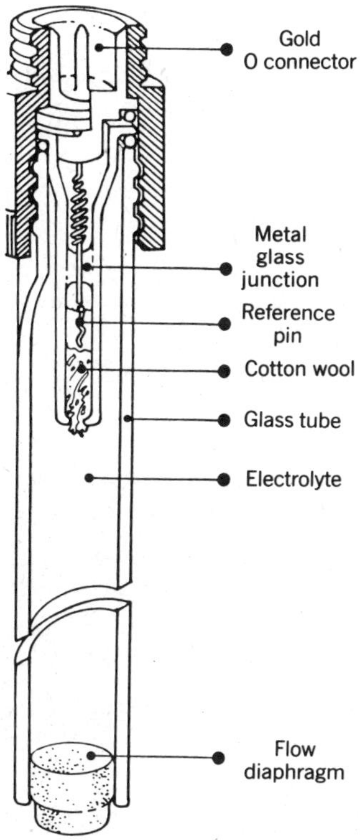

FIGURE 13.2 The reference electrode. (Courtesy of Yokagawa-Electrofact)

Since the concentration of all the components from the metal element to the KCl solution is fixed, the potential from the metal to the KCl is also fixed. The overall potential of the reference electrode is thus essentially constant at a given temperature, and virtually independent of process liquid composition changes. The potential of the silver/silver chloride electrode is 0.222 V at 25°C, referred to the normal hydrogen electrode.

Another example is the calomel electrode. This is probably the most widely used reference electrode in pH measurement, since it is easy to construct, is stable, and requires very little maintenance. The electrode consists of a platinum element in contact with mercury, mercurous chloride (calomel), and potassium chloride solution of known concentration. Normally saturated potassium chloride solution is used, under which conditions the electrode potential is 0.2444 V at 25°C, referred to the normal hydrogen electrode. For pH measurements at higher temperatures, an alternative reference electrode is the tamal electrode, which consists of thallium amalgam in contact with thallium chloride and 3.3 molal potassium chloride.

To provide electrical continuity, the reference electrode must have a *liquid junction* where the salt solution and the process liquid come into direct contact. For satisfactory operation, there must be a continuous, albeit very small, flow of salt solution across the junction. This flow prevents process liquid from entering the electrode and contaminating it.

One of the most difficult problems has been to design a reproducible liquid junction with a small enough leakage rate so that only infrequent replenishment of the salt solution is required. A large variety of different junctions are available today, including ceramic, glass sleeve capillary, PTFE, and wood plug types. Often the electrolyte is in the form of a gel, which flows at a very small rate into the process liquid (e.g., 0.01 ml/day). Reference electrodes are also available with a double junction. They are useful for particularly difficult applications, where contamination is a major problem.

Some manufacturers can also supply combination electrode systems, consisting of the glass electrode and reference electrode mounted together in one unit. Although this system is convenient and easy to service, it has the disadvantage that it is not possible to provide the wide selection of pH-sensitive glass types or design variations required for industrial applications.

(e) Temperature Compensation

Although pH is one of the more commonly measured industrial parameters, the temperature compensation associated with pH measuring instruments has been widely misunderstood. Measurement of pH is temperature dependent as a result of two different effects:

Solution Temperature Effects. The pH of pure water is 7.00 only at 25°C. As a result of changes in ionization with temperature, pure water has a pH of 7.47 at 0°C and 6.63 at 50°C. Further complications occur when dissolved ionic materials (e.g., hydrazine and ammonia) alter the ionization equilibrium and its temperature dependence. Note that this is a true pH change with temperature and is not compensated for by conventional pH instrumentation. Since it is a property of the sample itself (sample composition), it can vary from one sample to another.

Nernstian Effects. The slope of the Nernst equation, which describes the output of all ion-sensitive electrodes, is expressed in mV/pH, as shown in Fig. 13.3, and varies directly with absolute temperature, as evidenced by the term RT/F. This well-documented effect is handled by manual or automatic temperature compensation on virtually all process pH instruments. The various isothermal lines (plots of mV against pH at a given temperature) should intersect at one point (S) called the isothermal point. The intersection point depends on the pH of the buffer solution used in the glass electrode (usually 7.0). It is of the greatest importance that the isothermal lines intersect at one point. Only if this is achieved, normally requiring good electrode design and manufacture, will it be possible to make accurate temperature compensation within the pH amplifier.

The solution temperature effect just described is an actual change in hydrogen ion concentration (or activity). In many applications, compensa-

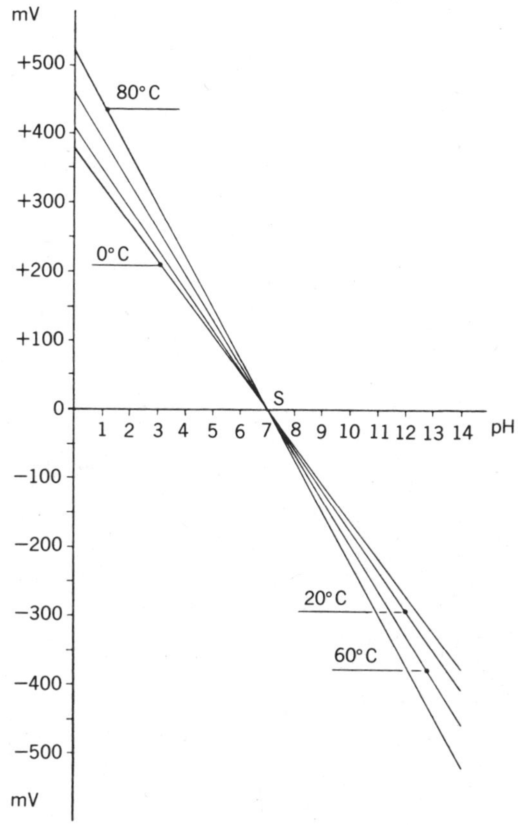

FIGURE 13.3 Temperature compensation (Nernst effect).

tion for solution temperature coefficient is desirable, particularly in the utilities industry (e.g., boiler plants). Often it is necessary to make this correlation to account for differences between process analyzer and laboratory readings. Various methods of solution temperature coefficient compensation, both manual and automatic, have been tried in industry. The advent of microprocessor-based pH instrumentation has brought increasing flexibility in handling pH and temperature signals and the ability to deal with both Nernstian and solution temperature coefficient compensation together in one instrument.

(f) Measuring Equipment Design

The measuring system using a glass electrode and a reference electrode is shown in Fig. 13.4, together with its equivalent circuit. The potentials which affect the final potential (E_f) between the two electrodes are as follows:

E_1—the potential difference between the outer surface of the glass membrane and the process liquid.

E_2—the potential difference between the electrolyte in the glass electrode and the inner surface of the glass membrane.

E_3—the potential difference between the electrode connector and the electrolyte in the glass electrode.

E_4—the potential difference between the electrolyte and the electrode connector in the reference electrode.

E_5—the potential difference between the electrolyte in the reference electrode and the process liquid, at their interface (across the liquid junction).

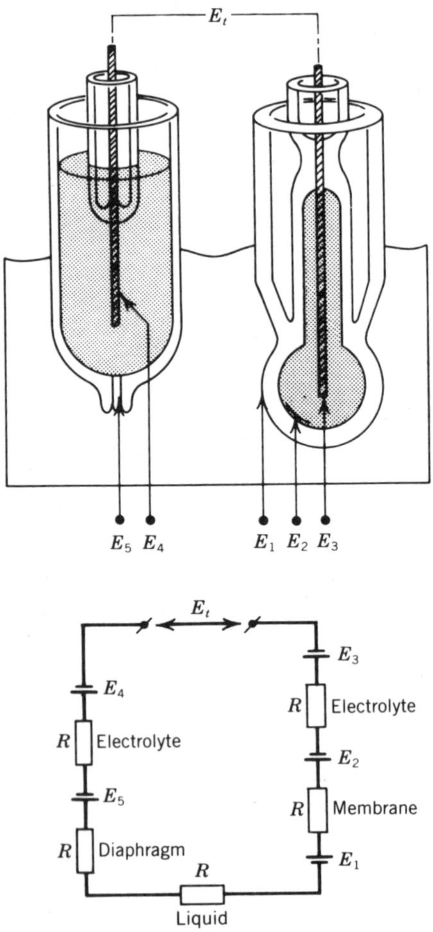

FIGURE 13.4 Typical pH measurement system.

It is the sum of these potential differences that is measured by the signal converter, although we are interested only in the potential difference between the pH glass membrane and the process liquid (E_1); the remaining potentials must be compensated for so that they do not affect the pH measurement.

If the reference systems in the glass and reference electrodes are identical and at the same temperature, E_3 is equal and opposite to E_4. With correct selection of the electrolyte used in the reference electrode and a good flow across the liquid junction, E_5 can be neglected, so that

$$E_t = E_1 + E_2$$

Since E_1 and E_2 in the measuring loop are of opposite polarities:

$$E_t = E_2 - E_1$$

or

$$E_t = 0.05916\ (\text{pH}_{inner} - \text{pH}_{outer})$$

The potential RT/F. pH_{inner} can be kept constant by filling the electrode with an electrolyte with good buffer properties. Thus, the measured potential E_f will be dependent only on the potential difference between the glass electrode and the process liquid.

These ideal conditions cannot always be achieved in practice and a small potential difference may exist when the glass electrode and the reference electrode are both immersed in a liquid with similar properties and pH value as the electrolyte. This potential difference is known as the *asymmetric potential* and may be caused by the liquid diffusion potential (E_5) not being zero, potentials $E_3 + E_4$ not being equal to zero, and by differences in texture between the inner and outer surfaces of the glass membrane. However, the errors resulting from asymmetric potential are compensated for during calibration using buffer solutions.

The pH-sensitive glass membrane has a high resistance (typically between 25 and 500 Mohm); therefore, the signal converter must have a high impedance, at least a factor of 1000 higher than the membrane resistance. Insulation and screening of cables and connections must also be of the highest integrity. The converter should be installed as near to the electrode system as possible and all connections kept dry to avoid any decrease in insulation resistance. Insulation of the reference electrode is less critical (resistance of 10 Mohm is usually satisfactory).

To achieve measurement accuracy of ± 0.02 pH, the limit of most secondary buffer solutions, the converter must be sensitive to ± 1.2 mV. Negligible current must be drawn during measurement, otherwise a voltage drop will occur across the pH electrode system, resulting in considerable error in measured pH. The rather extreme example that follows illustrates this problem. Consider a pH electrode of internal resistance 500 Mohm (R_e) and a true potential of 2.00 V. If the measuring circuit has a resistance of 10^9 ohms (R_m), the current flowing in the circuit is given by:

$$i = \frac{E}{(R_e + R_m)} = 2.00/(5 \times 10^8) + 10^9$$
$$= 1.33 \times 10^{-9} \text{ A}$$

Therefore, the voltage drop across the pH electrode is:

$$iR = (1.33 \times 10^{-9})(5 \times 10^8) = 0.665 \text{ V}$$

This voltage drop is in opposition to the cell potential; therefore, the measuring circuit will actually see only 2.00 − 0.665 = 1.335 V, resulting in an error of 33%! In general, it is necessary to reduce this error to 0.1% or less, requiring a measurement circuit resistance of at least 10^{12} ohms.

13.4.4 Redox Measurement

The measurement of pH and redox are discussed together since they both involve essentially the same type of measurement, namely that of the potential difference between a measuring electrode and a reference electrode, immersed in the solution under test. It should be noted that the terms redox and ORP (oxidation-reduction potential) are frequently used interchangeably.

Unlike pH and other ion-selective electrodes, which are highly specific in nature, redox electrodes are inert elements that measure the ratio of the activities of the oxidized and reduced forms of various ions in solution.

The term "oxidation" was originally defined as the reaction of a substance with oxygen. This has now been expanded to apply to any material that has *lost* one or more electrons in a chemical reaction. By definition, reduction is the opposite of oxidation, and thus applies to any material that has *gained* one or more electrons in a chemical reaction. Many elements can exist in two or more oxidation states when in solution; iron and chromium are typical examples. Iron can exist in solution in the divalent (Fe^{2+}) and trivalent (Fe^{3+}) state; chromium can exist in the divalent (Cr^{2+}), trivalent (Cr^{3+}), and hexavalent (Cr^{6+}) state.

(a) Oxidation-Reduction Potential

Let us consider the reduction of a metallic ion such as iron:

$$Fe^{3+} + e = Fe^{2+}$$

The electrons required for such a reduction may be supplied from an electrode consisting of a noble metal, such as gold or platinum, immersed in the solution under test, which is a mixture of the oxidized and unoxidized substances (e.g., ferrous and ferric chlorides). If an oxidation-reduction equilibrium exists in the solution, the metal electrode will assume a definite potential depending on the relative concentrations (activities) of the oxidized and reduced forms. The potential of the redox half-cell is given by the Nernst equation:

$$E = E_0 + \frac{RT}{nF} \cdot \ln \frac{a_{ox}}{a_{red}}$$

where E_0 is the standard electrode potential
R is the universal gas constant
T is the absolute temperature
n is the number of electrons taking part in the electrode reaction
F is the Faraday constant
a_{ox} and a_{red} are the activities of the oxidized and reduced forms respectively.

By substituting numerical values for constants, concentrations for activities, common logarithms for natural logarithms, and assuming a temperature of 25°C, the Nernst equation becomes:

$$E = E_0 + \frac{0.0591}{n} \log \frac{[\text{ox}]}{[\text{red}]}$$

The redox half-cell may be combined with a standard reference electrode, such as the silver/silver chloride, calomel, or tamal electrodes described earlier, and the oxidation-reduction potential may be measured accurately in the same way as pH.

Let us now consider the reduction of ferric iron to ferrous iron by potassium iodide:

$$2FeCl_3 + 2KI = 2KCl + 2FeCl_2 + I_2$$

Note that in this reaction there is a corresponding oxidation of iodide ion to iodine. The ionic equations for this reaction are:

$$2Fe^{3+} + 2e \rightarrow 2Fe^{2+}$$

$$2I^- - 2e \rightarrow I_2$$

This example demonstrates an important principle of redox reactions. When an excess of the oxidant is present (in this case $FeCl_3$), the potential of the platinum-reference electrode couple is wholly dependent on the activities of the reduced and oxidized forms of the oxidant, or:

$$E = E_0 + \frac{2.303\,RT}{nF} \cdot \log \frac{[Fe^{3+}]}{[Fe^{2+}]} - E_{ref}$$

where $n = +1$ and $FeCl_3$ is in excess. When potassium iodide, the reductant, is in excess, the potential is wholly dependent on the oxidized and reduced

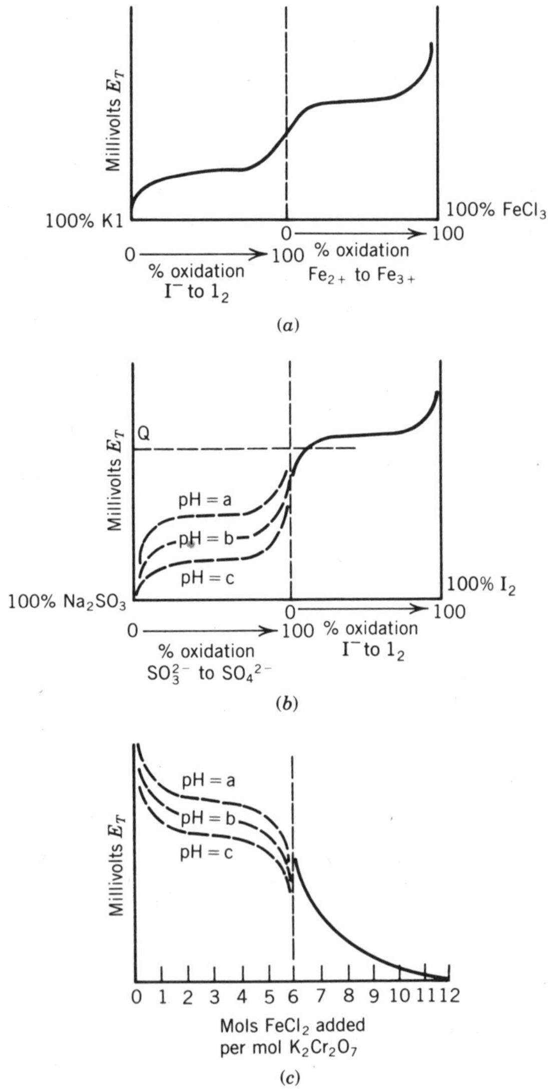

FIGURE 13.5 Potentiometric curve for REDOX reaction. (Courtesy of The Foxboro Company)

forms of the reductant, or:

$$E = E + \frac{2.303RT}{nF} \cdot \log \frac{[I_2]}{[I^-]^2} - E_{ref}$$

where $n = +2$ and KI is in excess. Since all solids and gases, such as I_2, are automatically defined as having unit activity, this equation becomes:

$$E = E_0 + \frac{2.303RT}{nF} \cdot \log \frac{1}{[I^-]^2} - E_{ref}$$

The potentiometric curve for the complete reaction is shown in Fig. 13.5. The curve is in reality two curves, but they are joined to make a smooth curve from 100% KI to 100% $FeCl_3$.

(b) Effect of pH

One common industrial application of redox measurement and control is the reduction of toxic hexavalent chromium (in the form of the dichromate radical $Cr_2O_7^{2-}$) to the non-toxic trivalent ion Cr^{3+}. This is a good example of the dependency on pH shown by many REDOX reactions. The ionic equation of the reaction is

$$Cr_2O_7^{2-} + 14H^+ + 6e = 2Cr^{3+} + 7H_2O$$

and the Nernst equation for this reaction is

$$E = E_0 + \frac{2.303RT}{nF} \cdot \log \frac{[Cr_2O_7^{2-}][H^+]^{14}}{[Cr^{3+}]^2}$$

where $n = +6$ and the term $[H^+]$ shows that the potential is dependent upon the solution pH. As long as the pH of the solution is constant, the pH term can be included with E_0, so that the variable potential is truly a function of the redox state of the chromium ions. In some cases, the pH effect is present in both the oxidation and the reduction reactions; in other cases, pH variations affect the potential of only one reaction.

(c) Effect of Temperature

As with pH measurement, temperature has two independent effects on the redox measurement system, the solution effect and the Nernstian effect. These effects were described earlier and act essentially the same as for the pH electrode.

13.4.5 On-Line Process pH and Redox Analyzers

While the basic concepts of continuous pH measurement are fairly straightforward, the practical application of the technique to on-line industrial measurement has not always been simple. Innovative technology has in fact been largely responsible for the reliability and convenience of industrial pH measurement today. Some of the more common problems with industrial installations are discussed next.

(a) Glass Electrode Fouling

Fouling of the glass membrane is the most common cause of error in any pH measuring system. As the measured potential is generated on the surface of the pH sensitive glass membrane, it is of the greatest importance that this membrane surface remains clean at all times. In certain industrial applications, fouling cannot be entirely prevented without periodic cleaning. Various cleaning methods may be used:

- Mechanical brushing. A brush activated by an electrical or pneumatic mechanism is passed periodically across the membrane.
- Ultrasonic cleaning. With this method, the liquid surrounding the electrodes is vibrated, thus preventing and/or removing any deposit formed on the electrodes. Optimal cleaning depends upon the vibration frequency and the velocity of the process liquid past the electrodes.
- Chemical cleaning. It is often possible to keep the surface of the glass electrode clean by periodically spraying it with a cleanser. The cleanser used, its concentration, and the cleaning frequency will depend upon the particular application.
- Flow impingement. A reduction in the rate of deposition on the electrodes can be achieved by increasing the flowrate of process liquid past the electrodes. Certain pH cells are designed to provide such an increased flowrate. This method is particularly suitable when measuring pH in liquids containing solids that settle out quickly.
- Sterilization. It is often necessary to sterilize the electrodes when measuring in processes where the formation of bacteria is critical. Chemical or steam sterilization may be applied.
- Manual cleaning. If it is necessary to hand clean the electrode periodically, suitable cleansers such as dilute hydrochloric acid may be used, followed by thorough rinsing with clean water. This treatment is often sufficient for many applications.

It is important to select the correct electrode for any particular application, since some electrodes are not suitable for cleaning by these methods. The maximum operating pressure of the reference electrode is also important.

Minute cracks in the glass membrane are not always visible to the naked eye. They may be caused by frequent shocks to the electrode and will result in significant measurement error. Usually this will show as a measurement of pH 7 (i.e., an input of 0 mV).

(b) Reference Electrode Fouling

It is equally important to keep the junction of the reference electrode properly clean, since fouling of the junction will interrupt the contact between the reference system and the process liquid. This will result in a reduction in the speed of response and may result in unstable measurement or complete failure.

Clogging or fouling of the reference electrode can be prevented or reduced by:

- Correct choice of junction type for a particular application.
- Increasing the flow of electrolyte.
- Ultrasonic cleaning.

(c) Poisoning of the Reference Electrode

Poisoning of the reference electrode will occur when the flow of electrolyte is insufficient to prevent process liquid from entering the electrode. Any change in composition of the electrolyte in the reference chamber will cause a measurement error. This can occur as a result of higher than expected process pressure or short-term, high-pressure fluctuations. Often the best solution to this problem is to use a double-junction reference electrode. If the process liquid temperature is significantly lower than the inner saturated KCl solution in the reference electrode, considerable electrolyte crystallization may occur. This will restrict electrolyte flow and allow process liquid penetration.

(d) Two-Wire Transmitters

One fairly recent innovation in industrial pH measurement is the two-wire pH transmitter. This device was developed to meet the needs of the chemical process industries (CPI) for an inexpensive transmitter that will produce a current output signal proportional to pH or redox, with explosion-proof features and intrinsically safe operation. This transmitter will look exactly like the conventional flow or pressure transmitters used in industrial instrumentation systems.

The following are examples of the very wide range of industrial pH measuring systems commercially available today. Manufacturers are usually able to provide a wide variety of pH/redox sensors and mounting assemblies, two-wire transmitters, monitors, controllers, and microprocessor-based instruments. For a more complete listing, see Table 13.3 on page 541–543.

(a) Beckman Industrial Inc. (United States)

The 19000 Series Reference Electrodes include the LAZARAN range, which are manufactured from a polymer composite material, are virtually unbreakable, and are constructed without a flowing electrolyte junction. They do not require replenishment or external pressurization and are highly resistant to coating or chemical attack. Maximum operating conditions are 150 psig and 100°C.

Flowing junction reference electrodes are also available, using three types of junctions. The *fiber junction* is generally the preferred choice, since it is inert. However, it is not suitable for pressurized applications. The *palladium junction* has a very small semi-annular opening between the glass body and a fine palladium wire, which provides a slow but predictable electrolyte flow. The *sleeve junction* provides leakage of electrolyte between a ground glass taper on the electrode body and a matched ground glass sleeve. It provides a very fast flowing junction and is used in highly viscous solutions, heavy slurries, or solutions with components that might form a precipitate with the electrolyte solution. Maximum operating conditions for these systems are 23–212°F and 0–150 psig.

The 19000 Series Glass pH and Redox Electrodes include low-resistance types for general use, high-temperature types with low sodium-ion error, and metallic electrodes in gold or platinum for redox measurement, all designed for use in Beckman flow chambers. Thermo-compensators are also available for automatic temperature compensation. Both pH and redox combination electrodes are offered, most of which include a thermocompensator.

Electrode mounting assemblies include flow-through, submersion, and inline versions. These may be constructed of 316 stainless steel or Ryton R-4, an industrial thermoplastic resin. Stainless steel provides extra strength for high pressure and severe handling services, whereas Ryton R-4 has high thermal stability and excellent chemical resistance. Typical examples are shown in Fig. 13.6. In all configurations, a dome snaps over the electrode gland to protect the electrode connections and usually contains a sealed-in amplifier/transmitter.

Both flow-through and submersion assemblies in 316 stainless steel can be fitted with an ultrasonic electrode cleaning system. An ultrasonic transducer is mounted directly to the electrode assembly and is connected to an electronic exciter unit, housed in a NEMA 4 waterproof enclosure. The exciter unit provides an nominal output of 40 kHz to drive the transducer. The energy level of excitation is adjustable from the exciter unit, which may also be equipped with a 15-min timer that is adjustable from 0–100% of the basic period. Power consumption is 50 W nominal.

The Models 942 pH/REDOX Monitor and 943 pH/redox Monitor/Controller, an example of which is shown in Fig. 13.7, are microprocessor-based units designed to simplify standardization procedure, provide accurate temperature compensation, and transmit digitized information up to 3000 ft. Optical coupling ensures complete isolation between electrodes, preamplifier, and the monitor. Power for the preamplifier is transmitted via the same two wires (a shielded, twisted pair) that carry digitized data from the preamplifier. This virtually eliminates noise errors and allows greater separation

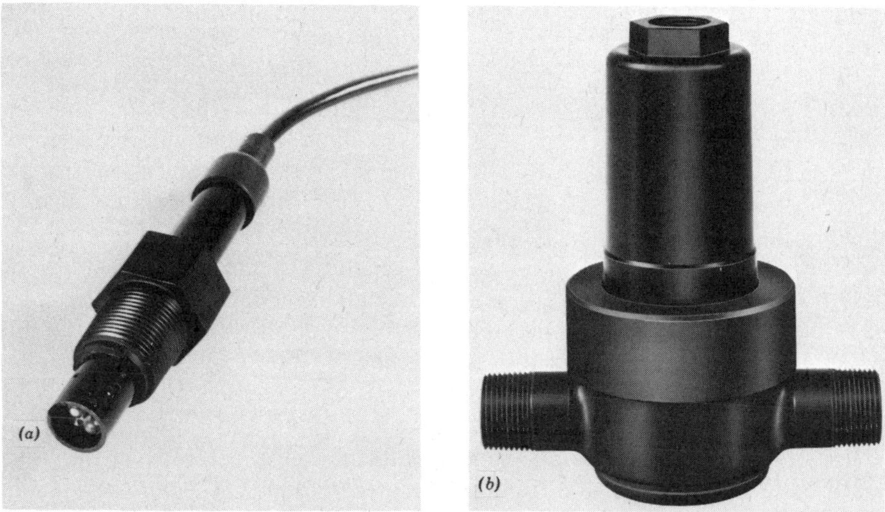

FIGURE 13.6 Beckman electrode mounting assemblies. (Courtesy of Beckman Industrial Inc.)

between preamplifier and monitor. A lightning protection module is available for applications where the remote preamplifier is located in a lightning-prone area. As an option, the preamplifier can be mounted integral with the monitor. A pH simulation function is provided, allowing verification of any pH-dependent function such as alarms, recorder or computer inputs, and recorder calibration. An additional feature is the ability to display temperature

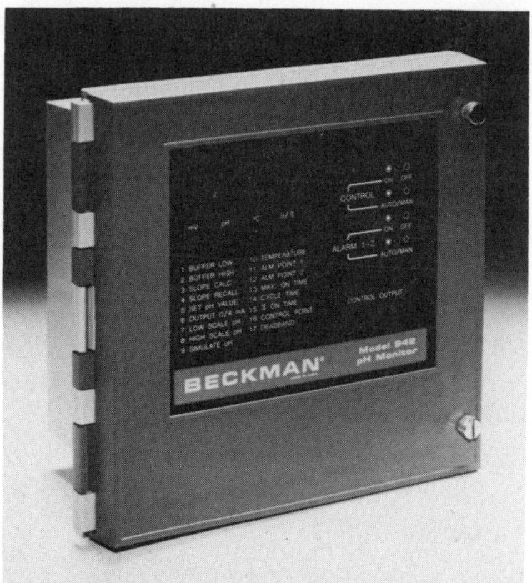

FIGURE 13.7 Model 942 pH/REDOX monitor. (Courtesy of Beckman Industrial Inc.)

at the electrodes, which can be very useful during calibration or evaluating process conditions.

Electronic circuitry of the pH or redox monitor is contained on three circuit boards—the power supply, controller and display/keyboard boards. A six-digit readout is provided, four digits for direct readout of the measured variable (0.00–14.00 pH or −1400 to +1400 mV redox) in the "monitoring" mode. All six digits are used when in the "input" mode. An isolated current (0–20 or 4–20 mA dc) output is provided for remote recording, programmable from the keyboard for any span of 2 pH units or greater, or 200 mV or greater in the case of the redox monitor. Dual alarm capability is available, with setpoints programmable from the keyboard and adjustable over the full range. The keyboard is accessible by opening the front cover of the unit, on which are listed the 17 addressable functions of the keyboard. All mode and status LEDs and the digital display are visible through the unit front cover.

For control applications, all control information is entered via the keyboard. The five addressable functions are Maximum On Time (adjustable from 0–1999 sec), Cycle Time (adjustable from 3–1000 sec), Percent On Time for the control output (adjustable 0–100% of cycle time) Control Point (adjustable 0.00–14.00 pH), and Deadband. A control auto/manual key and status light and a control output status light are provided.

(b) Foxboro Analytical (United States)

The Series 222 pH and ORP Sensors are available in Ryton or CPVC body material, with interchangeable plug-in sensing electrodes. Two types of pH electrode are offered, the standard glass electrode and an antimony electrode suitable for abrasive solutions or those containing hydrofluoric acid or other chemicals that attack glass. For ORP measurement, platinum and gold electrodes are offered. The gold electrode is usually specified for cyanide reduction systems. The electrode systems are fitted with an integral preamplifier and temperature compensator (thermistor or platinum RTD).

The standard reference electrode offered is a non-flowing silver/silver chloride half-cell immersed in saturated KCl, with a ceramic liquid junction. Body material may be Ryton or PTFE. The filling solution and ceramic junction are readily field-replaceable.

The above sensing electrodes may be mounted in various configurations, examples of which are shown in Fig. 13.8. An in-line ball valve assembly allows the sensor assembly to be inserted or removed from a process line or tank without the need to depressurize the system; a purgeable version is also available. Other alternatives include flow chambers in 316 stainless steel and PVC, with a twist-lock connection for the sensing electrodes, bushings for direct in-line mounting, and fabricated tees.

Two types of electrode cleaning systems are available. The Series ECS-1 uses a rotating wiper and may be mounted to operate as either a submersion or a flow-through unit. Two types of wiper are offered, nylon bristles and

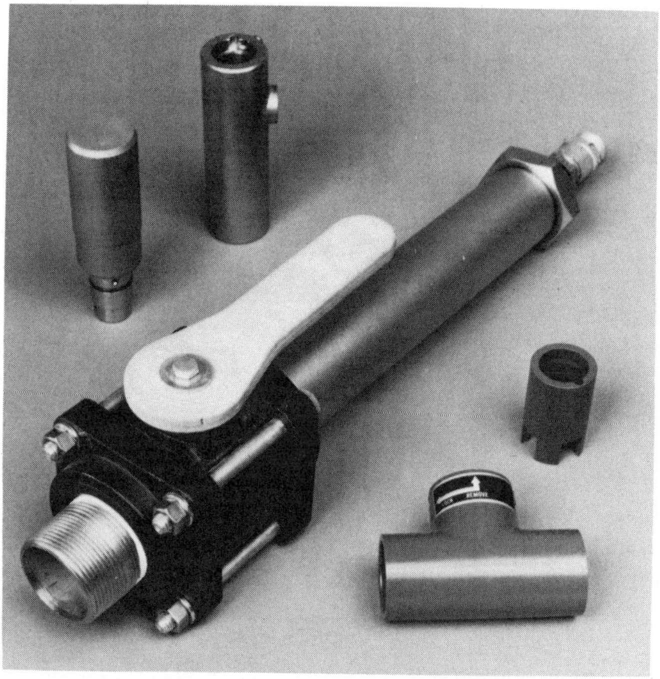

FIGURE 13.8 Foxboro electrode mounting assemblies. (Courtesy of The Foxboro Company.)

flexible ceramic fabric brushes, depending on the electrode used and the nature of the fouling medium. These are suitable for use with any of the 222-style electrodes. A hard spring-loaded ceramic block is available for use with antimony electrodes. The Ultrasonic Electrode Cleaner consists of two components, the ultrasonic transducer mounted as an integral part of the sensing electrode system, and the oscillator power supply unit, which is suitable for NEC Class 1, Group D, Division 2 hazardous locations and should be mounted within 25 ft of the transducer.

The Series 2220 pH/ORP Monitors are designed for use with the Series 222 sensing systems and cover several spans within the range 0–14 pH or 0–2000 mV dc ORP as specified. The monitors are solid-state, modular units suitable for panel or surface mounting in non-hazardous locations. Sensor and monitor may be separated by up to 1000 ft. Standard output is 0–10 V dc; optional current outputs (4–20 or 10–50 mA dc) are also available.

The Series E91 and E99 Electronic Transmitters are designed to receive a dc millivolt signal from pH and ORP sensors and convert it to a signal suitable for transmission to remote devices such as recorders, controllers, and so on. The high-impedance input module may be located either integral with or remote from the transmitter. Standard current outputs are 4–20 and 10–50 mA dc and are transformer-isolated. Certification for these transmitters has been obtained from CSA, FM, and PTB.

The Series 871 pH Sensors and Accessories are basically the same as the Series 222 described earlier, but designed specifically for use with the Model 870 pH Transmitters. These are two-wire transmitters, suitable for surface or pipe mounting, with integral dc milliammeter and 4–20 mA dc output. The transmitter has RFI protection and is certified as intrinsically safe for NEC Class 1, Groups A, B, C, and D, Division 1 and Class 2, Groups E, F, and G, Division 1 hazardous locations, when used with Foxboro Spec 200 and UIO I.S. equipment. A wide variety of different standard spans for pH and ORP are available, as well as custom spans. Range limits are −2 to +16 pH and −1400 to +1400 mV ORP. Repeatability is quoted as +−0.1% of span.

(c) Great Lakes Instruments Inc. (United States)

Whereas the conventional pH systems use a pH-sensitive glass electrode and a reference electrode (absolute measurement), the GLI pH systems operate on a patented differential principle, shown schematically in Fig. 13.9. Two glass electrodes are used to make the measurement differentially with respect to a third metal electrode. The standard electrode consists of a

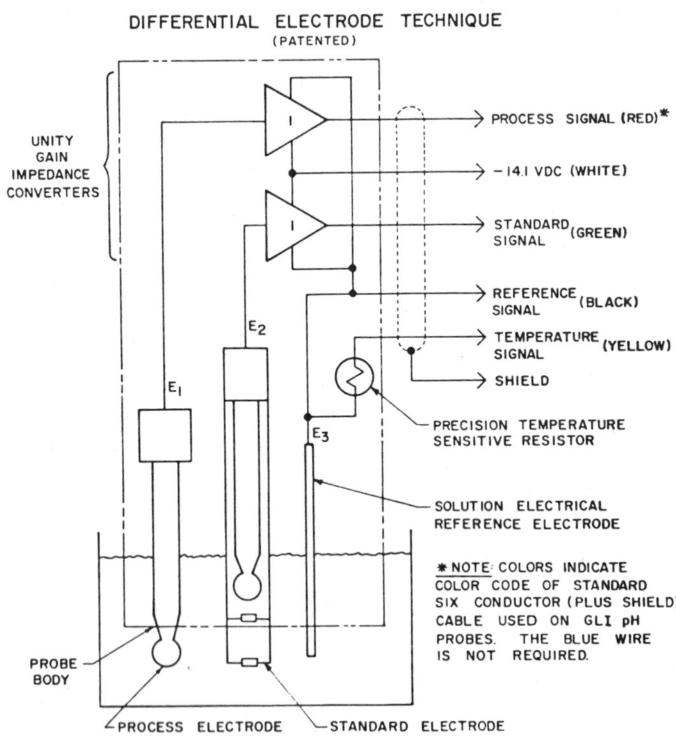

FIGURE 13.9 Great Lakes Instruments Inc. pH measurement systems: operating principle. (Courtesy of Great Lakes Instruments Inc.)

glass electrode contained in an inner chamber filled with a pH 7.0 buffer solution. The inner chamber makes contact with a "bridge" chamber through a ceramic plug and this chamber, which contains a protective intermediate solution, makes contact with the process liquid through a second ceramic plug.

This standard electrode maintains a constant pH level regardless of substantial dilution and, because of the protective intermediate solution, works under fouling conditions. This electrode requires no solution replenishment and is reputed to have a life of 2–5 years in harsh applications. The two glass electrodes are compared differentially with the third (metal electrode) and a unity-gain amplifier is associated with each electrode as shown. The circuitry of the monitor subtracts the outputs of the two unity-gain amplifiers; in so doing, it eliminates the potential of the third electrode, giving a measurement of process solution pH without the use of a conventional reference electrode. A precision temperature-sensitive resistor is incorporated into the measurement system for temperature compensation purposes.

The Model P60 Probe Assembly may be used in conjunction with the Models 60 and 70 pH transmitters; with the addition of an interface module it can be used with other manufacturers' analyzers. The assembly is completely encapsulated; wetted materials are Epoxy or Derakane (a vinyl ester). An antimony pH electrode is also available for specific applications. Measurement range is 0–14 pH. The Model R60 ORP Probe uses a platinum or gold electrode. Also available for food processing and dairy applications is the Model P60S Sanitary pH Probe. In the Model 68R pH Probes, the wetted parts of the probe body are made of Ryton. A selection of pH probes is shown in Fig. 13.10a.

A variety of mounting hardware is available, covering flow-through, submersion, and direct insertion types. Construction materials are PVC, CPVC,

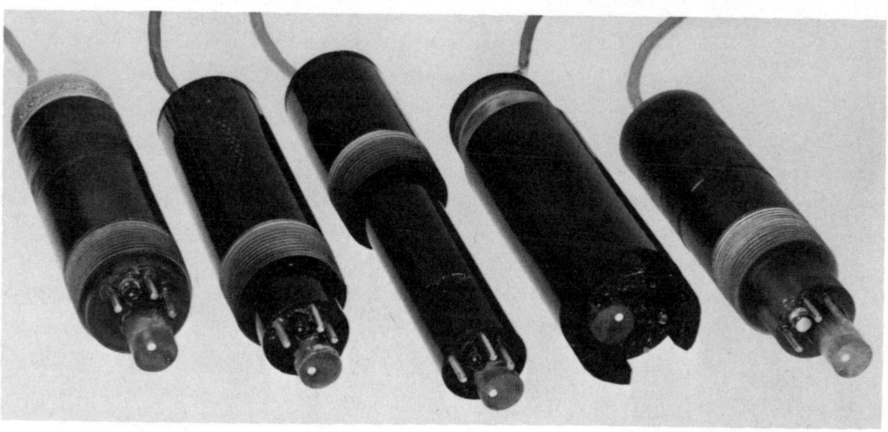

FIGURE 13.10 GLI pH measurement systems. (*a*) Typical pH probes. (Courtesy of Great Lakes Instruments Inc.)

and stainless steel. Mounting hardware assemblies include 50 ft of interconnecting cable and a junction box with terminal strip. The latter is suitable for pipe or surface mounting. The Model 223 Ultrasonic Cleaners are available for applications where sensor coating may be a problem and consist of an electronic generator unit mounted in a NEMA 4 fiberglass enclosure, a transducer, and mounting hardware. The transducer may be fitted to flow-through and submersion-type electrode assemblies.

The Model 60102 Steam Sterilizable pH Electrode Assembly uses a conventional electrode system with a large reservoir. The electrode housing has glass ports to allow visual inspection of the reservoir and may be pressurized to 30 psig. The assembly is constructed of stainless steel and interfaces to GLI transmitters via a preamplifier and 6 ft of cable.

The Model 95 Digital pH Controller is microprocessor based and can take up to four inputs—the pH sensor input, a temperature input, a 4–20 mA dc input, and a contact closure input. Up to six outputs are available, two voltage and four current. The unit has a 16-button keyboard and digital indication and is compatible with all GLI sensor probes. The controller has up to three modes and remote setpoint capability. One or two controller outputs (4–20 mA dc) are available, the latter having an adjustable dead

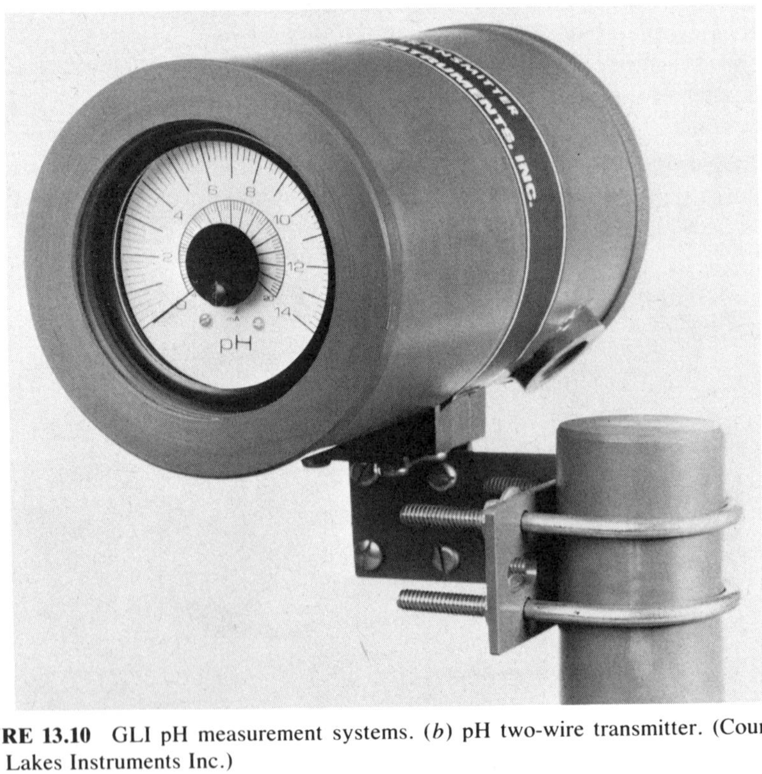

FIGURE 13.10 GLI pH measurement systems. (*b*) pH two-wire transmitter. (Courtesy of Great Lakes Instruments Inc.)

zone. A pulsed output is available for control of variable-speed motors. Repeatability is quoted as 0.01 pH.

The Model 90 Analyzer/Controller is an analog unit suitable for panel, surface, or pipe mounting. Standard outputs are 0–5 V dc and 0–1 mA dc, with options of 4–20, 0–20, or 10–50 mA dc. Analog or digital indication is available. The controller is two-mode with one or two outputs.

The Model 60 pH/ORP Transmitter is FM-approved for hazardous locations using a zener barrier and is suitable for panel or surface mounting. Modular, all solid-state circuitry is used. A two-mode controller is available, together with dual alarm capability. Standard pH ranges are 2–12, 0–10, 0–4, 4–10, and 4–14 pH; for ORP measurement, ranges are 0–1000 and −1000 to +1000 mV. Voltage or current outputs are available. The Model 60 is also available in a dual-channel version, with common indicating meter and channel switch.

The Model 70 pH/ORP Analyzers are available in general purpose, Division 2 or Division 1 versions, covering the 0–14 pH and −2000 to +2000 mV ranges and suitable for panel or surface mounting. A variety of controller options are offered. Voltage or current outputs are available.

The Model 690 Two-Wire pH/ORP Transmitter, shown in Fig. 13.10(b) is mounted in a cylindrical PVC enclosure and is of intrinsically safe design (Division 2 or 1) when used with a suitable zener barrier. The unit has a dual-scale meter, integral calibration controls, and may be separated from the sensor by to 3000 ft. Standard ranges are 0–14, 2–12, and 4–10 pH, and 0–500, 0–1000, and −500 to +500 mV ORP.

(c) Leeds and Northrup Instruments (United States)

L&N manufactures a wide range of pH and redox measurement systems for industrial applications. Glass and antimony electrodes are offered for pH measurement and redox electrodes are available in platinum, gold, and silver. The reference electrolyte is a non-flowing KCl gel or slurry. A replaceable or non-replaceable porous ceramic plug forms the liquid junction. Special double junction and wood junction electrodes are also available. The Meredian Electrodes are combination measurement/reference electrodes, designed for pH and redox measurement in a variety of applications including electroplating, waste treatment, fermentation, and electrolysis processes and for cooling tower and high-purity water applications.

A wide variety of mounting hardware is available. The Model 7773 Multipurpose Electrode/Preamplifier Mountings (see Fig. 13.11) are available in seven configurations covering flow-through and immersion types. Construction materials are stainless steel, polypropylene, or Ryton. A lightning protection option is also available. The Model 7779 Electrode and Transmitter Flow Mountings are of the self-cleaning, straight-through flow pattern, suitable for oily water solutions. The preamplifier, electrodes, temperature compensator, and terminal board are housed in a fiberglass NEMA 4 enclosure, with the all-plastic flow-through chamber extending below the enclosure to

FIGURE 13.11 Leeds & Northrup electrode mountings. (*a*) Model 7773 multi-purpose electrode/preamplifier mounting; (*b*) typical insertion type mounting; (*c*) insertion type mounting with valve. (Courtesy of Leeds & Northrup Company)

isolate the sample and moisture from the electrical connections and components.

The Model 7758 Pipeline Electrode Mountings shown in Fig. 13.12 are made of stainless steel or polypropylene, with up to three glands for mounting the measurement and reference electrodes (or Meredian electrodes) and the temperature compensator, and with the preamplifier provided in a separate enclosure. The Model 7776 Immersion Mountings are for Meredian pH or redox electrodes and are available in stainless steel or PVDC. Direct insertion mountings are also available for Meredian electrodes, with or without a valve assembly for removal (Models 7775 and 7774 respectively).

The Model 7073-33 REDOX Analyzer can be used with all L&N electrode mountings and consists of a measuring circuit, a voltage output circuit, and a 4-in. indicating meter mounted in a NEMA 4X case suitable for NEC Class 1, Groups C and D, Division 2 hazardous locations (using sealed relays). Standard measurement ranges are −500 to +500 and 0–1000 mV. Standard output is 0–1 V dc; isolated current outputs and alarms are obtained by the use of a separate module.

The Model 7073 Multichannel pH/REDOX Analyzer can accept from two to six independent measurement channels. The unit is suitable for NEC Class 1, Group C and D, Division 2 hazardous locations and mounted in a NEMA 4X plastic case. Various ranges are offered, with dual ranges available on pre-selected channels. Standard output is 0–10 V dc; isolated current outputs and alarms are obtained by the use of a separate module.

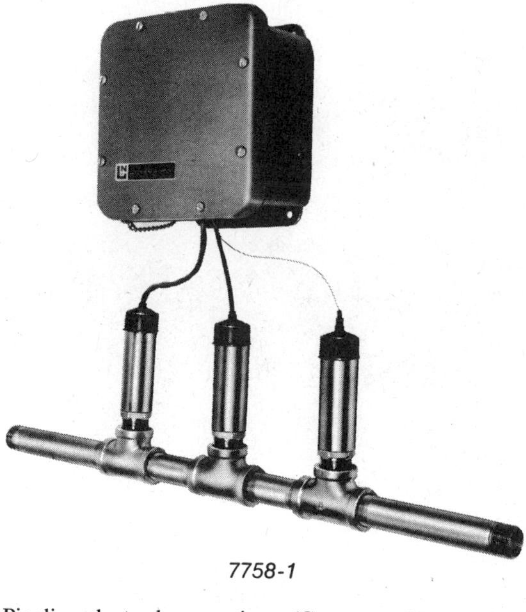

FIGURE 13.12 Pipeline electrode mountings. (Courtesy of Leeds & Northrup Company)

The Models 7075 and 7076 Analyzers are inexpensive, simplified instruments for pH and redox measurement respectively, and are suitable for panel, pipe, or surface mounting in NEC Class 1, Groups B, C, and D, Division 2 hazardous locations (using sealed relays). Analog or digital indication is available and high/low alarm set point knobs and lights are provided. Standard output is 0–1 or 0–10 V dc.

The Model 7079 Two-Wire pH/Redox transmitter provides an isolated 4–20 mA dc output and local digital indication. The unit is FM-approved intrinsically safe when used with zener barriers, and explosion-proof to NEC Class 1, Groups A, B, C, and D, Division 1. The cylindrical case is available in plastic or aluminum to NEMA 4X. Repeatability is quoted as 0.01 pH or 1 mV.

The Model 7083 pH Analyzer is a microprocessor-based instrument compatible with all L&N electrode mountings. The unit has a 24-button keyboard and digital (LCD) display. Solution temperature compensation is provided in addition to the conventional temperature compensation system. The unit has internal diagnostic capability as well as a special Electrolert detection system for electrode failure warning. Isolated voltage or current outputs are available.

The Model 7084 pH Analyzer/Controller shown in Fig. 13.13 is another microprocessor-based instrument capable of providing a wide variety of analysis and control functions. It is suitable for NEC Class 1, Groups B, C, and D, Division 2 hazardous locations and mounted in a NEMA 4X case.

FIGURE 13.13 Model 7084 pH analyzer/controller. (Courtesy of Leeds & Northrup Company)

The unit has non-volatile memory and extensive internal diagnostic programs. Up to three input signals can be accepted—pH/redox, temperature, and flowrate. Internal linearization is provided for the flow signal. The controller has field-selectable configurations of feedback, feedforward, or adaptive control. Controller output options are available are current adjusting, duration adjusting for solenoid valves, and pulse frequency for metering pumps.

A special feature allows user characterization of pH to match the process titration curve for more accurate proportioning control. Output range limits are digitally set anywhere within the range 0–14 pH and with 0.01 pH resolution. For redox measurement, range limits are −1350 to +1350 mV, with 1 mV resolution. The unit has a 24-button keyboard and two digital (LCD) displays, one for measurement data, one for control data. Two triacs and three reed relays are available for alarm and control functions. Isolated voltage and current outputs are available for remote recording. The Electrolert electrode failure detection system and solution temperature compensation are also provided.

(e) Rosemount-Uniloc (United States)

The Uniloc range of pH/ORP measurement systems include conventional and combination electrodes, flow-through, insertion, and submersion-type mounting assemblies, and a variety of transmitters, monitors, and controllers.

The Model 318 Combination pH/ORP Sensor has a rechargeable reference electrode with wood plug liquid junction and integral preamplifier. Body material is Kynar.

The Model 319 is a retractable pH sensor with ball valve and junction box containing the preamplifier. Various plug-in combination electrodes are available in general-purpose glass and high-pH glass. A large gel-filled, non-rechargeable reservoir is provided for reference electrolyte. Another version is made specifically for viscous, fibrous streams encountered in paper stock applications, for example.

The Model 320A/330A Flow-Through pH/ORP Sensors are available with the flow chamber of PVC, Kynar, Teflon, or stainless steel, and containing the measurement and reference electrodes and temperature compensator, together with a preamplifier junction box. The chamber cover contains the reference electrode, which uses saturated KCl gel and a large area liquid junction. A double junction reference electrode is also offered. The complete system is mounted on a plate suitable for surface mounting as shown in Fig. 13.14. The system is also available as a Division 2 explosion-proof version. Teflon balls may be used as a flow-powered cleaning mechanism (max. flow 4 gal/min) or an ultrasonic cleaner may be fitted.

In the Model 322/332 pH/ORP Submersion Sensors, the electrodes are housed in a Teflon body, with the preamplifier mounted in a separate NEMA

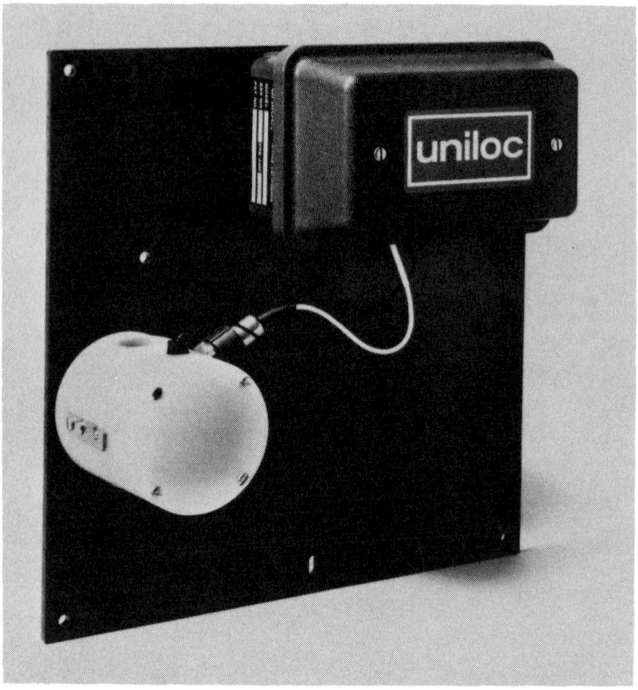

FIGURE 13.14 Model 320A/330A flow-through system. (Courtesy of Rosemount/Uniloc Division)

4X enclosure. The reference electrode has a refillable gel electrolyte chamber. A Division 1 version is also available. The Model 323/333 pH/ORP Insertion Type Sensors have an integral preamplifier, otherwise design is similar to the Model 322/332.

The Model 324/334 Flow-Through pH/ORP Sensors are designed for viscous liquid and slurry services and use a double junction reference electrode as standard. The flow chamber is constructed of Kynar or ABS plastic. The preamplifier is mounted in a separate NEMA 4X junction box. The complete assembly is installed on a plate suitable for surface mounting.

The Model 1003 pH Transmitter shown in Fig. 13.15 is housed in a corrosion-resistant, weatherproof, general-purpose enclosure and can be supplied with analog or digital display. Instrument range may be from 0.00 to 14.00 pH, with a minimum span of 2 pH. The transmitter may be located as far as 3 miles from the sensor. A variety of optically isolated current outputs are offered. Proportional plus reset control is available as an option and provides three field-selectable current output signals for two-mode control. Optional high-low alarms can also be provided. Repeatability is quoted as ±0.02 pH.

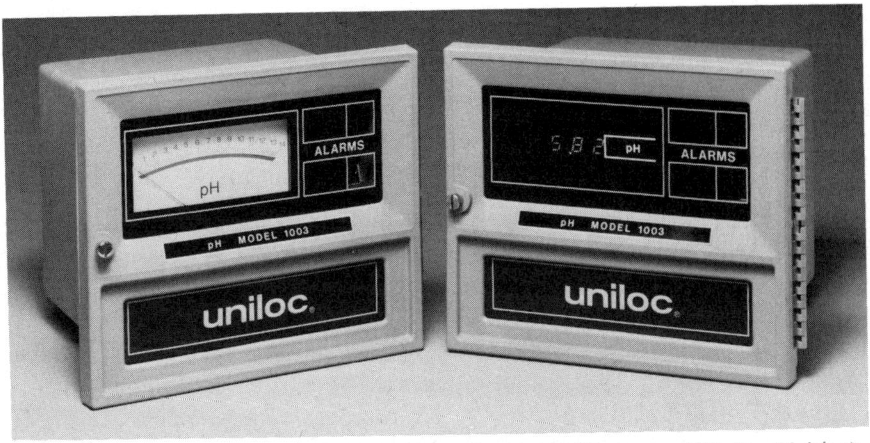

FIGURE 13.15 Model 1003 pH transmitter. (Courtesy of Rosemount/Uniloc Division)

The Model 1003-14 High Purity pH System uses an electronically modified Model 1003 transmitter and a Model 320A sensor designed specifically for pH measurement in aqueous solutions of very low dissolved solids content (below 50 μmho).

The Model 1023 ORP Transmitter has essentially the same design as the Model 1003. Instrument range is -1000 to $+1000$ mV, with minimum span of 200 mV.

The Model 1050A/1060A pH/ORP Transmitters are lower cost, general purpose units with analog indication. Standard output is 0–1 mA dc; other non-isolated current outputs are available as an option, as well as an isolated 4–20 mA dc. Models 1050C/1060C have CSA approval.

The Model 1505 pH Controller is housed in a general purpose enclosure suitable for panel mounting. Measurement and control circuitry, including manual and automatic temperature compensation, efficiency adjust, standardize, high/low alarm setpoints, and function switch are all contained on a single plug-in circuit board. Isolated current outputs are provided for remote recording. Two types of two-mode control signals are available, a non-isolated 4–20 mA dc output for PR control and a timed dry-contact closure for TP control, whose duty cycle is proportional to pH deviation from setpoint. The controller PCB features a direct/reverse logic switch which may be programmed for either acid or caustic control. Also available is a dual control, which provides both acid and caustic addition through the use of two controller circuits, each with individual setpoints, proportional band, and reset. Repeatability is quoted as ± 0.02 pH.

The Model 1181 pH/ORP Two-Wire Transmitters are housed in a weatherproof, explosion-proof enclosure and may be supplied with analog indication or blind. Output is 4–20 mA dc, isolated. The unit is CSA, FM, and

CENELEC approved intrinsically safe for Division 1 hazardous locations. Instrument range is 2–14 pH, minimum span 1 pH, or 200–2400 mV ORP, minimum span 200 mV. Repeatability is quoted as ±1% of full scale or ±2 mV full scale.

(f) Yokogawa-Electrofact (Netherlands)

Electrofact, recently acquired by Yokogawa (Japan), has been in the field of industrial process instrumentation and on-line liquid measurement for over 25 years, serving the refining, petrochemical, chemical, and power industries. A wide range of instrumentation is available, covering such measurements as pH and redox, conductivity, dissolved oxygen, sodium, chlorine, chloride, and sulfide. Electrofact equipment is certified intrinsically safe by CENELEC, PTB, and BASEEFA. The extensive range of pH and redox measuring instruments is described below.

The Type SM21 pH Electrodes are of all-glass construction, with Ryton R4 cap, and suitable for process pressures up to 140 psig. Various membrane types are offered to meet specific application requirements, including universal pH glass, high temperature, low resistance, and acid-resistant pH glasses. Process temperature range varies with selection from 0–135°C. The membrane is available in both spherical or dome-shaped construction, the latter being very strong, and in thicknesses of 0.1, 0.25, and 1.0 mm.

The Type SR20 Reference Electrodes are available in non-flow, refillable, and bellomatic versions. Reference systems are silver/silver chloride and Tamal. Liquid junctions include ceramic, porous PTFE, and sleeve types.

The Type SC21 Combination Electrodes are also available. For low conductivity applications a special reference ring is offered. A wide variety of mounting hardware is available covering immersion/submersion and flow-through configurations; special cables can also be provided. Electrode cleaning systems include ultrasonic and brush types, the latter being electrically or pneumatically driven. In addition, a chemical cleaning system is offered which periodically sprays a suitable cleaning fluid onto the surface of the glass membrane of an electrode.

The range of ELTRAN transmitters for pH and redox measurement are available in three versions. The "field" version is specially designed for mounting close to the sensor. The "indoor" version is a smaller unit designed for rack mounting in a remote location (e.g., control house). For two-wire transmission, the preamplifier is mounted in a "compact" housing designed for installation close to the sensor and for interfacing directly to a "field" or "indoor" transmitter. A variety of pH ranges is available with spans down to 2 pH minimum and covering the overall range 0–14 pH.

For redox measurement, electrodes are available in gold, silver, or platinum, having a pressure rating of 140 psig and suitable for process temperatures from 0–150°C. Reference electrodes, mounting hardware, and transmitters are the same types as used for pH measurement.

TABLE 13.3 On-Line Process pH and Redox Analyzers

Manufacturer	Country of Origin	pH Electrodes	Reference Electrodes		Redox Electrodes	Combination Electrodes	Electrode Cleaners	Mounting Assemblies	Temperature Compensation	Transmitters		Controllers
			Type	Liq. Junctions						Standard	Two-Wire	
1. Beckman Industrial Inc.	USA	pH-responsive Glass	Lazaran range	Palladium Frit Glass sleeve Fiber plug Double junction	Platinum Gold	Available for pH and ORP	Ultrasonic	Flow-thru Submersion In-line	Available as standard	Model 960B Model 942 (μP)[a]		Model 943 (μP)[a]
2. Exidyne Inc.	USA	pH-responsive Glass	Ag/AgCl	Annular wood	Platinum Gold	Available for pH and ORP		Flow-thru Submersion		Model 4110 Model 6110 (receiver)		
3. Foxboro Analytical	USA	pH-responsive Glass Antimony	Ag/AgCl	Ceramic plug	Platinum Gold		Ultrasonic Wiper-type	Flow-thru In-line Submersion	Available as standard	Series 2220 E91, E99, and E99S Series	Model 870	
4. Gam-Rad Inc.	USA	pH-responsive Glass	Ag/AgCl Calomel	Ceramic Teflon Nylon Double junction	Platinum		Ultrasonic Brush-type	Flow-thru Submersion	PHZ-1000 (μP)[a]	Model 501/1001		Model 1034
5. Great Lakes Instruments Inc.	USA	Uses patented differential Electrode system (see Section 13.4.5c)			Platinum Gold		Ultrasonic	Flow-thru In-line Submersion	Available as standard	Model 90 Model 570P Model 60 Model 70	Model 690	Model 95 (μP)[a] Model 90
6. Fischer & Porter Company	USA	pH-responsive Glass	Ag/AgCl	Ceramic plug	Platinum			Flow-thru Submersion		17RX1000		

TABLE 13.3 (*Continued*)

Manufacturer	Country of Origin	pH Electrodes	Reference Electrodes: Type	Reference Electrodes: Liq. Junctions	Redox Electrodes	Combination Electrodes	Electrode Cleaners	Mounting Assemblies	Temperature Compensation	Transmitters: Standard	Transmitters: Two-Wire	Controllers
7. Horiba Insts. Ltd.	Japan	pH-responsive Glass	Ag/AgCl	Polymer sleeve	Platinum Gold	Supplied as standard (nonrefill-able)	Ultrasonic Brush-type Jet wash-type	Flow-thru Submersion	Available as standard		Supplied as standard for K-7 system	
8. Kent Industrial Instruments Ltd.	Great Britain	pH-responsive Glass	Ag/AgCl	Glass sleeve Polypropylene	Platinum Antimony Gold	Available for pH and ORP	Ultrasonic Wiper-type	Flow-thru In-line Submersion	Available as standard	Model 9170 Model 2800 Series (IS)		Model 9140 Model 9160
9. Leeds & Northrup Company	USA	pH-responsive Glass Antimony	Ag/AgCl	Wood plug Ceramic plug Double junction	Platinum Gold Silver	Meridian range		Flow-thru In-line Submersion	Available as standard	Model 7073 Model 7075 Model 7076 Model 7083 (μP)[a]	Model 7079	Model 7084 (μP)[a]
10. Monitek Inc.	USA	pH-responsive Glass	Ag/AgCl	Peripheral junction	Platinum			Flow-thru In-line Submersion				Model 2015
11. Rosemount/ Uniloc Inc.	USA	pH-responsive Glass	Ag/AgCl	Wood plug	Platinum	Available for pH and ORP		Flow-thru In-line Submersion	Available as standard	Model 1003 (pH) Model 1023 (ORP) Model 1050A (pH) Model 1060A (ORP)	Model 1181	Model 1505

12. Sensorex Inc.	USA	pH-responsive Glass	Ag/AgCl		Platinum	Model S200C (pH) Model S500C (ORP)		In-line Submersion		Electrodes can be adapted for use with Beckman and Rosemount/Uniloc measuring instruments
13. Siemens AG	West Germany	pH-responsive Glass		Ceramic plug		Available for pH and ORP		Flow-thru In-line Submersion		Model 54110
14. Van London Inc.	USA	pH-responsive Glass	Ag/AgCl	Ceramic plug Glass sleeve Wood plug Double junction	Platinum	Available for pH and ORP	Ultrasonic	Flow-thru In-line Submersion	Available as standard	Model 604 Model 540A (pH) Model 541A (ORP) Model 514 (pH) Model 524 (ORP) Electrodes can be adapted for use with Leeds & Northrup and Rosemount/Uniloc measuring instruments.
15. Wallace & Tiernan	USA	pH-responsive Glass		Porous sleeve	Platinum		Ultrasonic	Submersion Flow-thru (Series 81-015)	Available as standard	Series E99 (cabinet-mounted electronics package)
16. Xertex-Delta Analytical	USA	pH-responsive Glass		Porous plug (polymer)		Available for pH and ORP (standard supply)		Flow-thru (self-cleaning) Submersion		Model 8212 (pH or ORP)
17. Yokagawa-Electrofact	Holland	pH-responsive Glass	Ag/AgCl Calomel Tamal	Ceramic plug Glass sleeve PTFE plug Reference ring (low conductivity)	Platinum Gold Silver	Available for pH and ORP	Ultrasonic Brush-type Chemical type	Flow-thru Submersion		ELTRAN Range: "Field" "Indoor" "Compact"

[a] μP: microprocessor-based instrument.

13.5 CONDUCTIVITY MEASUREMENT

13.5.1 Historical Background

The measurement of *electrolytic conductivity* is one of the oldest and most precise of electrochemical measurements and has been used in a wide variety of industrial applications for many years. Consequently, there is a wealth of academic and industrial literature on the subject.

The electrolytic conductivity of a solution is defined as its ability to pass an electric current. The most common electrolytic conductors are those in which the conductance results from the mixture of a solvent, which is usually in large excess (and may itself be only a poor conductor), and a second component, the electrolyte; all the ordinary solutions of salts in water fall into this category. In some cases, an electrolytic solution is produced when two substances, both of which are poor conductors in the pure state, are mixed together; solutions of acids in water are examples.

The conduction of electricity in electrolytic solutions occurs as a result of migration of the positively and negatively charged ions to the electrode of opposite charge. The conductivity of an electrolyte is therefore the sum of the contributions made by anions and cations and is simply related to the mobilities of these ions.

Faraday's classical work on the mechanism of electrolysis established what are known as Faraday's laws of electrolysis, namely, that the amount of decomposition is proportional to the quantity of electricity passed and that the amounts of the various substances generated at the electrodes are proportional to their chemical equivalents. Faraday introduced the terms electrode, anode (positive electrode), cathode (negative electrode), and ions (the particles that move when current passes). The fact that electrolytes obey these classical laws accounts for the accuracy of measurement available.

Although at the time of Faraday's investigations, the mechanism of electrolytic conduction was a matter of speculation, by the 1850s the view that electrolytic conduction takes place by the migration of charged ions in an electric field appears to have been generally accepted. Arrhenius, in his classical paper postulating the theory of electrolytic dissociation, proposed the fundamental hypothesis that an electrolyte in solution is dissociated into ions that conduct electricity and that the degree of dissociation may be determined by measurement of the conductance of the solution.

Extensive studies by Kohlrausch of the conductivity of electrolytes showed that Ohm's law is obeyed with great accuracy. He showed also that the conductivity of salt solutions in water, if they are sufficiently dilute, is the sum of the separate contributions from cations and anions. These accurate studies provided extensive data for the theoretical developments of electrolytic dissociation that came later. It is remarkable that this work has been improved on very little by subsequent investigators. Although there have been significant advances in technology in the area of measuring equipment and in refinement of techniques, differences in the data are relatively small. The extensive data and literature on conductivity values resulting from the earlier theoretical developments and subsequent investigations have provided the basis for practical, industrial applications of conductivity measurement.

13.5.2 Theoretical Basis

As shown by Kohlrausch, electrolytic solutions, like metallic conductors, obey Ohm's law except at very high voltages or high-frequency currents. That is, for an applied potential difference of V volts across two electrodes immersed in an electrolytic solution, the current I flowing through the electrolyte is inversely proportional to the resistance R of the electrolyte. The reciprocal of the resistance ($1/R$) is defined as the *conductance* (K) and is expressed in units known as reciprocal ohms (or mhos), or siemens (the SI unit of conductance).

The resistance R of a homologous substance of uniform cross-sectional area A and length l is given by the equation:

$$\mathrm{R} = \frac{rl}{A}$$

where r is defined as the *specific resistance* (*or resistivity*) of the substance and is characteristic of that substance under a given set of physical conditions.

Likewise, the standard unit of conductance is the *specific conductance* (k), defined as the reciprocal of the resistance in ohms of a 1-cm cube of electrolyte at a specified temperature. The observed conductance (K) of a solution is inversely proportional to the distance l between the electrodes and directly proportional to their area A, as given by the equation:

$$K = \frac{kA}{l}$$

The unit of specific conductance is the reciprocal ohm-cm (or mho/cm) or the siemens/cm. The specific conductances of some typical substances are shown in Table 13.4. The conductance of a solution is the sum of contributions from all the ions present and depends on the number of ions per unit volume of solution and on the velocities at which these ions move under the influence of the potential (i.e., their mobilities).

Most conductivity data was collected in studies of the properties of electrolytes to establish a relationship between degree of ionization and conductivity. The units quoted are those of *equivalent conductance* (Λ), which may be derived from the specific conductance (k) and the concentration (c) of the

TABLE 13.4 Specific Conductances of Typical Waters

Water Quality	Specific Conductance (micromho/cm)
Ultrapure water	0.05–0.1
Demineralized water (industrial)	0.1–0.9
Distilled water (in glass)	1.0–8.0
Boiler feedwater	1.2–12.0
Marine evaporated distillate	5.0–50
Raw water	90–900
River and stream waters	90–1000
Rinses (e.g., electroplating)	20–60 or 700–3000
Mineral waters (e.g., sodas)	900–2000
Circulated coolant	900–9000

solution (in equivalents per liter) by the expression:

$$\Lambda = \frac{K}{c}$$

At infinite dilution, the ions are theoretically independent of one another and each ion contributes its part to the total conductance. That is:

$$\Lambda_\infty = (\lambda_+) + (\lambda_-)$$

where Λ_∞ is the equivalent conductance at infinite dilution and λ_+ and λ_- are the ionic conductances of cations and anions respectively. The ionic conductance is constant for each ion in a given solvent at a given temperature.

As mentioned in Section 13.4, although pH and conductivity are often considered to be similar, it can be seen that they are completely different. Whereas the conductivity of a solution is related to the concentration of all ions present in the solution, the pH of a solution is related only to the concentration of hydrogen ions. Measurement of conductivity is basically a resistance measurement, whereas pH is determined by measurement of the potential of a glass electrode compared with a reference electrode potential.

13.5.3 Factors Affecting Conductivity

The conductivity of an electrolyte is determined by the number and species of ions present in the solution and their mobilities. These factors are, in turn, influenced by the composition and concentration of the solute, the temperature of measurement, and secondary factors such as the dielectric constant and fluidity of the solvent. In some cases, the conductivity of the solvent may contribute to the overall conductivity of the solution; examples are aqueous solutions where the solute concentration is very low, a few ppm or a fraction of a ppm (e.g., high-purity boiler feedwater and condensate).

The conductances of dilute aqueous solutions of inorganic salts increase nearly linearly with their concentration. The conductance-concentration curves of almost all inorganic acids exhibit maxima; sulfuric acid, in fact shows, several inflexion points. Although the ionization of strong electrolytes, such as inorganic salts, the common acids, and alkalis, decreases with temperature, the increased mobility of the ions at higher temperatures results in an overall increase in solution conductivity.

13.5.4 Methods of Measurement

Measurement of the conductance of a solution usually involves determination of thc resistance of a column of the solution by using some variation of the four-arm Wheatstone bridge. Passage of direct current through a solution not only produces changes in the chemical composition of the solution but gives rise to development of spurious potentials as a result of gas formation at the electrodes, a phenomenon known as *polarization*. Therefore, it is necessary to use alternating current in the measurement of conductance in order to reduce polarization effects to a minimum. These effects can be further reduced by using platinum electrodes coated with platinum black. These are particularly useful for measurement in high-conductance (low-resistance) solutions. However, platinization of electrodes is not considered essential for low-conductance measurements requiring only moderate accuracy.

Industrial conductivity measurements generally use an ac current within the range 50–600 Hz, whereas precise laboratory measurements are conducted at frequencies of 1000 Hz or higher. The normal range of resistances measured by high-accuracy conductivity bridges is from 100 to about 60,000 ohms. In on-line measurements, where a lower accuracy may be acceptable, the upper limit of resistance may be much higher. However, it is advisable to impose a lower limit of 100 ohms since below this value polarization and related effects become more pronounced.

The classical theory of the Wheatstone bridge for resistance measurements applies fundamentally to the use of direct current only. When ac current is used, the conductivity may be measured directly only when the ac resistance of the cell is the same as the dc resistance. Since the conductivity cell has electrodes separated by a dielectric material, it has appreciable capacitance. The adjustable arm of the bridge usually has a capacitor in parallel with it, to balance out the effects of cell capacitance. When the bridge is balanced, the impedance being measured is equivalent (though not identical) to the impedance represented by the capacitive and resistive units in the reference arm of the bridge. In a well-designed conductivity bridge, all arms should operate at close to pure resistance.

13.5.5 Conductivity Cell Design

The desired sensitivity and accuracy of electrolytic conductance measurements impose practical limits on the measured resistance of the solution, the optimum resistance range for a ±1% accuracy being approximately 500–10,000 ohms. For a given conductance cell, with electrodes of area A cm^2 separated by a distance of l cm, the specific conductance is given by:

$$k = \frac{1}{R} \cdot \frac{l}{A} = \frac{\theta}{R}$$

where the ratio l/A is a constant, known as the *cell constant* (θ), with dimensions of cm^{-1}.

As the dimensions of the cell are changed, the cell constant varies as the ratio of l to A. A cell of the dimensions given will have a cell constant of 1.00 and the conductance reading is numerically equal to the solution conductivity. For solutions of low conductivity, the electrodes can be placed closer together so that cell constants of 0.1 and 0.01 are obtained. This has the effect of raising the conductance as read between the electrodes and producing a value more easily handled by the measurement circuit. For solutions of high conductivity, the length l is increased, resulting in cell constants of 10 and sometimes 100; this also produces a more acceptable measured conductance (see Fig. 13.16*a*).

A conductivity cell is normally calibrated by measuring its resistance when the cell contains a standard solution of accurately known specific conductance. The cell constant may then be calculated from the above equation. The electrolyte invariably used for cell calibration is potassium chloride (KCl), for which there is a wealth of specific conductance data available. The procedure for obtaining very high accuracy data is covered in ASTM D1125-77.

Conductivity cells may be classified into two types, precision or laboratory-type cells and cells designed specifically for on-line industrial measurements. With laboratory-type cells, a wide range of designs and cell constants is possible, since there are few limitations on such factors as materials of construction and capacitance effects. Some typical precision conductivity cells of this type are shown in Fig. 13.16*b*. On the other hand, industrial on-line conductivity cells are usually required to withstand high temperatures, high pressures, and corrosive process streams, and to operate reliably over extended periods. The consequent limitations on materials of construction and basic design impose corresponding limitations on the range of cell constants available.

In high-conductivity applications, the cell is subject to fouling, by virtue of its required design (highly restricted sample passages). The Electrodeless conductivity system has been very successful in such applications. This approach uses a sensor consisting of two toroids in close proximity, as

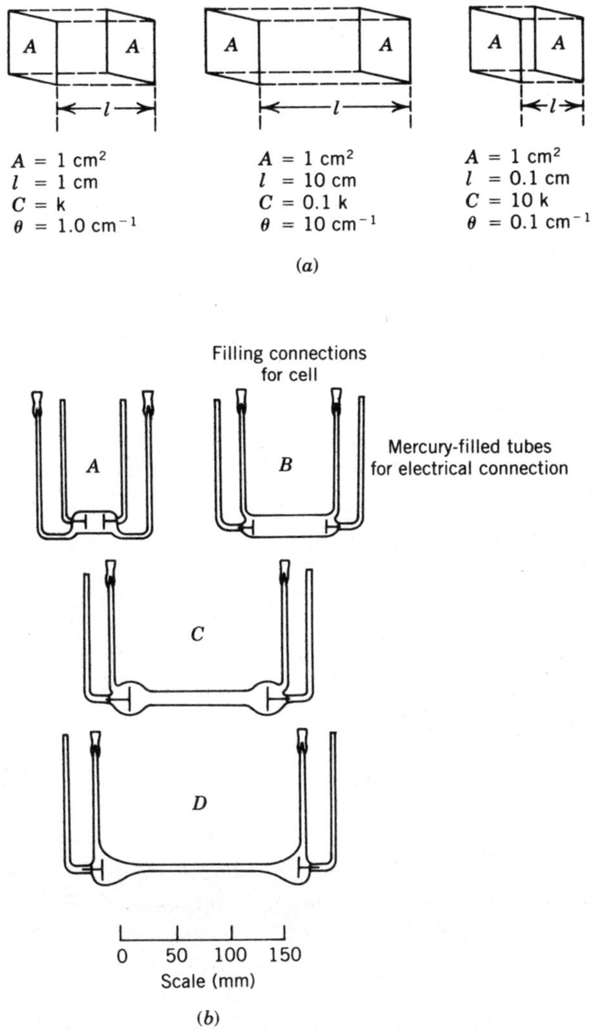

FIGURE 13.16 Conductivity measurement. (*a*) Specific conductance: cell design; (*b*) precision conductivity cells. Cell A, with electrodes close together, is suitable for solutions of low specific conductivity, cell D for high conductivity, cells B and C for intermediate conductivities.

shown in Fig. 13.17, one generating an electric field in the solution, the other acting as a receiver to pick up the small current from ions moving in a conducting loop of solution. The amount of magnetic energy sensed by the pick-up toroid is proportional to the conductivity of the medium between the toroids. The method was first described by Relis in 1951. With this system, there is no electrode contact with the process liquid (thus eliminating fouling or chemical abrasion), polarization is completely absent, and the sensor is relatively insensitive to coating.

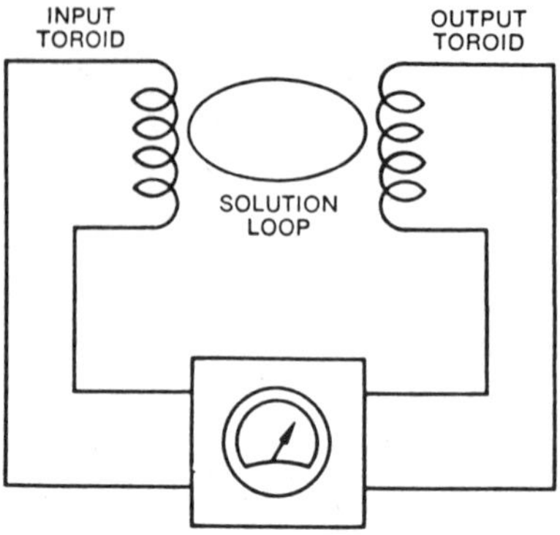

FIGURE 13.17 Electrodeless conductivity measurement.

13.5.6 On-Line Process Conductivity Analyzers

As with pH measurement, conductivity measurement is one of the most traditional of on-line analytical techniques, so a wide range of equipment is commercially available from a variety of manufacturers. The following are examples of this range; a more complete listing is given in Table 13.5 on page 556–557.

(a) Beckman Industrial Inc. (United States)

The Cedar Grove, NJ, operation of Beckman specializes in conductivity instrumentation and manufactures probably the widest range of industrial conductivity measuring equipment available in the United States.

The CEL-VS Cells are plastic, screw-in types with $\frac{3}{4}$-in. and 1-in. NPT connections. They are available with flexible conduit for explosion-proof requirements, have integral temperature compensator, and are suitable for 100 psig and 105°C maximum process conditions. Standard cell constants are 0.1, 0.2, 1.0, and 2.0 cm^{-1}.

The CEL-VY Cells have cell constants from 0.001 to 0.05 cm in seven versions. Maximum process conditions are 100 psig and 105°C. The CEL-VC Cells have graphite electrodes and integral temperature compensator and are ideal for contamination monitoring where plugging is a problem. Cell constants are 0.1, 0.2, and 0.5 cm^{-1} and maximum process conditions are 100 psig and 105°C.

The CEL-VK, VH and VKJ Cells are designed for medium to high conductivity measurements (salts, acids, and caustic), except where strong oxidizing agents are present. Cell constants range from 1.0 to 100 cm^{-1}. Rating is 100 psig and 105°C. The CEL-VH Cells have two parallel electrode chambers for high conductivity measurement, using platinum electrodes for acid and graphite electrodes for weak acid applications.

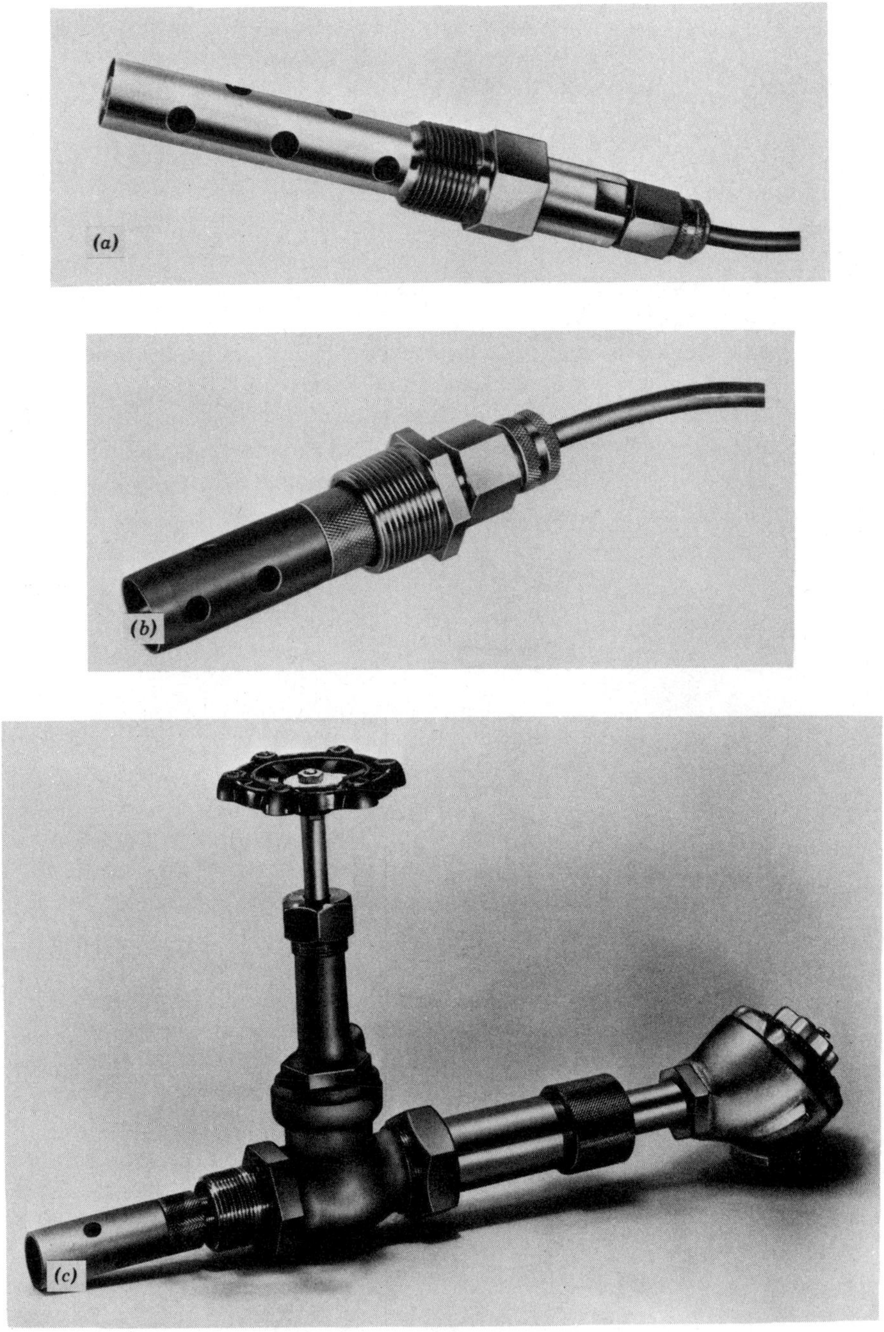

FIGURE 13.18 Beckman electrode type conductivity cells. (Courtesy of Beckman Industrial Inc.)

The CEL-L Cells are glass, screw-in types with temperature compensator and with cell constants from 0.01 to 2.0 cm^{-1} in 10 versions. Rating is 50 psig and 105°C. The CEL-K Cells are glass screw-in types for medium to high conductivity, provided with or without protection shield. Cell constants range from 0.1 to 50 cm^{-1} in 13 versions.

The CEL-H Cells are glass screw-in types designed for tank wall mounting and may be supplied with heavy-duty protective shield in 316, 304, or Carpenter 20 stainless steel. Rating is 50 psig and 105°C. The CEL-11 Series Cells are heavy-duty, screw-in types, designed for corrosive services. Standard rating is 500 psig and 105°C; rating up to 2500 psig is also available. Cell constants range from 0.001 to 100 cm^{-1}.

The CEL-Y Cells are designed for ultra-pure water applications and have cell constants ranging from 0.001 to 0.2 cm^{-1}. The CEL-1 Series Cells are designed for direct insertion with integral gate valve. Standard accessories include cell holders, sample coolers, temperature compensators, platinizing kits, and calibrating solutions. Examples of the range of conductivity cells are shown in Fig. 13.18.

FIGURE 13.19 Beckman electrodeless conductivity cells. (Courtesy of Beckman Industrial Inc.)

For difficult applications, Beckman offers a range of electrodeless conductivity cells, including dip-type and flow-through designs. The dip-type cells are available in epoxy, PVC, or polypropylene, and cylindrical or donut-shape design. They are fully immersed in the process stream, enabling the external fluid to provide the return path to close the loop. The flow-through cells feature bore sizes from $\frac{1}{2}$–3 in., with smooth interior surfaces. They are available in a wide selection of materials, end fittings and configurations. The toroid assembly is hermetically sealed and each cell is complete with 10 ft of interconnecting cable. Flexible explosion-proof conduit is available as an option. Typical examples of the range are shown in Fig. 13.19.

The range of conductivity measuring instrumentation offered includes the Solumeter series of indicators (analog and digital), recorders, controllers, and two-wire transmitters; the Type R15 servo-balanced indicators and accessories; the Type RA6 indicator/controllers; and the Solubridge series of

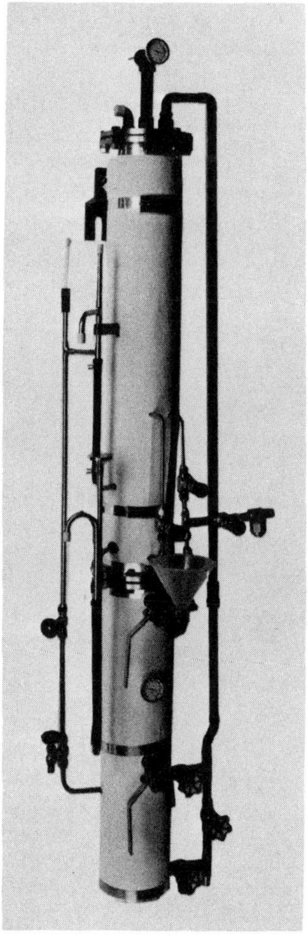

FIGURE 13.20 Larson-Lane stream purity analyzer. (Courtesy of Beckman Industrial Inc.)

indicators and controllers. One range of instruments is designed specifically for use with the electrodeless conductivity cells.

Beckman also manufactures a range of Larson-Lane Steam Purity Analyzers. The Model SC-19A shown in Fig. 13.20 employs cation exchange and degasification to provide a continuous measurement of steam purity. The steam sample is condensed at close to the boiling point by a water-cooled coil and passes to the bottom of the condensing chamber, where it is boiled to drive off the dissolved gases. The boiled condensate then passes through the first conductivity cell and down to the ion exchange chamber, where dissolved salts are converted to their corresponding mineral acids and ammonia/amines are eliminated. The sample is then reboiled to drive off dissolved gases and then passes through the second conductivity cell.

The conductivity as measured at the second cell is proportional to the mineral carryover content of the steam and is independent of carbon dioxide, hydroxides, carbonate, ammonia, or amines concentrations. Comparison of the two conductivity readings indicates the condition of the ion exchange resin. The Model SC-19C is similar to the 19A, except that it provides greater sensitivity. Conductivity cells used cover the range 0–1 to 0–20 mho/cm and have cell constants from 0.01 to 0.04 cm^{-1}.

(b) Foxboro Analytical (United States)

The Series 910 Conductivity Sensors have cell constants of 0.01, 0.1, 1.0, and 10. 0, and are available in seven mounting configurations with modular process connections. Electrodes are made of titanium or high-density graphite. The cells have integral temperature compensators and may be provided

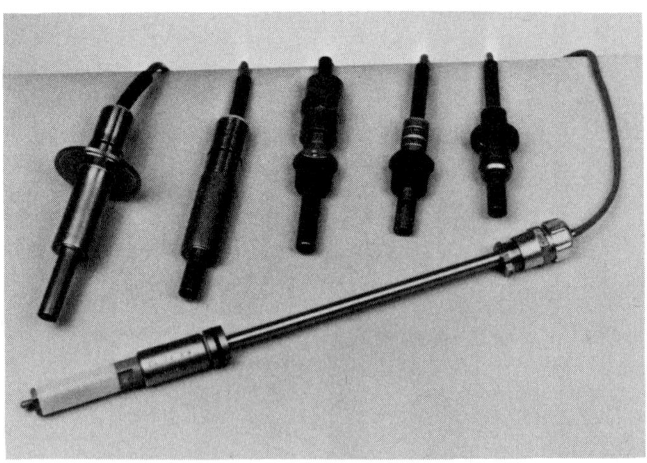

FIGURE 13.21 Foxboro electrode type conductivity cells. (Courtesy of The Foxboro Company)

with electrode guards. These sensors are designed for use with the Model 910M and 915H, M and N Monitors.

The Series 920 Sensors are for conductivity or resistivity measurement with electrodes of titanium or Monel. Sensor configurations include standard $\frac{3}{4}$-in. NPT, universal mount, sanitary fittings—twist-lock, dip type, and inertion type. High temperature versions are also offered. Integral temperature sensors are provided for temperature compensation. Typical examples of the range are shown in Fig. 13.21.

The Series 871CC Sensors are designed for use with the 870CC two-wire conductivity transmitters. There is a choice of two cell constants, 0.1 cm^{-1} with titanium electrodes and 10 cm^{-1} with high density graphite in a series of eight sensor configurations. An integral thermistor is provided for temperature compensation.

The Series 1210 Electrodeless Conductivity Sensors shown in Fig. 13.22 are designed for use with the Series 1210 Monitors. They are suitable for a wide variety of industrial applications such as environmental monitoring, acid service, slurries, and highly corrosive chemical solutions. Modular sensor mounts are offered, with integral thermistor for temperature compensation. There are six basic sensor configurations. The Series 871EC Electrodeless Sensors are the same as the Series 1210, except that they are specifically for use with the Model 870EC two-wire transmitter.

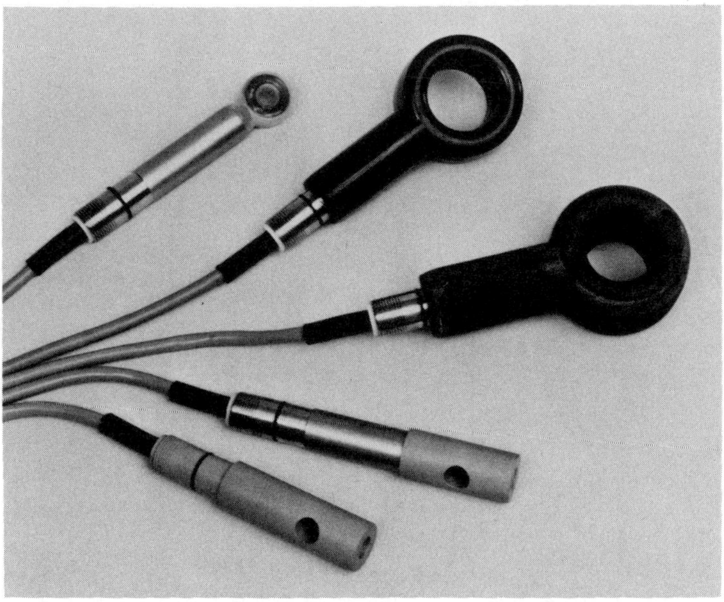

FIGURE 13.22 Foxboro electrodeless conductivity sensors. (Courtesy of The Foxboro Company)

TABLE 13.5 On-Line Process Conductivity Analyzers

Manufacturer	Country of Origin	Electrode Cells: Configurations	Electrode Cells: Cell Constants	Electrodeless Cells	Temperature Compensation	Transmitters: Standard	Transmitters: Two-Wire	Indicators & Controllers
1. Beckman Industrial Inc.	USA	Screw-in Dip type Flow-thru Removable (w/gate valve)	CEL series from 0.001 to 100	Screw-in Dip type Flow-thru Removable (w/valve)	Available as standard		Type SCT	"Solu Meter" range "Solu Bridge" range
2. Foxboro Analytical	USA	Universal Screw-in Sanitary flange Insertion Flow-thru Dip type	Model 920 series (0.01 & 0.1) Model 910 series (0.01, 0.1, 1.0, & 10) Model 871 series	Screw-in Flanged Dip type (Model 1210 & 871EC series)	Available as standard		Model 870CC (871 sensors) Model 870EC (871EC sensors)	Models 910, 915, 920, & 921
3. Gam-Rad Inc.	USA	Submersible In-line	Part of "Enviro-Monitor" range		Available as standard	Model 500 Series Model 1100 Series		
4. Great Lakes Instruments Inc.	USA	Submersion Flow-thru	0.05, 5.0, & 10	Submersion (Model 32014) Flow-thru (Model 33324)	Available as standard			Model 70 series Model 570C
5. Hartmann & Braun AG	West Germany	Dip type In-line	Model 2B1: 0.058 Model 10B1: 2.9 Model CFLD: 0.1 & 1.0 Model 4B2: 5.8 Model 5B2: 23.2		Available as standard	Model CFLG Model CFLZ (general purpose)		

6. Kent Industrial Measurement Ltd.	Great Britain	Screw-in Flow-thru Removable (w/valve)	Stainless steel: 0.01, 0.05, 0.1 Epoxy resin: 0.1, 1, 5, 10		Available as standard	Model 52 Series Model 5515 (multipoint)		Model 5123 Model 5122 Model 5756 "aqualarm" series
7. Leeds & Northrup Co.	USA	Dip type Insertion Submersion Flow-thru	0.01, 0.1, 1, 10 (4973) 0.01–50 (8 selections 4095) 0.001, 0.01, 0.1 (4902) 0.01–50 (8 selections 4800)		Available as standard	Model 7073 Model 7079	Model 7079-17	Models 7075, 7076, 7077, & 7078 Model 7085
8. Rosemount/ Uniloc Inc.	USA	Flow-thru Insertion Submersion	5, 10, 20, 50 (Model 110) 5, 20 (Model 112/113) 0.01, 0.1 (Model 142) 0.1, 1.0 (Model 150) 1.0, 2.0 (Model 160)	Submersion In-line Flow-thru	Available as standard	Model 723 (electrodeless) Model 733 Model 750		Model 1545
9. Siemens AG	West Germany	Flow-thru In-line Submersion	2-electrode & 4-electrode cells	Submersion	Available as standard	M54210 M54211		
10. Yokagawa-Electrofact	Holland	Flow-thru In-line Submersion	0.01, 0.1, 1.0, 10.0		Available as standard	"ELTRAN" range: "Field" "Indoor" "Compact"		

13.6 TURBIDITY/SUSPENDED SOLIDS MEASUREMENT

Turbidity is an important parameter used in determining water quality. The clarity of water has become associated with its purity, with the implication that departure from this condition indicates the presence of either dissolved material causing absorption; suspended material causing, primarily, the scattering of light, or both.

Thus, the term *turbidity* may be defined as one of the *optical* properties of a liquid, related to the presence, nature, and concentration of discrete aggregations of material different from the pure liquid "carrier." Being an appearance parameter, caused by the scattering of light energy passing through the liquid, turbidity can only be measured by optical techniques.

The degree of turbidity in a sample is dependent on the size, shape, refractive index, and color of the particulate matter; the liquid refractive index; the wavelength of light, and the viewing geometry. Thus, absolute mass concentration cannot be deduced from turbidity measurement unless these other parameters are held constant.

13.6.1 Theoretical Basis

The first successful attempts to establish the theory of the scattering of light by particulate matter were made in the second half of the 19th century by Rayleigh and Tyndall. The rigorous theory was developed by Lorentz, Mie, and Debye between 1890 and 1920. For the purposes of practical industrial measurements, such as filtration operations, the rigorous theory is usually simplified by making the following assumptions:

1. Particles are roughly spherical in shape.
2. Particle size is either within the range 0–0.5 micron or 0.5–100 micron.
3. Light polarization effects are ignored.
4. Particles do not absorb light energy at the measurement wavelength.
5. Incident light is within the visible spectrum (i.e., effective wavelength is assumed to be 0.5 micron).

For particles less than 0.05 micron, scattered light intensity is equal in the forward and backward directions and is reduced by 50% in the perpendicular direction (at 90° to the incident light beam axis). As particle size increases beyond 0.1 micron, forward scattering increases until it stabilizes at at least 100 times the perpendicular scattering. For both large and small particles, total scattered signal at a fixed mass particulate concentration varies with particle size, thus preventing the use of scattering measurement alone for mass concentration measurement. Thus, different instrument designs must

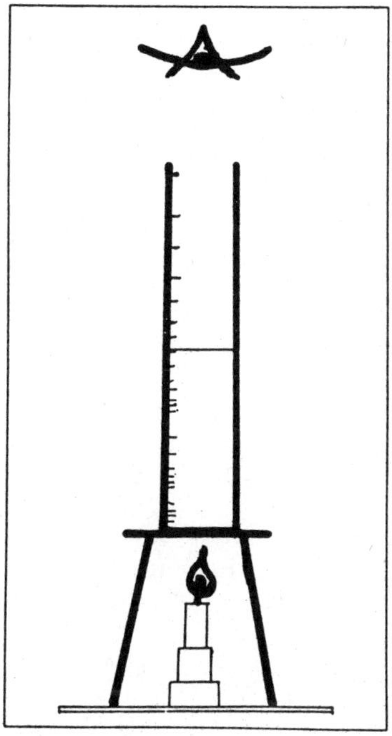

FIGURE 13.23 Jackson candle turbidimeter.

be used to measure small particle "hazes" and large particle phenomena, often known as "particulate turbidity."

The first attempt to standardize turbidity measurement resulted in the development of an instrument known as the *Jackson Candle Turbidimeter,* the forerunner of modern turbidity measuring instruments. As shown in Fig. 13.23, the instrument consists of a vertical glass cylinder graduated in units of height with a lighted candle placed beneath it. The sample liquid is poured into the cylinder until the candle flame is seen to disappear, leaving a uniform field of light, as viewed from above the cylinder. The height of liquid at which this occurs, which depends on the liquid turbidity, is converted by use of a standard table to Jackson candle units (JCU) or Jackson turbidity units (JTU). This technique depends upon narrow-angle, forward scattering and is extremely sensitive to particulate concentration, yet is not sensitive to light or moderate sample color. The technique is the basis of definition of the JTU scale of turbidity units, which is used extensively in some industries.

There are three types of instruments commonly used to measure the amount of light scattered by the presence of particulate matter in a liquid—the transmissiometer, the nephelometer (a side-scattering turbidimeter), and the forward-scattering turbidimeter.

In the *Transmissiometer,* turbidity is measured as the decrease in intensity due to loss of light from scattering and absorption by particles and the liquid in the path of a collimated beam of light, as the beam passes through the sample. The intensity of measured light is given by:

$$I = I_0(1 - S)(1 - A)$$

where I_0 = light intensity when there are no particulates or color in the sample
S = portion of scattered light
A = portion of absorbed light

The turbidity is given by:

$$T = kI$$

where k is a constant.

This type of instrument may be used to measure solids concentration (usually above 1% by weight) where S (scattered light) is high and change in A (absorbance due to sample color) is negligible. In addition to its sensitivity to the unwanted parameters mentioned, the transmitted signal is proportional only to the cross sectional area of the particles in the liquid. Thus, if we ignore the finer points of light scattering theory, the transmittance signal will vary as the square of the particle radius, while the mass of the particle varies as the cube of the radius. Transmittance measurements are usually expressed as "percent transmittance" for a definite light path distance in the sample liquid.

The *Nephelometer,* or side-scattering type turbidimeter, measures the light scattered at 90° to the incident light beam. With the addition of ratioing circuitry to measure the ratio of perpendicularly scattered light to direct transmitted light, the effects of color changes in the sample can be eliminated. Measured turbidity at 90° to the incident light beam is given by:

$$T_{90} = \frac{S_{90}}{I} = \frac{I_0 S_{90}(1 - A)}{I_0(1 - S_{90})(1 - A)}$$

or

$$T_{90} = \frac{S_{90}}{(1 - S_{90})}$$

Since for most materials, S_{90} is small, the denominator $(1 - S_{90})$ is unity for all practical purposes and the instrument gives a linear turbidity reading not sensitive to sample color variations.

Nephelometers are usually calibrated in FTU (Formazin turbidity units), NTU (Nephelos turbidity units), or TU (turbidity units). Formazin is a prescribed mixture of hexamethylene-tetramine and hydrazine sulfate, which forms a colloidal suspension suitable for use as a light scattering standard. Formazin calibration standards are covered in detail in *Standard Methods,* published by the APHA (American Public Health Association), 14th edition, 1975. In the United States the EPA has established nephelometric turbidity measurement as a standard for all drinking water.

The *Forward-Scattering Turbidimeter* is a relatively recent introduction. This instrument measures the amount of light scattered in the forward direction, usually about 25°, from a projected, collimated beam in the measuring chamber. Again, the addition of ratioing circuitry eliminates the effects of sample color variations and provides a direct measurement of the light scattered by particulates.

The intensity of the measured light is given by:

$$I = I_0(1 - S)(1 - A)$$

where I_0 = light intensity when there are no particulates or color in the sample
S = portion of scattered light
A = portion of absorbed light due to sample color

Scattered light intensity is given by:

$$S' = I_0 S(1 - A)$$

Measured turbidity is given by:

$$T = \frac{S'}{I} = \frac{I_0 S(1 - A)}{I_0(1 - S)(1 - A)}$$

or

$$T = \frac{S}{(1 - S)}$$

Since for most materials, S is very small, the denominator is almost unity. Therefore measured turbidity is linear and independent of light source intensity and sample color variations.

Forward-scattering turbidimeters use the units of JTU or ppm for a defined calibration material. As mentioned earlier, the Jackson Candle Turbidimeter provides the definition of JTU and the conversion table used with this instrument is such that the height of a column of prepared Fullers Earth suspension of 'x' parts per million reads 'x' JTU. This unit of turbidity is retained for calibration of forward-scattering continuous on-line turbidimeters used in the water industry. Calibration standards are suspensions of Johns Manville Filter-Cel (a diatomaceous earth) that is uncalcined or fluxed, having a particle size range of 2–5 micron.

13.6.2 On-Line Process Turbidimeters

(a) Great Lakes Instruments Inc. (United States)

The Model 80 Process Turbidimeter is built to the design criteria of EPA's Storet 00076 and consists of a sensor unit and a control unit. Twelve feet of interconnecting cable is supplied by the manufacturer as standard. However, by use of a separate power supply module, separation by distances up to 300 ft can be accomodated. The optical system shown in Fig. 13.24 combines 90° nephelometry with direct light transmittance to allow linear measurement from 0–0.1 to 0–1000 NTU (FTU) full scale. Signals from the three photocells are processed by solid-state circuitry in the control unit. Current outputs (0–20, 4–20, or 10–50 mA dc) are available for remote recording as well as dual alarm contacts. The sensor enclosure is designed to NEMA 12 or 4X and suitable for surface mounting. The sample cell is designed for sample flowrates of 0–1 gal/min at pressures from 0–60 psig. Wetted materials are PVC, Teflon (optional), glass, neoprene, and nylon. Repeatability is quoted as ±0.25% FS.

The Model 82 Process Turbidimeter consists of the sensor and transmitter, each housed in a NEMA 4X enclosure. The optical system uses 90° nephelometry, having two photocells, the outputs of which vary directly with the amount of scattered light caused by sample turbidity. Seven measuring ranges are available from 0–0.1 to 0–100 NTU (FTU). The transmitter is available with analog or digital indicator. Current outputs and dual alarm capability is the same as for the Model 80. Sensor/transmitter separation distance is 25 ft maximum, an optional power supply module will allow an increase to 300 ft. Repeatability is quoted as ±0.25% FS.

GLI also represents Sigrist Photometer in the United States and can supply a complete line of compact Sigrist Turbidimeters (called the 4800 Series). When coupled with the Model 48 Control Unit, these instruments can provide a variety of outputs and control functions. The Sigrist Turbidimeters are described later in this section.

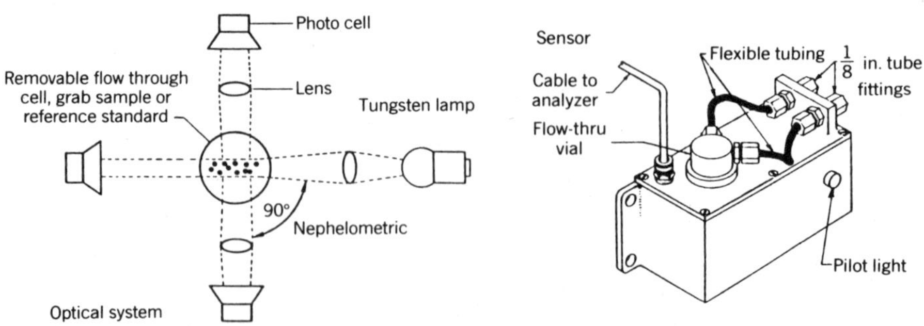

FIGURE 13.24 Model 80 process turbidimeter. (Courtesy of Great Lakes Instruments Inc.)

(b) Hach Chemical Company (United States)

The Model 1720A Low-range Turbidimeter is designed to meet the EPA standard for drinking water monitoring. The operating principle, as shown in Fig. 13.25, is based on 90° nephelometry. Standard ranges are 0–0.2, 1, 3, and 30 NTU (FTU), and a variety of voltage and current outputs are available for remote recording. The analyzer consists of a sensing unit and an indicator/transmitter. A bubble trap is also offered for elimination of air bubbles in the sample. Air bubbles can cause measurement errors by scattering light in the same way as turbidity. Sample requirements are 0.5–1.5 l/min at a pressure of 2–12 psig. An alarm relay module is also available in a single or dual version, with trip point(s) adjustable over the full analyzer span.

The Ratio 2000 Turbidimeter uses three geometrically positioned silicon photodiode detectors to measure 90° scattered light and transmitted light. The ratioed signals compensate for sample color, dust, and haze on the optics and temperature coefficients of detectors and amplifiers. Four standard ranges are offered, 0–2, 20, 200, and 2000 NTU (FTU), manually selectable. A 4–20 mA dc output is available for remote recording. Analog and digital indicators are available for use with this analyzer.

The Surface Scatter 5 Turbidimeter is designed for turbidity monitoring of most industrial process streams. The unusual operating principle is shown in Fig. 13.26. A light beam focuses on the sample surface at an angle of about 15° and light scattered by the sample is measured by a detector mounted immediately above the sample surface. The liquid surface produces a near-perfect optical surface, thereby eliminating the basic problems associated

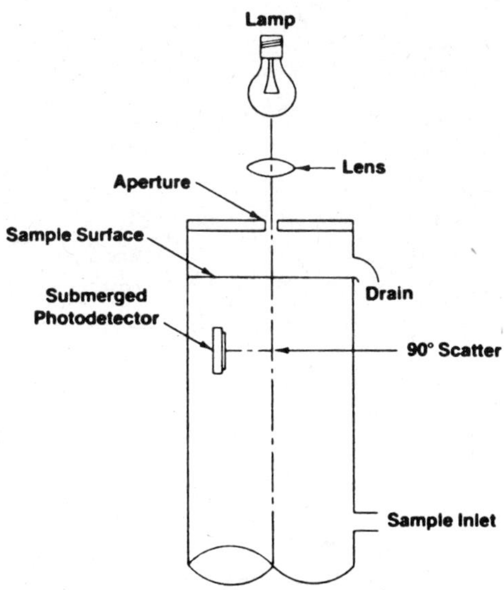

FIGURE 13.25 Model 1720A low-range turbidimeter. (Courtesy of Hach Chemical Company)

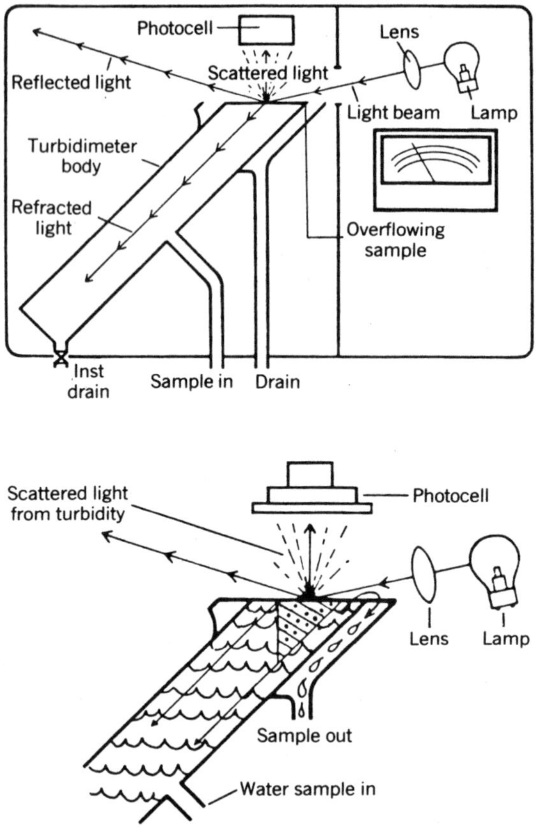

FIGURE 13.26 Surface scatter 5 turbidimeter. (Courtesy of Hach Chemical Company)

with "windowed" sample cells. The turbidimeter body is an inclined 2-in. diameter pipe about 12 in. long, with the bottom extended portion acting as a light trap. Sample flow is set at about 1–2 l/min; as it overflows to drain, it tends to carry away any floating materials such as scum. The sample container interior is painted an optical black, thus reducing stray light to a minimum. The analyzer unit is of NEMA 12 construction, suitable for wall or panel mounting. A variety of voltage and current outputs are available for remote recording.

(c) Monitek Inc. (United States)

Monitek manufactures a wide range of laboratory and on-line process turbidimeters capable of covering a variety of industrial applications for turbidity and suspended solids measurement over the range 0.01 to 50,000 ppm (JTU). Three different techniques are used, depending on the application. Forward scattered light measurement is used for turbidity (0.01 to 1000 ppm) and

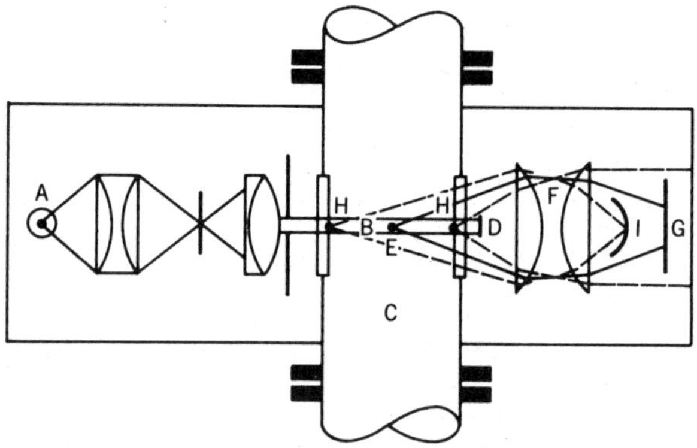

A Light source
B Light beam
C Process stream
D Direct beam detector
E Particulates
F Condensing lenses
G Scattered signal detector
H Sight glass
I Light trap

FIGURE 13.27 Monitek forward scatter system. (Courtesy of Monitek Inc.)

suspended solids (1000 to 50,000 ppm) measurements. A signal ratioing technique is used to provide accurate, repeatable measurements independent of sample color variations. Instruments employing direct light absorbance measurement are available for applications where the effect of sample color variations is negligible. Instruments are also available utilizing 90° nephelometry.

The spatial filtering system used in the forward scatter instruments is shown in Fig. 13.27. Condensing lenses send light scattered from the area closest to the left-hand window to the light trap and light from near to the right-hand window to infinity. Scattered light from the real area of interest falls on the detector. The signal from the detector is ratioed electronically to the signal generated by the direct beam detector. Monitek forward scatter turbidimeters also have an electronic "bubble reject" circuit, which eliminates the sharp peaks produced by the presence of air bubbles in the sample.

The Model 31 Turbidimeter is a 90° Nephelometer designed for potable water applications and has switch-selectable ranges of 0–2, 20, and 200 NTU. It is a single unit containing two modules, built into a NEMA 4 enclosure. Sample requirements are 0.5–5 l/min at a pressure of 1–45 psig. Response time is less than 1 sec full scale. A 4–20 mA dc output is available for remote recording.

The Model 160/131 Process Turbidimeter is a forward scatter instrument designed primarily for water and waste water treatment applications. Eleven switch-selectable ranges are offered within the overall range 0–2000 JTU.

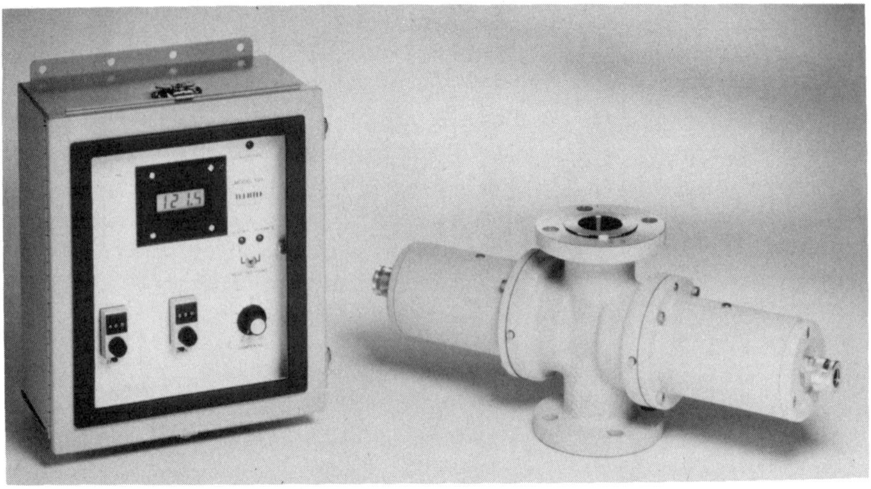

FIGURE 13.28 Model 230/134 suspended solids monitor. (Courtesy of Monitek Inc.)

The analyzer consists of the Model 160 sensor and Model 131 transmitter, each housed in an NEMA 4 enclosure. The sample cell is rated for 70°C and 90 psig maximum. Repeatability is quoted as ±0.5% full scale.

The Model 210 Process Turbidimeter is a high-sensitivity, fast response, forward scatter instrument that can be used with either the Model 130 or 134 transmitter. Standard range is 0.1–100 ppm. The transmitter is UL approved for NEC Division 1 hazardous locations. The sensor is of in-line design and rated for 120°C and 150 psig maximum sample conditions.

The Model 211 Process Turbidimeter is a forward scatter instrument designed primarily for use in filtration operations and is UL approved for NEC Division 1 hazardous locations. Range is 1–1000 ppm. The sensor is rated for 120°C and 150 psig maximum sample conditions.

The Model 230/134 Process Suspended Solids Monitor shown in Fig. 13.28 is a forward scatter instrument with three ranges, 0–1, 2, and 5% suspended solids. The sample cell has stainless steel, neoprene, and pyrex wetted parts, is rated for 120°C and 150 psig maximum sample conditions, and is available for in-line or slip-stream installation.

The Model 350/136 Process Turbidimeter is designed as a submersible unit, suitable for immersion at depths up to 200 ft. It is a forward scatter instrument designed for open, non-pressured applications in waste water treatment such as clarifying, thickening, and settling facilities. Instrument range is 0–1000 JTU.

(d) Sigrist Photometer (Switzerland)

Sigrist manufactures a range of absorption, fluorescence, and turbidity measuring instruments, based on a unique oscillating mirror principle and having

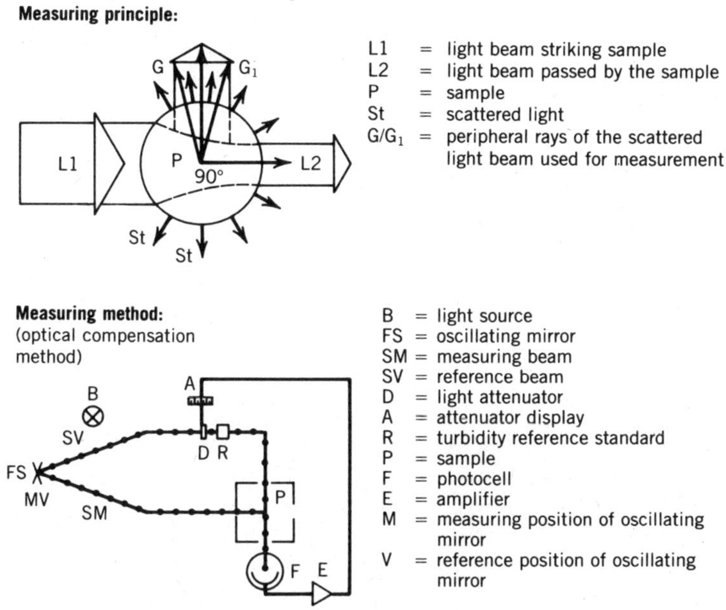

FIGURE 13.29 Model KT30S process turbidimeter. (Courtesy of Sigrist Photometer Ltd.)

application in a variety of industrial processes, including water and waste water treatment.

The Model KT30S Process Turbidimeter uses 90° nephelometry to measure turbidity in accordance with German (DIN 38404) and EPA standards. The operating principle is shown in Fig. 13.29. Light from the source is split into two beams by an oscillating mirror. The measuring beam passes through the sample cell and scattered light is measured by the photocell detector located at 90° to the measuring beam axis. The reference beam passes through an adjustable shutter, a standard cell, and the sample cell, impinging directly on the detector. The photocell detects the differences in light intensity of the two beams. The photocell output is amplified and used to drive the adjustable shutter until the reference beam is attenuated to the same intensity as the measuring beam.

At this equilibrium position, the shutter position indicates the turbidity value and may be calibrated in NTU, FTU, or EBC units (European Brewery Convention). Measurement range is from 0–2 to 0–100 NTU, with two to six ranges available, switchable manually from the control unit. Up to three alarm contacts can also be provided. The analyzer consists of the sensing unit containing the sample cell and optical system and a control/display unit. A current output (0–20 or 4–20 mA dc) is available for remote recording. The standard flow cell is made of stainless steel and suitable for 40–200°C and 140 psig maximum sample conditions. An encased version is available for cold sample applications.

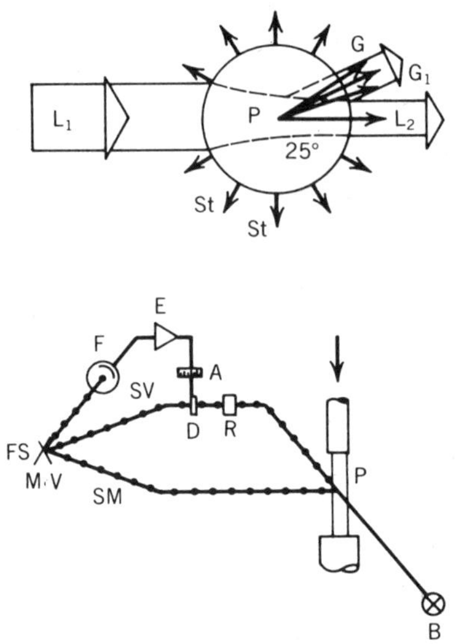

FIGURE 13.30 Model KTJ25 process turbidimeter. (Courtesy of Sigrist Photometer Ltd.)

The Model KT65 Process Turbidimeter uses the same operating principle as the Model KT30S, but is designed for applications requiring higher sensitivity. Measurement ranges are from 0–0.2 to 0–5 NTU. The flow cell is available in brass or 316 stainless steel and rated for 60°C and 90 psig maximum sample conditions.

The Model KTJ25 Process Turbidimeter shown in Fig. 13.30 uses forward scattered light at 25° for turbidity measurement. In this case, light scattered at 25° is directed onto the oscillating mirror as the measuring beam. The reference beam emerging undeflected from the sample cell is attenuated by sample color, a variable shutter, and the turbidity standard, and then directed onto the oscillating mirror. Alternately, the mirror directs the measuring and reference beams onto the photocell detector. Once again, an equilibrium position is obtained by attenuating the reference beam until the photocell detects no difference in intensity between the two beams. Measuring range of the analyzer is 0–1 to 0–200 NTU. Analyzer and control unit may be separated by up to 300 ft.

13.7 RESIDUAL CHLORINE MEASUREMENT

13.7.1 Background

Chlorination is the process in which chlorine gas or chlorine compounds (e.g., hypochlorites) are added to water or sewage, usually for the purpose of disinfection. Its principal use is to safeguard water supplies by killing disease-producing organisms.

In addition to their bactericidal properties, chlorine and chlorine compounds are used in the reduction or removal of objectionable tastes and odors from potable water supplies; the oxidation of iron, manganese, and hydrogen sulfide; and the oxidation, decoloration, and destruction of organic matter. Chlorination is also widely used in the control of slime, algae, and so on in cooling water systems and is injected into sewage and industrial wastes prior to discharge into streams.

In the clarification process, chlorine is often applied to raw water along with coagulant, one of the benefits being improved coagulation, probably due to chlorine's action on organic material present in the water. The addition of chlorine to partially or completely treated sewage aids in bacterial control and sterilization.

13.7.2 Theoretical Basis

Chlorine hydrolyzes in aqueous solution to form hypochlorous acid:

$$Cl_2 + H_2O \rightleftharpoons HOCl + H^+ + Cl^-$$

The hypochlorous acid in turn is ionized to form hydrogen and hypochlorite ions:

$$HOCl \rightleftharpoons H^+ + OCl^-$$

It is the hypochlorous acid produced in this hydrolysis that provides the disinfecting and oxidizing properties of chlorine solutions. The chemical reaction is dependent upon the pH of the solution; the lower the pH, the higher the concentration of hypochlorous acid present. Below pH 6.5, HOCl constitutes almost 100% of the free available chlorine, whereas at pH 9.0 and above, OCl^- accounts for practically all of the free available chlorine.

Since hypochlorous acid and not hypochlorite ion is the principal disinfectant in chlorine solutions, the efficiency of disinfection will be substantially greater at low pH. The free available residual chlorine of a treated water is thus considered to be that portion of the total residual chlorine present in the water as hypochlorous acid.

In the chlorination process, it is very difficult to predict accurately how much chlorine is required to effect complete sterilization. Sufficient chlorine is often added to ensure that an excess of chlorine remains in the water after sterilization has taken place, thereby providing a margin of safety. This excess of chlorine indicates that organisms have been destroyed.

Chlorine takes an electron from surrounding molecules, causing oxidation, undergoing reduction itself by gaining an electron. Other reducing agents present in the water, such as ferrous ions (Fe^{2+}), manganese ions (Mn^{2+}), nitrite ions (NO_2^-), and hydrogen sulfide, will also enter into an oxidation-reduction reaction with chlorine without contributing to organic cell destruction.

Total residual chlorine consists of both free and combined chlorine. Ammonia or amines are often present in natural waters as decomposition products of organic matter and react with hypochlorous acid to produce a combined form consisting predominantly of chloramines, as given by the following chemical equations:

$$NH_3 + HOCl \rightleftharpoons NH_2Cl + H_2O$$

$$NH_3 + 2HOCl \rightleftharpoons NHCl_2 + 2H_2O$$

$$NH_3 + 3HOCl \rightleftharpoons NCl_3 + 3H_2O$$

The pH of the treated water determines which product will be formed. At pH 8.5, monochloramine is the major product, whereas at pH 4.5, dichloramine is predominant. Mixtures of the two exist between these values. Below pH 4.5, nitrogen trichloride begins to be formed. The rate at which these reactions take place and the concentrations of the various chloramines present depend on the concentration of ammonia as well as the pH of the water. Where chloramines are present, it is necessary to provide longer contact time or increased chlorine residual for comparable bacterial reduction. The chloramines do have sterilizing power, but to a much lesser extent than hypochlorous acid or hypochlorite ion. However, in some cases, ammonia is added to chlorinated water to form chloramines, which act as a reserve disinfectant.

Chlorine may also oxidize and bleach organic matter, with the extent of oxidation dependent on time of contact and residual chlorine present. In any event, oxidation of inorganic and organic compounds requires the use of greater amounts of chlorine to satisfy the demand before a chlorine residual for disinfection can be established.

"Chlorine demand" may be defined as the difference between the amount of chlorine applied to a water and the amount of free combined or total residual chlorine remaining at the end of a specified contact period. This will vary for any water with the contact time, water quality, temperature, and pH. The "chlorine requirement" is the amount of chlorine that must be added to a water to achieve a specific objective (e.g., bacterial kill, H_2S removal, or oxidation of iron and manganese).

The residual chlorine requirement in treated water is usually established by bacteriological testing to determine the amount necessary for effective disinfection. Normally, a residual of 0.2–0.5 ppm is satisfactory after a 10-min minimum contact period at or above 20°C. Greater concentrations may be required if chlorine exists as combined residual or if the pH is greater than 7. In the food processing industries, chlorine may cause taste problems in canned and frozen foods or corrosion of metal containers. Thus, very critical limits are usually established for chlorine residuals.

In addition to gaseous chlorine, some commercially available hypochlorites are sometimes used for chlorination. The most widely used salt for water treatment is calcium hypochlorite $Ca(OCl)_2$, with 70% available chlorine, which ionizes in aqueous solution as follows:

$$Ca(OCl)_2 \rightleftharpoons Ca^{2+} + 2OCl^-$$

The hypochlorite ions establish equilibrium with the hydrogen ions depending on the pH, as does gaseous chlorine. Sodium hypochlorite (NaOCl) is available in several different forms, varying in available chlorine from 3% to as high as 20%. All hypochlorites are corrosive to some degree and require careful handling and the use of corrosion-resistant materials for storage and dispensing.

13.7.3 Process Analyzers for Residual Chlorine Measurement

(a) Capital Controls Company (United States)

The Advance Series 870 Residual Chlorine Analyzer shown in Fig. 13.31 is an amperometric instrument designed for the measurement of free or total chlorine residual in process waters.

The process sample is delivered to an inlet filtering chamber at a flowrate of about 500 ml/min, a fixed level being maintained in the chamber by an overflow weir. Sample passes through the filter screen, then flows at 100 ml/min to the entrance of the measuring cell, excess sample overflowing to drain. The measuring cell comprises a cylindrical copper electrode and a

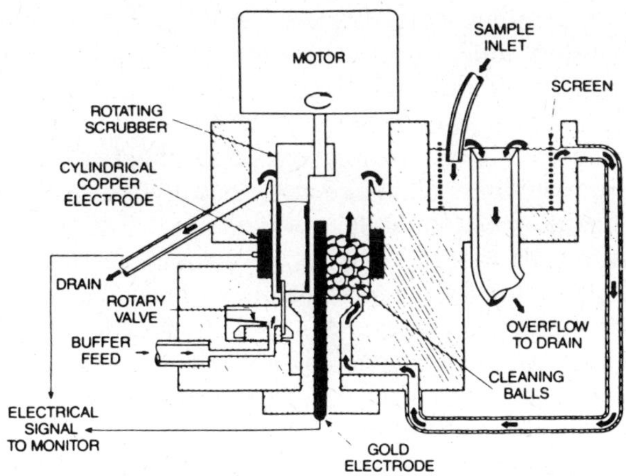

FIGURE 13.31 Series 870 residual chlorine analyzer. (Courtesy of Capital Controls Company)

fixed gold electrode. The surfaces of both electrodes are kept clean by the continuous action of PVC balls agitated by a motor-driven, rotating striker. Sample passes through the annular space between the electrodes and then flows over a weir to drain. The difference in weir heights between the inlet chamber and the cell chamber maintains a constant sample flowrate.

A small dc current is generated by the sensing electrode combination in proportion to the amount of chlorine present in the water. For measurement of free chlorine, the sample pH is adjusted to 4.0–4.5 by the addition of a buffer solution. For measurement of total chlorine, potassium iodide is added. This reacts with free and combined chlorine in the sample to liberate iodine (in direct proportion to the chlorine), which is measured in the sensing cell. Both reagents are stored in separate bottles (with capacity for seven days operation) and fed from constant head reservoirs through rotary valves located on either side of the measuring cell. These valves are actuated by an extension of the motor-driven striker to feed a precise amount of solution during each valve revolution.

The Model 871 (Free Chlorine) and Model 872 (Total Chlorine) Analyzers are general purpose instruments suitable for wall or panel mounting, or may be supplied as a complete analysis system in a free-standing cabinet. Seven measurement ranges are available from 0–0.5 to 0–20 ppm chlorine residual. The sensing cell output is 0–50 mV dc (0–25 mV dc for the 0–0.5 range). The analyzers can be supplied complete with local recorder or indicator fitted with a re-transmitter providing a 4–20 mA dc output for remote recording.

(b) Xertex, Delta Analytical (United States)

The Series 8000 Residual Chlorine Analyzers are designed to measure either total free chlorine or total residual chlorine in treated water or waste water. The analyzers use unique polarographic sensors capable of discriminating between the various ionic species of chlorine. One sensor is capable of determining HOCl with virtually no response to OCl^-, making it ideal for direct measurement of sterilizing power. The sensor is similar in operation to the polarographic oxygen sensors described in Chapter 7 and consists of a plastic structure containing an anode and electrolyte terminated by a membrane-covered noble metal cathode.

Each measurement system consists of a sample conditioning module, sensor module, and an analyzer/transmitter. The analyzer/transmitter is available in two versions. The Model 8224 is housed in a NEMA 1 general purpose enclosure, whereas the Model 8324 is housed in a NEMA 4 weatherproof enclosure. Air purge facilities can be provided as an option. A standard 4–20 mA dc output is provided for remote recording, as well as relay contacts for external alarm actuation. Standard ranges are 0–1, 5, and 10 ppm.

For total free chlorine measurement, the pH of the sample is adjusted to 5.0 by addition of a buffer acetate solution. The reaction converts all the

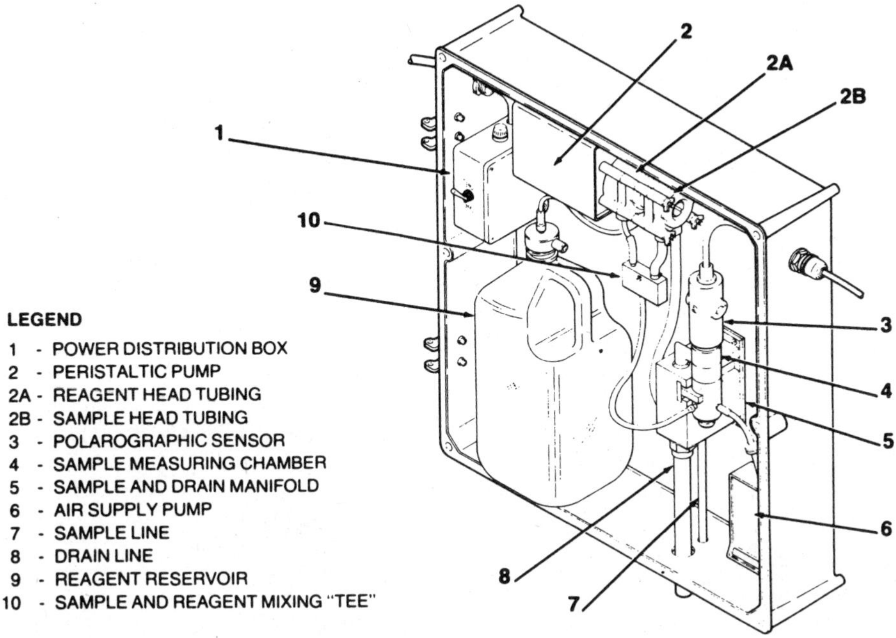

FIGURE 13.32 Series 8000 residual chlorine analyzer. (Courtesy of Xertex Corporation)

existing free chlorine to HOCl prior to measurement by the Model 82124 sensor. For total residual chlorine measurement, acidified potassium iodide is added to the sample. All forms of chlorine present in the sample oxidize the iodide to iodine, which is measured by the Model 82125 sensor.

A typical analysis system is shown in Fig. 13.32. Sample and reagent are precisely metered by peristaltic pumps. The outlets of the two pump heads empty via a mixing tee into the measuring chamber, where the mixture is stirred by the action of a small air pump. The sensor is mounted in the chamber so that its membrane is immersed in the sample. Sample continuously overflows from the measuring chamber to a common drain line. Reagent supply is provided for 30 days operation.

(c) Fischer and Porter Company (United States)

The Series 17B4200 Anachlor Residual Chlorine Analyzer shown in Fig. 13.33 is an amperometric device designed for continuous measurement of residual chlorine in water. Sample enters the instrument via a constant head tank mounted at the rear of the analyzer case. After passing through a filter element located in the tank, sample enters the flow control block and passes through the flow control valve to the head control block, where reagent (dilute acetic acid or buffer solution) is added by means of a special feed pump assembly. Reagent supply is suitable for 30-day operation.

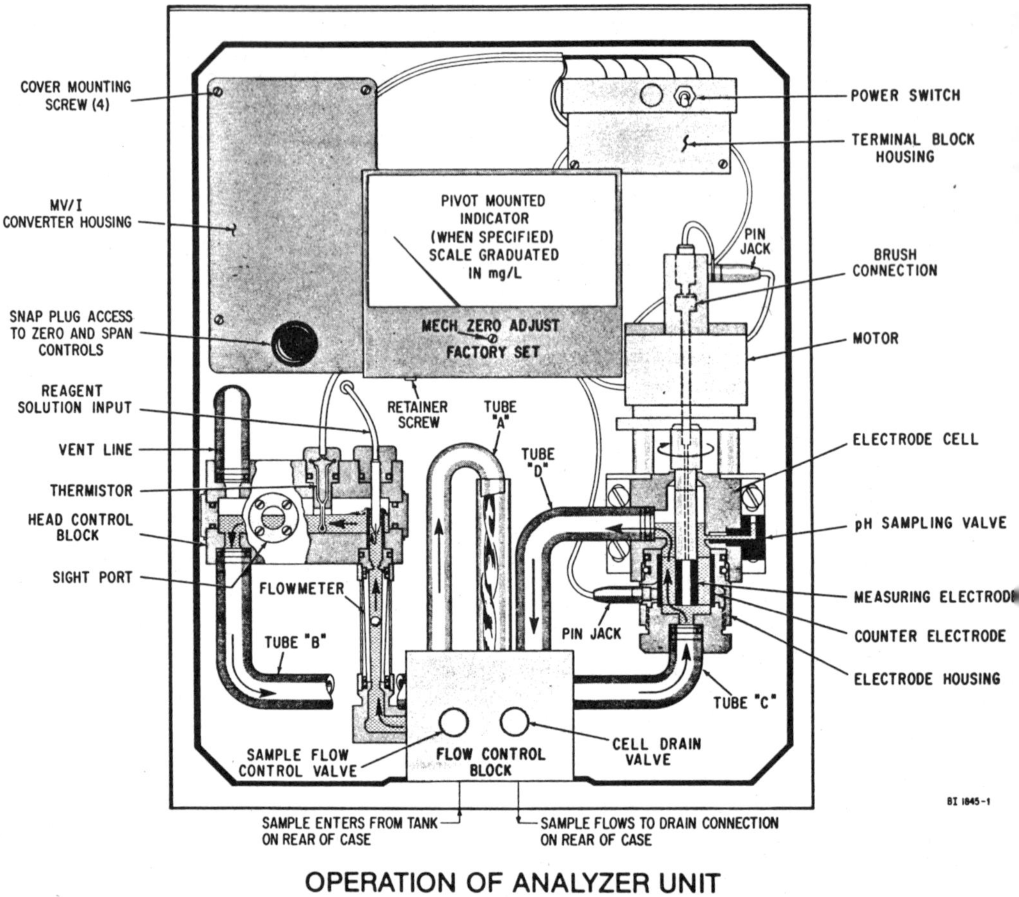

FIGURE 13.33 Series 17B4200 residual chlorine analyzer. (Courtesy of Fischer and Porter Company)

At a flowrate of 150 ml/min, the buffered sample flows via the flow control block to the bottom of the electrode cell assembly. In this assembly, sample flows up through the annular space between the concentrically mounted outer counter electrode (copper) and the inner measuring electrode (gold), then out to drain. The annular space contains plastic pellets that are continuously agitated by the sample and keep the electrodes free from contamination. The measuring electrode is rotated at 1550 rpm by a motor to maintain ideal, reproducible conditions. The electrical signal from the measuring electrode is picked up by a spring-loaded brush at the top of the motor.

The composition of the electrode is such that polarization prevents current flow in the absence of a strong oxidizing agent (e.g., chlorine). The amplitude of the self-generated depolarizing current produced by the presence of traces of chlorine is proportional to the chlorine concentration in the sample.

When free residual chlorine is measured, potassium bromide is added to the buffer and reacts with the free chlorine to liberate bromine, which in turn depolarizes the measuring cell and produces a current directly proportional to the free residual chlorine. When total residual chlorine is measured, potassium iodide and sodium hydroxide are added to the buffer. This reacts with free and combined chlorine to liberate iodine, which in turn depolarizes the electrode in the same manner as chlorine and bromine.

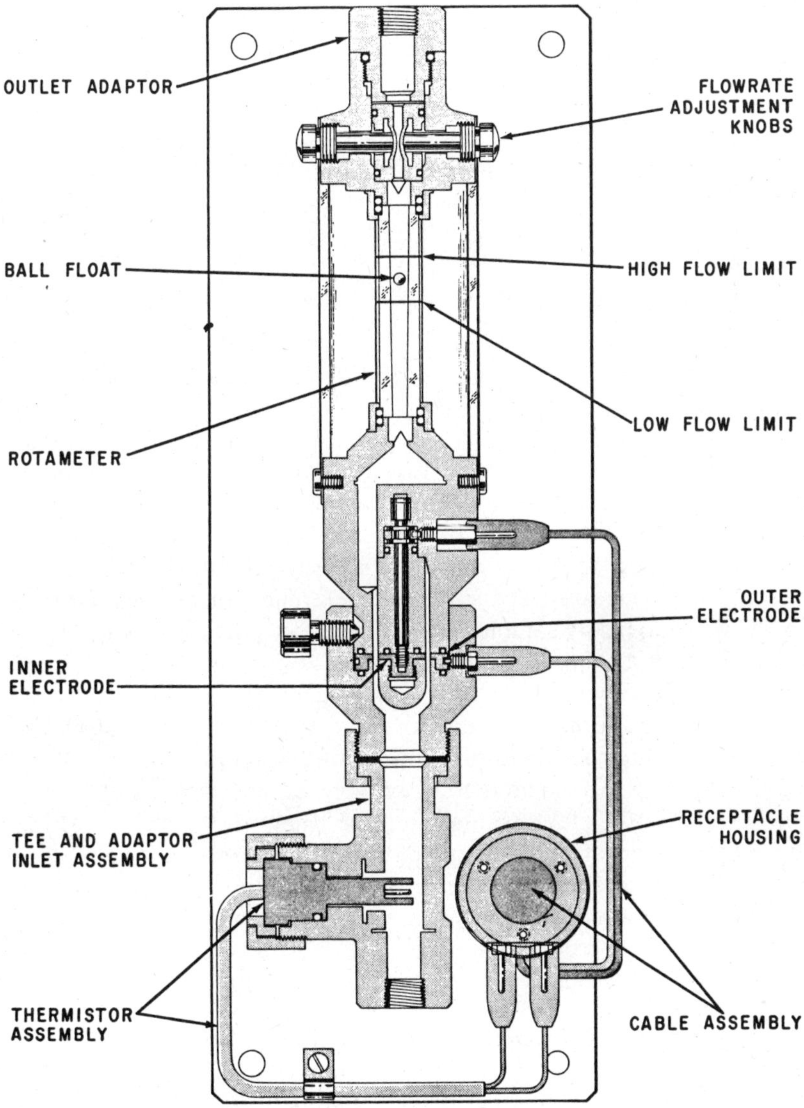

FIGURE 13.34 Chlortrol Series 17K1000 residual chlorine analyzer. (Courtesy of Fischer and Porter Company)

The generated current is amplified and signal-conditioned to produce a 4–20 mA dc output; temperature compensation is also provided. Operating ranges offered are 0–0.5, 1, 2, 3, 5, 10, and 20 ppm. Sample requirements are 2l/min at 0.5–2 psig. The analyzer is housed in an enclosure suitable for surface mounting.

The Chlortrol Series 17K1000 Free Residual Chlorine Analyzer shown in Fig. 13.34 is a low-cost amperometeric instrument designed specifically for drinking water, swimming pools, and other water treatment applications. Sample enters the analyzer through the bottom inlet and passes upward past the thermistor temperature compensator and between the concentrically mounted electrodes, then through the rotameter and flow control valve.

The small dc potential impressed across the electrodes is transmitted from the measuring circuit located in the indicator and polarizes the electrodes. The presence of free chlorine allows current to flow in the circuit and is measured by a series-connected microammeter located in the indicator. Zero and span adjustments are provided as standard. Measurement range is 0–2 ppm and response time is quoted as 5 sec full scale.

(d) Pennwalt-Wallace & Tiernan (United States)

Wallace & Tiernan specializes in chlorination systems for industrial water and waste water treatment processes and its process analyzers can be supplied separately or as an integral part of a complete chlorination system.

The Series A-773 Residual Chlorine Analyzer shown in Fig. 13.35 is designed for water treatment applications. Sample enters the instrument at between 5 and 60 psig, where it is filtered and flow-regulated to about 400 ml/min. Sample pH is stabilized at the regulator by constant CO gas or liquid buffer dosage. The buffered sample then flows to the measuring cell containing the platinum (measuring) and copper (reference) electrodes. Grit placed in the cell and carried by sample velocity bombards the electrodes to prevent contamination. For total residual chlorine measurement, potassium iodide is added to the flowing sample by a metering pump, liberating free iodine proportional to the total residual chlorine in the sample. The Model A-767 recorder or indicator used with the analyzer have standard ranges of 0–1, 2, 5, 10, and 20 ppm. An isolated 4–20 mA dc output is also available.

The Series A-792 Total Residual Chlorine Analyzer shown in Fig. 13.36 is designed for waste water treatment applications. Operation is basically similar to the A-773, except that provisions are made for dealing with the more stringent sampling requirements of waste water samples. A special PD pump is used to deliver as true a sample as possible for analysis and to prevent break-up of particles that would expose new organic matter and lower the residual chlorine.

A rotational plate-type filter is used in the bypass loop to filter a small quantity of the sample for analysis. The filtered sample passes to a multichamber, constant-level dilution box. (Dilution of the sample reduces reagent requirements and cleaning frequency.) Sample leaving the dilution box

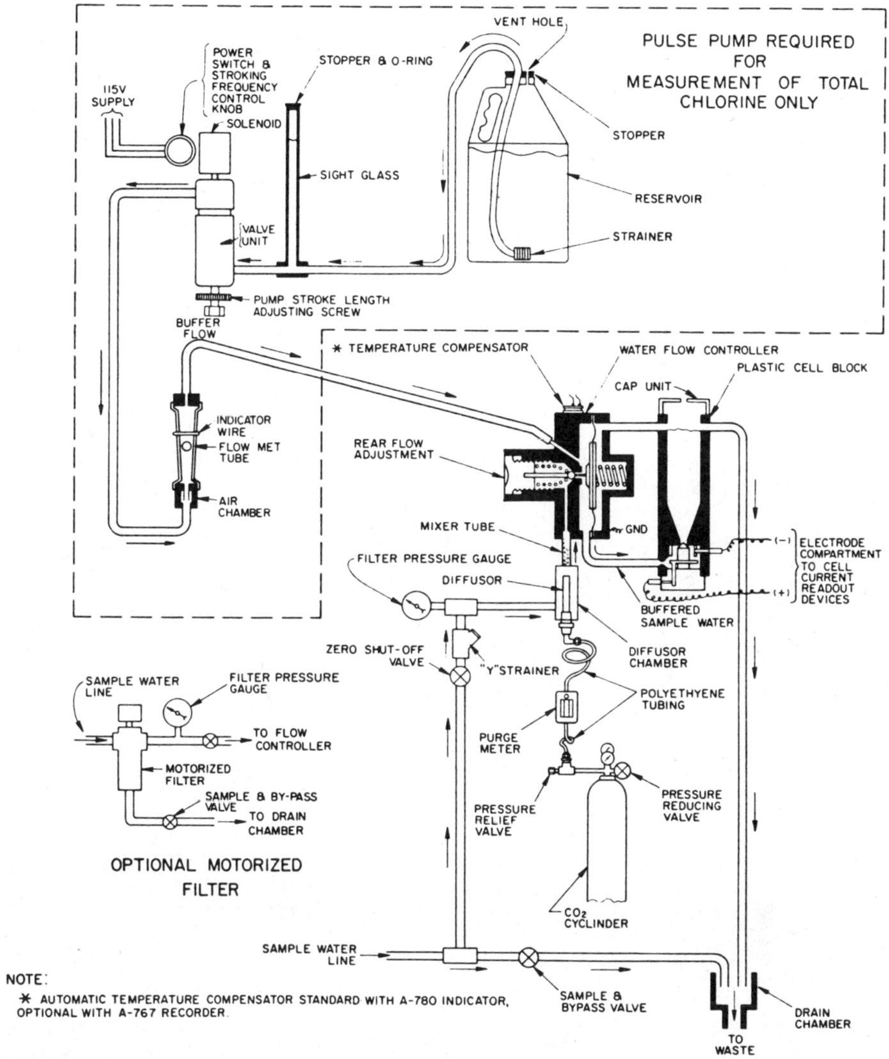

FIGURE 13.35 Series A-773 residual chlorine monitor. (Courtesy of Penwalt-Wallace & Tiernan)

has a constant amount of phosphoric acid/detergent solution added by a metering pump. An identical pump adds potassium iodide reagent, buffer solution, and detergent. Pump range is adjustable from 0.2 to 2.5 ml/min. Diluted and buffered sample then passes through an orifice to the measuring cell. The buffers and acid keep the sample pH at about 4.3, at which pH chloramines rapidly oxidize iodide to iodine. The free iodine produces a depolarizing current proportional to the total residual chlorine.

The Series 50-236 Residual Chlorine Analyzer shown in Fig. 13.37 is an

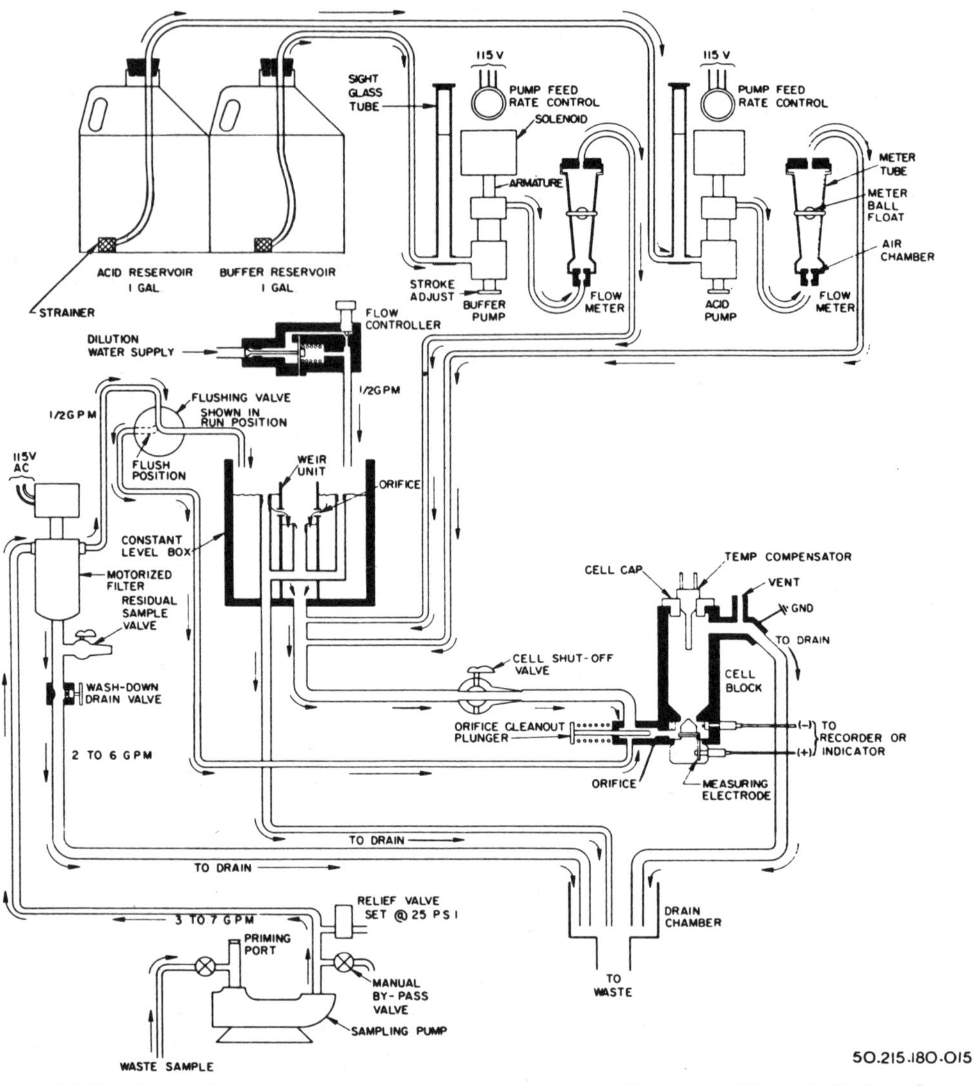

FIGURE 13.36 Series A-792 total residual clorine monitor. (Courtesy of Penwalt-Wallace & Tiernan)

electrochemical device designed for cooling water applications. Sample enters the analyzer at a flowrate of about 2l/min, is filtered, and flows to a constant head tank, where about 1l/min is diverted to the measuring cell; the remainder of the sample flows to drain. Sample enters the bottom of the cell through a venturi nozzle. A salt bridge maintains electrical contact between the measuring electrode and a reference electrode. The output of the electrode system is converted to a 10–50 mA dc signal, which operates an indicating meter and provides an output for remote recording.

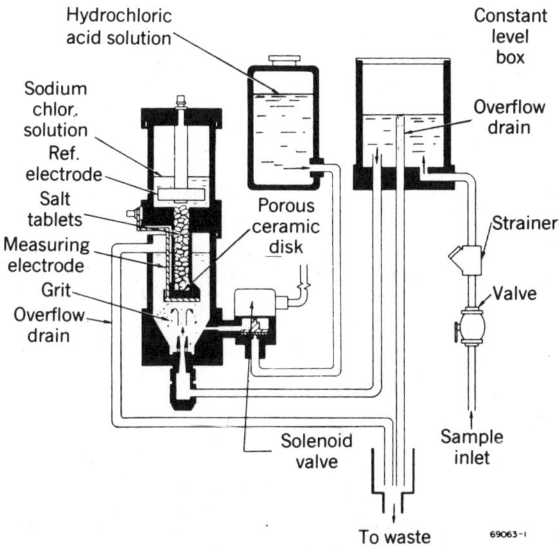

FIGURE 13.37 Series 50-236 total residual chlorine monitor. (Courtesy of Penwalt-Wallace & Tiernan)

Hydrochloric acid is used to clean the measuring electrode and is injected from a reservoir by a solenoid valve as shown, which is operated by a program timer/relay system, adjustable between 5 and 30 min. Acid is added, the electrode is cleaned and flushed, and cell output is stabilized automatically. When the analyzer is used as part of a chlorination control system, the acid cleaning cycle may be programmed with the chlorination cycles. Standard instrument range is 0–1 ppm, with a maximum of 0–2 ppm. Response time is quoted as 15 sec.

13.8 SPECIFIC ION MEASUREMENT

13.8.1 Theoretical Basis

As discussed in Section 13.4, electrode systems for pH and redox measurement are potentiometric devices that obey the Nernst equation and are specific to hydrogen ions and oxidation-reduction couples respectively. A recent introduction is the *ion-selective* (or *specific ion*) *electrode,* which is an electrochemical sensor that generates a potential, according to the Nernst equation, when immersed in a solution containing free ions of a certain specie. These electrodes are finding widespread application in industry as primary elements for process monitoring and control. As with the glass electrode, which develops a potential in response to the hydrogen ion activity in solution, the ion-selective electrode develops a potential in response to the activity of the ion for which the electrode is selective.

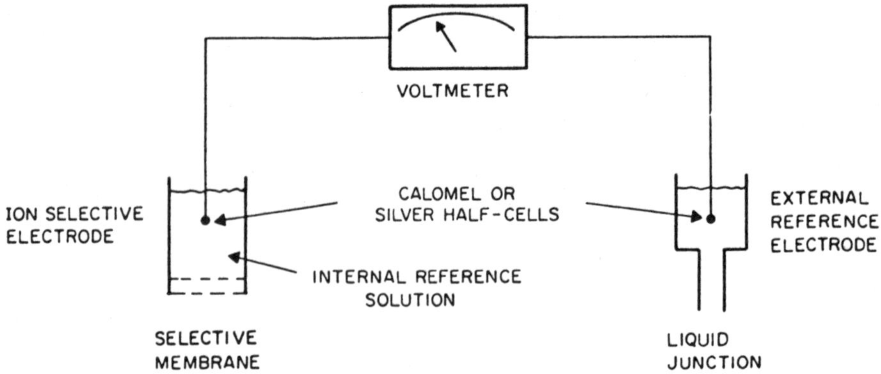

FIGURE 13.38 Ion selective electrode measurement system.

These electrodes have a sensing membrane that separates the process stream from an internal reference solution. When the electrode is immersed in a process stream, there is a momentary migration of ions across the membrane toward the solution containing the lower activity of the mobile ion. Eventually, an equilibrium is established in which the potential across the membrane is exactly that required to prevent further movement of ions. Changes in this potential may be measured by establishing electrical contact with the inner cell using a suitable reference element and at the same time contacting the process solution with an external reference electrode via a liquid junction. Such a system is shown in Fig. 13.38. Each electrode is designed to have a high degree of specificity for a single specie or class of ions in solution. Although the materials of construction of each electrode assembly vary, the principle of operation is basically the same.

The potential generated by the electrode is given by:

$$E = E_0 + 2.303 \frac{RT}{nF} \log \frac{a_1}{a_2}$$

where a_1 is the activity of the measured ion in the process stream
a_2 is the activity of the measured ion in the internal solution
E_0 is a constant for a given electrode at a given temperature

Normally, the electrode is constructed so that a_2 is constant; the equation thus becomes:

$$E = E_0 + 2.303 \frac{RT}{nF} \log a_1$$

The potential measured by the indicating meter in Fig. 13.38 is the sum of the potentials developed by the ion-selective electrode (E), the reference electrode (E_r), and the liquid junction (E_L):

$$E_t = E + E_L - E_r$$

The reference electrode potential and the liquid junction potential are assumed to be constant at a given operating temperature (although sometimes this is not the case).

As with pH and redox electrodes, temperature has two independent effects on ion-selective electrode measurements, the Nernst effect and the Solution effect, as described in Section 13.4. For precise interpretation of ion-selective measurement in terms of *concentration,* each of these temperature effects must be taken into consideration.

When making measurements with ion-selective electrodes, one is often faced with the question of how to report data, whether in terms of activity or concentration. Activity is related to the free (active) specie, whereas concentration is related to the total amount of specie present. With laboratory standard test methods involving titrimetric, gravimetric, or colorimetric techniques, results are usually reported in terms of concentration, which is useful for material balance calculations and pollution control.

However, for process control, reaction status is an important parameter; ion activity is the quantity that measures this status and ion-selective electrodes make this measurement directly. These electrodes can be used to obtain concentration data if calibration standards are made to match the process sample, if a reagent is added to the sample to control its ionic background, or if compensation is provided for the presence of other (interfering) ions.

13.8.2 On-Line Specific Ion Monitors

A wide variety of on-line specific ion measuring equipment is commercially available today and is used in a broad range of industrial applications. For meaningful measurement, it is frequently necessary to add a reagent to the sample in order to control its ionic background, so the technique is often used as the end measurement in wet chemical analyzers (see Section 13.9). The following are examples of the range of equipment available; a more detailed listing is given in Table 13.6 on pages 585–586.

(a) Foxboro Analytical (United States)

Foxboro manufactures a range of systems for the measurement of monovalent and divalent cations and anions in aqueous process streams, including sodium, cupric, silver, fluoride, chloride, bromide, cyanide, and sulfide. A typical system is shown in Fig. 13.39 and includes:

- An ion-selective electrode that generates a potential proportional to the logarithm of the ionic activity.

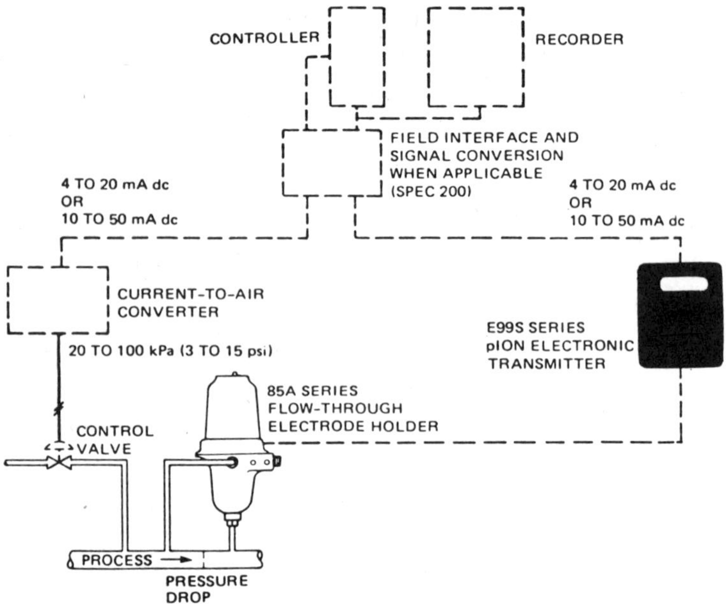

FIGURE 13.39 Foxboro specific ion measurement systems.

- A reference electrode.
- An electrode holder assembly to support and protect the electrodes.
- A transmitter that provides a current output proportional to the logarithm of the ionic activity.

Most ion-selective electrodes cover the range 10^{-6} to 1.0 molar ionic activity, and are capable of withstanding the pH range 0–14, although this range may be restricted depending on the type of measurement and interferences. For example, the sodium electrode has a measurement range of 0.023–23,000 ppm. However, hydrogen ions and silver ions affect electrode response, and high levels of lithium, potassium, cesium, and ammonium ions will interfere. Electrode temperature range is 0–80°C. Fluoride and sodium ion electrodes are rated at 25 psig maximum. Solid-state crystalline membranes are used in most systems, except the sodium ion electrode, which uses a sodium-sensitive glass membrane.

Reference electrodes are glass body, flowing types, or plastic body, nonflowing, with ceramic liquid junction. They are the same as used for pH and redox measurements (see Section 13.4).

The Series 85A Electrode Holders include reinforced plastic submersion/immersion types and a choice of flow-through types to meet most application requirements. For pressures up to 150 psig, a 316 stainless steel flow-through holder is available.

The Series E995 Transmitter, designed for use with the ion-selective electrode systems, is an all solid-state device which can be mounted at up to 1000 ft from the sensor assembly. The Hi-Z input module may be located remotely from or integral with the transmitter. Range limits are −1000 to +1000 mV dc, with spans from 50 to 840 mV dc. The transmitter is available with or without local indicator. Automatic temperature compensation (Nernst effect) is also offered. Single or dual alarms are adjustable over 5–90% of span (normally-closed contacts).

(b) Orion Research (United States)

The Series 1500 and 1700 Process Analyzers use ion-selective electrode systems for the measurement of a wide variety of ionic species in aqueous process streams.

The Series 1500 range includes analyzers for the measurement of sodium, chloride, hydrazine, and calcium hardness, and are all basically similar in design. The Model 151102 Sodium Monitor shown in Fig. 13.40 is designed to measure sodium in the range 0.1 ppb to 1 ppm at critical points within the steam/water circuit of power generation and water treatment plants.

After filtration and pressure reduction, process sample passes through a reagent diffusion bottle, where its pH is adjusted by addition of a reagent. It then flows to the measuring cell, which contains the sodium and reference electrodes and a thermistor for automatic temperature compensation. Spent sample then flows to an atmospheric drain. Zero sodium calibration sample can be introduced via a manual three-way valve. Reagent supply provides for 30 days of operation.

The Model 151101 Sodium Monitor is designed to measure sodium carryover in high-pressure steam with a standard range of 0.1–100%. Two sodium electrodes are used to monitor each of two process streams, boiler feed water and steam condensate. Design is otherwise similar to the Model 11502.

Sample requirements for both analyzers are a minimum flowrate of 50 ml/min at a temperature of 0–50°C and a pressure of 20–100 psig. Voltage or current outputs are available for remote recording. The transmitter has a local indicator with four-decade logarithmic scale. The analyzer system is housed in a front-access enclosure suitable for wall or panel mounting.

The Model 1518 Hydrazine Monitor is designed to measure hydrazine within the range 0.001 to 1 ppm, using an iodometric method in conjunction with an ion-selective (iodide) electrode system. After filtration and pressure reduction, process sample passes through a reagent diffusion bottle where pH adjustment and iodine addition takes place. Hydrazine present in the sample reacts with the iodine to produce iodide ions, which are measured by the iodide electrode system.

The Series 1700 range includes analyzers for the measurement of fluoride, alkalinity, sulfite, and chlorine (in cooling water); have the same basic de-

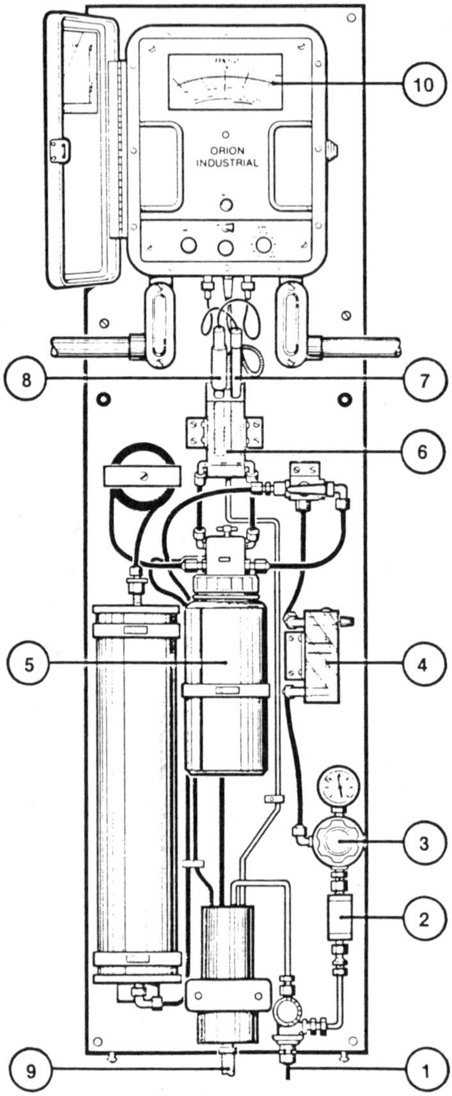

Sample stream enters the monitor at the sample inlet (**1**) and passes through a filter (**2**), a pressure regulator (**3**), and a flowmeter (**4**). Sample then passes through a reagent diffusion bottle (**5**), where pH adjustment takes place, and flows to the electrode holder (**6**). The holder contains sodium (**7**) and reference electrodes (**8**), as well as a thermistor for automatic temperature compensation. Sample then passes from the electrode holder to an atmospheric drain (**9**). Results are read directly on a four-decade logarithmic scale (**10**).

FIGURE 13.40 Model 151102 sodium monitor. (Courtesy of Orion Research)

sign; and have sample handling systems designed specifically to handle dirty water applications. The Model 1709 Fluoride Monitor shown in Fig. 13.41 has a range of 0.01 ppm to 10%, with user selection of any two or four decades within this range.

Process sample first flows past a Teflon inlet screen, where 13 ml/min is drawn into the analyzer by a sample pump. Sample bypass flow keeps the

TABLE 13.6 On-Line Specific Ion Monitors

Manufacturer	Country of Origin	Available Electrodes	Measurement Ranges	Operating pH Range
1. Foxboro Analytical	USA	Fluoride	10^{-6}–1.0 M	5–11
		Chloride	5×10^{-5}–1.0 M	0–14
		Bromide	5×10^{-6}–1.0 M	0–14
		Cyanide	10^{-6}–10^{-2} M	11–14
		Sulfide	10^{-7}–1.0 M	>13
		Copper 2+	10^{-8}–1.0 M	0–14
		Silver	10^{-7}–1.0 M	0–14
		Calcium	–	–
		Magnesium	–	–
		Sodium	10^{-6}–1.0 M	4–13
2. Horiba Instruments Ltd.	Japan	Sodium	0.1–10 ppb	
3. Kent Industrial Measurements Ltd.	Great Britain	Fluoride	10^{-6}–1.0 M	5–6
		Chloride	5×10^{-5}–1.0 M	4–5
		Sulfide	10^{-7}–1.0 M	>13 optimum
		Cyanide	5×10^{-7}–10^{-2} M	13 optimum
		Bromide	5×10^{-7}–1.0 M	4.5 optimum
		Iodide	5×10^{-7}–1.0 M	4.5 optimum
		Nitrate	10^{-6}–10^{-1} M	4–5 optimum
		Sodium	5×10^{-9}–10^{-1} M	pH > pNa + 3
		Calcium	5×10^{-6}–1.0 M	4–10
		Silver	10^{-7}–10^{-2} M	<8 optimum
4. Gam-Rad Inc.	USA	Chloride	$\begin{cases} 2.8 \times 10^{-7}\text{–}10^{-3}\text{ M} \\ 2.8 \times 10^{-6}\text{–}10^{-1}\text{ M} \end{cases}$	1–6
		Bromide	10^{-7}–10^{-3} M	1–12
		Iodide	3.9×10^{-8}–1.0 M	1–13
		Sulfide	3.1×10^{-8}–1.0 M	11–13
		Cyanide	1.9×10^{-7}–10^{-2} M	11–13
		Silver	4.6×10^{-8}–1.0 M	1–13
		Mercury	5×10^{-8}–1.0 M	1–5
		Copper 2+	3.2×10^{-8}–1.0 M	1–13
		Lead 2+	9.6×10^{-8}–1.0 M	1–13
		Cadmium	8.9×10^{-8}–10^{-1} M	1–13
		Sodium	10^{-8}–10^{-6} M	8–12
		Calcium		
		Magnesium		
5. Leeds & Northrup Co.	USA	Sodium	0.1 ppb to 100,000 ppm	11.3 (Buffered)
6. McNab Inc.	USA	Sulfide		9–11
		Cyanide		
		Sodium	0.1 ppb to 100 ppm	
		Potassium		

TABLE 13.6 (*Continued*)

Manufacturer	Country of Origin	Available Electrodes	Measurement Ranges	Operating pH Range
		Ammonium		
		Silver		
		Copper 2+		
		Lead 2+		
		Calcium		
		Magnesium		
7. Orion Research	USA	Sodium	0.1 to 1000 ppb	Buffered
		Sodium	0.1–100%	Buffered
		carryover	0.01 to 100,000 ppm	
		Fluoride	0–50 to 0–500 ppb	
		Chloride	0–10 ppm standard	
		Sulfite	0.005–10 ppm	
		Calcium		
		Cyanide		
		Alkalinity	0–50, 100, 500 ppm (as $CaCO_3$)	
8. Rosemount-Uniloc Inc.	USA	Fluoride	0.1–20,000 ppm	
		Chloride	10–35,000 ppm	
		Bromide	0.4–80,000 ppm	
		Iodide	0.001–130,000 ppm	
		Sulfide	0.003–35,000 ppm	
		Sodium	1 ppb to 25,000 ppm	
		Potassium	1.0–1000 ppm	
		Silver	0.01–110,000 ppm	
9. Wallace & Tiernan Inc.	USA	Fluoride	0.2–2.0 ppm standard	
10. Yokagawa-Electrofact	Netherlands	Chloride		
		Sulfide		
		Sodium		
		Calcium		

Teflon screen surface clean. Reagent is delivered from a reservoir by a reagent pump and mixed with the sample. An air pump is used to completely mix sample and reagent and provide agitation. Sample then flows through the measuring cell containing the fluoride electrode and out to drain. Calibration sample can be injected into the analyzer by means of a three-way valve. The transmitter provides local indication of fluoride ion concentration and voltage or current outputs for remote recording. Precision is quoted as ±5% of reading.

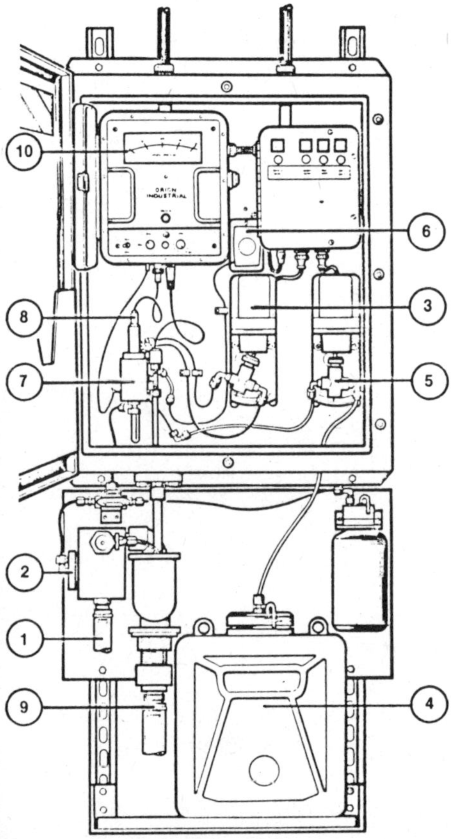

Sample is delivered to the sampling station (**1**) where it flows parallel to a Teflon inlet screen (**2**). The screen surface is kept clean by the bypass flow. Thirteen ml per minute of sample is drawn into the monitor by the sample pump (**3**). Reagent from a carboy (**4**) is delivered by a reagent pump (**5**) and mixed with sample. An air pump (**6**) is used to completely mix sample and reagent and to provide the agitation to keep plumbing lines and the flow cell (**7**) clean. Fluoride is detected in the turbulent flow by an electrode (**8**) after which the sample is discharged to a drain (**9**).

The fluoride electrode responds to the fluoride concentration and results are read directly on the meter scale (**10**).

FIGURE 13.41 Model 1709 fluoride monitor. (Courtesy of Orion Research)

13.9 ON-LINE WET CHEMICAL ANALYSIS

13.9.1 Background

As we saw earlier in this chapter, the majority of standard test methods used for laboratory measurement of water quality parameters use traditional wet chemical techniques. These are based on the development of intensely colored dyes or complexes by reaction of the sample with a selected reagent, followed by photometric or potentiometric measurement of the color intensity, which is proportional to the concentration of the component of interest. These tests are manpower-intensive and require accurate measurements by well-trained technicians and an adequately equipped laboratory.

Over the past twenty years, automatic analyzers have been developed and used successfully for standard tests in the laboratory. The first attempts to develop on-line wet chemical analyzers took the form of converting the automatic laboratory analyzers to process use. This was unsuccessful, as one would expect, since the laboratory analyzer was not capable of standing up to the rigorous environment in which it was placed.

With user and manufacturer experience gained over the years, coupled with significant advances in analyzer technology, the situation is different today. High-precision, automatic on-line analyzers are commercially available to carry out the standard tests as they are performed in the laboratory. Reagents are added in identical sequences, with mixing and time delay steps between additions, as the chemistry and laboratory procedures require. Where reactions are temperature dependent, a constant temperature bath is included as part of the analyzer hardware.

In general, three different techniques are applied for final measurement of component concentration:

- Coulometric measurement, which is the most suitable for trace components because of its high sensitivity.
- Volumetric measurement, mainly used for higher concentrations where direct titration is involved.
- Potentiometric measurement, where a specific ion electrode system is employed.

All these techniques involve precise metering of liquid and gaseous samples, reagents, and standard solutions, together with precise programming of reaction intervals, rinsing cycles, and so on.

If all the requirements described are satisfied, on-line wet chemical analyzers are extremely versatile and can be used to measure a wide range of components covering a variety of industrial applications, particularly those in water and waste water treatment processes. Some companies have specialized in the manufacture of this type of instrumentation.

13.9.2 Process Wet Chemical Analyzers

(a) Bran & Lubbe (West Germany)

B&L manufactures a wide range of on-line, wet chemical analyzers based on the colorimetric, titrimetric, or potentiometric principle of measurement and designed on the same modular basis. All instruments have current output and alarm facilities.

Colorimetric Analyzers. In these instruments, precisely metered amounts of reagent(s) are added to precisely metered amounts of process sample, resulting in a color-forming reaction. Color intensity is measured by a double-beam photometer. A typical program sequence, controlled by the analyzer program timer, includes the following steps:

1. Rinsing of the sample path.
2. Preparation of the zero solution.
3. Filling of the measuring cell.
4. Zero balance of the measuring bridge.
5. Preparation of the measuring solution.
6. Filling the measuring cell with measuring solution.
7. Measurement of the developed color intensity.
8. Recording of the measured value.

The following analyzers operate on the colorimetric principle.

The Series 270 Silikometer shown schematically in Fig. 13.42 operates on a 12-min cycle to measure silica in the range 0–20 ppb to 0–20 ppm, using the molybdenum blue standard method. The analyzer consists of an analysis unit with reagent storage and an electronics/programming unit with plug-in modules. Each unit is mounted in a separate compartment of a free-standing steel cabinet. The analyzer is capable of monitoring up to six sample streams. Measuring cell dimensions are chosen to meet the specific range requirements.

The Series 270 Phosphatometer is a special model designed to measure

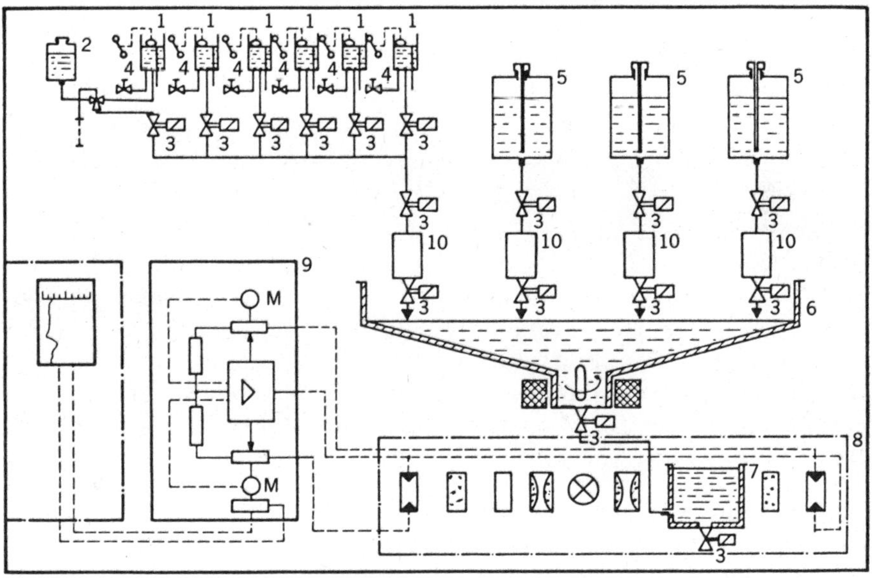

FIGURE 13.42 B & L Series 270 silikometer. KEY: 1, overflow vessels; 2, calibration solution vessel; 3, solenoid valves; 4, sample flow control valves; 5, reagent storage vessels with level switches; 6, reaction vessel; 7, measuring cuvette; 8, photometer; 9, compensator; 10, dispensing cylinders. (Courtesy of Bran & Lubbe GmbH)

ortho or total phosphates within the range 0–30 ppm, using the molybdate or vanadate/molybdate standard method. Cycle time is 12 or 30 min and up to three sample streams can be accommodated.

The Series 270 Kolorimeter is designed to automate any colorimetric method. The different programs available—metering volumes of 2–100 ml for sample and reagents and measuring cell thicknesses of 2–50 mm and a selective range of measuring wavelengths from 400–800 nm—make this a very versatile analyzer. Examples of measured components include aluminum, chlorine, chromium, iron, hydrazine, copper, manganese, hydrogen sulfide, and phenol.

Titrimetric/Volumetric Analyzers. In these instruments, a titrating reagent (titrant) of known concentration is added to a precisely metered sample until a measurable characteristic value of the mixture has reached a preselected point or suddenly changes. The pH value is used in many cases. The volume of titrant metered prior to the equivalence point is proportional to the measured component concentration. A typical program sequence, as controlled by the analyzer program timer, includes the following steps:

1. Rinsing of the sample path.
2. Emptying the titration vessel.
3. Metering of sample (and possibly reagent or dilution water) into the titration vessel.
4. Addition of titrant.
5. Stopping titrant addition by means of an indicator electrode.
6. Production of the measured signal.
7. Recording of the measured value.

The Titrometer U, (shown in Fig. 13.43*a*) is designed to carry out a classical titration over a 10-min cycle and has a wide range of industrial applications. The device can carry out both direct titration and back-titration, and is suitable for detection of several endpoints. Up to six sample streams can be monitored. Examples of measured components include calcium (hardness), chlorine, sodium hydroxide, hydrochloric acid, ammonia, sulfuric acid, acetic acid, and alkalinity. Minimum measurement ranges are from the ppm level (ammonia) up to percent level (acetic acid).

Potentiometric Analyzers. In these instruments, chemical treatment consists of adding reagent to the sample to prepare it for the electrode measurement. In most cases, this involves pH adjustment, stabilizing overall ionic strength, and eliminating possible interferences. The reaction product is transported to the measuring cell in which a measurable potential is produced by an ion-selective electrode system. This potential is proportional to the measured component concentration. A typical program sequence, as

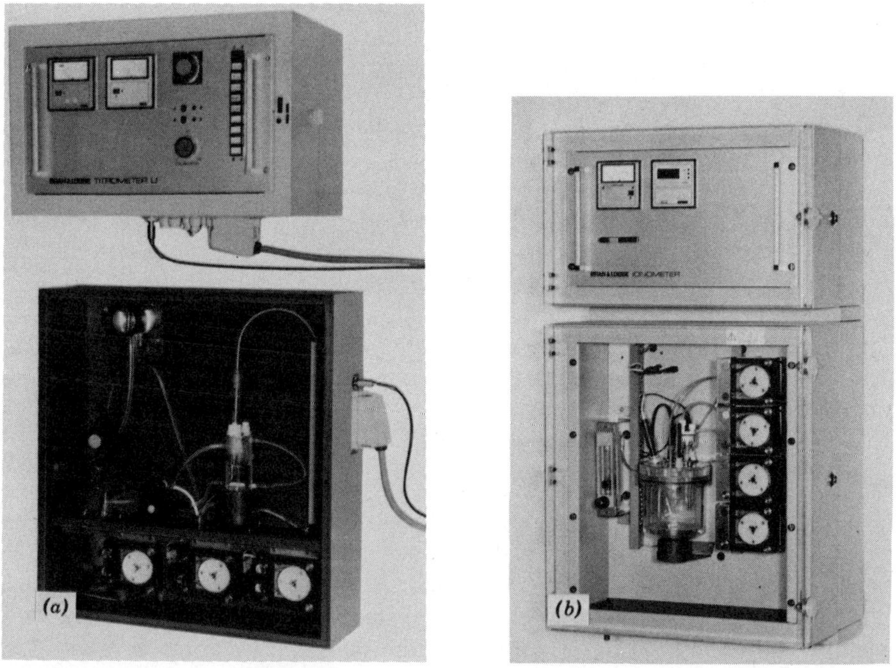

FIGURE 13.43 B & L water quality instruments. (*a*) B & L titrometer U; (*b*) B & L ionometer. (Courtesy of Bran & Lubbe GmbH)

controlled by the analyzer program timer, includes the following steps:

1. Metering of sample.
2. Metering of reagent (and possibly dilution water).
3. Mixing and conditioning of the measuring solution.
4. Flow of measuring solution into measuring cell.
5. Production and recording of measured value from the electrode system.

The following B&L analyzers operate on this principle.

The Ionometer (shown in Fig. 13.43*b*) is designed for measurement of ionic species in process, surface, and waste waters and operates on a continuous basis. Automatic calibration may be carried out on a preprogrammed basis between 1 and 60 hr and takes approximately 30 min. A continuous feeding system is provided for sample, calibration and buffer solutions using peristaltic pumps. Reagent storage capacity allows for at least 14 days operation. Examples of components measured are:

Ammonia/ammonium ion	0–0.1 to 0–1000 ppm
Chloride ion	0–5 to 0–5000 ppm

Fluoride ion	0–0.1 to 0–2000 ppm
Cyanide ion	0–0.1 to 0–100 ppm
Nitrate ion	0–2 to 0–1000 ppm
Sulfide ion	0–0.3 to 0–3000 ppm

The Natriometer is designed for measurement of sodium in steam, steam condensate, boiler feedwater, and treated water within the range 0.2–20 ppb to 0–1000 ppm and operates on a continuous basis. Peristaltic pumps are used for precise metering of sample, calibration, and buffer solutions. A gaseous buffer (ammonia) is used to avoid sodium contamination. The measuring cell has a magnetic stirrer and is thermostatically controlled, as in the Ionometer. Calibration is initiated by manual pushbutton. Up to six sample streams can be monitored.

(b) Hach Chemical Company (United States)

The Pump-Colorimeter Analyzer is designed to measure a variety of different water quality parameters based on standard colorimetric methods used in the laboratory. Basic mechanical operation and optical system design is shown in Fig. 13.44. The mechanical system consists of three parts—the sample transport pump, reagent pumps, and the flow control valve. The colorimeter cell through which the light passes also functions as the cylinder of the sample transport pump. All three parts are driven by an ac gear motor. Reagent and sample pistons are synchronized precisely for confluent reagent addition to the sample. Up to five reagent pumps can be accommodated.

The optical system is of single wavelength, dual beam design using a beam splitter, with the reference photocell located at 90° to the main light beam. Broad band or interference filters can be used, the wavelengths being chosen on the basis of measurement parameter and likely interference. Using this technique, compensation can be provided for moderate changes in sample color and turbidity.

A variety of voltage and current outputs are available for remote recording. High and low alarm contacts can be provided, with set points adjustable over the full analyzer range. Sample requirements are 10–100 ml/min at 10 psig minimum pressure and a temperature of 50–104°F. Reagent consumption is 0.95 l, suitable for 1 weeks operation.

The following analytical methods have been developed for the Pump-Colorimeter:

Parameter	Method	Range (ppm)
P-alkalinity	Thymol blue borate	0–50, 100*
T-alkalinity	Brown cresyl green acetate	0–50, 100, 500*
Chlorine dioxide	Chlorophenol red	0–0.5
Free chlorine	DPD	0–0.5, 1, 2, 5
Total chlorine	KI/DPD	0–0.5, 1, 2

Chromium	Diphenyl carbazide	0–0.2, 1
Hardness	Complexometric indicator	0–0.5, 1, 2, 5*
Manganese	KI/DPD	0–0.2
Molybdenum	Mercaptoacetic acid	0–10
Orthophosphate	Molybdovanadate	0–3, 10, 30
Permanganate	Direct colorimetry	0–5, 10, 20
Silica	Molybdenum blue	0–25, 50

* As calcium carbonate ($CaCO_3$).

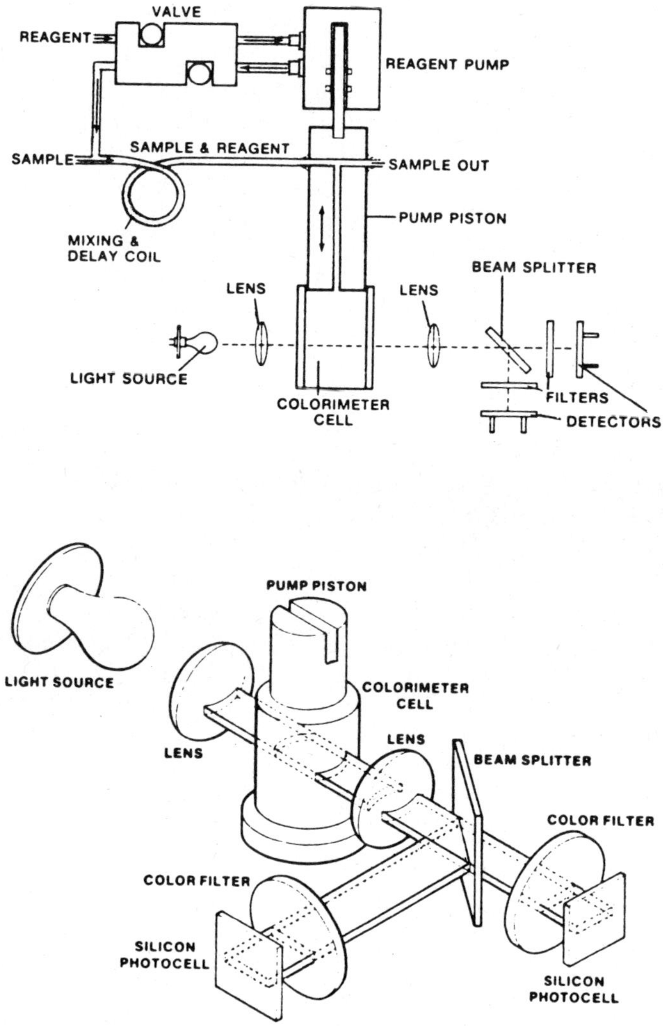

FIGURE 13.44 Pump-colorimeter analyzer. (Courtesy of Hach Chemical Company)

(c) *Ionics Inc. (United States)*

The Digichem 3000 Series of Programmable Chemical Analyzers are microprocessor-based instruments designed to duplicate a wide variety of laboratory analysis methods using a single physical configuration. The analyzer shown in Fig. 13.45 consists of four basic modules—the sample and reagent burettes, the rotary cell assembly, the detector assembly, and the microprocessor/electronics module.

The central feature of the system is the rotary cell assembly, which receives sample and reagent(s) from the burette modules and provides efficient mixing for rapid chemical reaction. Constructed of Teflon, the cell can rotate at up to 1500 rpm. Sample and reagent delivery is carried out by the burette modules, which can number up to five. They are driven by electrical stepping motors with a resolution of 2000 pulses per stroke. Fluid switching is performed by ceramic sliding-plate valves.

The system can accommodate one or more sensor techniques for monitoring reactions and determining endpoints. Electrochemical sensors such as pH, ORP, and selective ion types can be mounted directly in the reaction cell. For colorimetric endpoint, a non-contact colorimeter can be fitted. For direct colorimetric measurement, a flow-through colorimeter can be used.

The overall operation of the analyzer is supervised by a microprocessor, which has three basic functions:

1. Mechanical control of the rotary cell, burette drives, diverter, and rinsing valves.
2. Sensor monitoring.
3. Subsequent control of analysis routines and data handling.

The microprocessor also performs automatic calibration, data verification, and internal diagnostic routines. Up to 12 analog outputs are available, as well as a digital RS-232C serial interface with a 20 mA current loop. All data entry and operating functions are performed through a key-lock protected keyboard on the front panel of the instrument.

These are some of the typical analyses that can be carried out by the Digichem 3000:

Inorganics	Organics	Dissolved Metals
Boric acid	Amines	Aluminum
Carbonate	Formaldehyde	Cadmium
Chromate	Mercaptans	Copper
Cyanide	Hydrazine	Iron
Fluoride	Sulfonic acids	Manganese
Phosphate		Lead
Silica		

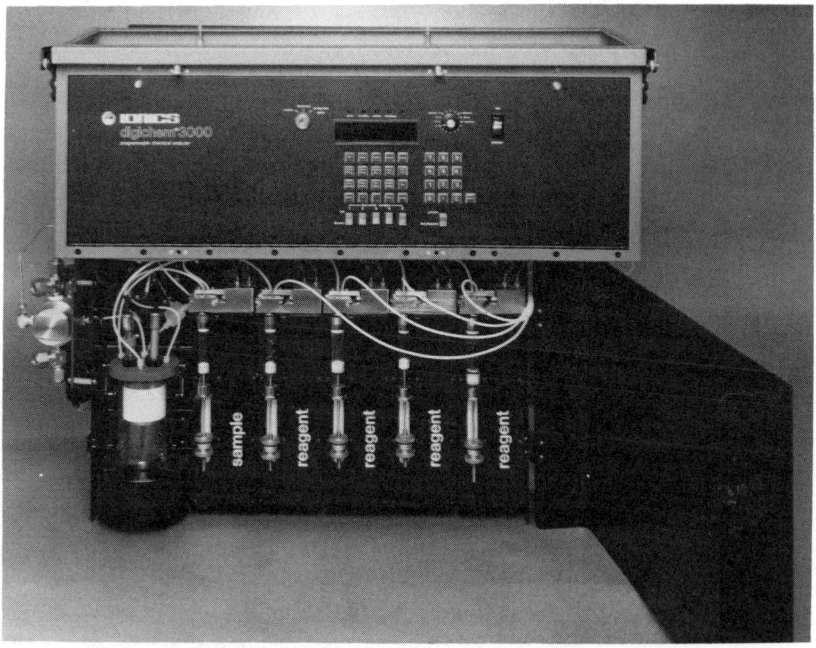

FIGURE 13.45 Digichem 3000 analyzer. (Courtesy of Ionics Inc.)

(d) Kent Industrial Measurements (Great Britain)

EIL Analytical Instruments, part of the Kent organization, manufactures a broad range of wet chemical analyzers for industrial on-line use, particularly in water and waste water treatment processes. These instruments have similar physical configurations, consisting of a modular electronics unit and an analysis unit, mounted in separate sections of a cabinet suitable for wall, bench or panel mounting.

The Model 8061 On-Line Silica Monitor shown in Fig. 13.46 is typical of the range and is designed for measurement of silica in aqueous process streams within the range 0–20 to 0–2000 ppb. Analytical chemistry is based on the molybdenum blue standard method. The instrument is mounted in a temperature controlled steel cabinet suitable for bench or surface mounting and divided into two sections, the lower analytical section and the upper section, which contains the rack mounted modular electronics.

Process sample and reagents are drawn into the analyzer by a multichannel peristaltic pump. Prior to the pump, the sample is divided into reference and measurement streams, the reference stream passing directly to the reference cuvette of the photometer. For the measurement stream, reagent addition is carried out sequentially in three mixing chambers. After each chamber, sample and reagent pass through a reaction/temperature equalization coil and finally flow to the measurement cuvette of the photometer.

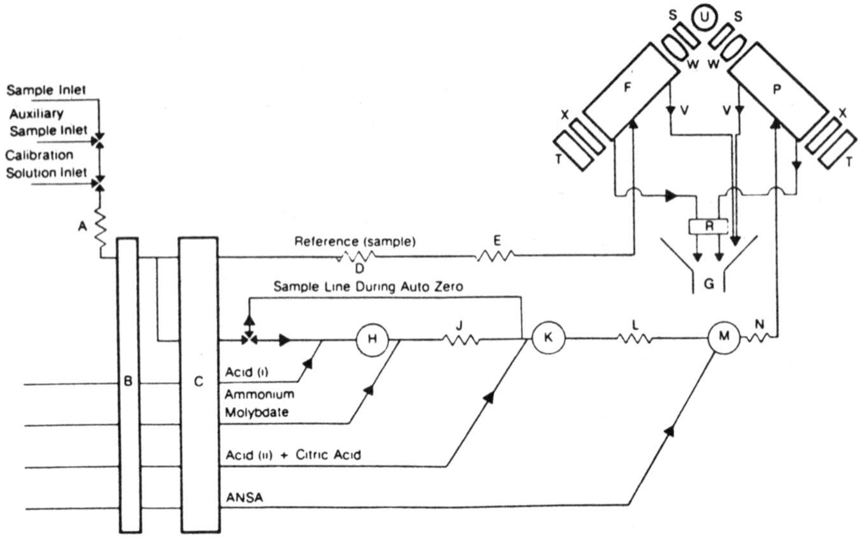

A. Heat Exchanger
B. Inlet Sensor Block
C. Multichannel Pump
D. Delay Coil
E. Temperature Equilibration Coil
F. Reference Cuvette
G. Main Drain
H. 1st Stage Mixer
J. Temperature Equilibration/Reaction Coil
K. 2nd Stage Mixer
L. Temperature Equilibration/Reaction Coil
M. 3rd Stage Mixer
N. Temperature Equilibration/Colour Development Coil
P. Sample Cuvette
R. Pinch Valve
S. Heat Filter
T. Photoelectric Cell
U. Light Source
V. Overflow
W. Lens
X. Cut-off Light Filter

FIGURE 13.46 Model 8061 silica monitor. (Courtesy of Kent Industrial Insts. Ltd.)

Although sample flows continuously, the cuvettes operate on a fill/overflow/drain cycle of 1 min, designed to eliminate interference by air bubbles during the measurement period. The use of reference and measurement cuvettes and an auto-zero facility eliminates errors due to reagent coloration, cuvette fouling, and reagent contamination by silica.

The analyzer may be zeroed manually or on a preprogrammed basis, adjustable between 3 hr and 8 days. Several analyzer malfunction alarms are provided to assist in troubleshooting. A special cabinet is available as an option for reagent storage and contains a 40-l supply of auxiliary sample sufficient for 4 days operation. A multistream selector is also available, enabling the analyzer to sample up to six process streams. A variety of voltage and current outputs are available for remote recording.

The Model 8023/8033 Sodium Monitor shown in Fig. 13.47 is designed for measurement of sodium in aqueous process streams within the ranges 0.1–10 ppb and 10–1000 ppb, using a sodium ion-selective electrode system. Construction and layout is similar to the Model 8061.

Process sample enters the instrument at a constant head tank and is drawn into the analysis unit by a peristaltic pump. Provisions are made for injection of primary calibration sample (for auto-standardization) and manual calibration sample. Adjustment to solution pH is made by diffusion of ammonia or

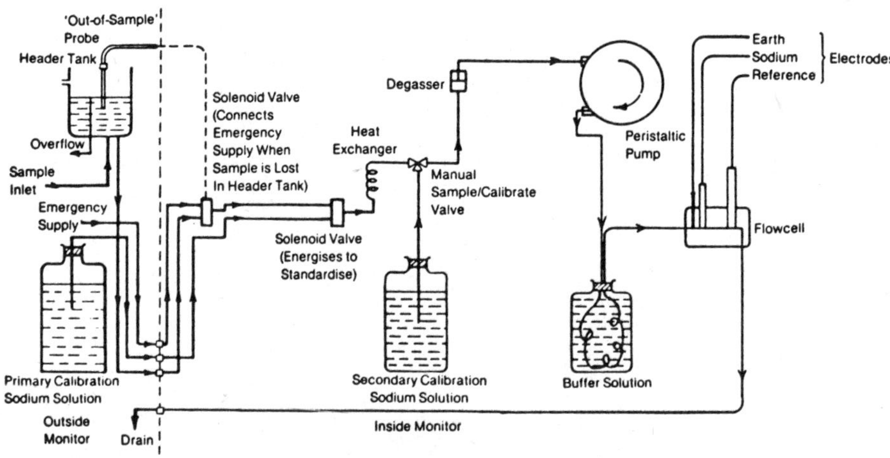

FIGURE 13.47 Model 8023/8033 sodium monitor. (Courtesy of Kent Industrial Insts. Ltd.)

diethylamine vapor through the walls of a loop of silicone tubing immersed in a reagent solution reservoir.

The electronics unit consists of seven modules, the primary functions of which are signal amplification, signal isolation and output, auto-standardization, alarm generation, power supply, analyzer function control, and digital readout. An optional analog display module is also available. A variety of current outputs are available for remote recording.

Other instruments are available for the measurement of total iron and copper, fluoride, ammonia, chloride, nitrate, and phosphate.

(e) Polymetron Ltd. (Switzerland)

Polymetron manufactures a range of microprocessor-based on-line wet chemical analyzers for measurement of water quality parameters.

The Silkostat shown in Fig. 13.48 is designed to measure silica in water within the range 0–20 to 0–500 ppb or 0–1 to 0–20 ppm, based on the molybdenum blue standard method. Sample and reagents are proportioned continuously by multi-channel peristaltic pumps and time delay steps are built into the program to allow adequate reaction time. The color intensity of the complex formed by the reaction is measured at 790 nm by a single-beam, dual-wavelength photometer. Analyzer cycle time is adjustable between 5 and 30 min. Up to six sample streams can be accommodated, with automatic switching. The individual sample streams are directed into overflow sample vessels equipped with level controls; solenoid-operated sampling valves are switched on a programmed basis by the microprocessor.

The microprocessor coordinates all the controlling functions of the analyzer, computes, monitors and displays critical parameters, and carries out internal self-diagnostic routines to ascertain correct analyzer operation.

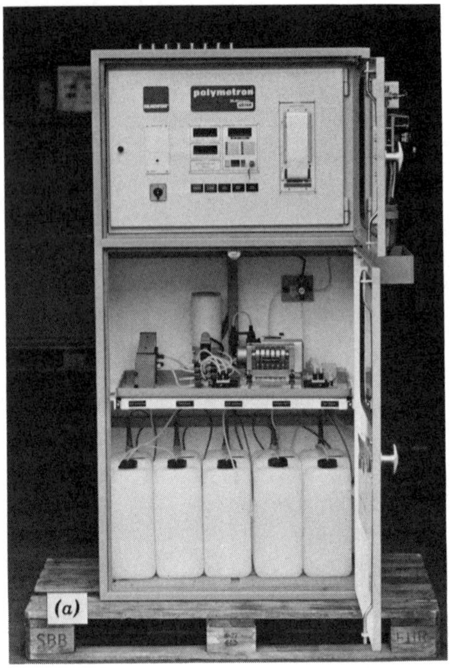

FIGURE 13.48 Polymetron Silkostat silica analyzer. (*a*) Analyzer cabinet. (Courtesy of Polymetron AG)

Flexible software allows a wide variety of programmable, specific features. Current outputs are available for remote recording, as well as alarm facilities. Response time is quoted as less than 10 min to 90% full scale at 20 ppb silica concentration.

The Calcimat is designed to measure total hardness (Ca + Mg) of industrial waters within the range 0–300 ppm. The Eriochromblack-T analysis method is used. Measurement of the color intensity is made at 645 nm by a single-beam, dual-wavelength photometer. Physical configuration is similar to the Silkostat.

The Sodimat is designed to measure sodium concentration in high-purity waters within the ranges 0–0.1, 1, 10, or 100 ppb or 0–0.1, 1, or 10 ppm, depending on the particular application. The analyzer uses a sodium-sensitive glass electrode as the sensor. Gaseous ammonia is used to adjust the pH of the sample before measurement to 10.3–10.5. The analyzer is microprocessor based and of similar physical layout to the Silkostat and Calcimat.

(f) Technicon Industrial Systems (United States)

The Monitor 650 Process Control Analyzer shown in Fig. 13.49 is a microprocessor-based wet chemical analyzer using a colorimetric detection sys-

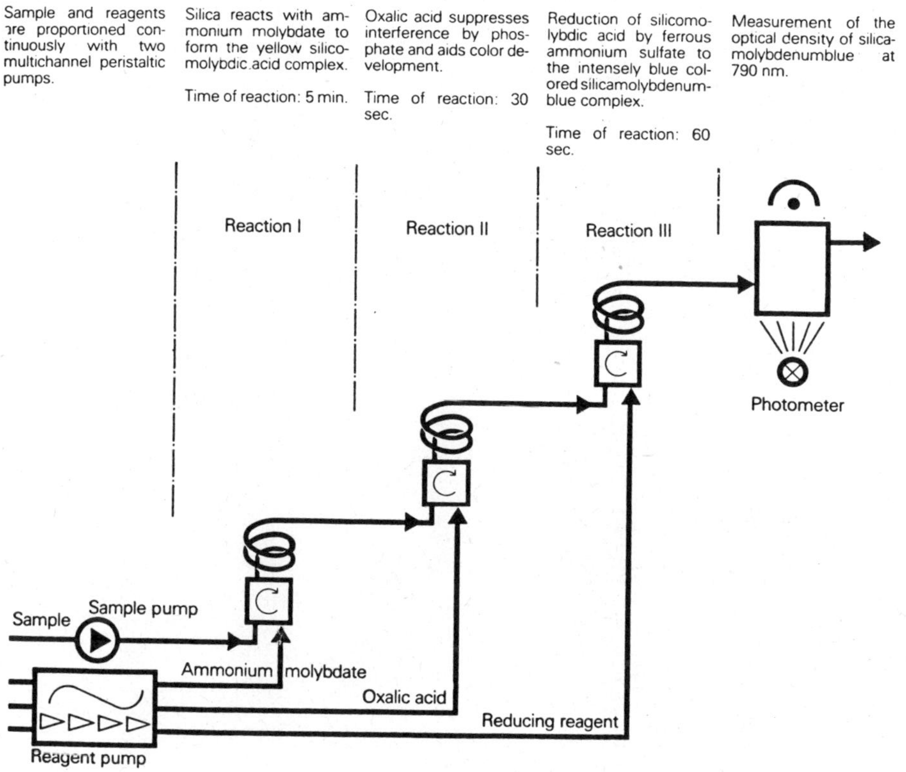

(b)

FIGURE 13.48 Polymetron Silkostat silica analyzer. (*b*) Method of operation. (Courtesy of Polymetron AG)

tem. The instrument is mounted in a NEMA 12 free-standing cabinet, which can be provided with Type Z purge to meet hazardous area installation requirements. The cabinet has three sections—the reagent storage section; the chemical section, which houses the overflow sampler, colorimeter, peristaltic pump(s), and manifold assembly; and the microprocessor/electronics section.

Reagent storage is suitable for 30 days operation. The colorimeter can accommodate flow cells from 10–75 mm path length and has a wavelength of 400–800 nm. A beam splitter/chopper configuration and filter optic system generate measurement and reference beams, which are alternately read by the detector. Dual and six-stream sampling capability is also offered.

The microprocessor controls all analyzer operating functions, processes output data, and provides alarms and diagnostics for the complete system. Automatic, preprogrammed calibration is also provided. In addition to voltage and current outputs, a host computer interface is available through an

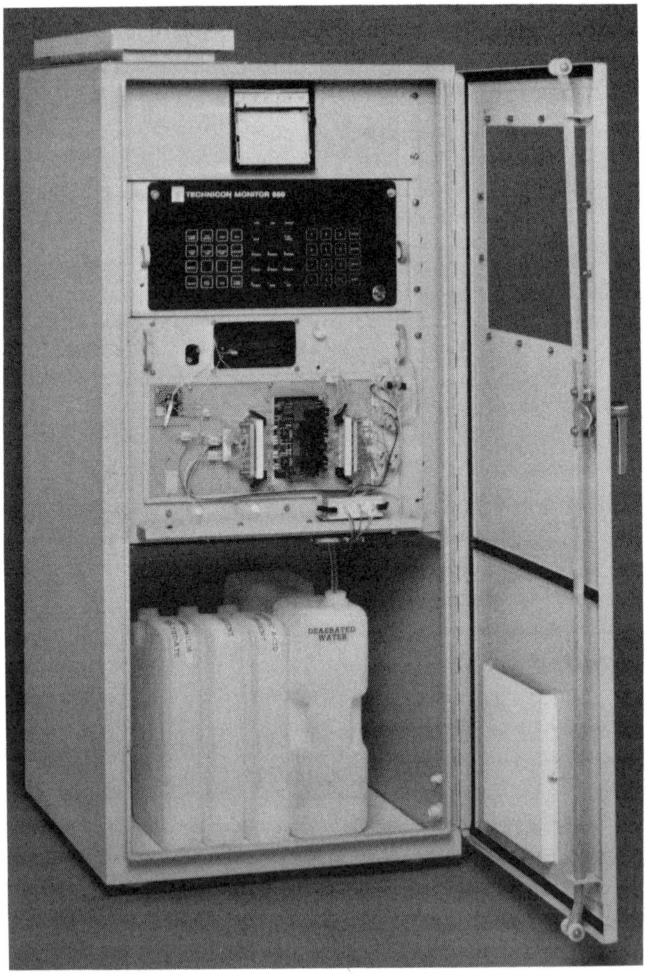

FIGURE 13.49 Monitor 650 analyzer: field cabinet. (Courtesy of Technicon Instruments Inc.)

optional RS-232C data link. A digital printer interface is also available. Over 200 analytical methods have been developed for the Monitor 650, covering a wide variety of measurement parameters. Technicon also offers to customize its analyzer to meet individual user requirements.

13.10 DISSOLVED OXYGEN MEASUREMENT

13.10.1 Background

Dissolved oxygen is present in all surface waters and rain waters by virtue of their contact with the atmosphere. If sufficient agitation has taken place in contact with the air, the dissolved oxygen content of the water will be equal to the saturation value at the temperature of the water. Mineral content of the water will also affect oxygen solubility to a minor extent. For instance, distilled water will absorb more oxygen than waters containing higher dissolved solids; sea water will dissolve less oxygen than fresh water; well water usually contains less oxygen than surface water, and water from deep wells in particular may contain no dissolved oxygen.

The major problem with dissolved oxygen in water used for industrial purposes is the corrosive effect it has on iron and steel equipment with which the water comes into contact. For example, failure of piping may occur when lines become blocked with the products of corrosion. Higher temperatures and low pH values accelerate oxygen attack. In boiler systems, corrosion can take place in feed lines, heaters, economizers, boilers, and steam and condensate return lines.

In cooling water systems, corrosion of heat exchangers, cooling water jackets on mechanical equipment, pumps, and piping can take place. Although corrosion can take place in the absence of oxygen, its presence is the major contributing factor to corrosion. The usual procedure in the investigation of any corrosion problem is to first obtain data on the dissolved oxygen content throughout the system.

When iron or steel is brought into contact with water, the iron goes into solution as ferrous hydroxide:

$$Fe + 2H_2O \rightarrow Fe(OH)_2 + 2H^+$$

If dissolved oxygen is present, it oxidizes the ferrous hydroxide to ferric hydroxide, which is insoluble:

$$4Fe(OH)_2 + O_2 + 2H_2O \rightarrow 4Fe(OH)_3$$

By removing ferrous hydroxide from the solution, the reaction equilibrium is upset and the reaction will proceed to completion until either the metal or the oxygen in the system is completely exhausted. The rate of solution of ferrous hydroxide is dependent on the pH of the water, the lower the pH the more rapid the rate of solution. If the water also contains carbon dioxide in solution as carbonic acid (H_2CO_3), this may lower the pH to 5.6–6.5, which is more conducive to the solution of ferrous hydroxide. Therefore it is important to remove carbon dioxide from the system and to increase the pH.

For boiler feedwater in particular, mechanical deaeration is commonly used to remove dissolved gases such as oxygen, carbon dioxide, and ammonia. This process involves raising the water to the saturation temperature for the particular operating pressure. Equipment is designed to provide intimate mixing of the water and scrubbing steam and to allow venting of released gases from the system.

Chemical deaeration may also be used for dissolved oxygen removal, usually accomplished by sodium sulfite, which reacts with oxygen to form sodium sulfate. At cold water temperatures, catalyzed sodium sulfite is required to provide a sufficiently rapid reaction.

Indirect methods may also be used to prevent corrosion. An example is the use of corrosion inhibitors such as alkaline chemicals and chromates. These form a protective scale film or oxide layer on the surface of the pipe, thus preventing contact of dissolved oxygen with the protected surface.

In the case of cooling water systems, aeration is often used to remove dissolved gases such as carbon dioxide and hydrogen sulfide. This is an efficient process for these gases, which are relatively soluble in water. However, the process results in saturation of the water with oxygen.

13.10.2 Process Analyzers for Dissolved Oxygen Measurement

A wide range of instrumentation is available for measurement of dissolved oxygen in industrial applications. Some items in this range are described next, and a more complete listing is given in Table 13.7 on pages 606–607.

(a) Beckman Industrial Inc. (United States)

The Model 7001 Dissolved Oxygen Monitor is designed to measure dissolved oxygen in high-purity waters within the range 0–50 to 0–200 ppb. The Model 7002 is designed to measure within the range 0–2 to 0–20 ppm in waste water, natural waters, and petroleum/petrochemical process streams. The heart of both analyzers is the Beckman polarographic oxygen sensor described in Chapter 7, Section 7.6.3. Various flow-through and submersion/immersion type configurations are available in PVC or Ryton for sensor mounting.

The analyzer/transmitters are housed in general purpose, NEMA 4 enclosures; a type Z purge kit is also available to meet hazardous area installation requirements. Maximum separation between sensor and transmitter is 1000 ft. Voltage and current outputs are available for remote recording, as well as single and dual alarm facilities.

(b) Leeds & Northrup (United States)

The Model 7931 Dissolved Oxygen Analyzer consists of a unique electrochemical sensor and an analyzer/transmitter mounted in a NEMA 4X weatherproof enclosure. The sensor shown schematically in Fig. 13.50 consumes oxygen at a multiplicity of cathodes in a similar fashion to conventional sensors. However, a like amount of oxygen is generated at a multiplicity of anodes. A balance exists so that there is no net consumption of oxygen (or other reactants) and no net generation of products. The anodes and cathodes are interspaced along an insulating support and covered with a rigid membrane. The reference electrode is located in the center of the support and immersed in the electrolyte. A small hole in the support allows electrolyte to bathe the anode/cathode area. A temperature compensation thermistor is mounted at the probe tip.

L&N Model 7931 dissolved oxygen analyzer

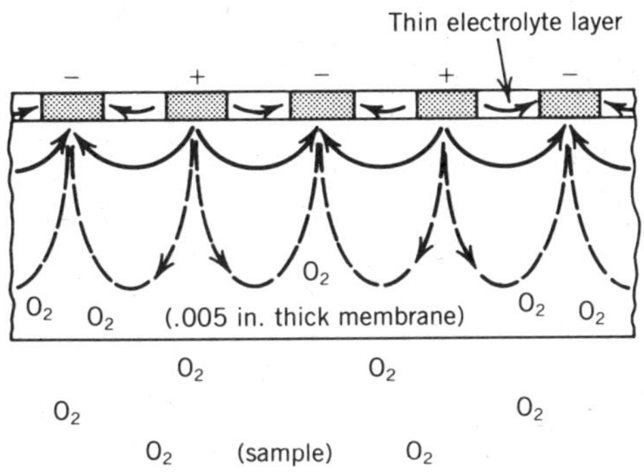

Schematic of the L & N oxygen sensor

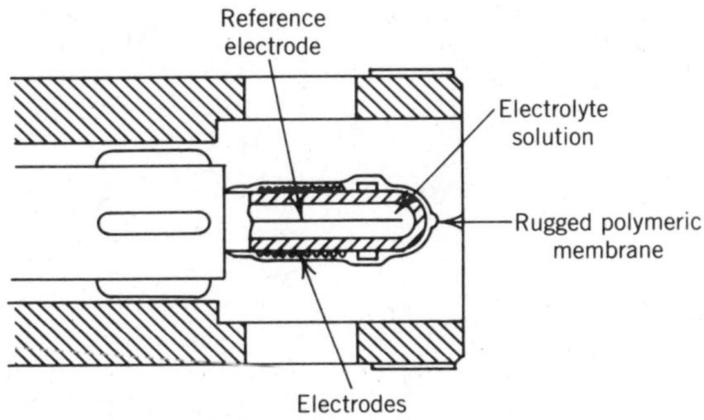

Diagram of L & N sensor construction

FIGURE 13.50 L & N dissolved oxygen sensor. (Courtesy of Leeds & Northrup Company)

When immersed in a process stream, oxygen will diffuse through the membrane until the oxygen tension on both sides of the membrane is equal and balanced. The current necessary to maintain the equilibrium is proportional to the dissolved oxygen concentration in the sample. The probe is housed in a rugged PVC enclosure and is supplied with 20 ft of waterproof cable. The main advantage of this sensor is that its operation is much less susceptible to fouling, making it ideally suited to waste water treatment applications. The balanced reaction of the sensor eliminates the problems of electrode oxidation and electrolyte depletion. In addition, the use of a heavy membrane covering extends the life of the sensor.

The analyzer/transmitter houses the measuring and output circuitry and an indicating meter and has switch-selectable ranges of 0–5 and 0–15 ppm. Isolated voltage or current outputs are available for remote recording, as well as optional high/low alarm facilities.

(c) Orbisphere Laboratories (United States)

Orbisphere manufactures a wide range of oxygen measurement systems, using polarographic sensors suitable for dissolved oxygen in aqueous process streams, oxygen partial pressures in process gases, as well as the Series 2727 systems designed specifically for low-level oxygen measurements in beer applications (e.g., bottle and can sampling). A few examples of the range of sensors and instrumentation are described next.

The Model 2120 Polarographic Sensor, which has a range of 0–2 to 0–200 ppm, features a novel method of attachment of the oxygen-permeable membrane to the sensor body by means of a metal ring. The membrane is lightly drawn but unstretched over the sensing face so that it always maintains its original thickness and its position relative to the cathode remains constant. The membrane is pressed tightly against a rubber O-ring on the cylindrical outer wall, creating a watertight seal isolating the interior from the exterior of the sensor. A specially formulated electrolyte can be used to make the sensor insensitive to hydrogen sulfide concentrations up to 3000 ppm. The sensor is thus ideally suitable for measurement in effluents, sewage plants, and natural waters where H_2S is present. The sensor is also insensitive to ammonia, sulfur dioxide, carbon dioxide, hydrazine, and suspended solids.

The Model 2120 sensor is available as a component of four Orbisphere analysis systems:

- System 2714. For dissolved O_2 measurement in flowing water systems.
- System 27141. For dissolved O_2 measurement in still waters.
- System 2715. For O_2 fugacity, tension, or partial pressure in gases of flowing liquids.
- System 27151. Similar measurement in still liquids.

The Model 2110 High-Sensitivity Polarographic Sensor shown in Fig. 13.51 is capable of dissolved O_2 measurement within a range from less than 1 ppb up to 20 ppm. The sensor incorporates a "guard ring" electrode, the function of which is to eliminate the contribution to the observed signal from oxygen originating from within the sensor. In conventional polarographic sensors, oxygen arrives at the cathode surface, not only by way of the permeable membrane, but also from within the sensor itself. This internal contribution is negligible provided that the rate of oxygen permeation through the membrane is high (e.g., dissolved O_2 concentrations greater than about 100 ppm), but is a dominant factor at very low concentrations.

The Model 2713 High-Sensitivity system is designed for dissolved O_2 measurement in boiler feedwater and consists of the Model 2110 Sensor, Model 2606 Indicator/Transmitter, and the Model 2950 Flow Chamber. Four standard ranges are offered, from 0–20 ppb up to 0–20 ppm. Digital indica-

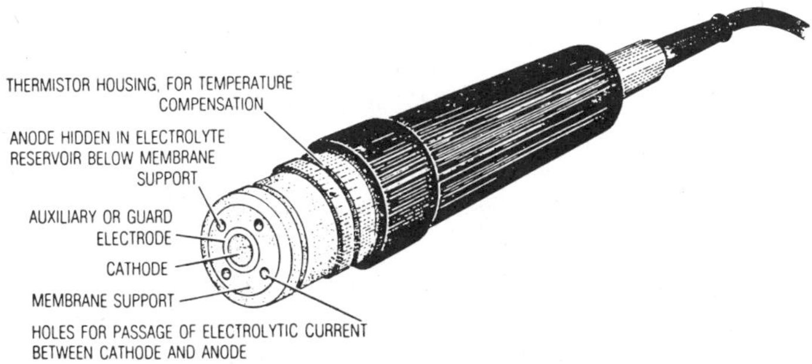

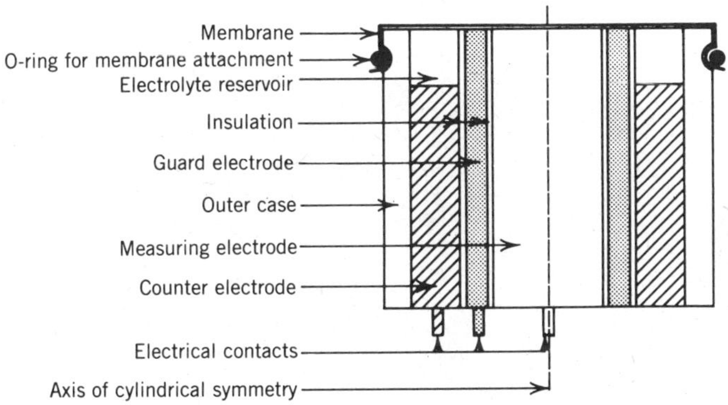

FIGURE 13.51 Model 2110 dissolved O_2 sensor. (Courtesy of Orbisphere Laboratories Inc.)

tion is provided and voltage or current outputs are available for remote recording. Response time is quoted as 10 sec to 90% full scale.

The Model 2714 System is designed for dissolved O_2 measurement in natural waters, effluents, and for BOD (biological oxygen demand) measurement. Standard ranges offered are 0–2, 20, and 200 ppm oxygen and 0–100°C temperature. Digital indication of oxygen and temperature are provided, as well as voltage outputs for remote recording of these parameters.

The Model 2609 Oxygen Indicator is a microprocessor-based device for measurement of oxygen partial pressure in gases, dissolved O_2 in fresh or saline waters, and dissolved O_2 fugacities in aqueous or non-aqueous liquids. The instrument, used in conjunction with the Orbisphere polarographic sensors, is mounted in a portable, waterproof case and can be mains or battery operated. A self-contained battery charger is fitted.

The Model 2610 Multi-Channel Indicator shown in Fig. 13.52 provides online measurement of gaseous or dissolved O_2 concentration for three to five sample streams. A variety of sensors and measurement ranges are available. The instrument provides digital indication of oxygen and temperature for each channel and may be located at up to 1500 ft from the sensors.

TABLE 13.7 On-Line Dissolved Oxygen Analyzers

Manufacturer	Country of Origin	Model No.	Measurement Principle	Mounting Configurations	Typical Measurement Ranges	Temperature Compensation	Response Time	Associated Instruments
1. Beckman Industrial Inc.	USA	7001 7002	Polarographic cell Polarographic cell	Flow-thru Submersion	0–50, 100, 200 ppb 0–2, 10, 20 ppm	Available as standard	90% in 20 sec	Transmitters (same model nos)
2. Exidyne Inc.	USA	20 (Probe) 6320 (Trnsm.)	Galvanic cell	Submersion	0–5, 10, 20 ppm	Available as standard	95% in 60 sec	Transmitter (6320) Series 6100 System
3. Fischer & Porter Co.	USA	Series 17DO1000	Amperometric cell	Submersion	0–2, 5, 10, 20 ppm	Available as standard (0–50°C)	90% in 30 sec	Transmitter (same model no.)
4. Great Lakes Insts. Inc.	USA	76 (Züllig)	Galvanic cell	Submersion	0–2, 10 ppm	Available as standard	99% in 3 sec	Model 176/376 transmitter; specifically designed for activated sludge applications
5. Hartmann & Braun AG	West Germany	CFOG (transm.)	Polarographic cell	Flow-thru Submersion	0–5, 15 ppm	Available as standard		Plug-in card system CFOZ-1 (multiparameter)
6. Horiba Insts. Ltd.	Japan	K-7 system	Polarographic cell	Flow-thru Submersion	0–20 ppm	Available as standard		Transmitter (9440) Controllers (9450 & 9455)
7. Kent Industrial Measurements Ltd.	Great Britain		Galvanic cell	Flow-thru Submersion	0–3, 10, 30 ppm	Available as standard		Transmitter (9440) Controller (9450) Controller (9455)
8. Leeds & Northrup Co.	USA	7931 7931-15	Equilibrium cell (multi-electrode)	Submersion	0–5, 15 ppm 0–20, 200 ppb	Available as standard	90% in 40 sec	Transmitter (same model no.)

9. Orbisphere Laboratories	USA	2120 (sensor) 2110 (sensor)	Polarographic cell (guard ring electrode)	Submersion	0.5 ppb–20 ppm 0.02–1000 ppm	Available as standard (0–75°C)	90% in 15 sec	Model 2714 Model 2715 Model 2609 (μP based) Model 2610 (multipoint) Model 2713
10. Philips GmbH	West Germany	PW9612	Polarographic cell	Flow-thru Submersion	0–30 ppm	Available as standard		Transmitter (KS201)
11. Rexnord Inc.	USA	62	Galvanic cell	Submersion	0–3, 15 ppm	Available as standard	95% in 40 sec	Model 3000 transmitter
12. Rosemount-Uniloc Inc.	USA	430	Polarographic cell	Flow-thru Submersion	0–5, 10, 20 ppm	Available as standard		Model 424 ''Sludge-Hammer'' Model 803/815 transmitters Model 1180D) two-wire transmitter
13. Siemens AG	West Germany	M54301	Polarographic cell with dual condy. measurement	Flow-thru Submersion	0–10 to 0–1000 ppb	Available as standard		M54302-A11 transmitter
14. Xertex Corp.	USA	Series 8210/8310	Polarographic cell	Submersion	0–5, 10, 20 ppm	Available as standard	90% in 20 sec	Also measures temp. simultaneously
15. Yokagawa-Electrofact	Netherlands	Type Z152/550	Polarographic cell	Submersion	0–5, 10, 20 ppm	Available as standard (0–50°C)		ELTRAN range of transmitters (''field,'' ''indoor,'' and ''compact'')
16. pHOX systems Ltd.	Great Britain	IC600	Galvanic cell	Submersion	0–2.5, 5, 10, 20 ppm	Available as standard	90% in 30 sec	65B/GS transmitter

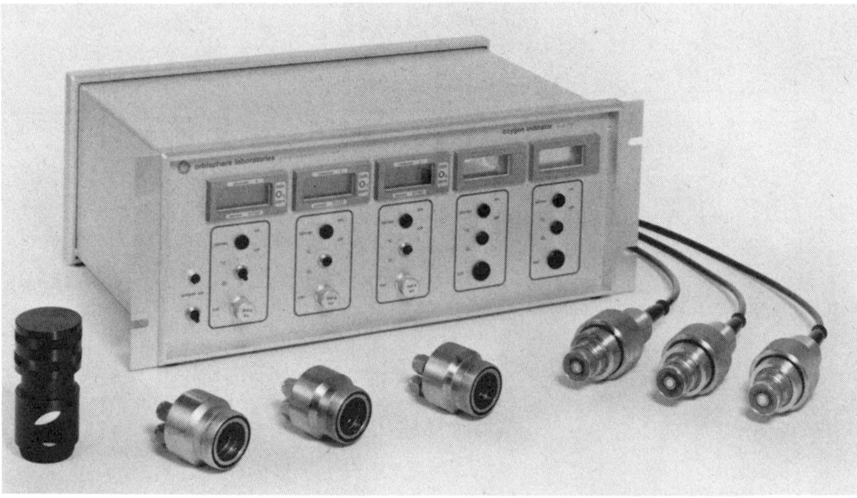

FIGURE 13.52 Model 2610 multichannel system. (Courtesy of Orbisphere Laboratories Inc.)

(d) Xertex-Delta Analytical (United States)

The Series 8210/8310 Dissolved Oxygen and Temperature Analysis System uses an in-situ, polarographic sensor for measurement of dissolved O_2 in aqueous process streams within the range 0–20 ppm. The analyzer/transmitter is available in a NEMA 1 general purpose (8210) or NEMA 4 weatherproof (8310) enclosure.

The sensor consists of an oxygen-permeable membrane, gold cathode, and silver anode and is filled with buffered potassium chloride electrolyte. Dissolved O passes through the membrane and is reduced at the cathode:

$$\text{Cathode reaction} \qquad O_2 + 2H_2O + 4e = 4OH^-$$

$$\text{Anode reaction} \qquad 4Ag + 4Cl^- = 4AgCl + 4e$$

A thermistor located in the sensor provides temperature compensation within the range −5 to +50°C. Dissolved oxygen and temperature are measured simultaneously. A variety of voltage or current outputs are available for remote recording, as well as high/low alarm capability.

13.11 TOTAL CARBON/ORGANIC CARBON MEASUREMENT

13.11.1 Background

Total carbon (TC) and total organic carbon (TOC) are two important water quality measurement parameters, the latter being particularly important in the petroleum and petrochemical industries. Continuous measurement of TOC in water is an excellent means of tracking and controlling organic spills, organic loading of influent streams to water treatment plants, and effluent discharges to the environment for regulatory compliance purposes. The classical biological oxygen demand (BOD) and chemical oxygen demand (COD) tests have long been used as indicators of water quality. The BOD analysis measures all molecules that exert an oxygen demand, but in addition to being a lengthy five or seven-day analysis it is often subject to significant variation. The COD analysis is shorter, but suffers from shortcomings in oxidation efficiencies. Direct correlation between BOD, COD, and TOC is usually not possible. However, with proper interpretation, the TOC can represent a rapid and quite accurate method of assessing pollution levels.

Ever-tightening regulations regarding waste emission often require that liquid effluents be "polished" by passing through an activated carbon bed before discharge to the environment. In addition, the process is frequently used for municipal water treatment before public use. The use of continuous TC/TOC analyzers can help reduce the cost of this expensive type of treatment through process optimization.

Another important application is in the monitoring of boiler steam/hot condensate recycle systems. Steam is often used in process heaters and heat exchangers. Leakage in this equipment will deposit hydrocarbons into the condensate, which will eventually find their way back to the boiler; if undetected, they can cause serious problems such as loss of heat transfer efficiency, corrosion, and tube burn-out. Measurement of TC is often preferred in these applications because of the possibility of inorganic carbonates build-up, which can also cause problems.

In a similar vein, many closed cooling water systems are susceptible to leakage from various process equipment, particularly process coolers (heat exchangers). Material lost into the cooling water system can result in reduced product yield, increased treatment costs for removal, loss of hydrocarbons to the atmosphere in cooling towers, and loss of cooling efficiency due to heat exchanger fouling. If the cooling system is open, that is, supplied from and returned to a natural water body, one is faced with the additional serious problem of major spills due to catastrophic heat exchanger failure.

Practical application of on-line TC/TOC analyzers began in the early 1960s, when Dow Chemical patented a combustion technique that was licensed to Beckman Instruments. In the early 1970s, Union Carbide introduced an analyzer to measure total carbon; subsequently, Ionics purchased the product line. Astro Resources introduced a patented non-catalytic combustion TOC analyzer in 1971. This company also manufactures a combustion type TOD (total oxygen demand) analyzer. The Dohrmann Division of Envirotech introduced its on-line TOC analyzer in 1975; Dohrmann is now part of Xertex.

The field performance of on-line TC/TOC analyzers is largely related to the cleanliness of the process stream being monitored and the treatment technique used to remove inorganic carbon (e.g., carbonates and bicarbonates). With clean process streams such as boiler feedwater and condensate recycle, which contain very little suspended or dissolved solids, performance in the past has been generally satisfactory. However, in waste water treatment and effluent applications, where high suspended solids contribute to organic loading, performance has been poor because of excessive maintenance requirements. Although IR absorption is probably the most specific, interference-free method available for measurement of CO_2, the major problem with these analyzers has been high maintenance associated with the IR detector.

In 1980, several manufacturers introduced on-line TOC analyzers using a low-temperature, UV-promoted chemical oxidation technique, which led to improved reliability, particularly in highly corrosive applications. There have also been some significant developments in IR detector design that have resulted in much lower maintenance requirements.

The combustion-type TOC analyzer will probably continue to be used in clean applications. The low temperature UV-promoted chemical oxidation type TOC analyzer generally avoids the problems associated with combustion techniques and is more suited to waste water treatment and effluent applications. These instruments can also be considered as on-line wet chemical analyzers, since reagent mixing and chemical treatment form part of the analyzer operation.

13.11.2 On-Line Process TC/TOC Analyzers

(a) Astro Resources Corporation (United States)

The Model 1800 LTO Analyzer shown in Fig. 13.53 is designed to measure TC or TOC in aqueous process streams within the range 0–2 to 0–10,000 ppm, using the UV-promoted chemical oxidation/IR detection technique. Process sample is pumped into the analyzer by a peristaltic pump (P1). Part of this flow is transferred by another pump to an inorganic carbon scrubbing assembly, the excess sample being bypassed to drain. Phosphoric acid is pumped by P5 from a reservoir and added to the sample before it reaches the

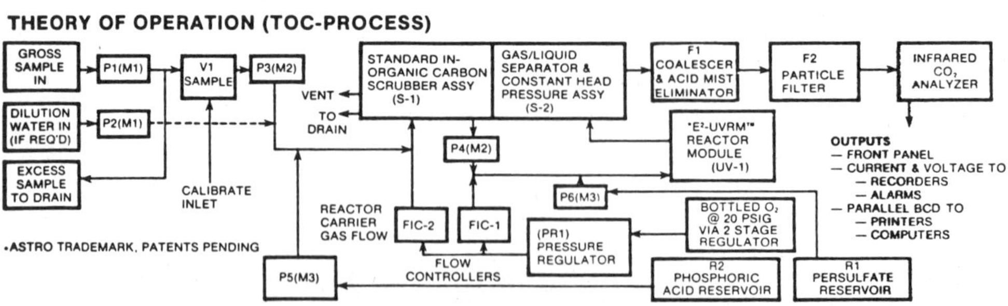

FIGURE 13.53 Model 1800LTO analyzer. (Courtesy of Astro Resources Inc.)

scrubbing assembly. All inorganic carbon is converted to CO_2, which is then scrubbed out of the solution.

Scrubber gas (oxygen or air) enters the scrubber under flow control (FIC-2). Pump P4 draws sample from the scrubber overflow, where it is mixed with persulfate solution from a reservoir (via P5) and delivered to the UV reactor module, where all organic carbon is converted to CO_2. Sample leaving the reactor is delivered to a gas/liquid separator and constant head assembly (S-2). Liquid is directed to drain, and vapor containing CO_2 and other byproducts leaves the top of S-2 and passes through a coalescer/acid mist eliminator and filter before entering the IR analyzer for CO_2 measurement.

The IR analyzer is designed to provide reliable operation in corrosive and dirty service. Compensating circuitry is provided for source aging and dirt accumulation on optical windows. Energy from the IR source passes alternately through two narrow-band interference filters, and the sample gas is then reflected back by a spherical mirror to a solid-state detector. Synchronous detection, dual-wavelength ratioing and digital processing of reference and sample signals are then carried out. Reference and measurement wavelengths are chosen to make the analysis specific to CO_2.

The Model 1800 is housed in a free-standing NEMA 4 cabinet suitable for ISA type Z purge. Alternatively, the unit can be supplied for surface or panel mounting. Bottled air or oxygen consumption is about 500 ml/min and reagent consumption is about 5 gal/month each of 10 vol% phosphoric acid and 1.0 molar persulfate solution. Voltage or current outputs are available for remote recording and repeatability is quoted as $\pm 1\%$ full scale. Multistream applications can be accommodated by use of the Model 800A Sequencer.

The Model 1800 is also available as a laboratory version (designated LTOL). For organic spill detection, the Model 1810 TOC Analyzer is offered. This instrument is identical to the Model 1800 described, except that the inorganic carbon removal step is omitted.

(c) Ionics Inc. (United States)

The Model 1218 TOC Analyzer, produced under patents from Union Carbide, is a microprocessor-based instrument designed for the measurement of TOC in natural waters, process waters, and waste waters, and is capable of handling particulate matter up to 200 m diameter. Standard ranges are 0–20 and 0–2000 ppm TOC full scale; other ranges are available as an option.

Analyzer operation is illustrated in Fig. 13.54. Inorganic carbon is first stripped from the sample by a purging system. Acid is added to the sample, converting inorganic carbon to CO_2, which is eliminated from the sample stream by a nitrogen purge. The sample is then transferred to an automatic sampling valve, which injects a precise amount of sample into a catalytic reactor maintained at 900°C. Sample injection size is 40 μl for the high-range analyzer and 100 μl for the low-range analyzer. All carbon is oxidized to CO_2

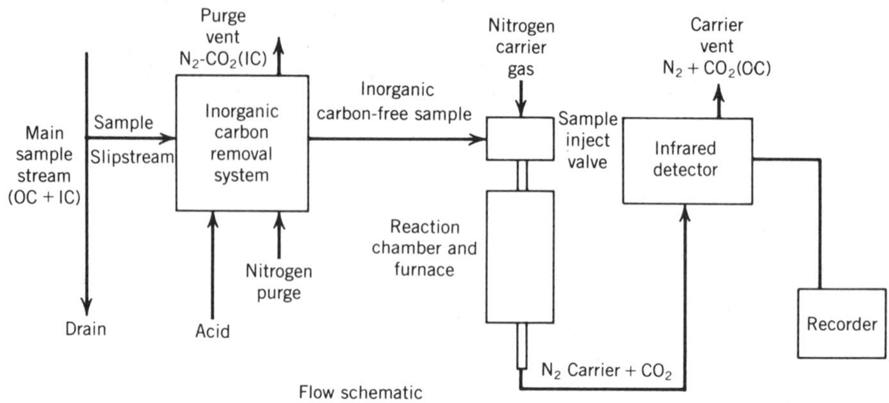

FIGURE 13.54 Model 1218 TOC analyzer. (Courtesy of Ionics Inc.)

and transferred to an IR detector for measurement. Typical cycle time is 5 min.

For measurement of total carbon (Model 1214A), the inorganic carbon removal step is omitted; otherwise, instrument physical configuration is identical to the Model 1218. Use of the Model 1265 Programmable Stream Selector allows sequencing of up to five process streams.

The analyzer is housed in a free-standing cabinet suitable for field installation, but requiring weather protection. Signal transmitters provide current outputs for single or multi-stream remote recording. The package includes a local recorder with alarm facilities.

The Model M-82 Organic Contamination Monitor shown in Fig. 13.55 is designed to respond to all organic chemical compounds in aqueous process streams at carbon levels ranging from 0–10 to 0–20,000 ppm. Analyzer cycle time is 2.5 min. Sample enters the analyzer and flows through an all-ceramic

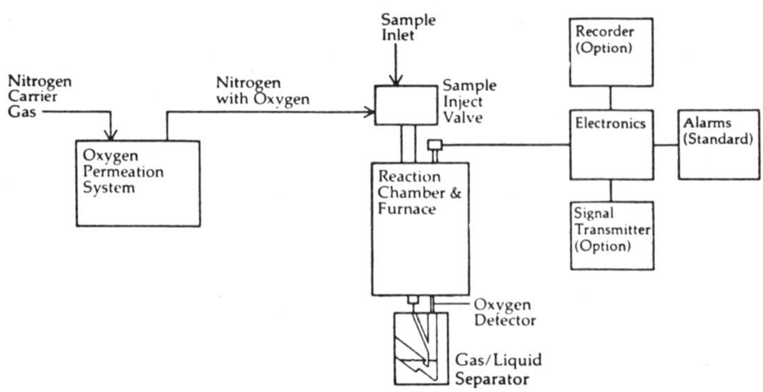

FIGURE 13.55 Model M-82 organic contamination monitor. (Courtesy of Ionics Inc.)

injection valve, which automatically injects a precise amount of sample into a reaction chamber, through which is flowing a nitrogen carrier stream. This stream contains a fixed concentration of oxygen, generated by an oxygen permeation system.

Oxidation of organic carbon in the reactor produces a reduction of carrier gas oxygen content, proportional to the carbon content of the sample, which is measured by a solid-state (zirconia) oxygen sensor. The analyzer is mounted in a splash-proof enclosure with front access and suitable for surface mounting. Weatherproofing and ISA Type Z purging options are also available. Voltage output is standard; current outputs are available as an option. Alarm facilities are also provided.

(d) Raytheon Company (United States)

The Model 2610 TOC Analyzer is designed to measure TC or TOC in process and waste water streams within the ranges 0–50 and 0–10,000 ppm and can be applied to process streams containing high concentrations of suspended biological and refractory solids. For these applications, Raytheon has developed a self-cleaning, in-line homogenizer (Model 2650), which reduces all solids to a uniform size without filtering out sample organics. The analyzer utilizes the combustion/IR detection principle.

To remove inorganic carbon, the sample is acidified and the CO_2 formed is stripped from the liquid by a high-velocity air stream. For TC measurement, this step is omitted. The heart of the instrument is the multi-finned, fixed-bed, non-catalytic reactor where, after mixing with a metered stream of oxygen, the sample vaporizes rapidly and all organic carbon is converted to CO_2. After water removal, the gas stream is filtered and passes to an IR analyzer.

Standard voltage and optional current outputs are available for remote recording, together with single or dual alarm capability. The instrument is mounted in a free-standing cabinet suitable for field installation, but requiring weather protection.

(e) Technicon Industrial Systems (United States)

The Monitor 650, described in Section 13.9.2 can also be used to determine TOC of plant effluents using a wet chemical/colorimetric technique. Inorganic carbon is removed from the process stream by acidification, followed by sparging with a high-velocity air stream. Treated sample is then mixed with acid/persulfate solution and subjected to UV irradiation. The resultant CO_2 generated is dialyzed through a silicone rubber membrane and reacted with a buffered phenolphthalein indicator solution, the decrease in color intensity being proportional to the original carbon concentration. The color is measured at 550 nm by a photometer. Minimum detection limit is quoted as 10 ppm.

(f) Xertex-Dohrmann Division (United States)

The Model DC-60 Process TOC Monitor is designed to measure either TC or TOC in aqueous process streams. Three standard ranges are available—high (0–2 to 0–100 ppm), medium (0–25 to 0–400 ppm), and low sensitivity (0–200 to 0–5000 ppm). Dual range capability is also offered. Cycle time is selectable from 5 to 20 min in six steps via a switch on the instrument front panel. The instrument uses the combustion/IR detection principle and is capable of handling suspended solids up to 0.1 in. in diameter.

Inorganic carbon is removed by phosphoric acid addition, followed by high-efficiency air sparging. The treated sample flows through the injection valve, where 0.125–2.5 ml is injected into the combustion tube. Here, the sample is vaporized and mixed with oxygen at 900°C and passed to the combustion zone, where all carbon is converted to CO_2. Following water removal, the vapor passes to an IR detector where the CO_2 concentration is measured. The analyzer signal is integrated and displayed on a local recorder. Voltage and current outputs are available for remote recording. The instrument is housed in a free-standing cabinet suitable for field installation, but requiring weather protection.

The Model DC-62 Process TOC Monitor uses the UV-promoted chemical oxidation/IR detection principle and is designed to handle process streams containing suspended solids up to 0.03 in. in diameter. Streams containing larger particles with little organic content should be filtered. Standard measurement ranges are 0–5 to 0–100 ppm for low-salt streams, 0–10 to 0–100 ppm for high-salt streams, and 0–100 to 0–1000 ppm for brines. An optional diluter can be used to handle high carbon levels and salt brines.

Sample is continuously pumped into the instrument, phosphoric acid is added, and inorganic carbon is removed as CO in a sparger and gas/liquid separator. Persulfate solution is added and the mixture is passed through a quartz reaction tube at 60°C where organic carbon is oxidized to CO_2 in the presence of UV light. After passing through another sparger and gas/liquid separator, the reaction products flow to an IR detector where the CO_2 concentration is measured. For high-salt applications, hydrazine sulfate is injected into the sample to protect the IR detector from corrosion.

The instrument consists of three modules—the electronics, reaction, and reagent storage modules—mounted on a free-standing framework. A local recorder is provided as part of the package and voltage or current outputs are available for remote recording.

The Model COA-1000 Continuous Organics Analyzer is a recent introduction by Xertex and was developed to satisfy the need for a highly reliable, low-maintenance organics analyzer. As illustrated in Fig. 13.56*a*, the instrument uses an entirely different operating principle from the other Xertex analyzers described above, that of partial oxidation by UV-irradiation and measurement of the change in conductivity resulting from absorption of the carbon dioxide generated.

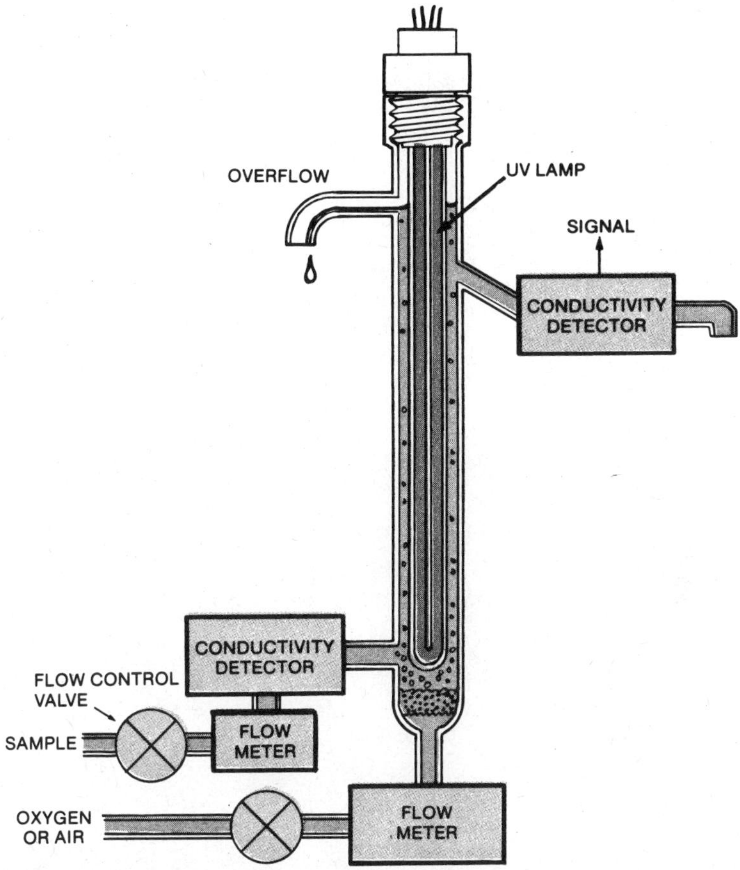

FIGURE 13.56 Model COA-1000 continuous organic analyzer. (*a*) Operating principle. (Courtesy of Xertex Corp., Dohrmann Division)

Process sample is metered at a flowrate of 1–10 ml/min, first into a conductivity cell and then into a reaction chamber in which a UV lamp is mounted. A flow of air or oxygen at 10–100 ml/min is also metered through a fritted glass section at the bottom of the reaction chamber. The reaction chamber is designed so that a thin film of liquid flows up through the chamber along the wall of the UV lamp.

Organics in the process sample are partially oxidized to carbon dioxide; gas and liquid separate in the chamber and the gas-free sample flows through a second conductivity cell. The increase in conductivity of the sample is a function of the organic content of the sample, the flowrate of the sample, and the availability of oxygen to the sample. At the specified flowrates, instrument response is quoted as 90 sec to full scale. Measurement range is 0–200 ppb to 0–300 ppm.

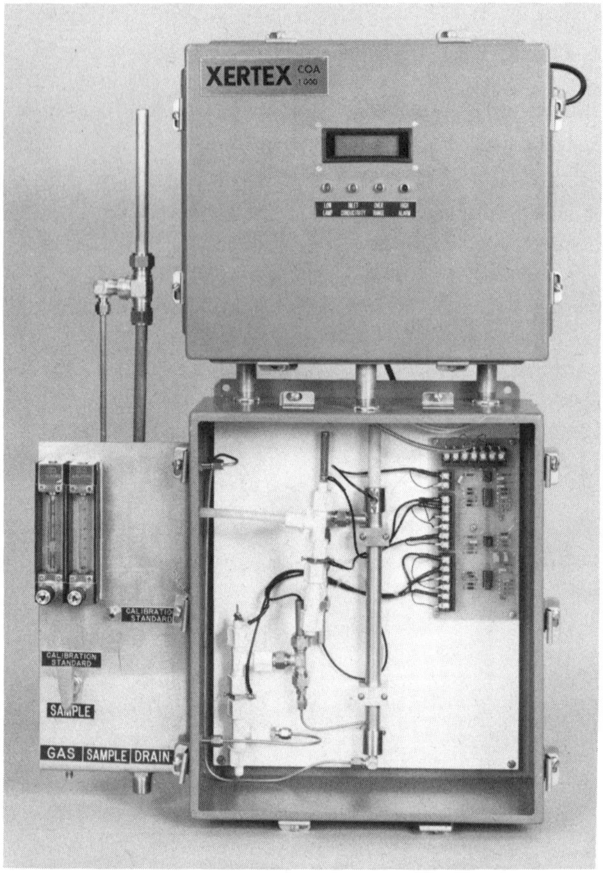

FIGURE 13.56 Model COA-1000 continuous organic analyzer. (*b*) Equipment layout. (Courtesy of Xertex Corp., Dohrmann Division)

The instrument consists of two modules as shown in Fig. 13.56*b:* the analysis module of NEMA 12 construction and the control module (NEMA 12 or 4X). The control module has a digital display and can provide a voltage or current output for remote recording. Alarm facilities are also available, as well as multi-stream capability for 4, 6, or 8 process streams.

13.12 MEASUREMENT OF OIL IN WATER

13.12.1 Background

In refineries and petrochemical plants in particular heat exchanger leaks will result primarily in the presence of light or heavy hydrocarbons (or both) in cooling water and steam condensate systems. An alternative approach to the on-line TC/TOC analyzers for these applications is the use of oil-in-water analyzers. Other important applications for these devices include the monitoring of effluents from offshore drilling and production platforms, and oil tanker ballast water and bilge water monitoring.

Oils floating on water prevent oxygenation and consequently have a catastrophic effect on marine life and vegetation. Oil in the water is equally undesirable since it contributes to BOD and can be toxic to aquatic life. The discharge of oily waters from tankers and other vessels has long been blamed as a major source of pollution. Current legislation now puts strict limitations on the amount of oil permitted to be discharged together with ballast, tank washing, and bilge water.

Therefore, environmental and legal guidelines, technological requirements, and financial considerations all contribute to the need for accurate on-line monitoring of oil content in water, wherever water is discharged to the environment or used in large quantities.

13.12.2 On-Line Oil-In-Water Analyzers

(a) Baird Corporation (United States)

The Petrolarm Oil-in-Water Monitor shown in Fig. 13.57 uses a molecular fluorescence technique for measurement of oil in water within the range 0–

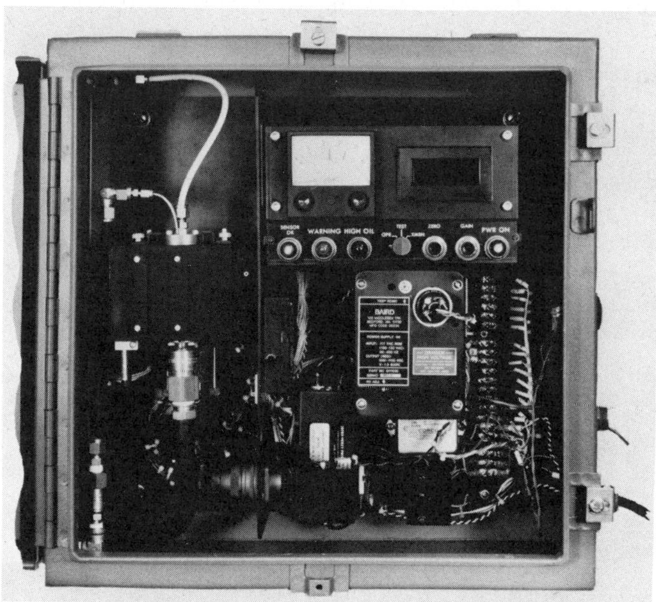

FIGURE 13.57 Petrolarm oil-in-water monitor. (Courtesy of Baird Corporation)

20 ppm. Sample is emulsified in the sample conditioning module (SCM) and pumped to the sample cell. Air supplied from the SCM is continuously forced through the sample tube to form a "curtain" around the walls and keep them free from deposits. UV energy from a lamp source passes through a narrow-band filter into the sample cell.

Oil molecules fluoresce when they absorb and are excited by the UV light. The fluorescence is isolated by a second optical filter and detected by a UV-sensitive photocell located at 90° to the main UV beam. A second UV-sensitive detector is located to detect direct transmission of UV light through the sample cell. This detector will actuate an alarm if light transmission intensity drops off to a preselected level due to lamp failure, contamination of the sample cell, or 100% oil flow.

The fluorescence detector output is processed by the electronics to produce a direct ppm reading on the display meter and to drive alarms. Repeatability is quoted as ±5% of full scale and response time is less than 10 sec to 90% of final value. The unit is housed in a NEMA 4 waterproof enclosure suitable for surface mounting. The sample conditioning module, available as an option, is housed in a matching enclosure and contains the sample pump, sample emulsifier, and air compressor.

(b) Biospherics Inc. (United States)

The Model OS100M Oil Sentry shown in Fig. 13.58 is designed to measure oil in water over the range 0–30 or 0–150 ppm, using the principle of forward light scattering and transmittance. The analyzer consists of a self-cleaning sensing unit designed for direct, in-line mounting and a solid-state control unit housed in a splash-proof enclosure suitable for surface or panel mounting. A dispenser is supplied for installation upstream of the sensor, the function of which is to break up larger oil droplets and ensure a homogeneous sample.

FIGURE 13.58 Model OS100M oil sentry. (Courtesy of Biospherics Inc.)

Every 8 sec, a 3 ml sample is drawn by a piston into the cylindrical measuring chamber, which is illuminated from the side by a small light source. Two photodetectors are located on the opposite side of the chamber. One detects a reference level of light (attenuated by solids and transmission losses through the sample), the other is obscured from the source by an occluding rod and receives only scattered light.

The position of the scattering detector is selected so that it responds to light scattered from oil rather than from solids. The ratio of scattered to transmitted light is calculated and displayed as ppm oil content on a LED meter. A recorder is also built into the control unit. Two switch selectable ranges of 0–30 and 0–150 ppm are available, as well as alarm facilities. Repeatability is quoted as ±2% of full scale.

(c) Bull & Roberts Inc. (United States)

The Model 240 Aqualert Oil Detector is designed to measure oil in water within the range 0–50 ppm (heavy oils) or 0–100 ppm (light oils). Operation is based on the measurement of UV and visible light transmission through the sample using a dual-channel photometric principle. The analyzer consists of an analysis unit and a solid-state control unit, which can be separated by up to 250 ft. The analysis unit contains the mercury vapor UV lamp, sample flow chamber, and detectors, as shown in Fig. 13.59.

The detectors monitor two segments of the emitted by the source. The primary portion is a UV wavelength specific to, and selectively absorbed by, hydrocarbon molecules; the other is a visible wavelength used as a reference to compensate for the presence of non-UV light-absorbing materials in the sample, lamp aging, sample chamber contamination, and so on. The two

FIGURE 13.59 Bull & Roberts Model 240 Aqualert. (*a*) Sensing unit. (Courtesy of Bull & Roberts Inc.)

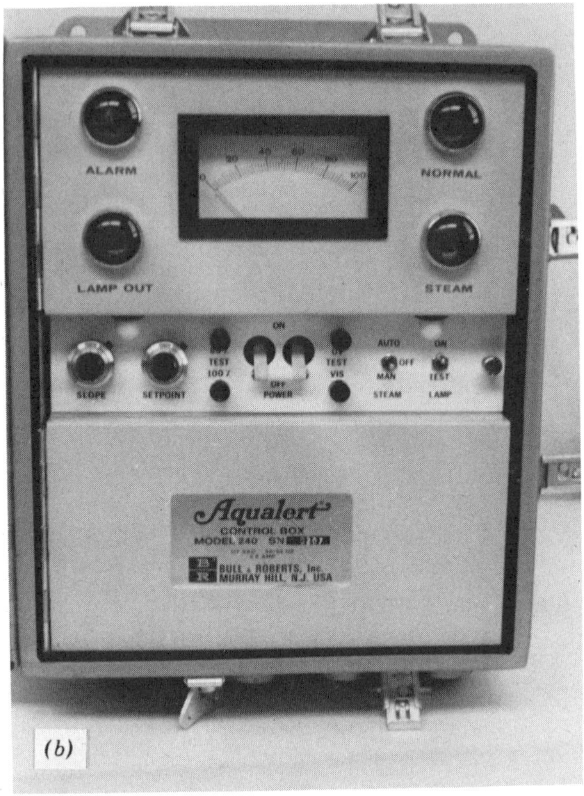

FIGURE 13.59 Bull & Roberts Model 240 Aqualert. (*b*) Control unit. (Courtesy of Bull & Roberts Inc.)

detector outputs are processed in the control unit electronics, which computes the logarithm of the ratio of visible to UV transmissivity. The output is calibrated as ppm oil and displayed on the local meter. Built-in test circuitry allows the meter to be used for troubleshooting the system. Voltage or current outputs are available for remote recording and alarm facilities.

The sample chamber is provided with steam cleaning connections and is rated for sample flowrates up to 30 gpm, pressures up to 225 psig, and temperatures up to 125°F. Sample calibration is obtained for a particular type of oil, either from a laboratory curve or by running a standard sample with a known concentration of a particular oil. Repeatability is quoted as ±2% of full scale.

(d) Horiba Instruments Inc. (Japan)

The Model OCMA-25 Oil Content Monitor is designed to measure oil in water using the principle of solvent extraction and IR detection (Horiba IR

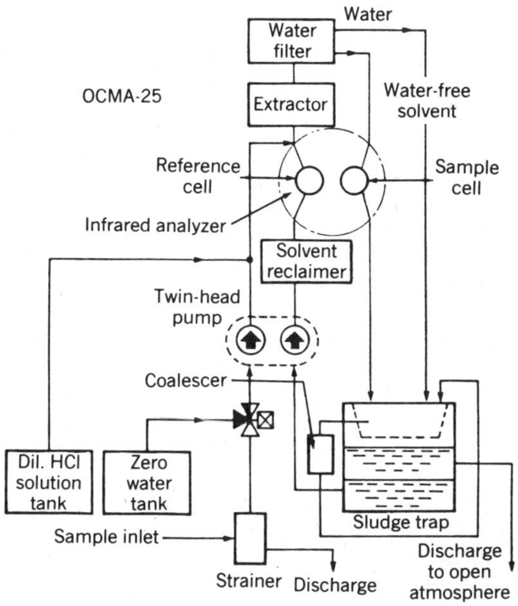

(a)

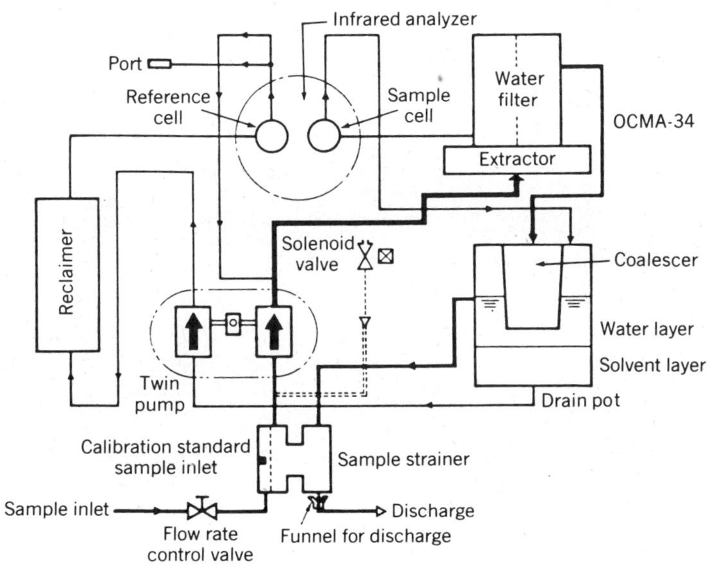

(b)

FIGURE 13.60 Horiba oil content monitors. (*a*) OCMA-25; (*b*) OCMA-34 (for shipboard applications). (Courtesy of Horiba Instruments Inc.)

analyzers are discussed in Chapter 10). Standard measurement ranges are 0–5, 10, 20, and 100 ppm. A schematic flow diagram of the analyzer is shown in Fig. 13.60a.

The oil is extracted from the water using Freon 316 as solvent, and the oil concentration is determined by IR absorption measurement at 3.4–3.5 μm, an absorption region common to all hydrocarbons but not to the solvent. The analyzer is therefore sensitive to different kinds of oil but unaffected by foreign substances, sample color, and so on. The analyzer has a built-in automatic calibration (zero and span) function, which can be preprogrammed from 1–24 hr. Alarms are provided for high oil concentration, out-of-calibration, filter failure, and solvent failure.

The Model OCMA-34 Oil Content Monitor operates on the same principle as the OCMA-25 and is designed to operate in a shipboard environment for monitoring bilge water discharge (see Fig. 13.60b).

(e) Monitek Inc. (United States)

The Model 160/829 Oil-in-Water Monitor shown in Fig. 13.61 uses the ratioed forward light scattering principle (see Section 13.6) to measure minute oil particles in aqueous process streams. A collimated beam of white light is passed through the sample chamber of the Model 160 sensor and its intensity is measured by the direct beam detector. Light scattered in a forward direction by oil particles in the sample is measured by the scattered beam detector. The two signals are then ratioed to provide a linear oil content measurement that is unaffected by sample color or light source intensity.

The Model 829 control unit is housed in a NEMA 4 enclosure suitable for surface mounting. Measurement range is 0–100 ppm, normally factory set at

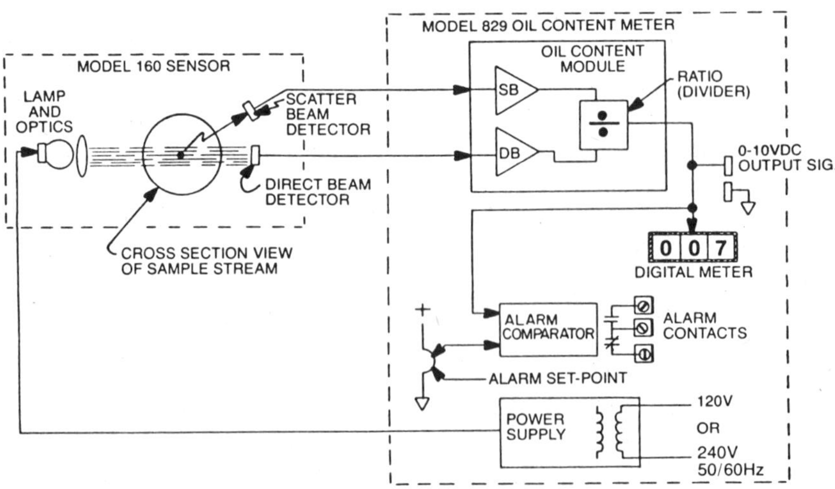

FIGURE 13.61 Model 160/829 oil-in-water monitor. (Courtesy of Monitek Inc.)

15 ppm. Voltage output (0–10 V dc) is standard. The sensing unit is rated for 90 psig and 158°F maximum sample conditions.

(f) Salen & Wicander AB (Sweden)

The Salwico Oil Content Monitor is designed for the continuous monitoring of oil content in water and was developed specifically for shipboard monitoring of oily water discharges. The equipment meets the requirements of the IMCO 1973 International Convention for the Prevention of Pollution from Ships and the 1978 Marpol Protocol, as specified in IMCO Res. A.393(X) and Res. A.496(XII).

The method of measurement is based on the empirical fact that oil color and content of volatile hydrocarbons appear with reasonable accuracy in opposite proportions and the two signals complement each other. By adding the two signals, the resultant signal is representative of the oil content regardless of the type of oil.

The monitor consists of an analysis unit and a microprocessor-based control panel with built-in printer. The analysis unit shown schematically in Fig. 13.62 is divided into chemical and electrical compartments by a gas-tight

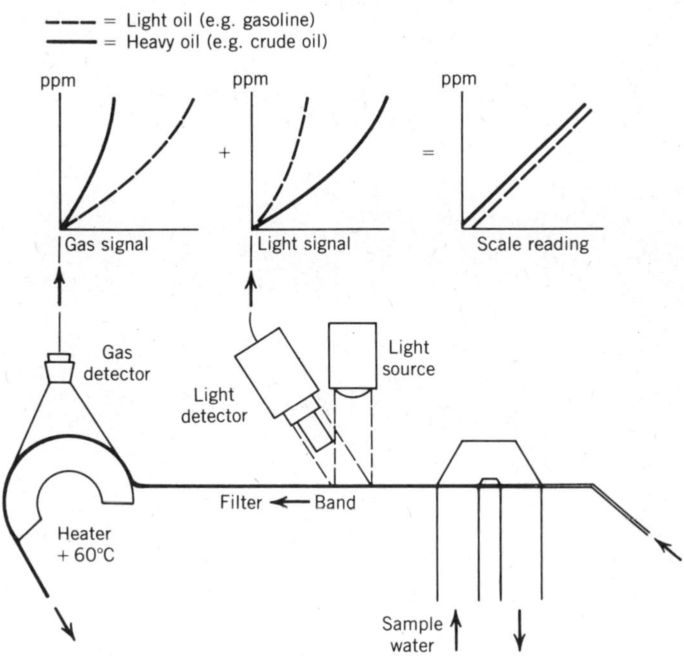

(*a*)

FIGURE 13.62 Salwico oil content monitor. (*a*) Operating principle. (Courtesy of Salen & Wicander)

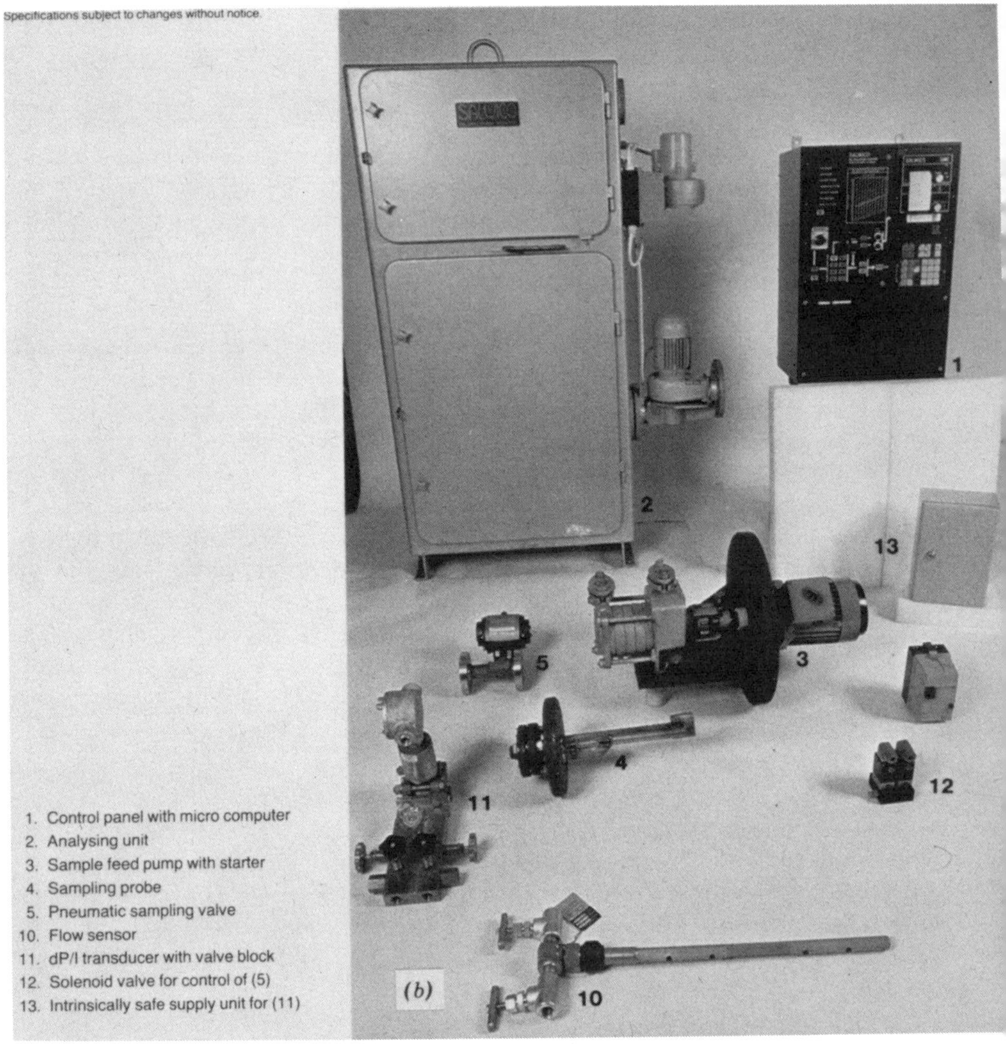

FIGURE 13.62 Salwico oil content monitor. (*b*) Component parts of system. (Courtesy of Salen & Wicander)

dividing plate. A cálibrated flow of sample water passes through a strip of oil-absorbing filter band, which moves at constant speed through the measuring chamber. The oil density of the filter band is in direct proportion to the oil content in the sample.

The filter band is subsequently scanned by an optical detector (which determines the discoloration) and by a gas detector cell (measuring the gas evaporation from the band after slight heating). The two signals are processed in the control unit to produce an output signal calibrated in ppm oil.

Two different measurement ranges are provided by using two selected sample water flowrates.

The control panel consists of three sections—the upper left section contains the alarm system display, control buttons, and potentiometers for zero adjustment; the upper right section contains the microprocessor and printer; and the lower section is a terminal compartment. The microprocessor calculates the oil discharge in liters per nautical mile and the accumulated total and displays them on the printer. The vessel speed and overboard discharge flowrate may be entered automatically or manually into the computer.

(g) Teledyne Analytical Instruments Inc. (United States)

The TAI Model 660 Oil-in-Water Analyzer shown in Fig. 13.63 operates on the principle of UV absorption to measure oil concentration in aqueous

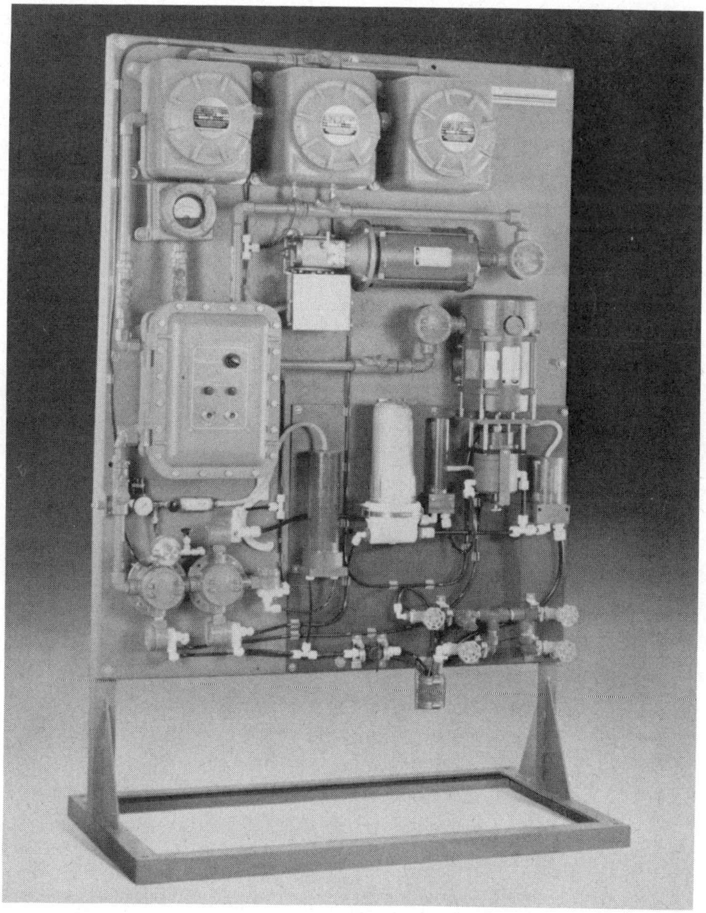

FIGURE 13.63 Model 660 oil-in-water analyzer. (Courtesy of Teledyne Analytical Insts. Inc.)

process streams, using a chopped beam, dual wavelength photometric analyzer.

The instrument uses optical filters to make measurements at two specific wavelength bands, the UV narrow-band measuring wavelength centered at 254 nm and a reference wavelength where oils do not absorb significantly. An automatic zeroing feature allows correction for changes in background absorption. The zero cycle is initiated on an hourly basis by a timer in the control unit. The zero sample is prepared by filtering and air sparging the process sample so that oil is removed, but the sample still contains compounds contributing to background absorption.

The TAI system incorporates conditioning components to prepare the process sample so that both dissolved and undissolved oil are detected and measured. A high-speed high-shear homogenizer mechanically disperses suspended oil droplets and oil adsorbed onto foreign matter to a minimum size so that they appear to be in true solution. When the system is calibrated with a standard containing oil identical to that being measured, reproducibility is quoted as ±1%. When the analyzer range is 0–10 ppm, changes of 0.1 ppm can be detected. The system can be housed in a general purpose or explosion-proof enclosure. Voltage or current outputs are available for remote recording; alarm facilities can also be provided.

BIBLIOGRAPHY

Acid rain: A rapidly shifting scene. *Envir. Sci. & Tech.*, **17,** No. 9 (1983).

Analyzer uses microprocessor to measure free chlorine. *Chem. Proc.* (mid-Nov 1984).

Arthur, R. M. On-line measurements improve activated sludge process control. *InTech* (Sept 1980).

Balazs, M. K., and Walker, S. Counting and identifying particles in high-purity water. *Semiconductor Intl.* (April 1982).

Bobeck, R. F. Electrochemical sensors for oxygen analysis. *Chem. Eng.*, **87,** 113 (1980).

Butler, J. W. *Ionic Equilibrium*. Addison-Wesley, Reading, PA (1964).

Calvert, J. G., and Stockwell, W. R. Acid generation in the troposphere by gas-phase chemistry. *Envir. Sci. & Tech.*, **17,** No. 9 (1983).

Camp, T. R. *Water and Its Impurities*. Reinhold, New York (1963).

A chemical engineer's guide to groundwater contamination. *Chem. Eng.* (Nov 26, 1984).

Chlorine Detector Saves a Life. *Publ. Works Mag.* (March 1978).

Clack, P. J., and Taylor, R. M. A flow and fouling independent dissolved oxygen monitor. *Proc. ISA Conf.* (Oct 1980).

Clack, P. J., Miskimon, W., and Stevenson, R. G. In-line dissolved oxygen control in the brewing industry. *Proc. ISA Conf.* (Oct 1981).

Condrashoff, G. Turbidity measurements in water treatment plants. *ISA Conf.*, Anaheim, CA (1981).

Conductivity of caustic processes. *InTech* (April 1980).

Cucchiara, O. The measurement of dissolved oxygen in organic liquids using a hygrometer. *Anal. Instr.* **15,** ISA (1977).

Davis, E. M. BOD vs COD vs TOC vs TOD. *Water & Wastes Eng.* (Feb 1971).

Dick, J. G. *Analytical Chemistry*. McGraw-Hill, New York (1973).

DuCross, M. J. F. Automated methods for assessing water quality come of age. *Env. Sci. & Tech.* (Oct 1975).

Gair, G. M., and Gayer, J. C. *Water Supply and Wastewater Disposal*. Wiley, New York (1954).

Gray, A. C., Paul, P. E., and Roberts, H. D. Evaluation of operation and maintenance factors limiting biological waste treatment plant performance. *EPA 600/2-79-078* (July 1979).

Gray, D. M. Microprocessor characterizes pH ahead of controller for easy tuning. *Contr. Eng.* (Jan. 1980).

Gray, D. M. Characterized feedback and feedforward pH control. *ISA Trans.*, **20,** *No. 2* (1981).

Gray, D. M., and Cahoone, R. A. Compensate your pH measurements for solution temperature. *Power* (July 1982).

Hach, C. C. Basic turbidity instrumentation. *Insts. & Contr. Sys.* (Dec 1968).

Hach, C. Non-reactive silica can be monitored on-line. *Power,* **128,** No. 1 (1984).

Hale, J. M., and Hitchman, M. L. Some considerations of the steady-state and transient behavior of membrane covered dissolved oxygen detectors. *J. Electroanal. Chem.*, **107,** 281 (1980).

Hamer, P., Jackson, J., and Thurston, E. F. *Industrial Water Treatment Practice*. ICI-Butterworths, London (1961).

Hitchman, M. L. *Measurement of Dissolved Oxygen*. Wiley, New York (1978).

Hoover, W. P., and Wickersham, C. pH sensor transmitter saves $40,000 annually in parts costs. *Chem. Proc.* (March 1985).

Horn, W., and Cunningham, O. H. Computerized system monitors petrochem waste stream. *Poll. Eng.* (April 1982).

Inghram, C. W. Keeping tabs on excess chlorine. *Water Eng. & Mngmnt.* (April 1983).

In-line system detects particle contamination in ultrapure liquids. *Chem. Eng.* (March 1984).

Kefford, F. D. Automated Analysis as an Analytical Tool. *2nd. Ann. Water Tech. Conf.*, Dallas, TX (Dec 1974).

Kehoe, T. J. Determining TOC in waters. *Env. Sci. & Tech.*, **11,** No. 2 (Feb 1977).

Kerker, M. *The Scattering of Light and Other Electromagnetic Radiation*. Academic Press, New York (1969).

Kidder, R. J., and Rosenthal, R. Advances in conductivity measurement. *Insts. & Contr. Sys.* (April 1965).

Krigman, A. Turbidity instrumentation: Clearing muddy waters. *InTech* (Feb 1984).

Kulin, G., and Schuk, W. W. Evaluation of a dissolved oxygen field test protocol. *EPA 78-D-X0024-1*.

Latimer, W. M. *Oxidation Potentials*. Prentice-Hall, Englewood Cliffs, NJ (1952).

Lilley, M. D., Story, J. B., and Raible, R. W. The chronoamperometric determination of dissolved oxygen using membrane electrodes. *J. Electroanal. Chem.*, **23,** 425 (1969).

Loehe, J., and Wickersham, C. P. Efficient pH probe saves $600,000 in one year. *Chem. Proc.* (Oct 1982).

McMillan, G. K. pH control: A magical mystery tour. *InTech* (Sept 1984).

Molvar, A. E. Instrumentation and automation experiences in wastewater treatment facilities. *EPA 600/2-76-198* (Oct 1976).

Moore, R. L. Neutralization of wastewater by pH control. *ISA Trans.* (1978).

Nachod, F. C. (Ed.). *Ion Exchange Theory and Application*. Academic Press, New York (1949).

Piovoso, M. J., and Williams, J. M. Self-tuning pH control: A difficult problem, An effective solution. *InTech* (May 1985).

Powell, S. T. *Water Conditioning for Industry,* 1st Ed. McGraw-Hill, New York (1954).

Riley, C. D. Jr. Synergetic design of ultrapure water systems for microelectronics applications. *Microelect. Manuf. & Testing* (July 1983).

Ryan, W. J. *Water Treatment and Purification,* 2nd Ed. McGraw-Hill, New York (1946).

Sirany, R., and Yitref, M. Batch fermentation pH measurement. *Proc. ISA Symp.,* Des Plaines, ILL (May 1984).

Standard Methods for the Examination of Water and Wastewater, 14th Ed. APHA, AWNA & WPCF, New York (1975).

Strauss, S. D., and Puckorius, P. R. Cooling water treatment for control of scaling, fouling, corrosion. *Power* (June 1984).

Tolliver, D. L., and Schroeder, H. G. Particle control in semiconductor process streams. *Microcontamination* (June, July 1983).

Toy, D. A., and Bonady, F. M. Guide to wastewater analyzers. *Chem. Proc.* (Sept 1984).

Trevathon, V. L. Advanced control of pH. Paper No. 839, *Proc. ISA Conf.,* Philadelphia, PA (1978).

14

Measurement Of Heating Value

14.1 INTRODUCTION AND BACKGROUND

In the early days of gas distribution, the main function was to provide light, and only one parameter—illuminating power—was important from a control standpoint. As gas found increasing use as an industrial and domestic heating medium, this parameter was replaced by *calorific value*. Coal gas, a major fuel until recent times, has fairly constant flame speed and density characteristics.

With the advent of gas mixing processes, the gas grid system, and the use of natural gas for industrial and domestic consumption, the vastly different properties of fuel gases produced many problems in combustion control.

With today's gas shortages, it is necessary to make efficient use of every available source of combustible gas, and composition variations are common and considerable. It is essential to measure and control the heating characteristics of a fuel gas to obtain optimum combustion control. In addition, to meet the objectives of energy conservation and smoothly functioning heating control systems, fuel gases must be blended to provide uniform calorific value or variations must be compensated for in the control system.

The familiar home appliance, the gas oven with calibrated dials, is designed to release a uniform heat flow at a given setting. The concept was introduced by an Italian, G. Wobbe, in 1926 and was subsequently adopted by gas utility distributors to provide the consumer with a fuel gas of uniform heating characteristics.

Thus, the important concept of uniform heat flow has been a part of our everyday lives for many years, but it was not effectively applied to industrial processes until quite recently.

The concept is more important to industry, where heating values can vary widely, depending on the source of fuel gas. For instance, in a refinery the off-gas from many process units may be mixed to supply fuel gas for firing furnaces and boilers. The fuel gas composition, and hence its heating value, may vary widely, depending upon the number and types of process units in operation. Consequently, control of heating value is essential for efficient combustion control and optimization.

Combustion efficiency is covered in detail in Chapter 18, which describes the use of analytical instrumentation dealt with in this chapter and in Chapters 7 and 12.

14.2 DEFINITIONS

The traditional scientific thermochemical unit is the *calorie*, which may be defined in several different ways.

- The international table calorie is the amount of heat required to raise the temperature of 1 g of water from 0° to 1°C.
- The 15° calorie is the amount of heat required to raise the temperature of 1 g of water from 14.5° to 15.5°C.
- The SI (Systeme Internationale) unit is the *joule* (*J*) and the thermochemical calorie is equal to 4.184 J.

The *calorific value* (or heating value) of a gas is a measure of the heat produced by burning unit mass or volume of the gas. Typical industrial units of measurement are cal/g, BTU/lb, and BTU/SCF. In the gas industry, the *British Thermal Unit* (BTU) and the *Centigrade Heat Unit* (CHU) are commonly used units of measurement. The BTU is defined as the amount of heat required to raise the temperature of 1 lb of water from 60° to 61°F. The CHU is defined as the amount of heat required to raise the temperature of 1 lb of water from 15° to 16°C.

One frequently hears the terms *gross calorific value* (H_g) and *net calorific value* (H_n) of a fuel. Calculation of H_g is based on condensation of the water produced in burning the fuel, so that the latent heat of vaporization of water is recovered. In practice, this recovery is often uneconomical and the net calorific value must be calculated. The relative difference between these two units ranges from about 2% for some coals to 10% for natural gas.

14.3 CONTROL OF HEATING VALUE

The object of determining and/or controlling the available heat flow from any fuel gas is to achieve a constant or predictable heating performance from burning the gas. The available heat flow is a function of the gas flowrate and the heating value of the gas:

$$Q = V_0 H_n \tag{1}$$

where V_0 is the *volumetric flowrate* of the gas at standard conditions (14.7 psia and 60°F) and H_n is the net calorific value at the same standard conditions. The net calorific value is defined as the gross calorific value generated by combustion, less the latent heat carried by the products of combustion. For one type of natural gas, the gross value (H_g) is 967 BTU/SCF and the net value (H_n) is 873 BTU/SCF.

For a stream of gas flowing from a burner, the volumetric flowrate, corrected to STP, is given by:

$$V_0 = KaB \sqrt{\frac{2gh}{\rho} \cdot \left[\frac{(P_0)(520)}{(\mathrm{T}_0)(14.7)}\right]} \tag{2}$$

where K = constant
a = orifice discharge coefficient
B = burner nozzle orifice area
g = gravitational constant
h = differential pressure across nozzle
ρ = gas density at flowing conditions
P_0 = absolute pressure at burner outlet
T_0 = absolute temperature at burner outlet

The discharge coefficient (a) is affected by gas viscosity, temperature, and pressure. However, the effect is small and can be considered as constant for a given set of operating conditions. If the burner outlet pressure is considered atmospheric, then the differential pressure across the nozzle is equal to the gas supply pressure ($h = P_1$). Equation 2 then reduces to:

$$V_0 = K' \sqrt{\frac{P_1}{\rho}} \tag{3}$$

Expressing the density in the customary gas industry terms of specific gravity (G_s) and substituting in equation 1, we get:

$$Q = K'' \sqrt{\frac{P_1}{G_s}} H_n \tag{4}$$

From equation 4, we can see that the factors to be measured or controlled in a fuel gas supply to provide a constant rate of heat flow, are:

1. Gas supply pressure.
2. Gas specific gravity.
3. Gas net calorific value.

The well-known *Wobbe index* or *Wobbe number* (W_0) is defined as:

$$W_0 = \frac{H_n}{G_s}$$

where H_n is the net calorific value
G_s is the specific gravity

thus embodying the two widely different qualities of calorific value and specific gravity in a single term. It can be seen from the preceding that both Q and W_0 express the heat flow per unit time (e.g., BTU/hr).

The significance of the Wobbe index is that it relates fuel gas heating characteristics in a manner that is useful for blending fuel gases or obtaining a constant heat flow from a gas of varying composition. It should be noted, however, that the Wobbe index does not relate to such factors as flame temperature, heat transfer coefficients, or temperature gradients.

Table 14.1 lists the heating characteristics of some typical fuel gases. Note that the Wobbe index generally groups these gases in a more usable format than classification by calorific value alone. Thus, for wide variations in specific gravity and calorific value due to composition variations, the Wobbe index varies over a much narrower range, providing a more realistic value by which to control the heating values.

While the use of the Wobbe index has advantages from a combustion control standpoint, calorific value (or heat of combustion) is still widely used as a quality control and specification parameter for gas sold on a custody transfer basis. The *mass* heat of combustion of a fuel is particularly important to weight-limited vehicles such as airplanes as the distance these craft can travel on a given weight of fuel is a direct function of the fuel's mass heat of combustion and its density. The *volume* heat of combustion is important to volume-limited craft, such as automobiles and ships, as it is directly related to the distance travelled between refuellings.

TABLE 14.1 Heating Characteristics of Typical Fuel Gases

Gas	Specific Gravity	Calorific Value	Wobbe Index
Blast furnace gas	1.000	100	100
Producer gas	0.878	129	138
Coal gas	0.431	527	803
Reformed gas (refinery)	0.497	530	752
Reformed gas (natural)	0.472	530	771
Natural gas (I)	0.635	1013	1271
Natural gas (II)	0.650	1145	1420
Refinery gas (high olefin)	0.890	1468	1556
Refinery gas (high paraffin)	0.990	1644	1652
Propane (commercial)	1.523	2550	2066
Butane (commercial)	1.950	3200	2292

The *extended Wobbe index* or *heat release* (eW_0), developed by Dr. O. Dommer, is sometimes used as a measurement parameter and is defined as:

$$eW_0 = W_0 \sqrt{P}$$

where P is the gas pressure.

14.4 LABORATORY METHODS OF MEASUREMENT

The methods used in the laboratory for the measurement of calorific value or heat of combustion of fuels generally fall into two categories:

- Methods using continuous recording calorimeters.
- Methods using the bomb calorimeter.

The ASTM Standard Test Method D1826-77 covers measurement of the calorific value of gases in the natural gas range by continuous recording calorimeter. The recommended instrument is the Cutler-Hammer recording gas calorimeter. This is essentially a process analyzer and is described later in this chapter. An extensive investigation of the accuracy of this instrument, when used with gases of high heating value, was conducted by the National Bureau of Standards in 1957 under a research project sponsored by the American Gas Association (AGA).

ASTM Method D2382-76 covers the measurement of the heat of combustion of hydrocarbon fuels by bomb calorimeter, and is designed specifically for use with aviation turbine fuels, where the permissible difference between duplicate determinations is in the order of 0.1%. It can be used for a wide range of volatile and non-volatile materials where slightly greater differences in precision can be tolerated.

The heat of combustion is determined by burning a known weight of sample in an oxygen-bomb calorimeter under controlled conditions. The temperature is measured by a platinum resistance thermometer and the heat of combustion calculated from temperature observations before, during, and after combustion, with proper allowance made for thermochemical and heat transfer corrections. This is a high-precision test, requiring attention to detail and strict adherence to all the procedures laid down in the ASTM test method. Details of the required apparatus are covered in the annex to the test method. The following equipment is recommended:

- For isothermal operation—the Precision Scientific Co. Model 63090 calorimeter or the Parr Instrument Co. Model 1230 calorimeter with Model 2651 controller.
- For adiabatic operation—the Parr Instrument Co. Series 1210, 1220 or 1230 calorimeter with Model 2601 controller.

14.5 PROCESS ANALYZERS FOR HEATING VALUE MEASUREMENT

The different techniques developed for on-line measurement of the heating value of gaseous fuels may be categorized as follows:

1. Traditional gas calorimeters, such as those manufactured by Cutler-Hammer in the United States. These instruments measure calorific value directly.
2. Open-flame-type analyzers, such as those manufactured by J. H. Reineke (Germany), Union A. G (Germany) and Fluid Data Inc. (United States). These instruments can be adapted to measure calorific value, Wobbe index, or extended Wobbe index.
3. Analyzers using correlative techniques (e.g., IR absorption).
4. Gas chromatographic analysis coupled with computation of heating value from total stream composition and pure component heating values.

14.5.1 Traditional Gas Calorimeters

These instruments, which are rather bulky, measure calorific value directly, and have been used as standards for custody transfer in the gas industry for many years. The Cambridge-Thomas calorimeter, until quite recently the standard for the British gas industry, is no longer in production and has been replaced by the Cutler-Hammer calorimeter, which has been the standard for the U.S. gas industry for some time. These two instruments operate on the same basic principle.

A prototype calorimeter, built by Analysis Automation Ltd. in England, was recently submitted to the Department of Energy and is awaiting approval for use as a standard.

(a) Fluid Data Inc. (United States)

The Cutler-Hammer Recording Gas Calorimeter, manufactured by Fluid Data is shown schematically in Fig. 14.1 and consists of a tank unit (the measuring unit) and a locally mounted recorder, which translates heat measurements into actual heating value of the gas sample being measured.

The sample gas is burned under controlled conditions and the heat generated is imparted to a stream of heat-absorbing air. The increase in temperature of the air is directly proportional to the calorific value of the sample gas. Flowrates of sample gas, combustion air, and heat-absorbing air are maintained in a fixed volumetric ratio by three separate wet-gas meters that are motor-driven, geared together, and work in a common water seal. Readings are independent of atmospheric pressure, temperature, and humidity variations, since metering of gas and air streams is carried out under identical conditions. Nevertheless, it is recommended that the instrument be installed in a temperature controlled environment.

The pressure of sample gas entering the instrument is reduced to atmospheric pressure by a governor and orifice, operating in conjunction with a bleeder burner. After metering, the sample is mixed with primary air (metered by the combustion air meter) and the gas-air mixture then passes to the burner unit. To ensure complete combustion, a small quantity of secondary air is supplied directly to the flame at the point of combustion. The ratio of primary to secondary air is set by means of orifices in the air distribution chamber. The heat generated by combustion of the sample is passed to the heat absorbing air (which flows in the opposite direction to the combustion gases) and the difference in temperature is measured by two resistance thermometers.

The gas meter, heat absorbing air meter, and combustion air meter are accurately machined, operating like the conventional water-sealed gas meter. The meter drums are driven, through suitable gearing, by an electric motor. The air supply for the air meters is taken directly from the atmosphere above the common water seal. Water level in the main tank is maintained constant by a pump, which draws water from a secondary (reservoir) tank, excess water being returned to the reservoir tank by an overflow weir. This eliminates the need for a permanent water supply, although periodic topping up is required to replace water lost by evaporation. The temperature measuring devices are calibrated and the measuring circuit designed so that the recording instrument can be calibrated directly in calorific value units (e.g., BTU/SCF at 30 in. Hg and 60°F saturated).

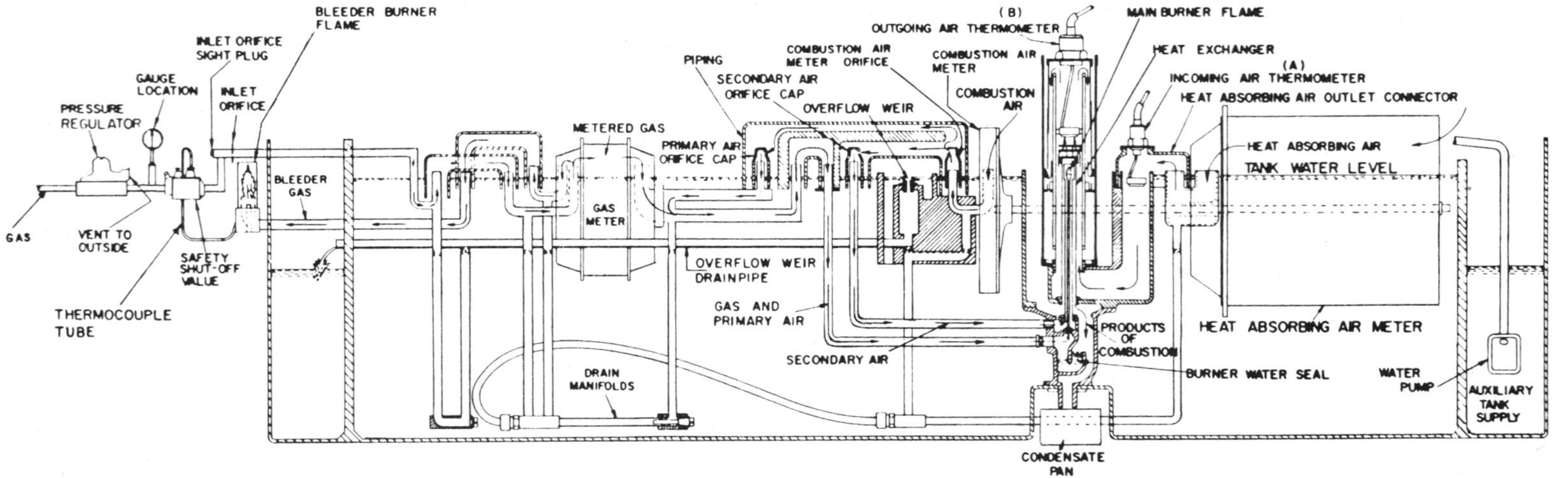

FIGURE 14.1 Cutler-Hammer recording gas calorimeter. (Courtesy of Fluid Data Inc.)

A variety of different measurement spans are available within the overall range 75–3600 BTU/SCF. Repeatability of the instrument is quoted as ±0.5% of full scale when room temperature and tank temperature are stable between 60 and 85°F (or preferably between 72 and 77 deg F ±1°F). The tank unit should be mounted on a floor free of vibration.

The standard method for signal transmission to a remote location is by retransmitting slidewire (0–100 mV) in the master (local) recorder. Pneumatic (3–15 psig) and electronic (1–5, 4–20, or 10–50 mA dc) options are also available. Precautions to be taken for installation are as for any open-flame device. A temperature-controlled and forced-draft ventilated enclosure is required. The air used for ventilation of the enclosure must be free from combustible gases.

14.5.2 Open-Flame-Type Analyzers

(a) Fluid Data Inc. (United States)

The FLO-CAL High-Speed Calorimeter shown schematically in Fig. 14.2 has the same basic design as the analyzer manufactured by J. H. Reineke, West Germany, and is manufactured by Fluid Data in the United States under exclusive agreement with Reineke.

The analyzer is designed to measure the heating value of gaseous fuels either as calorific value at standard conditions or as Wobbe index. The operating principle is based on measuring the change in heat released by continuously burning the sample fuel and varying the flow of air to control the temperature of the mixture of air and combustion products to a specified point of equilibrium. Since the air flow varies to maintain the specified equilibrium produced by heat generation, measurement of the air flow is directly related to the gas heating value.

Sample gas enters the instrument through a precision gas regulator, which maintains gas pressure constant as it flows to a calibrated orifice in the burner unit. Air is delivered to the instrument through a control valve and air orifice and then to the burner assembly, where a portion of it is used for combustion air; the remainder surrounds the burner chamber, serving as a heat exchange medium. The mixing baffle at the top of the burner chamber mixes the heat exchange air with the products of combustion. The compensation tubes adjust for incoming air temperature changes. The heated air, with the products of combustion, passes over a thermal expansion element, the length of which varies directly as a function of temperature.

The thermal-expansion element/compensation tubes assembly actuates a flapper/nozzle assembly that modulates the air flow through the air control valve to maintain the specified equilibrium. The net effect is a constant differential temperature between incoming air and heated air. The air flow through the air orifice is measured by a DP transmitter, which provides a continuous measurement directly proportional to the Wobbe index.

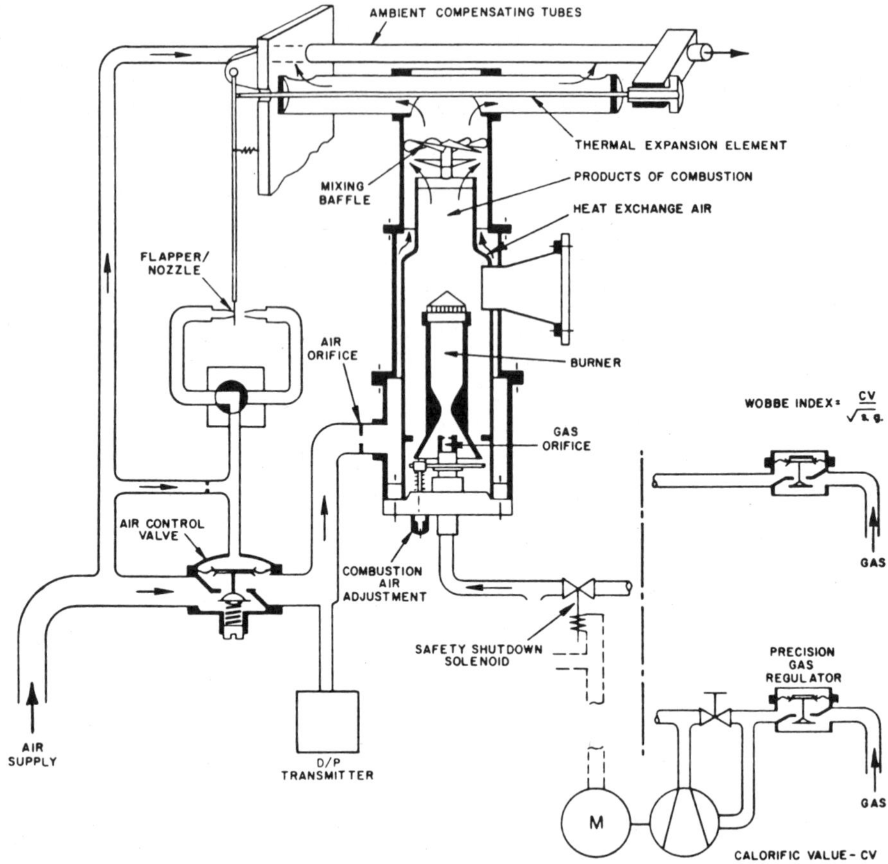

FIGURE 14.2 FLO-CAL high-speed calorimeter. (Courtesy of Fluid Data Inc.)

With a constant-volume gas pump replacing the gas governor, the system compensates for variations in gas specific gravity, enabling the instrument to be calibrated to measure calorific value.

Overall range of the analyzer is 65–3000 BTU/SCF or 650–30,000 kcal/m^3. Typical ranges offered are 75–150, 300–600, 600–1200, and 900–1800 BTU/SCF. Repeatability is quoted as $\pm 0.3\%$ of upper range value. Instrument response time is quoted as 6–8 sec to 90% of final reading (not including sample transport time). The fast response of this analyzer makes it ideal for control applications.

The MICRO-CAL High-Speed Calorimeter shown in Fig. 14.3 is a recent development from Fluid Data. It is a microprocessor-based instrument operating on the same basic principle as the FLO-CAL. The microprocessor measures incoming air temperature and heated air temperature. The change

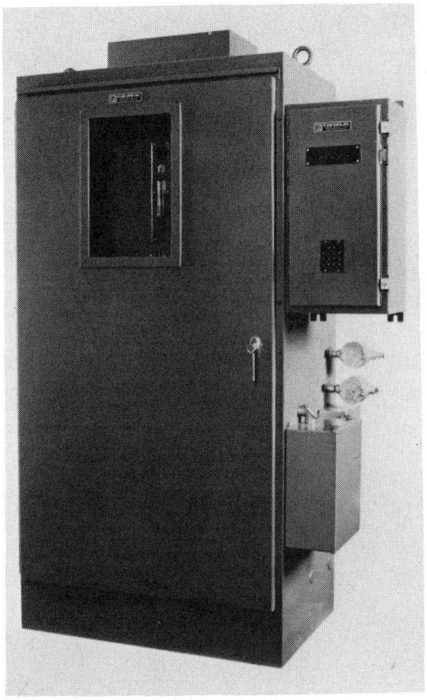

FIGURE 14.3 MICRO-CAL high-speed calorimeter. (Courtesy of Fluid Data Inc.)

in temperature is calculated and used to control a stepping motor, which actuates the air control valve. In addition, the microprocessor measures ambient air temperature, inlet gas temperature, and differential pressure across the air orifice. All of these measurements are used to calculate the heating value of the sample fuel, and a 4–20 mA dc output is provided, proportional to the calculated value.

The operator interface with the equipment is via a simple keyboard, through which changes to settings are made and calibration data is entered. Measured variables are stored in non-volatile memory. Alarm facilities and an auto-calibration system can also be provided as options.

Standard ranges offered are 600–1200, 700–1400, 800–1200, 800–1600, 900–1200, and 1000–2000 BTU/SCF. Repeatability is quoted as $\pm 0.2\%$ of upper range value. Response time is the same as for the FLO-CAL. Sample flowrate is 0.5–5.0 SCF/hr, depending on the orifice size selected. Inlet sample pressure should be 5.5 ± 0.5 in. WG.

(b) Union Apparatebau (West Germany)

The Union Quick-Response Instrument, distributed in the United States by Cosa Corporation, was developed in the late 1960s from Dr. O. Dommer's design to meet the demands of the European gas industry for gas heating value measurement. In 1971 an improved design was introduced, incorporat-

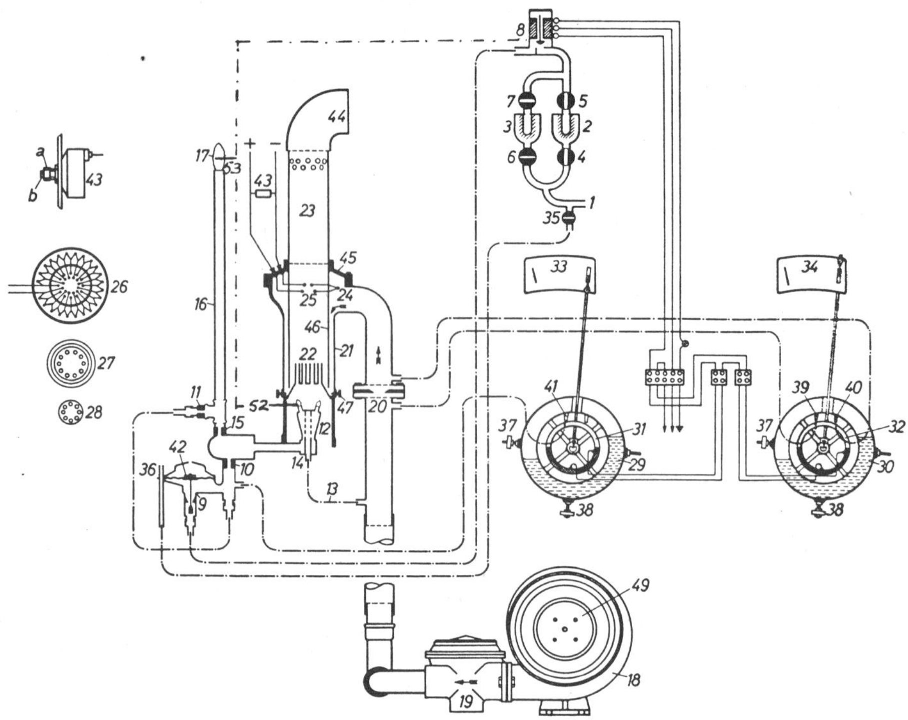

FIGURE 14.4 Union quick-response calorimeter. (Courtesy of COSA Corporation)

ing compensation for specific gravity, barometric pressure, and ambient temperature variations. Approximately 3500 units are now in industrial use worldwide.

The instrument is shown schematically in Fig. 14.4. Sample enters the analyzer via a filter system and solenoid shut-off valve (8) to the precision gas regulator (9). Gas then flows through an orifice (10), sized for the required measurement range, to an expansion chamber, then to the burner (12) and heat exchanger (22). Cooling air is supplied by a fan blower (18) through an air filter (49) and regulator (19) to an orifice (20), at which point the differential pressure is measured by a ring balance (30). Air flow continues to the cold junction of the thermopiles (24) and down the air duct-casing (21) to the heat exchanger (22).

The heat exchanger mixes hot flue gases together with the cooling air stream, resulting in a flue temperature about 15°C above that of the cooling air itself. The flue gas/air mixture passes over the hot junction of the thermopiles and then to atmosphere via the flue (23) and bend (44). The thermopile output of 0–30 mV is directly proportional to the calorific value of the sample gas. Converters for pneumatic or electronic output are available.

Compensation for gas density changes is provided by the compensation tube (16), pilot burner (17), and orifices (15) and (11). Orifice (15) has a cross-section ratio of 4 : 1 compared with range orifice (10). Constant flow of gas to the main burner is maintained by the interaction of orifices (10) and (15) when the gas density changes. For Wobbe index measurement, compensation for gas density changes is not required. In this case, orifice (15) is replaced by a blank plug, and gas to the pilot burner (17) is maintained by orifice (11). A flow regulator for oxygen is needed to maintain combustion at the main burner at low calorific value levels. It is used instead of the compensating tube/burner assembly.

Standard safety features include flame-out sensing of main and compensation burners, and monitoring of gas sample pressure and cooling air pressure at fixed set points by means of ring balances. Flame-out, or deviation from the set points, will result in equipment shutdown. Analyzer response time is quoted as 70 sec for 95% of final reading, making it ideal for control applications.

(c) Honeywell Inc. (United States)

The Heating Value Transmitter Model HVT100, a recent development from Honeywell, is a microprocessor-based instrument that measures indirectly the heating value of natural gas blends by a unique principle, measurement of the air/fuel ratio at stoichiometric combustion conditions.

It can be shown that for paraffinic hydrocarbons, the main constituents of natural gas, there is a linear relationship between heating value (BTU/SCF) and the air/fuel ratio (K) at stoichiometric conditions, represented by the equation:

$$H = \frac{Z_a}{Z_f} K_{AF} m + b \qquad (5)$$

where Z_a = compressibility factor of air
Z_f = compressibility factor of sample gas
H = gross calorific value at 60°F and 1 atm
m = constant (105.062)
b = constant (2.41853)
K_{AF} = air/fuel ratio at stoichiometric conditions

Thus, if the air/fuel ratio at stoichiometric conditions can be calculated, the heating value (H) may also be calculated from equation 5. Combustion of non-paraffinic hydrocarbons and other compounds will introduce errors into the calculation. However, it has been demonstrated that relatively large concentrations of these compounds must be present in the sample to cause a significant error.

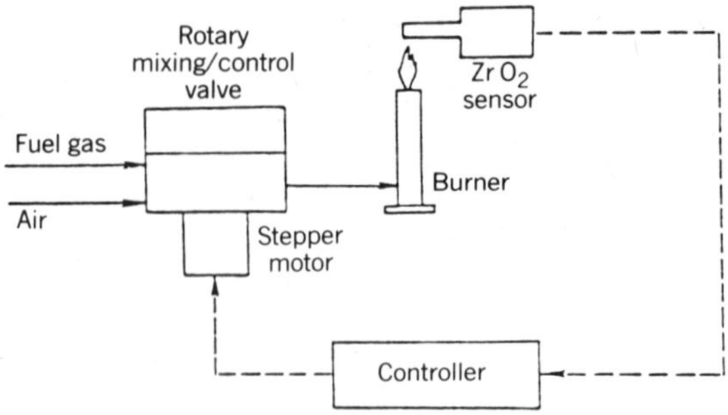

Stoichiometric combustion control scheme

FIGURE 14.5 Honeywell heating value transmitter HVT-100: operating principle. (Courtesy of Honeywell Inc.)

The measuring principle of the HVT100 is shown schematically in Fig. 14.5. A combustion control loop, incorporating a precision rotary mixing valve, is used for accurate measurement of the air/fuel ratio. The rotary valve is controlled by a microprocessor-based electronics unit. Oxygen concentration in the exhaust gases after combustion is measured by a zirconia sensor (see Chapter 7), whose output is monitored by the control unit, which in turn adjusts the rotary valve speed via a stepping motor to maintain the oxygen level at less than 0.005%.

The rotary valve consists of two parts, a manifold and a rotor. The manifold contains two pairs of concentric slots interconnected by 12 chambers located in the rotor. When the rotor is stationary, a constant air flow is routed through one pair of concentric slots and rotor chambers to the measurement burner. Likewise, the sample fuel is routed through the other set of slots and rotor chambers to a flare burner. When the rotor turns, sample fuel trapped in the rotor chambers is carried into the air-flow stream.

Similarly, for each volume of fuel transferred to the air stream, an equal volume of air is transferred to the sample fuel stream. The rotor therefore replaces volumes of air with sample fuel at a rate determined by rotary valve speed. The stepping motor attached to the rotary valve permits precise speed control of the rotor and therefore precise control of the air/fuel ratio of the mixture exiting the rotary valve and feeding the measurement burner.

In this way, the combustion control loop maintains the measuring burner flame at the stoichiometric point, where the speed of the rotary valve is inversely proportional to the air/fuel ratio, as given by the following equation:

$$K_{AF} = \left(\frac{Q_a}{K_3 f} - 1\right) \frac{K_1}{K_2} \qquad (6)$$

where K_{AF} = air/fuel ratio
Q_a = volumetric air flowrate
f = rotary valve speed
K_3 = rotary valve transfer constant proportional to the volume of the transfer chambers

$$K_1 = \frac{P_a C_0}{RT_a Z_a}$$

$$K_2 = \frac{P_f C_D}{RT_f Z_f}$$

and P_a and P_f are the absolute pressures of air and sample fuel respectively in the rotor chambers, T_a and T_f are the absolute temperatures of air and sample respectively in the rotor chambers, R is the universal gas constant, Z_a and Z_f are the compressibility factors of air and sample fuel respectively, and C_0 and C_D are the concentrations of oxygen in the air supply and dry air respectively.

A constant air flowrate to the rotary valve is supplied by maintaining a constant differential pressure across a precision jewelled orifice. The air flowrate (Q_a) is related to the orifice differential pressure in accordance with the classical flow equation:

$$Q_a = C\sqrt{\frac{P\Delta P}{T_a}} \tag{7}$$

where C is the orifice flow constant
ΔP is the differential pressure
P is the absolute upstream pressure
T_a is the absolute temperature

The air is passed through a regenerative drier to remove water vapor and hydrocarbons, so that $C_0 = C_D$. The temperature of the air flowing through the orifice is held constant at 158°F by means of a heating element and thermocouple attached to the valve manifold, and a controller in the electronics unit that operates in an on-off cycling mode.

By substitution for Q_a, K_1, and K_2 in equation 6, followed by suitable manipulation, the following expression can be obtained:

$$K_{AF}\frac{Z_a}{Z_f} = \frac{C}{K_3 f}\sqrt{\frac{P\Delta P}{618}} - 1 \tag{8}$$

The only unknown in this equation except K_{AF} is the constant C/K_3. This is determined by using calibration fuels by precisely known calorific values and calculation by equations 5 and 8. Once C/K_3 is determined by calibration, the air/fuel ratio can be determined from the motor drive frequency (f) by equation 8 and the calorific value of the sample fuel from equation 5.

In addition to the control unit's mathematical computation ability, based on these theoretical calculations, the microprocessor-based control unit has several other features:

- Data logging capability via a digital printer interface, customized for use with the Datel APP-20 thermal printer.
- Serial data link to a host computer (RS-422 and 20 mA current loops) with an adjustable baud rate (300–9600 baud).
- Statistical computations of hourly, daily, weekly, or monthly averages.
- A self-diagnostic and test feature that allows continuous on-line monitoring of analyzer performance. Battery back-up is provided for memory retention during power outages of up to two weeks. An analog current output (0–20 or 4–20 mA dc) and high/low alarm facilities are also available.

FIGURE 14.6 HVT-100 equipment layout. (Courtesy of Honeywell Inc.)

The transmitter shown in Fig. 14.6 is housed in a rugged, free-standing steel cabinet with gasketed door and is programmed and visually monitored from an operator's interface panel located on the door. Pertinent operating data is entered by the user via a membrane-type keyboard containing 18 active keyboard positions. Two instrument ranges are offered, 300–800 and 800–1400 BTU/SCF. Repeatability is quoted as ±2.0 BTU/SCF. Automatic calibration facilities are provided, allowing the user to program instrument calibration at time intervals between 10 min and 4 hr. The HVT100 is designed to operate continuously and unattended for up to 30 days.

(d) Herbert Controls and Instruments Ltd. (Great Britain)

The Sigma calorific value and Wobbe index analyzers have been used in the United States for more than 25 years and for longer in Europe. Both instruments, marketed in the United States by COSA Corporation, are simple in operation and are easy to maintain.

The Sigma Mark II Recording Calorimeter shown schematically in Fig. 14.7 consists of two components, a two-stage gas regulator assembly and an expansion tube sensing device linked mechanically to a recorder. The two-stage regulator system controls the flow of sample gas to the burner of the sensing mechanism so that the actual flow corresponds to a constant flow at 60°F and 30 in. Hg pressure.

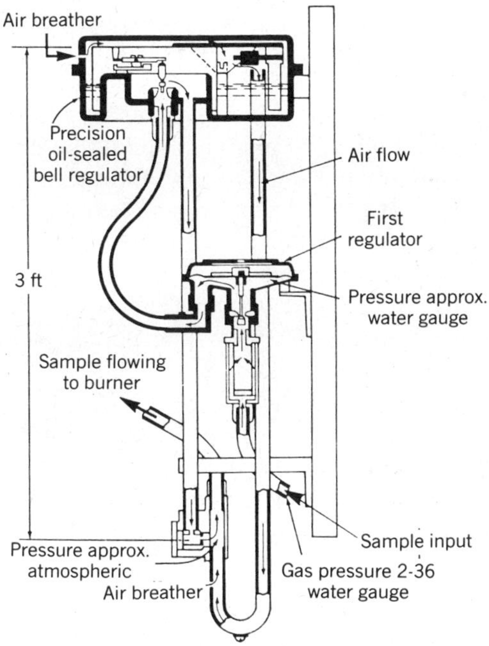

FIGURE 14.7 Sigma Mark II recording calorimeter. (Courtesy of COSA Corporation)

Sample gas, at a pressure of between 2 and 30 in. WG, enters the first regulator, where its pressure is reduced to 1 in. WG. This stage includes a sample shut-off feature for low gas pressure and a hand reset lever. The second-stage regulator is an oil sealed bell-type device, which gives precise sample pressure regulation. Sample gas leaves the bell and flows down a tube to the control orifice positioned 36 in. below the top of the bell. This system provides compensation for sample density variations.

From the control orifice, the sample gas feeds the burner of the differential expansion tube assembly, which consists of two concentric tubes of 1 and 2 in. diameter. The products of combustion pass up through the inner tube, causing it to expand differentially with respect to the outer tube. The longitudinal expansion is converted by a mechanical linkage into a movement of the recorder pen. This system provides compensation for ambient temperature changes, since both tubes respond equally.

Suction from the burner is used to induce a fresh air flow, via a breather hole, over the bell of the second stage regulator to maintain correct balance. A second breather hole is provided near the control orifice outlet to maintain a constant pressure irrespective of burner suction variations due to flue temperature changes. Secondary air for combustion enters the tube assembly via an outer jacket and is preheated as it passes through a baffle before flowing down to the burner. Secondary air intake is controlled to provide barometric pressure compensation.

The Sigma Wobbe Index Recorder operates on the same basic principle and with identical construction, except that sample density compensation is not provided. In this case, the sample control orifice is located directly at the outlet of the bell regulator and sample gas is fed directly to the burner from that point.

The two components of the instrument are usually mounted together and connected by flexible tubing. The analyzer should be installed in an enclosure to protect it from ambient temperature and pressure variations. A repeatability in the order of $\pm 0.5\%$ can be obtained under conditions where the barometric pressure is held between 29 and 30 in. Hg and the ambient temperature is held within the range 65–75°F. Pneumatic (3–15 psig) and electronic (0–20, 4–20, or 10–50 mA dc) outputs are available, in addition to millivolt outputs (0–500 mV, 0–5, or 0–10 V dc). Alarm facilities can also be provided.

The electronic outputs are available by using the TAM 35 solid-state telemeter amplifier, which takes its input from a variable resistance transducer mounted on the recorder pen spindle. A wide variety of different chart ranges are available to suit customer requirements. The local recorder may be supplied with electric drive, or spring-wound for two-week or one-month duration.

(e) Precision Measurement Inc. (United States)

The Therm-Titrator II was first introduced in 1978. It was used by Colorado Interstate Gas Company in an extensive evaluation program. A radical im-

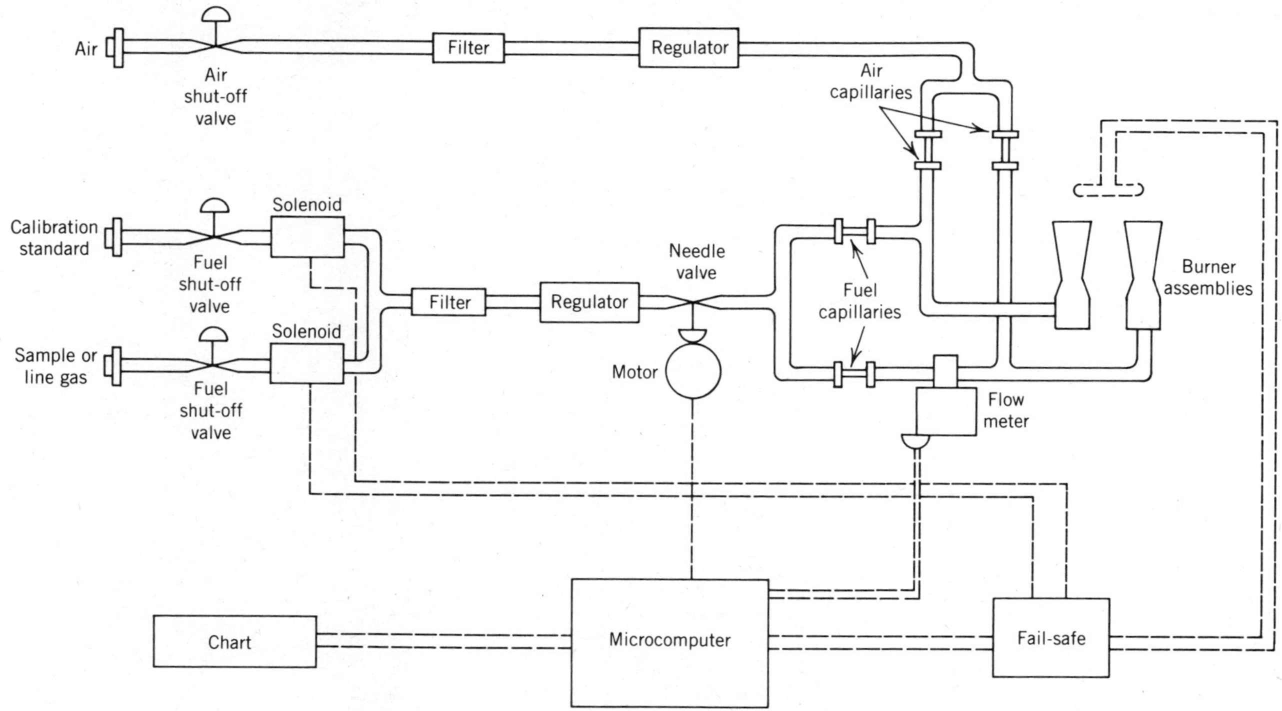

FIGURE 14.8 Therm-Titrator II. (Courtesy of Precision Measurements Inc.)

provement to instrument operation was made in 1980 with the introduction of a new gas volume measuring device. The successful experience of CIGC with the analyzer was summarized at the 60th annual Gas Producers Association Meeting in March 1981 in San Antonio, Texas.

Operation of the analyzer is based on the principle that there is an optimum air-gas mixture that will produce maximum combustion heat. If the mixture is too rich or too lean, burner temperatures will be less than maximum. The lower the gas heating value, the more gas is required to give maximum burner temperature when mixed with a constant air volume, and vice versa. If the correct air and gas volumes can be measured, the heating value can be calculated.

The Therm-Titrator II is shown schematically in Fig. 14.8. The heart of this microprocessor-based analyzer is a patented, solid-state flowmeter that measures gas flowrate, while holding the air volume constant. Maximum burner temperature is monitored by using two burners, one run slightly rich, the other slightly lean in air-gas mixture. A thermocouple in each flame measures the flame temperature. The microprocessor monitors the air and gas parameters and operates a regulator, increasing or decreasing gas volume until maximum burner temperature is found and maintained. The computer, using the measured air-gas ratio, calculates the gas heating value, which is displayed on a recorder.

Repeatability is quoted as ±1.0 BTU/SCF, and analyzer response time is less than 90 sec. Analyzer range is 800–1200 BTU/SCF. In 1981, CIGC had 14 Therm-Titrators in operation and reported that the maintenance requirement for all of them is less than that necessary for one calorimeter of the Cutler-Hammer type.

14.5.3 Measurement by Gas Chromatography

Since the introduction of microprocessor-based gas chromatograph systems, on-line measurement of the heating value of gas mixtures by gas chromatography has been made easier by virtue of the microprocessor's computational ability. The technique involves separation and measurement of the concentration of all components present in the sample, normalization, and computation of the average heating value from the pure component mole fractions and heating values.

For relatively simple gas mixtures such as natural gas blends this technique is satisfactory. However, for more complex mixtures such as typical refinery fuel gases, which have widely varying compositions, there are several difficulties. First, since all components in the process stream must be separated and measured, the more components there are the more complex the column configuration required to carry out the analysis. Second, the column system must be designed to take account of wide variations in composition and the effect this will have on component separation. Third, the more complex the analysis required, the longer the cycle time. In most refining or chemical plants, measurement of fuel gas heating value usually forms part of a BTU control system for furnace and boiler efficiency, where long cycle times are unacceptable. This has the effect of making the chromatographic technique uncompetitive with alternatives such as the open-flame analyzers described.

Because of these factors, the process gas chromatograph has found only limited application in refinery and chemical plant fuel gas applications. However, the technique has been used extensively in the natural gas industry, and some instruments have been designed specifically for this purpose, as described next. Of course, natural gas heating value measurement is also an analysis that has been offered by the major manufacturers of microprocessor-based chromatograph systems, such as Applied Automation, Beckman, Combustion Engineering, Foxboro, and Siemens.

(a) Daniel Industries Inc. (United States)

The Model 500-BTU Computer-Based Chromatograph is designed for natural gas pipeline and processing plant applications. It consists of three components—the sample handling system, the analyzer unit, and the controller, shown in Fig. 14.9. The sample handling system and analyzer unit are designed for installation in NEC Class 1, Group D, Division 1 hazardous locations. The control unit is of general purpose design.

The basic sample handling system is designed to meet the requirements for pipeline quality natural gas and consists typically of pressure regulator, filter, flow indicator, and the necessary valving to allow automatic selection of one of four possible process samples or the calibration gas. It is located directly below the analysis unit and is suitable for field mounting at or near the sample point.

The Model 500 chromatograph uses micro-packed columns for high resolution, a "backflush-to-measure" technique, and a data normalization procedure giving a quoted repeatability of ± 1.0 BTU in one thousand. The chromatographic oven operates on the "heat sink" principle, which allows installation in remote areas since instrument air is not required for its operation. The thermal conductivity detector uses two measurement and two reference thermistors in a four-element, modified Wheatstone bridge.

The Model 2250 Controller is microprocessor-based and uses the same computer used in the Daniel Model 2230 Series Flow Computers. Its functions include highly accurate timing of chromatograph operations, automatic preprogrammed calibration, precision calculations, self diagnostics, report generation, and interfacing to other devices via its analog outputs or by direct digital link to a printer and/or host computer. The operator interface is a 24-key keypad and LED display, through which configuration data can be entered.

Three analog inputs accept 4–20 mA dc signals, one from the GC preamplifier, the other two from other analyzers (e.g., moisture and sulfur analyzers). Three analog outputs are available for remote trend recording (4–20 mA

FIGURE 14.9 Model 500 BTU chromatograph system showing field unit with sample handling system. (Courtesy of Daniel Industries Inc.)

dc or 0–10 V dc). The printer interface is a standard RS232C compatible output with selectable transmission rates from 300 to 9600 baud. The digital link requires an external prompt to initiate data transmission and has 300 to 9600 baud selectable transmission rates.

A report is normally generated after each analysis on a standard 80-column printer, but printing may be skipped for up to 40 analyses. The actual content of the report may be decided by the operator during configuration. Reports can include computation of mol %, weight %, average BTU, average specific gravity, compressibility factor, component ratios, and Wobbe index, together with 24-hr averages and active alarms.

(b) Electronic Associates Inc. (United States)

The ENCAL Gas Energy Information System, another computer-based chromatograph system, is designed for on-line measurement of natural gas composition, heating value, and specific gravity. It is one of a range of ENCAL information systems developed by EAI, based on extensive experience in computer-based data acquisition and control systems.

The system shown in Fig. 14.10*a* consists of two main components, the chromatographic analyzer unit, designed for installation at or near the sample point, and the microprocessor-based control unit. The analyzer unit is made up of three subassemblies—the chromatograph itself, an electronic interface unit, and the sample handling system. The chromatograph uses microcolumn technology, high-sensitivity thermal conductivity detectors, and diaphragm-type switching values, all well-proven components.

The chromatograph is designed to separate individual hydrocarbon components in the sample from methane to pentanes, as well as nitrogen and carbon dioxide. A backflush feature is used to measure C_6 hydrocarbons and heavier. Helium is used as the carrier gas and the cycle time is approximately 13 min. The sample handling system incorporates solenoid valves for switching up to three sample streams or calibration gas, and includes sample flow control and pressure regulation facilities.

The electronic interface unit is the communications link between the control unit and the chromatograph and sample handling system. Functions of this unit include detector balancing, detector signal conditioning, amplification, and analog-to-digital conversion. Instructions from the control unit are decoded and commands issued for stream switching, sample injection, and column backflush. Interconnection between the interface unit and control unit is provided by two twisted, shielded cable pairs, providing full-duplex communication.

The control unit, shown in Fig. 14.10*b* and designed for installation in non-hazardous locations only, performs automatic unattended operations such as chromatograph control and sequencing, system calibration, computations for heating value and specific gravity, and stream switching. Battery back-up is provided to preserve memory in the event of a power failure. The system shuts down in an orderly manner and start-up is automatic when power is restored.

Operator displays and function switches are provided on the front panel of the unit. LEDs indicate chromatograph operational status (e.g., sample inject, backflush, and calibrate) and stream identification. Switches allow manual stream selection, calibration, and raw detector output recording. A strip printer is also provided.

Configuration data for the system is entered from a hand-held communications module, which is connected to the control unit via a plug-in cable. These data include analytical parameters, report headings, content and frequency, calibration gas composition, factors for BTU and specific gravity computation, peak area limits, and analyzer output ranges.

System calibration is normally carried out every 24 hr or on demand by the operator. The normal ENCAL report may be generated once per cycle, every 30 min, every hour or as configured, showing stream identification, date, time of day, heating value, and SG. Wobbe index and compressibility

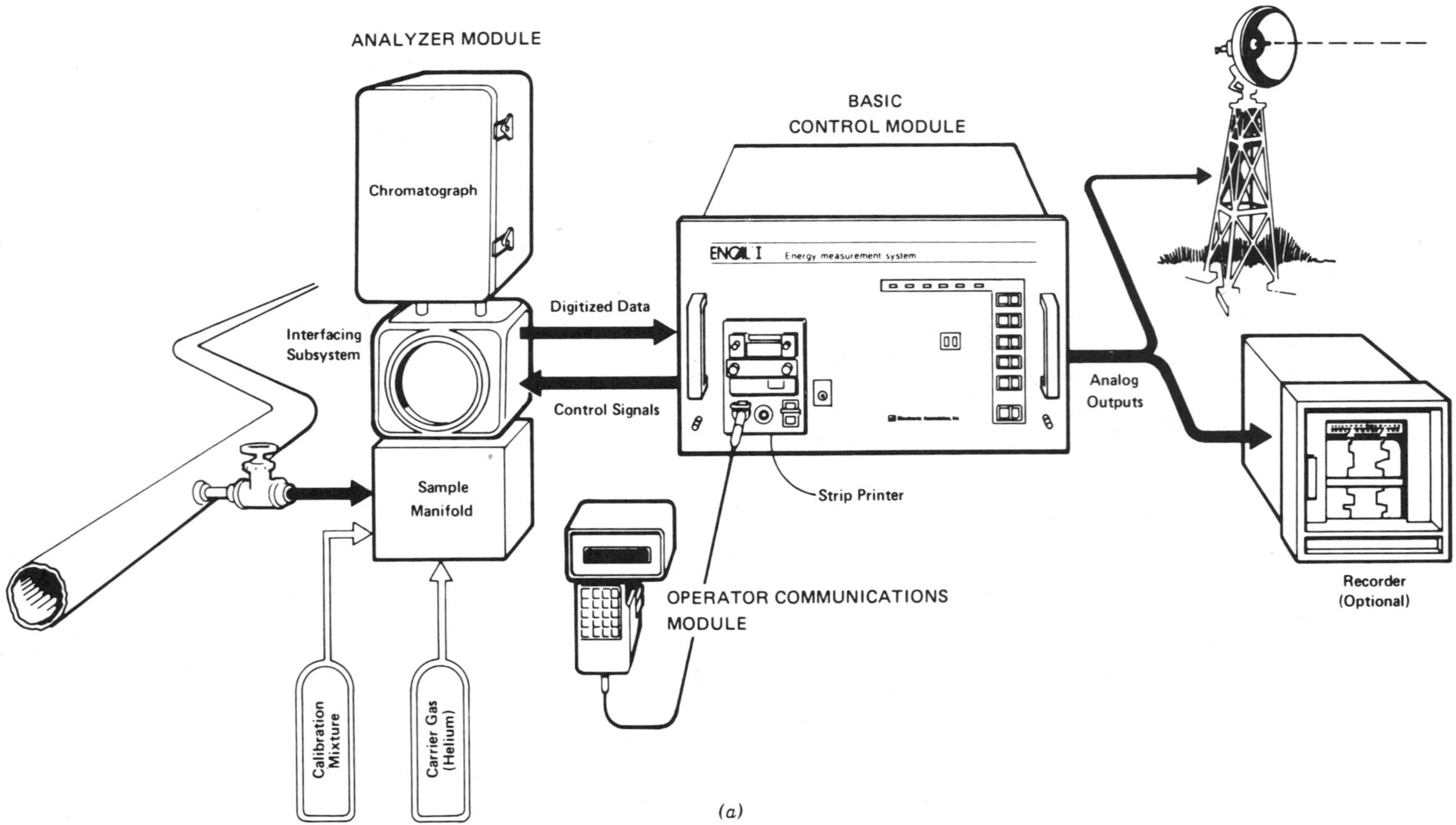

(*a*)

FIGURE 14.10 Electronics Associates ENCAL system. (*a*) System overview. (Courtesy of Electronics Associates Inc.)

FIGURE 14.10 Electronics Associates ENCAL system. (*b*) Control unit. (Courtesy of Electronics Associates Inc.)

factor can also be printed as a user option. A 24-hr average report can be generated at any time of day as configured. A gas analysis report is generated immediately after automatic calibration or in the event of an alarm. Three analog outputs are provided for remote recording. System reproducibility is quoted as ±0.1% (±1.0 BTU in 1000).

(c) *Fluid Data Inc. (United States)*

The COMPU-CAL Energy Measurement System shown in Fig. 14.11 is a microprocessor-based chromatographic analysis system designed for on-line measurement of heating value and specific gravity of natural gas, to meet the needs of natural gas production, processing, pipeline transmission, and distribution companies. Sampling system designs are based on Fluid Data's extensive experience with the FLO-CAL and MICRO-CAL high-speed calorimeters, described earlier, while the analyzer section and microprocessor-based control unit use hardware and application technology developed by a major process chromatograph manufacturer.

The chromatograph is a compact temperature-controlled oven containing switching valves, columns, and thermal conductivity detector, which provides separation of individual natural gas components in a 10-min cycle. Electrical components are located in junction boxes on each side of the oven, making the unit suitable for installation in NEC Class 1, Group D, Division 1 hazardous locations.

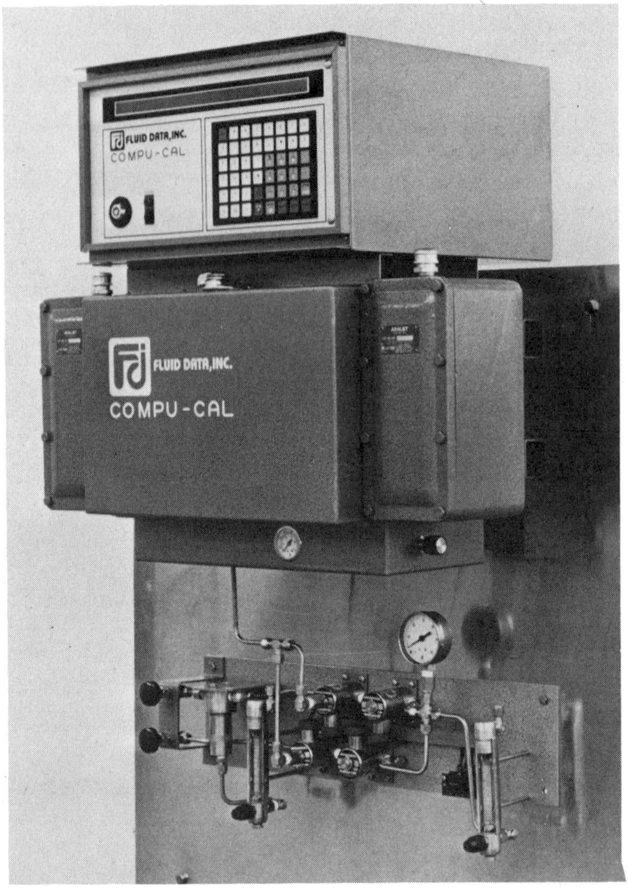

FIGURE 14.11 COMPU-CAL system showing chromatograph, controller, and sample handling system. (Courtesy of Fluid Data Inc.)

The sample handling system consists of two parts, pressure reduction and condensibles removal facilities installed adjacent to the sample point, and final sample conditioning components installed as an integral part of the analyzer unit. Auto-calibration, multi-stream manifolding, and bypass loops are included in the sample conditioning system.

The microprocessor control unit, suitable for installation in NEC Class 1, Group D, Division 2 hazardous locations, controls all operations of the chromatograph and sample handling system; provides validity checking of all data; computes stream composition, calorific value, and specific gravity; is capable of transmitting digital data for report generation; and carries out internal diagnostic routines. Battery back-up maintains all factory application information without data re-entry for up to six months after shipment and protects the COMPU-CAL programs in the event of a power failure.

Stream calorific value and specific gravity reports are generated every 10 min, with the additional option of Wobbe index and compressibility factor. Automatic calibration checks may be carried out at 8, 12, or 24 hr intervals, operator selectable. Short, long, and 24 hr average reports are available, along with alarm printouts. Analog outputs (4–20 mA dc) are provided for BTU/SCF and SG, with up to 14 additional analog output capability. The system reproducibility is quoted as ±0.1% (±1.0 BTU in 1000).

14.5.4 Measurement by IR Absorption

IR absorption is covered in detail in Chapter 10, together with other spectrometric methods of analysis. Traditionally, this technique has been used extensively in industry for many years in applications involving the measurement of a single component in a gaseous or liquid mixture. The success of the technique depends on the correct choice of measurement and reference absorption wavelengths. The reference wavelength is selected so that there is little or no absorption of IR energy by the component of interest, whereas there is maximum absorption at the measurement wavelength.

Thus, selectivity is an important consideration and, because of interference problems and background composition variations, IR absorption is seldom used for on-line multi-component analysis.

Nevertheless, there are instruments operating on the IR absorption principle that are designed for multi-component measurement. One such instrument is described next.

(a) Thermo Electron Corporation (United States)

The CALSENSOR Infrared Analyzer is a microprocessor-based instrument designed specifically to measure the dry heating value of natural gas on a continuous basis without combusting the sample. First, the gas composition is determined and then the heating value computed by the microprocessor. The analyzer was developed from research carried out in conjunction with Southern California Gas Company during the mid 1970s and was first introduced in 1978. It is an extension of the Thermo Electron PROSENSOR Gas Analyzer, an IR single-component analyzer introduced in 1974.

The CALSENSOR makes two absorption measurements, and a third (reference) wavelength is chosen so that no absorption occurs at this wavelength. One measurement wavelength is used to determine the concentration of methane in the sample (the major component of natural gas); the second measurement wavelength is used to determine the concentration of the non-methane components. Based on research data, the second wavelength is chosen so that each of the non-methane components absorbs in proportion to its heating value. Thus, the total absorption produces an effective ethane concentration with the same total heating value as the sum of all the actual components in their actual concentrations.

The operating principle of the instrument is shown schematically in Fig. 14.12. The output of a broad-band IR source is divided into three narrow

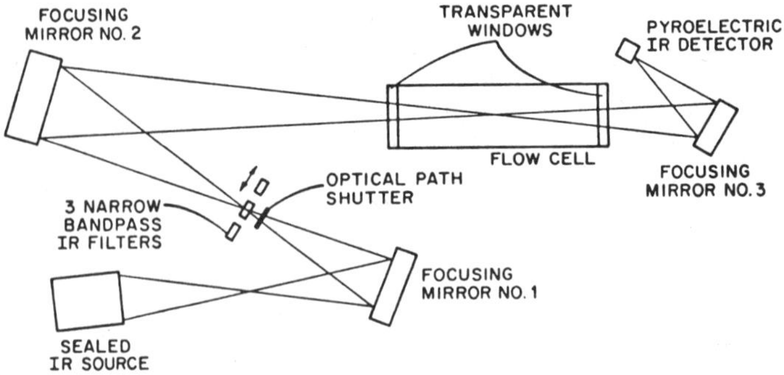

FIGURE 14.12 CALSENSOR system: operating principle. (Courtesy of Thermo Electron Corporation)

wavebands using precision optical filters. Three focusing mirrors direct the IR energy from the source through one of the optical filters, then through the sample cell, which contains the gas under analysis, and finally to a deuterated triglycine sulfate (DTGS) pyroelectric detector.

The optical system also includes a tuning fork shutter that interrupts the source radiation signal to generate a nearly sinusoidal radiation signal at the detector. The three filters are mounted on a positioning mechanism that alternately positions each filter in the beam. The sample cell is 15 cm long, chosen to optimize the absorption characteristics of each filter in terms of sensitivity.

The detector output is an ac signal, the amplitude of which depends upon which filter is in position and the sample gas composition. The microprocessor electronics unit converts the detector output to a digital signal, computes the ratios of methane output and non-methane output to the reference output, and computes the heating value from these ratios. The heating value is digitally displayed and also converted to an analog signal for remote recording.

The Calsensor is mounted in a two-compartment NEMA-4 style enclosure, as shown in Fig. 14.13. The upper section houses the electronics systems; the lower section contains the optical bench, flow controls, and temperature stabilization circuitry. Six sets of thumbwheel switches are provided. The five sets below the digital display are used to calibrate the analyzer, and provide the constants for curve fitting the methane and non-methane absorption/concentration curves. The two above the display control the display mode. Absorption ratios and other diagnostic readings can be displayed for calibration and maintenance purposes.

Precise control of sample pressure in the flow cell is achieved by two precision regulators, one a constant differential pressure regulator (set at 3 psid), the other a back pressure regulator referenced to atmospheric pressure. Sample flowrate is approximately 200 ml/min.

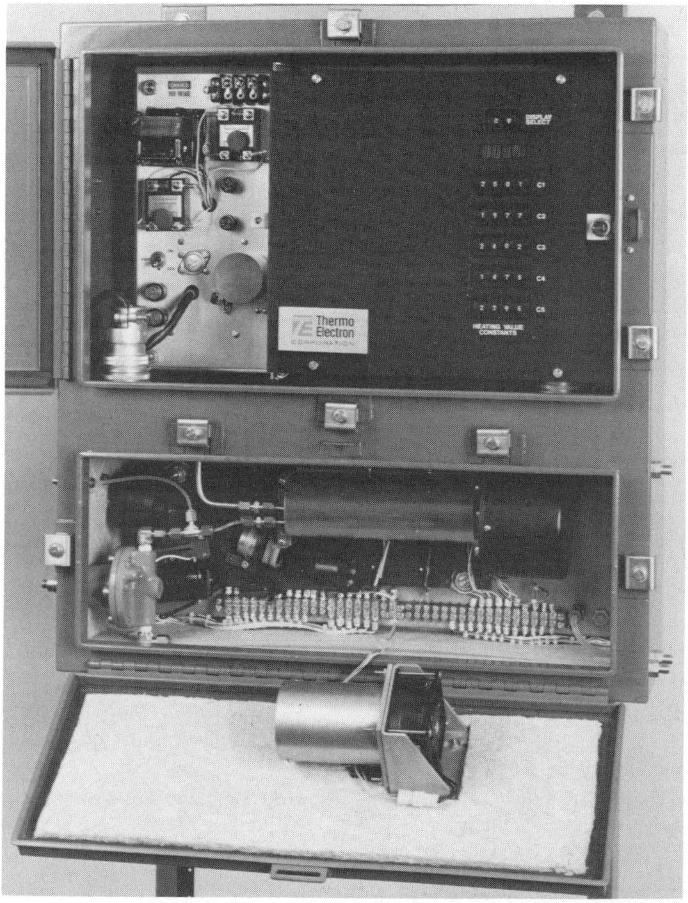

FIGURE 14.13 CALSENSOR system: equipment layout. (Courtesy of Thermoelectron Corporation)

Three heaters are used to stabilize the measurement components. The lower compartment is controlled at 40°C by a heater/blower combination. The sample cell is controlled by a separate heater to ensure constant gas sample temperature. A third heater maintains the three optical filters at 55 ±0.1°C, required because of the large temperature dependence of the components.

The standard range of the Calsensor is 750–1200 BTU/SCF (dry) resulting from 75–100% by volume of methane and 0–20% by volume of non-methane hydrocarbons. Hydrogen and carbon dioxide heating values are not measured. Output span normally corresponds to either 950–1150 or 800–1200 BTU/SCF (dry). Normal instrument response is 10–12 min for 99% of final reading at a sample flowrate of 200 ml/min; fast response is 1 min for 90% of final reading at a sample flowrate of 1000 ml/min. Sample pressure should be 6–100 psig and water vapor content must be less than 100 ppm, since water interferes with the measurement of the non-methane components.

As mentioned earlier, this analyzer is designed specifically for heating value measurement of natural gas. It cannot be used in services where the composition of the gas differs significantly from typical natural gases. Diatomic gases such as hydrogen and nitrogen do not absorb in the IR portion of the electromagnetic spectrum and therefore cannot be measured.

Two optional accessories are available to complement the basic CALSENSOR. An integral recorder (Honeywell Electronik 112) can be supplied. A complete dry air supply system is also available, consisting of an oilless compressor, heatless dryer, and associated flow and pressure indicators, all mounted in a self-contained enclosure.

BIBLIOGRAPHY

Bradley, J. N. *Flame and Combustion Phenomena*. Methuen, London (1969).

Broadwater, S. R. Columbia Gas moves toward heating value measurement. *Oil & Gas J.* (Aug 25, 1980)

Calorie counter provides a better boiler diet. *Chem. Week* (June 4, 1980).

Clingman, W. H. Jr. New approach to the continuous measurement of calorific value of gaseous fuels. *AGA Oper. Sect. Proc.* (1972).

Clingman, W. H. Jr., Kennedy, L. R., and Gleason, D. L. Therm-titrator round-robin testing and evaluation. *AGA Oper. Sect. Proc.* (1978).

Dofasco fine tunes its cokemaking act with feedforward combustion control. *Metal Prod.* (July 1980).

DuPont improves ethylene yield with feedforward fuel control. *Oil & Gas J.* (June 1, 1981).

Feedback loop improves BTU control of mixed gases. *Gas. Ind.* (April 1981).

Foundos, A. P. Measuring heat release rate from fuel gases. *Instr. Tech.*, ISA (1977).

Foundos, A. P. On-line gas calorimeter applications. Session 3, *Instr. Asia '82*, Singapore (Nov 1982).

Fraim, F., Curran, R., and Paquin, M. Natural gas heating value determination using an infrared calorimeter. *Proc. AGA Trans. Conf.*, Montreal, Canada (May 1978).

Goodyear, E. M. *Hydrocarbon Fluids*. Macmillan, London (1975).

Jasper, G. *Gas Service Technology*. Benn, London (1978).

Johnson, S. R. Gas analyzer ensures efficiency in cogeneration facility: Helps save more than $2 million/yr. *Chem. Proc.* (Feb 1985).

Light, J. The therm-titrator, a comparison with you-know-what. *60th Ann. GPA Conv.*, San Antonio, TX (March 1981).

Rosko, G., Wilson, W., and Downer, S. W. More accurate gas energy measurement using advanced microelectronics and chromatography. *ISA Natl. Conf.*, Chicago, ILL (1979).

Tests prove that calorimeters work with control loops. *Plnt. Energy Mgmnt.* (April 1981).

Wood, J., and Fraim, F. New instrument finds diverse application in gas industry. *Gas. Ind. Mag.* (Oct 1980).

Yazhari, F. Control ammonia process with varying natural gas LHV. *Hcbn. Proc.* (May 1982).

15

Measurement of Octane Number

15.1 INTRODUCTION AND DEFINITIONS

For fifty years, the petroleum industry has used standardized ASTM laboratory procedures to evaluate the expected performance of gasoline products in internal combustion engines. The octane quality of a fuel is defined as the fuel's resistance to knock and is expressed as octane number (ON). The *octane rating system* for motor gasolines measures the resistance of the fuel to *knock*.

Knocking makes itself evident as a metallic clacking noise, often called "pinking," which occurs when the engine is subjected to a heavy load, (e.g., rapid pulling uphill in high gear). It occurs when the octane quality of the fuel is too low for the engine compression ratio and other combustion conditions.

Thus, octane quality is directly related to knock resistance; the higher the octane number, the more resistant the fuel is to knock and the better its performance in high-compression engines.

Knock occurs late in the combustion cycle and results in the generation of high-frequency, high-pressure waves that bounce off the piston top and cylinder walls of the engine, producing audible noise.

The laboratory octane rating system uses a special internal combustion engine with adjustable operating parameters. The octane rating of an unknown gasoline is determined by comparing its knock intensity (KI) with the KI of a known mixture of iso-octane and n-heptane. These two pure hydrocarbons are used to define the octane scale. Iso-octane, which has a high resistance to knock, is assigned the octane number 100. Normal heptane, which has a low resistance to knock, is assigned the octane number zero.

Thus, the octane number of a mixture of these two compounds is, by definition, the percentage of iso-octane in the blend. The mixtures are known as *primary reference fuels* or PRFs. The octane rating of the unknown gasoline is obtained by a technique known as "Bracketing," in which the unknown gasoline KI is bracketed by two PRFs of accurately known octane number and the result determined by extrapolation.

15.2 HISTORICAL BACKGROUND

The first studies of the anti-knock quality of motor gasolines with respect to road performance in an internal combustion engine were made in 1932 in the United States. They led to the introduction of the *Motor Method Test* (*MON*), carried out in a special test engine built by Waukesha Motor Company, Waukesha, Wisconsin, and known as the cooperative fuel research (CFR) engine. The method remains substantially unchanged to the present day.

During World War II the only fuel available was of very poor quality compared with the high-octane fuels of the late 60s and 70s. During the latter period, the anti-knock quality of motor gasolines increased steadily, made possible by new developments in refining and gasoline production technology. The same period saw an increase in automobile engine compression ratios, made possible by improvements in engine design technology.

With these changes and improvements in design, it became evident that a single test could no longer predict road performance, so a second test, the *Research Octane Number Rating* (*RON*) was introduced. The RON test represents engine operation under mild conditions (low speed, low temperature), whereas the MON test represents operation under more severe conditions (high speed, high temperature). Together, these two laboratory ratings can be used to estimate the performance of a fuel in a car on the road.

15.3 PRODUCTION OF MOTOR GASOLINES

Motor gasolines are produced in a refinery by blending various component streams, at the same time introducing anti-knock and other additives. To manufacture a good final product, it is essential that the octane number be distributed evenly over the entire boiling range of the gasoline. If this is not done—for instance, if the low boiling range components have a lower octane number—component segregation can occur at the engine intake manifold, resulting in loss of power when starting from cold.

Tetra-ethyl lead (TEL), once universally used as an anti-knock additive, has a boiling point of 392°F (the approximate end point of gasoline) and therefore aggravates the fuel segregation problem. This led to the introduction of tetra-methyl lead (TML), which has a boiling point of 240°F. In many cases, a blend of these two compounds has been found to give the best solution.

With the introduction in the 1970s of federal legislation to reduce or eliminate the use of lead compounds as gasoline additives, there has been a gradual decrease in the octane number of motor gasolines and a search for alternative anti-knock additives. One additive that has recently received government approval is methyl tertiary-butyl ether (MTBE), leading to a spate of activity in the construction of plants to manufacture it.

Refinery processes producing feedstocks for motor gasoline blending may be classified into three categories:

- Cracking processes.
- Catalytic reforming processes.
- Alkylation processes.

Cracking processes are generally used to produce medium-quality gasoline from heavy middle-distillate feedstocks. *Thermal cracking* was used extensively prior to World War II, but has been largely replaced by *catalytic cracking*. However, two forms of thermal cracking are still used in modern refineries.

Visbreaking is a process in which crude oil residue is cracked under conditions minimizing coke formation. In *coking,* the time of cracking is extended to allow coke formation. This process upgrades heavy fuel oils to gasoline, with solid coke, gas oil, and gas as by-products.

Catalytic cracking is a process of major importance world-wide. The vaporized feedstock is cracked at a temperature of 1000–1070°F in the presence of a fluidized catalyst. Reaction products and catalyst are separated in cyclones at the top of the reactor and catalyst is regenerated by steam stripping and coke burn-off. Reaction products are further refined by fractionation. *Hydrocracking* may be regarded as a combination of hydrogenation and catalytic cracking. This process is very flexible, since it can handle a wide-range product distribution depending upon operating conditions.

Steam cracking is used primarily for the production of petrochemical feedstocks such as ethylene, propylene, and butadiene, but the process also produces significant quantities of good-quality gasoline blendstock.

The *Catalytic reforming* process, which developed from thermal cracking, began to be widely used for high quality gasoline feedstock production in the early 1950s. Since then, there has been a tremendous increase in catalytic reforming capacity world-wide. The process produces high-octane compounds (aromatics and branched-chain paraffinic hydrocarbons) from a naphtha feedstock. Three main reactions take place:

1. *Dehydrogenation* of naphthenic hydrocarbons to form aromatics:

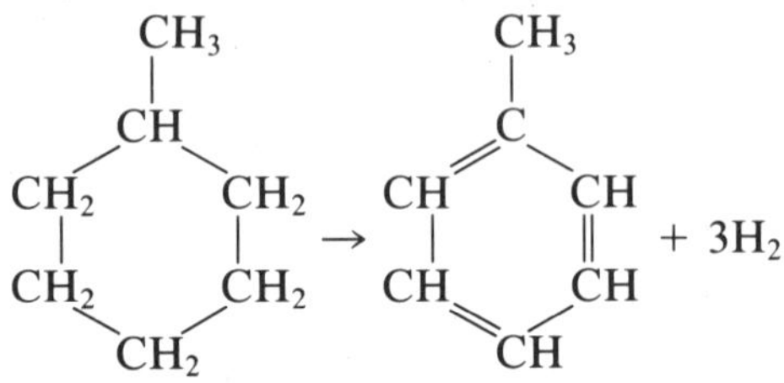

methyl cyclohexane — toluene

2. *Dehydrocyclization* of paraffinic hydrocarbons to form aromatics:

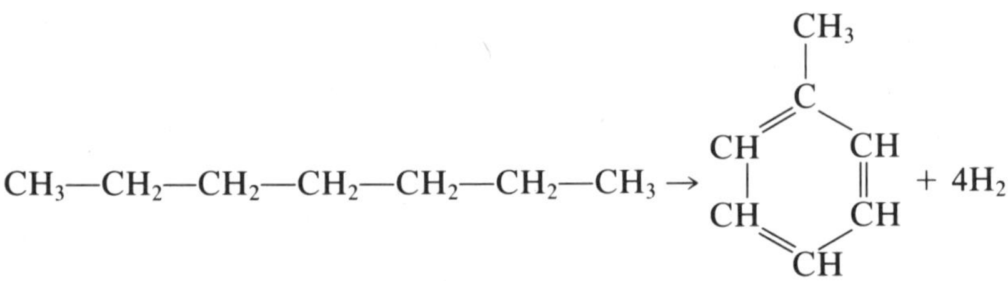

n-Heptane — toluene

3. *Isomerization* of paraffinic hydrocarbons to produce branched-chain compounds:

$$CH_3{-}CH_2{-}CH_2{-}CH_2{-}CH_2{-}CH_3 \rightarrow CH_3{-}\overset{\overset{\displaystyle CH_3}{|}}{CH}{-}CH_2{-}CH_2{-}CH_3$$

n-hexane — 2-methyl pentane

Hydrogen is produced in the first two reactions. The process is therefore a net producer of this valuable by-product, which is recycled in the process to maintain reactor pressure and suppress coke formation.

Alkylation is a process in which light olefins and isoparaffins are combined to produce high-quality gasoline feedstock. This process was used extensively in the United States during World War II to produce aviation gasoline blendstock. The majority of alkylation processes are designed around the combination of isobutane and butylene, but propylene and other olefins may also be used. The process is important in gasoline blending, since the alkylate produced has excellent gasoline characteristics.

15.4 THE ROLE OF ON-LINE OCTANE ANALYZERS

For many years, refineries producing motor gasoline and gasoline blending components have relied on the standardized laboratory knock-rating procedures (ASTM Methods) to evaluate the expected performance of their products in internal combustion engines. In the early days when production and blending processes were mostly batch-type, these laboratory methods were adequate for control of product quality.

The situation today is quite different. As the demand for high-quality gasoline increased, together with the need for more efficient and economic production, the batch-type process was replaced by the in-line blending process. This provides for simultaneous pumping and blending of all blendstocks, including anti-knock and other additives, to give a finished product for shipment to a final destination with little or no intermediate tankage or product hold-up.

Such a system has very little margin for error and cannot be operated efficiently unless product quality data is made available rapidly. The standard laboratory knock-rating methods are no longer adequate, particularly for the control of such a critical and elusive characteristic as octane number. This applies not only to octane number but to other motor gasoline quality parameters such as RVP, boiling point, flash point, and so on.

With the introduction of on-line octane analyzers in the mid 1960s, it was possible to closely control the in-line blending process. At the same time, great savings were possible from a reduction in octane give-away. In addition, there are other savings, varying from refinery to refinery, depending upon such factors as total production requirements, number of grades of finished gasoline produced, and methods of dispatch, all of which may be considerable.

15.5 LABORATORY METHODS OF MEASUREMENT

15.5.1 The ASTM-CFR Engine

The ASTM Manual of Engine Test Methods for Rating Fuels was first published in 1948. Since then, there have been numerous revisions to the manual, including supplementary publications. The Manual contains a number of appendices with valuable information to users on installation, operation, and maintenance of the test equipment; details of the equipment itself; and data on reference materials and accessories for use in blending PRFs and other standards. The current edition covers the latest ASTM-IP procedures for measurement in the laboratory. The following ASTM standards are included:

- ASTM D2699-80/IP237-69 Knock Characteristics of Motor Fuels by the Research Method.
- ASTM D2700-80/IP236-69 Knock Characteristics of Motor and Aviation Fuels by the Motor Method.
- ASTM D2623-80/IP238-68. Knock Characteristics of LP Gases by the Motor (LP) Method.
- ASTM D2885-79 Research and Motor Method Octane Ratings Using On-line Analyzers.
- ASTM D2886-79 Knock Characteristics of Motor Fuels by the Distribution Octane Number (DON) Method.
- ASTM D909-79/IP119-81 Knock Characteristics of Aviation Fuels by the Supercharge Method.

Of these methods, D2699, D2700, and D2885 are the most important in relation to process analyzers. The first two are the traditional laboratory tests referred to earlier. The third is evidence of the importance the ASTM attributes to the use of continuous on-line octane measurement and the acceptance of the technique in refinery gasoline blending operations.

For these methods, the test equipment, known as the CFR engine, consists of a single-cylinder, 4-cycle engine with a variable compression ratio. The engine is of very heavy-duty construction so that it can withstand the stress of continuous operation under heavy knocking conditions. Compression ratio adjustment is accomplished by moving the cylinder head up or down with respect to the piston and crankcase. This allows the engine to be operated under knocking conditions with fuels covering a wide range of octane numbers.

The complete system, shown in Fig. 15.1, is manufactured by Waukesha Motor Company, Waukesha, Wisconsin, the sole authorized supplier. A console type panel is supplied with the unit, complete with all instrumentation and wiring. Instruments include knock meter, oil pressure gage, oil temperature gage, hours meter, crankcase vacuum gage, air intake temperature controller, ammeter for intake air heater, and ammeter for mixture heater (motor method only). The complete system is installed on a specially prepared anti-vibration foundation in the laboratory engine room. Recommendations for installation are specified in the appropriate ASTM standard.

Compression ratio is adjustable between the limits 4.0 and 18.0. A micrometer mounted on the cylinder sleeve of the engine (or a remote mounted digital counter) is used to indicate the position of the cylinder in relation to the piston. The cylinder height is adjusted by means of a worm gear drive operated by hand crank or by a panel-mounted gear motor. The ignition system is of the capacitor discharge type with coil and distributor. When the

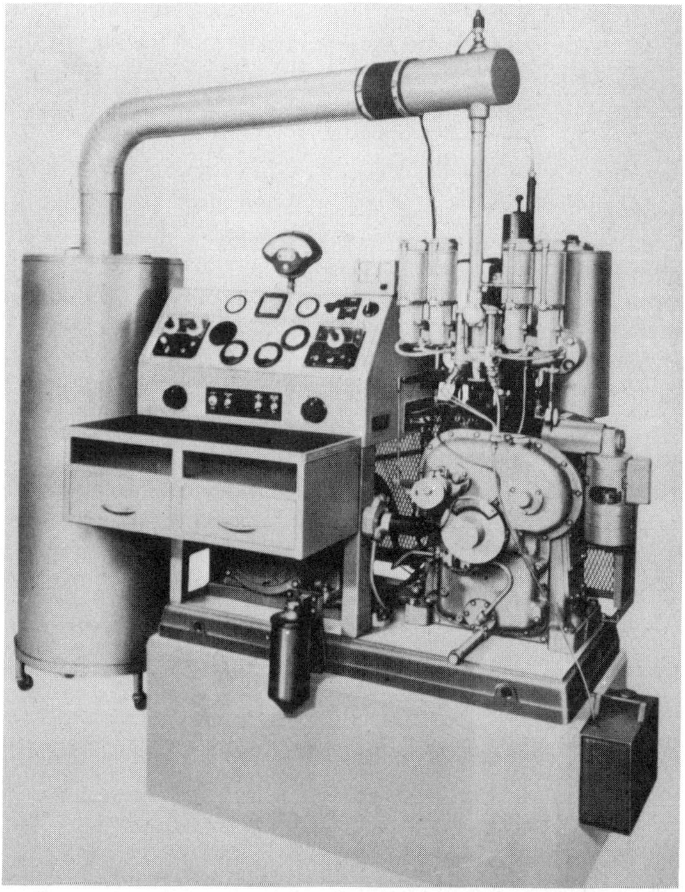

FIGURE 15.1 CFR (cooperative fuel research) engine. (Courtesy of DuPont Company)

engine is running, the spark advance is indicated in degrees on a scale on the engine housing, marked by a flashing neon light. For research method units, the spark advance is set at 13° BTDC. For motor method units, the spark advance is varied automatically with variation in compression ratio, in accordance with requirements listed in the ASTM manual.

The engine is connected to a reluctance-type synchronous motor capable of starting the engine, absorbing the power developed by it, and maintaining the specified engine speed within close tolerances (see Table 15.1). The engine exhaust port is connected by a flexible exhaust pipe to a surge tank. The discharge pipe from the surge tank should be designed to maintain an exhaust back pressure of between 0 and 10 in. WG.

The engine is equipped with an adjustable three-bowl carburetor and fuel containers complete with all piping. A four-bowl carburetor is also available.

The fuel-air ratio is varied by adjusting each float bowl individually, the fuel level for a given fuel-air ratio being dependent upon the size of the metering jet. The air intake surge pipe is fitted with a heater to maintain the intake air at the specified temperature (see Table 15.1).

Since air humidity has a noticeable effect on the performance of knock test engines, the ASTM standard specifies control of the intake air humidity at between 25 and 50 grains of water per pound of dry air (0.36–0.72% by weight). This is accomplished using an ice tower consisting of an insulated cylindrical tank designed to pass the intake air through an ice bed so that it emerges saturated at 32°F and with the required moisture level (26–28 grains of water per pound of dry air).

For motor method units, the intake mixture temperature must be controlled at 300 ±2°F. This is accomplished by a heater assembly consisting of an intake manifold, an electric immersion heater, voltage controller, and a thermometer. Automatic control is provided by the Waukesha temperature controller, the sensing element of which is a thermistor mounted in a boss on the intake manifold. The heater assembly is mounted on the engine immediately opposite the carburetor inlet. For research method units, the heater assembly is omitted.

TABLE 15.1 ASTM/CFR Engine Operating Conditions

Parameter	Research Method	Motor Method
1. Engine speed (rpm)	600 ± 6	900 ± 9
2. Crankcase oil	SAE 30	SAE 30
3. Oil pressure (lb/in.^{2})	25–30	25–30
4. Oil temperature (°F)	135 ± 15	135 ± 15
5. Coolant temperature:		
Range (°F)	212 ± 3	212 ± 3
Accuracy (°F)	±1	±1
6. Intake air humidity (grains/lb of dry air)	25–30	25–50
7. Intake air temperature (°F)	[a]	100 ± 5
8. Mixture temperature (°F)	—	300 ± 2
9. Spark advance (°BTDC)	13	[b]
10. Break point gap (in.)	0.02	0.02
11. Valve clearances[c]		
Inlet	0.008	0.008
Outlet	0.008	0.008
12. Fuel/air ratio	Adjusted for maximum knock intensity	

[a] Adjusted to within ±2°F of the temperature specified in the ASTM procedure.

[b] Varies automatically with compression ratio.

[c] Measured with engine hot and running under standard operating conditions on a reference fuel of 100 octane number.

Knock intensity is measured by a combination of three units—the detonation pickup (mounted in the cylinder head), the detonation meter (panel or desk mounted), and the knock intensity meter (usually mounted at the top of the operating console). Knock in the engine gives rise to pressure fluctuations, which are detected by the detonation pickup. The resulting output signal is modified by the detonation meter and transmitted to the knock meter, the readings of which indicate the relative degree of knock intensity. These instruments should have a stabilized power supply to avoid the effect of mains power fluctuations.

The operating conditions for research and motor method engines are summarized in Table 15.1, from which the main differences between the two methods are immediately apparent. Physically, the CFR engines are the same for both test methods, the differences being in the test conditions only. By virtue of the higher speed, preheating of the fuel/air mixture, and advanced timing, the motor method test stresses the fuel more than the research method test. In general terms, the motor methods tends to give a lower octane rating than the research method for the same gasoline.

15.5.2 Engine Operating Procedure

Detailed coverage of engine operating procedure is outside the scope of this book; should the reader be interested, the ASTM manual will give all the necessary information. The following summary will give the reader some idea of the requirements for carrying out an octane rating in the laboratory and the relative complexity of the procedures.

Engine operating procedure may be split arbitrarily into three stages:

1. Engine warmup.
2. Engine standardization.
3. Sample fuel rating.

During the warmup period, the various operating parameters such as pressures, temperatures, spark advance, valve clearances, and so on are checked to ensure that they do not exceed those specified in Table 15.1. The warmup fuel should be of sufficiently high octane number to avoid heavy knocking and the engine should be run at a compression ratio less than standard knock intensity (defined later).

When the engine is thoroughly warmed up, the detonation meter and knock meter may be zeroed and engine standardization can then be carried out. This basically concerns adjustment for standard knock intensity and fuel/air ratio adjustment for maximum knock meter indication.

The knock meter is zeroed initially with the main power supply and operational switches of the detonation meter in the "off" position. After a 30-min warmup, the knock meter is re-zeroed with the operational switch in the "off" position and the mains supply switch "on."

The detonation meter is also zeroed and calibrated as follows. The detonation meter has "meter reading" and "spread" controls, each with coarse and fine adjustment knobs. The "meter reading" control is adjusted to give a mid-scale knock meter reading, then the fuel/air ratio is adjusted for maximum knock by varying the carburetor float level. If the fuel level for maximum knock is not well-defined, this is an indication of low sensitivity. The sensitivity is increased by adjusting the "spread" control to give a knock meter reading of 100 and then bringing this reading back to mid-scale with the "meter reading" control.

The fuel/air ratio is then determined for a second fuel differing from the first by not more than two octane numbers and the spread determined by comparing the maximum knock meter readings for the two fuels. Ideally, the spread for a difference of two octane numbers should be 30 knock meter divisions, but it may be necessary to use a smaller spread, although this should never be less than six knock meter divisions.

For fuels above 100 ON, 30 divisions per octane number is a desirable spread. The spread may be increased by adjusting the "spread" control and then adjusting the "meter reading" control to restore the original knock meter reading.

Note that all octane ratings by research or motor methods are carried out at the same degree of knock, termed *standard knock intensity,* regardless of the octane number of the fuel. Thus, a 90 octane fuel is rated at a higher compression ratio than an 80 octane fuel.

Standard knock intensity and maximum-knock fuel/air ratio are established for the test sample as follows. A quantity of the sample fuel under test is poured into one of the carburetor tanks and the fuel level set at the estimated maximum knock position. After allowing time for the engine to reach equilibrium conditions, the cylinder height is adjusted to give a knock meter reading of 50–52. The fuel level for maximum knock is then established by raising the fuel level by increments of 0.1 until the knock meter reading has decreased by at least five divisions from the maximum reading obtained. The fuel level is then lowered following the same procedure. The fuel level is then adjusted to the position of maximum knock meter reading. The cylinder height is then adjusted to give a knock meter reading of 55, and this cylinder height is used for the remainder of the test. The engine is now adjusted for standard knock intensity and maximum-knock air/fuel ratio for the sample fuel under test.

The procedure for rating the sample fuel involves "bracketing" or matching the knocking tendency of the sample with the knocking tendency of two reference fuels (PRFs) of accurately known octane number. With the cylinder height determined on the test sample as described, the engine is operated on the first PRF, which is chosen to have an octane number near the estimated octane number of the sample fuel. After the engine has reached equilibrium, the fuel level is adjusted to give maximum knock meter reading. A similar procedure is then followed with the second PRF. This PRF is chosen so that it differs by no more than two octane numbers from the first PRF, and the two PRF's are expected to bracket the knock meter reading of the sample fuel. If the sample fuel knock meter reading is not bracketed, a third PRF that should differ by no more than two octane numbers from the second PRF is required.

Assume that the sample fuel knock meter reading is now bracketed by the two PRFs and all carburetor bowls are set to give maximum knock. A second series of readings is then taken in the order sample, second PRF, first PRF. The average knock meter readings are then determined for the sample and the two PRFs. Assume that the two PRFs have octane numbers of 96 and 98 and give knock meter readings of A and B respectively. Assume that the sample fuel gives a knock meter reading of C. The octane number $(ON)_s$ of the sample fuel is given by:

$$(ON)_s = 96 + 2\frac{(A - C)}{(A - B)}$$

Note that the knock meter reading for the low-octane PRF is higher than that for the high-octane PRF. The value obtained is reported to the nearest 0.1 octane number. Any octane number ending in exactly 5 in the second decimal place is rounded off to the nearest *even* tenth (e.g., 95.55 and 95.65 are both reported as 95.6).

15.6 CONTINUOUS ON-LINE OCTANE ANALYZERS

The detonation of gasoline in internal combustion engines is an unstable phenomenon and produces an erratic knocking signal, even in the specially designed ASTM-CFR engine used for octane measurement in the laboratory. So the development of on-line octane analyzers presented quite a challenge. This development proceeded along two different routes, analyzers based on the ASTM-CFR engine and analyzers based on correlative techniques.

Three companies chose the first route and developed octane comparators for use in conjunction with the ASTM-CFR engine:

- Ethyl Corporation (United States)
- DuPont Company (United States)
- Associated Octel Ltd. (England)

Two companies chose the second route and, after considerable research, developed on-line analyzers based on correlative techniques:

- Universal Oil Products Inc. (United States)
- Arcas—now Foxboro Analytical (United States)

15.6.1 Analyzers Using the Automated ASTM-CFR Engine

The laboratory test engine described earlier (and its associated operating procedures) is satisfactory for batch-type testing, but when applied to continuous in-line blending processes (requiring continuous octane monitoring), several problems must be overcome.

First, it is found that the octane rating of the same fuel in the same engine decreases with time over a period of hours or days due to change in engine characteristics. To eliminate this problem, the knocking tendency of the sample fuel is compared with that of a prototype fuel of acccurately known octane number. Note that primary reference fuels are not used. Any changes in engine characteristics beyond the operator's control, affect the proto and sample fuels in the same way and to the same extent. In the comparator-type system, relative (delta) octane ratings are made, rather than absolute octane ratings. The actual octane number of the sample fuel is calculated from the ΔON and the actual octane number of the proto fuel.

Second, the knock intensity for each firing of the engine is not constant, but varies widely over a period of seconds. This problem is well known in laboratory test engines and is reduced to a manageable level by increasing the electronic dampening of the detonation meter.

Third, the knock intensity also varies erratically on a time scale of minutes due to engine drift. This problem is reduced by using a short test cycle of 3–4 min. The comparator systems therefore allow for programmed switching between proto and sample fuels at or near to this frequency.

(a) Ethyl Corporation (United States)

The Ethyl Model 8021 Octane Analyzer is designed for automatic on-line measurement of the octane number of in-line blended gasolines or process streams used as gasoline blendstocks. The instrument is used with an ASTM-CFR engine and determines the difference in octane number between the sample stream and a proto fuel of accurately known octane number. Changes in compression ratio (CR) at constant knock intensity are related to octane number difference, which is indicated and recorded as delta octane number (ΔON). The comparator unit of the analyzer is shown in Fig. 15.2.

The octane comparator equipment consists of five basic assemblies:

- Analyzer Console—a pedestal-mounted enclosure housing display instruments, operational switches, controls, sequencing devices, and analog signal handling circuits. The console is installed to the right of the engine and connected to it by electrical cables.
- Engine Interconnection Unit—an enclosure usually mounted below the engine desk assembly. It contains a relay chassis, control unit for automatic CR adjustment, electrical connectors, and terminal strips.

FIGURE 15.2 Model 8021 octane analyzer: comparator console. (Courtesy of Ethyl Corporation)

- Fuel Panel—designed for wall mounting and containing all components required to condition and deliver fuels (sample and proto) to the engine.
- ASTM Engine Accessories:
  1. A converter assembly to produce an electrical output proportional to CR, and a CR safety limit switch. Both devices are mounted on the engine cylinder and sleeve assembly.
  2. A carburetor modification kit to achieve conversion to the "flow-through" principle that allows large volume sample fuel flow. Carburetor devices include a special selector valve, a fuel suction tube for removal of excess sample fuel, a sample fuel rotameter, and suitable carburetor cooling and fuel overflow safety devices.
- Fuel Cooling Unit—a mechanical water chiller and circulating pump for installation behind the engine for cooling sample and proto fuels.

Measurement of ΔON is accomplished by operating the ASTM-CFR engine on proto and then sample fuel in an automatic, alternating sequence. Typically, 5 min of proto testing is followed by 25 min of sample testing. The five assemblies described form three subsystems that carry out fuel conditioning, engine control, and operational sequencing respectively.

The fuel conditioning system handles both sample and proto fuels. Sample is filtered, passed through a coalescer, then pressure regulated for delivery through the "flow through" carburetor assembly to a solenoid-operated selector valve, which alternately supplies proto and sample fuels to the engine. Proto fuel is delivered directly to a conventional ASTM engine carburetor float reservoir and then to the selector valve. The fuel cooling unit circulates chilled water at 40°F through heat exchangers to cool both proto and sample fuels. Cooling is also provided at the conventional ASTM float reservoir, the selector valve, and the carburetor body.

The "flow-through" carburetor assembly uses a fuel level sight glass into which sample fuel is introduced at greater than ten times the rate required by the engine, thereby significantly reducing the response time of the system. Excess sample is removed by a suction tube, which projects down into the sight glass. Maximum knock fuel/air ratio is obtained using the conventional reservoir level adjustment device.

Constant knock intensity-CR control is continuous in the automatic mode. The KI signal produced by the standard ASTM detonation meter is the input to a control unit, which compares it with a preset standard KI once every few seconds. The control unit produces a timed electrical pulse to drive the CR motor in the proper direction and in the amount proportional to the error. (e.g., small KI deviations produce short pulses and small CR changes).

CR measurement and ΔON signal generation are handled by a series of electronic devices. A CR converter assembly mounted on the engine produces a signal proportional to the cylinder position. Spanning of this signal to accomodate various octane levels is selected by a six-position switch. The resulting ΔON signal is the input to either of two "track and hold" function modules, one for sample, one for proto. An "add-subtract" module calculates the ΔON by summing these values along with a signal representing the offset ON (offset ON is the difference between the proto fuel ON and the desired target ON). A selector switch permits selection of offset values between -1.0 and $+1.0$ ON in 0.1 increments. The ΔON value is displayed on an indicating meter and recorder mounted on the analyzer console.

Alarm contacts are provided for remote annunciation of engine shutdown, sample shutoff, CR high limit, and CR low limit. In addition, safety actuators are provided in conjunction with these alarm conditions. Power is removed from the analyzer console if the engine shuts down; sample fuel is shut off and the operation is locked in proto mode if any of several fuel conditions are abnormal.

The octane analyzer meets all requirements of ASTM D2885. The sample octane number as measured, will have 95% probability confidence limits as follows:

Octane Number	+− Confidence Limits	RON MON
90	0.11	0.15
95	0.09	0.18
100	0.11	0.20
101	0.12	0.19
102	0.13	0.18

The Comp-Con III Computer Octane Analysis System shown in Fig 15.3 is designed to accommodate from one to three in-line gasoline blending units, with both research and motor octane engines servicing each blender. The engines are provided with suitable interface equipment, such as KI and CR transducers, safety alarm sensors, fuel conditioning, and cooling equipment, as described previously. The computer automatically controls the knock intensity of each engine by CR adjustment, sequentially operates the engines on prototype and product fuels, and computes road octane number from delta RON and delta MON input data.

A single-bay steel enclosure houses the computer and computer interface equipment, consisting of the following:

- DEC PDP-8 computer, with programming panel and dual floppy disk storage
- Operator's panel
- Computer interface equipment, which includes power supplies, input/output logic, analog-to-digital converter, and I/O termination facilities

The unit is interfaced with a DECwriter hard copy terminal for report printing and data entry.

The master analysis program is permanently stored on floppy disk and can be loaded into the computer when required. Such subroutines as integrated RVP and V/L Ratio calculations can be incorporated as options, based on input data from other analyzers.

Initial set-up of each engine is carried out from the operator's panel in the MANUAL mode, including maximum knock air/fuel ratio selection for prototype and product fuels, detonation meter signal level adjustment, and proper selection of compression ratio for the specific prototype fuel calibrated octane number. When the RON and MON engines are switched to AUTO at the interface unit, the STANDBY mode is activated, allowing operation on prototype fuel only. From the standby mode the system is

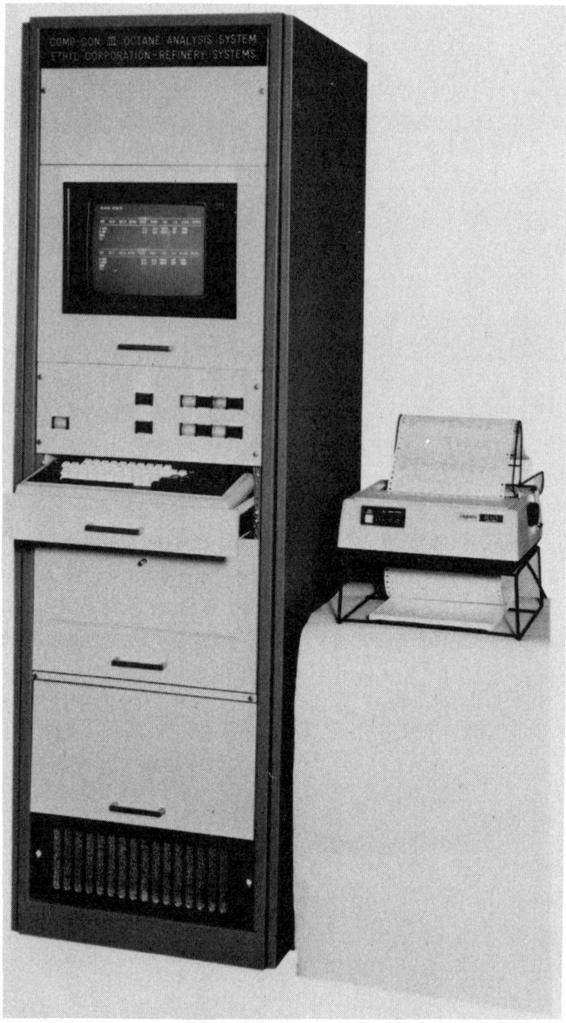

FIGURE 15.3 COMP-CON III computer octane analysis system. (Courtesy of Ethyl Corporation)

switched to full on-line analysis by pressing the RUN switch. The run mode is inhibited unless certain interlocking conditions are satisfied:

- Both RON and MON interface units are in the AUTO mode.
- The computer program is running correctly.
- The product fuel systems are operative.

If during a run, any interlock condition fails, the system automatically reverts to standby. If any engine fail-safe devices cause engine shutdown, the computer is disconnected from that engine, since the engine will revert to the manual mode. Printout of blend quality information can be programmed to occur on a predetermined time basis. Such data includes:

- Instantaneous ΔRON, ΔMON, and ΔRoad ON
- Instantaneous RON, MON, and Road ON
- Flow weighted averages for RON, MON and Road ON

(b) Dupont Company (United States)

DuPont developed its first octane comparator over 15 years ago and now has over 150 units in service. The Mark IV is the latest version. The comparator calculates delta octane (ΔON) from the direct measurement of the difference in KI between prototype and product fuels, while running the engine at a constant compression ratio setting. The calculation is based on a calibration factor (knockmeter units per octane number) previously determined using primary reference fuels, similar to the "bracketing" procedure used in the traditional laboratory test method.

The comparator is designed to eliminate the problems of engine cyclic instability and drift mentioned previously. A 4-min test cycle is used on each fuel with step-wise computer logic. A typical comparator installation is shown in Fig. 15.4; comparator console functions are shown schematically in Fig. 15.5. The complete comparator installation consists of the following components:

- A console unit for mounting directly on the engine chassis.
- A specially designed fuel selection system including a refrigerant-type cooling unit.
- A specially designed single "spill" carburetor.

The complete carburetor and 501-T detonation meter are included as part of the octane comparator. No ice tower is required since the comparator compensates for long-term atmospheric humidity changes. However, in an air-conditioned engine room, some means of humidity regulation may be required to eliminate rapid humidity fluctuations.

The fuel selection system consists of a junction box where the individual fuels are selected. The selected fuels then pass through a heat exchanger, which chills them, through the flow control valves to the spill carburetor, then to the engine. At the junction box, line fuel from the sample fast loop is pressure-reduced to about 25 psig by a regulator. Proto/line fuel selection is by means of solenoid valves.

The spill carburetor is designed for precise setting of the air/fuel ratio. A bidirectional stepping motor positions the appropriate nozzle over the center section of the carburetor. The spill-over of gasoline keeps the level in the center section constant, allowing precise air/fuel ratio setting by a mi-

FIGURE 15.4 Mark IV octane comparator showing console mounted on CFR engine. (Courtesy of DuPont Company)

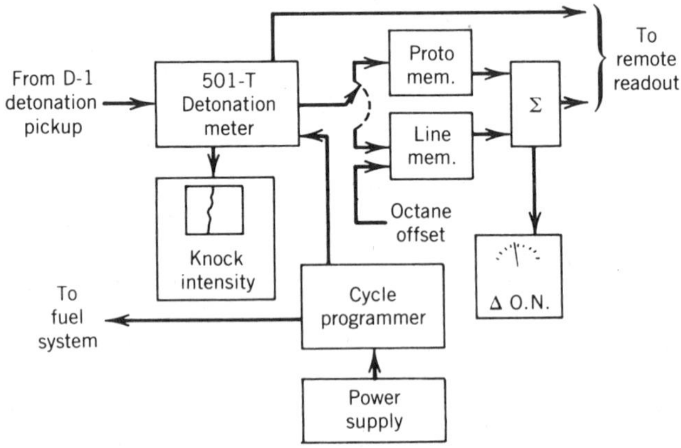

FIGURE 15.5 Mark IV octane comparator: functional diagram. (Courtesy of DuPont Company)

crovalve. The bottom drain is set to drop very slowly, keeping the bottom of the spill tower free of water and particulates. The fuel cooling system is a self-contained and self-controlled refrigeration unit feeding the fuel heat exchanger and the carburetor and allowing satisfactory operation on high RVP fuels.

The comparator consists of a fixed upper panel and a slide-mounted lower panel and chassis. The 501-T detonation meter, knock intensity recorder, and ΔON meter are mounted on the upper panel. The lower panel contains switches and associated hardware for the ON computer, cycle programmer, and power supply sections located on the attached chassis. The knock signal from the engine is fed to the 501-T detonation meter, which averages the KI signals and transmits them to the KI recorder on the comparator console, the ΔON computer, and a remote readout device (optional).

The ΔON computer takes the knock signal of each fuel at the end of each 4-min cycle and converts knock difference to ΔON. Octane offset is provided in the event that product target ON is different from proto fuel ON. The cycle programmer directs all functions of the comparator, switches fuels, turns the pumps on and off, switches memory banks, and tells the computer when to sample data and when to compute. The power supply furnishes power to the entire console and fuel system.

A number of accessories are available with the Mark IV Octane Comparator:

1. An Engine Trouble Alarm Package, which consists of four sensors—oil pressure, water level, oil temperature, and water pressure, and their associated readout lamps and contacts. Readout lamps and contacts are also provided for environmental sensors (e.g., fire, hydrocarbon, and CO monitors). Auxiliary contacts are provided to permit engine shutdown, activate annuciators, and so on.
2. A Cumulative Octane Barrel Indicator (COBI), which takes the ΔON error signal, multiplies it by the flowrate signal from a turbine meter in the blend header and totals it as ΔON barrels of error from the target value.
3. A Road Octane Computer (ROC), which combines research and motor ON and calculates the road ON, according to the following equation:

$$a \times \text{ON research} + b \times \text{ON motor} + \text{offset} = \text{ON road}$$

   where a and b are coefficients (typically 0.5).

Normal range of the Mark IV Octane Comparator is 70–110 ON and standard deviation is quoted as ±0.06 for RON and ±0.12 for MON. Span is normally ±2.0 ON fullscale; ±1.0 and ±5.0 is also available as an option.

The COMOC II Computer Octane Comparator System is a development of the Mark IV Octane Comparator using a PDP 8 Series computer. This system is designed to handle data from up to four engines. Normally, two engines, one research method and one motor method, monitor a blend stream. Therefore, COMOC can monitor and output control signals for two gasoline blenders simultaneously.

The local COMOC console is mounted on each engine stand and contains the electronics necessary to interface the engine with the COMOC central computer. The local console includes a strip chart recorder to provide a continuous record of KI, the 501-T Detonation Meter, and switches for controlling engine operation. Manual modes may be selected by switches located in the center of the console panel.

The COMOC central computer shown in Fig. 15.6 receives averaged KI information from the engine consoles via a mutiplexer, which digitizes the

FIGURE 15.6 COMOC II computer octane comparator system. (Courtesy of DuPont Company)

data and passes it to the computer one engine at a time. The computer then calculates both absolute and delta octanes for the product, as well as total blend octane error (COBI), obtained from blender flow data.

All data is printed on an associated teleprinter. The teleprinter keyboard is used to give instructions to the computer in simple conversational language. An analog error signal is provided for control purposes. Relay drivers provide commands for operating equipment in the engine room. KI meters are also provided on the computer panel.

Octane reports are available in four formats:

1. Format 1—ΔON, proto ON, target ON, product ON, and COBI.
2. Format 2—ΔON and COBI.
3. Format 3—product ON only.
4. Format 4—ΔON only.

At the end of a blend, the system can provide a printout of average ΔON, average absolute ON, the total octane error (COBI) for research, motor, and road, and the number of barrels in the blend. In addition, the computer performs diagnostic routines to assess system performance, looking for evidence of engine malfunction, high or low knock intensity, compression ratio changes, and non-repeatability on proto KI.

(c) Associated Octel Ltd. (Great Britain)

The Mark 5 Octane Comparator is the latest in the line of octane comparators manufactured by Associated Octel and is a microprocessor-based system with a CRT/keyboard operator interface. The system is much easier to operate than previous units and gives a wider range of information. The comparator console and CRT are shown in Fig. 15.7.

The comparator, in conjunction with an ASTM/CFR engine, measures the difference in maximum knock intensity between a known reference fuel and the product fuel by taking samples on each in turn. From this information, the ΔON is calculated and thus the absolute ON of the product fuel. The comparator system consists of the following components:

- A comparator console, which houses the micro-processor, controls the supply of fuels to the engine, and calculates the octane number data.
- A CRT/keyboard interface, which displays all relevant information concerning operational settings of the system, fuel octane number data, and status of all alarm circuits. Information is entered via the keyboard. Graphic display capability is also provided.
- A junction box mounted on the engine for all interconnections.
- A fuel bowl and sight glass assembly for the control of reference and product fuels to the engine.

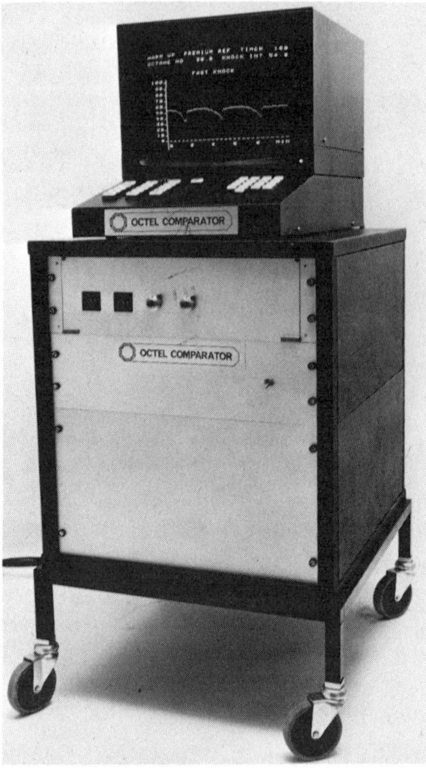

FIGURE 15.7 Mark 5 octane comparator system showing comparator unit and CRT-based operator station. (Courtesy of Associated Octel Ltd.)

- A modified detonation meter to provide knock intensity signals to the comparator.
- An optional microprocessor-based unit for remote mounting in the control room and connected to the comparator by a four-core cable (maximum distance 2 km). This unit provides analog output signals for recording, digital output signals for host computer/printer interface, and alarm and/or status signals.

There are five modes of operation:

1. Run—for normal operation.
2. Warmup—for engine checking and warmup prior to the start of blending operations.
3. Calibrate—to set up and check engine and detonation meter settings.
4. Test—to check comparator without the engine in operation.
5. Manual—where operator can control fuel valves for setting up purposes.

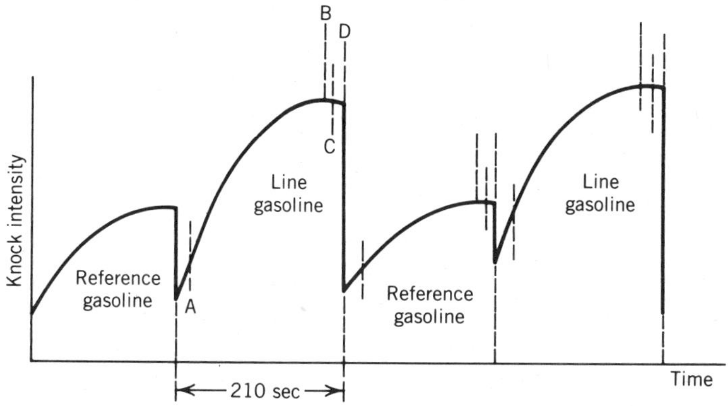

FIGURE 15.8 Mark 5 octane comparator: operating sequence. (Courtesy of Associated Octel Ltd.)

The comparator meets all of the accuracy requirements of the ASTM test method D2885-79. A typical operating sequence for the system is shown in Fig. 15.8. Just before point A in the sequence, the fuel bowl is filling with sample fuel. When the liquid level reaches the tip of the optical dipstick, a light beam (directed onto a photocell) is interrupted, a relay is energized to close the sample fuel solenoid valve, and the programmer is reset. After a 2–3 sec delay, the programmer starts. Fuel level in the fuel bowl falls as the engine runs on sample. A voltage proportional to the KI signal now builds up on a capacitor; a peak-picking diode ensures that the maximum signal is retained.

At point B in the sequence, the maximum value is read into the comparator. At point C, the read-in is complete. At point D, the fuel bowl drain valve is opened. At point E, the drain valve is closed and the reference fuel solenoid valve is opened, allowing the fuel bowl to fill with reference fuel. The sequence is then repeated with the engine running on reference fuel. The cycle time for each fuel is 210 sec. In the warm-up mode, the engine operates on reference fuel only.

The range of the Mark 5 Octane Comparator is normally ±2.0 octane number full scale output span; other spans are available as an option. Standard deviation is quoted as better than 0.05 between 85 and 100 RON (ASTM D2699).

In addition to the Mark 5 Octane Comparator, Octel can also provide complete packages for octane monitoring. A typical system would include two engine/comparator systems, one for research, one for motor, including reference fuel storage tanks, sample preparation and associated equipment, all mounted in a prefabricated shelter to customer's requirements. Also available are process analyzer packages for gasoline blending applications, including boiling point, RVP, density, and lead-in-gasoline analyzers, all mounted in a prefabricated shelter.

15.6.2 Analyzers Based on Correlative Techniques

The conventional and automated knock test engines described earlier in this chapter are mechanically complex, requiring frequent and extensive maintenance, and use an artificially dampened device to indicate knock intensity. The equipment is also expensive to purchase and install.

It can be demonstrated, by listening, that not all fuels knock alike in the same engine, even though they have the same octane rating. If a knock engine is adjusted to maximum knock on a predominantly paraffinic fuel and then switched to a predominantly aromatic fuel with the same knock rating, the sound of the knock is different in both pitch and intensity. Oscilloscope traces of engine pressure are appreciably different, demonstrating that there is a difference in behavior of the fuel during combustion. However, the knock meter readout will not reflect this difference. Considerations such as these led Universal Oil Products (UOP) in the early 1960s to conduct research into finding an alternative technique for measuring octane number.

Although the mechanism by which an internal combustion engine knocks was not precisely known, it was established that knock is associated with pre-ignition of "end-products," defined as the part of the fuel-air mixture in the cylinder that has not yet been ignited by the flame front. Hydrocarbon fuel and air are mixed, subjected to a system of increasing temperature and pressure in the engine, and then ignited by the spark. The fuel reacts spontaneously with oxygen (present in the air) and proceeds through a series of reactions that finally produce CO, CO_2, water, and other compounds.

With increasing temperatures and pressures, a point is reached where reactions become violent enough to generate the high-frequency, high-pressure waves recognized as knock. It was considered possible, therefore, that somewhere in the chain of reactions preceding knock, there are one or more reactions, occurring under milder conditions, that can be correlated with octane rating as measured by an ASTM/CFR engine. Investigation confirmed that the reactions related to octane number are partial oxidations that proceed at elevated temperatures, without the presence of an ignition source or catalyst.

At about the same time as the UOP investigations, Gulf Research and Development was conducting extensive research on partial oxidations and their relationship to octane number (the so-called "cool flame" technology). This work led to the licensing of Anacon Inc. by Gulf to produce instruments for installation at Gulf facilities worldwide. The Anacon Model 80-P octane analyzer replicated the original Gulf units. Combining this successful Anacon experience with the use of microprocessor technology led to the introduction of the Arcas Model 81-P Octane Analyzer, which is now marketed by Foxboro Analytical.

(a) Universal Oil Products Inc. (United States)

The UOP Monirex Octane Analyzer, first introduced in 1966, is designed to monitor the reactions that are the precursors to knock. The instrument is shown schematically in Fig. 15.9.

The heart of the system is a reactor tube maintained at an elevated temperature. The partial oxidations are exothermic and result in a fairly well defined peak temperature at a certain distance from the reactor inlet. After the initial temperature rise, the reactions either stop or continue at a much slower rate, resulting in a fall in reactant stream temperature. As the octane number of the sample fuel *increases,* the temperature peak is displaced *away* from the reactor inlet, and vice versa. Furthermore, the temperature peak is displaced *toward* the reactor inlet by *increasing* the reactor pressure, and vice versa. This fundamental discovery led to the basis of the measurement, which is maintaining the temperature peak in a fixed position by varying the reactor pressure.

The temperature peak is located in the reactor by two thermocouples mounted 1 in. apart, as shown in Fig. 15.10. Note that although there are several locations within the reactor where the temperatures are equal there is only one location where they are equal and exactly 1 in. apart.

The thermocouples are connected to a differential temperature controller, the output of which operates a control valve in the reactor outlet line. When the sample octane number increases, the temperature peak is displaced away from the reactor inlet, resulting in a temperature difference between the two thermocouples. The controller acts to close the reactor outlet valve to increase reactor pressure and move the temperature peak back to the position of zero temperature difference. The reactor pressure is measured by an

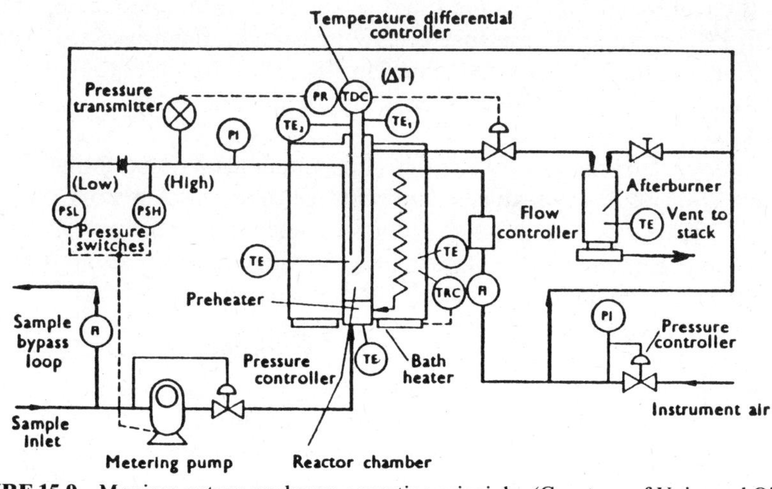

FIGURE 15.9 Monirex octane analyzer: operating principle. (Courtesy of Universal Oil Products Company)

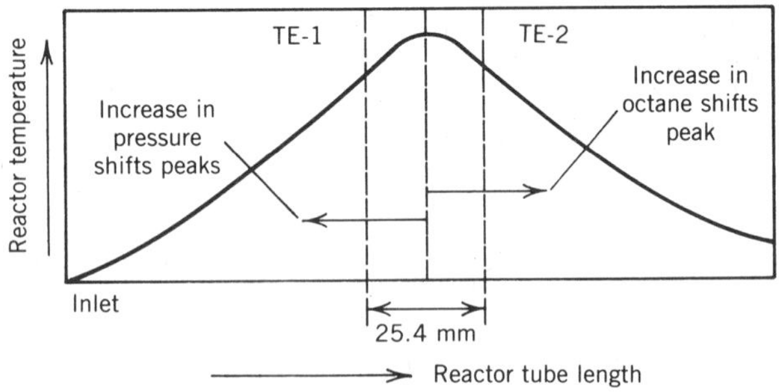

FIGURE 15.10 Monirex: thermocouple location.

independent transmitter, the output of which is calibrated in units of octane number.

Process sample is metered to the reactor chamber at 1.0 ml/min by a metering pump. Excess sample is bypassed via a flowmeter. A pressure control valve in the pump outlet maintains constant differential pressure across the pump. Clean, oil-free air enters the unit at 45 psig via a pressure controller. Primary reaction air is metered and passes through a heater section in the reactor bath into a preheater, where it is mixed with the incoming sample. Reaction products are a mixture of partially oxidized compounds (e.g., aldehydes, ketones, alcohols, etc.) which are foul smelling and have a tendency to deposit as "varnish." An afterburner disposes of these products. Additional air is fed to the afterburner to carry the oxidation reactions to completion, resulting in an effluent of essentially carbon dioxide, nitrogen, water, and excess air, which is discharged to atmosphere via a small stack.

Two pressure switches are provided to shut down the metering pump in the event of high reactor pressure or low instrument air pressure. Various thermocouples are provided, as shown in Fig. 15.9, for operational checking. These are connected to terminations at the analyzer unit for connection to a portable recorder.

The analyzer consists of two basic units, the analysis unit shown in Fig. 15.11 and suitable for installation in NEC Class 1, Group D, Division 1 hazardous locations, and a control unit designed for remote mounting in non-hazardous locations. The control unit contains the differential temperature controller, span and zero controls, and power supply switches for mains supply, heater, sample pump, and control system.

Calibration of the instrument involves establishing the response curve slope, that is, the curve representing the relationship between octane number and reactor pressure. Three primary reference fuels and a prototype fuel are required. Two of the PRFs should have octane ratings that 'bracket' the expected octane range of the process stream and the third with the same octane rating as the prototype fuel. Reactor temperature is varied by adjusting the bath temperature until the same pressure readings are obtained for

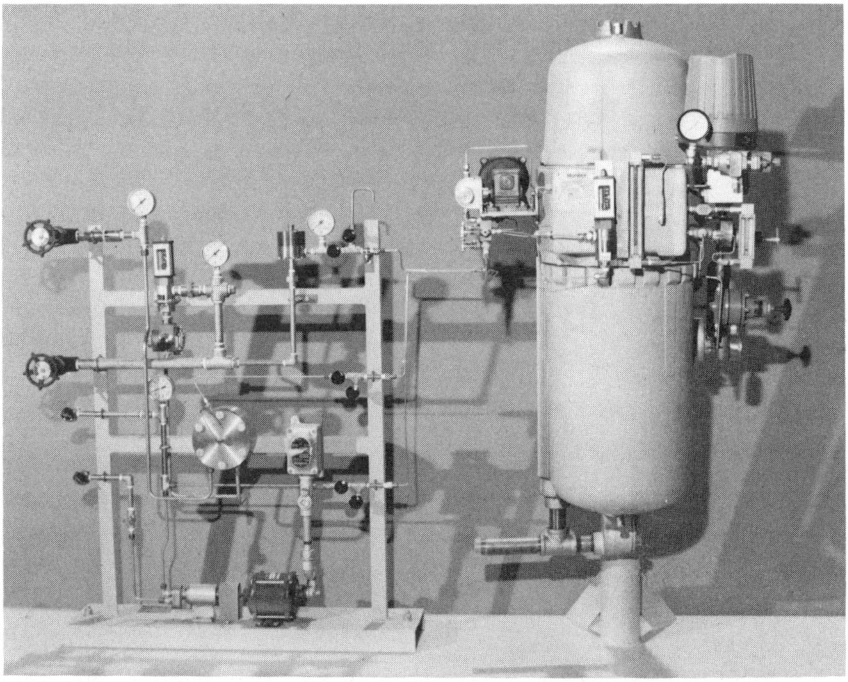

FIGURE 15.11 Monirex octane analyzer showing field unit and sample handling system. (Courtesy of Universal Oil Products Company)

prototype fuel and the PRF with the same octane rating. The three PRFs are then run to establish the slope of the response curve. Thereafter, it is unnecessary to run the PRFs unless there is a major shutdown, but the prototype fuel should be run periodically as a check on calibration.

(b) Foxboro Analytical (United States)

The Arcas Model 81-P Process Octane Analyzer uses a simple thermal oxidation, known as a "cool flame" reaction, to simulate the preliminary reactions that occur during engine knock measurements. Certain parameters of these reactions correlate with the octane number of the fuel.

The sample fuel is injected into a reaction chamber at elevated temperature and the magnitude of the exothermic reaction that occurs is measured by a thermocouple. The severity of the partial oxidations in the cool flame region shows good correlation with octane number, the reactions becoming more severe as octane number decreases, as shown in Fig. 15.12. While reactions have been observed in the reactor temperature range 270–330°C, most practical systems operate within the range 300–310°C.

A layout of the reactor system is shown in Fig. 15.13. The reactor housing contains an insulated stainless steel block with a spherical chamber in the center. The housing is heated to 70°C with a proportional temperature con-

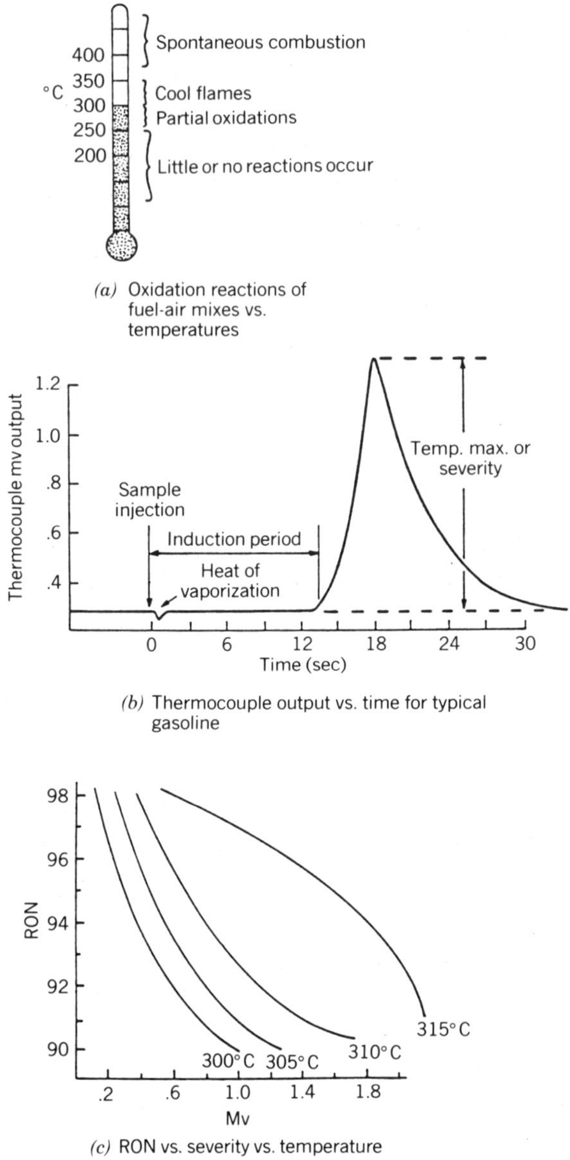

FIGURE 15.12 Cool flame reactions vs. octane number.

troller. The reactor block itself is maintained at 300 ±0.1°C by a second proportional controller. The sample volume injected into the reaction chamber is critical in this measurement; therefore, a specially-modified chromatographic-type liquid injection valve is used. A continuous flow of air also passes through this valve into the chamber. Every 5 min, a 12-μl sample is injected into the chamber and the low-mass thermocouple measures the

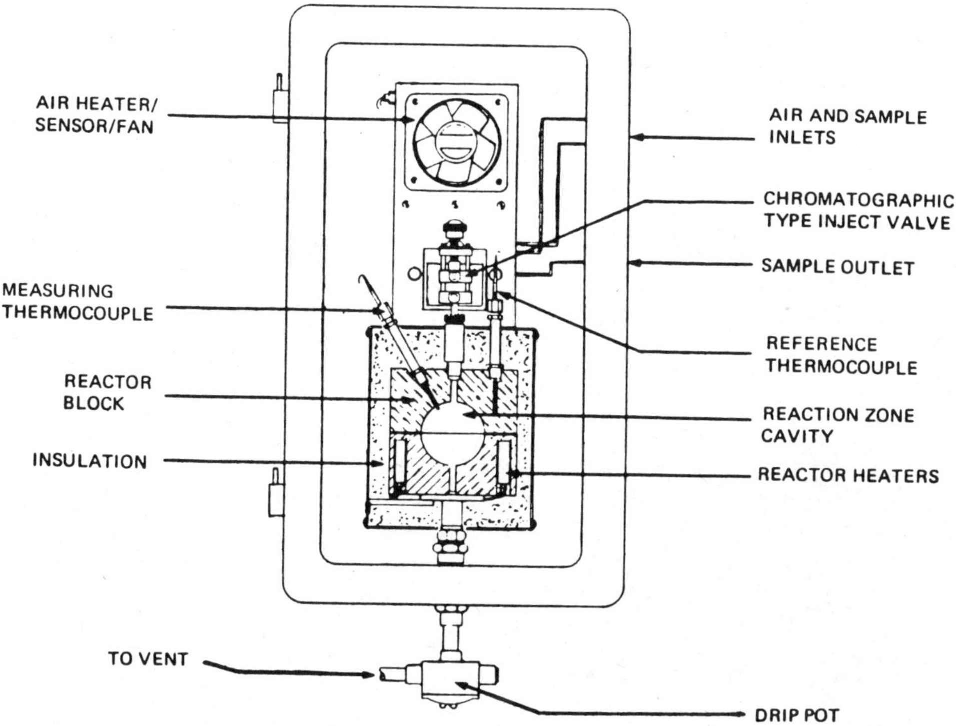

FIGURE 15.13 Model 81-P octane analyzer: reactor system. (Courtesy of The Foxboro Company)

temperature of the exothermic reaction. The reaction products are vented to atmosphere through a flame arrestor connected to a bottom exit port.

The analyzer unit shown in Fig. 15.14 consists of a free-standing frame on which are mounted three explosion-proof housings containing the reactor system, the microprocessor-based controller, and the digital display unit respectively. Also mounted on the frame are the sample conditioning equipment and associated controls, the air flow control equipment, and the sample pump (located at the base).

All timing and data conversion functions for the analyzer are handled and controlled by the microprocessor unit. The display unit allows complete manual local control of the analyzer. The analyzer is designed to interface with a remote terminal unit such as the Texas Instruments Silent 700, using an ASCII asynchronous serial data link. The terminal may be used for both data readout and to address the analyzer via a three-wire communication link. In addition, the analyzer can be connected to a remote recorder if required.

Every 5 min, the Silent 700 printer gives date and time, reactor temperature, octane number and a bargraph calibrated to 0.1 ON. Information on

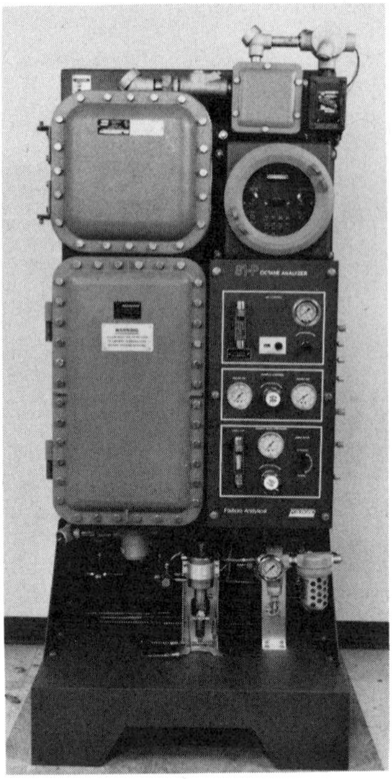

FIGURE 15.14 Model 81-P: equipment layout. (Courtesy of The Foxboro Company)

analyzer malfunction is also displayed, using the diagnostic capability of the microprocessor control unit.

Analyzer range is field-adjustable over a 5 ON span. Repeatability is quoted as 0.1 ON, depending on the accuracy of the engine-rated calibration standards used. Instrument air is required at 60–100 psig and 1 standard ft^3/min. Sample requirements are approximately 300 ml/hr at a maximum pressure of 300 psig and within the temperature limits 10–40°C. The analyzer should be protected from ambient conditions since the housings and condulets are not weatherproof.

15.6.3 Octane Analyzer Applications

There are two major applications in petroleum refining for measurement of octane number:

1. Control of gasoline blending operations.
2. Control of process units producing feedstock for gasoline blending (e.g., catalytic reformers).

In spite of the disadvantages of high total installed cost and extensive maintenance requirements, most users still favor the ASTM/CFR engine-octane comparator approach described in Section 15.6.1 for gasoline blending applications. This is almost certainly related to difficulties in using the correlative-type analyzers in an application where considerable composition variations can affect the correlation with octane number.

However, the continuous correlative-type analyzers have been used successfully in monitoring and controlling the octane number of catalytic reformer product. In semi-regenerative reformer start-up, target octane operation can be established with minimum offgrade production, minimizing dependency on laboratory analysis. In swing reactor reformer operation, continuous octane readout from a correlative-type analyzer allows reactor temperature control to minimize octane swing and the time to establish target octane. In many refineries, the reformer stabilizer handles other streams in addition to the reformate. Such an operation makes octane control of the reformer difficult because of octane changes in extraneous streams. In such cases, or where time lags in the stabilizer make control difficult, the correlative-type octane analyzer can sample the reformer separator liquid (a high vapor pressure sample), allowing better process control.

15.6.4 Other Measurement Techniques

Although the two measurement techniques described in Sections 15.6.1 and 15.6.2 are used in most refinery applications for octane number measurement, two other techniques worthy of mention have been used.

(a) Refractive Index Measurement

Anacon Inc. has used its Model 31S Differential Refractometer (see Chapter 17) for measurement of octane improvement in the hydroforming process. In this process, the octane number of low-grade naphtha is improved principally by conversion of paraffins and naphthenes to aromatics. The refractive indices of aromatics are much higher than those of corresponding naphthenes and paraffinic compounds.

With two naphthas of different cut points or from different crude sources, an absolute correlation between refractive index and octane number cannot be made. However, the improvement in octane number of a particular crude may be measured and a correlation obtained by running samples of different octane number through a refractometer and plotting a curve.

When a number of different feedstocks are used, the correlation curves for all feedstocks tend to converge at the higher octane numbers. It is sometimes possible to use one basic calibration curve for all feedstocks if the desired end product is high-octane gasoline. One approach to the analysis of octane improvement is to monitor the feedstock, followed by the improved gasoline product, each for a 5–10 min period. Both readings may be displayed on the same recorder for easy comparison. Continuous monitoring in this manner will allow the operator to maintain the proper cut point on non-aromatic fractions.

(b) Chromatographic Measurement

At the ISA Analysis Instrumentation Symposium in 1973 (1), Applied Automation presented a paper demonstrating that it is possible to measure re-

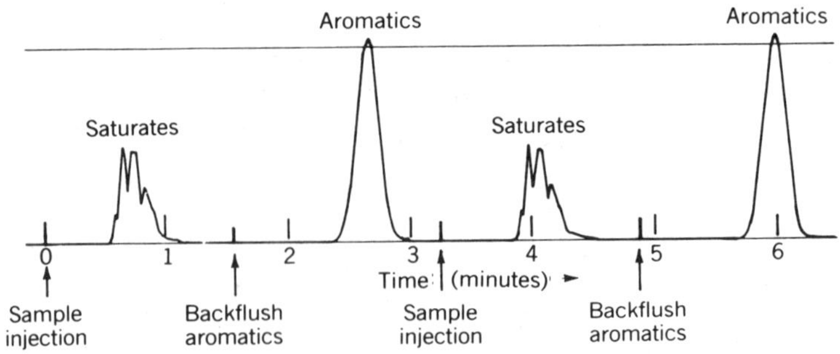

FIGURE 15.15 Total aromatics separation by chromatograph. (Courtesy of Applied Automation Inc.)

former product octane number using chromatographic analysis. The technique is based on the measurement of percent total aromatics in the sample. Data was collected in conjunction with Phillips Petroleum Company.

From the experimental data obtained, it was established that the octane number was related to total aromatics by the following expression:

$$\mathrm{ON} = \mathrm{A}\ (\%\ \text{aromatics}) + \mathrm{B}$$

where A and B are constants. It was also found that 100A + B is essentially constant and equal to 120.3. A single laboratory knock engine measurement, coupled with a chromatographic measurement of total aromatics, was used to give the value of A, and hence the value of B from the expression.

The AA Model 102 chromatograph (see Chapter 1) was fitted with multiple capillary columns with a highly polar stationary phase to provide the necessary separation of aromatics from saturates. A typical chromatogram is shown in Fig. 15.15. To obtain maximum reproducibility of the aromatics concentration, a normalization procedure was used.

Using this technique, it was shown that for a given feed composition, measurement of total aromatics will allow reformate octane number to be calculated with a reproducibility of better than ±0.1 ON.

REFERENCES

1. McCoy, R. D. Catalytic reformer product octane measurement via total aromatics. *19th Ann. ISA Symp.*, St. Louis, MS (April 1973).

BIBLIOGRAPHY

McLaughlin, J. H., and Bajek, W. A. An instrument to continuously monitor the octane quality of gasoline. *Proc. ISA Conf.*, Houston, TX (Oct 1969).

16

Thermal Conductivity Measurement

16.1 BACKGROUND AND DEFINITIONS

All gases possess the ability to conduct heat to varying degrees. The rate of conduction of heat depends upon a property of the gas known as the *thermal conductivity* (λ). This is defined classically as the quantity of heat in calories transferred in 1 sec in a gas between two surfaces of area 1 cm^2 placed 1 cm apart when the temperature difference between the two surfaces is 1°C.

Continuous measurement of thermal conductivity has been used for many years as a method of gas analysis and is the simplest and one of the oldest of on-line analytical techniques. The gas sample is passed over an electrically heated wire filament. The extent of cooling that takes place, which results in a change in resistance of the filament, depends on the gas thermal conductivity and is compared with a "reference" condition. The sample filament and a reference filament are connected into a Wheatstone bridge circuit; the imbalance of the circuit is calibrated in terms of gas concentration.

The difference in the ability of a gas to conduct heat may thus be used to determine quantitatively the composition of gaseous mixtures. However,

TABLE 16.1 Relative Thermal Conductivities of Typical Gases at 100°C[a]

Acetone	7.8	Hydrogen	100.0
Acetylene	12.9	*n*-Hexane	9.5
Air	14.3	Hydrogen sulfide	7.2
Argon	10.4	Methane	20.8
Benzene	8.2	Methyl alcohol	10.2
n-Butane	10.6	Methyl chloride	7.5
Carbon dioxide	9.8	Neon	26.4
Carbon monoxide	13.7	Nitrogen	14.2
Cyclohexane	8.2	Nitrous oxide	10.4
Ethane	13.9	Oxygen	15.1
Ethyl alcohol	10.0	*n*-Pentane	10.1
Ethylene	13.2	Sulfur dioxide	5.4
Helium	83.5	Water vapor	11.1

[a] Hydrogen = 100.

the technique has limited selectivity, since all components present contribute to the overall thermal conductivity of the mixture. The technique is best applied to the analysis of binary mixtures where the two components have significantly different thermal conductivities. More complex mixtures can be analyzed adequately if all components in the mixture except the components of interest have similar thermal conductivities. Hydrogen has a much higher thermal conductivity than hydrocarbons and other gases except helium. Table 16.1 lists the thermal conductivities of various gases at 100°C relative to hydrogen (=100.0).

In view of this considerable difference, one of the most common industrial applications of the on-line thermal conductivity analyzer is the measurement of hydrogen concentration in hydrocarbons and other gases, or measurement of the purity of hydrogen-rich process streams. Undoubtedly, the most common use of the thermal conductivity *detector* (TCD) is in laboratory and process gas chromatographs. Since it was the only detector available prior to 1960, it dictated the use of hydrogen and helium as carrier gases for GC applications. The TCD is still the most popular choice of GC detector in view of its sensitivity to all sample components except the carrier gas, its simplicity, and its low cost. For more information on GC thermal conductivity detectors, see Chapter 1.

The thermal conductivity of a pure gas is given approximately by the equation:

$$\lambda = K\left(C_v + \frac{9R}{4}\right)\sqrt{\frac{T}{M} \cdot \frac{1}{\sigma^2}}$$

where C_v is the molar specific heat at constant volume (cal/mol °C)
R is the universal gas constant
T is the absolute temperature (°K)

M is the molecular weight of the gas
σ is the collision diameter of the molecule (Å)
λ is the thermal conductivity of the gas (cal · cm^{-1} · °C)
K is a proportionality constant (2.67×10^{-5})

One equation for the thermal conductivity of binary mixtures, which is based on elementary kinetic theory, has been moderately successful in satisfying experimental data:

$$\lambda_m = \frac{\lambda_1}{\left(1 + \frac{AX_2}{X_1}\right)} + \frac{\lambda_2}{\left(1 + \frac{BX_1}{X_2}\right)}$$

where

$$A = \frac{(\sigma_1 + \sigma_2)^2}{2\sigma_1} \cdot \frac{(M_1 + M_2)^{1/2}}{2M_2}$$

and

$$B = \frac{(\sigma_1 + \sigma_2)^2}{2\sigma_2} \cdot \frac{(M_1 + M_2)^{1/2}}{2M_1}$$

and X_1 and X_2 are the mole fractions of carrier gas and vapor, σ_1 and σ_2 are the molecular diameters of carrier and vapor molecules, M_1 and M_2 are the molecular weights of carrier and vapor, λ_1 and λ_2 are the thermal conductivities of carrier and vapor, and λ_m is the thermal conductivity of the mixture.

On-line measurement of thermal conductivity, as mentioned earlier, is used primarily to *infer* the composition of binary or pseudo-binary mixtures of gases. Generally, composition measurements in the laboratory are made by gas chromatographic analysis, including the measurement of hydrogen, and in most cases using the thermal conductivity detector. Consequently, calibration of on-line thermal conductivity analyzers must usually be carried out by correlation with chromatographic analysis of the process stream.

16.2 PRINCIPAL COMPONENTS

A thermal conductivity analyzer consists of three major components—the measuring cell, a Wheatstone bridge circuit with regulated power supply, and a temperature-controlled enclosure, or "oven," in which the cell is located.

The measuring cell must, of course, be compatible with the process sample and consists of a relatively large mass of metal to provide a stable heat sink and, at the same time, have a high thermal-conductivity coefficient.

Normally, stainless steel is used. The metal block is machined to provide flow passages for the sample and the reference gas and recessed cavities for the sensing elements. These may be hot-wire filaments or glass-bead thermistors.

Since just about *any* change in operating conditions will result in a change in temperature of the sensing element, it is important to maintain a constant sample flowrate and install the measuring cell in a temperature-controlled enclosure. Early developments of the thermal conductivity analyzer in Great Britain led to the term "katharometer" to describe instruments of this type; the term is still widely used in Europe today.

If hot-wire filaments are used as sensing elements, they are usually tungsten or platinum alloy materials. They may be installed directly in the flow path (flow-through design) or recessed in cavities. The flow-through design gives fast response but is generally sensitive to flowrate variations, whereas the recessed design is not flow-sensitive but generally has a relatively slow response.

Thermistor sensors were popular prior to 1965, but suffered from frequent failure in hydrogen-rich sample streams due to cracking of the thin glass coating with excess heat and reduction of the metal oxide by hydrogen, causing drift. Considerable improvements in the design of hot-wire filaments have taken place over the years and they are now the preferred choice.

The hot-wire filaments are used in pairs, one located in the sample stream, the other in a reference gas. One or two pairs are normally used, but some cell designs include up to eight pairs to improve sensitivity. Best results are obtained when the reference cells are exposed to a flowing reference gas that contains all components in the sample except the component of interest. However, because of the operating costs incurred by having a flowing reference, often the reference cell(s) is merely filled and sealed with air.

The Wheatstone bridge, the basic sensing system, includes a high-quality regulated power supply to provide maximum stability of operation. Many circuit designs are in use today, and some of them are discussed next in relation to a particular manufacturer's equipment.

16.3 PROCESS THERMAL CONDUCTIVITY ANALYZERS

In view of the wide range of thermal conductivity analyzers commercially available, all operating on essentially the same principle, typical examples are described in detail in this section and the remainder summarized in Table 16.2 on page 704–705.

16.3.1 Beckman Industrial Inc. (United States)

The Model 7C Thermal Conductivity Analyzer is well suited to the analysis of two-component gas streams, but analysis of multi-component streams is

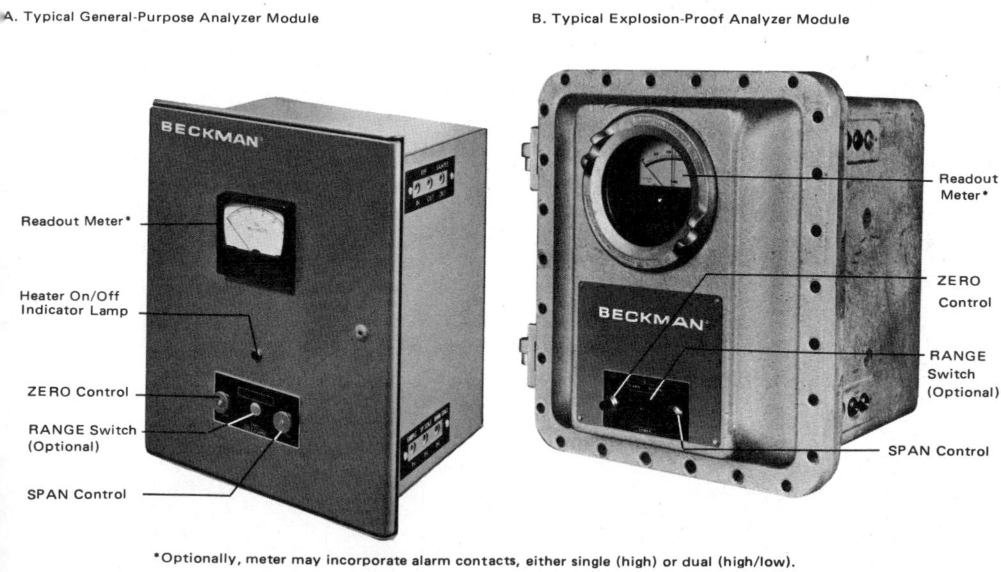

FIGURE 16.1 Model 7C thermal conductivity analyzer. (Courtesy of Beckman Industrial Inc.)

also practical, provided that the various components of the background gas occur in relatively constant ratio or have similar thermal conductivities.

Depending on the particular application, the analyzer module may be supplied in either general purpose or explosion-proof design as shown in Fig. 16.1. The thermal conductivity cell shown in Fig. 16.2 is a 316 stainless steel block with separate passages for sample and reference gases. In all applications, the sample passage receives a constant flow of sample gas. The reference passage may either have the reference gas sealed within it or may receive a flowing reference gas, depending on the particular application. The sample passage contains a matched pair of temperature-sensitive tungsten filaments; the reference passage contains a similar matched pair. Corrosion-resistant filaments are also available.

The filaments are connected to form the arms of a Wheatstone bridge circuit, which receives power from an internal voltage-regulated power supply (Fig. 16.3). The interior of the analyzer case is maintained at 130°F by a heater and solid-state temperature controller. The analyzer module also contains solid-state circuitry to condition the bridge unbalance signal as required to provide a readout on the front panel meter and voltage or current output for remote recording.

Normally, a calibration curve is provided for all standard ranges; however, a linearizer circuit can be supplied as an option. Also available is a gas selector panel, which allows selection, flow adjustment, and measurement for sample, flowing reference, and span and zero gases. Reproducibility is quoted as ±0.5% of full scale. Cell response time is 30 sec for a 95% re-

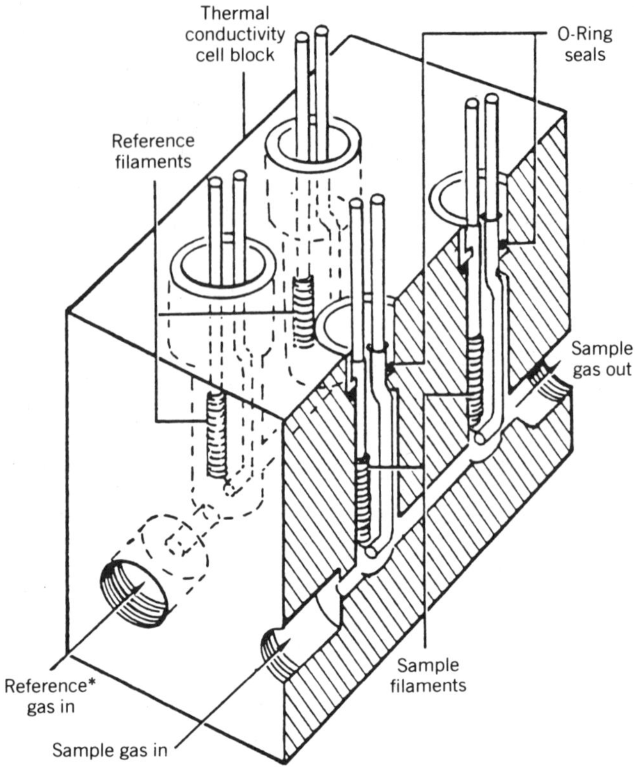

FIGURE 16.2 Model 7C thermal conductivity cell design. (Courtesy of Beckman Industrial Inc.)

sponse at a sample flowrate of 250 ml/min. Single range is standard, but a switch-selectable dual or triple range option is also available.

16.3.2 Hartmann & Braun (West Germany)

The Caldos 4T and 4T-Ex Thermal Conductivity Gas Analyzers are designed for measurement of a single gas component in a binary or quasi-binary gas mixture. Specific components measured include hydrogen, carbon dioxide, sulfur dioxide, ammonia, chlorine, hydrogen chloride, and argon.

The measuring unit consists of four resistance elements mounted in a gas chamber system; two elements are exposed to the process sample, the other two to the reference gas. The resistance elements are protected by thin-walled glass capillaries to prevent corrosion and catalytic ignition. The measuring unit is installed in a temperature-controlled oven. The resistance

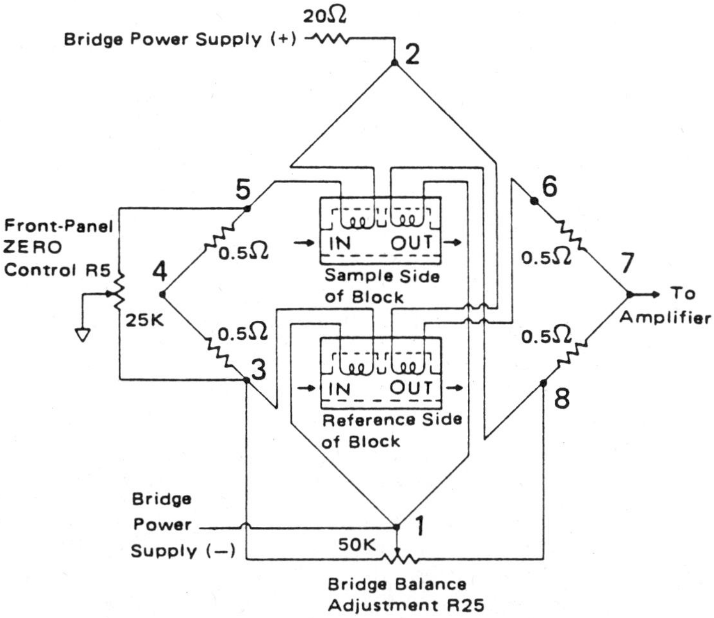

FIGURE 16.3 Model 7C Wheatstone bridge circuit. (Courtesy of Beckman Industrial Inc.)

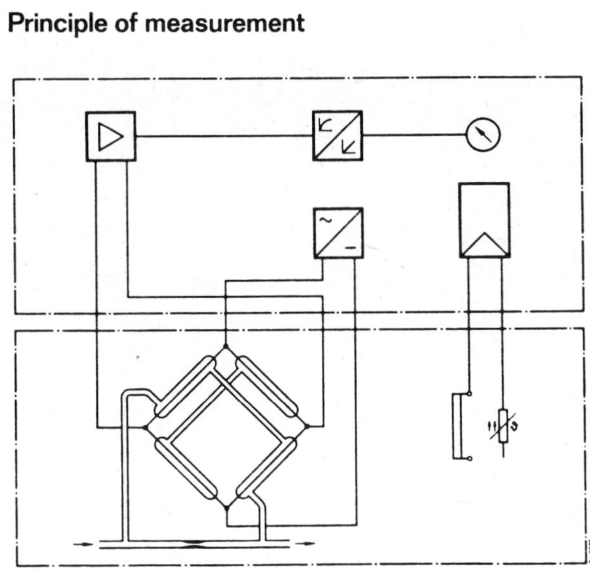

FIGURE 16.4 Caldos 4T-Ex. (Courtesy of Hartmann & Braun)

elements form the four arms of a Wheatstone bridge; the imbalance of the bridge is proportional to the concentration of the measured component.

The most sensitive measurement range of the instrument depends on the component of interest, varying from 0–0.5 to 0–5.0 vol%. A variety of reference gases can be used, depending on the application. Reference gas is normally sealed into the reference chamber, although a flowing reference version is also available. Dual measurement range is available, as well as suppressed zero ranges.

The Model 4T is a general purpose instrument, suitable for surface or panel mounting in non-hazardous locations. The Model 4T-EX shown schematically in Fig. 16.4 is suitable for surface mounting in Zone 1 hazardous locations by virtue of its intrinsically safe design to PTB standards. The analyzer has an integral indicating meter and provides a current output (0–20 or 4–20 mA dc) for remote recording.

16.3.3 Kent Industrial Measurements (Great Britain)

Kent has manufactured thermal conductivity analyzers (katharometers) for many years, based on a 65-year-old design for measuring hydrogen purity in gas. The latest instruments, the 6517 Series Katharometers, have been applied in a wide range of applications, spanning several industries.

The katharometer consists of four filaments constructed of platinum wire, finely coated with glass. All four filaments are matched for electrical and thermal characteristics. They are installed in four separate cells in a metal block and form the four arms of a Wheatstone bridge, as shown in Fig. 16.5. A general view of a typical katharometer block is shown in Fig. 16.6. All katharometers in the 6517 Series are housed in a steel case of common size and presentation and are usually supplied mounted on a steel back-plate also holding the associated sample conditioning accessories. A variety of katha-

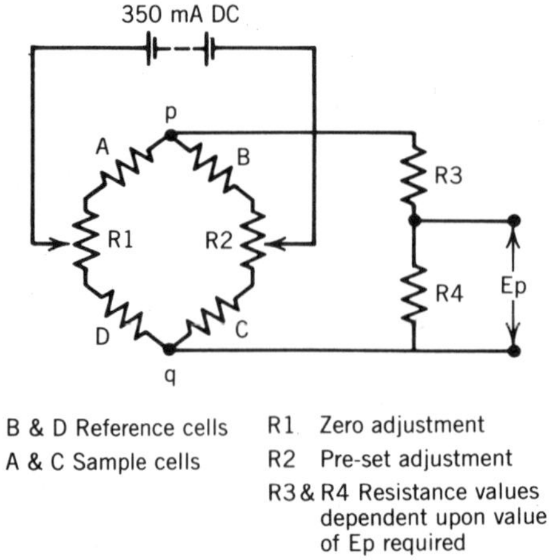

FIGURE 16.5 Katharometer Wheatstone bridge circuit. (Courtesy of Kent Industrial Insts. Ltd.)

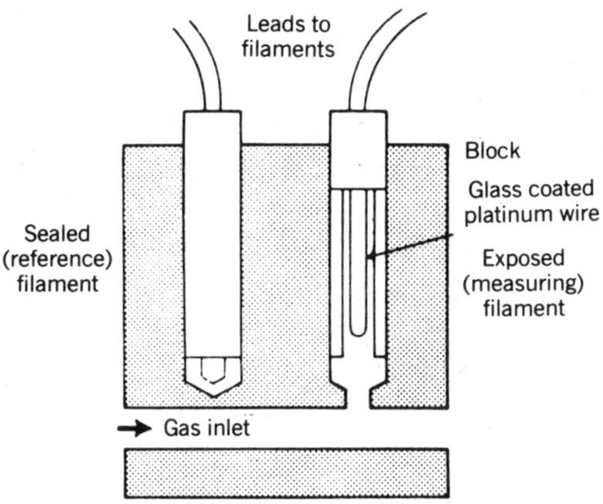

FIGURE 16.6 Katharometer block: typical. (Courtesy of Kent Industrial Insts. Ltd.)

rometer types is available, suitable for measurement in both non-corrosive and corrosive gases.

In its simplest form, a complete gas analyzer for a particular application consists of the katharometer itself, a constant current power supply, and a potentiometric indicator. Amplifiers are available from the Kent range to provide a current (4–20 mA dc) output for remote recording. Katharometer outputs are 0–1, 0–5, and 0–50 mV, with 0–5 mV as the preferred standard.

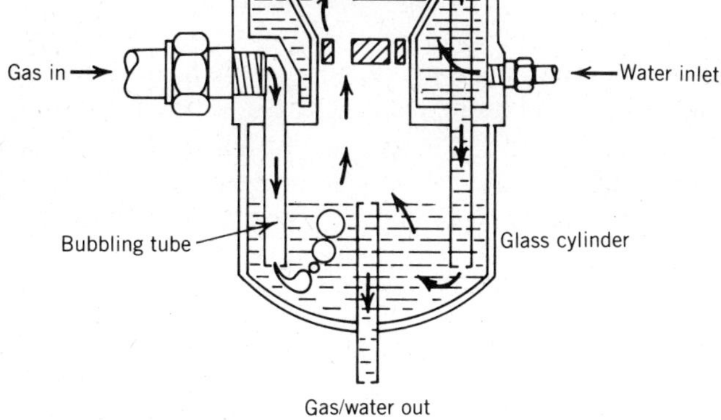

FIGURE 16.7 Katharometer designed for measurement of carbon dioxide in flue gas. (Courtesy of Kent Industrial Insts. Ltd.)

For monitoring flammable gas mixtures, a special power supply unit and shunt diode safety barriers are required to meet BASEEFA certification requirements. The approximate volume of the katharometer is 5 ml and optimum sample flowrate is 50–200 ml/min. Repeatability is quoted as better than 1% of full scale.

Figure 16.7 shows a special Kent Katharometer-type measuring unit that has been used for measurement of carbon dioxide in flue gases. The upper part of the unit consists of the katharometer surrounded by a water cooling jacket. The lower part of the unit is a bubbler/aspirator that draws a sample continuously from the flue and cleans, cools, and saturates it prior to passing it to the katharometer.

16.3.4 Leeds and Northrup Instruments (United States)

Leeds and Northrup manufactures a family of three gas analyzers with common design features but with sensing unit capabilities tailored to the specific analysis:

- The Model 7863 Thermomagnetic Oxygen Analyzer.
- The Model 7866 Thermal Conductivity Analyzer.
- The Model 7867 Combustibles Analyzer.

The Model 7866 Thermal Conductivity Analyzer shown in Fig. 16.8 is designed primarily for the analysis of binary mixtures. It may also be calibrated to measure the concentration of a single component in a multi-component mixture, provided that the background gases constitute a stable mixture (e.g., air) or have approximately the same thermal conductivity.

The sensing unit consists of two sections, the cell block assembly and the electronics assembly. The cell block assembly is of stainless steel construction with two identical internal cells, as shown in Fig. 16.9. A highly stable thermistor is mounted in each cell. The thermistors, which are a matched pair, form the active arms of a Wheatstone bridge circuit. The cells in which the thermistors are mounted are dead-ended so that the sample gas enters only by diffusion, minimizing the effect of sample flow variations. The entire cell block assembly is maintained at a constant optimum temperature by two heaters and a control thermistor located in the cell block assembly, and a temperature controller located in the electronics assembly.

The reference gas chamber may be open or sealed depending on the application. For all zero-based standard ranges and the 20–50% hydrogen range, the reference cell is air-filled and sealed. For hydrogen ranges starting above 50% and for the 98–100% oxygen range, a flowing reference is used.

The electronics assembly includes the bridge current regulator, the cell block temperature controller, and the voltage/current converter amplifier, the output of which is transmitted to the remote mounted control unit. The

FIGURE 16.8 Model 7866 thermal conductivity analyzer. (Courtesy of Leeds & Northrup Company)

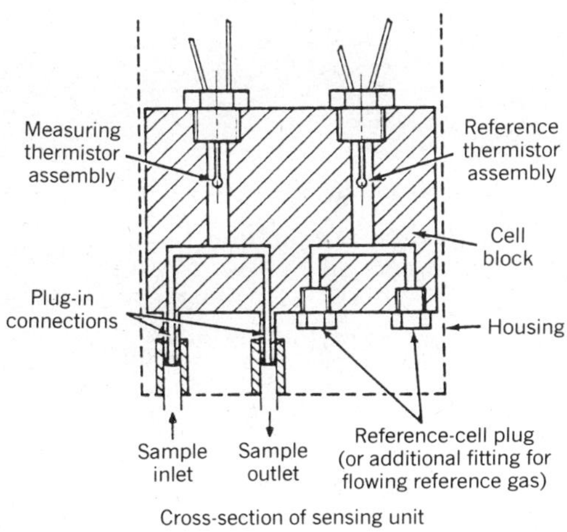

FIGURE 16.9 Model 7866 thermal conductivity cell design. (Courtesy of Leeds & Northrup Company)

control unit, which may be mounted at up to 1000 ft from the sensing unit, houses the remaining measuring and power supply circuits, zero and span calibration adjustments, indicating meter, alarm set-point thumbwheels and indicating lamps, and an auto calibrate switch. A variety of voltage and current outputs is available for remote recording. Sample requirements are 100–2000 ml/min flowrate at a minimum pressure of 20 in. water. Many standard ranges are available. Repeatability is quoted as ±1% of span.

16.3.5 Mine Safety Appliances Company (United States)

The MSA Thermatron Model T-3 Thermal Conductivity Analyzer is designed primarily for analysis of a single component in simple gas mixtures and consists of two units, a detector unit and a control unit. These may be mounted together as a single unit or the control unit may be mounted remotely at up to 500 ft. Connection between the units is by a four-wire shielded cable. Both units are available in general purpose or explosion-proof versions. The explosion-proof version is shown in Fig. 16.10.

The detector unit is insulated, heated, and temperature-controlled at 120–130°F. A system of two detector cells is used in which the effects of thermal conduction and thermal convection are different. Thermal convection de-

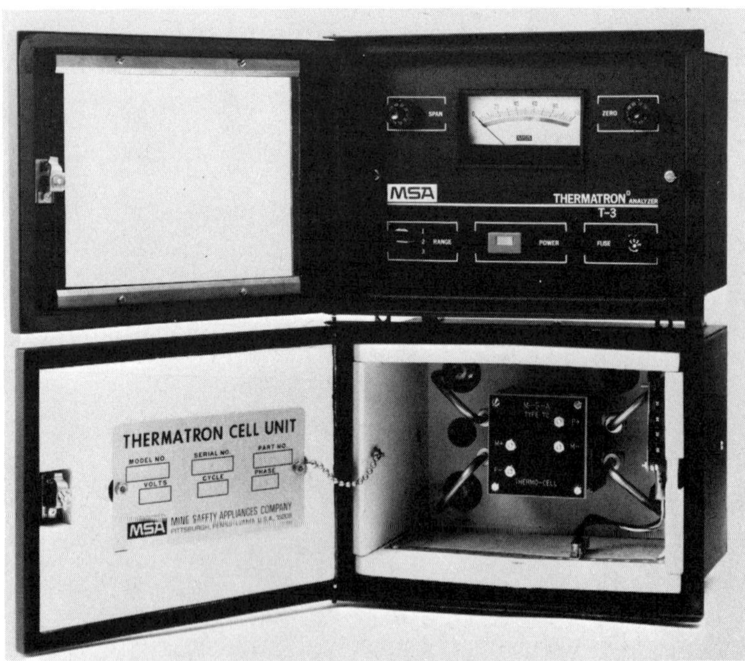

FIGURE 16.10 MSA Thermatron Model T-3. (Courtesy of Mine Safety Appliances Company)

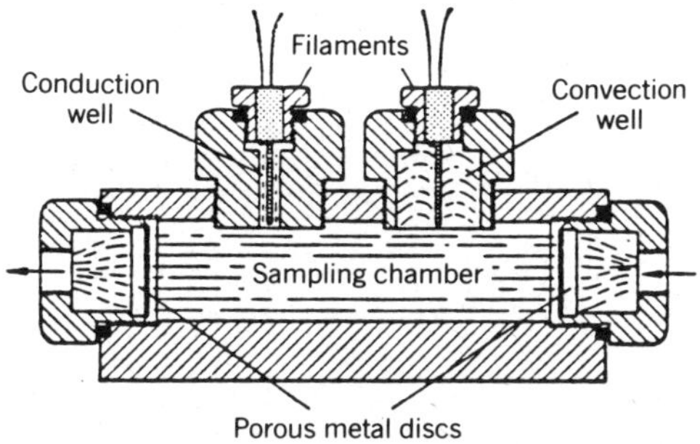

FIGURE 16.11 Thermatron Type M flowcell. (Courtesy of Mine Safety Appliances Company)

pends upon gas density and viscosity as well as thermal conductivity. According to the manufacturer, by careful selection of the cells the effects of variations in the concentration of background gases can be eliminated with this design.

The type M flow cell (sample side) is shown in cross-section in Fig. 16.11. Each chamber contains a conduction well and a convection well. The flow block itself is designed to eliminate turbulence, which interferes with the natural convection currents in the large well and tends to upset thermal conductance in the small well. The small well has a $\frac{3}{16}$-in. diameter, where heat loss is mainly due to thermal conduction. The large well has a $\frac{3}{4}$-in. diameter, where heat loss is due to convection as well as conduction. The heat loss of the filament in the large cell is therefore faster than that in the small cell.

To illustrate the operation of this system, consider the case of hydrogen measurement. Hydrogen contributes little to convection because of its high diffusion rate; convection currents in the large cell do not, therefore, have a chance to become established. However, carbon dioxide diffuses more slowly, thus convection currents are set up in the large cell. The temperature difference between the filaments in the small and large cells is very small in the case of hydrogen; in the case of carbon dioxide, the filament in the large cell cools measurably faster than the filament in the small cell. By careful choice of cell dimensions and filament temperatures, it is possible to measure, for example, hydrogen in nitrogen containing a variable concentration of carbon dioxide.

The four filaments are connected as four arms of a Wheatstone bridge; the imbalance of the bridge is proportional to the concentration of the component of interest. The control unit contains a transistor-controlled power supply for the bridge, a range switch (three ranges standard), span and zero controls, indicating meter, and power supply switch and indicating lamp.

Speed of response is quoted as 60 sec for 90% of final reading. Sample requirements are 50–2000 ml/min flowrate at a pressure of 10 in. water

TABLE 16.2 On-Line Process Thermal Conductivity Analyzers

Manufacturer	Country of Origin	Model No.	No. of Filaments: Reference	No. of Filaments: Sample	Typical Flowrates (ml/min): Sample Gas	Typical Flowrates (ml/min): Ref. Gas	Ref. Cell Configuration	Response Time	Remarks
1. Beckman Industrial Inc.	USA	7C	2	2	50–300	5–10	Sealed or flowing	95% in 30 sec	Gas selector panel available
2. Comsip-Delphi Inc.	USA	B	2	2	150	?	Flowing	90% in 30 sec (% ranges) 90% in 60 sec (ppm ranges)	
3. Hartmann & Braun AG	West Germany	Caldos 4T & 4T-Ex	2	2	500–1000	?	Sealed or flowing	90% in 6–20 sec	
4. Horiba Instruments Ltd.	Japan	TCA-31/31F	2	2	300–700	?	Sealed or flowing	90% in 40 sec	
5. Kent Industrial Measurements Ltd.	Great Britain	6608 6609 6553 6421	2 2 2 2	2 2 2 2	100–2000	10–100	Sealed Sealed Sealed Sealed	5 min to 90%	All use the 6517 series katharometers

6. Leeds & Northrup Co.	USA	7866	1	1	100–2000	10–100	Sealed or flowing	90% in 23 sec
7. H. Maihak AG	West Germany	Thermor 6N	2	2	60–1600		Sealed	90% in 8 sec
8. Mine Safety Appliances Co.	USA	Thermatron T-3	2	2	50–2000	10–100	Sealed or flowing	90% in 60 sec
9. Schlumberger-Sereg	France	Type HCD 5	1	1			Flowing	90% in 50 sec
10. Siemens AG	West Germany	Type TCA	2	2	300–1000	300	Sealed or flowing	90% in 35 sec (sealed ref.) 90% in 51 sec (flowing ref.)
11. Teledyne Analytical Insts.	USA	225	2	2	40–200	?	Flowing	90% in 50 sec
12. Telsec Process Analyzers Ltd.	Great Britain	TPA 410/411	2	2	250–500	?	Flowing	

minimum. Both flowing reference and sealed reference cells are available, the selection depending on the application and range of measurement desired.

BIBLIOGRAPHY

Tye, R. P. The art of measuring thermal conductivity. *InTech* (March 1969).

17

Measurement of Refractive Index

17.1 THEORETICAL BASIS AND DEFINITIONS

When light strikes a medium other than the one through which it is transmitted, a portion of the incident light is *reflected* and a portion *refracted* into the new medium.

Many effects in everyday life are caused by refraction. A pond or swimming pool appears much shallower than it really is; a straight stick appears bent when partially immersed in water; the landscape "shimmers" on a hot summer's day. Mirages are usually associated with hot deserts. The traveller in the desert often sees what appears to be a sheet of water a short distance ahead. It can never be reached because it is an optical illusion, created by progressive refraction of the light rays as they pass through warmer layers of air of decreasing refractive index to reach the eye.

17.1.1 Laws of Refraction

The law of reflection was known to Euclid. He was the first to observe that a ray of light bends toward the normal when it passes from a medium such as air to a more dense medium such as water or glass. Ptolemy, an Egyptian astronomer who lived in the second century A.D., made measurements of the angle between a ray of light in air and the normal to the plane refracting surface, and the corresponding angle in water. He concluded that these angles, called the angles of inclination in air and water respectively, were proportional to one another. This conclusion was supported by his measurements for small angles, but was quite wrong for large angles.

Although many scientists worked on the problem, the law governing the refraction of light when it passes from one substance to another resisted discovery for centuries. It was not until 1621 that Willebrord Snell, Professor of Mathematics at Leyden University, discovered the exact relationship between the angles of incidence and refraction. He may have been led to his discovery by the fact that a pond appears shallower than it really is. Snell measured the ratio BC/CD in Fig. 17.1, assuming that the image of the point B at the bottom of the pond was at point C on the normal to the bottom. The constancy of this ratio for small angles of incidence may have led him to consider measurements of the ratio AB/AC, which he also found to be constant for all angles of incidence:

$$\frac{\mathrm{AB}}{\mathrm{AD}} \div \frac{\mathrm{AC}}{\mathrm{AD}} = \text{constant}$$

or

$$\frac{\operatorname{cosec} r}{\operatorname{cosec} i} = \text{constant}$$

where i and r are the angles of inclination in air and water respectively. Or,

$$\frac{\sin i}{\sin r} = n$$

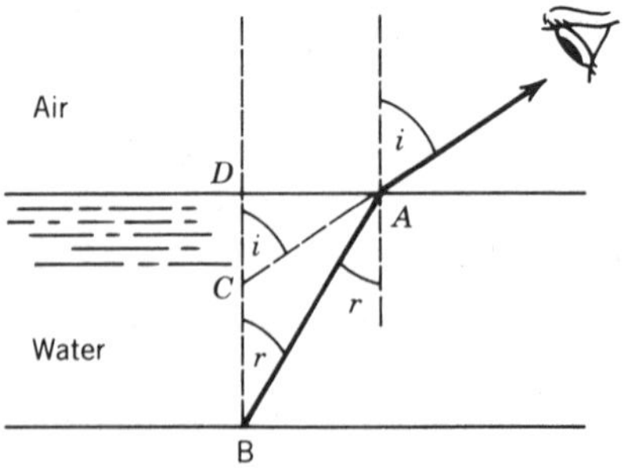

FIGURE 17.1 The laws of refraction.

The factor n is a constant for the two given media and is called the *refractive index* (RI) of water relative to air. Since refractive indices are usually measured relative to air, the value n is normally referred to as the refractive index of water, which is 1.33.

Strictly speaking, the *absolute refractive index* of a medium is the value of n when the first medium is a vacuum. The refractive index of a material may also be defined as the ratio of the velocity of light in a vacuum to the velocity of light in the material.

The laws of refraction can be stated as follows:

Law 1. The incident and refracted rays are on the opposite sides of the normal at the point of incidence and all three are in the same plane.

Law 2. (also known as Snell's Law). The ratio of the sine of the angle of incidence to the sine of the angle of refraction is a constant.

These laws may be extended to the refraction of light between any two media. Referring to Fig. 17.2, if a ray of light is incident on a plane surface separating two media of refractive indices n_1 and n_2 respectively, the ray BC in the second medium will bend toward the normal if n_2 is greater than n_1. Measurements of the angles of inclination i and r in the two media establish that the results satisfy the law

$$\frac{\sin i}{\sin r} = \frac{n_2}{n_1}$$

or

$$n_1 \sin i = n_2 \sin r$$

which reduces to Snell's Law when the first medium is assumed to be air ($n_1 = 1.0$)

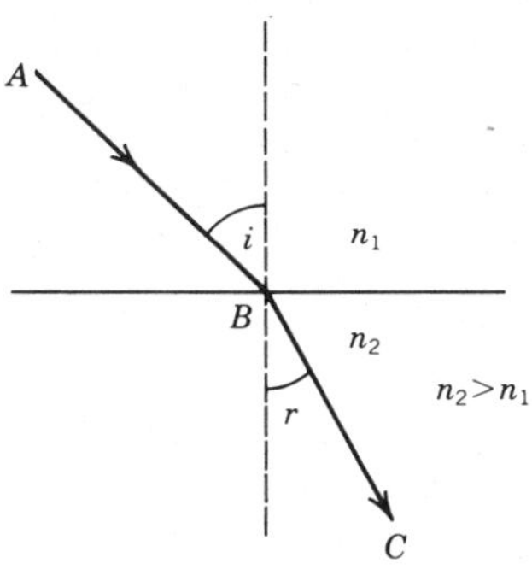

FIGURE 17.2 Refraction of light between any two media.

17.1.2 Total Internal Reflection and Critical Angle

When light passes from one medium to a more optically dense medium, there will be both reflection and refraction for all angles of incidence. This is not always so when light passes from one medium to a *less* optically dense medium. Referring to Fig. 17.3, consider a ray of light passing from glass to air. Starting with a small angle of incidence, we get a weak internally reflected ray and a strong refracted ray. As the angle of incidence increases, the angle of refraction also increases; at the same time, the intensity of the reflected ray gets stronger and that of the refracted ray weaker. Finally, at a certain *critical angle* of incidence (c), the angle of refraction becomes 90°. Since it is impossible to have an angle of refraction greater than 90°, it follows that for all angles of incidence greater than the critical angle c, the light undergoes what is known as *total internal reflection*. From this discussion,

$$\frac{\sin c}{\sin 90°} = \frac{1}{n} \text{ (air to glass)}$$

But sin 90° = 1, therefore

$$\text{n} = \frac{1}{\sin c}$$

For crown glass of refractive index $n = 1.5$, the critical angle is about 42°. For water (n = 1.33), it is about 49°, as calculated from the preceding equation. One interesting effect of the critical angle/total internal reflection concept is that a fish at any depth in water will have a full view of everything above the water, all contained within a cone of angle about 98° (twice the critical angle for water).

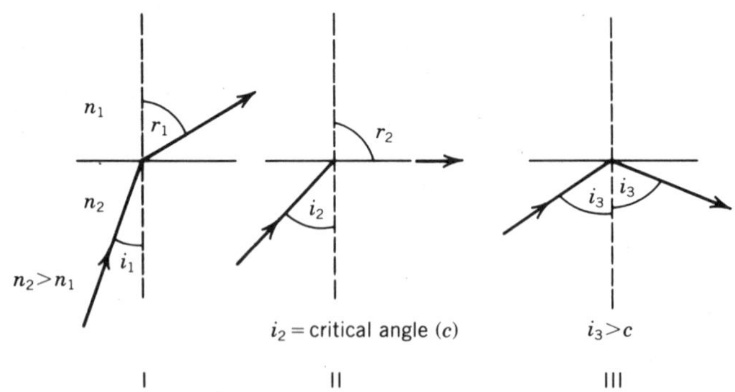

FIGURE 17.3 Critical angle and total internal reflection.

17.2 LABORATORY METHODS OF MEASUREMENT

Measurement of the refractive index of liquids in the laboratory involves the use of commercially available precision refractometers. The operating principle of these instruments is similar to the process refractometers described later in this chapter.

The ASTM has established a number of standard test methods for industrial use. Two that specifically apply to the petroleum industry are described next.

ASTM Method D1218-61 (re-approved in 1977) covers measurement of the refractive index of transparent and light-colored hydrocarbon liquids with an RI in the range 1.33–1.50 at temperatures from 68 to 86°F. This method is not applicable (within the stated accuracy) for liquids with colors darker than ASTM #4 (Method D1500). The refractive index is measured by the critical angle method with a Bausch & Lomb Precision Refractometer, using monochromatic light. The expected measurement accuracy is 0.00006 RI units.

ASTM Method D1747-62 (re-approved in 1978) covers measurement of the refractive index of transparent and light-colored, viscous hydrocarbon liquids and melted solids with RI in the range 1.33–1.60 at temperature from 176 to 212°F. Again, the method is not applicable (within the stated accuracy) to liquids with colors darker than ASTM #4 (Method D1500). The refractive index is measured by an Abbe Precision Refractometer with a measurement range of 1.3–1.63 RI units. Expected measurement accuracy is 0.0002 RI units.

17.3 PROCESS REFRACTOMETERS

17.3.1 Anacon Inc. (United States)

Anacon manufactures three process refractometers—Model 31, Model 47, and Model 70—the last of which is designed specifically for the food industry.

The Model 31 Differential Refractometer shown in Fig. 17.4 measures the amount a light beam is bent when the refractive index of a liquid sample changes. A light beam passes through a slit, which in effect narrows the beam, and then through a lens, which focuses the light through the sample cell. After the light passes through the cell, a mirror reflects the refracted light back through the cell for a second traversal, thereby optimizing sensitivity. The refracted light beam then passes through a null glass deflector to a beam splitter, which directs half the beam to each of two photocells, A and B, connected to form two arms of a Wheatstone bridge balancing circuit.

As the sample RI changes, the amount of light falling on one photocell

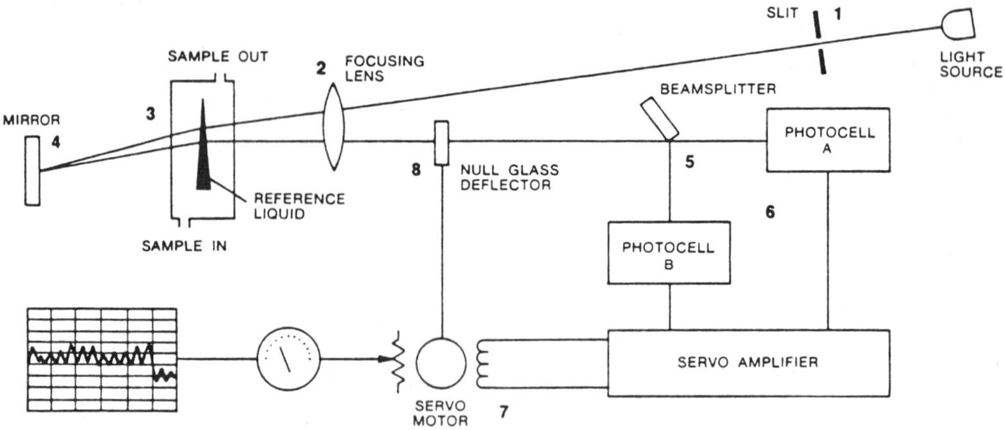

FIGURE 17.4 Model 31 differential refractometer. (Courtesy of Anacon Inc.)

changes with respect to the other, unbalancing the bridge circuit. The resultant signal is amplified and drives a servomotor, which alters the position of the null glass deflector to reduce the unbalance to zero. At the same time, the servomotor drives a helipot in proportion. The helipot output drives a meter and provides an output proportional to sample RI for remote recording.

Two types of sample cell are available. One uses a small, sealed reference cell around which the process sample flows. The other uses a flowing reference. Normally, the sealed cell is used in applications where a permanent reference standard is required. The flowing reference cell is normally used in applications where a secondary (background) component is present in the process stream, which could vary, thus preventing a true reading of the individual sample. A heat exchanger and proportional temperature controller maintain both sample and reference liquids to within $\pm 0.05\%$ of the heat exchanger temperature within the range 20–150°C. Sample flow control facilities are also available as an option.

Both general purpose and explosion-proof designs are available. In the general purpose design, the measurement and electronics sections are mounted together in the same enclosure. In the explosion-proof design, the two sections are mounted in separate enclosures on a backplate suitable for surface mounting. Instrument span is from 0.00005 to 0.1 RI units full scale. In addition to voltage (0–100 mV) and current (1–5, 4–20, or 10–50 mA dc) outputs, a BCD output is also available for computer or printer interface. Dual alarm contacts can also be provided.

The Model 47 Process Refractometer operates on the critical angle principle and therefore requires no depth of penetration of the sample by the light beam. Two types of sensing unit are available, as illustrated in Fig. 17.5, a weir-type valve sensing head and a probe-type head. Both are in-line units.

A spinel prism is mounted in the process line in contact with the sample.

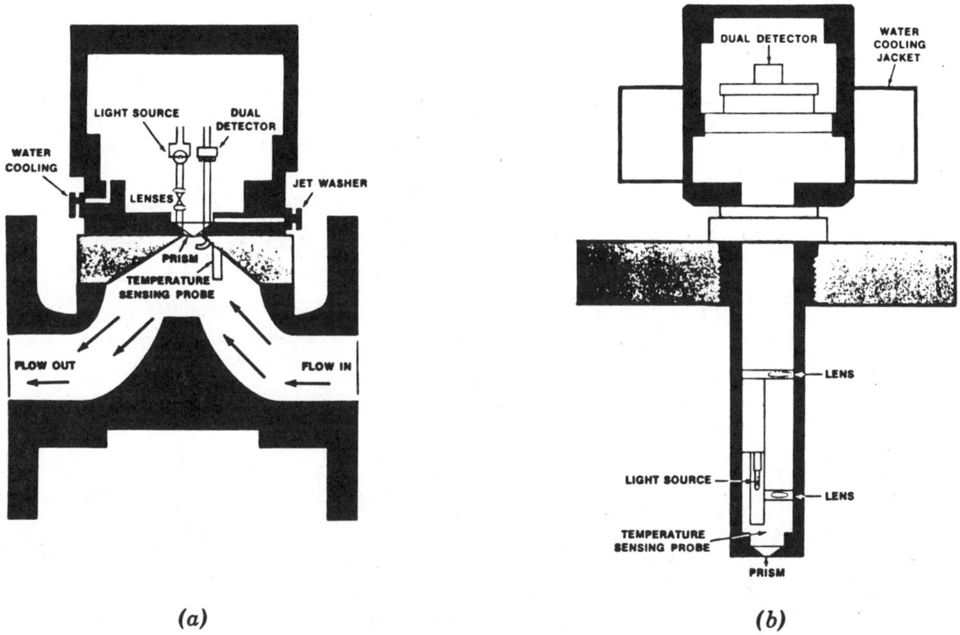

FIGURE 17.5 Model 47 process refractometer; sensing unit designs. (*a*) Valve body type design; (*b*) flanged, in-line type design. (Courtesy of Anacon Inc.)

A conical beam of light is directed through the prism at the interface of the prism with the process liquid. A portion of the light is transmitted into the liquid and a portion is reflected, as determined by the critical angle. The reflected light is directed to a dual-element solid-state photodetector. One element is positioned in the constant light area (reference), the other at the light-to-dark edge (sample), as shown in Fig. 17.6.

As the process stream RI changes, the critical range and the amount of light reflected change, all proportionately. The shadow edge on the sample detector element moves, changing the resistance of this element. The reference and sample detector elements are connected in the feedback loop of an

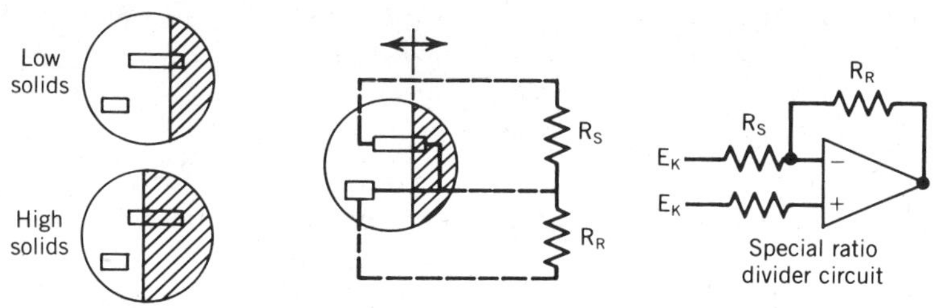

FIGURE 17.6 Model 47: operating principle. (Courtesy of Anacon Inc.)

operational amplifier. This input stage divides a stable, constant voltage exactly by the ratio of the dual-element photodetector resistance, thereby converting an optical signal into a directly related linear electrical signal.

A silicon-junction temperature sensing probe is mounted in the sensing head adjacent to the prism to provide automatic temperature compensation. The probe is inherently linear over the range −25 to 150°C. A wide variety of outputs is available, including a digital display meter and BCD output for computer or printer interface. Also offered is a fully automatic prism washing system. With this system, an auto-hold function is provided in the electronics, which holds and maintains meter indication and all outputs at the reading prior to initiation of the wash cycle, for its entire duration. Dual alarm contacts are available as an option.

The instrument consists of the sensing unit and the control unit, which may be separated by up to 500 ft. Sensing heads are constructed in stainless steel as standard, although other materials are available. Valve type bodies can be supplied with a wide variety of linings to suit the application, and are available for pipe sizes from $\frac{1}{4}$ in. up to 12 in. The sensing head requires a cooling water supply to provide a cooling medium for the photodetector. Sensing units are available in general purpose and explosion-proof designs.

Instrument range is 1.25–1.65 RI units, with minimum and maximum spans of 0.003 and 0.18 RI units respectively. Repeatability is quoted as ±0.5% of full scale. Process sample temperature should be within the range −25 to 220°C and at a pressure up to 1000 psig maximum.

The Model 70 Instream Refractometer, designed specifically for the food and beverage industry, is a continuous direct reading instrument for the determination of total soluble solids in a variety of liquids, including turbid and dark-colored concentrates such as jellies, pastes, juices, and syrups. It operates on the critical angle principle and is available in three models, each with a different Brix range (0–25%, 12–72%, and 36–82% total soluble solids). Other ranges are also available on special order.

The RI of the sample is determined by how obliquely light is reflected back into the instrument. This angle shows up as a well-defined transition line between dark and light areas on an internal graduated Brix scale on which readings are taken directly. Suspended solid matter or air bubbles do not affect the measurement.

All models can be used successfully in a connecting pipeline or mounted in the wall of a process vessel. When mounted by a coupling flange, the instrument is designed to operate at pressures up to 40 psig and temperatures up to 250°F. Instruments are factory-calibrated at 68°F and may be recalibrated on-line by a simple external set-screw adjustment.

17.3.2 Electron Machine Corporation (United States)

This company introduced its first critical angle refractometer in the late 1950s. Its latest process analyzer, the Model SSR-72 Process Critical Angle

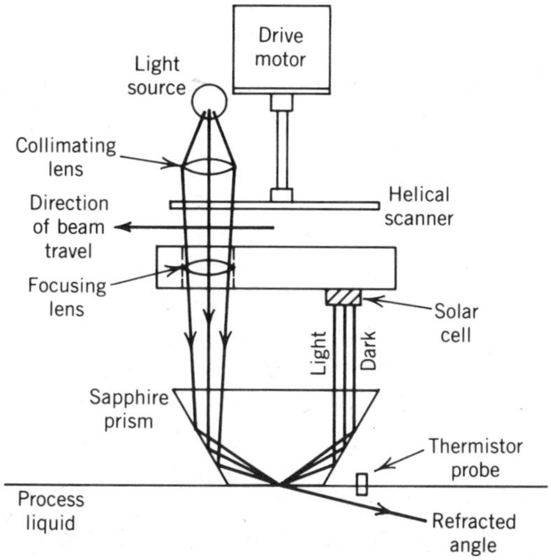

FIGURE 17.7 Model SSR-72 refractometer. (Courtesy of Electron Machine Corporation)

Refractometer, is a fourth-generation instrument, using a unique approach to RI measurement. The optical system and a typical probe/control unit are shown in Fig. 17.7.

The probe unit is designed for installation in a bypass loop or directly in the process piping or vessel wall. A rotating helical scanner driven by an electric motor is used to sweep a beam of light through a series of angles at the interface of a sapphire prism and the process liquid. Thermal isolation of the prism ensures that its temperature is the same as that of the process liquid. An exceptionally large prism is used to minimize measurement errors due to the presence of solid particles in the process liquid or partial coating buildup on the prism. Depending on the RI of the sample, light will be reflected onto a single solar cell, producing a digital signal unaffected by entrained air or opaqueness of the process liquid.

Indication is based solely on the time measurement between a light and a dark pulse, essentially a square wave with a maximum point occurring when the sweeping ray is reflected and a minimum when it is refracted. If the RI of the process liquid changes, there will be a proportional change in the critical angle, resulting in more or less time given to refraction versus reflection and the relative widths of the light and dark pulses. Sixty measurements per second are taken, with a change in the process detected by the solar cell within 16 msec.

An automatic gain control circuit maintains the light source at the degree of illumination irrespective of power supply variations, component aging, and so on. Useful life of the lamp and scanning motor are quoted in excess of 50,000 hr. Automatic temperature compensation is provided, using a thermistor probe mounted adjacent to the prism.

The sensing head is NEMA 4 design and may be located up to 500 ft from the control unit. The latter contains modular solid-state components housed in a NEMA 12 enclosure. Automatic prism wash facilities can also be provided. The cleaning medium may be water, steam, caustic, or other acceptable solution, depending on the application. Ultrasonic cleaning may also be employed (manual only). The sensing head requires a cooling water supply to provide cooling for the detector (solar cell).

Instrument spans range from 0.0015 RI units minimum to 0.130 RI units maximum. Repeatability is quoted as ±0.5% of span. The sensing unit is suitable for process temperatures up to 400°F and pressures up to 300 psig standard. Units for more stringent process requirements are available on request. Sensing units are also available in explosion-proof design.

17.3.3 Liquid Solids Control Company (United States)

The LSC Model 829 Process Refractometer is a recent addition to the marketplace, having been in production for about four years. Most applications to date appear to be in the food, beverage, and pulp/paper industries. The instrument is a compact, solid-state system consisting of three units—a sensing unit, a field junction box, and a remotely mounted control unit. The control unit is shown in Fig. 17.8.

The analyzer operates on the critical angle principle. A light source directs a focused beam of light to the base of the sensing (prism) window, which is wetted by the process liquid. The focused beam consists of rays with angle of incidence both above and below the critical angle. As a result, the light pattern on the output side of the sensing window will indicate a distinct light-to-dark division, the position of which is a direct function of the process liquid RI.

The prism is a synthetic sapphire cut in a hemispherical shape. This material has high density and superior hardness and is highly resistant to wear and abrasion. The detector is a specially cut silicon with high output and the ability to operate stably and dependably at temperatures from −25 to 200°C. Therefore, a cooling water supply is not required for the sensing unit. The temperature compensation element is an active silicon device installed in the mounting plate adjacent to the sensing window.

The output signals from the sensing unit are brought to a field-mounted junction box and then to the control unit. The control unit has span, zero, and temperature compensation adjustments, a liquid crystal display of the measured variable, a process temperature indicator, and a prism wash pushbutton. The unit houses up to six printed circuit boards, three of which are optional accessory boards.

Standard material of construction for all wetted parts of the sensing unit is 316 stainless steel; other materials are also available on request, depending on the application requirements. The sensing unit shown in Fig. 17.9 is a Teflon type designed for highly corrosive acid streams. In this case, only Teflon and sapphire are in direct contact with the process liquid. The sensing

FIGURE 17.8 Model 829 process refractometer: control unit. (Courtesy of Liquid Solid Controls Company)

unit may be installed in virtually any piping from small tubing up to 12-in. pipe. It is also possible to mount the sensing unit in the walls of vessels or tanks. A wide variety of different installations are available.

Analyzer range is 1.3000 to 1.6000 RI units, with minimum and maximum spans of 0.0015 and 0.1100 respectively. Process sample temperature range is −25 to 200°C at operating pressures up to 1000 psig maximum. Sensing unit and control unit may be separated by up to 1000 ft. Voltage (0–10 V dc or 0–100 mV dc) and current (1–5, 4–20, or 10–50 mA dc) outputs are available for remote recording.

17.3.4 Maselli Misure S.p.A. (Italy and United States)

Maselli manufactures a range of process refractometers, all operating on the critical angle principle. In these refractometers, a beam of light converges through an optical focusing lens, then passes through a prism to the interface

FIGURE 17.9 Model 829 sensing unit: typical. (Courtesy of Liquid Solid Controls Company)

between the prism and the process liquid. Rays with a lower deflection than the critical angle are refracted into the liquid; rays with a higher deflection are reflected. The critical angle coincides with the interface between illuminated and dark zones and is detected by selenium photocells. There are two types of detection units available.

In the FF Series Refractometers, a fixed photocell arrangement is used, as shown in Fig. 17.10. One photocell is always located in the light zone (reference) and one in the light-to-dark interface. A third photocell is incorporated to compensate for liquid turbidity, turbulence, and so on. A change in RI of the process liquid will change the critical angle and thus the position of the light-to-dark interface on the measuring photocell. The corresponding bridge imbalance produces an output proportional to sample RI.

In the FM Series Refractometers shown in Fig. 17.11, the fixed reference photocell is always located in the light zone. The measuring photocell is mounted on a moving arm and always searches for the light-to-dark interface. The movable arm is driven by an amplifier, motor, and cam, so that

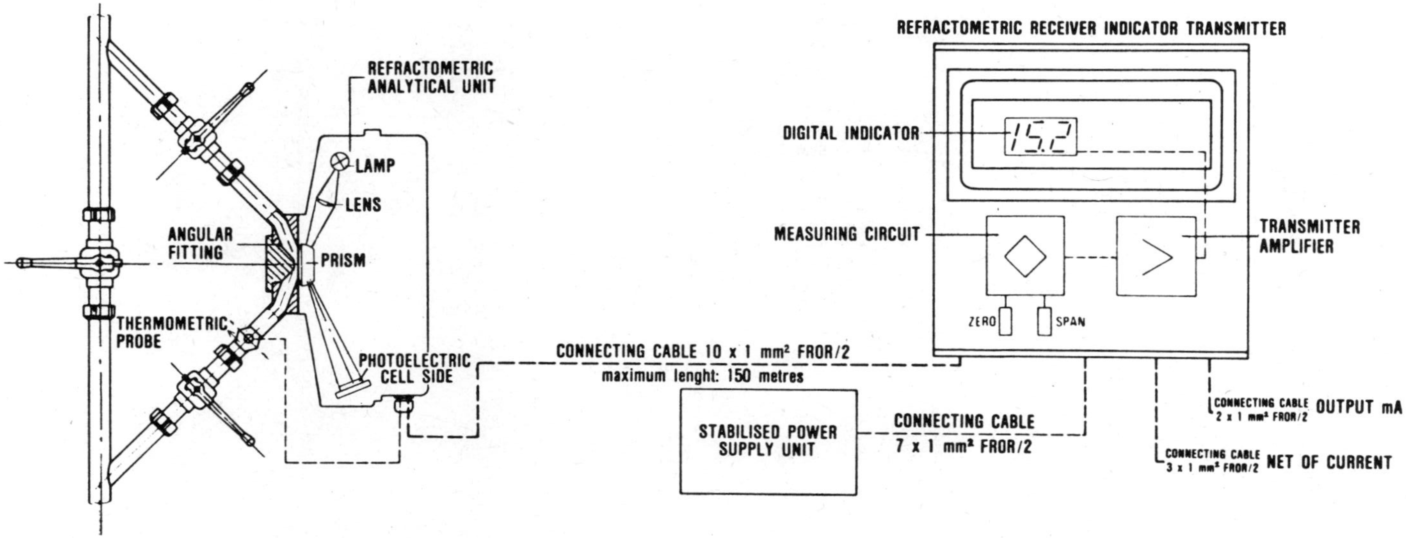

FIGURE 17.10 FF Series refractometer. (Courtesy of Maselli Measurements Inc.)

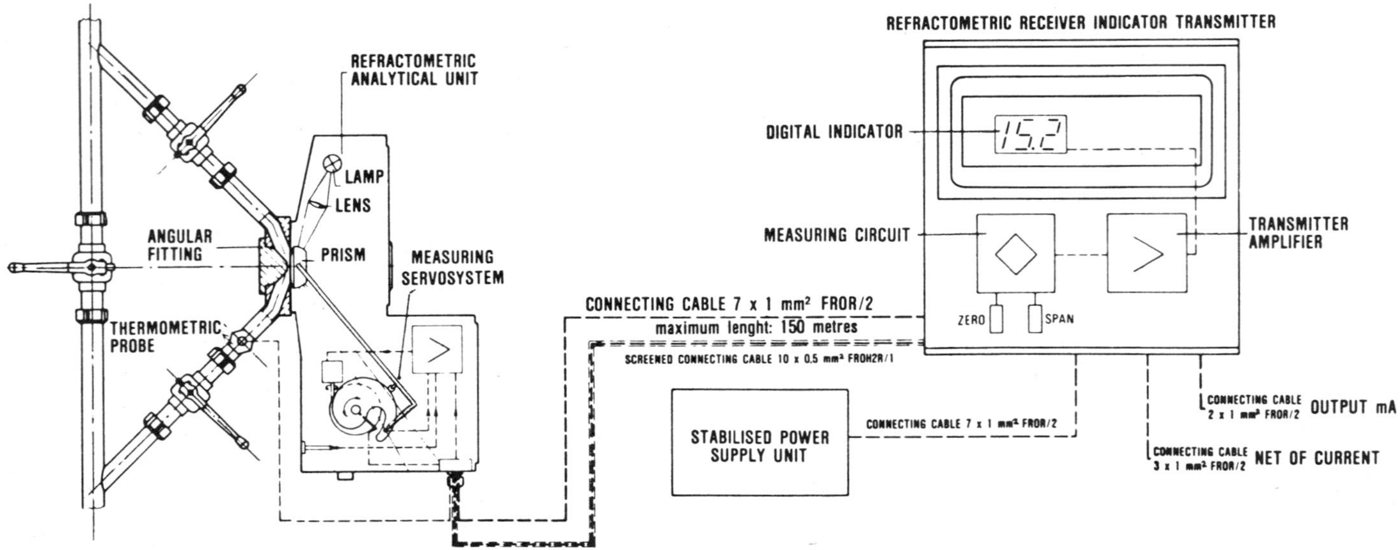

FIGURE 17.11 FM series refractometer. (Courtesy of Maselli Measurements Inc.)

when the system balances, the measuring (mobile) photocell is located exactly at the interface. The movement necessary to achieve balance is transmitted to the control unit by a slide wire.

The complete measurement system consists of three units—the sensing unit, the control unit, and a stabilized power supply unit. Two sensing units are available, the Models UR1 and UR6. Both use a special angular fitting to provide self-cleaning of the prism and facilitate maintenance. Typical examples are shown in Fig. 17.12. The body is of cast aluminum and of waterproof construction. Wetted parts are of 316 stainless steel or glass, with Viton or Teflon gaskets. The Model UR6 is also available in explosion-proof construction. The sensing unit is suitable for process sample temperatures from 2 to 120°C and pressures up to 180 psig.

The same solid-state control unit, Model RIT-2-600, is used with either sensing unit. This is suitable for installation in non-hazardous locations only. The control unit provides digital display of sample RI and current output

FIGURE 17.12 (*a*) Typical Maselli sensing units. (Courtesy of Maselli Measurements Inc.)

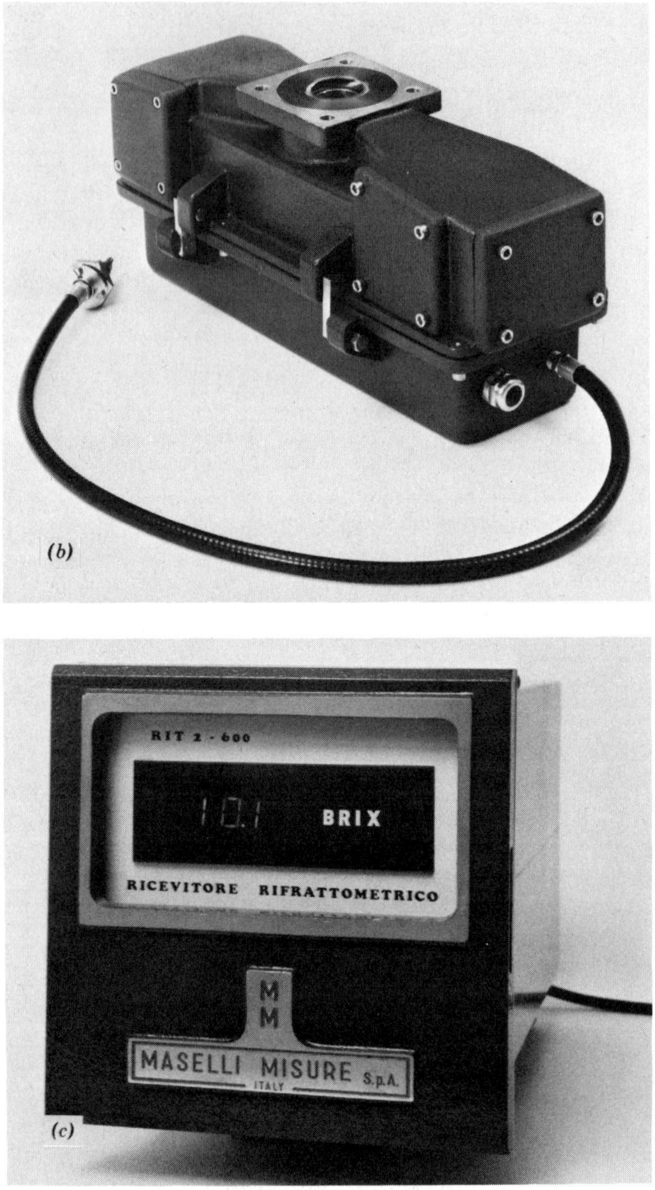

FIGURE 17.12 (*b*) Typical Maselli sensing unit. (*c*) Refractometer control unit. (Courtesy of Maselli Measurements Inc.)

(1–5 or 4–20 mA dc) for remote recording. Automatic temperature compensation is available, with a standard span of ±10°C of sample temperature; larger spans are also available. Measurement range is 1.3000 to 1.5500 RI units. For the Model UR1, minimum span is 0.0300; for the Model UR6, minimum span is 0.0200. The sensing unit in the Model VR6 is used for applications requiring high accuracy or very large spans. Repeatability is quoted as ±0.25% of span.

A recent introduction is the Model UR8 Monitor In-Line Refractometer. This instrument is designed for applications where indication of RI is not required. The monitor has two alarm setpoints, normally used for maximum and minimum values, to provide a visual alarm when the RI value reaches either setting. The setpoints are individually adjustable over the entire instrument range. The analyzer uses the same operating principle as the other Maselli refractometers. Measurement range is 1.3250 to 1.5500 RI units, with a repeatability quoted as ±0.5% of span.

17.3.5 Pier-Electronic GmbH (West Germany)

The Pier Process Refractometer operates on the critical angle principle and uses a novel detection method as illustrated in Fig. 17.13. Light from a tungsten lamp is focused by a lens through the glass prism onto the prism-liquid interface. Directly after the lens is a double-apertured diaphragm, which allows reference and measurement light beams through, one to the left and one to the right of the optical axis. These beams impinge on the interface at different angles of incidence. The angle of incidence of the measurement beam is less than the critical angle, while that for the reference beam is greater. The reflected reference beam and the reflected portion of the mea-

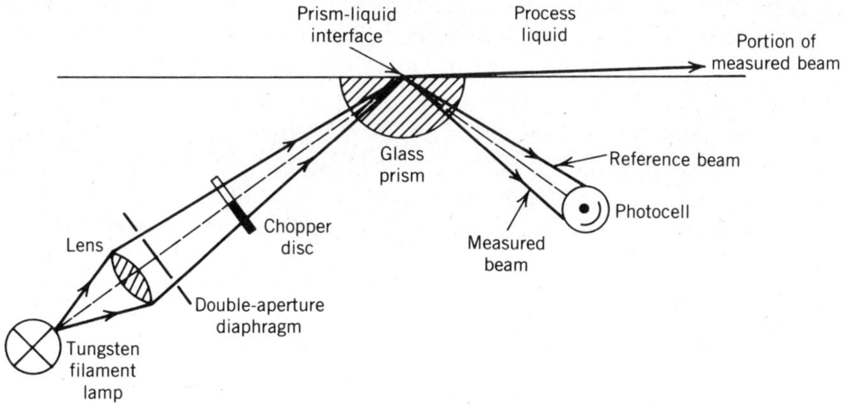

FIGURE 17.13 Pier process refractometer: operating principle. (Courtesy of Pier Electronic GmbH)

FIGURE 17.14 Typical refractometer sensing unit. (Courtesy of Pier Electronic GmbH)

surement beam fall onto a photocell detector. A chopper disc, located after the diaphragm, allows reference and measurement beams to pass alternatively to the photocell. As the RI of the process liquid changes, there is a proportional decrease in intensity of the measurement beam.

The system consists of two components, the sensing unit and the control unit, which may be separated by up to 300 feet. The sensing unit shown in Fig. 17.14 is designed for in-line mounting by means of a flanged connection. All wetted parts are made of stainless steel, glass, or Viton. The unit is suitable for process sample temperatures up to 60°C and pressures up to 75 psig. At higher sample temperatures, it is necessary to water-cool the measuring head.

The control unit incorporates zero, sensitivity, and temperature compensations adjustments and an analog indication of measured variable, and provides voltage (0–1 V dc) or current (0–20 or 4–20 mA dc) outputs for remote recording. The unit is suitable for panel or rack mounting in non-hazardous locations. Measurement range at maximum sensitivity is 0.001 RI units and precision is quoted as ±1% of full scale. The temperature compensation system incorporates a platinum resistance thermometer, installed in the sensing unit, and an amplifier stage in the control unit. The system will compensate for sample temperature changes of ±15°C from normal operating conditions.

17.4 INDUSTRIAL APPLICATIONS

We rarely measure the refractive index of a process sample to obtain a knowledge of that parameter. Usually, we use changes in refractive index to *infer* changes in composition. The refractometer is therefore a non-selective instrument, in the same way as the thermal conductivity analyzer (see Chapter 16) since it measures the refractive index of the entire sample. However, process refractometers are very sensitive instruments, as evidence by the ranges and spans quoted in Section 17.3.

The RI of a mixture may be calculated from the following equation:

$$N_m = C_a N_a + C_b N_b + C_c N_c + \cdots C_n N_n$$

where C_a, C_b, C_c, etc. are the mole fractions of components a, b, c, etc. and N_a, N_b, N_c etc. are the RIs of the pure components. Thus, when used as a process analyzer, the refractometer is ideally limited to binary mixtures, although more complex mixtures can be measured, provided that the component(s) of interest have RIs significantly different from the remainder of the sample components.

In spite of some limitations, there are numerous applications where on-line refractometers are used in industry. One major area is in the food and beverage industry, where RI is used for the measurement of sucrose concentration in jams, preserves, sauces, fruit juices, milks and sugars, coffee, and soft drinks. The weight percent concentration is known as degrees Brix. Another important application is in the measurement of iodine values of edible oils during the hydrogenation cycle.

In the petroleum industry, major applications include furfural extraction of lube oil, lube oil dewaxing (MEK/toluene ratio), gasoline octane rating (although correlation is still a major problem, see Chapter 15), phenol manufacture, aromatics production, and styrene and formaldehyde manufacture.

The process refractometer is used in the chemicals industry for monitoring phosphoric acid, sulfuric acid, nitric acid, urea/ammonium nitrate, brine concentration, glycol and glycerine, and in caustic soda production.

The brewing industry uses on-line refractometers for monitoring alcohol mixtures. They are also used in the pulp and paper industry for the measurement of percent solids in back liquor.

BIBLIOGRAPHY

Downie, R. A. Practical applications involving critical angle process refractometers: Parameters to consider for automatic monitoring and control. *Proc. ISA Symp.*, Des Plaines, ILL (May 1984).

Maley, L. E. Analysis by in-line refractometer. *ISA J.* (Oct 1962).

18

Combustion Efficiency Monitoring and Control

18.1 INTRODUCTION

Combustion efficiency control and optimization of fired heaters and steam generation facilities (boilers) has become a subject of major importance in industrial process plant operations. Spiralling fuel costs over the past decade have made industry look toward state-of-the-art technology for this purpose. With the accent today on energy conservation and economic operation, substantial savings in fuel and utilities costs can be achieved by operating equipment at maximum efficiency. This is particularly relevant to large industrial complexes, where what appear to be relatively small increases in efficiency represent very large savings in operating costs. Energy costs will no doubt continue to form a growing percentage of product cost; thus, conservation as an activity will occupy an increasingly important place in corporate strategies.

One method of improving efficiency is to reduce the amount of excess air used for combustion of the fuel, thereby approaching the requirement for stoichiometric combustion. Various programs to reduce excess air have been conducted by industry over the past ten years, with a relatively high measure of success. These programs have been enhanced by the use of process analyzer technology. This chapter outlines the ways in which various process analyzers can be used in this important application.

18.2 OPERATION AT OPTIMUM EXCESS AIR

Theoretically, perfect (or stoichiometric) fuel combustion occurs when fuel molecules react with oxygen molecules and there are no unburned fuel molecules or unreacted oxygen molecules left over. Why would we want to operate at this point?

First, there will be obvious savings in fuel costs. Second, the air contains not only the oxygen needed for combustion of the fuel but about 79% of nitrogen and other inert gases that all absorb heat. These air molecules and the other molecules such as sulfur dioxide, carbon dioxide, water, and nitrogen oxides formed in the combustion process carry heat out of the stack in the form of unused energy. Therefore, any air in excess of that required for combustion causes an increase in the loss of heat up the stack and requires additional fuel to offset the increased heat loss. In addition, for forced-draft

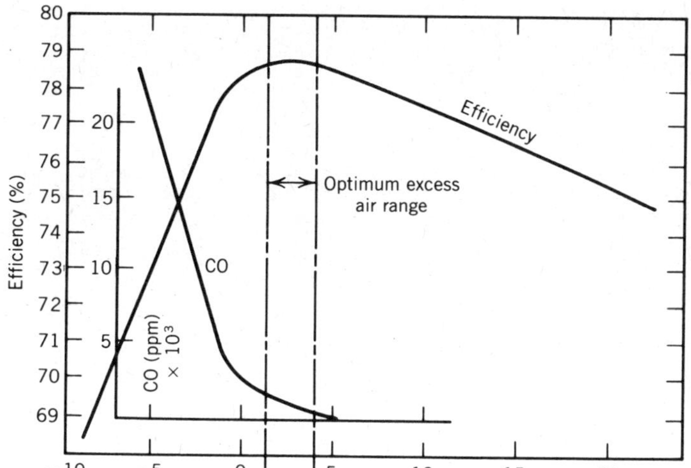

FIGURE 18.1 Relationship between combustion efficiency and excess air.

heaters and boilers, combustion air is provided by an electric motor-driven fan, another source of energy use.

The relationship between fired heater or boiler efficiency and excess air usage is shown in Fig. 18.1. Ideally, the efficiency peak occurs at near the stoichiometric point; in reality it occurs at somewhat higher excess air levels. Note that operation at too low, as well as too high, excess air levels will result in an efficiency loss. The efficiency loss to the right of the peak results from the increased flow of gases up the stack and the increased temperature of these stack gases that accompanies higher excess air. The efficiency loss to the left of the peak results from incomplete combustion of the fuel.

Also shown in Fig. 18.1 is the relationship between excess air and stack CO concentration. The peak efficiency usually occurs with a stack CO content generally below about 300 ppm. A stack CO content of about 1000 ppm will result in an efficiency loss of around 0.5%. Since the rate of combustibles emission increases so rapidly with reduction in excess air, efficiency can drop dramatically as excess air is reduced to below the optimum level.

The standard method of measuring excess air is to measure oxygen concentration in the stack flue gas. As mentioned in Chapter 7, this has long been a major industrial application for oxygen analyzers. The relationship between excess air and stack flue gas oxygen content is shown in Fig. 18.2.

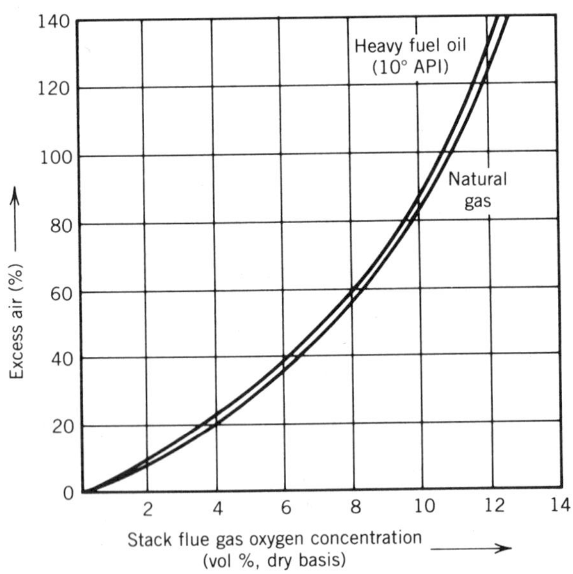

FIGURE 18.2 Excess air vs. stack flue gas oxygen content.

18.3 CONSTRAINTS ON EXCESS AIR REDUCTION

Excess air reduction programs conducted by industry have shown that it should be possible to reduce excess air to about 5% for gas-fired equipment and to about 10% for oil-fired equipment. However, several constraints must be overcome to achieve these goals:

- Air leakage, a major problem, particularly with older equipment.
- Inadequate draft control.
- Inadequate combustion air flow control.
- Inadequate instrumentation.
- Process constraints such as emissions, where appearance of stack CO or smoke may limit excess air reduction.
- Large swings in firing rate.
- Poor burner performance.

Note that the point of minimum heat loss does not necessarily occur at low excess air. For example, air leakage into equipment, turndown operation, or poor burner performance can result in combustibles emission at quite high excess air levels. Therefore, the target excess air may be different for every fired heater in a plant and this target may also be a function of heater firing rate.

It is therefore necessary to establish, by testing procedures, the excess air target for every heater. The test procedures involve measurement of stack oxygen and combustibles as excess air is reduced.

For gas-fired heaters and boilers, it has been found that combustibles emission is primarily in the form of CO. In the case of oil-fired heaters, the combustion chemistry is more complex. It has been demonstrated that CO constraint control sometimes works well, while at other times heavy smoking (high opacity level) occurs without CO appearing in the flue gas. In addition, hydrocarbon level in the flue gas sometimes shows a significant increase, while the opacity and CO levels remain relatively low. The characteristic shape of the CO/excess air curve shown in Fig. 18.1 is similar for opacity and hydrocarbons. It also appears that oil viscosity and temperature have an effect on whether CO, opacity, or hydrocarbons will be the limiting constraint on excess air reduction.

In summary, for gas-fired equipment, the limiting constraint on excess air reduction is the appearance of CO in the flue gas. For oil-fired equipment, any one of the parameters CO, opacity, and hydrocarbons can exert a limiting influence on excess air reduction, depending on the operating conditions. While the reasons for this are not completely understood, there seems to be universal agreement that the combustion process is not simply the direct formation of carbon dioxide and water.

18.4 PROCESS ANALYZERS FOR COMBUSTION CONTROL

18.4.1 Historical Background

Over the years, a variety of schemes have evolved for controlling excess air in fired heaters and boilers. Prior to the 1960s, no instrumentation was used; excess air was set by visual observation, was entirely manual, and almost always resulted in very high excess air in order to ensure complete combustion of the fuel.

The first stack-flue-gas oxygen analyzers used were of the paramagnetic or electrochemical type (see Chapter 7), requiring a sample handling system to take a sample of the stack gas, condition it, and transport it to the analyzer for measurement. These systems were plagued with maintenance problems because of the wet and corrosive nature of the sample. Consequently, closed-loop control was rarely possible with these systems. Nevertheless, in view of the importance of the application, much work was done by users and manufacturers in an attempt to improve sample handling systems from a maintenance standpoint. A certain measure of success was achieved, and the extractive sample type of oxygen analyzer system is still used today for flue gas analysis. An example of such a system is shown in Fig. 18.3.

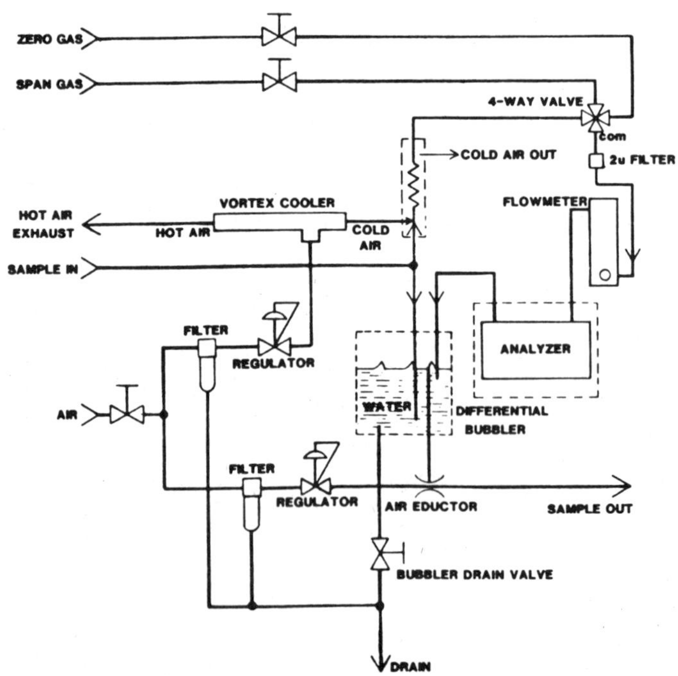

FIGURE 18.3 Typical extractive sampling system for stack flue gas analysis.

The early 1970s saw the introduction of the zirconia oxygen sensor, which removed the need for a complicated sample handling system. Maintenance requirements were reduced and system reliability increased, opening the way for closed-loop control. The technique was most applicable to forced-draft (FD) heaters and boilers, where the oxygen analyzer output signal could be used to control the position of the FD fan inlet vanes. For natural-draft equipment there was a problem; to what should the controller be connected? Initially, the oxygen analyzer was used to control the position of the stack damper. However, this did not achieve fully automatic control, since the operator was still required to manually set burner air registers.

The late 1970s saw the first use of CO analyzers to rationally set excess air targets. Data from the CO analyzer revealed that too low excess air resulted in combustibles emission and that different excess air targets were needed at different firing rates to achieve optimum efficiency. As a result of this initial test work, it became clear that control could be best achieved by using a CO analyzer to reset (or constrain) the excess air target for gas-fired equipment.

As mentioned earlier, in the case of oil-fired heaters and boilers, smoke (opacity) and hydrocarbon limits are often encountered before a CO limit is reached. Therefore the use of CO analyzers may not be sufficient to determine the optimum excess air level in these cases.

18.4.2 Process Analyzer Selection

While there are many commercially available oxygen, carbon monoxide, opacity, and hydrocarbon analyzers for the analyzer engineer to choose from, in practice the selection is based primarily on economics. A key factor that should be considered in this economic decision is the cost of maintaining the instruments. In addition, the "total system" approach, as opposed to the individual analyzer approach, can have an impact on the decision.

(a) Oxygen Analyzers

As indicated in Chapter 7, 30 or more manufacturers are active in the oxygen analyzer market today, giving the analyzer engineer a difficult task in selection the right instrument for the job. For combustion control applications, these may be divided into two main groups:

- Analyzers requiring an extractive sample system and using paramagnetic or electrochemical measurement techniques.
- In-stack or on-stack analyzers based on the zirconia (electrochemical) measurement technique.

The in-stack or on-stack types are generally preferred, since they eliminate problems with sample handling. However, if sweet fuels are used, the extractive sample type analyzers may still be applicable, especially if a conditioned sample is required for a CO analyzer. The reader is referred to Chapter 7 for a detailed coverage of available oxygen analyzers.

(b) Carbon Monoxide (CO) Analyzers

Commercially available CO analyzers are all based on the principle of IR absorption, as discussed in detail in Chapter 10. Once again, CO analyzers may be categorized into two groups:

- Conventional IR analyzers requiring an extractive sampling system.
- In-stack IR analyzers, not requiring a sampling system.

The extractive IR analyzers are covered in detail in Chapter 10. In Chapter 12, one example of the in-stack CO analyzer (Land Combustion Inc.) is described. As for oxygen analyzers, there is a fairly wide market for CO analyzers, as shown in Table 18.1 on page 747. Some of these instruments are described next.

The in-stack CO analyzers are a relatively recent introduction, so there is not much operating experience with them. However, they appear to be generally preferred for the same reason namely, elimination of the sampling system. More use of these analyzers is anticipated in the future as operating experience and reliability increases.

Dynatron Inc. (United States). The Model 3100M In-Stack CO Monitoring System shown in Fig. 18.4 is designed to measure CO in flue gas within the range 0–500 or 0–1000 ppm. The system consists of a stack- or duct-mounted sensing unit and a microprocessor-based, remote-mounted control unit.

The sensing assembly consists of two parts, the IR source section and the detection section, mounted on opposite sides of the stack or flue duct. An analog meter is provided as standard on the IR source section to facilitate optical alignment of the system. Combination sensor/air purge blower enclosures provide weather protection for the sensor assemblies. The regenerative blower provides a source of clean filtered air to protect the optics from dirt accumulation.

The collimated optical system is designed for path lengths from 1.6 to 32 ft., maximum path length being a function of required CO measurement range. The ratio measurement technique used provides compensation for IR source intensity variations, dirt accumulation on optical surfaces, detector aging, and flue gas opacity variations.

The microprocessor-based control unit, designed for panel mounting, provides digital signal processing, averaging, and smoothing capability and performs self-diagnostic routines on the complete system. A system status display is provided as standard. An automatic zero/span check system can be programmed for any time period from 1 to 99 hr, or can be actuated manually on demand. During the zero/span check, a "data hold" circuit maintains the

FIGURE 18.4 Model 3100M CO monitoring system. (*a*) Field mounted unit; (*b*) remote mounted control unit. (Courtesy of Dynatron Inc.)

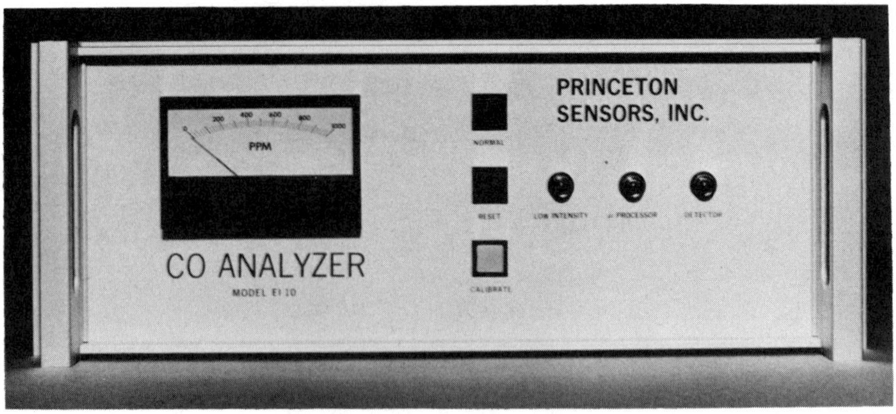

FIGURE 18.5 Model EI-10 CO analyzer control unit. (Courtesy of Princeton Sensors Inc.)

analog output at its previous value prior to initiation of the check. A total of four SPST contact closures are provided as standard for external alarm actuation.

Princeton Sensors Inc. (United States). The Model EI-10 In Situ CO Analyzer shown in Fig. 18.5 is designed to measure CO in flue gases within the range 0–1000 ppm. The analyzer consists of two air-purged field units (the source and receiver unit, which are mounted on either side of the stack or flue duct) and a microprocessor-based control unit. A beam of IR light is transmitted from the source unit across the stack into the receiver, which uses high-resolution absorption spectroscopy to find the differences between the intensity of the light transmitted and the light received at the detector.

The detector output signal is transmitted to the control unit, where it is processed to provide a signal proportional to CO concentration, which is used to drive a local indicator and provide an output for remote recording. Self-diagnostic routines are carried out by the microprocessor to monitor the system status, which is indicated on the front panel of the control unit. Calibration checking can also be carried out from the control unit. Voltage or current outputs are available and repeatability is quoted as ±1% of full scale.

Syconex Corporation (United States). The Supervisor 5000/6000 Series Combustion Analysis System is an in-situ analyzer, designed to monitor up to two components in flue gases using a gas filter correlation IR absorption measurement system, microprocessor-based signal processing, and a novel calibration technique designed to eliminate all interferences due to changes in sample composition and temperature, optical system variations, and so on.

The analyzer consists of two field units, a transceiver unit and retroreflector unit, mounted on either side of the stack or duct and a remote mounted control unit. The optical system is shown in Fig. 18.6. A beam of IR light

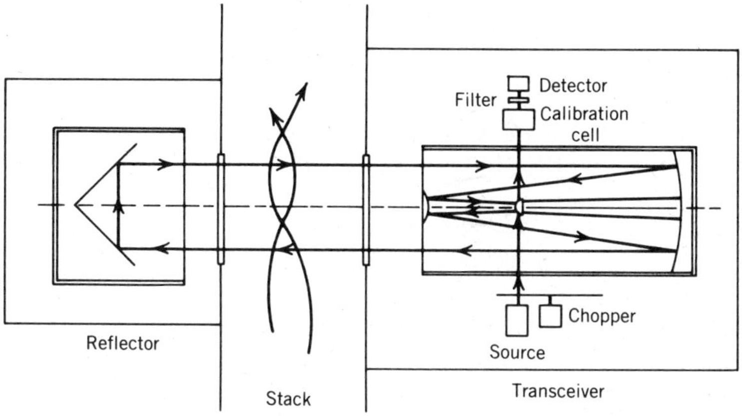

FIGURE 18.6 Model 5000 CO analyzer. (Courtesy of Syconex Corporation)

from the source is chopped at 900 Hz and, after collimating and focusing, passes across the stack through the sample to a retroreflector. The beam is inverted and returns across the stack to the transceiver, where it is collected by the primary mirror, focused on the secondary mirror and prism, passes through a gas filter correlation cell and narrow bandpass filter, and finally impinges on the detector. The optical elements are all contained within a prealigned and sealed telescope assembly.

When measuring CO, the gas correlation filter contains a high concentration of CO and the narrow bandpass filter allows only radiation within the band 4.5–4.7 μm to pass to the detector. The 2-in. diameter gas filter wheel is designed to accommodate up to eight cells (1-in. long and 0.5 in. in diameter), only four of which are used for CO measurement. These four cells are:

- Reference cell, filled with nitrogen.
- Correlation cell, filled with a known concentration of CO.
- Calibration standard #1 cell (90% of span).
- Calibration standard #2 cell (20% of span).

In normal operation, a digitally controlled stepper motor moves the gas filter wheel from the "index" position to the reference cell position and dwells there long enough for the stack-mounted electronics to make an intensity measurement, digitize the signal, and transmit it to the control unit over an RS422 link. The gas filter wheel is then stepped to the correlation cell position and a reading is taken, digitized, and transmitted. The stepper then dwells for a time equivalent to one step, a second reading is taken, and it then steps back to the reference cell and another reading is taken. The cycle of reading each cell twice takes about 1 sec.

Instrument response time is selectable from the electronics unit and is the number of individual measurements that are averaged. For instance, 10 measurements averaged give a 5-sec response time. Thus, in all cases the average is updated every 0.5 sec, so that changes are seen almost instantaneously.

The microprocessor subtracts the correlation cell signal (background absorption only) from the reference cell signal (background and sample absorption) and divides by the correlation cell signal, thus:

$$\frac{(S + B) - B}{B} = \frac{S}{B}$$

which is directly proportional to CO concentration.

In the internal calibration mode, which is operator-selectable or available on demand, the internal instrument zero and span are set and calibration cell concentrations are validated. When this mode is selected, a calibration retroreflector moves into the path of the IR beam so that it is reflected directly back into the telescope, never leaving the transceiver unit. Since there is no CO sample in the IR beam, the reference cell has no absorption; this is therefore the instrument zero. The stepper now moves to Calibration cell #1, which gives the instrument span setting. The microprocessor computes:

$$\frac{(\text{span} + B) - B}{B} = \frac{\text{span}}{B}$$

Finally the stepper moves to calibration cell #2. The microprocessor takes the three intensity measurements and computes the ratio of each pair—corr./cal #1, corr./cal #2, and cal #1/cal #2—and compares the results with the known ratios in its memory, based on the known concentrations of cal #1 and cal #2. If there is a leak in any of the cells, it will show as an error in the computations.

The unit is also capable of carrying out a complete system calibration. This mode can be initiated at a frequency selectable from the control unit. In this mode, the stepper motor steps so that cal cell #1 then cal cell #2 is placed in the path of the IR beam. The IR beam has passed through the sample and the cal cell in series; therefore the microprocessor receives three signals equivalent to S, (S + cal #1) and (S + cal #2). The known calibration values are subtracted to give:

$$(S + \text{cal } \#1) - \text{cal } \#1 = S_1$$
$$(S + \text{cal } \#2) - \text{cal } \#2 = S_2$$

and a check is made to see that $S = S_1 = S_2$. If it isn't the microprocessor adjusts zero and/or span to give the equality.

This calibration technique corrects for any and all interferences on a cumulative basis and makes measurements in 0.5 sec, thus measuring changes in real time without the use of flowing standard gases. It also uses exactly the same components used to measure sample concentration, since the IR beam is passing through the sample during the calibration mode.

(c) Opacity Analyzers

These devices have been used extensively in the utilities industry for many years to comply with EPA regulations and are well proven in this application. The reader is referred to Chapter 12 for detailed coverage of the analyzers.

(d) Multi-Function Analyzers

Analyzers that are designed to measure more than one flue gas parameter are also commercially available. Some examples are given here.

Ametek-Thermox (United States). The Model WDG-IIIC Oxygen and Combustibles Analyzer shown in Fig. 18.7 combines a new combustibles monitor with the field-proven WDG-III zirconia oxygen analyzer (described in Chapter 7). The physical layout and construction of the system is the same as the WDG-III, except that the combustibles sensor is installed immediately before the temperature controlled furnace containing the zirconia sensor. The combustibles sensor consists of two sensing elements, a catalytically active element and a reference element, maintained at a constant temperature. The current decrease that occurs when combustible gases burn on the catalytic element provides the output signal to the control unit. Local indi-

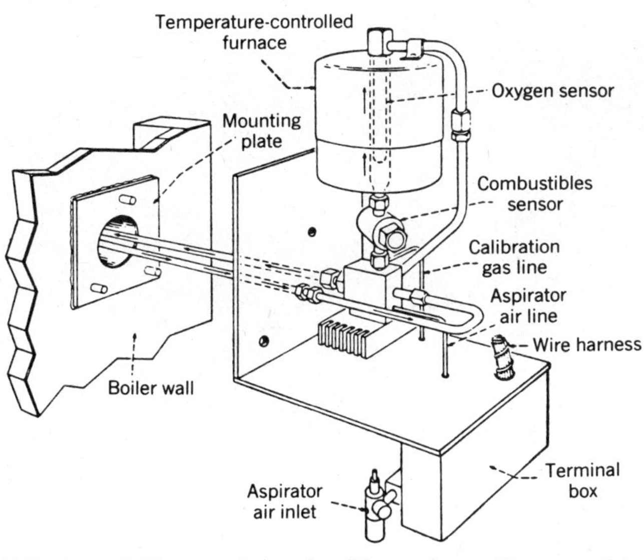

FIGURE 18.7 Ametek-Thermox O_2/combustibles analyzer. (Courtesy of Ametek Inc.)

cating meters are provided on the control unit for oxygen and combustibles. Analyzer repeatability is quoted as ±0.2% of reading.

Lear Siegler Inc. (United States). The Model EX4700 ENERGY EXPERT In-Stack Analyzer uses a narrow-band non-dispersive IR absorption technique to measure CO and CO_2 in flue gases within the ranges 0–1000, 2000, or 5000 vol ppm (CO) and 0–20% v/v wet basis (CO_2). The system consists of a probe/transducer unit attached to the duct or stack by a mounting flange, a locally mounted junction box, and a remote-mounted microprocessor-based control unit. The optical system shown schematically in Fig. 18.8 is a single-

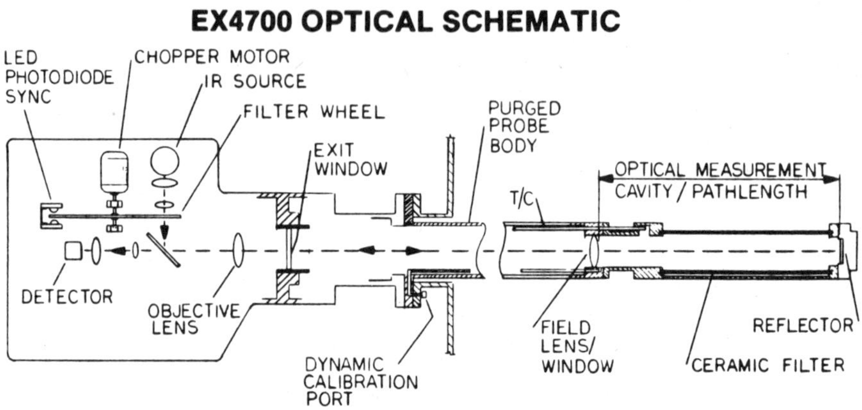

(a)

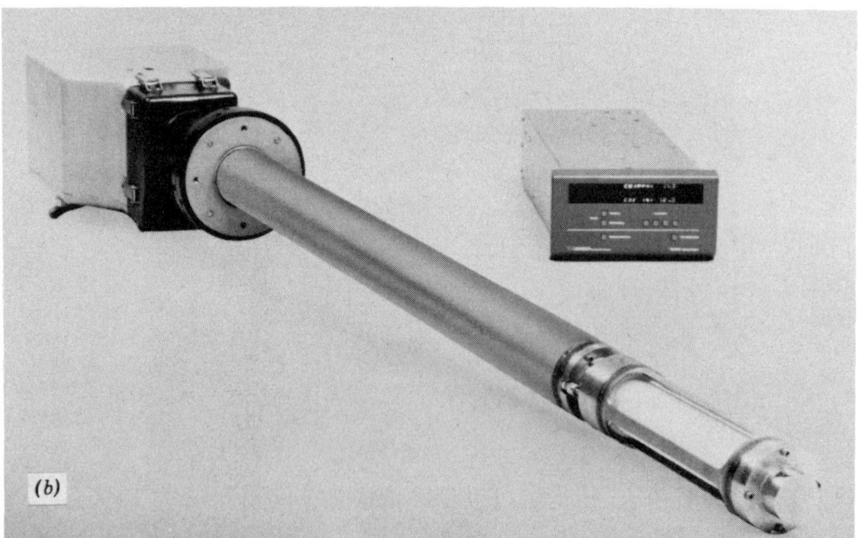

(b)

FIGURE 18.8 Lear Siegler Model EX4700 CO/CO analyzer. (*a*) Operating principle; (*b*) probe and control unit. (Courtesy of Lear Siegler Inc.)

beam, chopped, multiple-wavelength design that measures the IR absorption of each of three gases, CO, CO_2, and water vapor. Separate wavelengths are used for each measured component. The transceiver contains basic signal processing circuitry to provide current outputs to the control unit representative of each component absorption measurement.

The probe has a ceramic diffusion filter that screens out particulates before the flue gas diffuses into the measurement cavity. Zero calibration is accomplished by manually or automatically actuating a solenoid valve to allow instrument air to flood the measurement cavity. Optional span-calibration solenoid valves provide for automatic calibration with bottled gases of known concentration.

The junction box, connected to the transceiver by 6 ft of cable, provides access to all transceiver functions and monitors sensor response. A diagnostic panel is available as an option for sensor performance verification. Primary power supplies and power-consuming circuits are located in the junction box.

The control unit provides linearization of CO and CO_2 signals, corrected for temperature, pressure, and moisture content. It also monitors all diagnostic signals from the transceiver and provides alarms, operator access to system data, and verifies/corrects calibration of the system.

18.4.3 Complete Combustion Control Systems

Several manufacturers offer complete systems for combustion efficiency control utilizing analytical sensor technology.

(a) R.J. Bambeck Company (United States)

The Model CO-3000 System is a flue-gas carbon monoxide air-fuel ratio trim control system designed primarily for industrial boiler applications. The system consists of an in-stack CO analyzer and a panel-mounted control unit.

The CO analyzer shown in Fig. 18.9 consists of two units, a transmitter and a receiver, mounted on opposite sides of the flue duct or stack. A single beam of collimated IR energy leaves the transmitter and passes through the flue gas to the receiver, where it is focused through a gas cell onto a thermoelectric detector.

The gas cell has two compartments, one filled with a high concentration of CO and the other empty. The cell is rotated at 7 cycles/sec and positioned so that the IR beam is interrupted alternately by each side of the gas cell. Windows provide a positive seal between the flue gas and the source and receiver units. Purge air is continuously blown between the windows and the mounting hardware opening. The windows are mounted in sliders to allow rapid access for cleaning. Measurement range is 0–1000 ppm CO.

The CO-3000 control unit, also shown in Fig. 18.9, is microprocessor-based and performs all supervisory and monitoring functions for the system, including self-diagnostic routines and CO analyzer calibration. Combustion

FIGURE 18.9 Model CO-3000 system showing sensor (2 units) and remote mounted control unit. (Courtesy of R. J. Bambeck Inc.)

control strategy uses the CO analyzer signal as the primary control with auxiliary oxygen and opacity control (provided as options). Manual control is also available.

Remote setpoints are available with bumpless transfer between control modes. Continuous digital readout is provided for CO, CO_2, opacity, and controller output, and for alarm facilities. The system has the capability for remote setpoint control and can interface with a process computer system. Current (4–20 mA dc) outputs are available for all parameters.

(b) Econics Corporation (United States)

The Model 700 Fuel Optimization System consists of an in-stack analysis unit designed for simultaneous measurement of CO, CO_2, opacity, and hydrocarbons in flue gases, and a remote mounted operator/control console. Both units are microprocessor-based and data transmission between them is by means of a single-conductor fiber optic cable.

The analysis unit mounts directly on the stack or flue duct. Modular electronics and optics are mounted outside the duct in a NEMA 12 enclosure, the measurement tube projecting inside the duct. The optical system shown in Fig. 18.10 consists of two parts, the IR absorption system for measurement of CO, CO_2, and hydrocarbons, and a visible light folded-beam transmissionmeter for measurement of opacity.

Light from the two sources passes through the measurement tube to a retroreflector and back to the IR and visible light detectors, using a chopped, beam-splitting principle. The sensing unit and retro-reflector are separated from the flue gas by sapphire windows. The measurement tube, when open,

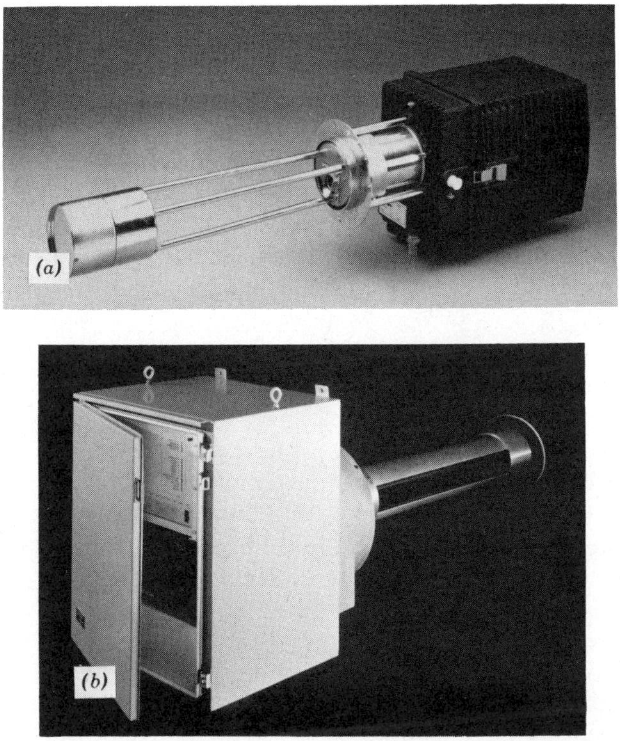

FIGURE 18.10 Econics Model 700 system: typical sensing unit designs. (Courtesy of Econics Corporation)

allows flue gas to pass through it perpendicular to the measurement beam path.

For zero standardization, an outer tube is rotated to form an air foil. This converts the kinetic energy of the flue gas into a pressure differential, which scavenges flue gas out of the measurement chamber and floods it with outside air. The microprocessor performs measurement calculations, standardization, and calibration functions.

The operator/control console uses the same 16-bit microprocessor as the analysis unit (except for memory board) and includes a 19-in. CRT and operator keyboard through which all operational data is entered. Various displays are available for combustion data, alarms, and production history. Extensive built-in self-diagnostic routines are carried out for the complete system.

The Model 700 is designed to interface with existing control systems. Standard measurement ranges are:

Carbon monoxide	0–10,000 ppm
Carbon dioxide	0–20%

Hydrocarbons	0–15,000 ppm
Opacity	0–100%

(c) Measurex Corporation (United States and Great Britain)

Measurex has manufactured complete computer-based fuel optimization systems for industrial power and cogeneration plants for over 15 years; a recent example is the Series 2001 system. An important component of the 2001 system is the Model 2224 Flue Gas Analyzer, shown in Fig. 18.11. This in-situ instrument is designed to measure simultaneously up to five components (CO, CO_2, NO, SO_2, and opacity) in flue gases, in order to meet federal requirements for emission monitoring equipment.

The analyzer consists of light sources, detectors with associated electronics, and mounting hardware, located directly on the stack or flue gas duct. Detector output signals are transmitted to the computer system, which processes the data and calculates the concentration of each measured component.

Measurement of CO and CO_2 is by IR absorption. For CO measurement, light from an IR source passes through a rapidly rotating wheel containing two cells, one filled with nitrogen (reference), the other filled with CO (measurement), then impinges on the lead selenide detector. The computer calculates the CO concentration by comparing the two detector output signals. For CO_2 measurement, filters located on the same rotating filter wheel alternatively present to the detector a wavelength that is absorbed by CO_2 and a reference wavelength that is unaffected by CO_2. Computer comparison of the two detector signals yields a measurement of CO_2 concentration.

Measurement of NO and SO_2 is by UV absorption. The beam from a UV source is collimated and passes to a high-resolution diffraction grating that separates the beam into its component reference and measurement wavelengths. Photomultiplier detectors then measure the UV absorption characteristics. The computer analyzes the raw detector signals and calculates the NO and SO_2 concentrations.

Opacity is measured in the visible region of the spectrum using a double-pass design. Visible light from a source lamp is chopped, collimated, and sent through the flue gas to a retroreflector, which reflects the light back to the detector unit. The computer compares the detector output signal with a stored calibration curve and calculates the flue gas opacity.

For an in-situ zero check, the sensor alignment tube within the stack is sealed off and purged of flue gases. Calibration of the IR and UV channels is carried out using sealed test cells or equivalent filters of known component concentration. For the opacity sensor calibration, a reflector is automatically inserted into the light path to return the beam to the detector (bypassing the stack), thus establishing the instrument zero. A 100% absorption shutter is then inserted in front of the retroreflector to complete the standardization check.

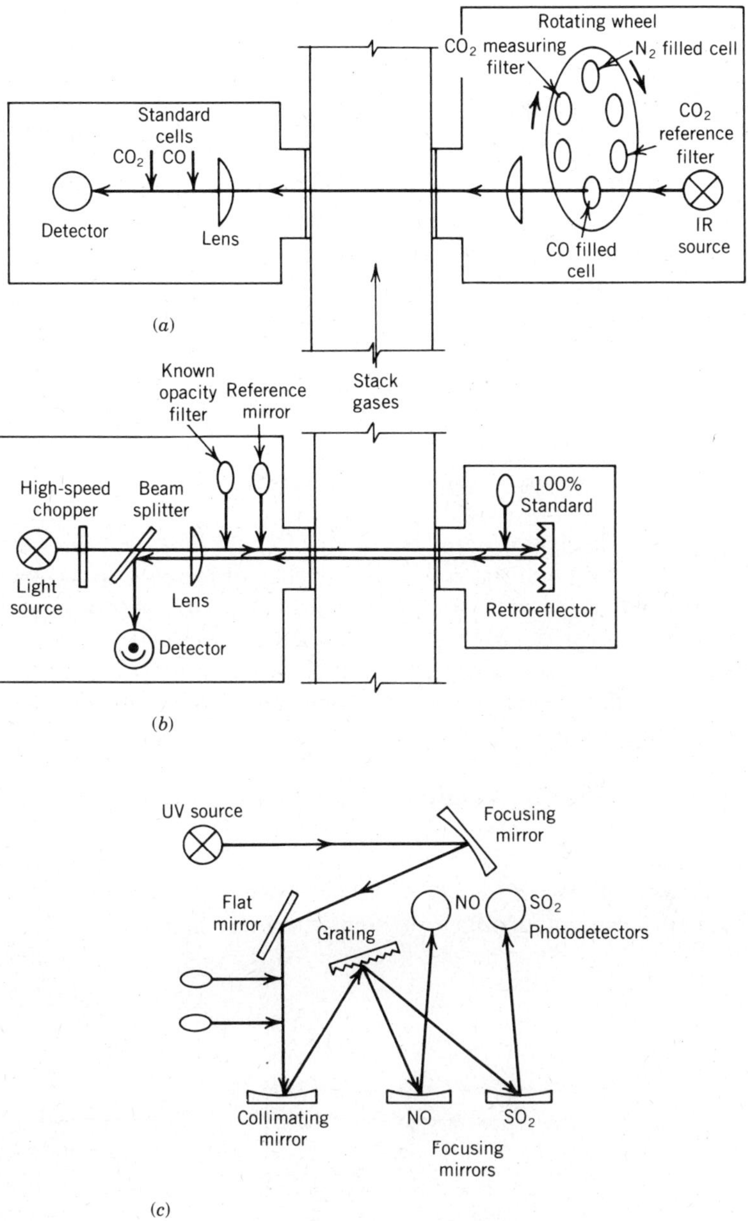

FIGURE 18.11 Model 2224 flue gas analysis system. (*a*) CO and CO_2 measurement; (*b*) opacity measurement; (*c*) NO and SO_2 measurement. (Courtesy of Measurex Inc.)

The measurement ranges offered and the quoted component repeatabilities are as follows:

Carbon monoxide	0–2000 ppm	±4% of reading
Carbon dioxide	0–30%	±5% of reading
Nitric oxide	0–1000 ppm	±4% of reading
Sulfur dioxide	0–1500 ppm	±4% of reading
Opacity	0–100%	±2%

Instrument response time is quoted as less than 1 sec.

18.5 CONTROL OF FIRING RATE

18.5.1 Objectives

So far we have discussed efficiency control of fired heaters and boilers by excess air reduction and the role various types of process analyzers play in the overall control system. Another important factor in overall combustion efficiency is control of firing rate. The primary objective of a firing rate control system is to maintain the coil outlet temperature (COT) of the heater or boiler constant irrespective of variations in feed flowrate and variations in fuel gas heating value.

18.5.2 Feedforward Control Systems

To compensate for feed flowrate variations, a feed-forward control system, such as the one shown in Fig. 18.12, is often used. Here, the feed flowmeter signal is linearized and changes are shaped against time to match the equip-

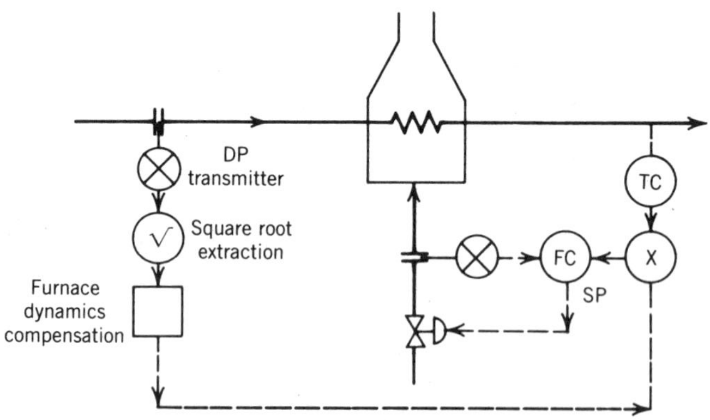

FIGURE 18.12 Typical feedforward control system for feed flowrate compensation.

ment dynamics (by use of a "lead-lag" algorithm). The COT controller output signal is then multiplied by the shaped feed-flowrate signal and the resultant output used as the setpoint to the fuel gas flow controller. If equipment dynamics are not taken into account, the change to fuel gas flowrate will take place too quickly, resulting in a "bump" in coil outlet temperature in the opposite direction to the "bump" that will occur if feedforward control is not used. The advent of digital instrumentation systems and computer control has made the implementation of feedforward control easy compared with a few years ago when such a system had to be designed using analog electronic or pneumatic instrumentation.

18.5.3 COT Control Systems

Three types of coil-outlet-temperature control systems are commonly used in industry. They are illustrated in Fig. 18.13 and form part of the overall heater or boiler control system. One should assume at this point that the heating value of the fuel gas is constant.

Direct temperature control (A) is frequently used with liquid fuels. With gaseous fuels, where supply pressure changes are slow compared with COT control loop response, the choice of control method is open. When pressure changes are more rapid, a cascade system is required to avoid COT upsets. Either pressure (B) or flow (C) cascade will prevent upsets, but flow is preferred, since constant flow is more positive in maintaining constant heat input. However, when the number of burners in gas service varies widely, the limited rangeability of the orifice meter may make pressure cascade preferable.

18.5.4 Heating Value (BTU) Control Systems

The COT control systems described are ideal because the assumption is made that the fuel gas heating value is constant. This is rarely true for most

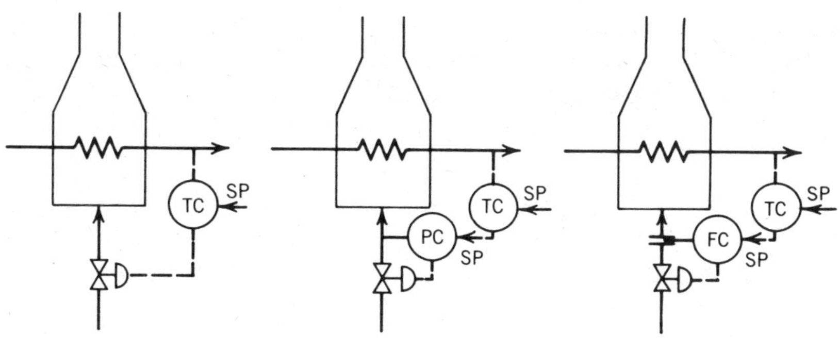

(*a*) (*b*) (*c*)

FIGURE 18.13 Typical COT control systems.

industrial complexes, particularly refineries and petrochemical plants, where the fuel gas heating value varies widely depending on the number and type of units providing feed to the fuel gas system at any time. Should a unit shut down, the fuel gas composition will change considerably. Therefore, additional efforts are required in the overall control system to maintain constant heat input.

Three methods are in common industrial use. The choice of method depends on the assumptions made about fuel gas composition; the more limited the assumption, the cheaper, faster, and simpler is the correction method.

(a) Method 1

The assumption here is that the fuel gas contains hydrogen and hydrocarbons, with less than 2% variation in interfering gases (such as non-combustibles, H_2S, CO_2, etc.) and the average molecular weight is greater than about 20. In this case, it can be shown that the heat content of the fuel gas (in BTU/lb) will be practically constant.

To provide control, an in-line density analyzer is used to measure absolute density at line conditions, the fuel gas flow transmitter is calibrated to give volume flowrate at line conditions, and the two signals are multiplied to give mass flowrate. Constant mass flowrate gives constant heat input to the burners. This is the fastest, cheapest, and simplest method of control.

(b) Method 2

The assumption here is that the fuel gas molecular weight varies between 2 and 60 and the interfering gas concentration is practically constant. In this case, it can be demonstrated that the heat content of the fuel gas (in BTU/SCF) will vary linearly with the molecular weight.

To provide control, the molecular weight of the fuel gas relative to air is measured by a specific gravity analyzer and the fuel gas flow controller is calibrated to give volume flowrate at STP (60° F and 14.7 psia). Constant heat input is maintained by manipulating the flowrate to maintain the product SCF × K (MW) constant, where K is a constant.

The system is somewhat more complicated than Method 1, since a sampling system is required for the SG analyzer, introducing dead-time into the control system and increasing the overall cost.

(c) Method 3

This method makes no assumptions regarding fuel gas quality. Variations in concentrations of interfering gases exceed ±2%; there may be sudden changes in CO and non-combustibles such as nitrogen. Under these conditions, control of heat input can only be accomplished by using an analyzer that measures fuel gas heating value directly (e.g., a Wobbe index analyzer or a GC). Again, a sample handling system is required, introducing dead-

TABLE 18.1 Flue Gas Carbon Monoxide Analyzers

Manufacturer	Country of Origin	Model No.	Measurement Principle	Micro-processor-Based	Sampling Configuration	Typical Ranges	Remarks
1. Anarad Inc.	USA	AR-200 series	NDIR	No	Extractive		Also measures O_2
		AR-500 series	NDIR	No	Extractive		Single component
		AR-600 series	NDIR	No	Extractive		Dual component
		AR-800 series	NDIR	No	Extractive		Multipoint systems
		AR-1500 series	NDIR	No	Extractive		Single component
		AR-1800	NDIR	Yes	Extractive		Computer-based, multistream system (24 max)
2. Anatek Insts. Ltd.	Great Britain	PSA-402	NDIR	No	Extractive	0–100 ppm to 0–100%	Single-beam design
		PSA-	NDIR	No	Extractive		Gen. purpose only
3. R. J Bambeck Company	USA	CO-3000	NDIR	Yes	Cross-stack	0–1000 ppm	Complete combustion efficiency contr. syst.
4. Beckman Industrial	USA	864/865	NDIR	No	Extractive	0–1000 ppm	Ex-proof & gen. purpose versions available
5. Combustion Eng. Process Analytics	USA	501	NDIR	No	Extractive	0–500 ppm; 0–15%	Digital control unit
6. Contraves-Goertz Corp.	USA	100	NDIR	No	Cross-stack	0–1000 ppm	Part of a complete comb. eff. monitoring & control system

TABLE 18.1 (*Continued*)

Manufacturer	Country of Origin	Model No.	Measurement Principle	Micro-processor-Based	Sampling Configuration	Typical Ranges	Remarks
7. Datatest Inc.	USA	EDIT 500 (EPA compliance)	NDIR	No	Extractive	0–1000 ppm	Also measures SO_3 NO_x, opacity, & O_2.
8. Dynatron Inc.	USA	3100M	NDIR	Yes	Cross-stack	0–1000 ppm; 0–500 ppm	Gas filter correlation
9. Econics Corp.	USA	7220	NDIR	Yes	Cross-stack	0–2000 ppm	Also measures Hcbns, opacity, & temp.
10. Foxboro Analytical	USA		NDIR		Extractive		
11. Hartmann & Braun	West Germany	URAS 2T/2T-Ex URAS 3G URAS 3E	NDIR NDIR NDIR	No No No	Extractive Extractive Extractive	Min 0–100 ppm Min 0–100 ppm Min 0–100 ppm	
12. Horiba Insts. Ltd	Japan	EIA-31/31F TIA-31/31F CFA-32/32F	NDIR NDIR NDIR	No No No	Extractive Extractive Extractive	0–2000 ppm 0–50 ppm 0–10 ppm	
13. Infrared Industries Inc.	USA	IR-700 series	NDIR	No	Extractive	0–1000 ppm	

14. Kent Industrial Instruments Ltd.	Great Britain	Infragas 200/300 Infragas 400	NDIR NDIR	No No	Extractive Extractive	Min 0–5000 ppm Min 0–50 ppm	
15. Land Combustion Ltd.	Great Britain	8000	NDIR	No	Cross-stack	0–400; 0–1000 ppm	
16. Lear Siegler Inc.	USA	EX 4700	NDIR	Yes	In situ	0–1000, 2000, 5000 ppm	
17. Leybold-Heraeus	West Germany	BINOS Range	NDIR	No	Extractive		
18. Measurex Inc.	USA	2224	NDIR	Yes	Cross-stack	0–2000 ppm	Part of a complete comb. eff. monitoring & control system; also measures CO_2, NO, SO_2, and opacity
19. M.S.A. Company	USA	LIRA 202/204 LIRA 3000	NDIR (Luft) NDIR	No Yes	Extractive Extractive		
20. Princeton Sensors Inc.	USA	EI-10	NDIR	Yes	Cross-stack	0–1000 ppm	
21. Rosemount/ Uniloc Inc.	USA	5100	NDIR	Yes	Cross-stack	0–1000 ppm	

TABLE 18.1 ***(Continued)***

Manufacturer	Country of Origin	Model No.	Measurement Principle	Micro-processor-Based	Sampling Configuration	Typical Ranges	Remarks
22. Siemens AG	West Germany	Ultramat 1	NDIR	No	Extractive	Min 0–300 ppm	
		Ultramat 2	NDIR	No	Extractive	Min 0–500 ppm	
		Ultramat 3	NDIR	No	Extractive	Min 0–50 ppm	
		Ultramat 4	NDIR	No	Extractive	Min 0–20 ppm	
		Ultramat 31	NDIR	No	Extractive	Min 0–500 ppm	
		Ultramat 32	NDIR	No	Extractive	Min 0–50 ppm	
23. Taylor Analytical Instruments Inc.	USA	PSA-402	NDIR	No	Extractive	0–100 ppm to 0–100%	See Anatek Insts. Ltd.
24. Syconex Corp.	USA	Series 5000	NDIR	Yes	Cross-stack	0–1000 ppm	Complete system calibration facility; also measures other gases
25. Teledyne Analytical Insts.	USA	700 (also 9300)	NDIR	No	Extractive	0–500 ppm	Can also measure CO_2 (typical 0–20%)
26. Thermoelectron Corp.	USA	48	NDIR	No	Extractive	0–10 to 0–200 ppm	
27. Xertex Corp.	USA	OP-3300	NDIR	No	Cross-stack	0–500 ppm to 0–5%	
28. Sereg (Schlumberger)	France	Type ANIR	NDIR	No	Extractive	0–100 ppm	

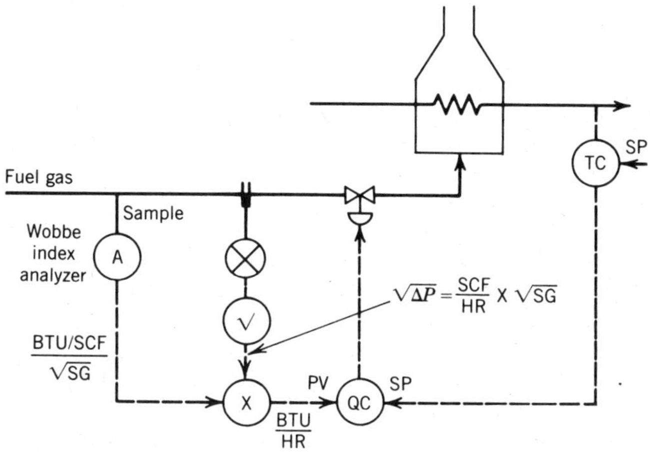

FIGURE 18.14 Typical heating value (BTU) control system.

time into the control system. In addition, for a GC one must consider the dead-time in the analyzer itself (cycle time). Although the Wobbe index analyzer can have a fast response time, it is an open-flame device, so additional safety protection is needed for the equipment. This system is therefore relatively complex and expensive, but does the job admirably.

The use of a Wobbe index analyzer allows direct determination of the heat input and provides an output proportional to calorific value (BTU/SCF) divided by the square root of the relative density (SG). Multiplying the Wobbe index analyzer signal by the linearized signal from the fuel gas flow transmitter results in an output proportional to heat input in BTU/hr (or kW).

This type of heating value (BTU) control system is illustrated in Fig. 18.14. For more detailed information on density/SG analyzers and Wobbe index analyzers, the reader is referred to Chapters 8 and 14 respectively.

BIBLIOGRAPHY

Annino, R., Cheney, M. C., Connelly, J. P., and Fletcher, K. S. A stack probe for the analysis of carbon monoxide and oxygen. *Proc. ISA Symp.*, Boston, MA (May 1985).

Anson, D., Clark, W. H. N., Cunningham, A. T. S., and Todd, P. Carbon monoxide as a combustion control parameter. *Combustion* (March 1972).

Boiler control system cuts fuel costs. *Chem. & Eng. News* (Oct 8, 1979).

Branham, R. L., Pritchard, J. J., Gilmour, W. A., and Pregler, D. N. Continuous measurement of carbon monoxide improves combustion efficiency of CO boilers. *IECTC 3rd. Ann. Conf.*, Houston, TX (April 1981).

Gilbert, L. F. Precise combustion control saves fuel and power. *Chem. Eng.* (June 21, 1976).

Gilbert, L. F., CO control heightens furnace efficiency. *Chem. Eng.*, (July 1980).

Gilmour, W. A., and Kaye, M. J. The application of a flue gas carbon monoxide sensor in the chemical process industries. *ISA Conf.*, Anaheim, CA (Oct 1981).

Gomes, V. G. Controlling fired heaters. *Chem. Eng.* (Jan 7, 1985).

Kirsch, R. P. Project 'O' saves energy and money for union camp. *J. Paper Trade* (Sept 1978).

Leffler, N., and Bargh, D. J. Dedicated computer systems increase co-generation efficiency. *Inst. Fuel Conf.*, Birmingham, England (Feb 1978).

Leonard, D. J., and Kehoe, T. J. Automatic control ups heater combustion. *Oil & Gas J.* (Sept 21, 1981).

Lord, H. C. CO_2 measurements can correct for stack gas dilution. *Chem. Eng.* (Jan 31, 1977).

Lord, H. C. In-stack monitoring of gaseous pollutants. *Env. Sci. & Tech.* (March 1978).

Lord, H. C., and Maier, J. A new NDIR analyzer incorporating a complete system calibration. *Proc. ISA Symp.*, Des Plaines, ILL (May 1984).

Mangin, H. V., and Bautz, R. R. Improving the efficiency of industrial boilers. *ISA Conf.*, St. Louis, MO (March 1981).

McFadden, R. W. Multi-parameter trim: It pays in combustion control. *InTech* (May 1984).

Measuring stack pollutants on the spot. *Bus. Week* (June 1, 1974).

Miller, J. G. A multivariable digital analyzer for combustion optimization. *Proc. ISA Symp.*, Boston, MA (May 1985).

Opel, A. E., and Gilbert, L. G. Closed loop source monitoring saves energy and money. *ASME Ann. Winter Mtg.*, San Francisco, CA (Dec 1978).

Presepe, P. W., Mangin, H. V., and Barnett, W. K. CO/O_2 combustion control strategy provides fuel dollar savings. *ISA Cap. Cities Contr. Conf.* (May 1980).

Presser, C., Semerjian, H. G., and Juberts, M. Evaluation of O_2 and Co monitoring systems for combustion controls. *Adv. in Instr., ISA Intl. Conf.*, Philadelphia, PA (Oct 1982).

Pruce, L. M. Maximize boiler efficiency and save fuel with CO monitoring. *Power* (Jan 1980).

Putman, R. E. J. A comprehensive approach to plant energy management. *7th Ann. Energy Tech. Conf.*

Schwartz, J. R., Fisher, H. J., and Gilmour, W. A. Carbon monoxide monitoring: A successful application. *ISA Conf.*, Portland, OR (April 1982).

Swanson, K. An advanced combustion control system. *AICh. E, Plnt. Oper. Prog.* (Jan 1983).

19

Process Analyzer Data Systems

19.1 BACKGROUND

Although Chapter 1 mentions the available on-line systems for acquisition and presentation of chromatographic data, the subject of analyzer data systems in general is worthy of more detailed discussion in view of its importance in today's world of process control.

The process analyzer has become an important component of advanced control systems, and its role as compositional sensor has added a significant dimension to overall process control strategy. Not only have reliable compositional data improved regulatory control, but have made a significant contribution to process optimization in terms of improved quality and other advanced control applications.

There is, therefore, a very real justification for providing reliable, up-to-date analytical information for use in process control. This chapter discusses the historical development of process analyzer data presentation from the late 1950s to the present and highlights the considerable effect the microprocessor has had on analyzer data system design since its introduction in the early 1970s.

19.2 HISTORICAL DEVELOPMENTS

In the late 1950s, when most process control systems were of the pneumatic variety, the process analyzer was considered an unusual device since it provided an electrical output, normally a voltage or millivoltage (the thermocouple is another such device). The process chromatograph, as discussed in Chapter 1, was considered even more unusual. Its primary output, the "chromatogram" (from which column performance can be assessed) was often displayed directly on a recorder or in the so-called "bargraph" format shown in Fig. 19.1.

In this format, the components of interest are displayed on the recorder as a series of vertical lines about $\frac{1}{8}$ in. apart, the height of each line representing the concentration of the particular component. When the component "gate" opens as the peak begins to elute, the chart drive motor remains de-energized. As the detector output signal increases to a maximum and then decreases, the recorder pen traces this movement without the chart itself moving. When the component "gate" closes at the end of the peak, it energizes the chart drive motor for a few seconds via a time delay relay.

The problem with bargraph presentation is that it is difficult to read, especially if a large number of components are being measured. More important, it cannot be used for process control. Nevertheless, the bargraph pre-

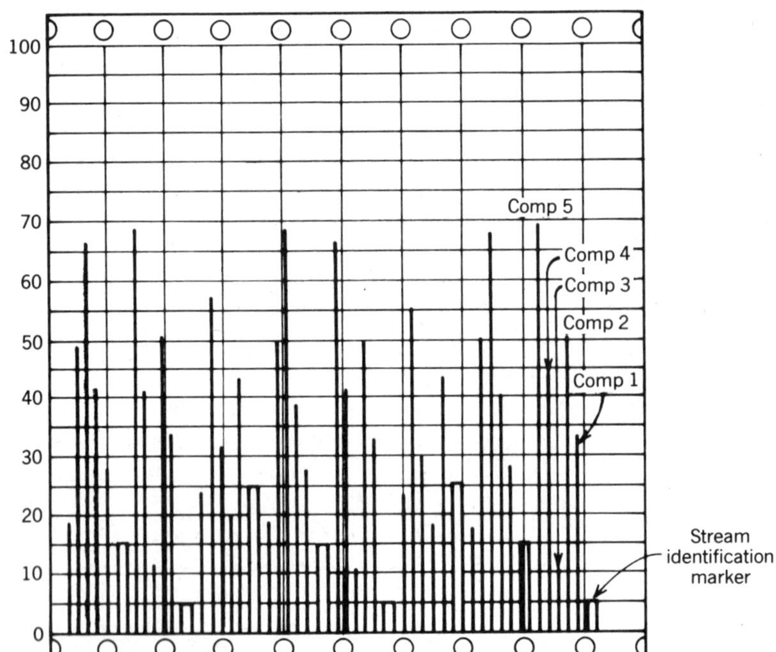

FIGURE 19.1 "Bargraph" format for chromatograph data.

sentation is still offered by some chromatograph manufacturers as one of their data presentation options.

For continuous analyzers, whose output is a continuous trend representing the measured variable (e.g., pH, conductivity, oxygen, etc.), there is no problem in recording the signal or using it in a process control scheme. Some process analyzers were even designed to provide pneumatic outputs so that they could easily be interfaced with pneumatic control systems, but they were the exception rather than the rule.

In view of its importance as an on-line quality measuring device, the process gas chromatograph has been the focal point of data presentation development over the past two decades. As we shall see, this development is still proceeding apace today.

As a result of advances in electronic circuit technology, the first major step forward in GC data presentation was the introduction of the "peak picker," or memory unit, usually located in the GC programmer. This device measures the maximum peak height of the component of interest during the analysis cycle, presents this value to the recorder, and stores the signal until the next cycle. The stored signal is erased just before the component gate opens, allowing maximum peak height to be measured again. The recorder pen therefore traces a stepwise, continuous line on the chart (trend recording), the amplitude of which is representative of component concentration. A typical example is shown in Fig. 19.2. Trend recording can also be achieved using pneumatic circuitry. This enabled the GC to be used with conventional pneumatic or electronic control systems, enhancing its versatility. Since a separate memory unit was required for each component of interest, the first GC programmers using trend facilities were rather bulky.

The advent of solid-state electronic circuits had a major effect on process analyzer design and on instrumentation and control systems in general. In addition, there was a trend away from pneumatic instrumentation and toward the use of electronic instrumentation, together with a standardization to current transmission (4–20 or 10–50 mA dc signals). As a result, by the late 1960s most process analyzers used solid-state electronics wherever possible and offered the option of current as well as voltage outputs. In particular, GC programmers used solid-state timing circuitry in place of the traditional cam-type timers, increasing the precision of timing functions. Trend output was accomplished with solid-state memory circuits.

The result was that process analyzer control units became more versatile, precise, and compact, and that process analyzer data could be transmitted in the same fashion as the main process variables (flow, level, pressure), namely, as 4–20 mA dc current signals.

Some manufacturers of pH and conductivity analyzers went a step further with the introduction of the so-called two-wire transmitters, described in Chapter 13. These transmitters contain a preamplifier that provides a 4–20 mA dc current output proportional to the measured variable and that looks from the instrumentation system end exactly like a conventional field flow or

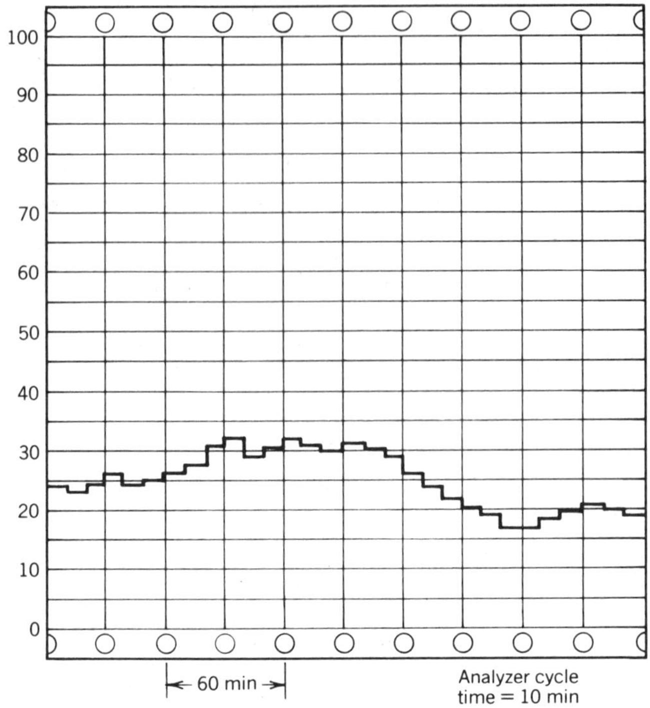

FIGURE 19.2 "Trend" recording of chromatograph data.

pressure transmitter. The current loop may be powered by a 24 V dc supply in the instrumentation system, eliminating the need for a separate power supply at the transmitter.

19.3 MICROPROCESSOR-BASED ANALYZER SYSTEMS

The advent of the microprocessor in the mid 1970s had a profound effect on the development of instrumentation systems and on the philosophy of process control. Previous chapters illustrate the influence this remarkable invention has had on the design and development of process analyzer systems—influence that is still being exerted today and will almost certainly continue into the future. In fact, the pace of development in recent years has been so rapid that it is difficult to keep up-to-date on new microprocessor-based process analyzers entering the marketplace.

Once again, one of the major focal areas has been the development of microprocessor-based GC systems, covering single GC controllers and systems capable of supporting multiple GCs. The principal advantages of the microprocessor in chromatographic operations are discussed in detail in Chapter 1, together with descriptions of some of the systems commercially available today. Other areas of note include the development of new microprocessor-based boiling point analyzers (Chapter 3), moisture analyzers (Chapter 11), pH analyzers (Chapter 13), BTU analyzers (Chapter 14), in-situ CO analyzers (Chapter 18), and oxygen analyzers (Chapter 7).

A recent major advance has been the introduction of process analyzer computer systems, as described in Chapter 1, capable of supporting not only GCs but other discontinuous types such as wet chemical analyzers as well as continuous analyzers such as oxygen, moisture, IR, pH, and conductivity. So, for the first time, the analyzer maintenance technician is able to troubleshoot the complete process analyzer system for a process plant from one central location.

In addition, the system's capability for internal self-diagnosis and report generation is a powerful tool for improving the efficiency and reliability of process analyzer systems and an aid to more efficient maintenance operations. This subject is discussed further in Chapter 23.

19.4 ANALYZER DATA LINKS

In the past decade there has been a marked increase in the use of computers for direct digital control (DDC) of industrial processes. The savings attributable to advanced process control very often hinge on the availability of reliable process-stream quality data from on-line process analyzers. In fact, some major processes are unable to operate efficiently without analyzers. Therefore, the need has arisen for a reliable and cost-effective means of transmitting process analyzer data to the process control computer.

Traditionally, continuous analyzer data are transmitted directly to the instrumentation system as hard-wired 4–20 mA dc signals. Process GC data (component concentrations) are converted at the GC programmer to 4–20 mA dc trend outputs, which are also hard-wired to the instrumentation system. Although the analyzer signals are treated in the same way as other field transmitter signals, additional costs must be incurred for signal cable purchase and installation just to provide analyzer data to the instrumentation system.

The introduction of the microprocessor-based GC system provided the first opportunity to transmit data from either a single GC or multiple GCs over a single serial digital data link to the process control computer (often called the host computer or PCC).

19.4.1 Analyzer Computer/Host Computer Data Links

Data links from the analyzer computer system to the process control computer are generally interrupt-driven links, the connection being through a single serial communication port and based on either 20-mA current loops or a subset of the EIA RS-232C interface.

However, the RS-232C interface only defines the type of connection, not the format of the message transmitted. Consequently, analyzer computer system manufacturers have had to provide many different types of data links, based on the customer's specified message format; there has been little standardization in this respect. The following discussion is therefore limited to typical examples of the analyzer computer/host computer data links currently used in industry.

(a) Functionality

Although many types of serial data links are available from analyzer computer system manufacturers, they all have the same functions in common. The functional objectives of a serial data link may be summarized as follows.

1. Checksums, positive acknowledgment, and error recovery procedures are used to ensure the integrity and availability of data.
2. Extensive data validation techniques are used to qualify data as acceptable for process control. Such validation may be carried out in the host computer or the analyzer computer.
3. The data format must allow easy cross-referencing to the appropriate PV tag in the host computer database.
4. Compositional data should be presented in engineering units and in a format compatible with wide-dynamic-range data.
5. Data format should be compatible with process GCs and other process analyzers.
6. The data link should be able to report not only operating data but calibration results in terms of response factors and/or scaling factors.
7. The host computer and the analyzer computer should be able to recognize transmission or link failures.
8. The data link should be able to report analyzer status change information to the host computer, with a distinction made between analyzers taken out of service for maintenance and analyzers taken out of service on a *temporary* basis for manual or automatic calibration.
9. The host computer should be capable of supporting up to two serial data link channels to take advantage of analyzer computer systems where redundancy is built in to improve analyzer data availability (see Chapter 1, Section 1.8).
10. The host computer and the analyzer computer should be able to verify the integrity and availability of any one or both data link channels.

TABLE 19.1 USASCII Character Set

	000[b]	001	010	011	100	101	110	111
0000[a]	NUL	DLE	SP	0	@	P		p
0001	SCH	DC1		1	A	Q	a	q
0010	STX	DC2	"	2	B	R	b	r
0011	ETX	DC3	#	3	C	S	c	s
0100	EOT	DC4	$	4	D	T	d	t
0101	ENQ	NAK	%	5	E	U	e	u
0110	ACK	SYN	&	6	F	V	f	v
0111	BEL	ETB	,	7	G	W	g	w
1000	BS	CAN	(	8	H	X	h	x
1001	HT	EM	)	9	I	Y	i	y
1010	LF	SUB	*	:	J	Z	j	z
1011	VT	ESC	+	;	K	[	k	
1100	FF	FS	,	<	L	\	l	l
1101	CR	GS	–	=	M	]	m	
1110	SO	RS	.	>	N	↑	n	~
1111	S1	VS	/	?	O	←	o	DEL

[a] Bits are shown in transmission sequence 4, 3, 2, 1.

[b] Bits are shown in transmission sequence 7, 6, 5.

11. The hardware requirements for the data link interface should be as simple as possible.

(b) Hardware Requirements

Serial data links are usually based on either RS-232C or 20 mA current loops. In the latter case, it is usual to support optical isolation for receivers in both the analyzer computer and the host computer. Transmission rates for each serial data link are usually field programmable or jumper selectable; data rates include 300, 1200, 2400, 4800, and 9600 baud. The standard USASCII 10-bit character formats are normally used (i.e., one start bit, seven data bits, one parity bit (usually even), and one stop bit). The standard ASCII character set is shown in Table 19.1.

Although most data links are physically installed as "full duplex" (i.e., communication can be bidirectional), they are normally operated in the "half duplex" mode (message transmission from the analyzer computer to the host computer only).

(c) Communications Protocol

As mentioned, most serial data links have common function objectives and hardware requirements; the main differences between them are in their communications protocol and in the extent of the data contained in the particular message transmitted. These functions are usually developed for specific user

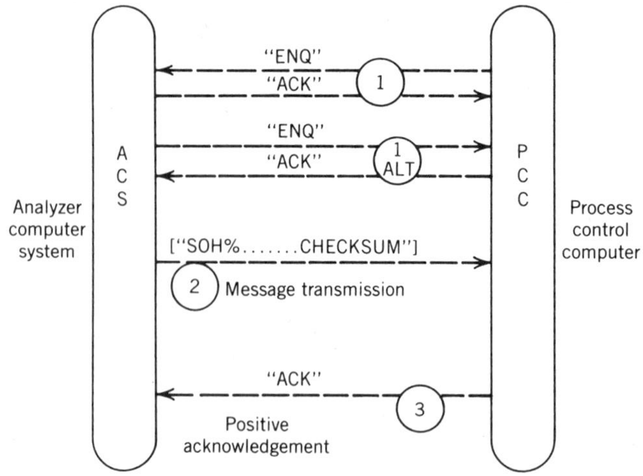

FIGURE 19.3 Typical data link protocol.

requirements. In this discussion, and in order to illustrate the basic concepts, two typical examples are described here.

The protocol is usually based on three distinct functions:

1. Data link establishment to verify either link integrity or I/O buffer availability in the host computer.
2. Message transmission, including checksum.
3. Error recovery procedure (usually positive acknowledgment or retransmission).

The protocol is therefore "interrupt-driven," where the analyzer computer has control of the data link once the link is established. However, either the analyzer computer or the host computer may attempt to establish the link. Figure 19.3 shows a typical detailed flow chart of protocol implementation.

Link Establishment. Referring to Fig. 19.3, the analyzer computer (ACS) attempts to establish the data link upon system initialization, whenever an analysis report becomes available, or when a change in analyzer or stream status is to be communicated. To establish the link, the ACS transmits a single "ENQ." The PCC responds with a single "ACK," upon which the ACS resumes control of the link and proceeds to transmit the message. If any character other than "ACK," such as "NAK" or no response, is received by the ACS, two more attempts are made. After three unsuccessful attempts, link establishment is aborted and the message is diverted for printing on the analyzer computer system printer (if so desired).

Often, a "data link status flag" is incorporated, such that the flag can be set up or down by the ACS, depending on the success or failure of the link establishment procedure. In addition, link establishment or failure can be communicated to the ACS printer, with the time and date.

The PCC may also attempt to establish the link upon system initialization or on a periodic basis to detect if no messages have been received from the ACS for a certain period, say 5 min, or as set by a timer. If an "ACK" is received from the ACS, no further action is taken. If "NAK" or no response is detected within, say, 1–2 sec, the PCC will indicate a failure to poll the ACS.

Message Transmission. Once the data link has been established, the ACS proceeds with transmission of the appropriate message. Messages usually have the following format:

Byte#	Code	Description
1	"SOH"	Start of header control character
2	Message ID	Message ID character
3 to n−1	Message text	
n	"ETX"	End of text control character
n+1	Checksum	Checksum byte calculated by ACS

The message ID is a single character indicating the type of message transmitted. Usually two types of messages are used, the "ANALYSIS REPORT" and "STATUS REPORT." In the following example the ASCII codes for the message IDs are "%" and "#" respectively.

A typical ANALYSIS REPORT message is shown in Table 19.2. This message is usually generated after the completion of an analysis cycle (GCs), at the end of a signal sampling and averaging time for continuous analyzers supported by a continuous analyzer controller, or optionally after a calibration cycle, where new response factors (GC) or scaling factors (CA) are

TABLE 19.2 Analysis Report Message Format (Example 1)

Byte #	Code	Description
1	"SOH"	Start of header
2	%	Message ID—analysis data
3–4	99	Analyzer number (01–99) (in continuous analyzer systems CAC Number)
5–6	99	Stream number (01–99)
7–10	9999	Analyzer cycle time (0001–9999 seconds)
11–15	99:99	Sample inject time (GC) (HR:MN)
16–24	.9999E 9	Component concentration/value
25	9	Validity status code
n	"ETX"	End-of-text
n+1	CHECKSUM	

transmitted in lieu of component concentration. In some cases, the analysis report may be generated as a short-form or a full report.

Referring to Table 19.2, note that the format provides for the following data to be transmitted:

Analyzer number	Stream number
Analyzer cycle time	Sample inject time
Component concentration	Validity status codes

This example allows for an analyzer system capable of supporting up to 99 analyzers and 99 streams per analyzer. This, of course, will depend on the particular analyzer computer system capabilities.

The component concentration (or value) field is presented as ASCII representation of floating point numbers, the format of the decimal number being similar to the "E" format used in Fortran. Where continuous analyzers are interfaced to the analyzer system via a continuous analyzer controller (CAC), the analyzer number will be the CAC number; also, the cycle time will be the CAC update time. Bytes 16–25 are repeated for every component of interest up to a maximum of 17 components. The maximum number of bytes that may be transmitted in each message will depend on the input buffer capacity of the PCC.

Validity status codes can be presented in a variety of ways. In the following example, there are four such codes, with a capability of reporting up to nine:

Status Code	Description
0	Calculated component value is valid and acceptable for process control.
1	This is a warning condition. Data are still considered valid, but the validation criteria implies pending failure. Analyzer maintenance should be scheduled.
2	Indicates analyzer is temporarily unavailable due to autocalibration. Only applies to systems with continuous analyzers other than GCs, capable of autocalibration (e.g., infrared).
3	Component value is invalid or unavailable and the PV-tag should be flagged "bad".

In this example, extensive data validation is carried out in the analyzer computer system. The analyzer computer system manufacturer will usually provide a complete list of alarm codes that can be generated by monitoring detector signals, analyzer status sensors (e.g., carrier gas pressure, oven temperature, etc.), concentration limits, and analyzer computer internal di-

TABLE 19.3 Status Report Message Format (Example 1)

Byte #	Code	Description
1	"SOH"	Start of header
2	#	Message ID—status change
3–4	99	Analyzer number (01–99)
5–6	99	Stream number (01–99)
7	9	Status code
8	"ETX"	End-of-text
9	CHECKSUM	

agnostics. However, data validation is often carried out in the process control computer.

The checksum is computed by the ACS for all messages transmitted to the PCC. This is equal to the least significant seven bits of the arithmetic sum of all characters from "SOH" through "ETX." The PCC computes the checksum correspondingly on all appropriate characters received from the ACS. After transmitting "ETX" and "CHECKSUM," the ACS ceases transmission; the PCC is expected to respond with a single "ACK" if the PCC detected neither a checksum nor any parity errors.

A typical STATUS message is shown in Table 19.3. This type of message is usually transmitted when one of three status conditions occur, represented by the following codes:

Status Code 0. "Off-Line" message is generated whenever:

- Any process GC is taken off-line for maintenance or manual calibration.
- Any continuous analyzer controller is taken off-line for maintenance.
- Any process stream is bypassed or skipped because of sample system maintenance or when the process is shut down.

Note that with this format it is not necessary to have a status change message if one continuous analyzer is taken off-line; the unavailability of the analyzer will be indicated by Validity Status Code 2 in the ANALYSIS REPORT message (see Table 19.2).

Status Code 1. "On-Line" message is generated:

- Whenever any process GC or continuous analyzer controller resumes the normal reporting mode to the PCC subsequent to a Status Code 0 condition.
- Whenever any process stream is reactivated subsequent to being bypassed or skipped (Code 0 condition).
- After completion of automatic calibration cycles by a process GC subsequent to a Code 2 condition.

TABLE 19.4 Analysis Report Message Format (Example 2)

Byte #	Code	Description
1	"SOH"	Start of header
2	"R"	Message ID
3–5		Analyzer number (001–128)
6–7		Stream number (01–16)
8		Status code
9	">" or "<"	Result in or out of limits
10–19	.99999E 99	Analysis result
n	"ETX"	End-of-text
n+1	CHECKSUM	

Status Code 2. "Temporarily Off-Line" message is generated:

- After any process GC initiates an automatic calibration analysis cycle(s), but not until immediately before the first calibration analysis report is generated.

Referring to Table 19.3, this example provides the following status information:

Analyzer number   Stream number
Status code

Note that if the stream number is reported as 00, the status of the *analyzer* has changed (e.g., a GC).

Another typical message format is shown in Table 19.4. In this example, the analyzer system is capable of supporting up to 128 analyzers and up to 16 streams per analyzer. Status is reported as one of three ASCII codes:

1. No Errors—data are valid.
2. Warning Alarm—data are still valid, but analyzer is approaching the need for maintenance.
3. Fault Alarm—analyzer unable to complete an analysis; previous good data transmitted.

Byte 9 indicates whether the result is in or out of preset limits. The result data is again in floating point ASCII representation, but this time includes 10 characters. Once again, bytes 9–19 are repeated for each component of interest, up to a maximum of 16 components per stream.

19.4.2 Disadvantages of the ACS/PCC Data Link

The serial data links described have proved to be an effective way of transmitting analyzer data to the process control computer for use in advanced process control schemes; however, they do have some shortcomings.

To illustrate them, take a look at a specific situation, illustrated in Fig. 19.4. In this process control system, regulatory control is exercized by a digital instrumentation system with a CRT-based console as the operator interface to the process; advanced control, using analyzer data, is carried out by the process control computer. The analyzer computer system is connected directly to the PCC via a serial data link such as those described. Conventional field sensors such as flow, pressure, level, and temperature are hard-wired into the instrumentation system and are available for display on the operator's console CRT and available to the PCC via a data hiway. As long as the PCC is running, analyzer data are available for display on the computer CRT, also located on the operator's console. However, if a PCC failure occurs, analyzer data are no longer available to the operator.

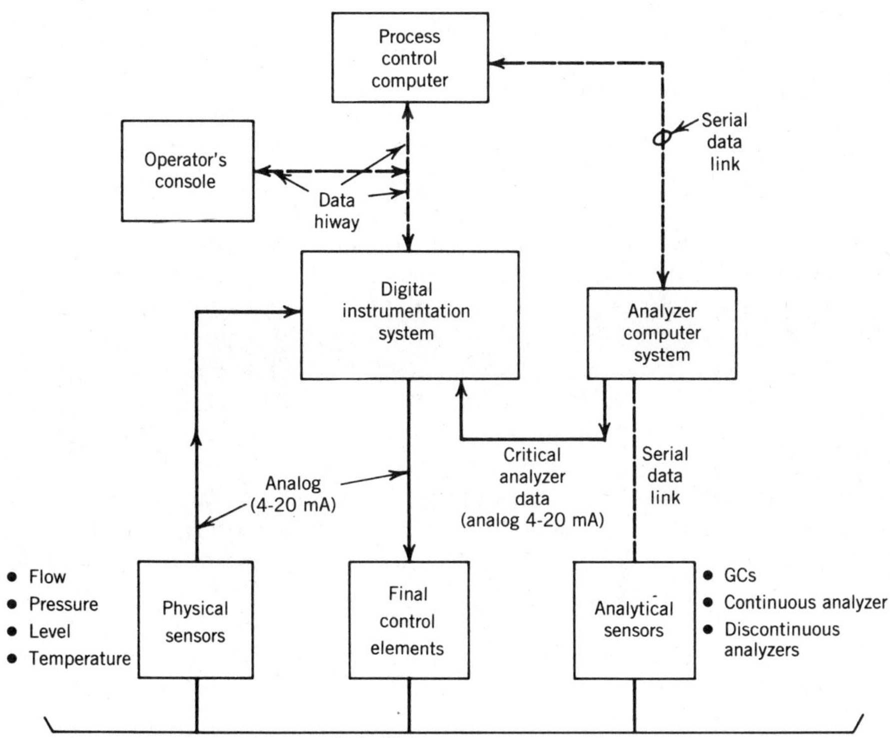

FIGURE 19.4 Typical digital instrumentation system with process control computer.

Unfortunately, this is a time when analyzer data would be most useful to the operator in maintaining the process under control in the basic regulatory mode. Therefore, it is essential to provide an alternative path for analyzer data to reach the instrumentation system. This is usually done by providing analog output (4–20 mA dc) facilities at the analyzer computer system for critical analyzer data and connecting them directly to the digital instrumentation system via hard-wire cables. There are two problems with this approach:

1. An agreement has to be reached with process operations as to what analyzer data are considered critical. This can often be a difficult decision to make.
2. One has to seriously consider the cost of purchasing and installing the necessary cables.

19.4.3 Other Types of Data Links

The process analyzer computer system may be looked on as a peripheral system from the control system point of view. There are many other types of peripheral systems used in process control, examples of which are tank gaging systems, digital temperature indication systems (DTI's) and, in particular, programmable logic controllers (PCs, not to be confused with the Personal Computer). The PC has been used for some time in industry for relay logic replacement (primarily sequence control), emergency shutdown systems, and so on. Traditionally, these have been stand-alone devices, with no interface as such to the process control system.

Recently, process control system manufacturers such as Honeywell, Foxboro, Beckman, and others have been looking seriously at ways of interfacing such peripheral devices with their digital instrumentation systems. In view of its very widespread use in industry, the PC was the first such device to be considered. Devices are now available that allow the PC to be interfaced to a digital instrumentation system; examples are the Honeywell Data Hiway Port (DHP) and the Foxboro Foxnet Device Gateway (FDG).

The data link between these devices is entirely different from that described earlier. First, the link is not interrupt-based but scan-based; second, the data link protocol is not ASCII-based but PC-based (the data base is composed of "coils" and "registers" as in a PC). From the control system end the peripheral system will look exactly like a PC.

There is much interest now in trying to interface process analyzer computer systems with these scan-driven devices. The major advantage, of course, is that it would no longer be necessary to provide hard-wired inputs to the digital instrumentation system; the data would already be available directly via a digital data link. In addition, there would be no need to provide a data link to the process control computer, since analyzer data would be available to the PCC via the instrument system data hiway.

Therefore, if the PCC should fail, all analyzer data would still be available to the operator, although it would not be used for advanced control. The major problem with this approach is that analyzer computer system manufacturers will have to reorganize the analyzer computer system database to look exactly like that of a PC.

Most work in this area is in the preliminary stage at present, so it is too early to say what direction will be taken. Nevertheless, it is anticipated that there will be considerable progress in the next few years as process computer control systems themselves develop. We have recently seen one such development enter the market place; Honeywell's TDC 3000 system, which was first exhibited at the ISA Show in Houston in October 1984.

BIBLIOGRAPHY

Baker, C. W. Design and architectural considerations for a distributed analyzer and control system. *Proc. ISA Symp.*, Boston, MA (May 1985).

Roberts, W. C. A unique microcomputer-based controller for managing and supervising continuous data from on-line sensors. *Proc. ISA Symp.*, Boston, MA (May 1985).

20

Sample Handling Systems

20.1 INTRODUCTION

Process analyzers are installed and operated to provide up-to-date and reliable information on critical process stream properties so that the processes involved can be controlled at optimum efficiency to produce on-grade products meeting the required specifications. From more than twenty-five years of experience in analyzer installations, operation of the analyzer itself presents few problems. The main problem lies in obtaining a clean sample, continuously representative of the process stream being analyzed, and ensuring that the sample is in a condition fully compatible with analyzer operation.

An analyzer that is properly installed with a well-designed sample handling system will usually give economical and trouble-free service, whereas a badly designed system will produce long-term maintenance and operating problems, resulting in a loss of confidence in the application by the user. This can have a serious effect on confidence in analyzer installations generally. As is true in many other fields, reputations are built up over long periods but are lost very quickly. This is as true today as it was twenty-five years ago. It has been reported from many sources in the process industries that of the problems occurring with analyzer installations, over 80% are related to the sample handling system. In most of these cases, the problem may well have been avoided if due attention had been paid to the design of the sample handling system.

Because of their inherent complexity in the analytical sense, analyzer installations are usually costly. Successful application is a product of painstaking attention to detail and, above all, technical expertise in recognizing potential problem areas and therefore a broad knowledge of the field. It is safe to say that there is no single solution to the sample handling problem. Each application must be considered on its own merits. Many points must be taken into consideration and each application studied individually before a satisfactory system can be developed. However, this is not to say that there are no fundamental principles on which to base sample system design. It is the intention of this chapter to review some of the basic guidelines, take a look at the functions of sample system components, and understand what information is necessary to design and build efficient and reliable sample handling systems.

The sample handling system is made up of three component parts—the sample tap and return point, the sample transport system (or fast loop), and the sample conditioning system. Where there is insufficient pressure to return analyzer spent sample to the process, a sample recovery system may be required. This should also be considered as part of the sample handling system. Individual attention must be given to every component in order to design an efficient and reliable system.

20.2 DESIGN OBJECTIVES

The basic requirements of a sample handling system are to extract a sample of the process stream and deliver it to the process analyzer in such a way that:

1. The sample measured by the analyzer is *representative* of the process stream with respect to the measured variable(s).
2. The sample is presented to the analyzer in a state and condition *compatible* with the measurement technique used by the analyzer.
3. An *acceptable* time delay occurs between a change in process conditions affecting the measured variable(s) and the response of the analyzer to that change.
4. The sample system must be *highly reliable,* with a minimum of maintenance attention required.
5. The *total cost* of the system is kept to a minimum, subject to requirements 1 through 4.
6. The system is *safe*.

20.2.1 Representative Sampling

The sample reaching the analyzer should be identical in composition to the process stream or differ in composition in a known and predictable way. Known and acceptable differences may be (1) the removal of certain components from the process stream by filtration, condensation, and so on, (2) as a result of dilution, dosing, or other planned sample treatment. Provided the effect of these treatments has been carefully evaluated, the analysis will still be valid for the purpose desired.

Unrepresentative sampling may be due, for example, to a poor choice of sample point, bad sampling at the probe, or to inadvertent changes in the sample before it reaches the analyzer. Unfortunately, it is rather difficult to check whether a sample is truly representative, since the true process composition is unknown (except for infrequent lab testing). Therefore, there is a very strong reliance on *design*.

The principles of good design are well known, but are often ignored in practice in the belief that "it's really not important." As a result, many analyzers are producing, reliable and repeatable results bearing little relationship to what is really happening in the process. Obtaining good representation is vitally important. Without this, one may as well take the analyzer out of service before confidence is lost.

20.2.2 Compatibility

For ideally representative sampling, the sample would be presented to the analyzer in exactly the same condition as it exists in the process. Most analyzers, however, require that the sample be modified in some way to make it compatible with the technique of measurement (e.g., cooling, pressure reduction, and condensate removal). Any such change in the sample to meet compatibility requirements must be carefully engineered to ensure that no additional problems are created. For example, condensation of water may increase the measured concentration (by removal of part of the sample) or may reduce it (by dissolving it in the condensate). Obtaining both representative and compatible samples are the primary objectives of the sampling system.

20.2.3 Response Time

Even with perfectly representative and compatible samples there is a delay before samples from the process reach the analyzer. There may be a further delay before the analyzer responds to any change in the measurement. These measurement lags must be minimized for effective open- or closed-loop control.

Sample transport time depends upon the total volume of pipework and fittings between sample tap and analyzer and on the sample flowrate. These parameters determine the *distance-velocity lag*. Additional delays, known as *first-order lag,* occur because of exponential mixing of sample in sample conditioning devices and pipework. First-order lag can be quite significant, and it may be well worthwhile to eliminate or reduce the size of larger vessels and unpurged volumes where mixing can occur. A good design, often no more expensive than a bad one, provides substantial benefits in performance.

20.2.4 Reliability and Cost

Unfortunately, sample handling systems have earned a poor reputation for reliability over the years. As pointed out earlier, most analyzer failures have been attributed to the sampling system. One principal reason for this is the lack of attention given to the initial design of the system. Much care is exercised in the selection of analyzers, but the sampling system remains an afterthought. One must bear in mind that good sampling system designs are expensive. Don't be surprised to have to pay as much for the sampling system as for the analyzer. However, often the lowest cost job is accepted. The effect of over-zealous cost cutting is to simplify the hardware to such an extent that the system cannot cope with even minor plant disturbances.

Design decisions must often be made on the basis of incomplete data. This is the "art" of sampling system design. Many trade-offs must be made and there is no specific "best" design for a particular application. However, there are some firm design principles that, if followed, will help to avoid known "traps." More important, existing plant sample handling systems may be examined in detail using these principles; perhaps with only minor changes, their performance can be substantially improved.

20.2.5 Safety of Analyzer Installations

When you are considering the safety of analyzer installations, take into account the effect of a potential hazard on equipment, personnel, and the environment.

Ensure that all components of the sampling system are accessible, preferably at ground level. Adequate escape routes should be available to personnel should a hazard arise. Personnel should be protected from coming into contact with toxic or corrosive substances, hot or cold materials. Attention must be paid to insulation of piping.

All equipment must be rated for the pressures and temperatures likely to be encountered. In addition, the equipment must meet the required electrical classification codes and "T" ratings for hot surfaces.

If toxic substances are released to atmosphere, the environmental impact must be known and acceptable from a legal standpoint. The system design should allow for dangerous material to be flushed and pressure to be released before breaking lines.

Above all, people involved with the installation should understand the potential hazards and know what to do in case of an emergency.

20.3 SAMPLE HANDLING SYSTEM TERMINOLOGY

To avoid confusion, the terminology used to describe the various parts of a sample handling system is summarized here:

- **Sample Tap (or Point).** Location of the source of the sample (the process connection).
- **Sample Return Point.** Place where the sample is returned to the process.
- **Fast Loop.** Direct flow path between sample tap and return point.
- **Transport Line.** Tubing (or pipe) with fittings, used to construct the fast loop.
- **Analyzer Take-Off.** Point where the sample leaves the fast loop and flows to the analyzer, via the sample conditioning system. Usually at a self-cleaning filter.
- **Supply Line.** Pipework system from the sample tap to the analyzer take-off point.
- **Sample Supply Pressure (Temperature).** The pressure (temperature) available at the sample tap. Pressure may be reduced or increased from process pressure (e.g., by a regulator or a pump).
- **Return Line.** Pipework system from analyzer take-off to sample return point.
- **Return Pressure (Temperature).** Pressure (temperature) available at the return point. Pressure may be reduced or increased from return point process pressure (e.g., by a sample return pump).
- **Pipe.** Term used to mean either pipe or tube. What is used depends on the process application.
- **Sample Conditioning System.** The part of the sample handling system, downstream of the analyzer take-off point, where the sample is conditioned to make it compatible with the analyzer requirements.

- **Sample Recovery System.** System designed to take spent sample from one or more analyzers and return the sample to the process. Usually required when analyzers operate at low pressure.

20.4 SAMPLING OPERATIONS AND HARDWARE

This section outlines the various operations that may take place in the sample handling system and discusses the physical equipment used to perform these operations. To give the reader an idea of the relative complexity of the sample handling system, the types of equipment that may be required in such systems are listed here:

Sample probes	Rotameters
Pumps	Relief valves
Pressure regulators	Check valves
Coolers	Flow switches
Vaporizers	Pressure gages
Filters	Temperature gages
Heat tracing	Shut-off valves
Line insulation	Needle valves
Coalescers	Absorbers
Knock-out pots	Permeable membranes
Driers	Reactors
Scrubbers	

Analyzer engineers need to know (and maintenance technicians need to understand) when a particular component is required, the capabilities and limitations of the device, and how to design the total system correctly.

20.4.1 The Process Interface

(a) Sample Taps

With relatively clean process streams, a simple sample tap, as shown in Fig. 20.1, may be sufficient. When gases or vapors are sampled, the tap should be located at the top of the process line to help exclude suspended solids. When liquids are sampled, the tap should be located at the side of the process line. This will provide the best protection from solids and from gas bubbles in the process liquid. Never install the tap at the bottom of the line. If you do, you will create a sampling problem before the sample is even taken from the process.

When installing the sample tap, the line should be bored with the same size hole as the bore of the nozzle to be used and the internal edge radiused

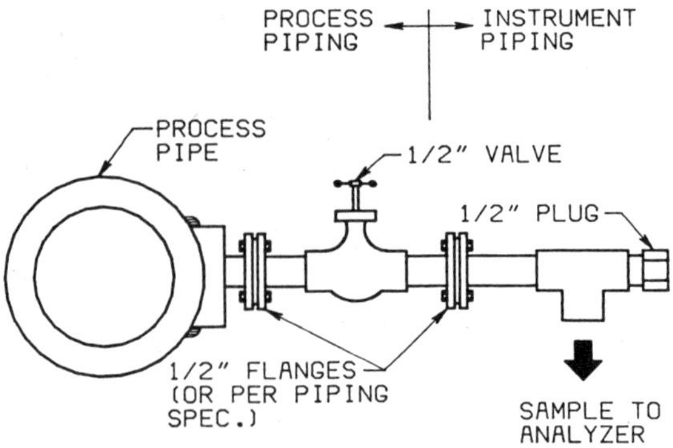

FIGURE 20.1 Typical simple sample tap.

to remove burrs. If the nozzle is to penetrate the pipe wall, it should be set so that its end is in line with the inside pipe wall.

These precautions will help reduce the accumulation of deposits that tend to build up at sharp edges in the sample path and that might eventually block the tap.

The tap must be fitted with a shut-off valve. A good choice would be a gate or ball valve; then, if a sample probe is found to be necessary, it can be inserted through the valve without any difficulty. Following the valve should be a pipe tee with the connection opposite the probe capped off. This will allow the sample tap to be rodded out when necessary.

(b) Sample Probes

The purpose of having a probe rather than just a simple tap is to ensure representative sampling and to assist with sample conditioning. The first advantage of using a probe is that the higher velocity fluid at the center of the pipe may be sampled. This layer is likely to be more representative by virtue of turbulent mixing and will provide more rapid response to process changes. The sample probe should extend into the process line to at least one-third of the pipe internal diameter.

The second advantage is the elimination of some of the suspended solids from the sample, providing some in-situ conditioning. This is accomplished by two mechanisms. First, large particles tend to travel near to the pipe walls and are eliminated by a correctly installed probe. Second, probes, which by virtue of their design, force the sample to turn through 180° at the point of entry, tend to preclude particles that cannot make the turn because of their inertia.

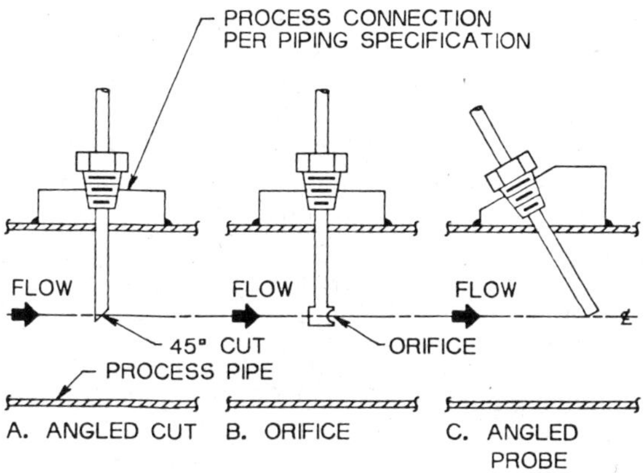

FIGURE 20.2 Momentum-assisted probes.

(c) Open Probes

Various designs of momentum-assisted probes are shown in Fig. 20.2. Clearly these must be installed with the correct orientation to the process flow to be effective. The sample velocity in the probe should be significantly less than the velocity of the process fluid in the pipe. By virtue of the vacuum effect of process fluid streaming across the mouth of the probe, the sample will emerge at a slightly lower pressure than that of the process fluid. A similar probe at the sample return point will enhance the differential pressure available. In really difficult situations this probe can be designed specifically for maximum vacuum effect.

When it is desired to sample at various points across the process pipe or vessel diameter to increase the degree of representative sampling, a probe with multiple entry ports can be used, as shown in Fig. 20.3. The use of such

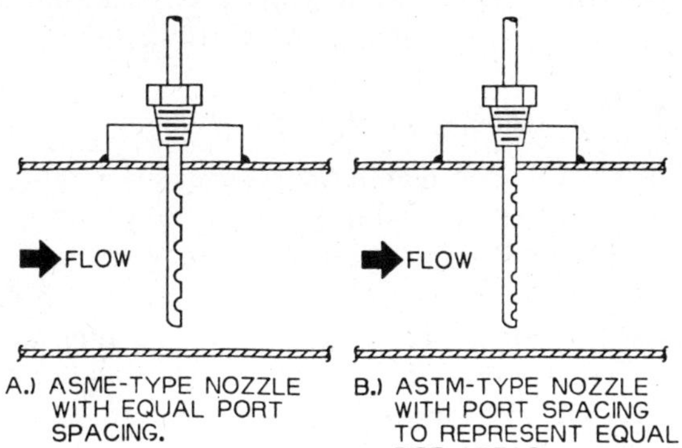

FIGURE 20.3 Multiple-port entry probes.

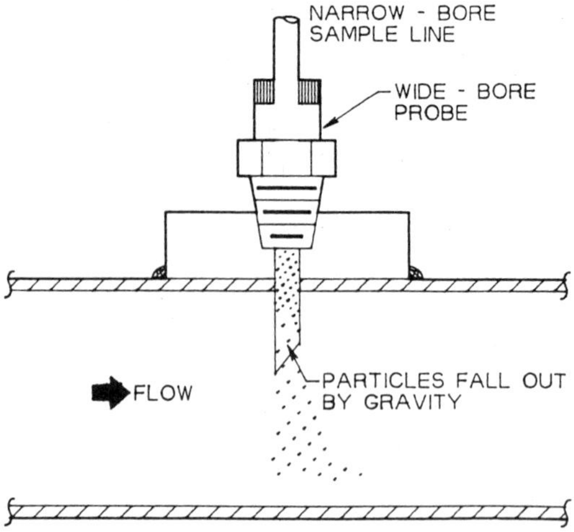

FIGURE 20.4 Stokes' law probe.

a probe immediately raises questions concerning the distribution of process fluid across the pipe diameter. Probes with equally spaced ports will extract samples at regular intervals across the pipe. If samples from equal cross-sectional areas are required, the port spacing can be reduced toward the pipe wall. More complex spacing can be devised to account for area and velocity differences at different points across the diameter, but such complexity is rather academic and generally not of practical value.

The Stokes' law probe shown in Fig. 20.4 uses gravity in addition to momentum to separate solid particles. Since the bore of the probe is large compared with the sample line, the vertical sample velocity is low and allows solid particles to drop back into the process stream. Such a probe must be very carefully designed to be effective. One other problem is the additional response time caused by the probe volume, but this may be acceptable if the solids removal benefits are worthwhile.

(d) Water-Cooled Probes

When the process sample is at high temperature, say 1500–3200°F, a water-cooled probe similar to that shown in Fig. 20.5 may be a suitable solution. For stack sampling, when there is a severe dust loading, a water-washed probe (see Fig. 20.6) will both cool the probe and wash out most of the solids. Water-washed probes are limited by the flashing of wash water to steam at higher temperatures. The graph shown in Fig. 20.6 defines the usable range of the probe in terms of its length. For operating conditions above the curve, a probe combining the features of water washing and water cooling will be necessary.

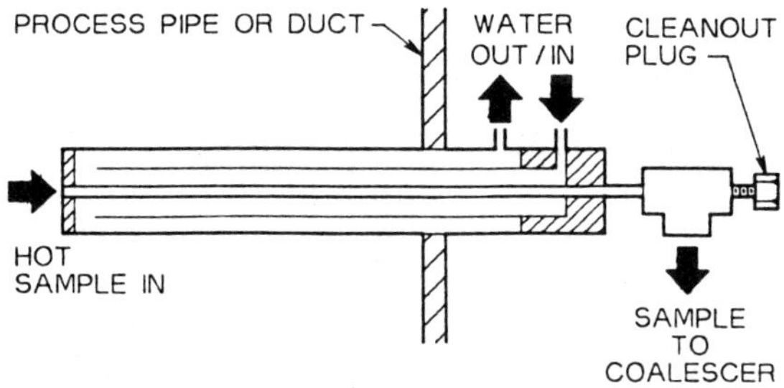

FIGURE 20.5 Water-cooled probe.

A very high water-vapor content in a gaseous process stream can be condensed and dropped back into the process line using the system shown in Fig. 20.7. Some judgement is required here. If the solids content is low, the condensate will also wash solids back into the line and assist in cleaning the sample. However, very high solids content may cause blockage. In this case, it is better to use a heated filter with automatic blowdown, followed by a water condenser.

(e) Ejector Probes

The water ejector probe shown in Fig. 20.8 sucks sample into a pressurized water flow. A water/gas separator is then used to separate water from the

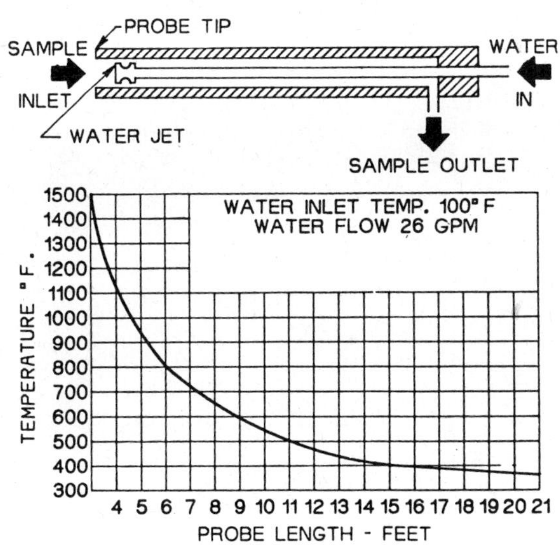

FIGURE 20.6 Water-washed probe.

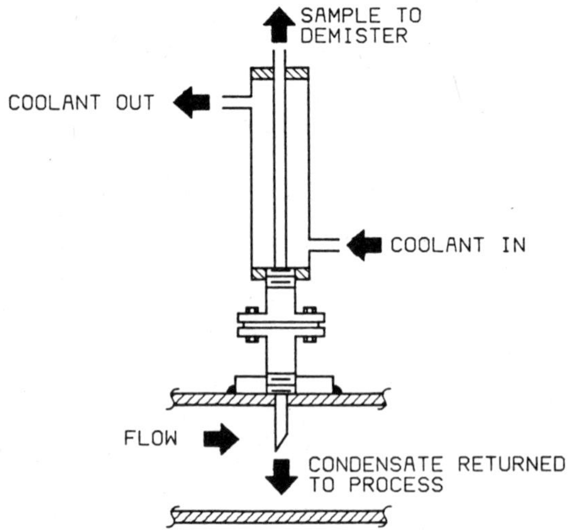

FIGURE 20.7 Condenser probe.

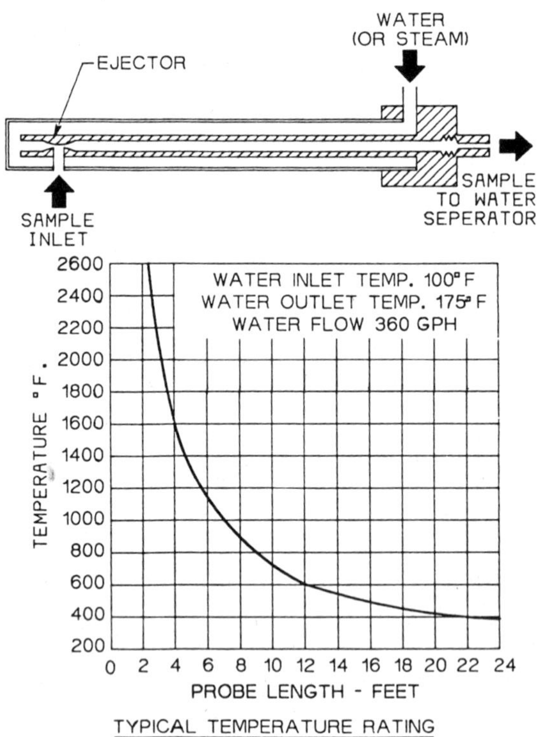

FIGURE 20.8 Water ejector probe.

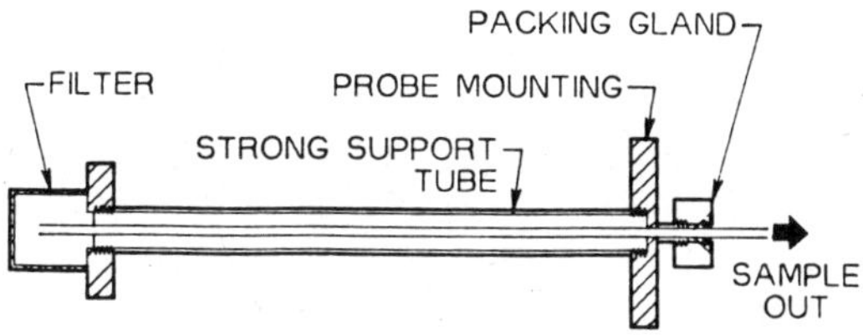

FIGURE 20.9 Filter probe.

gas sample. Most of the solids will remain in the water and be washed to drain. When less cooling (or heating) is required, a steam-ejector probe of similar design may be used. Steam-aspirated air probes are preferred to water ejectors whenever possible. However, they are limited to relatively low levels of dust, since dust and condensing steam tend to form "mud" in the sample system. For all water-washed sample systems, it is important that the sample gases are not significantly soluble in water at the separator temperature.

(f) Filter Probes

For very high dust loads, a filter probe such as the type shown in Fig. 20.9 may be the only solution. These probes are available in an assortment of designs and construction materials. Abrasion can be reduced by using a suitable shield to deflect particles, but it is never completely effective. Filter probes are much bulkier than simple probes. They are generally used with an aspirator sample pump, and in all cases automatic blowback with air, nitrogen, or steam is mandatory. Therefore, the filter probe is usually a rather expensive solution to a problem.

(g) Isokinetic Sampling

Sometimes it is necessary to sample the whole of the two-phase process stream. In this case, isokinetic sampling will be required. The basis of the technique is the collection of sample at the same velocity as the process fluid, thus ensuring that the sample contains both phases in the same proportion as they occur in the process. This is more likely to be achieved by sampling from a vertical pipe, preferably flowing in a downward direction. Under these conditions, gravity will have a minimal effect on the distribution of the suspended phase.

Two methods for isokinetic sampling are shown in Fig. 20.10. It is important to equalize the velocities; if the velocity of the sample at the probe entrance is different from that of the process stream, momentum effects will produce a non-representative sample.

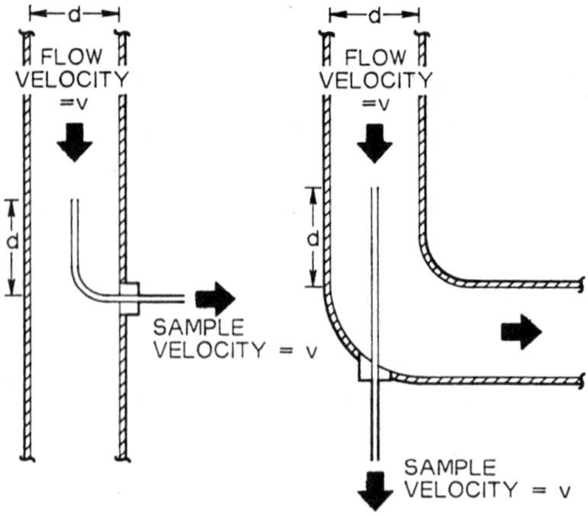

FIGURE 20.10 Isokinetic probes.

(h) Installation of Probes

Probes should be installed through a gate valve. The fittings can often be arranged so that the probe can be withdrawn through the valve and the valve closed before breaking any joints, as shown in Fig. 20.11. The valve should be tagged to prevent someone closing it before the probe has been withdrawn; this may damage both probe and valve. The probe stem should be marked to ensure that the probe nozzle is correctly oriented with respect to the process flow and is at the correct depth in the process line.

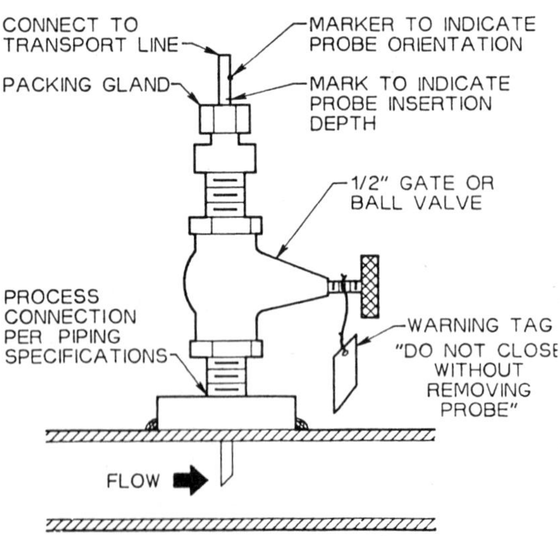

FIGURE 20.11 Typical sample probe installation.

(i) In-Line Sensors

Ideally, the need for a sampling system is eliminated by having the sensor make a direct measurement in the process line. In recent years we have seen this technique used more frequently with success. Examples are the zirconia probe for oxygen measurement, X-ray-type analyzers for density measurement and, more recently, the in-situ IR analyzer for carbon monoxide measurement.

However, making a measurement in-line does not mean neglecting the basic principles of effective sampling. For instances, the zirconia probe measures oxygen at one point in a stack, but it may not necessarily be a point truly representative of the excess oxygen level in the combustion chamber of the furnace.

Another problem with in-situ devices is calibration. Without some very special in-line equipment, most sensors must be removed for calibration. However, this changes the operating conditions and a careful correlation between "calibration" and in-line operation is needed. This is not a severe problem with a simple pH electrode, but might be difficult with an in-line IR analyzer.

In many cases, this attractive concept cannot be used because the process sample must be conditioned to make it compatible with analyzer operating requirements.

20.4.2 Sample Pressure Control

(a) Sample Pumps

Special-purpose sample pumps should be avoided whenever possible. They need a lot of maintenance and contribute to low system reliability, mainly because of their special nature. It is far better to use process differential pressure to drive the sample through the fast loop, since the process pumps are well maintained and sample pressure will always be available when the process is operating. If a sample pump cannot be avoided, many different

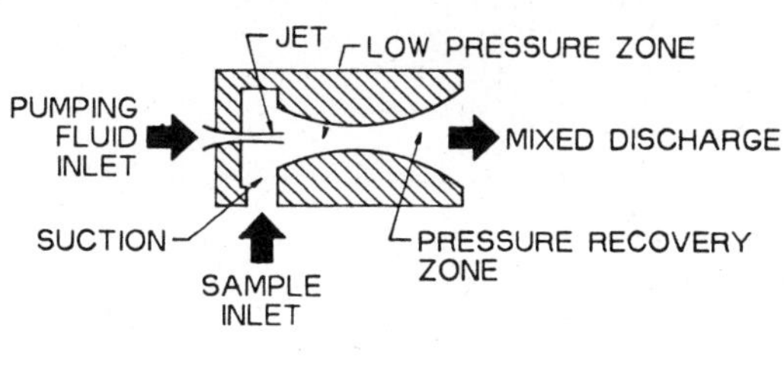

NOTE: EJECTOR PUMP SUCTION IS LOWER PRESSURE THAN EITHER FEED OR DISCHARGE

FIGURE 20.12 Ejector pump: operating principle.

types are available, depending on the application. The following will be helpful in deciding which type to select.

Ejector Pump. This is the pump of choice whenever practical (see Fig. 20.12). It has no moving parts and tends to be inexpensive and reliable. The source of motive power is usually plant air, steam, or occasionally water. Ejector pumps are particularly useful when the process pressure is just

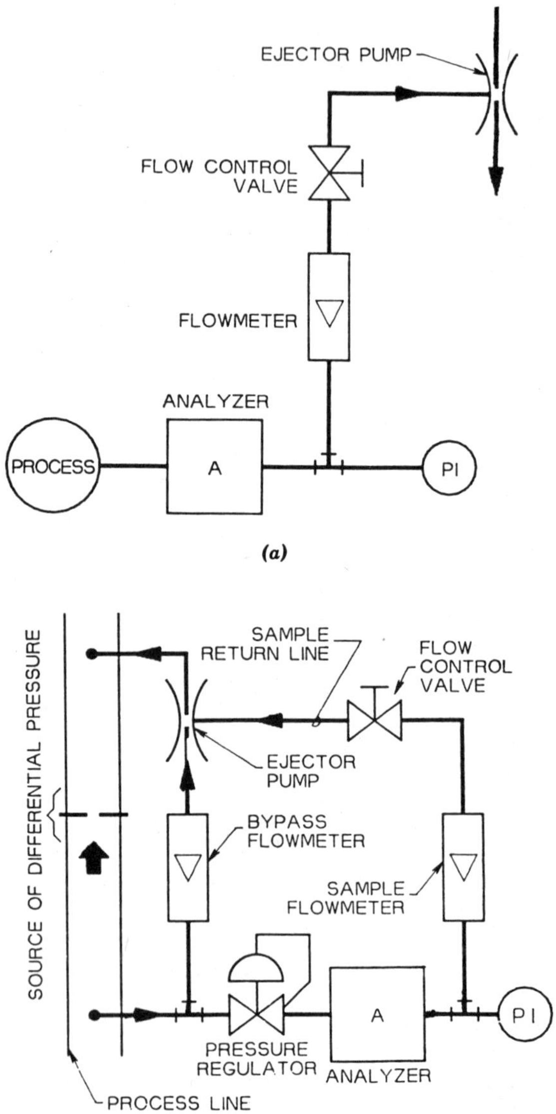

(b)

FIGURE 20.13 Typical ejector pump applications.

above or just below atmospheric. Of course, the analyzer must be capable of operating with suction on the sample vent. By restricting the flow after the analyzer, as shown is Fig. 20.13a, the analyzer will work at process pressure.

Another interesting application of the ejector pump is shown in Fig. 20.13b, where the sample return point pressure is too high for the analyzer but the spent sample has to be returned to the process. Here the ejector pump uses the fast-loop flowrate to generate the driving force to pull sample from the analyzer into the return line.

The biggest problem with ejector pumps is getting them to work efficiently and consistently. They require a considerable flow to provide the necessary driving force.

Centrifugal Pump. This uses a rotating impeller in a circular housing as, shown in Fig. 20.14. The impeller accelerates the fluid toward the discharge port. Normally of simple construction, the only parts that wear are the shaft, seal, and bearings. This type of pump provides good capacity at low pressure. Pumps in liquid service should be primed to ensure a flooded suction. Maximum motor loading occurs at minimum head. As the outlet pressure increases, the flow drops off rapidly and motor loading decreases.

Centrifugal pumps are not affected by small quantities of suspended solids but are not suitable for slurries or viscous liquids.

Gear Pump. The gear pump consists essentially of two meshed spur gears which, as they rotate, trap a small quantity of liquid between the teeth and the housing, transferring it from the low-pressure inlet to the high-pressure discharge port (see Fig. 20.15). They provide higher pressure (up to 100 psig) and lower flow than centrifugal pumps. Because of the close tolerances, they are suitable for clean, non-abrasive liquids only, but are able to handle more viscous liquids. Like the centrifugal pump, they provide a smooth, non-pulsating flow.

Reciprocating Pump. There are two main types, diaphragm and piston. The diaphragm pump consists of a flexible diaphragm that moves up and down in a chamber, creating suction and pressure alternatively. As the dia-

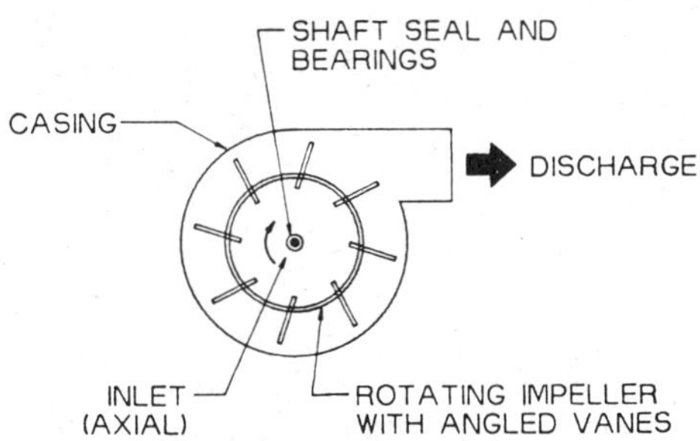

FIGURE 20.14 Centrifugal pump.

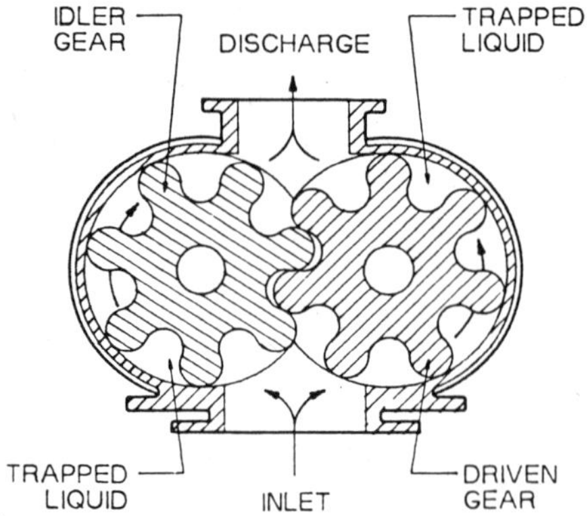

FIGURE 20.15 Gear pump.

phragm moves up, it draws fluid into the chamber via a non-return valve, as shown in Fig. 20.16. As the diaphragm is forced down, the suction valve closes and fluid is forced out of the discharge valve.

The reciprocating pump can handle fluids containing a high percentage of solids such as mud, silt, and sludge. Piston pumps work in the same way and are familiar because of their mechanical similarity to the internal combustion engine. They are normally used when exact quantities of fluid at very high pressure must be watered.

Other Sample Pumps There are other types of pump in which the mechanism traps a volume of fluid and physically transfers it to the discharge. These include flexible impeller, roller, vane, rotary screw, and peristaltic pumps, some of which are shown in Fig. 20.17. In the flexible impeller type, the impeller rotates in an eccentric housing. The volume of the spaces between the vanes decreases as the fluid is transferred from inlet to discharge.

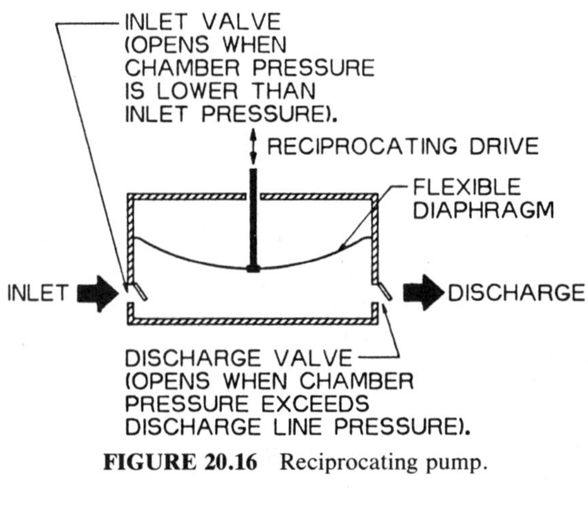

FIGURE 20.16 Reciprocating pump.

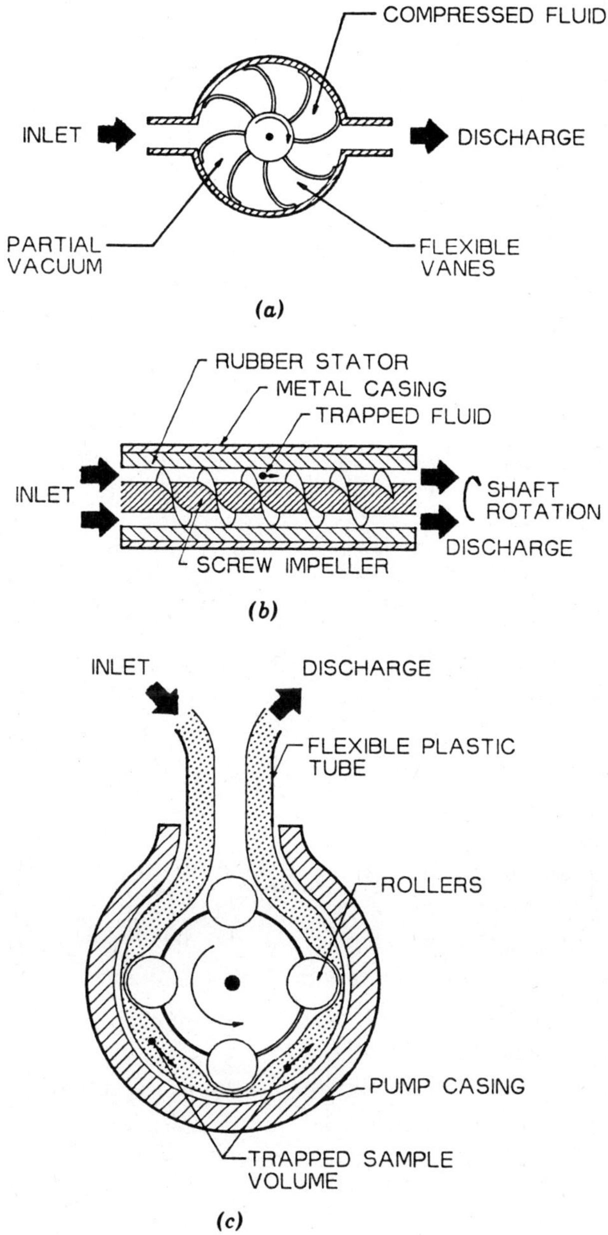

FIGURE 20.17 Other types of sample pump. (*a*) Flexible impeller type; (*b*) screw type; (*c*) peristaltic type.

This compresses the fluid and creates the pumping action. Pressures up to 30 psig can be generated. The roller or vane pumps are similar in construction, but can develop pressures up to 200 psig.

The rotary screw pump has a screw-shaped rotor turning in a flexible stator housing. Cavities between screw and housing transfer the fluid between suction and discharge. Typically, this pump will develop pressures up to 75 psig, and at lower speed can handle slurries or samples containing abrasive solids.

The peristaltic pump works by pinching a flexible tube between a roller and a fixed support. When the roller is moved, it drives the fluid along the tube. This pump suffers from continual tube wear and requires regular maintenance. It is ideal for laboratory-type work as it provides a reliable method of metering a constant flowrate at low volume. It is not suitable for continuous on-line service because of the high maintenance commitment required, although this has improved in recent years. Peristaltic pumps have been used in the past in on-line wet-chemical type analyzers but without great success for this reason.

Selection of Sample Pumps. For those who are unlucky enough to have to specify a sample pump, the procedure outlined below should be followed:

1. Determine the required capacity from the transport line volume and the response time required.
2. Select the materials of construction for the sample-wetted parts of the pump, considering chemical compatibility with the fluid being pumped.
3. Ensure that the materials selected are compatible with the physical condition of the sample, including its temperature and solids content.
4. Remember that high-viscosity samples must be pumped at low rpm.
5. Select the type and size of pump from manufacturer's specifications.
6. From specifications, deduce the motor horsepower required to drive the pump.
7. Determine whether the pump can run at the motor rpm or, if not, what reduction ratio will be required.
8. For direct drive, select the appropriate coupling and shims to align both shafts.
9. For belt drive, select the appropriate pulleys to achieve the reduction in rpm and/or increase in horsepower required.
10. Ensure that the pump is safe. It must meet the classification for the hazardous area in which it is to be installed and have mechanical guards for the belt or coupling drive.

(b) Sample Pressure Regulation

Most analyzers require a constant pressure sample to give repeatable measurements. For gas analyzers, the measurements are usually directly proportional to pressure and very good regulation is required. For liquids, pressure is less critical but must be kept reasonably constant to prevent gassing and to maintain constant flow.

The most common procedure is to analyze the sample at atmospheric pressure, although atmospheric pressure is not constant and may vary by as much as ±3% from the mean. For all practical purposes it is generally accepted as constant. Sample pressure regulation is thus used to control sample flow and the actual sample for analysis is taken at atmospheric pressure. Even if the sample for analysis is pressure controlled, it should be noted that conventional pressure regulators are referenced to atmosphere.

The Lute(or Bubbler). This is the simplest means of pressure regulation; though seldom used, it is effective if sample can be vented to atmosphere. The use of a lute to maintain small positive or negative pressures is illustrated in Fig. 20.18. In both cases, process pressure must be higher than lute pressure.

Diaphragm Regulator. This is the most common device used for sample pressure regulation in sample handling systems. Its function is to maintain the discharge pressure at the required value to counterbalance the spring force on the other side of the diaphragm. Since atmospheric pressure also acts on the diaphragm, the regulated pressure is "gage," not absolute. Absolute pressure regulators have the spring side of the diaphragm in a sealed evacuated enclosure. Differential pressure regulators have a connection to the

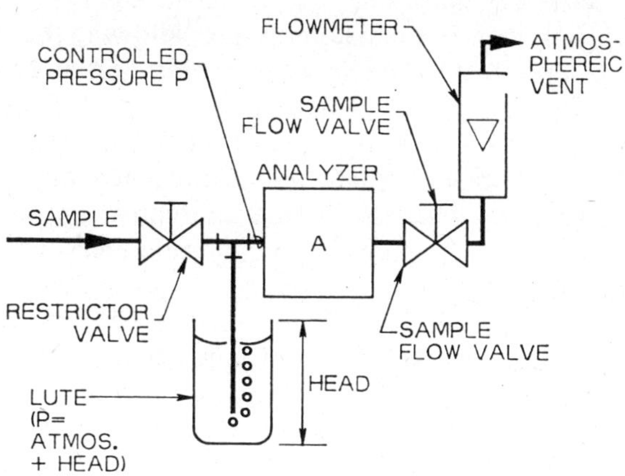

(a)

FIGURE 20.18 Use of the lute for sample pressure control. (*a*) Positive pressure.

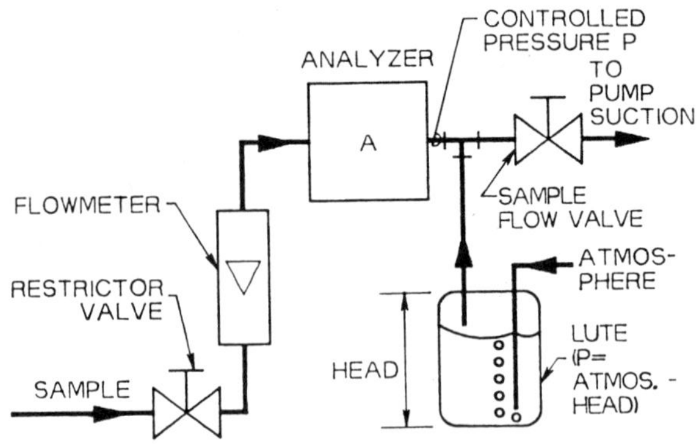

(*b*)

FIGURE 20.18 Use of the lute for sample pressure control. (*b*) Negative pressure.

spring chamber for applying the reference pressure. The spring may be retained to provide a bias or be removed completely.

Regulator valve seats are generally not designed for tight shut-off. When a regulator is closed, a residual flow proportional to the pressure differential across the valve seat is obtained. The "non-bleed" type of regulator is preferred for toxic or corrosive samples. It cannot control output pressure unless the flow is greater than the leak rate of the valve seat. Also, if the flow is shut off downstream of the regulator, pressure will increase on this side unless provision is made to vent the leak flow. The alternative "bleed" type is capable of bleeding the excess downstream pressure to atmosphere. A typical non-bleed regulator is shown in Fig. 20.19.

For high pressure drops, a two-stage regulator is often used. This is essentially two series regulators in a single body. The first stage has a fixed pressure reduction to an intermediate pressure, allowing the second stage to output a constant pressure independent of supply pressure variations.

With all types of diaphragm regulators, it is important to specify a diaphragm that will not adsorb components of the sample(rubber is useless in this respect). Stainless steel is usually suitable for most applications, except for corrosive substances when a suitable alternative must be found. This mistake is often made with regulators for calibration gas cylinders, resulting in inaccurate calibration.

Pressure regulators should never be installed in the fast loop, as the heavy duty will invite early failure. This also requires that the entire fast loop be filtered, which is certainly undesirable from a maintenance standpoint. Pressure reduction at the sample point is sometimes required to satisfy minimum lag time requirements, but should be avoided where possible.

Atmospheric Balancing. This is a technique for adjusting sample pressure to atmospheric and is used mostly for discontinuous analyzers, when variations are expected from the sample disposal system (e.g., flare header) or

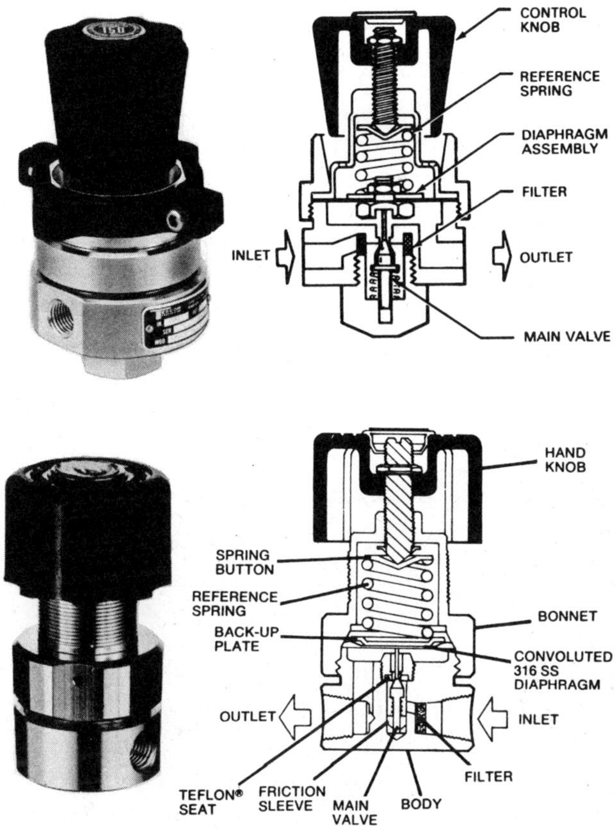

FIGURE 20.19 Typical non-bleed regulator. (Courtesy of Kent Industrial Insts. Ltd.)

because of sample flow variations. It is particularly valuable for samples at near atmospheric pressure drawn through the analyzer by vent suction.

The basic system is shown in Fig. 20.20. Two solenoid valves isolate the sample from the supply and vent disposal systems. The normally closed port of the three-way valve allows the sample to come to atmospheric pressure prior to sample injection. Typically, the sample flow is blocked about 3–8 sec before sample injection and released about 5 sec after. This system is commonly used in process GCs.

Pressure Gauge. A pressure gauge is normally installed downstream of a pressure regulator to enable the regulator to be set correctly. Bear in mind, however, that a pressure gauge such as the bourdon type is an unpurged volume (the stem and the bourdon tube) connected to the sample line. The sample trapped in this volume will slowly diffuse into the fresh sample flowing to the analyzer, causing a slow response to sample quality changes. It is better to put the pressure gauge in the sample bypass, as shown in Fig.

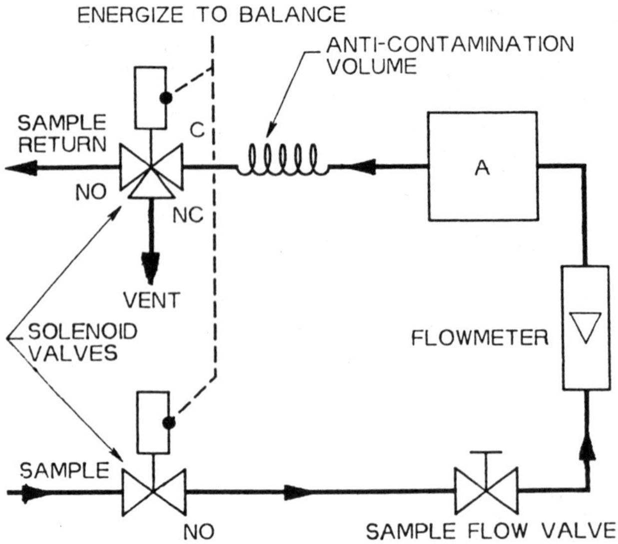

FIGURE 20.20 Principle of atmospheric balancing.

20.21*a* than to locate it as shown in Fig. 20.21*b*, which is a quite common but very poor design. Often, the gauge and flowmeter can be mounted after the analyzer (see Fig. 20.21*c*), which also solves the problem.

Pressure Relief Devices. Spring-loaded relief valves and rupture discs should be installed when process pressure is high and when using glass-tube rotameters. To be effective, these devices must be installed correctly so that they protect the relevant sampling equipment and the analyzer itself if required. It is often thought that because rupture discs and relief valves burst or open at a preset pressure, the line pressure should fall to this value.

Note that a relief device has some flow resistance and will generate a backpressure proportional to the vent flow. If the vent flow is not restricted upstream of the relief valve and the line/components it is protecting, the total pressure will be applied across the device itself, resulting in very high flow and no overpressure protection for sample system components or analyzer.

The principles are illustrated in Fig. 20.22. Failure of the regulator in Fig. 20.22*a* will apply 500 psig to the relief valve. When the valve opens, a high flow is vented, but the only restrictions are the regulator and relief valve. The pressure in the protected line will therefore stabilize at an intermediate value, say 400 psig, depending on the relative restriction of the failed regulator and the open relief valve. Adding an extra restriction, negligible at normal flow but high compared with the open relief valve, as shown in Fig. 20.22*b*, ensures that the pressure is reduced in the protected line and in the vent flow under relief conditions. The resistance can be a filter or narrow-bore sample tubing, so it accomplishes two purposes.

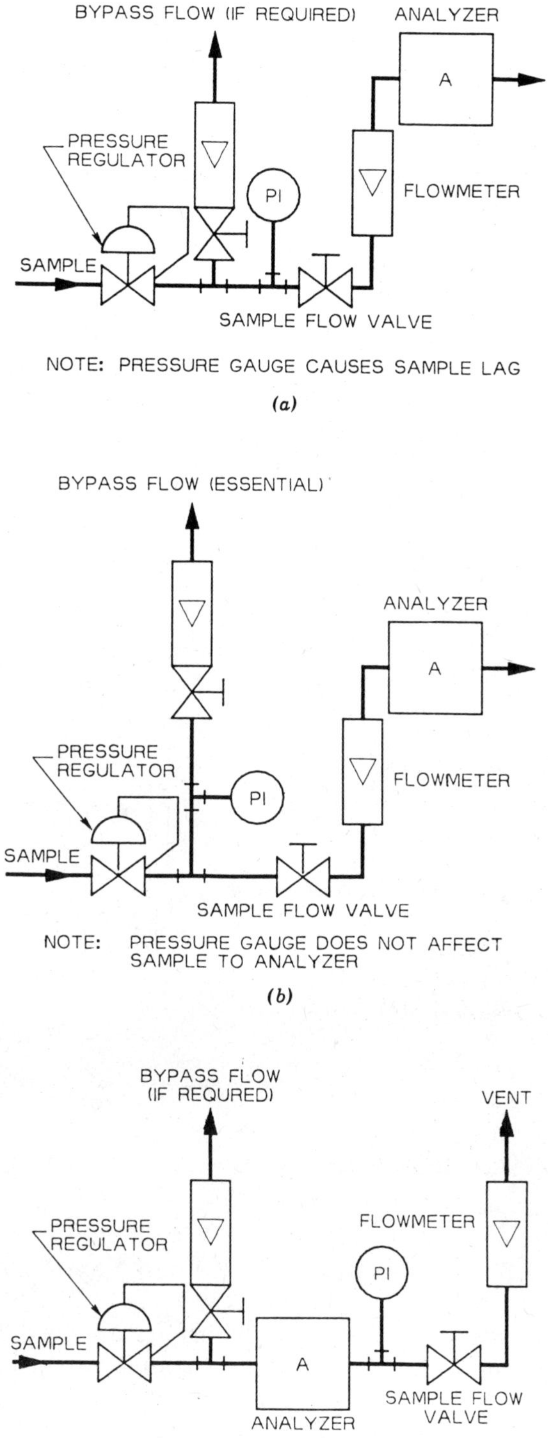

FIGURE 20.21 Pressure gage installation. (*a*) poor design; (*b*) pressure gage in bypass line; (*c*) pressure gage after analyzer.

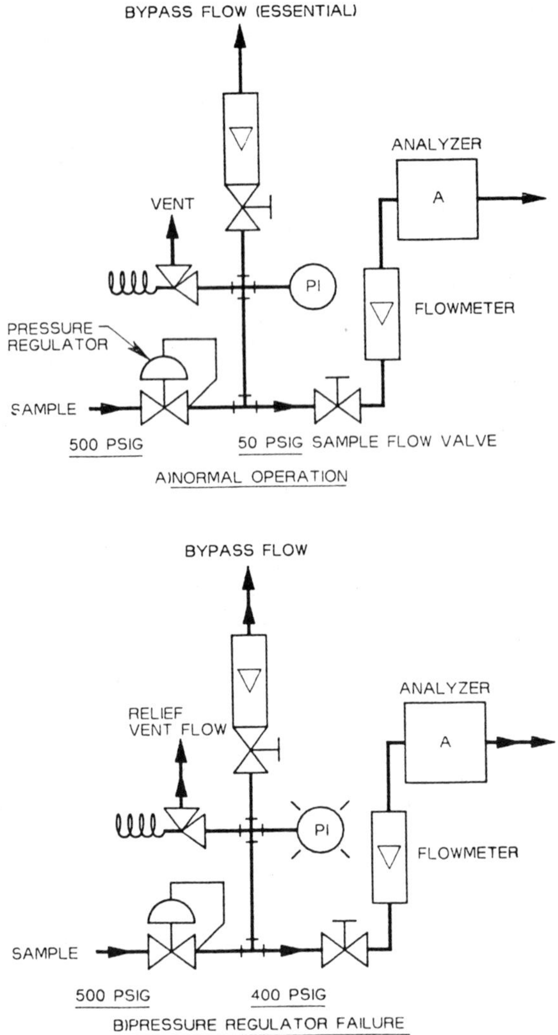

(a)

FIGURE 20.22 Relief valve installation. (*a*) Poor design.

20.4.3 Temperature Control and Vaporization

Three important sampling system operations are considered here, namely, sample cooling, sample heating, and liquid sample vaporization.

(a) Sample Cooling

Small-bore sample lines lose heat so rapidly that a gas sample may be cooled sufficiently without any special equipment. However, it may be necessary to

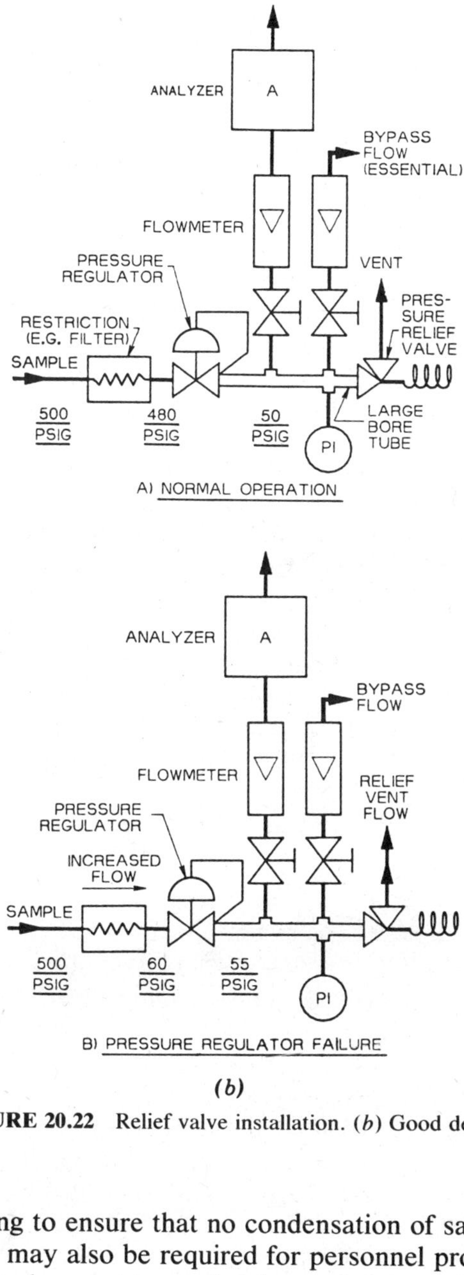

(b)

FIGURE 20.22 Relief valve installation. (*b*) Good design.

provide heat tracing to ensure that no condensation of sample components occurs. Insulation may also be required for personnel protection.

When liquid samples must be cooled down close to ambient temperature, a conventional tube-in-tube sample cooler using water or refrigerant coolant is necessary. In most cases, an air cooler may be adequate. This can be a simple coiled tube with a sheet-metal cylindrical safety guard around it.

For cooling to subambient temperatures, a cryogenic exchanger is generally used, but the problem of icing must be considered. Electrical cooling by solid-state Peltier effect elements can be controlled more closely. Also useful is the Hilsch vortex tube, which is a cyclone-like device creating separate streams of hot and cold air from a supply of dry instrument air or nitrogen.

(b) Sample Heating

For most gas samples, heating is applied to sampling equipment to maintain the sample above its condensation temperature (dew point). The two common techniques used are electrical heating tape and steam tracing. Electrical tape heaters are available for use in hazardous areas and have the advantage of being controllable to a desired temperature. Steam tracing is less versatile, although some control is possible by pressure adjustment. Where steam is used as the heating medium, control valves that operate from a temperature-sensing bulb and filled thermal system are available.

Steam tracing relies on the transfer of heat from the steam tube to the sample line, which are bound and insulated together. Commercially available composite tube bundles are designed specifically for this purpose and offer savings in installation time and manpower. For samples condensing at temperatures close to the steam supply temperature, it may be necessary to use a steam-jacketed sample line (tube-in-tube construction); this is done infrequently because of the expense.

When using steam tracing, the system should be fitted with a trap and fresh steam inlet at intervals of about 100 ft. The sample heating and cooling requirements will vary from site to site depending on local ambient conditions; therefore, each application must be considered individually.

(c) Vaporization of Liquid Samples

To vaporize a liquid,

1. Reduce the pressure, or increase the temperature, to provide conditions under which the sample will exist as a vapor.
2. Supply heat energy at a sufficient rate to provide the latent heat of vaporization of the sample liquid.

To obtain representative samples, *all* of the liquid must be converted to vapor. With relatively pure liquids this is usually not a problem and a heated pressure drop may suffice (e.g., a steam-jacketed capillary tube or pressure regulator). The device must be heated to a temperature high enough to maintain the sample in the vapor state at the exit pressure. In practice, a significantly higher temperature will be needed to provide the latent heat of vaporization.

For samples containing significant levels of impurities, this method will not work because these samples do not vaporize at a single temperature but have a boiling range. In a simple vaporizer, the lighter components flash off as the sample approaches the heated region; often, the large quantity of vapor produced drives the liquid back down the feed tube to a cooler region. Here the vapor condenses, allowing liquid to once more surge into the heated region. This produces an oscillatory effect and results in unrepresentative sampling. It will occur even in expensive vaporizers if the operating conditions are incorrectly specified.

The secret of effective vaporization is to start with the liquid at a pressure well above its own vapor pressure. This liquid is passed into a vaporizer that is hot, but not hot enough to reach the initial boiling point of the liquid. The liquid then passes through a narrow orifice in the vaporizer that suddenly reduces its pressure, causing complete vaporization. The vaporizer chamber and exit port are maintained at a pressure low enough to prevent condensation. These basic conditions for good vaporization are shown in Fig. 20.23*a*.

In addition to the pressure-temperature relationship, the latent heat of vaporization of the liquid must be supplied. The vaporizer design will usually arrange for the liquid exiting the pressure-reduction stage to spray onto a metal plate close to the heater, thereby allowing good heat flux. In addition, heat is available in the form of liquid superheat. It is therefore desirable to increase the vaporizer temperature above that needed to maintain the vapor phase. As the temperature is increased, the initial boiling point of the liquid becomes the critical factor. In some applications, the sample pressure must actually be increased to prevent premature vaporization of the sample!

When designing a vaporizer application, start with the minimum sample vapor pressure needed for reliable operation of the sample transport system and the analyzer. The vaporizer and sample system must be maintained at a temperature higher than the dew point at the minimum sample vapor pressure. The sample initial boiling point at the process pressure will determine the highest workable vaporizer temperature. The vaporizer temperature should be set somewhere between these extremes.

The flow of liquid must then be optimized. In good vaporizer design, liquid flow is determined solely by the pressure drop across the vaporizing orifice. The flow therefore depends on the differential between the inlet and outlet pressures. Excess vapor is usually vented by a built-in backpressure regulator. The inlet pressure must also be optimized at a level high enough to prevent premature flashing of the liquid, but not high enough to overload the flow capacity of the vaporizer.

One vital feature of vaporizer design, often ignored in practice, is the volume change that accompanies the change of state. For hydrocarbons, the vapor volume at typical sample transport temperatures and pressures is between 100 and 400 times the volume of liquid vaporized. This can give rise to substantial time delays in the sampling system if a bypass liquid flow loop is not used.

The design of the vaporizer inlet is critical here. The vaporizer must *always* be connected by a tube tee screwed into its input. This fitting should be parallel-thread with a bottom sealing gasket, or have a shaped end providing metal-to-metal sealing at the bottom of the hole, as shown in Fig. 20.23*b*.

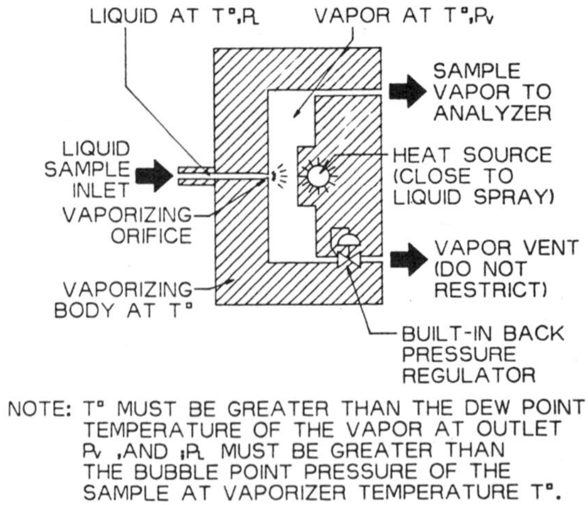

(a)

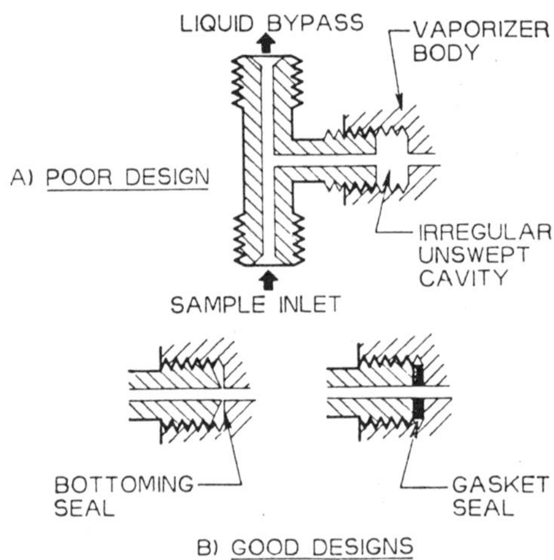

(b)

FIGURE 20.23 Principles of good vaporization.

This feature is almost completely ignored in practice, even by some vaporizer manufacturers, but it must be counted as important when 1 cm^3 may be sufficient to cause a 60-sec time delay! Figure 20.24 shows a typical commercially available vaporizer built into a sampling system.

20.4.4 Sample Flow Control

Sample flowrate is an important parameter in most applications. Although few analyzers are directly affected by flow, most require a constant flow to

FIGURE 20.24 Typical sample vaporizer. (Courtesy of Applied Automation Inc.)

keep the sample pressure constant and to ensure predictable transport delays.

In practice, flow is generally regulated by a combination of pressure regulator and restrictor; flow controllers are used less frequently. There are, however, two important weaknesses in the conventional regulator/needle valve method. First, small needle valves are quite unreliable and tend to shut off the sample flow after a period of time; flow regulators are likely to have improved setpoint stability. Second, it is often desired to limit sample flow to a safe rate in the event of a tube rupture. Flow regulators will maintain the preset rate (within certain limits) irrespective of downstream restriction. The regulator/needle valve setup will satisfy neither requirement.

With automatic calibration of process analyzers becoming more commonplace, it is important to match the flowrate of calibration sample to the normal process sample flowrate. Pressure differences between process and calibration can introduce substantial errors. Pressure differences between process streams may cause similar errors in multi-stream systems. Flow regulation has, in principle, the potential to overcome these problems.

The main disadvantage of flow regulators is their cost. The needle valve is considerably less expensive, so is a major factor in vendor systems. However, even when the regulator is made of stainless steel or other special material, its benefits may outweigh the capital cost. Flow regulators, like needle valves, are susceptible to blockage and should be protected by an upstream filter of about 5–10 microns.

(a) Rotameters

For analyzer sampling systems, the ubiquitous flow measurement device is the "rotameter" or variable area flowmeter, a typical example of which is shown in Fig. 20.25. This device consists of a ball or float mounted in a tapered tube. The float is often shaped to induce spin, thus reducing adhesion to the tube walls. Tubes are usually made of glass (sometimes armored), but metal tubes with magnetic followers are common in industrial use.

Rotameters perform rather badly in practice. The float will often end up at the top or bottom of the tube, either because of inadequate flow control, as discussed earlier, or because of liquid condensation in the tube or accumulation of solid material. When these problems occur, they should be taken seriously, since the rotameter tube is a window into the internal passages of the sample system.

For instance, in gas service, liquid or solid particles may indicate the same situation throughout the analytical system, a clear sign that the sample system is inadequate and that the analyzer is being subjected to contamination that may eventually cause it to fail. Conversely, a clean tube and freely moving float indicate a healthy and reliable system.

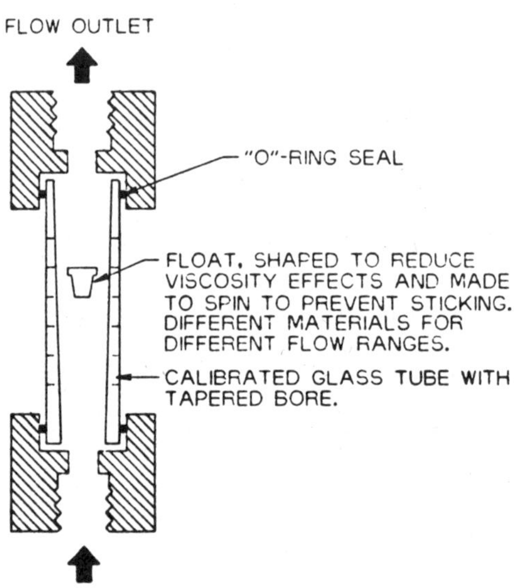

(a)

FIGURE 20.25 Rotameters for sample flow measurement. (*a*) Typical design; (*b*) rotameters installed in a sample handling system (Courtesy of The Foxboro Company.)

Rotameter readings are proportional to the square root of the fluid density and are also affected by fluid viscosity. Calibration therefore varies with sample composition, temperature, and pressure. Normally, this is not a problem, since the device is used to indicate flow rather than to make a precise measurement. If accurate measurement is required, the reading must be corrected for the actual flowing conditions. Viscosity effects are minimized by float design and can usually be ignored.

Most rotameters are calibrated with arbitrary fluids such as air and water at a specific temperature and pressure and the calibration data is marked on the tube. To calculate the actual flow, the observed reading must be corrected for fluid density, according to the following expression:

$$\text{actual flow} = \frac{\text{observed reading} \times \text{calibrated density}}{\text{actual density}}$$

where calibrated density is the density of the fluid used to calibrate the rotameter at the specified temperature and actual density is the density of the process sample at the flowing temperature. For liquids, no pressure correction is required. For gases, actual density may be calculated from the average molecular weight, temperature, and pressure of the process sample.

(b) Flow Indicators

A number of commercially available devices give a visual indication of flow in a system. Two typical examples are shown in Fig. 20.26. The rotary-type unit uses a multi-blade rotor to indicate flow; the other is a bubble sight-glass indicator. Another type uses a hinged flapper raised by the flowing fluid.

(c) The Soap-Film Flowmeter

This device is frequently used to measure carrier gas flowrates in a chromatograph and is probably the most accurate and convenient way to do it. The device is a calibrated tube with a tee at the lower end, coupled to a soap

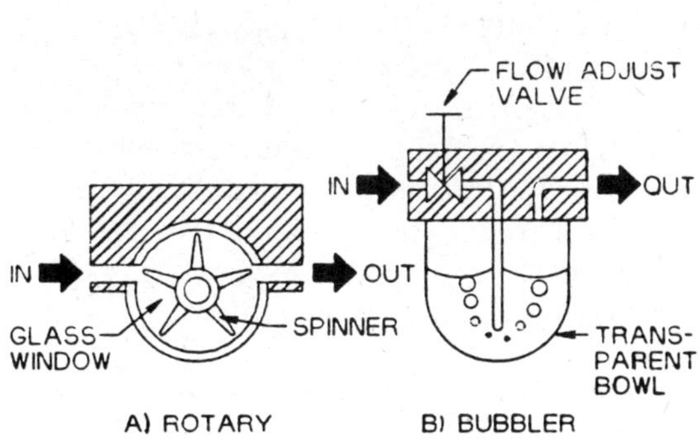

FIGURE 20.26 Other types of flow indicators. (*a*) Rotary type; (*b*) bubbler type.

solution reservoir. Gas enters the side arm of the tee and carries soap bubbles up the tube. By timing the bubbles, quite accurate volumetric flow measurements are possible, except for hydrogen, which tends to diffuse through the bubbles.

(d) Shut-Off Valves

Various types of shut-off valve are used in sampling systems. Ball valves and gate valves provide an unobstructed opening and are usually used at the sample tap. The advantages are that a sample probe can be installed through the open valve and the tap can be rodded out if it becomes blocked. These types of valves are not suitable for adjusting flowrate; they are designed for open/shut service only.

When using valves to switch streams, simple shut-off valves should *not* be used since they cause "dead legs" in the sample flow passage. A three-way ball valve will eliminate this problem. For multi-stream switching it is best to use a double-block-and-bleed system, which avoids cross-contamination. The subject of stream switching will be discussed in more detail later.

20.4.5 Sample Filtration

Most process samples require some filtration, depending on the process and the analytical technique to be used. For example, aqueous samples for electrochemical measurement need only a coarse screening, whereas samples for gas chromatography or mass spectrometry must be filtered down to about 1 micrometer or less.

Strainers and screen filters are rated by their mesh size and finer filters by their pore size or by their actual performance. A single-number rating of, say, 5 microns may be interpreted as an average pore size. When a range of, say, 2–5 microns is given, one can assume that the filter will stop particles of the larger size and let through most particles smaller than the stated range. The best filter specifications will quote actual performance. The retentive efficiency is different for gases than for liquids. Generally, ratings for gases are based on the percentage retention of a particular particle size, whereas for liquids it is easier to quote what size particles are retained.

Bear in mind that a filter is designed for a purpose—that of filtration. A blocked filter is not a problem; it has protected the analyzer from a worse fate. Filters are designed to block eventually. If element replacement becomes too frequent, or too much downtime is caused by the blockage, a better filtration system is required. Operating the system without the filter, or substituting a coarser size, is usually a step in the wrong direction. Temporary relief from filter blockages will most likely be followed by expensive and troublesome analyzer failure.

Filters fall into two categories, devices to catch occasional particles and working devices to remove particles on a continuous basis. Strainers and instrument filters are in the first category and should not be expected to cope with continuous solids removal unless the necessary maintenance time is acceptable. For continuous filtering, some kind of self-cleaning filter should be used.

(a) Filter Media

Various types of filter media and their characteristics are summarized in Table 20.1. It is important to select the right element for the particular duty. Specifying too fine a filter will result in increased expense and excessive maintenance. For liquid samples, filtration to 8 microns will protect pumps and orifices; down to 2 microns will protect optical systems; down to 1 micron will be adequate for a GC sample valve. Gas filters are usually sized for 90% removal of 0.6-micron particles.

Fibrous media provide better filtration for liquid samples. For liquid samples a coarser filter than for gas samples is used, although this does not apply to sintered stainless steel and membrane media. These trap particles at the surface rather than inside the filter medium and work equally well for both liquids and gases. When selecting materials, it should be borne in mind that a

TABLE 20.1 Characteristics of Various Filter Media

	Sintered Stainless	Sintered Ceramic	Glass Microfiber Tubes	Glass Microfiber Paper	Teflon Membrane	Cellulosic
Physical form	Tubular cartridge	Cylinder	Cylinder	Paper disc	Thin disc	Thin disc
Field-handling characteristics	Excellent	Somewhat fragile	Good	Fragile	Fragile	Fragile
Filtration efficiencies (micrometers)						
Gas filtration	5	1	0.1	0.3	0.3	0.2
Liquid filtration	5	0.5	0.3	0.5	0.3	0.2
Separate liquid from gas	Ineffective	Fair–poor	Excellent	Ineffective	Ineffective	Ineffective
Separate liquid from liquid	Ineffective	Varies	Varies	Varies	Ineffective	Ineffective
Separate gas from liquid	Ineffective	Fair	Excellent	Fair	Ineffective	Ineffective
Life (solids capacity)	Fair	Fair	Good	Fair	Poor	Poor
Flowrate per unit area	Fair	Fair	Excellent	Good	Poor	Poor
Maximum temperature	815°C	815°C	540°C	540°C	120°C	93°C
Chemical resistance						
Acids	Good, except HCl, HF	Excellent, except HF	Excellent, except HF	Excellent, except HF	Excellent	Poor
Bases	Excellent	Poor	Fair	Fair	Excellent	Poor
Solvents	Excellent	Excellent	Good	Good	Good	Poor

Source: Strauss, R., "Sample Filters for Analyzers", *Instrumentation Technology*, pp. 53–56 (August, 1977).

filter element has a high surface area and will be more susceptible to corrosion than the solid material. Chemical attack will reduce element life and may contaminate the sample. Seals and gaskets must, of course, be suitable for the proposed duty.

Filter element life will depend on the particulate content of the sample, the pore size, and the type of medium used. Media that block particles mainly at the surface (e.g., sintered stainless steel) have less capacity than a matrix-type element (e.g., glass fiber). In addition, sintered materials are difficult to clean and may be fragile, especially the ceramic variety.

(b) Protective Filters

Strainers. Y-type strainers should always be fitted in the sample supply line at a convenient point for maintenance. The device is a major maintenance item and should be easily accessible. A typical example is shown in Fig. 20.27*a*. For most applications, a screen size of about 100 mesh is suitable.

Instrument Protection Filters. These are available in two varieties, cartridge and in-line. The bowl-type cartridge filter used in instrument air systems has too much internal volume for sample line application, but is extensively used in the bypass mode (see Section 20.4.5c).

For final instrument protection, a small in-line filter like the example in Fig. 20.27*b* is suitable. The element should be replacable without disconnecting fittings. Bypass valving should not be installed; it is better to shut off the sample flow than to risk letting particles flow into the analyzer.

(c) Self-Cleaning Filters

Back-Purged Filter Probes. The use of a filter in or on the probe is best avoided because of the difficulty of access. All installations using filter probes should incorporate some form of backwashing, preferably automatically sequenced, to clean the filter. Filter probes are used when the process sample contains a heavy concentration of solids and it is impractical to

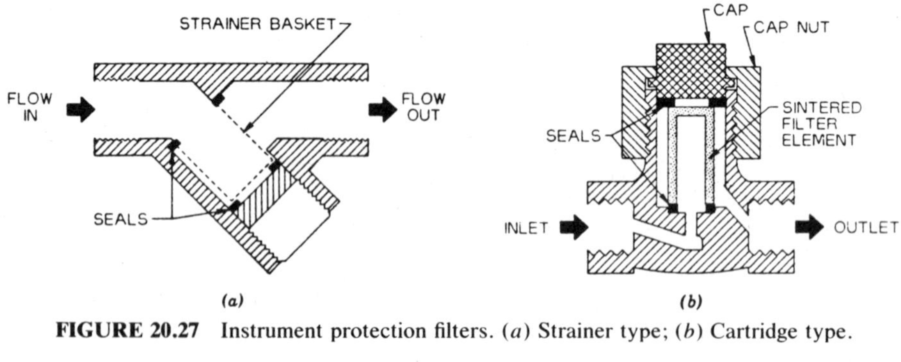

FIGURE 20.27 Instrument protection filters. (*a*) Strainer type; (*b*) Cartridge type.

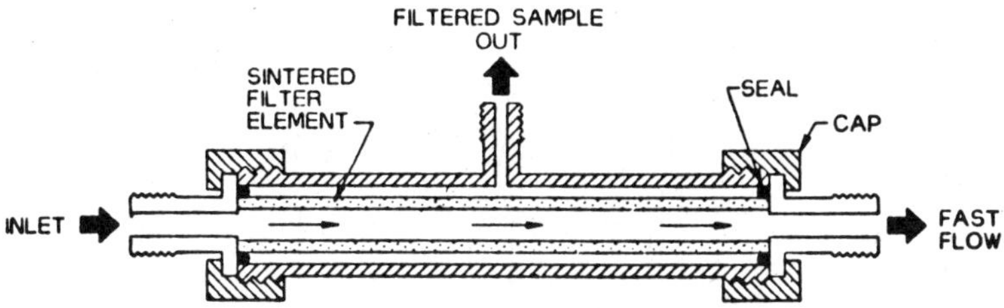

FIGURE 20.28 Cylindrical bypass filter.

withdraw the sample through a pipe. As an alternative, a separate, heated filter mounted close to the probe can be considered. An example of such an application is blast furnace top gas, which is extremely dirty and wet.

One successful sampling method is to use a momentum-selective probe with a steam-heated filter mounted directly on top. Duplicate probe-filter assemblies are automatically switched and cleaned in turn using a high-pressure nitrogen blowdown. However, having the filters accessible for maintenance is much preferred to an in-situ filter probe.

Cylindrical Bypass Filters. This in-line type of filter uses a cylindrical filter element, usually of stainless steel or ceramic, mounted in a cylindrical steel body. A typical example is shown in Fig. 20.28. This type is commonly used in sample fast loops. The major sample flow passes axially through the filter unimpeded. Filtered sample exits at a side connection. The main flow provides a scrubbing action to keep the inside of the tube clean. By ensuring that the major flow is over ten times the filtered sample flow, particles are made to impact the filter element at a shallow angle. This, together with the momentum effect of the fast stream, helps to increase the efficacy of filtration without clogging the element.

There are two problems with this type of filter. First, the annular volume of filtered sample within the housing provides space for exponential mixing to occur, generating a capacitive time delay. If a liquid sample is to be vaporized after filtration, it may be necessary to bypass some of the filtered liquid to achieve an acceptable response time. Second, filter elements are difficult to change and hard to clean; the ceramic elements are also fragile and easily broken. Filters and replacement elements are moderately expensive.

Cartridge Bypass Filters. These have bowl-type design. The bowl may be transparent or made of a solid material, depending on the service. Typically, for sample systems, the body is stainless steel. Two designs are shown in Fig. 20.29. The bowl-type filter has too much volume for use without a bypass, but is frequently used in the bypass mode since it is inexpensive and the element easy to replace. For low particulate loadings, the miniature version is cost-effective and suitable for direct coupling to a vaporizer.

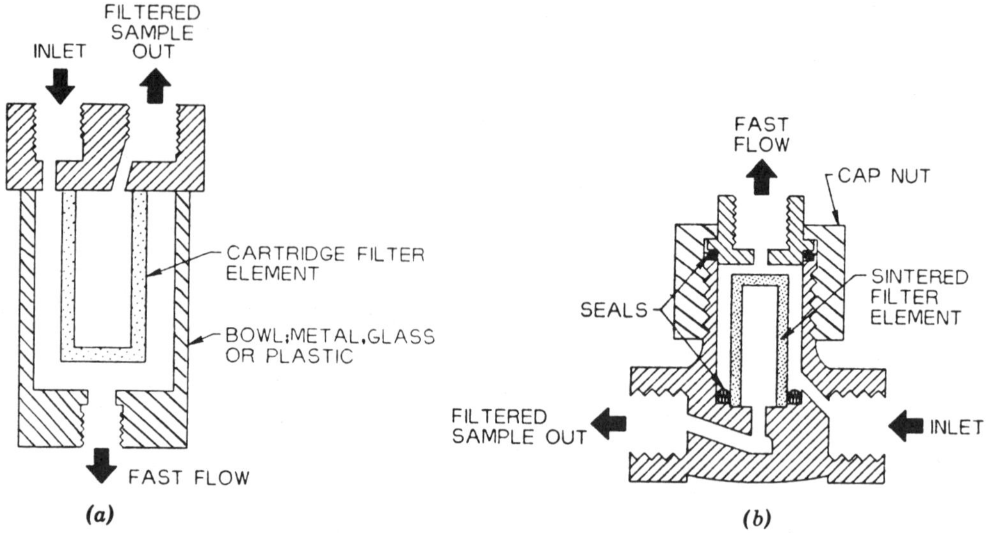

FIGURE 20.29 Cartridge bypass filters.

Swirl-Type Bypass Filter. With this type, the fast loop flow enters the circular filter body tangentially, creating a swirling action. The velocity of the swirling fluid keeps the element clean, and filtration to 0.5 micron is possible using membrane filter elements. The swirling action is created to provide a high velocity across the face of the element, not for inertial separation of solids. However, the filter will also coalesce and remove water droplets in gas and liquid samples. A special hydrophobic membrane element is used for this duty. For filtering liquids prior to vaporization, the single-sided version is used, to allow close coupling between filter and vaporizer. A typical example is shown in Fig. 20.30.

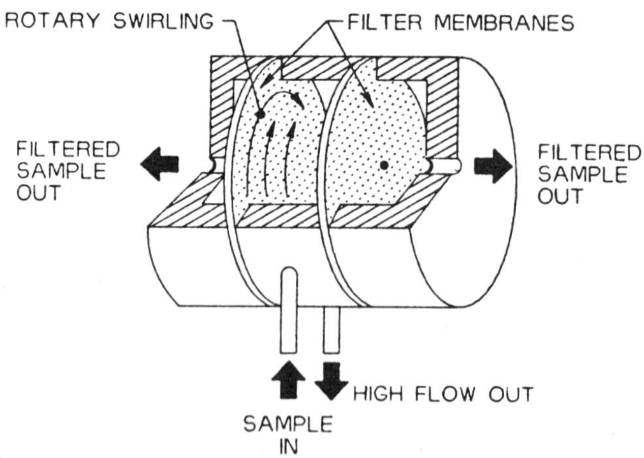

FIGURE 20.30 Swirl-type bypass filter.

Edge-Type Filters. The edge-type filter consists of a stack of steel or stainless steel discs with precision spacers between to generate the filter matrix, as shown in Fig. 20.31. The thickness of the spacer determines the size of particles retained by the filter. Spacing is generally between 90 and 200 microns. For finer filtration, a model using alternate discs of different diameters is available. The clearance relies on the precision dimensions of the discs and is about 40 microns. The filter may be cleaned without disassembly. The cleaner blades are fixed and extend into the openings between the discs. By rotating the disc and spacer assembly one complete revolution, all particles lodged on or between the discs are removed.

For relatively clean samples, the cleaning can be done on a manual, routine basis, the model used having a handle for manual operation. For process streams containing high particle loads, continuous cleaning is possible using a model with a motor to rotate the cartridge at about 2 rev/min.

The flow pattern of an edge-type filter allows a bypass flow to be taken from the bottom (sump) connection. It is advisable to use this configuration to reduce dead time and return the rejected solids to the process. It is essential that the filter be installed so that the flow is in the correct direction. The filter body will be clearly marked with inlet and outlet connections.

If an element becomes plugged, it is inadvisable to use force, since this will result in damage to the precision discs and spacers. Excessive thermal shock can also damage these filters. When used in hot services, care should be taken to ensure that heating and cooling occur slowly and that the motor drive (if provided) is not operated until the temperature has stabilized.

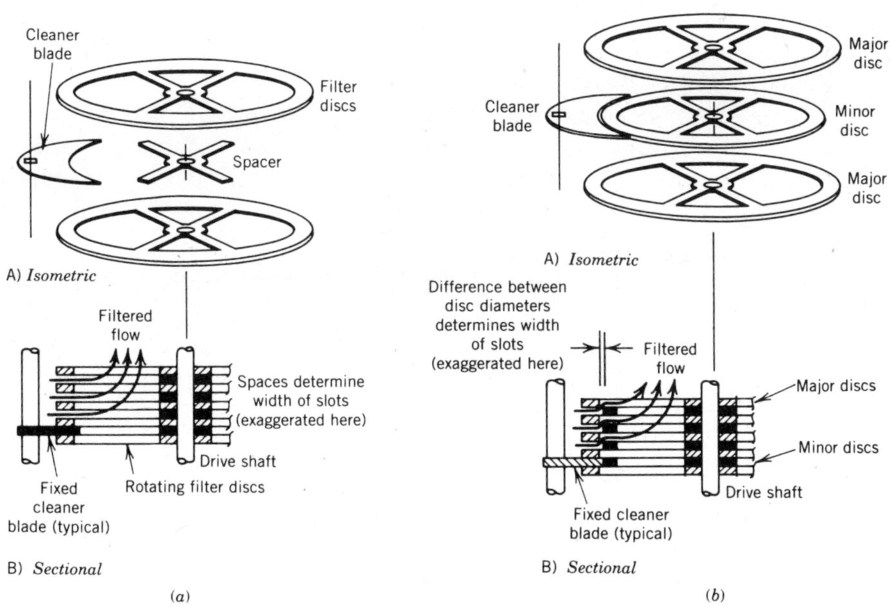

FIGURE 20.31 Edge-type filters. (*a*) Spacer; (*b*) Dual diameter.

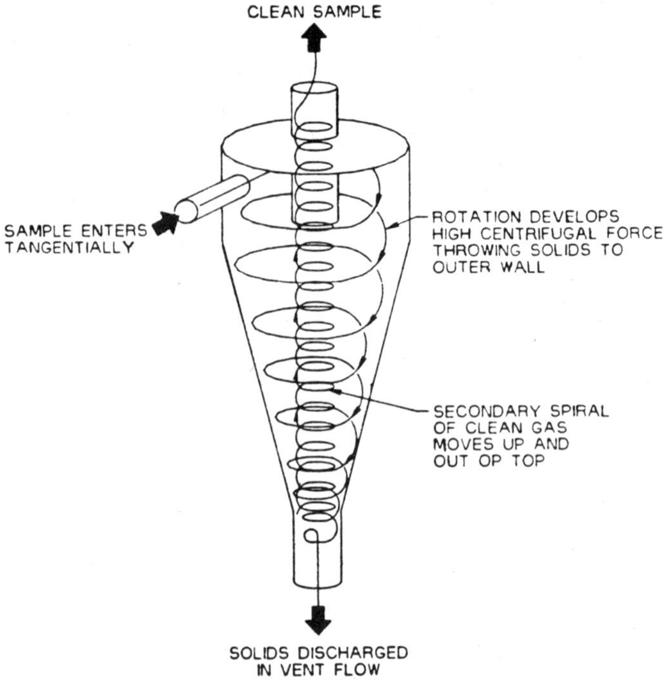

FIGURE 20.32 Miniature cyclone.

(d) Miniature Cyclones

A cyclone is not a media filter, but uses inertia to separate particles from a vapor stream. The cyclone body is cylindrical with a conical bottom section, as shown in Fig. 20.32. Gas enters near the top, tangentially to the diameter. The entry nozzle is designed to produce a high-velocity jet, which causes the gas to spin at a high rate. This spinning action throws particles to the outside, where they fall down the smooth sides and exit at the bottom. A bypass flow is taken from the bottom connection. Filtered sample exits from an axial port protruding from the top and extending to a level below the inlet jet.

Cyclones are very effective for removing the larger-sized particles from medium dust loads and will also coalesce and remove suspended liquid droplets.

20.4.6 Coalescers

The purpose of a coalescer is to force finely divided liquid droplets to conglomerate into drops large enough to separate by gravity. Since some of the sample will be removed by this process, the affect of this operation on the analysis must be evaluated.

The general principle of coalescing is to make the sample fluid take a tortuous path through a porous obstruction. Momentum causes the heavier droplets to impact the matrix material, where successive impacts conglomerate the droplets.

The simplest coalescer is a bundle of glass, stainless steel, or polymer "wire wool." A packed bed of inert granules can be just as effective. Provision must be made for removal of the coalesced liquid; this is often done by arranging the packed bed in a vertical plane. Sample gas flows up through the

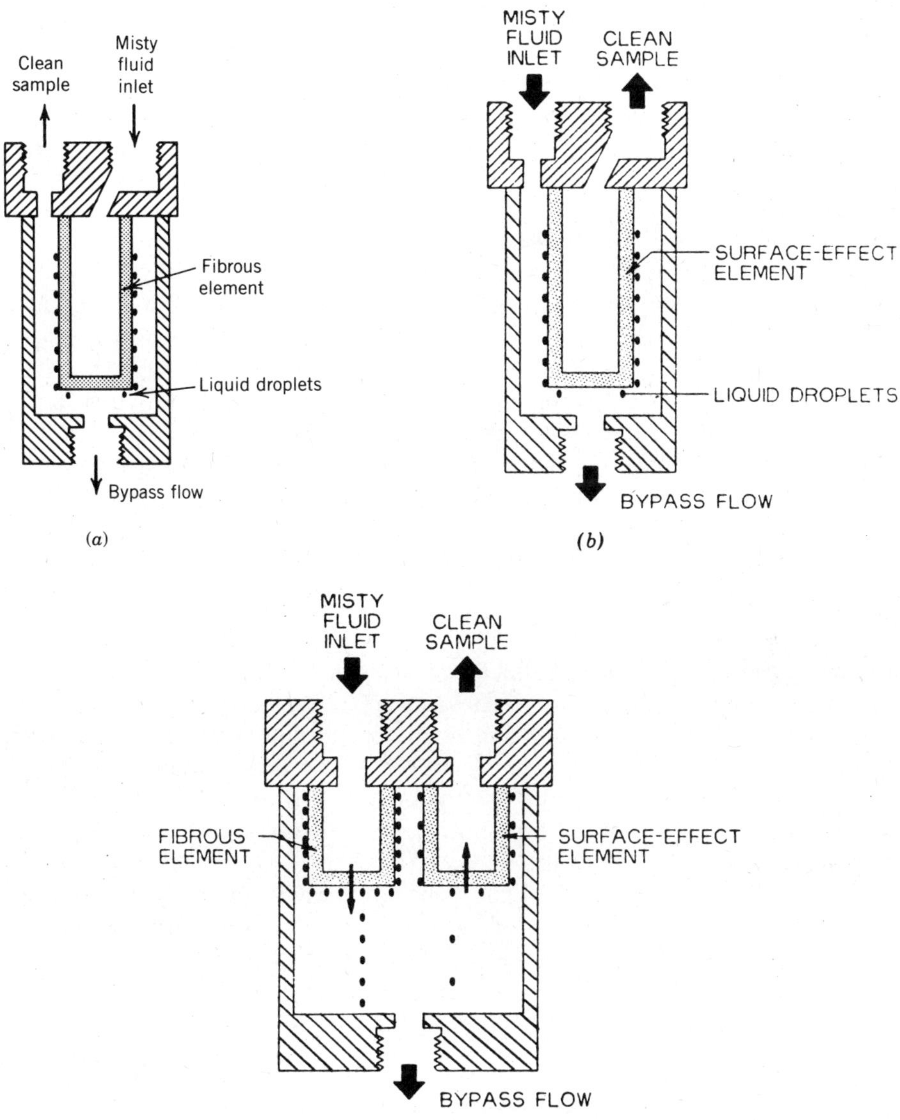

FIGURE 20.33 Typical filter-coalescers. (*a*) Fibrous element type; (*b*) Surface-effect element type; (*c*) Combination type.

bed, countercurrent to the liquid percolating down. This arrangement keeps the bed wet, increasing coalescing efficiency.

For less arduous duties, a filter-coalescer can be used. Essentially, this is a bypass cartridge filter working in reverse to its normal configuration. A typical example is shown in Fig. 20.33*a*. Sample enters inside the cartridge and flows to the outside. Liquid coalesces in the filter matrix and runs down the outside of the element into the bowl, where it is continuously removed as a bypass flow in order to reduce sample system dead time.

The second type of element, used mainly for coalescing liquid emulsions, allows a passage of one phase while blocking the other. Blocking may occur because the diameter of the capillary openings in the medium are too small to admit the molecules or, more commonly, because one liquid wets the surface of the medium and the other does not. Thus, a hydrophobic material such as sintered PTFE will not allow water to pass. With this type of coalescer, the coalesced liquid separates on the upstream side, as shown in Fig. 20.33*b*. A combination of both types of coalescing element, shown in Fig. 20.33*c*, will provide a greater degree of phase separation.

When the process stream is very dirty, the solids should be removed first to avoid blockage of the coalescing membrane. For vapor samples, a heated filter will remove solids and vaporize liquid mist (assuming it is volatile). The clean sample may then be cooled and coalesced. If only small amounts of solids are expected, the coalescer alone will filter out solids and remove liquid mist. Be prepared to replace the element at regular intervals.

To remove trace quantities of aerosols from gases, a filter 99.95% efficient at 0.6 micron should be specified. Finer filters will block more frequently and are less effective at coalescing aerosols.

20.4.7 Knock-Out Pots

Knock-out pots are used to collect liquid separated from vapor samples to protect the analytical equipment downstream. They work by reducing the velocity of the vapor and allowing liquid droplets to separate by gravity. Normally, the liquid is removed by an automatic drain facility. The main problem with knock-out pots is their volume and the effect this has on the response time of the system. For this reason, a knock-out pot should never be placed in a direct line to the analyzer.

The separating function of a knock-out pot in the bypass line is illustrated in Fig. 20.34. Note that the gas-liquid separation occurs at the tee by inertia and gravity. If the knock-out pot is fitted with a manual drain only, a reliable maintenance schedule will be required to ensure regular draining. Unfortunately, one cannot predict when a problem will occur involving liquid carryover, so it is best to provide an automatic drain facility.

20.4.8 Other Sample Conditioning Techniques

The sample handling procedures described so far (with a few exceptions) have been physical in nature and care has been taken not to change the

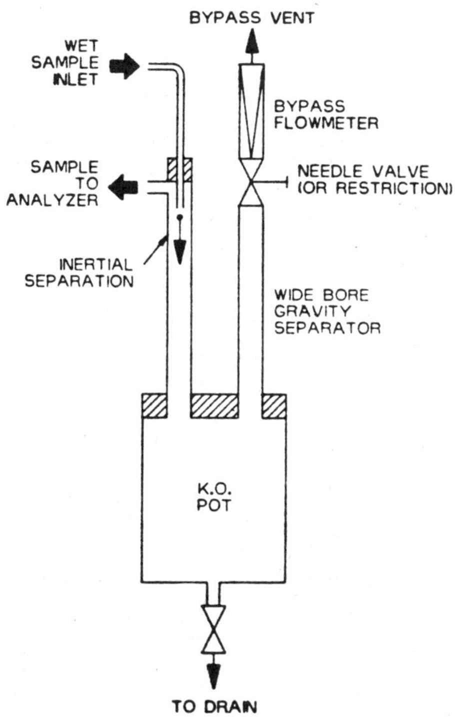

FIGURE 20.34 Typical knock-out pot designed for a bypass system.

concentration of the measured components. In this section, techniques are described that change the sample to make it more amenable to analysis but may also change the measured variable. Thus it is essential to evaluate the effects these methods have on the measurement. It is important that the original stream concentration can be inferred from the measurement result.

These techniques are useful when the sample contains corrosive substances, interfering components are present, or the sample would partially condense, freeze, or vaporize at the analyzer temperature.

Corrosion normally occurs when acidic gases or their solutions are present in the sample together with moisture. Sometimes, thoroughly drying the sample is sufficient to solve this problem. Otherwise the techniques described next may be helpful. Interference occurs when other components in the sample affect the analysis. There are two ways this can happen:

1. The other component generates a response that is indistinguishable from the desired measurement. A good example is the effect of sodium ions on a pH electrode. The electrode is designed to measure hydrogen ion concentration, but responds also to sodium ions in the sample. The ability of an analytical method to resist responding to other components is termed the *selectivity* of the method.

2. The other component may interact with the measured component. An example is the effect of pH on the fluoride electrode. At low pH, some of the fluoride ions combine with the hydrogen ions to form HF molecules, which cannot be detected by the electrode. Although the electrode does not respond to hydrogen ions, their presence prevents the electrode from accurately measuring fluoride ions. This is called *interference*.

All methods of removing fractions from or changing the sample have some effect on the measurement. The simplest way is to increase the percentage of the measured component. Often, an analysis made on this basis is quite adequate. When water is removed, this is called a "dry basis" analysis. More serious is the partial loss of a measured component during treatment. For example, carbon dioxide may dissolve in the condensate when water is removed from the sample.

Most of the methods described use a physical or chemical device that gradually becomes saturated with the component being removed and must be replaced or regenerated on a regular schedule or the reliability of the system will be unacceptable. It is always best to use the simplest method that will provide adequate measurements. For some applications, these more complex conditioning techniques may be the only satisfactory solution to the problem. Each of them has been used to solve a particularly difficult sampling problem.

(a) Condensers

Condensers are used to remove water or heavy hydrocarbons from the process sample. The simplest form is the air-cooled type, suitable for high-temperature samples. Most samples, however will need forced air, water, or refrigerant cooled condensers. A single tube-in-tube construction is normally adequate. Dimensions will depend on the required duty and require careful evaluation.

When condensing heavy hydrocarbons from a complex mixture, a simple condenser may not give sufficient control over the fraction removed. To get a better split, the condenser should be in the form of a packed tower, with the sample entering at the bottom and stripped sample exiting from the top. The packing can be selected from laboratory-type distillation column packings. A typical example is shown in Fig. 20.35. This fractionating sample condenser has been used successfully in furnace effluent sampling and is mounted directly above the process line so that condensate runs back into the line. In effect, it is a small distillation tower, with the process line acting as the reboiler. By careful adjustment of the column temperature and the overhead flowrate, a good cut-off can be obtained at any desired boiling point.

All condensers must have an outlet for the condensate, which may be the process line or a knock-out pot. If a pot is used, it should be fitted with an automatic drain, kept cool, and situated out of the flow path to the analyzer.

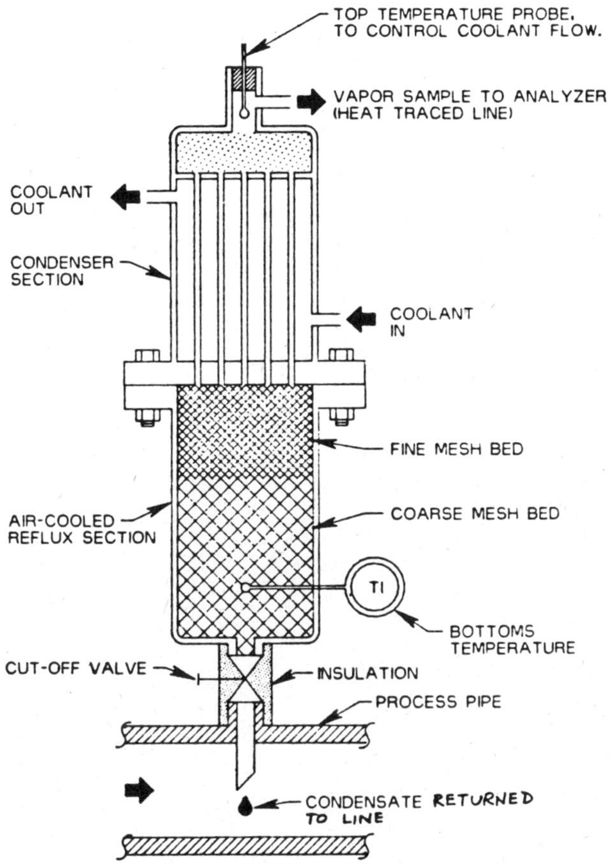

FIGURE 20.35 Fractionating sample condenser.

The sample leaves the condenser saturated with condensate at the operating temperature. To avoid further condensation, the sample must be immediately reheated and maintained in this condition throughout the analytical system. Steam heating is the usual method, using prefabricated tube bundles manufactured specifically for this purpose. If better temperature control is required, electrical heating tape should be used. If very cold refrigeration is used in the condenser, heating the first portion of the sample line to prevent icing may be all that is required.

It is, however, not unusual to experience liquid mist in the sample and occasional carry-over of liquid at times of plant upset or condenser malfunction. Therefore, whenever a condenser is used, it is essential to use a coalescing filter in the sample system downstream of the condenser. The preferred place for this is at the analyzer location so that condensate formed in the transport line may be removed, provided that the transport line slopes downward to the analyzer.

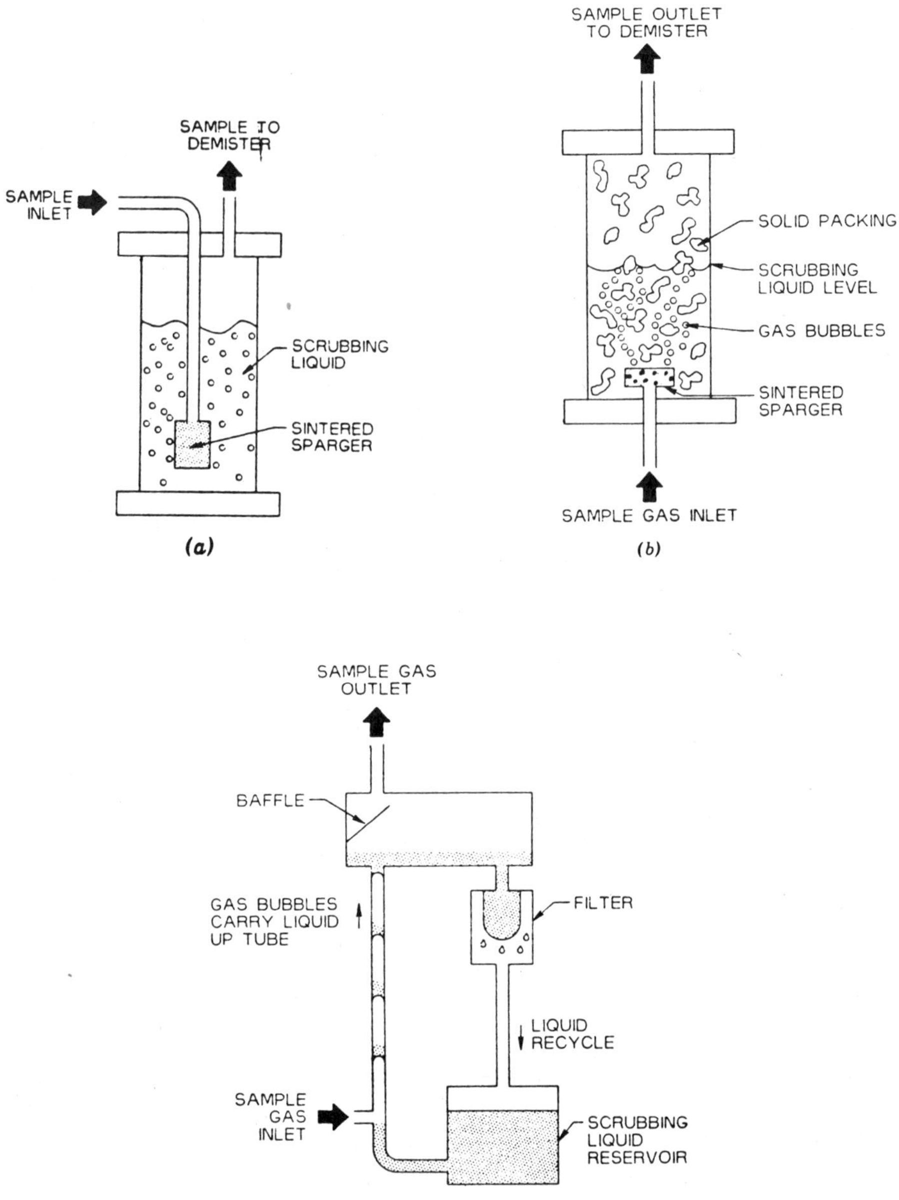

FIGURE 20.36 Typical scrubbers. (*a*) Using a scrubbing liquid; (*b*) using a scrubbing liquid with a solid packing; (*c*) gas-lift type scrubber.

(b) Scrubbers

These are devices that remove unwanted components from gaseous samples by contact with a liquid or solution. The simplest type allows the sample gas to bubble through the chosen liquid. Often, a sparger is used to provide good gas-liquid contact. A typical example is shown in Fig. 20.36*a*. Another type commonly used is the packed-bed type of scrubber shown in Fig. 20.36*b*, where the packing can be rings, beads or some other inert object that provides a large surface area.

The gas-lift scrubber shown in Fig. 20.36*c* is a special design that incorporates a self-cleaning action for the scrubbing liquid. It is particularly valuable when the sample gas contains solid particles or reacts with the scrubbing liquid to form a precipitate. Sample gas enters the gas lift and breaks up into bubbles, each lifting a little liquid ahead of it. As it travels up the lift tube, each bubble is in contact with fresh liquid on the walls and at the bubble surface. At the top of the lift tube, the liquid separates and returns through the filter to be used again. The liquid used for scrubbing may be a pure liquid or a solution. Typical examples are:

- Caustic soda solution—acid gases and carbon dioxide.
- Dry alcohol—moisture.
- Acetic acid—alkaline gases (ammonia).
- Methyldiethanolamine—hydrogen sulfide.
- Alkaline pyrogallol solution—oxygen.

All scrubbers should be followed by a suitable entrainment separator that can handle the worst case, where all of the liquid is blown over. Most scrubbing liquids cause costly damage if they enter the analyzer. Scrubbing requires a low velocity in a relatively large vessel, but good dynamic response requires fast purging of the scrubber head volume and the downstream device. Getting the right balance calls for careful attention. In most cases, the greatest problem results from failing to recognize the time delay in the separator.

(c) Packed-Tube Absorbers

These perform the same function as scrubbers, but are packed with solid materials rather than liquids and are therefore preferred. Construction is simple, as illustrated in Fig. 20.37. The packing is retained by a plug of glass wool, a sintered disc, or a wire screen. The packing itself is a granular material such as molecular sieves, alumina, silica gel, charcoal, caustic soda pellets, or calcium chloride. True *absorbers* rely on chemical reaction between the impurities in the sample gas and the solid material; an example is the reaction between carbon dioxide and caustic soda pellets. Thus, absorbent material gradually deteriorates and must be replaced.

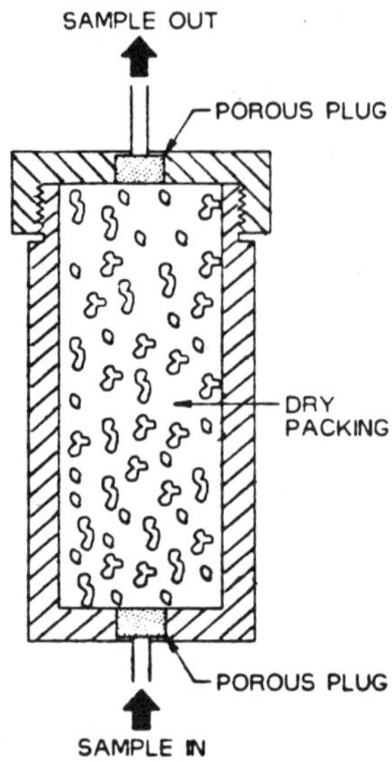

FIGURE 20.37 Typical packed adsorber/absorber.

The other type of mechanism is physical in nature and is preferred. Materials such as silica gel or molecular sieves are *adsorbers,* that is, they trap impurity molecules at the surface. They also deteriorate as they become saturated with trapped impurities, but they can be regenerated by heating while purging with dry nitrogen or helium. When selecting a solid material for a packed-tube adsorber, consider the following:

1. The efficacy of removing the unwanted component.
2. The expected life of the packing material.
3. The effect on other components in the sample.

No. 3 is one of the problems encountered with adsorbers. Many will retain components of interest, perhaps not completely, but enough to delay them, rather like a GC column. For example, a silica gel adsorber may pass hydrogen in 30 sec but retain carbon dioxide for up to 1 hr. The analyzer may show very good response to carbon dioxide but the response is an hour late! If in doubt, test the system by switching between sample gas and pure nitrogen; this will provide a practical measurement of the time lag caused by the adsorber for each component being measured. Finally, consider the effect of other components on the solid packing. If any of these are retained, it may not affect the analysis but it may deactivate the packing material.

(d) Permeable Membranes

These are materials that allow certain molecules to pass through while obstructing larger ones. In the gas phase, the rate of transfer is proportional to

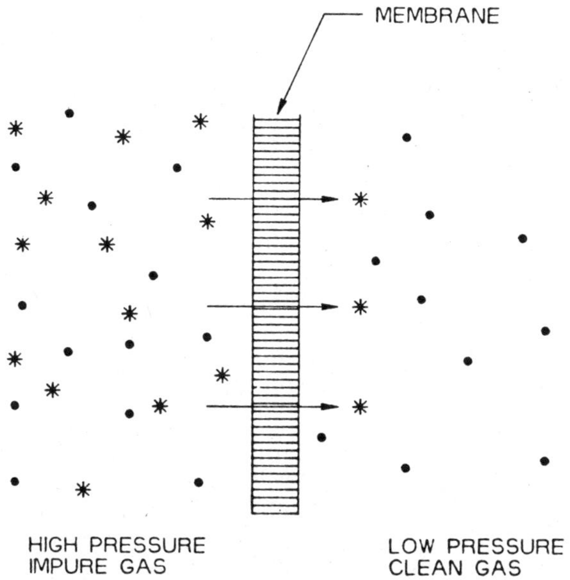

FIGURE 20.38 Diffusion through a membrane.

the partial pressure of the substance being transferred. Referring to Fig. 20.38, the rate of diffusion will be greatest when there are more molecules on the left-hand side than on the right. To optimize the left-hand side, the pressure should be high, the contact area and time as high as possible, and the depth of the gas space minimized. To optimize the right-hand side, the pressure should be low, the gas should not contain any of the diffusing molecules, and any molecules emerging from the membrane should be swept away as soon as possible.

The main use of permeable membranes in gas sampling systems is to remove water vapor from the sample. A typical example, the permeable-tube drier, is shown in Fig. 20.39. The concentric tube arrangement ensures that the sample gas is in intimate contact with the tube and that moisture is swept away from the outside surface. The direction of purge flow is countercurrent to the sample flow. As the sample passes through the tube, it becomes progressively drier. At the exit end the purge gas must be dry to ensure maximum water removal.

The efficiency of the permeable-tube drier is surprisingly high. It also provides a good lesson in the selection of tubing for sample systems and other analyzer gas supplies. Polymer tubes, such as polyethylene and nylon, are permeable to water vapor even though the pressure differential is opposed to it. Another feature of polymer tube is its ability to attract and hold certain molecules. If the sample contains acidic gases, the permeable-tube drier will probably not be suitable, particularly if any of these gases is to be measured.

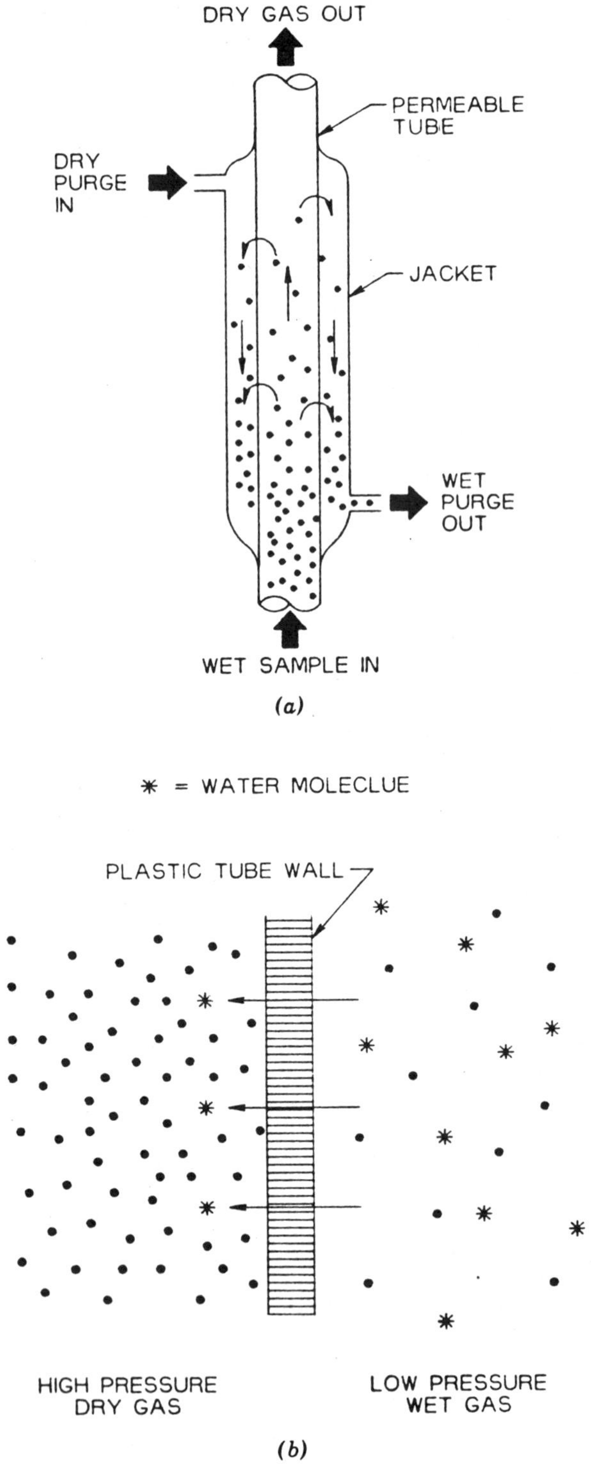

FIGURE 20.39 Permeation tube drier. (*a*) Basic design; (*b*) operating principle.

Another permeable gas membrane in use is the palladium hydrogen transfer tube. A palladium alloy heated to 300°C is permeable to hydrogen. The design is similar to Fig. 20.39 and allows hydrogen to be transferred from one flowing gas into another. This device is used as part of an analyzer rather than in the sampling system.

In liquid systems, membranes can be used to transfer dissolved solids from a concentrated solution to a dilute one, a process known as *dialysis*. It is frequently used in analyzers themselves but rarely in sampling systems.

(e) Reactors

Reactions that are useful for the removal of unwanted components in the sample are usually specific to the method of analysis. An example is the measurement of boiler feedwater conductivity, where an ion exchange resin is used to take ammonia out of solution and replace it with hydrogen ions. Simultaneously, the ion exchange resin converts all dissolved salts to free acids, which increases their sensitivity to conductivity measurement. This type of specific reaction is best considered as part of the analytical technique rather than a sample conditioning technique.

Another reactor commonly used in process GLC is the methanator. When carbon monoxide or carbon dioxide is to be measured at the ppm level, the flame ionization detector (FID) is the obvious choice. As mentioned previously, the FID does not respond to these components. It is well known that they and certain other compounds are converted quantitatively to methane at high temperature in the presence of a nickel catalyst.

The methanator consists of a short tube containing a granulated solid support (e.g., Chromosorb) on which nickel is impregnated. The tube is wound round a heater, which maintains the tube at about 500°C. The methanator is installed between the column and the detector. The column performs the component separation; as the carbon oxides elute they are selectively converted to methane and measured by the FID in the normal way. This is an elegant solution to what was a difficult measurement.

(f) Sample Dilution

Sample dilution can solve some difficult vapor sampling problems but is rarely used. The difficulty is to provide a constant, accurately known dilution factor. Without this, it is impossible to calculate the true stream composition.

Dilution is potentially useful for reducing the concentration of the measured component to bring it within the range of the measurement technique being used. Again, dilution is used more as part of the analytical method than in the sampling system. For instance, it is used in "wet chemical" analyzers, where the continuous dilution is sometimes carried out using a peristaltic pump. As mentioned earlier, this type of pump is a high maintenance item, probably why the technique is not often used.

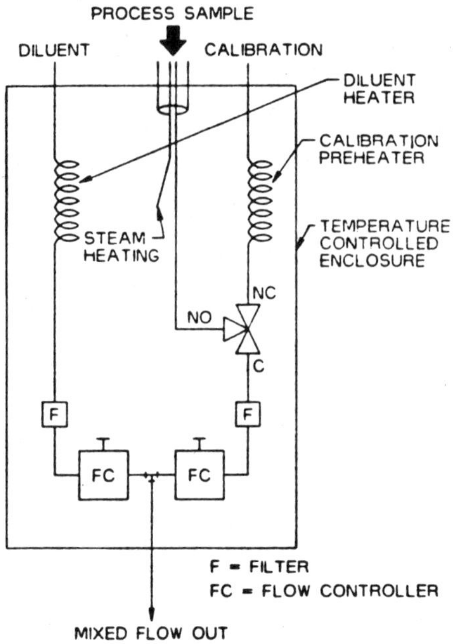

FIGURE 20.40 Sample dilution system.

With vapor samples, condensation must be avoided at all costs. When the whole sample is required for analysis and the sample cannot be held in the vapor state at the temperature and pressure limitations of the analyzer, sample dilution is the only method available. Even when other methods are possible, the inherent simplicity of dilution may provide the best solution.

When designing a dilution system, the emphasis must be on constancy. The sample vapor must be mixed in a constant ratio with diluent gas, which requires careful control of both flowrates using flow controllers for both sample and diluent. The whole system should be mounted in a temperature-controlled enclosure to prevent changes in the dilution factor. This is particularly critical for widely dissimilar flows. The second important design criterion is how to calibrate the dilution factor so that the true sample composition may be calculated. Measuring both flowrates is not the best way. Recognizing that the analyzer itself must be regularly calibrated, it is better to calibrate the dilution system and the analyzer concurrently. A typical complete system is shown in Fig. 20.40. In some applications, the calibration sample cannot be easily held in the vapor state, so a more complex design is necessary.

20.5 SAMPLE HANDLING SYSTEM DESIGN

Previous sections in this chapter cover the basic objectives of the sample handling system, a description of some of the physical equipment used and its function in the system, and the terminology used in identifying the various parts of the system. This section outlines the basic principles that should be followed when designing the sample handling system.

The most important requirement is that the sample handling system consist of a number of components, all of which play a significant role in achieving an efficient and reliable system. Attention must be given to each component.

20.5.1 Selection of the Sample Point (TAP)

When selecting a sample point location, there are three major considerations:

1. Providing an analysis that is useful for the desired purpose.
2. Minimizing the complexity of the sample handling system required.
3. Ensuring that the equipment is accessible for safe installation and maintenance.

In practice, conflicts often occur. For example, to improve the response time it may be desirable to sample further upstream, but this may result in a hotter, dirtier, or even a mixed-phase sample and therefore require a more complex sampling system. It is important that these judgements are made early in design and as part of the total system design. It is not sufficient to select a sample point merely to meet process requirements; conversely, it is inadequate to regard the sample point as "fixed," "unimportant," or "someone else's responsibility."

To be useful, the analysis result must reflect the conditions of the process being monitored at the point where control is to be effected. This may seem obvious, but there are four major sources of potential error.

First, be sure that the process stream composition has not changed between the point at which control is to be exerted and the sample point. Consider the case of a fractionator overhead circuit shown in Fig. 20.41. There are five possible sample points as shown. As to which one is right, there is no single correct answer, since none of the sample points are equivalent. The correct choice can only be made if the reason for making the analysis is known.

Second, the process lag time between the operation being controlled and the chosen sample point must be acceptable. Long delays may be caused by another process operation between the real process being controlled and the sample point. Sometimes, the intervening process is a clean-up operation—a scrubber or condenser. There are significant incentives for taking advantage of unit operations to clean the sample, as this can be expensive and troublesome if done in the sample system. However, the delay incurred in the process must be carefully evaluated. To get the speed of response required, it may be necessary to sample upstream and accept the increased complexity of the sampling handling incurred.

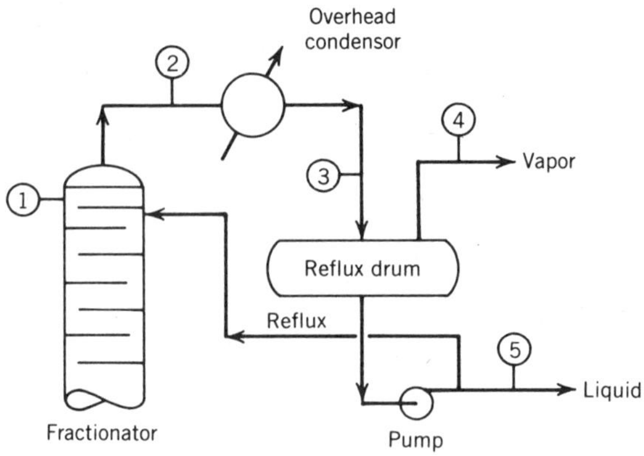

FIGURE 20.41 Fractionator overhead sample points.

Third, special care must be taken to ensure that the sample extracted from the process line is truly representative of the flowing material. This point cannot be emphasized too strongly.

Fourth, an analysis may not be useful because of the time lag in the sampling system itself. When selecting the sample point, the distance to the analyzer location must be considered. The sample point location is often dictated by other considerations and the analyzer location must be chosen to suit. Sometimes, as in the design of a grass-roots plant, there is a choice of positions for the probe and the analyzer.

Whenever possible, positions should be chosen to minimize the sample run length and allow a convenient analyzer location. Where large numbers of analyzers are to be installed, it is economical and convenient to group them together in shelters. This makes the selection of sample points and run lengths even more difficult and a compromise must be sought.

In conjunction with sample point selection, the sample return point must be selected. The objective should be to provide a source of constant differential pressure for the sample transport system. This will facilitate transport of the sample to the analyzer with a predictable time delay and obviate the need for a sample pump. If the sample is to be returned to the process, as is almost always the case today, the differential can be obtained across a process pump or heat exchanger.

Tapping across a control valve should be avoided, since the differential pressure will vary with valve opening and the sample flow bypassing the valve may affect its characteristics, especially when the valve is small. Never tap across an orifice, since this will negate the reason for the orifice in the first place—to measure the total flowrate in the process line.

All sample points should be accessible for safe installation and maintenance. Arranging for good accessibility is best done at the time of plant detailed engineering; often a process pipe can be run in a way that provides a good representative sample and good accessibility. The engineering model is an excellent vehicle for reviewing these criteria.

Special care must be exercised in the area of personnel safety when the process is at a very high pressure or temperature. In addition, other special precautions may be necessary for the handling and disposal of toxic samples.

20.5.2 Design of Sample Transport Systems

The sample transport system (or fast loop) is designed to transport the process sample to the vicinity of the analyzer and sample conditioning system. Its sole purpose is to reduce the time delay caused by having the analyzer distant from the sample tap. The most important design objective is therefore to obtain a satisfactory transport time. What is acceptable will depend on the purpose of the measurement.

It is essential to understand the application and to what extent it will be affected by time lags in the measurement. In the absence of more stringent requirements, a 60-sec maximum time delay in the fast loop is usually specified; an advanced control or safety application may require a faster response. When evaluating such an application, bear in mind that additional time delays occur in the sample conditioning system, in the analyzer itself, and in the process.

Although some processes demand the use of sample piping in accordance with the process piping specification, sample transport systems are normally run in tubing (e.g., $\frac{1}{4}$ in. OD stainless steel) and connections made using compression fittings. This is now the normal material for conventional field transmitter impulse lines. However, one should be conversant with the process piping specification and any ramifications regarding analyzer installations. The materials of construction must meet these specifications and be suitably corrosion-resistant to the sample. Some applications may demand the use of Monel, Hastelloy C or other exotic materials. Normally, the first valve at the sample tap must be to line specification.

Sharp bends and elbows should be avoided. The shortest practical route should be taken and, if possible, a downward slope of at least 1 : 12 be maintained, particularly for gas samples containing condensibles. All fittings should be accessible for maintenance.

Finally, once the initial design is completed, it should be reviewed in detail to see if there is any way to economize without prejudicing the effectiveness of the design.

(a) Available Pressure Differential

The sample tap and return point are selected initially to meet the process and application requirements. As a starting point, these define the available differential pressure. If problems arise, alternatives are often available, espe-

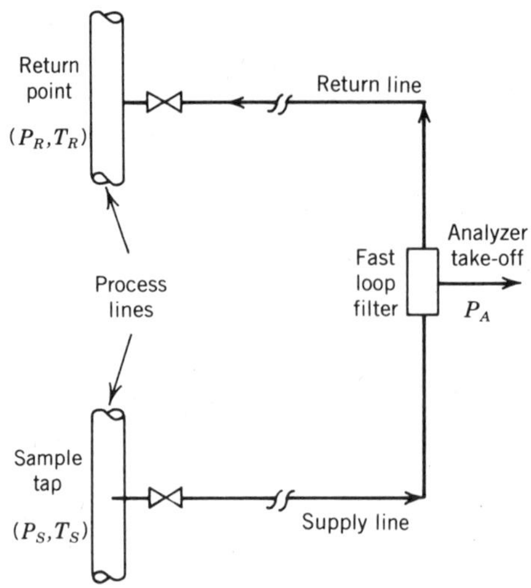

FIGURE 20.42 Typical fast loop system.

cially for the return point. It is important to follow up on these alternatives, rather than design a complex or inadequate system.

In Fig. 20.42 there are two sample transport lines, the supply line and the return line. The design of each depends on the line length and the differential pressure available. The pressures at the sample tap and the return point determine the maximum differential available. However, the nature of the sample and the pressure required at the analyzer take-off may limit the amount available across the supply or return pipework.

For liquid samples, the bubble point pressure must be known. When the sample is to be analyzed as a liquid, the take-off pressure must be kept high enough to prevent bubbles forming at the analyzer temperature. When the sample is to be vaporized, the take-off pressure must be kept high enough to prevent premature flashing in the vaporizer. Thus, the need to keep the pressure high at the analyzer take-off may reduce the available pressure drop in the supply line.

Vapor samples must not be allowed to condense in the lines. If the vapor at the sample tap is at a high temperature, it may be necessary to reduce the pressure at the sample tap and provide cooling to avoid having very hot sample transport lines. Pressure reduction at the sample tap will, of course, reduce the differential pressure available for the transport system.

In some applications, the return point pressure may be too high for the analyzer. In this situation, a sample return pump cannot be avoided. When conditions permit, an ejector pump can be used in the fast loop (see Fig. 20.13).

(b) Design Prerequisites

Before going ahead with the detailed design of the fast loop, it is necessary to define what other equipment will be contained in the loop. Each device will have a capacity and a flow resistance, which must be taken into account in the flow calculations.

Ideally, the fast loop should be a simple pipe run carefully sized to provide the desired flowrate. It is best not to include equipment such as filters, pressure regulators, and vaporizers unless absolutely necessary. There are some exceptions. For instance, some very dirty streams require special blowback filters or liquid separators at the sample point, as described earlier. In most cases, the sample is not filtered or separated in the fast loop. To preserve a single phase, the fast loop may be heated or cooled as necessary.

Pressure reduction at the sample point may be justified in order to:

1. Save the cost of special high-pressure pipework in the sample transport lines.
2. Prevent a vapor sample from condensing in the lines as it cools, saving the cost of heat-tracing the lines.
3. Reduce the mass flow of a gas sample. Since a fixed volume flowrate is required to purge the line, less gas will be used at the lower pressure.

Vaporization at the sample point should be carried out only if the return point must be vapor. In all other cases the sample should be transported as liquid and be vaporized at the analyzer take-off point. If pressure reduction or vaporization is to be done at the sample point, a problem arises because of the need to filter the process stream in order to protect the conditioning equipment. The filter is a necessity, but may introduce time delays. One solution is to install a local fast loop at the sample tap, which allows a self-cleaning filter to be used in front of the pressure reducer or vaporizer as shown in Fig. 20.43. The sample transport line then becomes a secondary fast loop.

Although the fast loop is usually not filtered, the analyzer take-off flow should always be filtered through a self-cleaning filter. This minimizes the volume of sample that must pass through the filter and uses the high velocity of the fast loop to keep the filter clean. Flow control of the fast loop is better

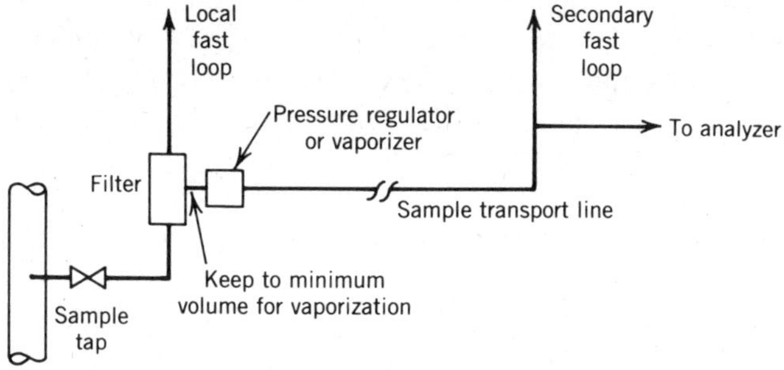

FIGURE 20.43 Use of a local fast loop.

done by line-sizing rather than by a valve. If a flow indicator (rotameter) is used, it should be installed on the downstream side of the analyzer take-off, as should a pressure indicator (if one is used). Although the time-lag effects are probably negligible in a fast-flowing system, it is best to keep these devices out of the direct path to the analyzer.

(c) Sizing of Sample Transport Lines

The flow of fluids through pipes and fittings has been extensively studied and is well documented. The precise methods for calculating the optimum line size, flowrate and pressure drop are fairly complex. For the limited range of conditions of interest in sample transport systems, these can usually be covered more simply on a step-by-step basis and will meet most requirements.

To calculate the size of the transport lines, complete data are required. Whether designing the system or specifying the job for others, the following information is essential:

1. The acceptable lag time in the supply line.
2. Operating pressure at the sample tap. Consider both normal and abnormal conditions and always use the worst case (minimum available pressure drop).
3. Operating pressure at the sample return point.
4. Pressure limitations at the analyzer take-off point, if any.
5. Pipe-run distance, sample tap to analyzer take-off.
6. Pipe-run distance, analyzer take-off to return point.
7. Process piping specifications, sizes, and materials.
8. Nature of the process fluid—vapor or liquid, composition, bubble point, dew point, density, viscosity, and so on. Density and viscosity are particularly important for pressure-drop calculations.
9. Analyzer requirements—pressure, temperature, flowrate, and so on.
10. A schematic drawing showing details of the sample transport system—valves, unions, elbows, vessels, and so on.

There are two types of sample transport system. The simple type consists only of piping/tubing and fittings. The complex type contains vessels where mixing can occur (e.g., pumps, filters, condensers, etc.). The complex system is less desirable but sometimes unavoidable.

The mixing volumes are important only when they are in the sample supply line (in the return line they do not matter). Usually, they have a negligible resistance to flow and may be ignored when calculating the pressure drop. However, the delay time they cause must be recognized when calculating the necessary fast loop flowrate.

Linear Velocity Required in the Fast Loop. For the fluid to travel from the sample point to the analyzer take-off within a given time, it must have sufficient velocity. The velocity may be calculated from the expression:

$$\text{velocity(ft/sec)} = \frac{\text{distance(ft)}}{\text{lag time(sec)}}$$

The distance is actual pipe-run distance of the supply line, ignoring fittings and other equipment in the line. The greater the distance, the faster the sample must travel to reach the analyzer take-off point in the specified time.

Equivalent Length. Sufficient pressure differential must be available to overcome the resistance to flow offered by pipe and fittings. One method in common use is to express the pressure drop in valves and fittings as an equivalent length of pipe. Probably the most useful document to refer to for this type of information is Publication No. 410 by the Crane Company. The equivalent length of each valve and fitting is added to the pipe run distance to find the total equivalent length of the supply line and the return line:

$$\begin{aligned}&\text{total equivalent length(ft)} = \\ &\quad \text{distance(ft)} + \text{equivalent length of valves and fittings(ft)}\end{aligned}$$

Calculation of Fast Loop Pressure Drop. After calculating the equivalent length of the supply line and the return line and the required sample fluid velocity to meet the response time requirement, the following step-by-step procedure can be used to carry out sample transport system design calculations to ensure that:

1. System response time is adequate for the application.
2. Sample pressure at the analyzer take-off point is adequate for analyzer operation.
3. Flowrate in the system is sufficient for analyzer operation.

Step 1. Table 20.2 gives the velocity(ft/sec) and the flowrate(cm^3/min) for various sizes of stainless steel tubing. A suitable line size should be selected, bearing in mind that the flowrate in the transport line must meet the needs of the analyzer. This depends on the type of analyzer.

Step 2. Calculate the Reynolds number and friction factor (f) for the supply line, from the expression:

$$\text{Reynolds number } (R_e) = 123.9\,\frac{dV\rho}{\eta}$$

where: d = line ID(in.)
V = velocity(ft/sec)
ρ = fluid density(lb/ft^3)
η = fluid viscosity(cP)

TABLE 20.2 Flowrate (cm^3/min) versus Linear Velocity (ft/sec) for Commercially Available Stainless Steel Tubing

Tubing Size		ID	X Sect.	Linear Velocity (ft/sec)													
OD (in.)	Wall (in.)	(in.)	Area (in.^{2})	0.5	0.75	1.0	1.5	2.0	2.5	3.0	3.5	4.0	4.5	5.0	5.5	6.0	6.5
	0.020[a]	0.085	0.0057	34	50	67	101	135	168	202	235	269	302	336	369	404	436
	0.025	0.075	0.0044	26	39	52	78	104	130	156	182	208	234	260	286	312	338
$\frac{1}{8}$	0.028[a]	0.069	0.0037	22	33	44	66	88	110	132	154	176	198	220	242	264	286
(0.125)	0.032	0.061	0.0029	17	26	34	51	68	85	102	119	136	153	170	187	204	221
	0.035[a]	0.055	0.0024	14	21	28	42	56	70	84	98	112	126	140	154	168	182
	0.049	0.027	0.0006	3.5	5.3	7.0	10.5	14.0	17.5	21.0	24.5	28.0	31.5	35.0	38.5	42.0	45.5
	0.020	0.210	0.0346	204	306	408	612	816	1020	1224	1428	1632	1836	2040	2244	2448	2652
	0.022	0.206	0.0333	196	295	393	590	786	983	1179	1376	1572	1769	1965	2162	2358	2555
	0.028	0.194	0.0295	174	261	348	522	696	870	1044	1218	1392	1566	1740	1914	2088	2262
$\frac{1}{4}$	0.035[a]	0.180	0.0254	150	225	300	450	600	750	900	1050	1200	1350	1500	1650	1800	1950
(0.a250)	0.049[a]	0.152	0.0181	107	160	214	321	428	535	642	749	856	963	1070	1177	1284	1391
	0.058	0.134	0.0141	83	125	166	249	332	415	498	581	664	747	830	913	996	1070
	0.065	0.120	0.0113	67	100	133	200	266	333	399	466	532	599	665	732	798	865

	0.020	0.335	0.0881	520	780	1039	1559	2078	2598	3117	3637	4156	4676	5195	5715	6234	6754
	0.022	0.331	0.0860	507	761	1015	1523	2030	2538	3045	3553	4060	4568	5075	5583	6090	6598
	0.028	0.319	0.0799	471	707	943	1415	1886	2358	2829	3301	3772	4244	4715	5187	5658	6130
$\frac{3}{8}$	0.035[a]	0.305	0.0731	431	647	862	1293	1724	2155	2586	3017	3448	3879	4310	4741	5172	5603
(0.375)	0.049[a]	0.277	0.0603	356	534	711	1067	1422	1778	2133	2489	2844	3200	3555	3911	4266	4622
	0.058	0.259	0.0527	311	466	622	933	1244	1555	1866	2177	2488	2799	3110	3421	3732	4043
	0.065[a]	0.245	0.0471	278	417	556	834	1112	1390	1668	1946	2224	2502	2780	3058	3336	3614
	0.083	0.209	0.0343	202	304	405	608	810	1013	1215	1418	1620	1823	2025	2228	2430	2633
	0.020	0.460	0.1662	980	1471	1961	2942	3922	4903	5883	6864	7844	8825	9805	10786	11766	12747
	0.028	0.444	0.1548	913	1370	1826	2739	3652	4565	5478	6391	7304	8217	9130	10043	10956	11869
	0.035[a]	0.430	0.1452	857	1285	1713	2570	3426	4283	5139	5996	6852	7709	8565	9422	10278	11135
$\frac{1}{2}$	0.042	0.416	0.1359	802	1203	1603	2405	3206	4008	4809	5611	6412	7214	8015	8817	9618	10420
(0.500)	0.049[a]	0.402	0.1269	749	1123	1497	2246	2994	3743	4491	5240	5988	6737	7485	8234	8982	9731
	0.058	0.384	0.1158	683	1025	1366	2049	2732	3415	4098	4781	5464	6147	6830	7513	8196	8879
	0.065[a]	0.370	0.1075	634	951	1268	1902	2536	3170	3804	4438	5072	5706	6340	6974	7608	8242
	0.083	0.334	0.0876	517	775	1034	1551	2068	2585	3102	3619	4136	4653	5170	5687	6204	6721

[a] Commonly used wall thicknesses for commercial SS tubing.

If the fluid specific gravity is given, multiply by 62.36 to get the density in lb/ft^3.

The friction factor may be obtained from Crane Publication No. 410. If the calculated Reynolds number is less than 4000, the friction factor may be calculated from $f = 16/R_e$ (critical or laminar flow conditions).

Step 3. Calculate the pressure drop in the supply line from the following equation:

$$P = 5.18 \times 10^{-3} \cdot \frac{fL\rho V^2}{d}$$

where: f = friction factor
L = equivalent length of supply line(ft)
ρ = fluid density(lb/ft^3)
V = line velocity(ft/sec)
d = line ID(in)

Step 4. Calculate the Reynolds number, friction factor, and pressure drop for the return line as in Steps 2 and 3. The analyzer flowrate should be subtracted from the supply line flowrate to get the return line flowrate. The return line velocity, which will be lower than the supply line velocity used in Steps 2 and 3, must then be calculated.

Step 5. The available pressure drop should be compared with the total pressure drop calculated in Steps 3 and 4. Check the analyzer take-off pressure to ensure that it is adequate for analyzer operation.

If the transport line pressure drop is too great, select a larger line size and repeat the calculations. Two alternatives may be considered:

1. Using larger tubing for the complete transport system.
2. Using larger tubing for the return line only.

   If the analyzer take-off pressure is adequate and calculated line pressure drop is not much greater than available pressure drop, 2 may be better than 1 and should be considered first. The decision on which alternative to use will be based on pressure drop *and* economic considerations.

Purge Time for Vessels in the Supply Line. As mentioned earlier, a complex transport system may contain vessels in which mixing can occur and which must be adequately purged. Once the volume(s) has been allowed for, the line-sizing calculations can be carried out as described.

Since the vessels must be purged, the fast loop flowrate is important. To calculate purge time, assume that three volumes will be required to fully purge the vessel. Once the internal volume of the vessel has been estimated, the purge time may be calculated from the following expression:

$$\text{purge time (sec)} = 3 \times \frac{\text{volume (liters)} \times 60}{\text{flowrate (liters/min)}}$$

The purge time must be subtracted from the acceptable lag time to obtain the time available for transporting the fluid from sample tap to analyzer take-off.

The effect of the additional volume is therefore to reduce the time available for sample transport. In many cases, however, the vessel volumes will turn out to have a negligible effect. The best procedure is to ignore the vessels and calculate pipe sizes and flowrate as outlined. Then calculate purge time and transport time. If available transport time is not enough for the line size chosen, the procedure is repeated using a greater velocity.

(d) Other Types of Sample Transport System

The main emphasis so far has been on fast loop systems, but other alternatives may be used.

When the analyzer is located adjacent or near to the sample point, it may be possible to use a single-line system as shown in Fig. 20.44*a*. This will be satisfactory if the 60-sec maximum time lag can be met and there is little or no sample conditioning required. The time lag will depend on the analyzer flowrate requirement.

If time lag is a problem, a bypass system as shown in Fig. 20.44*b* can be used. The problem with this system is that the sample bypass flow must either be vented or returned to a low-pressure point. Atmospheric venting of significant flows of process sample is usually not allowed in these days of energy conservation, so some alternative must be found.

The need for a local bypass when vaporizing at the sample point has already been discussed (Fig. 20.43).

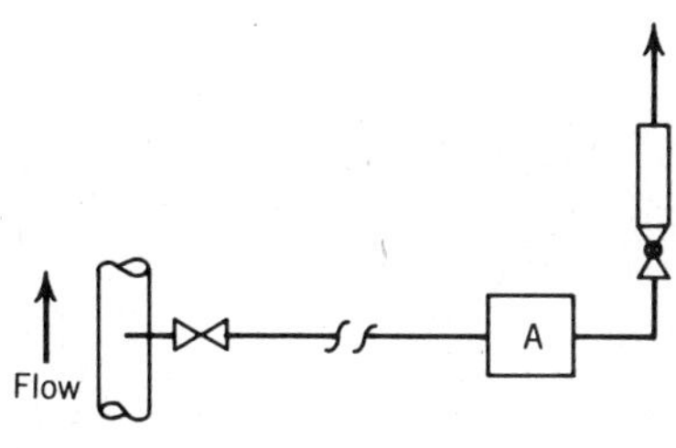

(*a*)

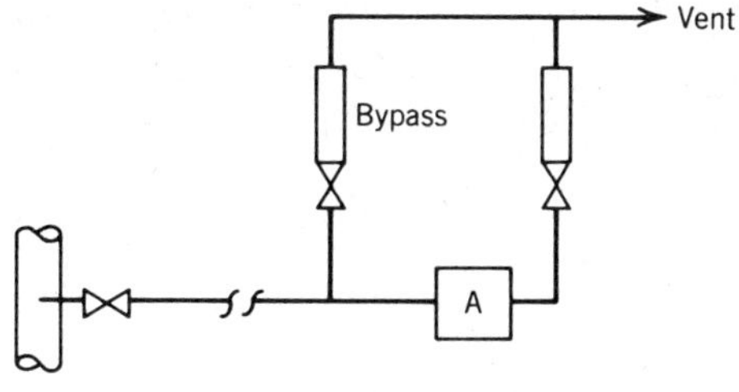

(*b*)

FIGURE 20.44 Single-line and bypass transport systems. (*a*) Single-line without bypass; (*b*) single-line with bypass.

20.5.3 Sample Conditioning System Design

The various components of the sample conditioning system have been described in Section 20.4 in relation to their function in the system. The question now is "how are these components put together to produce an efficient system?" A number of basic rules can be applied to answer this question, but bear in mind that they are only guidelines; unusual samples may require special treatment.

The crucial factor in sample conditioning system design is correct selection of the *order* in which certain operations are carried out, bearing in mind the nature of the process sample and what one is trying to achieve in the conditioning system. For each of the following examples, a sequence of operations is specified for a particular sample type:

(a) Gas/Vapor Sample, with Condensibles Retained

1. Filter hot
2. Reduce pressure hot
3. Cool, but not below the dewpoint
4. Scrubber or knock-out pot
5. Coalescer, always used after a knock-out pot or scrubber
6. Heat
7. Sample to Final Sequence

(b) Gas/Vapor Sample, with Condensibles Rejected

1. Filter hot (if necessary)
2. Cool to dew point required
3. Separate liquid in knock-out pot
4. Coalesce
5. Heat
6. Reduce pressure
7. Sample to final sequence

(c) Liquid Sample to be Vaporized

1. Cool, if required
2. Filter
3. Vaporize
4. Reduce pressure
5. Coalesce
6. Heat
7. Sample to final sequence

(d) Liquid Sample for Analysis

1. Filter
2. Cool
3. Reduce pressure
4. Separate Phases if necessary
5. Coalesce
6. Warm to analyzer temperature
7. Degas, if required
8. Sample to final sequence

In these examples, the last step is "sample to final sequence." This consists of any further conditioning of the sample required, multi-stream switching systems, calibration sample injection systems, final filtration, sample flow adjustment, actual analysis, sample disposal, and so on. As a general rule, the final sequence should be carried out in the following order:

1. Dry (if necessary).
2. Final filtration.
3. Regulate pressure.
4. Switch sample streams.
5. Switch calibration streams.
6. Adjust flow (unless flow controlled).
7. Match temperature to analyzer.
8. Analyze.
9. Dispose of sample.

There are naturally exceptions to this sequence, so each application should be considered individually.

20.5.4 Multi-Stream Switching Systems

Much has been written about multi-stream sampling systems and the precautions to be used in the many different designs in use today. Wherever possible, analyzer systems should be designed to monitor one sample stream only. However, a number of factors for and against multi-stream systems must be considered when making a decision.

First, the more complex the sampling system, the greater the chances of failure, with a consequent decrease in reliability and availability. Second, if there is an analyzer failure, the analysis on all streams is lost; therefore, the relative importance of each analysis is a prime factor for consideration. Third, maintenance personnel are often influenced by the relative simplicity of the equipment when establishing a maintenance schedule. This may result in the routine maintenance of a small group of single-stream analyzers being less than that of the combined sampling system of a multi-stream analyzer. Fourth, one must pay considerable attention to the accessibility of components for maintenance. This can often be a major problem with a complicated system.

Nevertheless, there are distinct advantages in using a multi-stream sampling system, not the least of which is the overall cost; analyzers are expensive instruments to purchase and install. For instance, in chromatographic applications, the cost of the GC alone may justify the use of a multi-stream system, but there are low-cost GCs on the market, suitable for straightforward analyses. For complete GC analysis of multiple streams, it is preferable that the streams be of similar composition, otherwise an expensive installation will be needed because of column resolution requirements.

Several fundamental principles should be observed when designing multi-stream sampling systems. Each stream should be treated as if it were the only sample for analysis, that is, it should have its own sample conditioning components (e.g., filter, coalescer, pressure regulator, etc.). Each stream should be manifolded to the analyzer in such a way that there is no possibility of cross-contamination between sample streams; always assume that valves will leak. Care should be taken to avoid "dead legs."

Performance of the system will depend entirely on the reliability of the switching valves. To keep the time lag in the system as small as possible, installation of a bypass to maintain continuous flow in the system may be needed. To ensure that individual samples are representative of the process stream, a continuous purge system may be necessary whereby the new sample scavenges the system containing the previous sample.

Many different designs of multi-stream systems are documented in manufacturers' literature, technical journals, and other publications. They may suggest preferences for individual system components, based on availability in the country of origin, operational experience and so on. These should be used as a reservoir of ideas by the potential designer.

The systems described next are examples and should be used as guidance only in demonstrating the basic principles. For instance, in these examples, solenoid-operated three-way valves are shown for stream switching. Pneumatically-operated three-way valves that receive air from solenoid valves actuated by a stream selector may also be used. In fact, there are advantages in the use of such a system, such as more positive seating, greater thrust, and so on. Some companies prefer to use GC column switching valves for multi-stream switching systems because of their reliability, precision, and low volume.

The final design of the system should always be decided on the basis of previous operating experience and familiarity with the available equipment. A designer with limited experience can only rely on the experience of others.

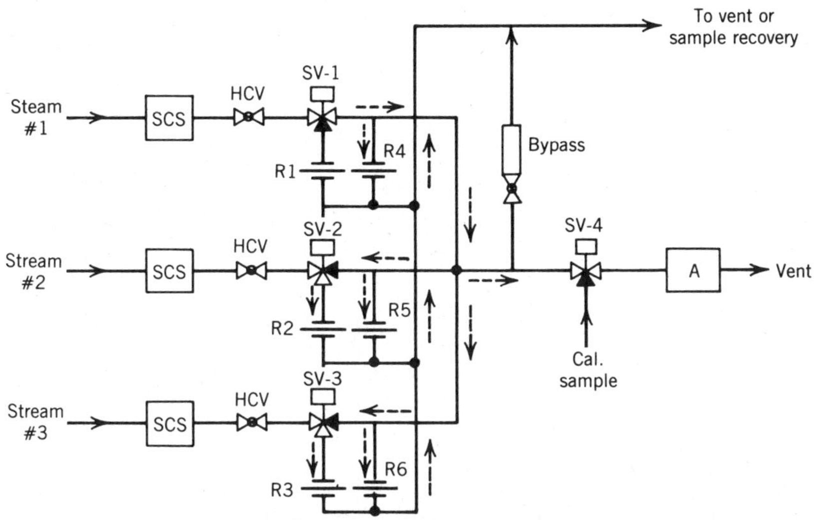

FIGURE 20.45 Single-valve system with back purge.

(a) Single Valve System with Back Purge

The main features of this system, shown in Fig. 20.45, are that a single three-way valve is used for each stream and that the sample manifold is back-purged by the sample under analysis, thereby preventing contamination. In addition, the sample streams not under analysis are bypassed, keeping all samples current. The main disadvantage of this system is that a considerable quantity of sample must be bypassed, which may be a problem if the sample must be returned to the process, as is almost always the case.

(b) Double Block and Bleed Systems

The purpose of double block and bleed is to eliminate the possibility of cross-contamination. There are several variations on this theme. The system shown in Fig. 20.46*a* uses two three-way solenoid valves for each sample stream. When the valves are de-energized, the tubing between them is vented to atmosphere, thereby preventing contamination. A secondary bypass is used to scavenge all sample lines, keeping all samples current. The system is shown with stream 2 supplying sample to the analyzer.

Another example is shown in Fig. 20.46*b,* commonly called double block with cascade. In this case, fewer solenoid valves are used and the manifold is only partially scavenged by the sample under analysis. Contamination is prevented by venting the space between solenoid valves, as indicated by the dotted lines. The system is shown with stream 2 supplying sample to the analyzer. However, when stream 1 is switched in, the sample in the line

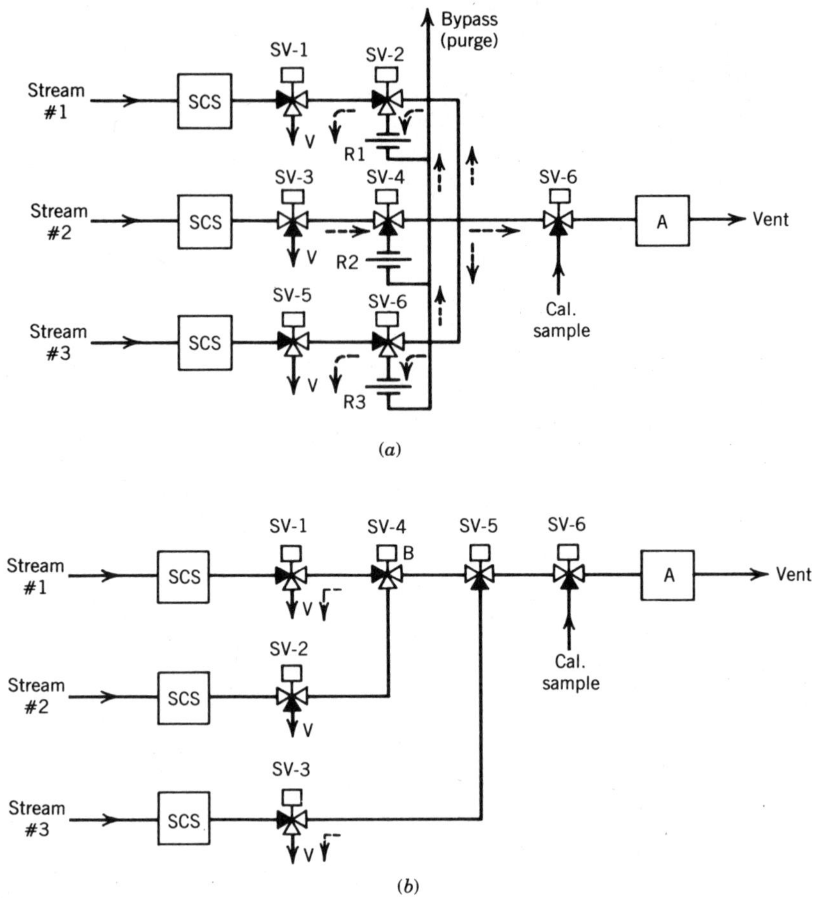

FIGURE 20.46 Double-block-and-bleed systems. (*a*) With bypass (purge); (*b*) with cascade.

between point B and the analyzer will not represent the sample flowing in stream 1 fast loop and the analysis will give incorrect information for a short time. The length of time will depend on the analyzer flowrate and the dead volume of the line. This system has been used successfully and proved to be an acceptable solution in most cases. If response time problems do occur, a secondary bypass can be considered.

(c) Loop Manifold System

The system shown in Fig. 20.47 is popular with some analyzer manufacturers because of its simplicity and low cost. One major problem with the system is that it does not completely remove "dead legs." One advantage of the loop manifold is that it can be constructed neatly and with good accessibility for maintenance.

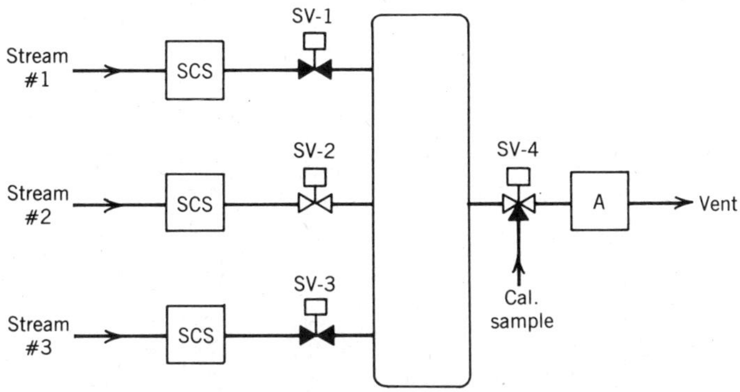

FIGURE 20.47 Loop manifold system.

20.5.5 Analyzer Calibration Systems

Calibration makes a very important contribution to the overall performance of an analyzer installation and its acceptance by operating personnel as part of the process control system. In view of this, the subject is covered in detail in Chapter 22.

The calibration facilities provided for an analyzer form part of that analyzer's sample handling system. The calibration sample is treated as another sample stream, so is usually included in the "final sequence" described before. Referring to Figs. 20.46*a* and *b*, the calibration sample is injected into the analyzer downstream of the multi-stream switching system, using another three-way solenoid valve. The same criteria apply for the calibration sample as for the process sample (e.g., cross-contamination, pressure regulation, filtration, etc.).

20.5.6 Sample Recovery Systems

When several liquid phase analyzers are installed together and are operating at near atmospheric pressure, the quantity of spent sample may be considerable. In addition, it may be difficult to return the sample to the process in the conventional manner. In many cases, the sample is a valuable product, so it is economical to recover the sample in some way. This is usually achieved with a sample recovery system.

Several manufacturers of physical property-type analyzers offer a sample recovery system as part of their overall package. A typical system is shown in Fig. 20.48.

The system consists of a collecting vessel of 30–40 gal capacity and a pump. The complete unit, together with all electrical switchgear, is mounted on a free-standing metal framework. Sample drains from the analyzers are connected to the top of the vessel, ensuring that the lines slope downward

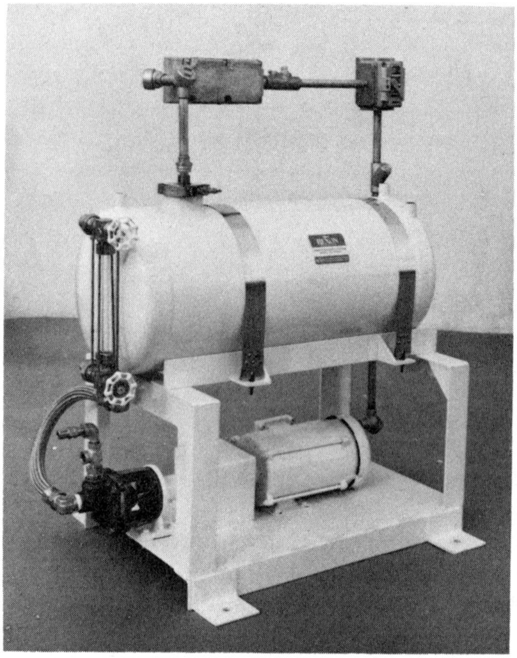

FIGURE 20.48 Typical sample recovery system. (Courtesy of Rexon Corporation)

and there are no bends to allow liquid to collect in the lines. The vessel is fitted with four level switches, two for pump on/off operation, one for low level alarm and one for high level alarm. The pump discharge is connected to a preselected point in the process.

When the level in the vessel reaches a predetermined high position (LS-1), the pump is started to return sample to the process. When the level reaches a predetermined low position (LS-2), the pump is stopped. The level switches LS-3 and LS-4 are for alarm at high and low level respectively and for vessel and pump protection. These switches are usually connected to the control house annunciator to bring the operator's attention to a potential problem. The vessel should be sized to provide about 30 min hold-up between the level switches operating the pump. The pump should have a minimum design capacity of 200% of sample in-flow rate.

The sample recovery system is a neat solution to the sample return problem when several analyzers are installed together. However, the system is expensive and cannot be justified for individual analyzers. The main problem with the sample recovery system is pump and level switch reliability. Failure of either may result in a major sample spillage. Although pump reliability has improved, there is still concern over the reliability of the level switches used by some manufacturers.

BIBLIOGRAPHY

Cooper, R. L. Industrial wastewater sampling. *Insts. & Contr. Syst.* (May 1980).

Cornish, D. C. *Sampling Systems for Process Analyzers.* Butterworths, London (1981).

Cortina, B. V. Sampling systems for dew point hygrometers. *InTech* (March 1985).

Cortina, B. V. Sampling systems for chilled mirror dew point hygrometers. *Proc. Intl. Moisture & Humidity Symp.,* Washington DC (April 1985).

Foundos, A. P. In-line sampling systems for pyrolysis gas and other hot gases. *Proc. ISA Symp.,* Des Plaines, ILL (May 1984).

Heisig, C. G. Filters and coalescers for wet gas sampling systems. *Proc. ISA Symp.,* Des Plaines, ILL (May 1984).

Houser, E. A. Principles of sample handling systems design for process analysis. *Proc. ISA Symp.,* Pittsburgh, PA (1977).

Strauss, R., Sample filters for analyzers. *Instrumentation Technology* (August 1977).

21

Design and Installation of Analyzer Systems

21.1 INTRODUCTION

The design of analyzer systems for industrial applications has generated considerable discussion among analyzer specialists. If a large analyzer design project were given to six different analyzer specialists, final outcome would probably be six different designs. Each would be workable and each would reflect the personal preferences of the designer. This leads us to the conclusion that there is no best solution to an analyzer system design. This is reasonable, since each analyzer specialist has learned the subject from field experience and will design on the basis of that experience. Nevertheless, some basic guidelines can be followed when designing analyzer systems. One of the aims of this chapter is to outline them.

The installation of analyzer systems is also a subject of great variety. The way installation is handled will depend upon the preferences of the user company and the type of organization involved. However, once again there are some basic guidelines that can be followed.

In the early 1960s, when analyzer technology was in its infancy, there was very little experience to draw on in either the user's or the analyzer manufacturers' organization. Therefore, analyzer installations suffered from poor design, unreliability, and so on. Today, with over 25 years of experience behind us, the situation is entirely different. We now know far more about sample handling systems and the problems associated with them, analyzers themselves are much more reliable, and we know how to handle major analyzer packages more effectively.

This chapter will cover the basic concepts to be applied to analyzer systems, from the conceptual design stage up to the precommissioning stage, just before process startup.

21.2 THE ANALYZER SYSTEM

The term "analyzer system" is used here to describe a number of component parts designed to take a representative sample of the process stream, transport it to the analyzer location, make required analysis, and provide the necessary information for monitoring or control of the process. Obviously, the system consists of far more than just the analyzer; typically it has the following components shown in Fig. 21.1:

- Sample take-off and return points.
- Sample transport system (fast loop).
- Sample conditioning system.
- Analyzer (sensing) unit.
- Analyzer control unit.
- Analyzer calibration facilities.

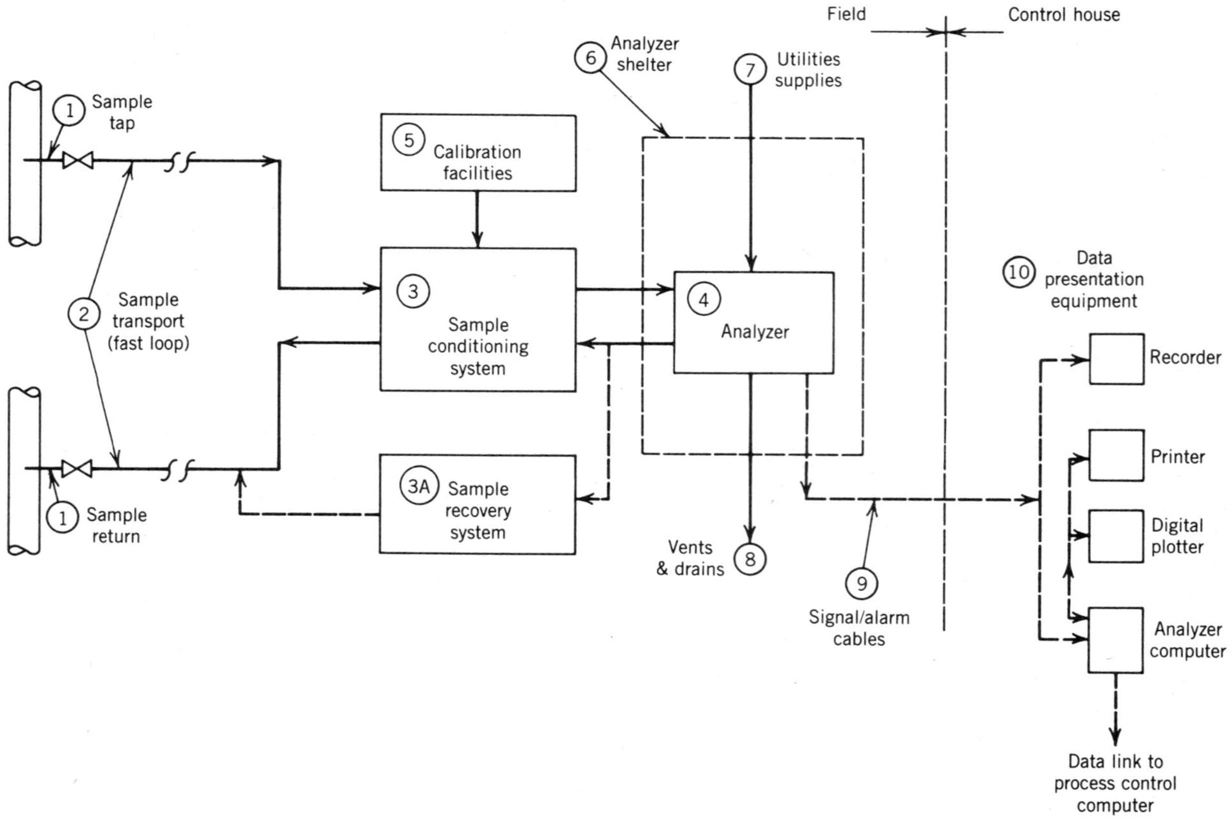

FIGURE 21.1 The analyzer system: component parts.

- Utilities supplies, such as air, water, power, steam, carrier gas, and purging facilities.
- Analyzer shelter or enclosure to protect the analyzer and its associated components from ambient conditions.
- Signal transmission, alarm, and power supply cables.
- Data recording and presentation equipment.

Each one of these components must receive due consideration during the design stage if the system is to operate reliably and provide the benefits to justify its purchase and installation.

Experience has shown that the sample handling system is usually the part of the analyzer system where most of the problems occur. So it is important that the sample handling system be given proper attention during the design phase; many potential operating and maintenance problems can be prevented if due care is exercized here. Bear in mind that analyzer results are meaningless unless the analyzer receives a representative and reliable sample of the process stream. On the other hand, a certain amount of experimentation is always necessary with sample handling systems; one can rarely expect to solve the problem on paper because there are too many variables involved.

Analyzer systems can be relatively simple, highly complex, or anywhere in between, varying from the design and installation of a single analyzer on an existing process plant to the design and installation of a complete "analyzer package" consisting of 150 or more analyzers, sample handling systems, analyzer shelters, and utilities on a large, grass-roots plant. Nevertheless, some ground rules may be applied to both situations.

21.2.1 Basic Information Requirements

At the earliest stage of system design, it is essential for *every process analyzer application* that the process stream be thoroughly defined in terms of process operating conditions, component concentrations, contaminants, and so on. In addition to normal process operating conditions, upset conditions must be considered. Analyzer data will, in many cases, be useful to the process operator during plant upsets. However, these conditions, if not taken into account, may upset the operation of the sample conditioning system and the analyzer itself. Typical information required by the analyzer engineer is listed in Table 21.1.

21.2.2 Sample Handling System Design

Sample handling systems are covered extensively in Chapter 20, so it is unnecessary to go into the subject in detail here. Nevertheless, it is wise to reiterate some of the more important concepts in the context of overall system design.

TABLE 21.1 Typical Process Data Required for Analyzer System Design

1. **Process stream conditions at sample tap**
   Sample tap location/flowsheet no.
   Pressure: min/normal/max
   Temperature: min/normal max
   Phase: Note 2-phase sample is not acceptable.
   Viscosity at operating temperature
   Dew point at operating conditions
   Bubble point at operating conditions
   Total composition, including:
      Corrosive components
      Particulates
      Toxic materials (required special handling)
2. **Process stream conditions at sample return point**
   Return point location/flowsheet no.
   Pressure: min/normal/max
   Temperature: min/normal/max
3. **Application Details**
   Analyzer tag number
   Service
   Stream number (1 for a single-stream analyzer)
   Reason for analysis
   Measured components/analysis required:
      Range
      Minimum detection limit
      Reproducibility required
   Cycle time required/maximum acceptable
   Hazardous area classification at proposed location of analyzer
   Data presentation requirements
4. **Other Information**
   Sample handling requirements:
      In-line analyzer
      fast loop transport
      sample conditioning
      sample recovery system
   Calibration requirements:
      Calibration method
      Availability of calibration standards
      Plant sample available for factory check-out
   Environmental impact of sample release

(a) Sample Take-Off and Return Points

Selection of these points must ensure that there is sufficient pressure drop to drive sample through the fast loop at a flowrate consistent with the response time requirement of the particular application. The analyzer engineer must be familiar with the exact purpose of the analyzer installation (monitoring or advanced control) so that the *correct* sample take-off point can be selected in relation to the process being controlled.

Particularly in the case of vapor samples and where there are relatively few low-pressure points in the process, it may be worthwhile to consider gathering several analyzer fast loop returns together into one return line, running to one specified return point (e.g., a compressor suction).

(b) Sample Transport System

Design of the fast loop is based on the distance between the sample point and the analyzer unit, together with the response time requirement of the system. Usually, fast loops are designed for a maximum response time of 60 sec. However, there are exceptions where a faster response time is necessary for control, based on process dynamics. This is another area where the analyzer engineer must have full information on the application in order to design the system correctly. Chapter 20 gives full details on sizing of fast loop systems.

(c) Sample Conditioning System

At this point in the analyzer system, process sample is conditioned to meet the operating requirements of the analyzer itself. Operations carried out here and the equipment available to carry them out are covered in detail in Chapter 20. The most important point here is that a thorough knowledge of the process stream under both normal and upset conditions is essential in order to adequately design the sample conditioning system.

(d) Analyzer Calibration Facilities

Facilities for manual or automatic calibration of process analyzers usually form part of the sample handling facilities. In view of its great importance, the subject of process analyzer calibration is covered separately in Chapter 22, including facilities design.

21.2.3 Analyzer Selection

We have seen in preceding chapters that a wide choice of on-line process analyzers is available today for industrial applications. Nevertheless, it is important that the *correct* analyzer be selected for a particular application. One factor contributing to the low analyzer service factors reported has been the misapplication of analyzers. Most analyzer engineers select analyzers on the basis of their previous experience, but this is made difficult by the rapid advances in technology taking place and the number of new instruments appearing on the market. Maximum use must therefore be made by the analyzer engineer of expertise within his or her own organization and within the analyzer manufacturing organizations.

21.3 ANALYZER SHELTERS AND ENCLOSURES

Most process analyzers are designed for installation in hazardous locations, that is, they are weather-proof and explosion-proof or flame-proof, whereas their associated control units are usually of general purpose design (non-hazardous). Therefore, the process analyzer can in theory be mounted in the open without any further protection. Analyzers installed in the late 1950s and early 1960s were often mounted directly in the open or under sun/rain canopies adjacent to the sampling point.

As more analyzers were installed in existing or new plants during the 1960s and 1970s, grouping analyzers together in shelters or enclosures became a justifiable proposition. Not only did these facilities provide protection for the analyzers against adverse ambient conditions, but they afforded the analyzer maintenance technician a reasonably comfortable environment in which to work. Some companies found that grouping analyzers together in this manner resulted in a higher overall service factor for the grouped installations than for analyzers that were installed separately.

21.3.1 Types of Analyzer Shelter

Analyzer shelters come in a wide variety of shapes, sizes, constructions, and degree of complexity. They have been developed over the years by user companies and are usually custom-built to company standards and preferences.

Some walk-in analyzer shelters are brick-built or made of precast concrete blocks. They have the advantage that internal temperature gradients are reduced, but they are very expensive. The shelter must first be built on site; then all analyzers, sample handling systems, utilities and internal piping and wiring are installed.

The trend over the past few years is toward prefabricated analyzer shelters constructed of precast concrete block, reinforced glass fiber, insulated sheet steel, or aluminum. They are significantly cheaper than the brick-built or block-type shelters. They also have the great advantage that they may be prepiped and prewired with all analyzers and sample handling systems, utilities, and power supplies installed at the factory, and the complete assembly tested and shipped to site as a single package. In this way, site installation work is kept to a minimum.

Sometimes it is impossible to install an analyzer in a walk-in shelter because of excessive sample transport run lengths. In these cases, analyzers are installed individually in cubicles or enclosures. These are usually prefabricated and are tested at the factory in the same way as a large analyzer shelter.

21.3.2 Basic Design Concepts

In view of the wide variation in type, complexity, and design philosophy, it is difficult to lay down rules for shelter design. What may be suitable for installation in the southern United States or the Middle East will almost certainly be unacceptable for installation in Canada or northern Europe; the prevailing climate at a particular location will have a significant effect on the design. The guidelines that one can follow are based on sound engineering practices, taking into account the following considerations.

(a) Internal Dimensions and Layout

The internal dimensions and layout of an analyzer shelter are largely dependent on the number and types of analyzers installed in the shelter. To allow for safe transportation, shelter width is normally limited to a maximum of about 8 ft (2.5 m). Length can vary anywhere from 10 to 20 ft, or even greater, depending on the equipment installed. The layout of analyzers within a shelter should take into account the particular mounting arrangement. For instance, some analyzers are designed for wall mounting and require only front access for maintenance, whereas others are free-standing and require all-around access.

The clearance between individual analyzers mounted along the side walls of the shelter must be sufficient to allow adequate access to all components for maintenance. Accessibility criteria should also include all interconnecting piping, conduit/wiring, switches, terminal boxes, and so on, and is the prime factor for consideration in design. Clearance between rows of analyzers is usually set at about 4 ft to provide an unrestricted central area for safe

FIGURE 21.2 Typical analyzer shelter layout. (Courtesy of Ludlam-Sysco Ltd.)

exit in the event of an emergency. This space should be kept clear of obstacles. A typical analyzer shelter layout is shown in Fig. 21.2.

(b) Doors and Windows

Walk-in shelters should be designed with two doors, one for use as a main entrance, the other as an emergency exit. Each door should be fitted with a crash bar to provide rapid egress and be lockable from the outside only. Crash-bar operation must override the lock. The doors should be fitted with armored glass windows, which allow inspection of the entire shelter interior from the outside. Safety considerations usually dictate the maximum size of these windows and do not permit windows to be placed in locations other than the doors. It is also a good idea to have a small overhead canopy placed above each door for rain protection.

(c) Shelter Utilities, Vents, and Drains

One of the main advantages of the walk-in shelter concept is that utilities such as air, steam, water, and so on can be run in headers outside and inside the shelter, with take-offs to individual analyzers. Analyzers can also use common vent and drain systems.

Utilities and vent headers are usually run above the analyzers and the drain header below the analyzers. Care must be taken in the design of vent and drain systems to ensure that blockages cannot occur from the accumulation of liquids. A typical vent and drain system is shown in Fig. 21.3.

(d) Shelter Floors

The analyzer shelter is often installed on a concrete plinth, which forms the floor of the shelter. The concrete floor should be sealed with a fire-resistant, liquid-resistant, non-slip material to prevent absorption of sample liquids (particularly hydrocarbons). Care should be exercized in the floor design to ensure that, should a liquid spillage occur, the liquid flows naturally to a drain and puddles do not form on the floor. If a liquid analyzer is installed, it is good practice to fit a curb around the base of the analyzer to confine any liquid spillage. Sometimes the shelter has a steel checker-plate floor; the same considerations as for the concrete plinth apply to this design.

(e) Sample Conditioning Systems

The sample conditioning system, fast loop flowmeter, and filter are normally mounted on a panel or in a cabinet installed on the outside wall of the shelter and connected to the analyzer it is serving by the minimum length of small-bore sample line. The sampling cabinet can be heated if necessary and may be fitted with a window to allow internal inspection of flows, pressures, and temperatures without opening the door.

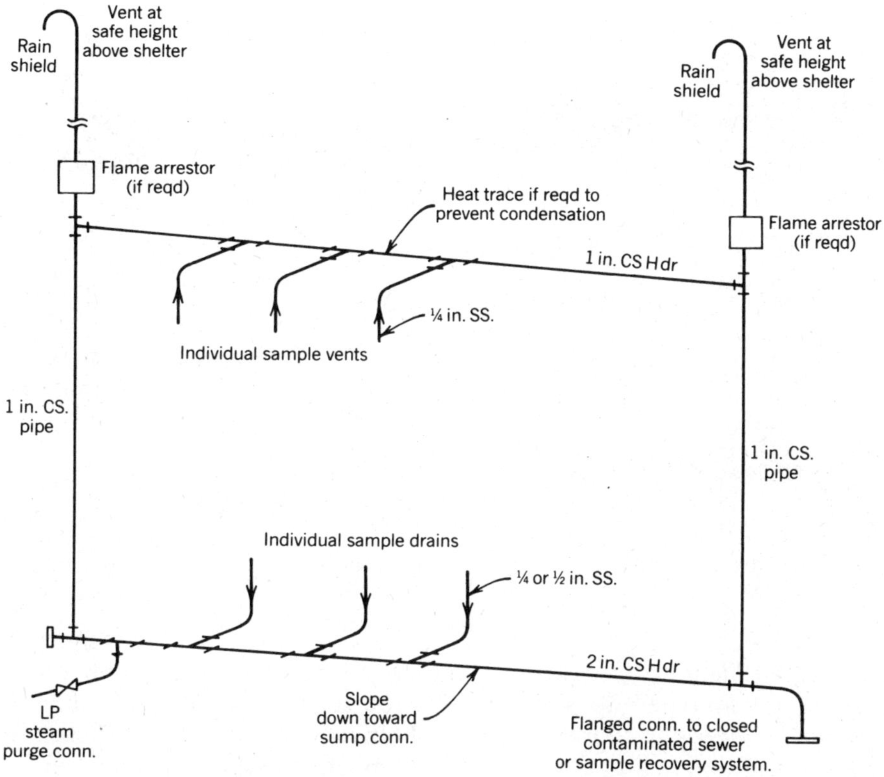

FIGURE 21.3 Typical vent and drain system.

Although this is a good method of mounting the sampling system, there are times when this approach is unacceptable; this is related to climate. For example, in locations with low ambient temperatures in winter, experience has shown that an external sampling system will receive very little maintenance under these conditions. Therefore it is sometimes necessary to mount sampling systems inside the shelter, which significantly affects shelter design from an electrical safety standpoint, as will be discussed later. Likewise, carrier gas cylinders and calibration standards are usually mounted outside the shelter. It is usual practice to fit external overhead canopies on the side walls to protect equipment and maintenance personnel from inclement weather. A typical arrangement is shown in Fig. 21.4.

(f) Shelter Lighting

Shelter internal lighting must be adequate in intensity to allow maintenance personnel to do their work properly. Fluorescent lighting, suitable for the area classification (see Section 21.3.3), has proved to be more than satisfac-

FIGURE 21.4 Typical overhead canopy arrangement. (Courtesy of Ludlam-Sysco Ltd.)

tory. If the shelter is fitted with external overhead canopies, lighting should also be provided here to permit work to be carried out safely on sampling systems, carrier gas cylinders, and calibration standards. The switches for internal and external lighting are usually mounted outside and adjacent to the main entrance.

(g) Heating, Ventilating, and Air Conditioning

As mentioned earlier, the functions of a walk-in analyzer shelter are to protect equipment and personnel from adverse ambient conditions and create a comfortable environment in which the analyzer maintenance technician can work. Therefore, heating and/or air conditioning are important factors to be considered. The subject of ventilation is dealt with in more detail later, since it is directly related to electrical safety requirements.

The type of system employed will depend on the climatic conditions at the site location. The main problem in very hot climates is to keep the shelter *cool* enough in summer; usually some form of air conditioning is required. In desert conditions, the shelter must be capable of withstanding sandstorms. Brick-built analyzer shelters are still fairly common in the Middle East, where summer day temperatures are very high, since this type of structure tends to reduce temperature variations within the shelter. However, the trend in recent years appears to be toward air-conditioned prefabricated shelters.

In cooler climates, the main problem is to keep the shelter *warm* enough in winter; therefore, heating is the most important factor and air conditioning is not usually provided unless there are extremes of temperature between winter and summer.

Steam is often used for heating, since it is readily available in most industrial plants and is cheap to run. One major disadvantage of steam as a heating medium is the lack of control one has over the shelter temperature. Electrical heating is more expensive but gives better temperature control.

When designing heating, ventilating, and air conditioning systems for analyzer shelters, bear in mind that the heat dissipated by the electrical equipment installed in the shelter will tend to heat the ventilation air. Conversely, the ventilation air will lose heat by conduction through the walls, roof, and floor of the shelter and in the air exhausted. To reduce heat loss, it is good practice to provide insulation between wall, roof, and door skins. If a ventilation fan system is installed, the shelter heater is often mounted in the fan ducting. This provides even distribution of heated air and good temperature control.

21.3.3 Electrical Classification and Safety

The electrical classification and safety of analyzer shelters is a subject of great importance, since it has a significant effect on shelter design and the type of protection that must be provided for equipment installed within the shelter. Bear in mind that an analyzer shelter is unique in that process samples are actually piped into the shelter and into the analyzers mounted therein. So there is always *one* potential source of ignition within the shelter, which requires more attention from an electrical safety point of view than most electrical equipment installed in industrial process plants. In addition, very few standards apply specifically to analyzer shelter installations. As a consequence, one finds varying opinions as to what level of safety is required, what standards apply, and to what extent.

(a) Hazardous Area Classification

The installation of electrical equipment in a potentially flammable atmosphere presents a fire or explosion hazard, with a risk to the safety of personnel and equipment. The subject of electrical safety, although covered here in relation to analyzer shelters, is applicable to *all* electrical equipment, including process analyzers, their associated control units, and other peripheral equipment.

The choice of proper equipment and protection methods initially involves determination of the nature and likelihood of the hazards at the location. In the United States, the National Electrical Code (NEC), Article 500, provides a method of classifying hazardous areas using three categories—Class, Group, and Division.

The Class defines the *physical nature* of the hazard:

- Class I—flammable gases and vapors.
- Class II—combustible dust.
- Class III—easily ignitable fibers or flyings.

Most analyzer shelters are located in Class I areas.

The group indicates the *chemical group* the material is in:

- Groups A, B, C and D—Class I chemicals.
- Groups E and F—Class II chemicals.
- Class III—no groups.

The Division is a measure of the likelihood that the hazard will exist in the area:

- Division 1—an area in which a flammable atmosphere is likely to exist during normal operation.
- Division 2—an area in which a flammable atmosphere is likely to exist only during normal abnormal operation.
- Non-hazardous—an area in which a flammable atmosphere is not likely to exist during normal or abnormal operation. This area is often known as a "general purpose" or "safe" area.

Area classification methods for Class I areas are also covered in recommended practices NFPA #497 and American Petroleum Institute (API) RP-500A.

In Europe, the Divisions listed above are now known as *Zones;* another Zone has been added:

- Zone 0—an area in which a flammable atmosphere is continuously present or present for long periods.

Because of these classifications, electrical equipment for installation and use in hazardous areas must be protected in some way and certified as suitable for use in the designated area. Various international organizations are responsible for providing certification of electrical equipment.

In the United States, Underwriter's Laboratories Inc. (UL) and Factory Mutual Research Association (FM) are responsible for certification of equipment in accordance with the applicable National Fire Protection Association (NFPA) and National Electrical Manufacturers Association (NEMA) standards and codes.

In Germany, the Physikalische Technische Bundesanstalt (PTB) certifies equipment in accordance with the requirements of the Verband Deuscher Electrotechniker (VDE).

In Great Britain, certification is provided by the British Approval Service for Electrical Equipment in Flammable Atmospheres (BASEEFA).

(b) Ventilation Requirements

The Class and Group designations are logical and straightforward, but often the Division designation can be difficult to determine. With regard to analyzer shelters, there is a problem when the shelter is located in a Division 2 area. Since the shelter is a confined space, recommended practice dictates that adequate ventilation must be provided to maintain the Division 2 classification. NFPA #497 defines adequate ventilation as that sufficient to ensure that the concentration of flammable gases does not exceed 25% of the lower explosive limit (LEL). In addition, the minimum ventilation rate for Class I combustible liquids in an industrial area is quoted as 1 cfm of fresh air per square foot of floor area.

If the leak rate is known, it can be used together with the volume of the shelter to calculate the required ventilation rate. Since transport of sample vapors or liquids into the shelter is normally via a closed system (stainless steel tubing), the leak rate is usually based on a tubing fracture. If the leak rate is unknown, it is recommended that 2 cfm of fresh air per square foot of floor area be used as the minimum criterion. For a typical analyzer shelter, say 8 ft high, this translates into 15 air changes an hour, a common rate used in industry. It should be emphasized that this is the *minimum* rate and implies proper ventilation system design to provide adequate mixing and ventilation as defined in NFPA #497.

(c) Equipment Protection

There are two potential electrical hazards to equipment installed in hazardous areas:

- The equipment can serve as a potential ignition source that could result in a fire or explosion.
- The equipment could be damaged by overcurrent.

Once the area classification has been determined, the equipment and wiring protection methods to meet this classification must be determined.

For Class I areas, the allowable methods are covered in NEC Articles 500 and 501. It is a general requirement that equipment and wiring must be approved for the specific Class, Group, and Division they are to be installed in, or fall under some of the exceptions allowed in NEC. NFPA #496 covers methods of purging and pressurization; NFPA #493 covers the use of intrinsically safe methods.

In Europe, the primary organization for electrical safety standards is the Comité Européen de Normalisation Electrotechnique (CENELEC), which has it headquarters in Brussels, Belgium. The members of CENELEC are the national electrotechnical committees of the following countries:

Austria	Italy	Belgium
Netherlands	Denmark	Norway
Finland	Portugal	France
Spain	Germany	Switzerland
Ireland	United Kingdom	

The standards issued by this organization are published in three languages—English, French, and German. A number of them were issued in 1977 under the general title "Electrical Apparatus for Potentially Explosive Atmospheres," each standard covering a particular method of protection, as listed below:

EN 50.014	General requirements
EN 50.015	Oil immersion 'o'
EN 50.016	Pressurized apparatus 'p'
EN 50.017	Powder filling 'q'
EN 50.018	Flameproof enclosure 'd'
EN 50.019	Increased safety 'e'
EN 50.020	Intrinsic safety 'i'

Some CENELEC standards are roughly equivalent to their American counterparts, but a little more stringent in their requirements. For instance, CENELEC standard 50.016, which covers methods of pressurization, is the equivalent in Europe of NFPA #496. Likewise, 50.020, which covers intrinsically safe equipment, is the equivalent of NFPA #493. The CENELEC standards also refer to other standards by the International Electrotechnical Committee (IEC) and the International Standards Organization (ISO). In Great Britain, the guiding standard for electrical equipment is BS 5345, which is relatively new. This standard covers the various protection methods given for CENELEC, with the addition of two more:

- Special protection—type 's' (e.g., specifically certified for Zone 0).
- Non-sparking—type 'N' or 'n'.

BS 5345 gives the Code of Practice; BS 5501 gives the requirements for Construction and Testing; BS 4683 gives the specifications for electrical apparatus.

(e) *Personnel Protection*

The question of personnel protection in analyzer shelters is equally important as it is in other industrial installations. The primary electrical hazard for personnel is electrical shock. What voltage level is safe for personnel? In view of the relatively small current levels involved (e.g., 80 mA causes ventricular fibrillation), many authorities such as the ISA recommend 30 V RMS as the maximum safe level.

One cannot really use the common guarding method "access by qualified personnel only," since the definition of a qualified person is open to dispute. It is recommended that any potentially hazardous source of electrical shock be clearly marked as a hazard and guarded by a physical barrier. Proper grounding of all electrical equipment is essential; approved methods are covered in NEC Article 500.

Mechanical hazards can be categorized into physical hazards, breathing hazards, and exposure to hazardous chemicals. All three can exist in an analyzer shelter. Physical hazards include sharp edges, points, tripping hazards, and so on; these should be eliminated during design and fabrication. Adequate working and operating space must be provided and no person should have to carry out an unsafe act in the course of his or her normal activities at the installation. Hot surfaces such as heat traced sample lines, sample conditioning systems and lines that are insulated but contain hot material, should be clearly identified.

Breathing hazards can be a major problem, since non-life sustaining materials are frequently brought into an analyzer shelter. Therefore, adequate safeguards must be provided to ensure that the ventilation system is working properly. Maximum leak size can be set by the use of flow limiting devices, and tubing volumes and run lengths kept to a minimum, as well as gas pressures.

Even more important is the potential exposure to hazardous chemicals. In addition to these precautions, specific detectors can be used to monitor build-up of hazardous gases.

21.3.4 Shelter Types By Electrical Classification

As mentioned earlier, analyzer shelter design is affected considerably by the electrical classification of the area in which the shelter is installed and the desired classification of the shelter interior. These factors, in turn, determine the extent of protection that must be provided for electrical equipment installed within the shelter. Section 21.3.3 shows how area classification may be downgraded from Division 1 to Division 2 or non-hazardous (as defined by NEC) or from Zone 1 to Zone 2 or non-hazardous (as defined by CENELEC) by the use of suitable protection methods.

This section describes some of the types of analyzer shelters used in industry (particularly petroleum and petrochemical), with emphasis on the extent of equipment protection and safety facilities that must be provided to meet the various electrical codes of practice.

(a) Location of Shelter

Analyzer shelters should be installed in Division 2 (Zone 2) or non-hazardous areas, never in Division 1 (Zone 1) areas. Since there are normally not many Division 1 areas in a process plant, it is not too difficult to find a suitable Division 2 location; however, finding a suitable non-hazardous area can be quite difficult, the major problem being to minimize sample run lengths in order to achieve acceptable response times. One exception, of course, is the water treatment plant, which is generally a non-hazardous area in any case. Therefore, it is quite common to find analyzer shelters installed on the edge of Division 2 areas, with part of the shelter located in a non-hazardous area, thus minimizing the potential hazard.

(b) Division 1 Shelters

Since the walk-in analyzer shelter is a confined space into which the process sample is conducted, the interior will be classified as Division 1, even though the shelter is located in a Division 2 area. Therefore, without further protection, all equipment installed within the shelter must be suitable for Division 1.

Although Division 1 shelters were fairly common at one time, they pose major problems from a maintenance standpoint. All electrical power must be removed from equipment before covers can be opened and servicing work carried out. In addition, a hot-work permit must be obtained and the area proved to be gas-free before work commences. It is possible to install non-Division 1 equipment in the shelter, using suitable protection methods, but unfortunately this does not eliminate the need for equipment de-energization and work permits to carry out maintenance.

(c) Division 2 Shelters

In view of these constraints, it is common practice to downgrade the area classification to Division 2 using some means of electrical protection as approved by NFPA or CENELEC. The usual method is a forced ventilation system, which provides a minimum of eight-to-ten air changes an hour, based on the volume of the shelter. In addition, all equipment within the shelter must be suitable for Division 2 minimum.

As we have seen from previous chapters, many process analyzers are designed for installation in Division 1 or Division 2 areas, whereas the associated control units are usually only suitable for installation in non-hazardous areas. This means that the control unit is usually installed far from the analyzer unit, creating an additional burden for the maintenance technician.

In addition to the forced ventilation system, it is recommended that a hydrocarbon (combustible gas) detection system be installed in the shelter, usually with sensors located at high and low levels. Should the ventilation system fail, or hydrocarbons be detected, the shelter interior classification will upgrade to Division 1. In this event, power must be removed from all equipment not suitable for Division 1, whereas Division 1 equipment may remain energized. One option is to use a type-Y purge system for Division 2 equipment, with alarm actuation in the event of a purge failure.

FIGURE 21.5 Division 2 analyzer shelter: example 1. (Courtesy of Combustion Eng. Process Analytics)

We see that downgrading the shelter interior to Division 2 results in increased cost of the analyzer shelter by virtue of the additional cost of protection facilities. Although the situation is improved from a maintenance standpoint, it is still common practice for a hot-work permit to be required in order to carry out maintenance work. Typical examples of Division 2 analyzer shelters are shown in Figs. 21.5 and 21.6.

(d) General Purpose Analyzer Shelters

The installation of more sophisticated process analyzer systems in recent years, together with the requirement for higher reliability, performance, and maintainability of these systems, has led to increased use of the general purpose analyzer shelter.

The interior of this type of shelter is classified as non-hazardous during normal operation, that is, it is downgraded from Division 1 (Zone 1) to non-hazardous by suitable protection methods. This permits the installation and use of general purpose equipment within the shelter and makes the task of servicing and maintenance easier. For instance, analyzer control units may be installed with their respective analyzers in the same shelter. In addition,

FIGURE 21.6 Division 2 analyzer shelter: example 2. (*a*) Exterior view; (*b*) Interior view showing shelter layout; (*c*) Division 2 analyzer shelter: exterior view showing power distribution, sample handling, and ventilation facilities. (Courtesy of Ludlam-Sysco Ltd.)

general purpose electronic test equipment may be used in the shelter. Provided that acceptable protection methods are used, work can be carried out in the shelter without conducting a gas-free test or obtaining a hot-work permit, normally the requirement for other shelter types.

Thus, the general purpose analyzer shelter's internal environment is conducive to efficient maintenance. Most companies using this approach feel that analyzers give better performance and have higher reliability and service factors when installed in a general purpose shelter. The additional cost of the shelter is usually justified on this basis.

A major contributor to the relatively high cost of this type of shelter is the rather sophisticated equipment protection and safety shutdown system required. The following is a description of the kind of protection typically required for general purpose analyzer shelters, based on the author's experience.

Ventilation System. Ventilation is provided by two identical fans, one in continuous operation, the other on standby. Each fan should be sized to provide a minimum of ten air changes an hour inside the shelter. An air low-flow switch and alarm is provided. The ventilation air must be taken from a *non-hazardous area* so that there is no likelihood of combustible gases entering the shelter via the ventilation air should there be an external release. The inside of the shelter is held at 0.1 in. WG by means of weight-loaded louvers located at the top and bottom of the shelter walls. Interior pressure is measured by a wall-mounted manometer.

Combustible Gas Detection. The combustible gas detection system should have two detection heads for each 150 square feet of floor area, one located at high level and one at low level. The system should be set to alarm at 10% LEL of the *least sensitive gas present.*

Sample Lines. Each exit and entry line carrying flammable, toxic, or asphyxiating fluids should be fitted with a restriction orifice, sized to allow not more than 200% of normal flow in the event of a line rupture. In the case of toxic fluids, it may also be necessary to install toxic gas detection equipment within the shelter. The length and size of sample lines from sampling systems to analyzers must be kept to an absolute minimum.

Sampling Systems. Sample fast loops and conditioning systems are mounted on the outside walls of the shelter under rain canopies and adjacent to the analyzers they are supplying.

Carrier Gases. Where hydrogen is used as a carrier gas for GCs, a solenoid valve must be installed in the carrier gas line *outside the shelter* so that all hydrogen to the shelter can be shut off in the event of an emergency.

Equipment Protection. All permanently installed equipment inside the shelter should be designed for installation in a Division 2 (Zone 2) environ-

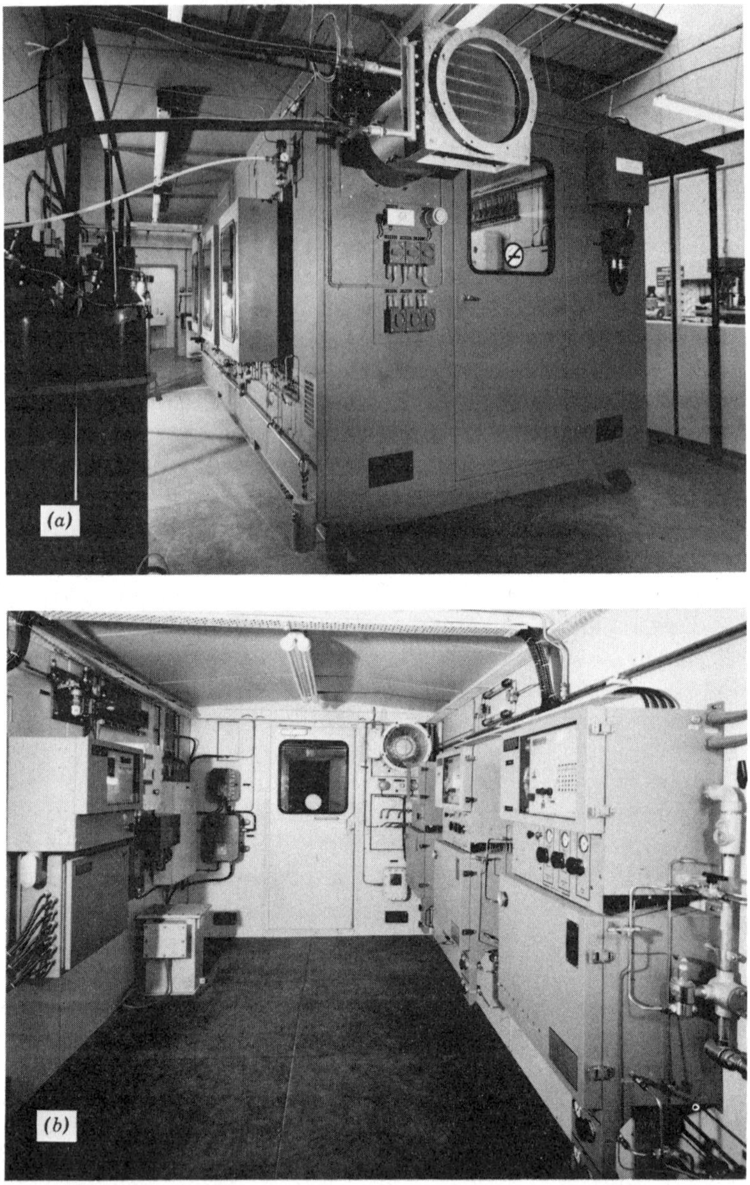

FIGURE 21.7 General purpose analyzer shelter. (*a*) Exterior view showing sample handling cabinets and ventilation system; (*b*) interior view showing shelter layout. (Courtesy of The Foxboro Company)

ment, as a minimum. However, the combustible gas detection system, alarm and shutdown system, and the shelter lighting must be designed for Division 1 (Zone 1) since this equipment has to remain operational during a shelter shutdown. Therefore, all general purpose equipment must be protected; the usual method is to use the type-Z purge in accordance with NFPA-496 or CENELEC EN50.016, with alarm on purge failure.

FIGURE 21.8 General purpose analyzer shelters. (*a*) During fabrication phase; (*b*) interior view showing layout of wall-mounted chromatographs. (Courtesy of Ludlam-Sysco Ltd.)

FIGURE 21.9 Dual-compartment shelter: exterior view showing utilities and carrier gas storage (zone I sampling room and general purpose analyzer room). (Courtesy of Ludlam-Sysco Ltd.)

Shutdown System. During normal operation (i.e., no alarms), the interior of the shelter is classified as non-hazardous. In the event of a ventilation failure, the interior classification is assumed to be Division 1 (Zone 1). Under this condition, all equipment not suitable for Division 2 (Zone 2) must be shut down automatically. Division 2 equipment will remain energized for up to 24 hr. If the ventilation system cannot be restored within this time, all Division 2 equipment must be shut down automatically.

After a ventilation system failure in which power to the equipment is switched off, power is not restored until the ventilation system is fully operational and at least ten air changes within the shelter have been completed.

This illustrates the level of complexity of the protection and safety shutdown systems required for general purposes shelters, and thus the main reason for the additional cost of the shelter. Typical examples illustrating the design and fabrication of such shelters are shown in Figs. 21.7, 21.8, and 21.9.

21.4 DESIGN AND FABRICATION—THE SYSTEMS APPROACH

21.4.1 General Considerations

In Section 21.2, the phrases "analyzer system" and "analyzer package" are used to describe the various components required to provide process stream quality data and some of the main points to be considered during system design. Advances in analyzer technology over the past decade or so, coupled with increased user and manufacturer experience in analyzer operation, have led to a significant increase in reliability and, at the same time, an increase in sophistication and complexity of analyzer systems. Recently, it has not been uncommon to see analyzer packages consisting of many different types of analyzers, sampling systems, shelters, and so on, specified for grass-roots plants or for existing plant expansions.

The main question to be addressed in such situations is how to handle design, fabrication, installation, and check-out of the complex systems. Some user companies have engineering departments capable of undertaking responsibility for an analyzer project from detailed design through to commissioning; however, they appear to be the exception. For other companies not quite so fortunate, the problem is how to define the various responsibilities and, at the same time, achieve a final installed package that meets all of the originally defined requirements.

There is no doubt that good overall coordination of effort is essential for the project to be a success, a criterion that applies equally well to all major project work. However, because of the fragmented nature of process analyzer project work, it is even more important here.

Ideally, it would be efficient and cost-effective for the user company if one organization were responsible for the complete project from conceptional design through to final commissioning of the system (the "turn-key" approach). In reality, it is extremely difficult to find an organization with all the necessary attributes and experience to do the complete job. The result is that one rarely achieves single responsibility; it is usually split between the user company and several other organizations.

One approach that has been taken by a number of user companies with considerable success is to employ a specialist subcontractor, often called the analyzer systems vendor (ASV) or system house, to design and fabricate the complete analyzer package and participate in installation and commissioning activities. This organization must have in-depth experience in the design of sampling systems, analyzer selection, analyzer shelter design, and fabrication to a variety of different standards (user defined), and in system installation, check-out, and calibration. Although there are still several problems to be overcome regarding division of responsibility, from the author's experience this approach is probably the nearest to ideal achieved to date.

There are several such systems houses, with varying amounts of experience in analyzer project engineering. In the United States, Xertex-United Controls, Johnson Controls, Texamation, and Pastech are all active in this area. In Europe, Comsip Entreprise (France), Saunier Duval (France), Intech (Italy), and Ludlam-Sysco (UK) are involved in analyzer project work. In particular, United Controls, Comsip, and Ludlam-Sysco have extensive experience world-wide. In addition, several analyzer manufacturers such as Foxboro, Applied Automation, and Combustion Engineering in the United States and Servomex in the UK offer complete analyzer package design and fabrication. Physical property analyzer manufacturers such as GCA-PSG in the United States also offer complete packages.

One key criterion for selection of a specialist subcontractor for analyzer systems is that the organization must be completely impartial about the choice of analyzers to be used. The reason is that most major companies are tied to a competitive bidding procedure for all equipment, including process analyzers. Therefore, it is rarely possible to preselect the type of analyzer. However, this situation may pose a problem for analyzer manufacturers bidding as systems houses, since they would obviously prefer to bid on a package that includes their own process analyzers.

21.4.2 Division of Responsibilities

For any major project in the process industries, there is usually a main contractor nominated to have overall responsibility, on a multi-discipline basis, for the detailed engineering and construction of the process plant and all associated equipment. The main contractor may subcontract to other companies for some portions of the work, especially those of a specialist nature (e.g., packaged units such as water treatment, steam generation facilities, process analyzer systems, etc.), but will still have overall responsibility for management of the subcontract. The user company (owner) will usually have a team assigned to the project whose function is to monitor and review the main contractor's work performance and progress during all phases of the project.

Large process analyzer packages generally form part of a major project. Therefore, one can lay down some reasonable guidelines regarding division of responsibility between the owner, main contractor, and ASV for handling such packages. The same basic philosophy can also be applied to smaller analyzer packages. In view of the close coordination needed, it is essential, based on the author's experience, that the owner nominate an experienced analyzer engineer to handle the project and act as overall coordinator. It is also advisable, and the owner should insist on it, that the main contractor nominate one person, preferably with considerable analyzer experience, to coordinate all analyzer work within the main contractor's office. Bearing in mind that the work here will involve several different disciplines, close coordination is essential. With this set-up in place, the project will be off to a good start.

(a) Initial Design Phase

This is an intensive period for both owner and main contractor, which will culminate in the production of a detailed design specification for the complete analyzer package and will involve the following work:

- Identification of all sample taps and return points on the mechanical flowsheets (P & IDs).
- Preparation of all process data for analyzers and sampling systems.
- Selection of the type of analyzer for each application.
- Establishment of the basic requirements for fast loops, sample conditioning systems, and so on.
- Deciding on the type of analyzer shelter to be used.
- Initial sizing of analyzer shelters, based on the number of analyzers to be installed in each.
- Locating analyzer shelters on plot plans.
- Deciding on the complexity of the package (e.g., the provision of an analyzer computer system).

These are all fundamental decisions that must be taken by the owner before work can commence in earnest on the analyzer package design specification. The owner and main contractor must also discuss and agree on the way the analyzer package is to be handled contractually. It is assumed, for the purposes of the following discussion, that a specialist subcontractor will be used.

It is the author's experience that preparation of the detailed design specification is best carried out by the owner. There are a number of reasons why this is the best approach, the most obvious being that the owner knows best what is required. Second, there can be considerable savings in time and manpower effort. If the DS is prepared with some thought given to the bidding process, the main contractor should have a minimum of work to obtain bids for analyzers, sampling systems, analyzer computer system, analyzer shelters, and so on or to obtain complete bid packages from the ASVs selected as potential bidders for the project.

The selection of potential bidders should not be treated lightly. Unless the owner's analyzer engineer is completely familiar with each ASV and up to date on their capabilities, it is recommended that visits be arranged so that current information can be obtained first-hand. In today's highly competitive environment, the situation can change significantly from month to month and it is difficult to keep up to date.

Unfortunately, there is usually a tight project schedule requiring analyzer package bids to be evaluated early in the project, often before detailed information such as plot plans, piping layouts, and so on are available. Therefore, the owner's and main contractor's analyzer engineers must be prepared to review all initial design work to ensure that changes are picked up and incorporated in the design specification. It is a fact of life that changes will occur on a continuous basis throughout the project.

During the bidding phase, it is essential that each bidder fully understands the requirements and is given plenty of time to prepare a comprehensive bid document. It is strongly recommended that at least one meeting be held with each bidder to answer questions. It is also important that bids are all on the same basis, so that the owner can compare "apples with apples." One approach that has worked quite well in practice is to divide the work into a number of "unit operations" and have each of these estimated by the bidder, giving total manhours for each operation and standard labor rates. The owner/main contractor can then use this information to estimate the total project cost. In this manner, should there be changes to the detailed design at a later date, the cost of the changes can be estimated quickly.

(b) Detailed Engineering Phase

Following the award of the subcontract for the process analyzer package to the selected systems house, the most important event will be the kick-off meeting. Here, the owner, main contractor, and ASV discuss the project in detail, review and agree on the scope of detailed engineering work, and finalize the project schedule and the various milestones within the schedule. The major responsibilities of each organization can be summarized as follows:

Analyzer Systems Vendor

- Detailed engineering of sample transport systems; sizing of fast loops, response time and fast loop flow rate calculations.
- Detailed engineering of sample conditioning systems.
- Detailed engineering of analyzer shelters.
- If applicable, purchase of all analyzers and the analyzer computer system.
- Preparation of all drawings and documentation for sampling systems, analyzer shelters, and so on.
- Gathering of all documentation from analyzer manufacturers, including installation drawings, operating manuals, spare parts lists, and so on.

Main Contractor

- Final location of all sample taps and return points on P & IDs and piping layouts.
- Final location of all analyzer shelters on plot plans.

- Design of concrete bases for analyzer shelters.
- Design of all utilities headers to analyzer shelters and enclosures (e.g., steam, cooling water, carrier gases, instrument air, nitrogen, etc.)
- Design of all power supply, signal and alarm cabling to analyzer shelters and enclosures.
- Design and layout of analyzer computer room, if applicable.
- Management of the analyzer package subcontract.

Owner

- Bid evaluation of analyzers and analyzer computer system.
- Review and approval for construction of all detailed engineering documentation (drawings and specifications) and project schedules from main contractor and ASV.
- Preparation of procedures for final inspection of analyzers and analyzer computer system at the vendor's factory and for the complete package at the ASV's factory prior to shipment.
- Preparation of requirements and procedures for training of owner's engineers and maintenance technicians. Normally the requirements will form part of the design specification.
- Establish requirements for calibration samples for all analyzers and prepare a schedule for procurement and delivery. Decide which organization will be responsible for this work. Once again, this will usually be covered in the design specification.

(c) Fabrication Phase

The ASV will have the main responsibility for this phase of the project. However, the owner and main contractor will need to monitor all aspects of the fabrication phase to ensure adherence to the agreed project schedule and to see that systems are built in accordance with the detailed engineering drawings. If slippage is going to occur in the schedule, this is the most likely place for it. Therefore, it is essential that adequate and qualified coverage be provided to closely monitor progress at this stage of the project. The ASV is often given the responsibility for procurement of all calibration samples. Careful monitoring is essential to ensure that they are available on site in good time for analyzer precommissioning.

21.5 INSPECTION REQUIREMENTS

21.5.1 General Considerations

Experience has shown that adequate inspection of most components of an analyzer package is fully justified prior to shipment of the package to the site. It is far less costly and time-consuming to have problems rectified at the vendor's factory than to have the work done at the site. Such a stipulation is usually written into the design specification.

When an analyzer systems vendor is given complete responsibility for an analyzer package, it is usual to have analyzers shipped to the ASV's factory, where they are installed in shelters, together with sample conditioning systems, utilities, and other peripheral equipment. For simple analyzers such as pH, conductivity, combustible gas detectors, and so on, inspection at the manufacturer's factory may often be waived. However, for more complex analyzers such as GCs, IR, moisture, wet chemical, and so on, as well as analyzer computer systems, it is advisable that these items be fully inspected prior to shipment to the ASV.

Following shipment of analyzers and other peripheral equipment to the ASV's factory and completion of final assembly of analyzer shelters, each shelter should be inspected as a complete operational system prior to shipment to the site. This will uncover any errors in installation that may be difficult and inconvenient to put right on site.

To conduct an effective inspection of individual analyzers or complete systems, it is essential that analyzer manufacturers and the ASV have full details of what is expected of them and a procedure for the inspection, well in advance of the inspection date. The author recalls numerous occasions where the manufacturer and/or ASV were ill-prepared for the inspection because of inadequate information. The final results were frustration on the part of the inspector and a lot of wasted time. In fact, on one occasion, the inspector insisted on returning at a later date, with the manufacturer agreeing to foot the bill for the additional 7,000 mile journey!

In addition to inspection procedures, the owner or main contractor should ensure that adequate samples are available for the check-out and inspection of analyzers. These samples can be synthetic samples prepared by the manufacturer, or preferably plant samples supplied by the owner. In any event, procurement and delivery of these samples should be organized well in advance of the inspection; often this point is overlooked, with disastrous consequences.

Any inspection of analyzer systems/packages consists of two separate parts:

- Physical inspection.
- Technical/operational inspection.

Some of the guidelines for each are given next.

21.5.2 Physical Inspection

The objective here is to ensure that the system has been fabricated in accordance with good engineering practices and with the detailed engineering drawings, and meets all the requirements laid down in the design specification. Items that require consideration during the physical inspection include the following:

- Equipment layout and accessibility for maintenance.
- Adequate identification of all piping connections (sample, carrier gas, air, hydrogen. nitrogen, cooling water, steam, etc.)
- Adequate identification of all terminations for power supply, signal and alarm cables.
- Provision of instrument nameplates.
- Compliance with manufacturer's drawings and documentation (e.g., connection size and type for piping, conduit, utilities, etc.)
- Neatness and tidiness for the installation (in accordance with good engineering practices).

With regard to the inspection of sample handling systems, it is rarely possible to conduct a complete operational inspection of this part of the analyzer system. Manufacturers and ASVs do not usually have adequate facilities for storage of and operation using large quantities of plant or synthetic samples, mainly because such operation would violate the factory safety standards. Therefore, factory inspection of sample handling systems is generally limited to a detailed physical inspection, including the following:

- Full compliance with detailed engineering drawings.
- Good workmanship and layout of components.
- Accessibility of all components for maintenance.
- Check on all components (e.g., tubing sizes, flowmeter sizes, relief valves, etc).

21.5.3 Technical/Operational Inspection

For this part of the inspection, the objective is to ensure that all analyzers/systems operate correctly and reliably, providing results that are repeatable within the limits set by the design specification. Items requiring consideration include the following.

For gas chromatographs, column performance is a critical parameter and must be evaluated carefully. Measured components should be adequately separated, elution times should be stable and repeatable, and cycle time should be within the limits set by the design specification. Column switching times should not be so tight as to be difficult to set up or troubleshoot. Selection of carrier gas and detectors should be such as to provide outputs linearly proportional to component concentration.

For all analyzers, it is recommended that a stability run of at least 8 hr be conducted. This can take place before the inspection, provided that it is fully documented and the results available for owner review at the final inspection. Repeatability of results should be within that listed in the design specification For any analyzers suspected of having non-linear outputs over the specified range, the manufacturer should provide a calibration curve, ensuring that an adequate number of points are taken to define the curve.

For analyzer computer systems, an extensive inspection is required. This should be conducted with all GCs and continuous analyzers (if applicable) connected to the system so that data transfer rates can be checked at various loadings. All functions of the computer system should be verified prior to shipment, since it is far easier to clear hardware and software "bugs" at the factory than on site. These functions include report generation and formats, program upload/download, analyzer/computer communications, start-up from cold, alarm functions and emergency shutdown, and so on.

An important function that should be checked and verified at the factory is the internal self-diagnostic program of the analyzer computer system. Another item that should be checked and verified if possible is the data link between the analyzer computer and the process control computer. Until quite recently, this was usually done on site, when both computer systems were available and the physical connection made between them. However, with modems (and good coordination between the two computer manufacturers), it may be possible to test the data link at the manufacturer's or ASV's factory prior to shipment. Several analyzer computer manufacturers have had success in this area.

For analyzer shelter inspections, it is recommended that all analyzer systems be operating on calibration standard samples or plant samples. Apart from sample handling systems, *all* components of the system should be checked operationally, including ventilation systems, purging systems, alarm and shutdown systems and so on. A good practice, if time can be allotted, is to start up the analyzer shelter from cold and, when fully operational, to create conditions for testing the alarm and shutdown system.

21.5.4 Documentation

As in all project-oriented work, the provision of good, clear documentation cannot be overemphasized. For process analyzers, good instruction manuals and drawings are also essential, since they will be used by technicians in the field for installation and troubleshooting. Although the quality of instruction manuals issued by manufacturers today has vastly improved in the last decade, there is still room for further improvement. Recent comments from the field still point to the fact that manuals are sometimes badly laid out, confusing to read, missing vital information required for installation and troubleshooting, and written without any appreciation of what it takes to maintain the equipment in the field.

For large analyzer packages, the ASV is often given the responsibility of providing complete documentation on the package. In addition to analyzer manuals/drawings obtained from individual manufacturers, the documentation should include operational instructions for the complete system (e.g., shelter startup/shutdown) and detailed drawings of all components, spare parts lists, and so on.

The requirements for package documentation are usually covered in detail in the design specification and procurement documents. For a large analyzer package, the gathering of documentation, drawings, and so on is a task that requires close coordination and monitoring by the owner. All documentation must be available at site prior to precommissioning; in fact, some of the relevant data is needed far earlier than this (e.g., wiring and installation drawings). In any event, full documentation should be available at the final inspection or shortly thereafter. It is good policy to carry out a detailed review of all documentation at the final inspection and develop a "punch list" for action by the manufacturer/ASV.

21.5.5 Technician Training

Technician training is an important component of any major analyzer project, so details of the owner's requirements should be included in the design specification. The final inspection of an analyzer package at the manufacturer's or ASV's factory is an ideal time for the owner's maintenance technicians to become familiar with the package and get hands-on experience in its operation. Not only are they involved with the package prior to its receipt on site but they have some ownership, which can be important from a psychological viewpoint. In addition, they will be adequately prepared when the equipment arrives on site.

If it can be arranged, it is a good idea to have analyzer manufacturers run short training courses on their analyzers during the fabrication phase, prior to the final inspection. However, this must be set up with the manufacturer well in advance.

21.6 INSTALLATION, CHECKOUT, AND PRE-COMMISSIONING

These phases of a major analyzer project are considered by many to be the most intense and critical of the project; understandably so, since it is during this period that the whole project begins to come together. One very soon finds out how good an engineering job was done and whether the complete package can be operated safely and reliably. Once again, the total site effort requires very close coordination by both owner and main contractor and the provision of good documentation for record purposes.

21.6.1 Installation

When a contract is placed with an ASV for complete package design and fabrication, considerable time can be saved during the installation phase, as mentioned earlier, provided that the complete package is inspected thoroughly prior to shipment to site. The main contractor (or possibly an installation subcontractor) will probably have been making progress on its part of the analyzer package work on site during the period prior to delivery of the analyzer shelters. This work will include installation of shelter plinths, sample probes (if used) at sample taps and return points, sample fast loop tubing, signal, alarm, and power supply cables, and all utilities piping. When the analyzer shelter arrives on site and is installed on its plinth, the remainder of installation work will be limited to connection of all piping, tubing, ventilation ducting, and all signal, alarm, and power cables. The system should then be ready for final checkout.

21.6.2 Final Checkout

The purpose of the final checkout is to ensure that all analyzers, associated equipment, piping, tubing, cables, and so on are installed in accordance with the design specification and detailed engineering drawings and that all components of the system are accessible for maintenance, are properly identified, and can be operated safely. Once again, a detailed inspection prior to delivery will have covered all equipment associated with the analyzer shelter, inside and outside.

It is essential that the final checkout be well documented. This will facilitate the identification of potential problems, "but"-listing the problems in terms of their priority and providing a record of all that is done.

A detailed loop checkout should be carried out on all cables running from analyzer shelters to control house and/or analyzer computer room, including all field junction boxes and recording/data presentation equipment. The main contractor should have standard checkout forms prepared for this purpose and will usually carry out the work with the owner witnessing. One of the most important items associated with loop checkout is a detailed review of all terminal and cable identification to ensure compliance with loop diagrams and cable routing drawings.

All sample tubing and utilities piping should be pressure tested to ensure that joints are tight and there is no potential for leakage or hazard to personnel.

21.6.3 Pre-Commissioning

The objective of the precommissioning phase is to demonstrate that the complete analyzer package can be brought up to full operation, with all analyzers calibrated and providing reliable data, prior to the startup of the process plant. Analyzers are usually pre-commissioned using standard samples of accurately known composition and calibrated on that basis. As mentioned previously, these standards must be organized well in advance so that they are available on site in good time.

An important point to bear in mind is that it is not usually possible to operationally test the sample handling systems during pre-commissioning. The real check occurs when the process sample is turned in after startup. It is then that one finds out whether all the data used for fast loop calculations, conditioning system design, gathered earlier, were correct.

Equally important, the process sample should *not* be turned into the analyzer system too early in the startup, since the sample handling system may not be able to cope with the unstable sample conditions. This could result in analyzer failure and costly downtime. The analyzer engineer may therefore need to use a little diplomacy in persuading process operations personnel to delay process sample turn-in until the process is reasonably stable.

In spite of these limitations, pre-commissioning is considered to be worthwhile for the following reasons:

- Maintenance technicians can become familiar with the complete system from an operational viewpoint (training).
- The analyzer systems are usually operated for a period of several days or even weeks prior to plant startup. This will uncover any potential problems and tend to improve reliability.
- Process operations are able to use the analyzers at the earliest opportunity, saving valuable time in getting the plant under control and producing on-grade product.

BIBLIOGRAPHY

API manual on installation of refinery instruments and control systems. Part II: Process stream analyzers.

Bajek, W. A., Kuchar, P. J., and Remec, A. A. B. Increased production of valuable products made directly through the use of full distillation range monitors. *Instr. Conf.*, Shanghai, Rep. of China (April 1983).

Berridge, S. A., Pearce, C. J., and White, E. N. Continous plant analyzers. *Proc. World Petr. Conf., Sect. 5,* Frankfurt, Germany (June 1963).

Burgess, D. The flammability limits of lean fuel-air mixtures. *Anal. Instr.*, **12,** ISA (1974).

Flow of Fluids in Valves, Fittings and Pipe. Crane Company, Publ. 410 (1969).

Frant, M.SA., and Weir, R. J. Putting process analyzers on line. *Insts. & Contr. Syst.* (Oct 1982).

Garside, R. Intrinsic safety for instrumentation: No reason to go anywhere else. *InTech* (Jan 1985).

Jones, E. B. On-Line analysis instrumentation. *Instrument Technology,* **2,** Newnes-Butterworth, London (1977).

Nootbar, R. F. An approach to the application of process stream analyzers. *API Div. of Refining,* **43** (1963).

Perry, J. H., and Chilton, C. H. *Chemical Engineer's Handbook,* 5th Ed. McGraw-Hill, New York (1973).

Villalobos, R. The application of on-stream analytical instrumentation in ethylene manufacture. *Beckman Tech. Bull. 5035.*

Villalobos, R. The application of on-stream process gas chromatographs in petroleum refineries. *Beckman Tech. Bull. 5039.*

Williams, A. O., and Green, P. W. Applying analyzers in the process industries. *Hcbn. Proc.* (June 1980).

22

Process Analyzer Calibration and Correlation

22.1 INTRODUCTION

Calibration of process analyzers and correlation of analyzer readings with laboratory standard tests has created more comment, discussion, and animated debate among analyzer specialists worldwide than any other topic in process analyzer technology. It is a subject of paramount importance, since the accuracy of the measurement made by a process analyzer is entirely dependent on the accuracy of the calibration standard used and the method of calibration.

User companies have recognized the importance of calibration from the time the first analyzers were installed and they have developed procedures based on their own field operating experiences. However, it was not until the early 1970s that attempts were made at standardization of these procedures. The Oil Companies Materials Association (OCMA) in Great Britain was the main driving force, producing OCMA Specification INP-3 in 1974. More recently, the Institute of Petroleum has issued "Codes of Practice for the Calibration of Process Analyzers," covering general principles and specific practices for viscosity, flash point, distillation, cloud point, and gas chromatographic analyzers. Not much work appears to have been done in the United States so far.

22.2 TERMS AND DEFINITIONS

Before proceeding to a description of the principles and procedures used to calibrate process analyzers, we will discuss some of the more common terms and definitions used and the relationship between them.

22.2.1 Repeatability, Reproducibility, and Accuracy

The standard laboratory tests described in previous chapters, particularly those used for physical property measurement, measure two types of property:

- An *absolute property,* where the measurement result is independent of the type of equipment and the method used for its measurement. In these cases, the results from a process analyzer and from the standard test should be identical. Viscosity and density are two examples.
- An *empirical property,* where measurement is only definable in terms of the standard test. The results from a process analyzer and the standard test may differ, and the analyzer must be calibrated to agree with the standard test. Flash point, pour point, cloud point, vapor pressure, and octane number are typical examples.

The precision of *laboratory standard test methods* may be quoted and defined by the following two terms:

(a) Repeatability

A quantitative measure of the variability associated with the same operator in a given laboratory obtaining successive, repeated results on identical test material using the same apparatus. It is the difference between two such single results that would be exceeded in only one case in twenty in the normal and correct operation of the test method (i.e., 95% confidence limit).

(b) Reproducibility

A quantitative measure of the variability associated with operators working in different laboratories, each obtaining single results on identical test material. It is the difference between two such single and independent test results that would be exceeded in only one case in twenty in the normal and correct operation of the test method.

The precision of *process analyzer measurements* may be defined in the same manner:

(c) Repeatability

A quantitative measure of the variability associated with the same analyzer in a given location obtaining successive, repeated results on identical test material. It is the difference between two such single results that would be exceeded in only one case in twenty in the normal and correct operation of the analyzer.

(d) Reproducibility

A quantitative measure of the variability associated with analyzers operating in different locations, each obtaining single results on identical test material. It is the difference between two such single and independent test results that would be exceeded in only one case in twenty in the normal and correct operation of the analyzers.

(e) Accuracy

In both cases, this is defined as the closeness to which the reading approaches the *true value* of the measured property of the sample.

Note the similarity in these definitions for both laboratory test method and process analyzer. Whereas both repeatability and reproducibility are well established precision criteria for the laboratory test method, reproducibility is not frequently used in relation to process analyzer precision, although there are some exceptions. Thus, the majority of analyzer manufacturers quote the *repeatability* of their instruments but not the *reproducibility* or the *accuracy*. Where accuracy is quoted, one is never quite sure exactly what the term means!

These terms and definitions pose some problems for the analyzer specialist. For example, let us consider again the case of physical property measurements, such as flash point, pour point, cloud point, and so on. The process analyzer has a far better precision in most cases than the corresponding laboratory standard test method. Unfortunately, routine test results are usually reported to the plant as a single result; no precision is included. Should the process analyzer reading be different from the test result, the tendency is to request an analyzer calibration or adjustment, but this may not be justified. This situation also effects the confidence process operations personnel have in the particular analyzer readings.

22.2.2 Measurement Errors

Any measurement signal—whether from a process analyzer or a conventional flow, pressure, level, or temperature instrument—is subject to a number of errors usually classified into three types—random, systematic, and gross errors.

Random errors cannot usually be assigned to a particular cause. They are normally evaluated by statistical methods beyond the scope of this book. The reader is referred to the excellent literature on this subject.

Systematic errors can usually be assigned to specific causes and may be reduced by calibration with standard reference samples. Normally, this type of error is unidirectional and may either be constant or vary with component concentration (or measured parameter) in the sample (e.g., non-linearity). The systematic error of a measurement is often known as the *bias*.

Gross errors usually result from catastrophic events.

Two other terms commonly used are absolute error and relative error. The *absolute error* (e_a) of a particular measurement is given by the expression:

$$e_a = x - x_t$$

where x_t is the true value of the measured parameter.

The *relative error* (e_r) of a particular measurement is given by the expression:

$$e_r = \frac{x - x_t}{x_t}$$

and is often expressed as a percentage.

The difficulty arises when one tries to use these two expressions; generally, the true value (x_t) of the measured parameter is not known. The usual procedure is to replace this with the *mean* ($\bar{x}$) of a number of readings, which is a reasonably good estimate of the true value provided that systematic errors have been eliminated.

22.2.3 Precision and Confidence Limits

The *precision* of a set of analyzer readings is related to the scatter or dispersion of the values about a central value, or *population mean* (μ), and normally follows the familiar *normal (or Gaussian) distribution* shown in Fig.

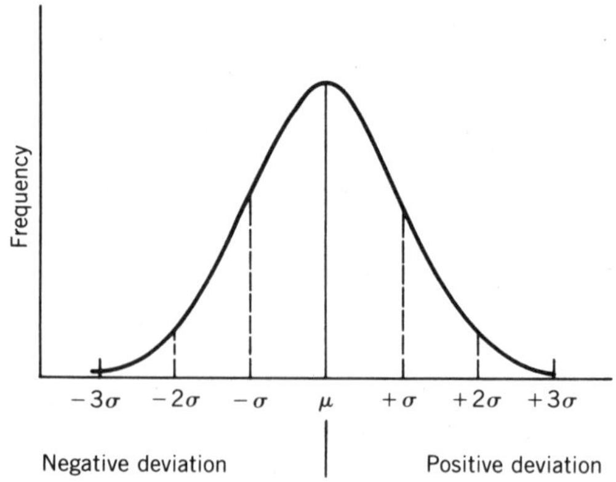

FIGURE 22.1 Normal (Gaussian) distribution curve.

22.1. For normal distribution, accuracy is best characterized by a term known as the *standard deviation* (σ), which is correctly defined for an infinite number of data points. However, for a set of n data points, a reasonably good estimate (s) of the value of σ is given by the expression:

$$s = \sqrt{\frac{\Sigma(x - \bar{x})^2}{(n - 1)}}$$

Standard deviation is frequently used to provide a measure of the precision by quoting *confidence limits*. Referring to Fig. 22.1, it can be demonstrated mathematically that the area beneath the curve from $(\mu - \sigma)$ to $(\mu + \sigma)$ is approximately 68.3% of the total area, from $(\mu - 2\sigma)$ to $(\mu + 2\sigma)$ it is about 95.5%, and from $(\mu - 3\sigma)$ to $(\mu + 3\sigma)$ it is about 99.7%. Therefore, in a large set of data points, 68.3% will lie within $\pm\sigma$ of the population mean (μ), 95.5% will lie within $\pm 2\sigma$, and 99.7% will lie within $\pm 3\sigma$. μ is approximately equal to ($\bar{x}$), defined earlier. For 95% confidence limits, a commonly used criterion, the measurement result is reported as equal to $\bar{x} \pm 2s$. The confidence limits used in the definitions of repeatability and reproducibility given previously were expressed as "one result in twenty," the 95% confidence limit.

22.3 CALIBRATION PRINCIPLES

The objective of calibrating a process analyzer are to provide a usable instrument giving meaningful results and to assure the user and other interested parties that the analyzer reading can be quoted with at least the same confidence as the result from the corresponding laboratory standard test method. Where the analyzer precision is better than that of the standard test method, calibration may be required to an accuracy sufficient to take advantage of that fact.

In theory, analyzer calibration is simple. The property of interest (e.g., component concentration, boiling point, etc.) is measured by the analyzer and by the corresponding standard test, on samples covering the range of the property and types of material of interest. The results are then compared. In practice, however, it is rarely simple to calibrate correctly and accurately. In addition, the manpower effort required for calibration is usually quite considerable. It should never be underestimated and attention should be given to minimizing the skills and effort required.

As we have already seen, process analyzers are fundamentally different from other process instruments—they measure the physical or chemical properties of the process stream rather than its state (e.g., temperature, pressure, and flowrate). No absolute standards are available for most properties measured. The precision of the reference test is frequently much lower than the precision of the process analyzer, contrary to the normal principles of calibration. Other important factors effecting calibration methods are the variety of different properties to be measured, the range of sample (product) types for which a particular property may need to be measured, and the wide range of analyzer systems required to meet industrial needs.

The following important factors have a bearing on analyzer calibration:

- The nature of the sample (e.g., gas, liquid, difficulty of handling).
- The nature of the property (specific, trace component, etc.).
- Whether it is an empirical or an absolute property.
- Continuous or cyclic measurement by the analyzer.
- Single, multi-stream or multi-component measurement per stream.
- Measurement principle and its precision relative to the standard test method.
- Availability of analyzer span and zero adjustments.
- Linear or non-linear analyzer calibration.

Each measurement by the standard test method and by the process analyzer is subject to error, as explained earlier. The test precision qualifies these errors in terms of repeatability, reproducibility, or both, and applies to the laboratory standard test method and the process analyzer. In the former case, data is usually well established for both criteria, whereas for process analyzers data is normally available on repeatability only.

Repeatability is usually better than reproducibility for the laboratory standard test method, since reproducibility takes into account variation associated with different laboratories, operators, and test apparatus. Repeatability for a process analyzer is usually better than for the corresponding laboratory standard test method, mainly because the analyzer is not subject to operator technique or judgment errors.

For calibration work, it is important to achieve the best accuracy possible. Obviously, the average value (mean) for a number of tests is a far better estimate of the value of a property than a single result. The accuracy of the mean improves in proportion to the square root of the number of tests used to calculate the mean. Therefore, any required degree of accuracy can be obtained by averaging the results of a sufficient number of tests performed under appropriate conditions. In practice, standard testing is very expensive in terms of the effort and skill required; therefore, the number of tests that can be recommended for calibration purposes must be limited. The general consensus seems to be that *four* standard tests for each calibration point is about the maximum that user laboratories would normally be prepared to accept.

22.4 BASIS OF CALIBRATION

The relationship between the result (y) of a standard test method and the corresponding result (x) from a process analyzer, is given by the expression:

$$y = A + Bx + Cx^2 + \cdots + Z$$

where A is the zero offset

B is the span or sensitivity

C is the curvature for non-linear calibration

Z is the sum of the errors associated with each measurement

Most analyzers have an approximately linear calibration, so the expression becomes:

$$y = A + Bx + Z$$

When the analyzer has facilities for zero and span adjustment, normally the case, by making $A = 0$ and $B = 1$, the calibration can be set up such that:

$$y = X + Z$$

If the errors (Z) are small, then we come close to the ideal relationship ($y = x$), that is, both standard test and process analyzer give identical results over the whole range of measurement. For the few non-linear analyzers, it is unusual to require terms in the expression beyond Cx^2.

Two points are sufficient to define a straight line; a third, intermediate, point can be used to confirm linearity. If non-linearity is suspected, up to five points may be necessary to define the curvature.

Therefore, the calibration procedure must establish accurately corresponding values of a property measured by the standard test method and the process analyzer at three (or possibly only two) points across the range of the analyzer. The best-fit line through the points can be fixed either by calculation or by a graph, the latter being more common and easier when the calibration is not linear.

22.5 PRECONDITIONS FOR ANALYZER CALIBRATION

Bear in mind that accurate calibration of a process analyzer is only worthwhile and justified if the following criteria have been satisfied:

- The analyzer has been correctly selected for the application.
- The analyzer has been properly installed and maintained.
- The analyzer is operating in accordance with the specified conditions.
- The analyzer is operating on a representative sample of the process stream.
- The analyzer is known to be capable of stable and reliable operation in the designated service.

This may sound straightforward, but a whole gamut of factors concerning design of the sample handling system and the analyzer installation itself must be considered.

22.5.1 Analyzer Selection

The development and design of most process analyzers referred to in this book have been improved largely as a result of industrial experience by users, followed by feedback to the manufacturers of information on operating problems. Therefore, the capabilities and limitations of the majority of analyzers are well known, and more than adequate information is available, from both users and manufacturers, for selecting process analyzers for particular applications.

But it would be unwise to expect performance better than that claimed by the manufacturer and due regard should be paid to any abnormal conditions that might reduce those claims. Dealing with analyzers employing new technology is an entirely different situation.

22.5.2 Factory Inspection and Testing

Analyzers and analyzer systems should be tested as fully as possible at the manufacturer's factory before shipment to the site. Corrections and modifications are far more easily handled at the factory than on site, and such testing will provide assurance to the user that the equipment will do the job it has been designed for. Test conditions and samples should simulate as closely as possible the expected site conditions. The analyzer should be calibrated using the procedures described later in this chapter. The subject of analyzer inspection is covered in more detail in Chapter 21.

22.5.3 Reliability and Stability

Factory calibration should be adequate in demonstrating that the analyzer is operating satisfactorily. This is essential before any calibration work is carried out on an installed analyzer.

In new plants (i.e., before the process stream is available) the analyzer can be run on a sample similar to the future process stream. Even in operating plants, there are advantages in testing first using a separate test sample before turning on the process stream. This will identify any specific analyzer problems from problems in the sample transport and conditioning system.

Reliability of the analyzer alone may be demonstrated by running on a test or process sample. However, reliability of the complete analyzer system can only be demonstrated if the sample transport and conditioning system is fully operational. On the other hand, *stability* of the analyzer can only be demonstrated on a constant-quality test sample.

Adequate reliability and stability are largely a matter of judgment and experience, but the consensus of opinion indicates that one should aim for routine attention at weekly or longer intervals. Any analyzer requiring attention more frequently than once a day is totally unacceptable. On this basis, satisfactory operation for at least one week is mandatory before attempting calibration.

22.6 CALIBRATION PROCEDURES

The general term "calibration," as applied to process analyzers, covers the initial calibration and any subsequent checking carried out during normal operation.

Initial calibration should accurately establish the relationship between the analyzer readings and the corresponding laboratory standard test results over the full range of the analyzer, including analyzer span and zero adjustments. Initial calibration may be carried out at the manufacturer's factory or on site before or after installation in the plant. It is often done in both places in order to correct any problems that may have occurred in transit.

After the analyzer has been commissioned, *in-service checking,* sometimes known as *validation,* is carried out at regular intervals or as required. Validation serves to identify any changes that may have occurred in the initial calibration. Usually only one point in the analyzer range is checked by validation.

A third type of calibration, often called *supervisory checking,* is a routine procedure using, for example, control charts for monitoring analyzer performance on a long-term basis. It is used largely to detect drift or other faults before they become serious.

It can be seen that analyzer calibration is not a single event, but a continuing process. The labor necessary to sustain and maintain confidence in analyzer results can be justified only by the benefits that accrue from using the analyzer. It is therefore important that analyzers be installed only in applications where worthwhile benefits are expected.

Calibration and Validation of process analyzers is usually accomplished by using either or both of the two methods described next. Each method has its advantages and limitations and the choice of which to use and when depends on many factors.

22.6.1 The Reference (or Standard) Sample Method

With this method, samples with accurately known measured properties, as established by the laboratory standard test method, are injected into the analyzer. The corresponding analyzer readings are obtained and compared with the laboratory test results. This method has the following advantages:

- It is the *only* method applicable when the normal process stream is not available. It is therefore used for factory checking and prestartup checking.
- The reference sample may be more easily handled and its reference value (or values) more accurately measurable than for the process stream.
- The whole range of the analyzer may be calibrated or tested at the user's convenience, whereas the alternative requires waiting for suitable values to occur in the process stream.
- The same reference samples may be used for initial calibration and for subsequent checking. They may also be used for testing different analyzers.

There are, however, some limitations:

- Reference samples are generally not identical in all respects to the process stream.
- The method whereby reference samples are injected into the analyzer does not generally use all of the sample conditioning system through which the process sample passes. Therefore the method may not show any effects of sample contamination, or check any factors affecting sample treatment.
- Sometimes it is not possible to provide suitably stable reference samples.

Therefore the choice of reference samples is usually dictated by the following criteria. First, the sample must be stable in bulk under the storage conditions prevailing at the plant. Often the user must take precautions to maintain stability (e.g., blanketing, heated enclosures). Second, the sample should be easily handled without deterioration of the measured property (e.g., mobile liquids or gases under pressure). Third, the sample material should be available at moderate cost. Process streams produced in the user's plant are frequently used for reference samples and afford this advantage to the user. However, the use of bought-in certified standards is becoming more widespread.

Reference samples provide an excellent means of validating a process analyzer at regular intervals or whenever its reading is questioned by process operations. The main advantage is that such validation can be carried out by maintenance personnel at any time and is independent of the availability of laboratory testing facilities. The result is also independent of the precision of the laboratory standard test, provided that the values of the measured parameters reported by the laboratory are accepted by all parties involved in analyzer calibration.

A satisfactory result by the reference sample method alone is not sufficient to check the complete analyzer system, since it generally cannot reveal errors arising from incorrect sampling and/or conditioning.

In summary, an incorrect result on a reference sample is a clear indication of analyzer malfunction. A correct result shows only that the analyzer has maintained its calibration for that reference sample. One or more of the reference samples used for initial calibration can be used for validation. It is advantageous to use the sample that corresponds most closely with the normal process stream value.

22.6.2 The Paired (or Grab) Sample Method

With this method, samples of the normal process stream being monitored by the analyzer are drawn off and tested by the laboratory standard test method. The corresponding analyzer readings and laboratory test results obtained are the calibration points. The method has the following advantages:

- Calibration takes account of sampling system effects on the analyzer reading. So the method is often specified for validation.
- The method always applies to analyzers in service, whereas the alternative procedure may not be practical.

The limitations of the method are:

- Initial calibration requires values of the measured property over at least a substantial part of the analyzer's range. Such a spread of values may occur in the process stream only over a long period, possibly weeks or even months, if at all. Therefore, initial calibration can rarely be done quickly or conveniently. This is perhaps the most severe practical limitation.

- The process stream property measured by the analyzer may be changing continuously in response to changing process conditions. Thus, it may not be possible to obtain, on any one sample, more than one analyzer reading from which to calculate an average analyzer value.
- The method is applicable only to analyzers in service (where the process stream is available). It cannot be used for factory or pre-startup checking.
- Testing in the laboratory is required. This may be a constraint and requires careful planning and scheduling.

Provided that the analyzer system is functioning satisfactorily and the plant is operating normally, the paired sample method checks the complete analyzer system. It therefore has significant advantages for validation.

The method uses a single analyzer reading because process stream quality cannot be assumed constant although, in reality, process stream quality during normal process operation does not fluctuate rapidly. Thus, any calibration work using the paired sample method should be timed to avoid periods of plant upset or deliberate operational changes. It is recommended that analyzer readings be taken before and after the validation value. These will serve to show the steadiness of the process during the test period and indicate the normality of the analyzer value taken. Supporting evidence of this nature will improve confidence in the results obtained.

22.7 DESIGN OF ANALYZER CALIBRATION FACILITIES

As we have seen in the previous section, there are basically two procedures in common use for analyzer calibration. Whichever method is used, some basic ground rules can be followed when designing facilities required for injecting standard samples into analyzers and for taking process stream samples for analysis by the laboratory standard test method.

22.7.1 Paired Sample Method

The basis of this calibration/validation method is the comparison of the result from laboratory analysis of a process stream sample, taken at a given time, with the corresponding analyzer reading at the same time. The method demands close coordination between maintenance and laboratory personnel to ensure correct timing of the various operations.

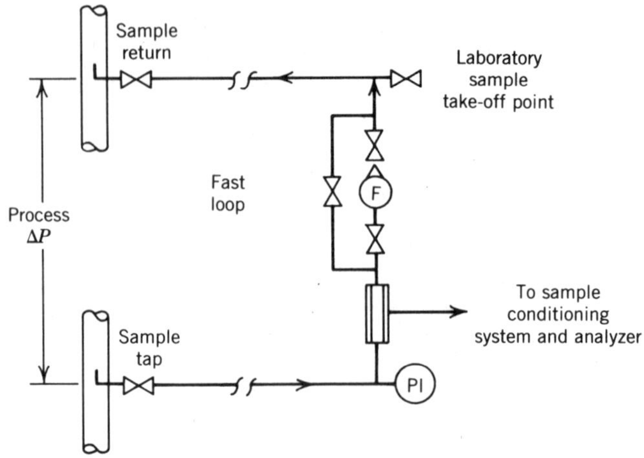

FIGURE 22.2 Location of laboratory sample point.

(a) Location of the Laboratory Sample Point

The laboratory sample point should be located such that the sample taken is representative of the process stream being monitored by the analyzer. This sounds relatively simple and logical, but there have been many instances where poor correlation between laboratory and analyzer results has occurred because the lab sample was taken from a different point in the process from that of the analyzer sample. The optimum location for the lab sample point is in the analyzer sample handling system. Where a fast loop is used, the lab sample point is best located downstream of the fast loop flowmeter, as shown in Fig. 22.2.

The lab sample point must be adequately identified. The identification should include the process stream being monitored and the tag number of the corresponding process analyzer. It is recommended that a schedule be produced, listing all laboratory sample points and their associated analyzers, together with their physical location on the plant. This will enable personnel responsible for taking samples to become fully familiar with lab sample point locations.

(b) Procedure for Taking Samples

Personnel responsible for taking samples must follow the correct procedure, as laid down by various organizations such as the ASTM or the IP. The method of taking the sample can have a very significant effect on the accuracy of the laboratory test result. There have been many instances where incorrect sampling procedure has created unnecessary problems.

(c) Effect on Analyzer Operation

Drawing laboratory samples from analyzer sampling systems must not affect the operation of the analyzer in any way (e.g., sample starvation). The sampling system should be designed so that there is an adequate sample flowrate in the fast loop to meet this requirement.

(d) Co-ordination with the Laboratory

Laboratory samples may be taken by process operations personnel or by laboratory personnel, depending on the plant organization. In either case, sample collection is usually a structured activity, that is, a sample is scheduled to be taken at a certain time each day, say 9.00 am. However, it is not always certain precisely at what time the sample is actually taken; for instance, it maybe any time between 8.30 and 9.30 am, depending on workload. If the process parameter being measured by the analyzer is changing at the time the lab sample is taken, this can have a detrimental effect on analyzer/lab correlation.

This is a difficult problem. One recommendation is to have the actual time of taking the sample noted on the lab sample ticket. Then we are faced with the problem of what the analyzer reading was at that time; one recommendation here would be to have the person taking the sample note the analyzer reading at that time.

With computer control of industrial processes becoming more commonplace, it is now possible to have the process control computer log analyzer readings at a specific time each day. Many users have adopted this procedure. Nevertheless, we are still faced with the problem of determining precisely when the lab sample is taken. In any event, the use of the paired sample method for analyzer calibration/validation presents problems of an organizational nature that can be solved in a number of ways, depending on the preferences of the user.

22.7.2 Reference Sample Method

This method is based on the injection of a known sample, accurately tested by the laboratory, into the analyzer and a comparison of the known value of the measured parameter with the corresponding analyzer reading. It is therefore possible to design facilities for injecting the standard sample into the analyzer as part of the analyzer sampling system.

(a) Design Criteria

When designing such systems, bear in mind that the calibration/validation sample is, in effect, another process stream. As in the design of multi-stream sampling systems, care must be taken to avoid cross-contamination. In addi-

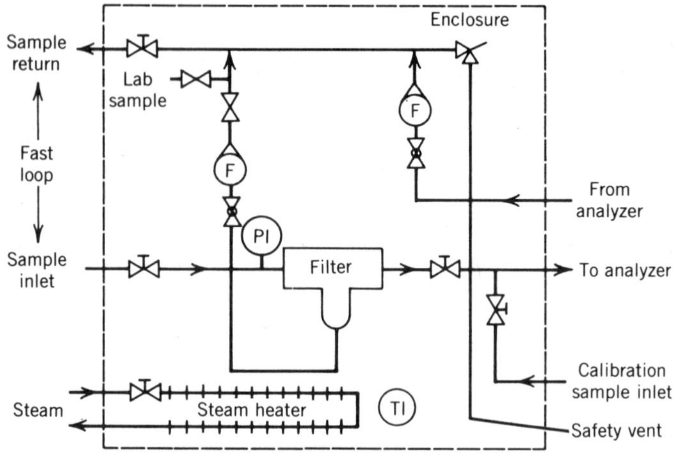

(*a*)

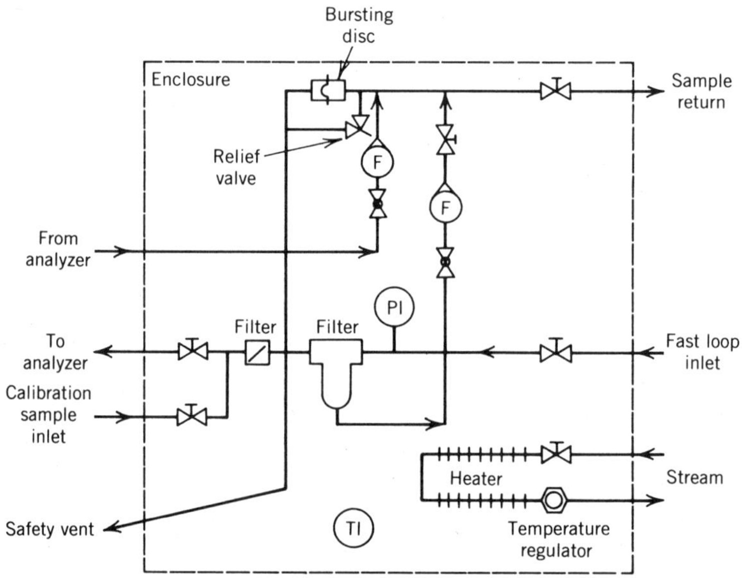

(*b*)

FIGURE 22.3 Location of the calibration sample inlet. (*a*) Typical liquid sample conditioning system; (*b*) typical liquid sample conditioning system for a gas chromatograph showing a manual calibration sample injection.

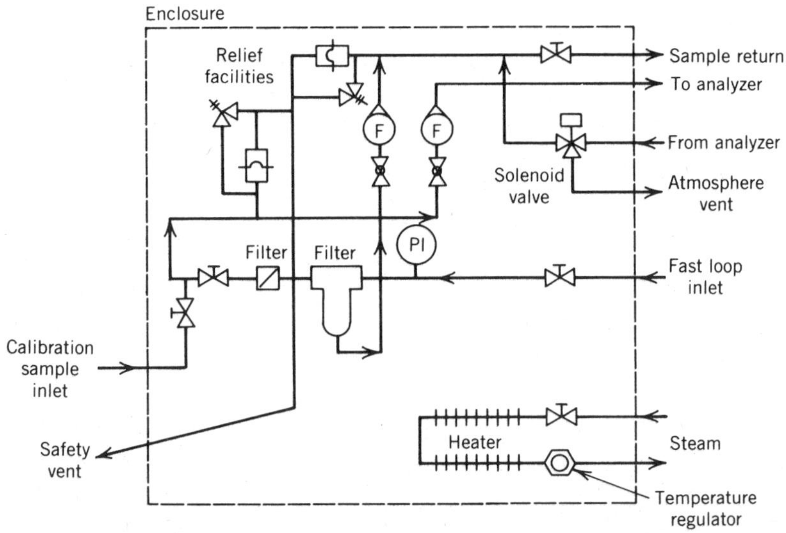

FIGURE 22.4 Typical vapor sample conditioning system for a gas chromatograph showing a manual calibration sample injection.

tion, introduction of the standard sample must be accomplished without disconnecting lines, changing settings, or making other adjustments.

In view of the extensive industrial use of the reference sample method, manufacturers supplying sample handling systems with their analyzers have for many years designed these systems with facilities for manual injection of calibration samples. Typical examples are shown in Figs. 22.3 and 22.4, for liquid and vapor samples respectively. The system shown in Fig. 22.5 is designed for a boiling point analyzer installation and allows filling of the calibration sample storage vessel from the analyzer fast loop. This sample can then be tested and validated by the laboratory standard test method and used as the certified standard for analyzer calibration or validation.

(b) Microprocessor-Based Analyzer Systems

Until microprocessor-based systems became available, analyzer calibration was done manually, that is, the standard sample was injected into the analyzer either locally (by manually switching the appropriate valves at the analyzer sampling system) or remotely (by pushing a button on the analyzer control unit). Adjustments to analyzer response factors were then made manually by the maintenance technician.

With the advent of microprocessor-based analyzer systems, particularly in process gas chromatography, there has been a trend toward automatic calibration. Again, standard sample injection facilities are provided as an

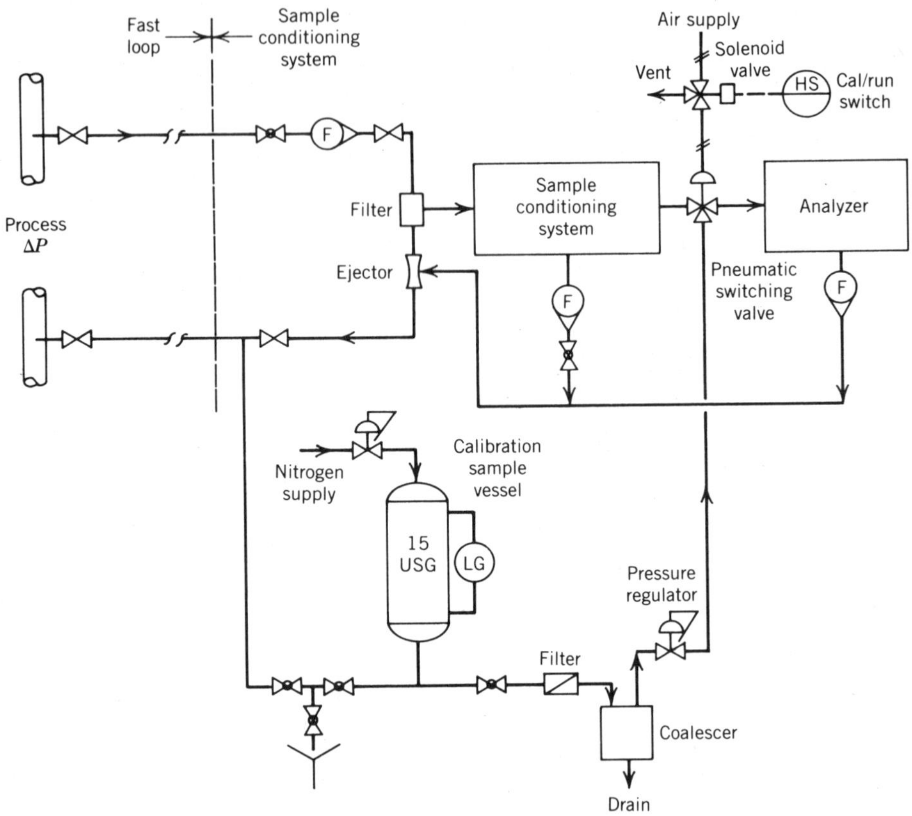

FIGURE 22.5 Calibration sample injection system for a boiling point analyzer.

integral part of the analyzer sample handling system, as described earlier. Using the power of the microprocessor, standard samples can be injected into the analyzer on command or on a preprogrammed time basis, say every 24 hours. This calibration/validation philosophy is also being extended to other analyzer types, such as IR analyzers, where zero and span calibrations can be carried out in the same manner.

One must be careful in defining the term "automatic calibration." With true automatic calibration, the standard sample is injected *automatically* into the analyzer and the analyzer response factor(s) are updated *automatically* by the analyzer computer. The disadvantage of this procedure is that analyzer response factors will be updated without the maintenance technician knowing it has occurred. Some constraints on this function are usually required. Some users set limits on the variation in response factor, say $\pm 2\%$; when the response factor is outside these limits, the situation is flagged in the computer and the response factor is not updated. Other users dispense with the automatic response factor updating function entirely. In any event, the choice is available in the analyzer computer.

The use of microprocessor-based GC computer systems and the recent trend to analyzer computer systems, where all process analyzers installed in a process plant are interfaced to this computer, makes it easy to control and document analyzer calibrations and highlight any major problems in analyzer reliability and stability.

22.8 SUPERVISORY CHECKING PROCEDURES

As we have seen, initial calibration is usually carried out prior to operation on the process stream and validation is carried out when required, at periodic intervals during normal operation. Supervisory checking, often known as correlation or control checking, is a routine procedure used for monitoring analyzer performance on a long-term basis.

The usual method is to use what are known as control charts. These are plots of process analyzer readings (x) against the corresponding laboratory test results (x_r), as shown in Fig. 22.6. The procedure is as follows. The analyzer readings and corresponding laboratory test results are noted in a log book, together with the date and time of the measurements. Pairs of values of x and x_r are plotted on a graph and the best straight line through the points is determined, represented by the equation:

$$x_r = mx + C$$

where m is the slope of the line and C is the intercept, often known as the *initial error* or *bias*.

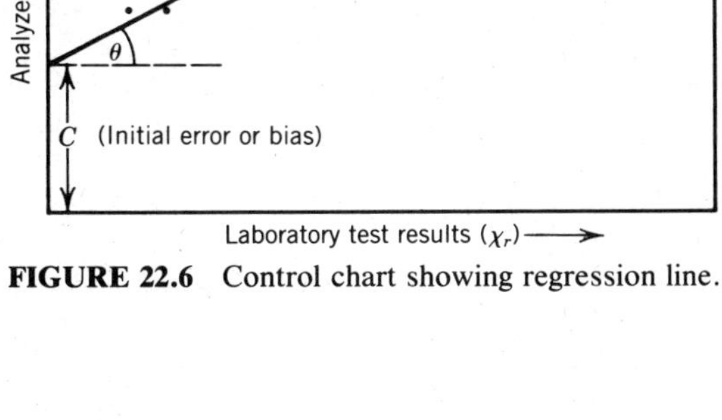

FIGURE 22.6 Control chart showing regression line.

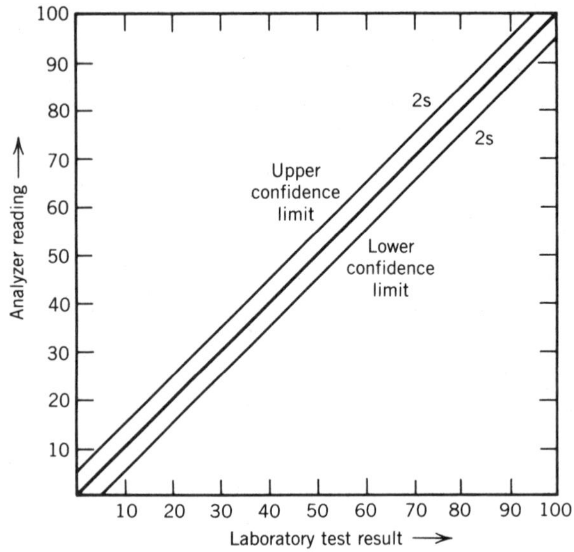

FIGURE 22.7 Control chart showing action lines.

Once the "regression line" has been established, two additional *action lines* are drawn above and below it, as shown in Fig. 22.7, such that 95% of all results lie between the action lines (±2s). The value of s used here is the same as that established for repeatability (see Section 22.2.3). If it is found that more than 5% of results fall outside the limits of the action lines, some remedial maintenance action is necessary.

Sometimes, in addition to plotting the 95% confidence limits, an extra pair of lines is plotted at ±3s, corresponding to 99.7% confidence limits. The ±2s lines are then known as "warning lines" and the ±3s lines as "action lines." Corrective maintenance action is usually taken if *one* point falls outside the action lines or if *two successive points* fall outside the warning lines.

BIBLIOGRAPHY

Bissell, A. F. Cusum techniques for quality control. *Appl. Statistics*, **18**, 1–30 (1969).

Converse, J. G. Internal reference introduction for performance verification, diagnostics and calibration validation. *Proc. ISA Symp.*, Boston, MA (May 1985).

Converse, J. G. Portable analyzers: They still have a place in the plant. *InTech* (June 1985).

Cornish, D. C., and Timmermans, F. A. G. Laboratory procedures for the measurement and evaluation of quality. *Proc. ISA Conf.*, Chicago, IL (Oct 1979).

Davies, O. L., and Goldsmith, P. L. *Statistical Methods in Research and Production*, 4th Ed. Oliver & Boyd, London (1972).

Fisher, R. S. Controlled saturators for accurate humidity generation. *Proc. Intl. Moisture & Humidity Symp.*, Washington DC (April 1985).

Fisher, R. S. An accurate flow-division humidity generator. *Proc. Intl. Moisture & Humidity Symp.*, Washington DC (April 1985).

Harding, J. C. Wide range humidity transfer standard. *Proc. Intl. Moisture & Humidity Symp.*, Washington DC (April 1985).

Hayward, A. T. J. Repeatability and accuracy. *Mech. Eng.*, London (1977).

Institute of Petroleum Code of Practice For Calibrating Analyzers:

IP 340/82: General Guide to Principles and Methods Used.

IP 341/82: Capillary-Type Viscometers.

IP 347/82: Cloud Point Analyzers.

IP 348/82: Distillation Analyzers.

IP 349/82: Flash point Analyzers.

IP 353/82: Gas Chromatography Analyzers.

Mator, R. J. Trace moisture analyzers and their calibration. *Anal. Instr.*, **12,** ISA (1975).

Oster, D. W., and Kowalski, B. R. Background detection and correction in multi-component analysis. *Anal. Chem.*, **57,** 908–917 (1985).

Scelzo, M., and Pierce, R. A high-precision saturation/dilution calibration system for water vapor in a carrier. *Proc. Intl. Moisture & Humidity Symp.*, Washington DC (April 1985).

23

Maintenance and Performance Evaluation

23.1 INTRODUCTION

With the present emphasis on computer control of industrial processes, maximum operating efficiency and optimization, high reliability, and service factor of process equipment in general have become important considerations in the eyes of plant management. Any cost-effective means by which reliability and service factor can be increased is considered justified. Since the process control computer relies on accurate, reliable data from field instrumentation such as flow, level, pressure, and temperature transmitters and on-line process analyzers, these instruments must receive adequate maintenance to achieve maximum reliability and service factor.

The process analyzer is significantly different from conventional field instrumentation in that it has an analytical chemical basis as well as an instrumental basis. In most cases, a sample handling system is required to provide a representative sample of the process stream. Many analyzers also require protection from adverse ambient conditions. Taking these factors into account, technicians responsible for analyzer maintenance need to be thoroughly familiar with the specific equipment and trained in all aspects of analyzer systems. It is a generally accepted fact, borne out by user experience over the years, that process analyzer maintenance should be carried out by experienced, dedicated analyzer technicians, responsible for analyzer maintenance only.

Analyzer maintenance always had to be a scheduled and coordinated activity of a preventive nature. In the early days of process analyzers, very little operational experience was available, so it was not possible to achieve this ideal. Until a pool of analyzer engineering and maintenance experience was built up in the user organizations, breakdown maintenance was the order of the day and probably contributed to the bad name that analyzers received in the eyes of process operations personnel in the early 1960s.

It is satisfying to note that the situation is entirely different today. A wealth of analyzer experience has been built up within the user organizations covering engineering, operation, and maintenance of analyzer systems; expertise and experience of analyzer manufacturers has increased significantly; and advances in technology over the years have resulted in more reliable equipment and improvements in analytical techniques. In addition to these improvements, we now have the power of the microprocessor to help us in diagnosing problems, as well as documenting and reporting on equipment maintenance and performance.

23.2 PROCESS ANALYZER MAINTENANCE

The success of a process analyzer maintenance activity depends on the organization, the caliber and training of maintenance personnel, and the size and complexity of the analyzer package installed at the particular plant. Analyzer system design can also play a very important part in this overall picture. For instance, as mentioned in Chapter 20, badly designed sample handling systems will require much more maintenance attention than well-designed systems.

23.2.1 Optimum Service Factor

An interesting item about analyzer systems is the term "optimum service factor" and what is meant by it. Experience by some companies indicates that by increasing the amount of preventive maintenance the number of analyzer failures can be reduced significantly, resulting in higher service factors. Unfortunately, a point is reached where there is very little further improvement in service factor to justify the additional manpower requirements and cost involved. Thus, there appears to be an "optimum" level of preventive maintenance required to achieve the maximum service factor for a particular analyzer.

Bear in mind that each type of analyzer may require a different level of preventive maintenance, based on the principle of operation. For example, a wet chemical analyzer will require more attention to keep it in tip-top condition than will a pH or conductivity analyzer.

How does one identify this "optimum"? By experimentation. First, increase preventive maintenance until little improvement in service factor is obtained, then gradually decrease it until the service factor starts to drop off. How can the optimum be increased? Here are some possibilities that might be considered:

- Select trouble-free sample handling systems.
- Obtain more reliable analyzer systems from the manufacturers.
- Extend the use of microprocessor-based analyzer systems, where improved fault diagnostic capability can be used to advantage.
- Conduct a careful analysis of repetitive failures for each type of analyzer and decide on systematic replacement of critical components.

These suggestions all depend on operational experience; continuous dialogue with the manufacturers to feed back user experience and thereby provide information to help them improve their products; use of state-of-the-art technology; and, above all, effective and efficient documentation systems.

23.2.2 Technician Training

As mentioned earlier, the training and experience of analyzer maintenance technicians will have a significant impact on the effectiveness of any maintenance activity.

For a specific analyzer, there is no doubt that the training course offered by the manufacturer is the best place to obtain an in-depth knowledge of that particular equipment. Most analyzer manufacturers are able to provide this facility. A good time to get hands on experience with it is during the factory check-out prior to shipment. This can be beneficial to both the analyzer engineer and the technician.

Some companies supplement the manufacturer's training course with in-house training courses of their own. These use the full extent of the company's expertise and experience in analyzer engineering and maintenance. In-house training has the great advantage that subjects pertinent to the company's operations can be discussed in detail, such as sample handling systems, applications, maintenance experience with a particular analyzer, and so on, information that no manufacturer's course can provide. In addition, the in-house course can be general rather than specific to particular equipment.

Until a few years ago, no general training courses in analyzer technology were available to industry. The Institute of Measurement and Control in England runs a School of Analytical Instrumentation at the University of Warwick every year. This has reportedly been a great success. In the United States, there has been little movement in this direction, although the Instrument Society of America (I.S.A.) is currently making a course in process analyzers available to industry.

23.2.3 Number of Analyzers per Technician

The number of analyzers to be maintained by one technician remains a subject of much discussion. The general consensus seems to be that an experienced technician should handle from 10 to 15 analyzers, but this rather simple statement requires further elaboration.

The relative complexity of analyzers must be considered. Some companies "weight" analyzers in terms of their relative complexity in order to assess maintenance manhours required and therefore number of technicians. A chromatograph, wet chemical analyzer, or moisture analyzer might be considered complex; an IR, oxygen, or density analyzer, of medium complexity and a pH or conductivity analyzer, of low complexity. An added complexity is the question of multi-stream analyzers. Again, "weighting" the analyzer in accordance with the number of process streams monitored appears to be the preferred approach.

23.2.4 Organizational Aspects

Process analyzer maintenance, as with all industrial equipment, needs to be a well-organized and coordinated activity. Management/supervision of the activity is best carried out by a person who is familiar with all aspects of process analyzer technology and with the equipment installed at his or her plant. In large industrial complexes (e.g., refineries, petrochemical plants, etc.) consisting of many different process units, there are several advantages with this arrangement:

- The supervisor can function as a central coordinating point for all analyzer maintenance activities. Process operations personnel know who to contact in the event of an emergency.
- The supervisor can act as a motivating influence on his or her technicians by being available to handle their problems and answer their questions.

- The coordination and allocation of maintenance manpower is far more effectively controlled by one supervisor.

For less experienced technicians, on-the-job training is an important factor in their career development and needs effective coordination. Once again, the analyzer maintenance supervisor is in a good position to provide such coordination.

For industrial plants where many different types of analyzers are installed, how to organize the maintenance activity has always been difficult to address. One approach would be to have each technician responsible for a given type of analyzer (e.g., chromatograph, viscometer, density analyzer). Over a period of time that technician would acquire an intimate knowledge of the particular equipment and the maintenance efficiency would increase. This arrangement has the disadvantage that the technician is not able to expand his or her overall knowledge of different types of analyzers; this may act as a demotivating influence, finally resulting in a reduction in efficiency. Obviously a compromise must be adopted, the objective being to achieve optimum maintenance efficiency with the available manpower.

The preceding discussion is directed primarily toward organizations that can draw upon in-depth analyzer expertise. What of those small plants that have little or no analyzer expertise? In this situation there are two possibilities:

- Build up analyzer expertise by initiation of a rigorous training program for selected maintenance technicians, including manufacturers training courses and in-house courses if available.
- Set up maintenance contracts with the analyzer manufacturers or companies that specialize in analyzer maintenance.

Unfortunately analyzer maintenance expertise is built up over a period of time, based on hands-on experience with the particular equipment. It cannot be obtained quickly.

Having the analyzer manufacturer carry out maintenance on a contract basis can be beneficial, since some in-house training of technicians can be conducted by the manufacturer's service engineer during the periodic visits. The success of such an arrangement depends on the extent of support available from the manufacturer at a particular location. For instance, a contract of this kind may work well in the United States or Europe but be an absolute disaster in the Far East. So it is vital for small operating companies with little analyzer expertise to find out how much manufacturer support is available at the plant location before an analyzer is purchased.

23.2.5 Calibration Schedules

As mentioned in Chapter 22, analyzer calibration has proved to be an important factor in maintaining high reliability and the confidence of operations personnel in analyzer installations.

The calibration frequency for a particular analyzer is usually established by operating experience. For example, after commissioning, an analyzer may be calibrated once a week or more frequently, dropping to once a month after experience has been gained in operation of the equipment and confidence in the readings has been built up.

Establishing a calibration schedule and monitoring the number of scheduled calibrations completed each month can be a very good way of maintaining a high level of confidence in analyzer operation and thus achieving high reliability and service factor.

With the increased use of computer-based analyzer systems, particularly GC systems, has come the possibility of conducting preprogrammed, automatic calibration of analyzers at the push of a button. It is believed that such systems will be used more and more in the future because of the contribution they can make to efficiency and confidence.

23.2.6 Spare Parts

To carry out efficient and timely maintenance on analyzer systems, an adequate supply of spare parts must be held in stock. The decision on which spares to hold for a particular analyzer is largely a matter of experience and the extent of standardization at a particular site location. Normally the manufacturer is requested, at the quotation stage, to provide lists of recommended spare parts for start-up and for two years operation of the equipment. These lists are then reviewed and revised (usually downward), based on previous site experience with the type of analyzer, and the spare parts are ordered with the analyzer.

Particularly in recent times, spare parts have proven to be an expensive proposition and contribute significantly to overall maintenance costs. With the accent now on cost reduction, some user companies have found it considerably cheaper to purchase spares directly from component suppliers rather than from the analyzer manufacturer.

In any event, spare parts costs must be included in the overall analyzer maintenance cost, when calculating analyzer payback (i.e., analyzer total installed cost plus operating cost vs process credits.

23.3 COORDINATION WITH THE PLANT LABORATORY

The plant laboratory has always been the governing body with regard to quality control in any industrial process plant and is likely to remain so. For many finished products, customer acceptance is based on specification tests carried out by the laboratory.

As the reliability of on-line process analyzers has increased over the past few years, more and more analyzers are being used as part of advanced control systems and for on-line monitoring of product quality specifications.

Nevertheless, from a traditional point of view, the laboratory is still looked upon as the "bible" as far as product quality is concerned. It is therefore logical that the laboratory should be intimately involved in the validation/ calibration of process analyzers. The question of how and to what extent this involvement should be has been the subject of animated debate for many years.

The methods of validation/calibration discussed in Chapter 22 rely on effective coordination between analyzer maintenance forces and the laboratory; in the preparation and testing of samples for use as analyzer calibration standards; or in the spot checking of plant samples for comparison with analyzer readings. In any event, there must be efficient and effective communication and mutual agreement reached as to whether an analyzer is giving correct data or not.

The fact that we are often dealing here with two completely separate organizations within the same company makes the job even more difficult. In the older industrial plants, process operations personnel are more likely to believe a laboratory test result than a process analyzer reading, even though the analyzer may be correct (one reason why scheduled calibrations are important).

Some companies have tackled the problem with a good measure of success by combining analyzer maintenance personnel and laboratory personnel into one group under the guidance of one supervisor. This arrangement has the advantage that even though problems still exist they are solved "under one roof" and an agreed position taken for the benefit of process operations.

The trend today appears to be toward increased use of analyzers for process control and decreased use of routine testing in the laboratory. In fact, some plants run largely on the basis of on-line quality measurement, with little routine laboratory testing. In such situations, the laboratory is deeply involved with analyzer validation/calibration.

In the future we may see a significant change in the role of the laboratory from that of a routine testing facility to that of a process analyzer calibration/ validation facility, together with the capability of conducting special testing when required. In such a situation, coordination between the analyzer maintenance and laboratory organizations will become even more important (if they exist as separate organizations).

23.4 SERVICE FACTOR AND PERFORMANCE

Earlier in this chapter, the term "service factor" was used several times. It is appropriate now to discuss some of the terms used in industry to define analyzer performance.

The *service factor* (or performance) of an analyzer installation is often reported using the term *availability,* which is calculated as a percentage from the following equation:

$$\text{availability (\%)} = \frac{\text{time analyzer is operating and calibrated}}{\text{time process unit is operating}} \times 100$$

Note that the time taken to calibrate the analyzer is built into the calculation of availability. The analyzer is not available for use in process control while it is being calibrated. The availability is not debited if the process unit is shut down.

Another term used to determine service factor is *utilization,* which is calculated as a percentage from the following equation:

$$\text{utilization (\%)} = \frac{\text{time analyzer is used by process operations}}{\text{time process unit is operating}} \times 100$$

Percent utilization is much more difficult to determine than percent availability, although in the long run percent utilization is more important. Utilization requires a knowledge of whether the analyzer is *used* by process operations, irrespective of whether it is considered *available* for use.

If a process control computer is used for advanced control of a process plant, the situation becomes a little easier. For analyzers used in advanced control loops, the time the analyzer is out of service for any reason (i.e., when the control loop is broken) can be logged by the computer. For monitoring-only applications, there is no easy answer. However, if analyzer availability is reported as 98% and its utilization as 25%, there is obviously something wrong!

A discussion with process operations will usually reveal the cause of the problem. For example, an analyzer might be highly reliable and give correct analyses when compared with lab results, but may be of little use to the process operator because it is taking a sample from the wrong point in the process.

Other terms related to analyzer performance are *reliability* and *maintainability,* both of which are a function of the time over which the parameters are measured.

$$\text{Reliability } (R) = e^{-\lambda t}$$

where λ is the failure rate (= 1/MTBF), t is the time duration, and *MTBF* is the mean time between failures.

$$\text{Maintainability } (M) = 1 - e^{-t_r/\text{MTTR}}$$

where t_r is the repair time and MTTR is the mean time to repair.

The repair time can be further divided into three categories:

- Time for corrective repair.
- Time for preventive maintenance.
- Time for out-of-service overhaul, should this be necessary.

For critical analyzer applications, where even a small amount of maintenance down-time is unacceptable, it is often the practice to install duplicate analyzer systems to achieve maximum availability. With such systems, it is essential that any failure be promptly recognized and corrected.

23.5 DATA GATHERING AND DOCUMENTATION

23.5.1 Basic Requirements and Objectives

It is apparent from the foregoing that for any analyzer maintenance or performance evaluation activity to be successful it must rely on an efficient and effective data gathering and documentation system. For companies having central engineering and support organizations, such a system can be extremely useful in helping the central organization keep up to date with the current situation at the different plant locations. This can be a difficult problem without adequate documentation.

The requirement for good documentation can be illustrated by reference to what industry is trying to achieve in terms of analyzer availability. Although several figures have been quoted in recent years, here is a reasonable set of objectives:

1. For computer control applications, 98% minimum.
2. For monitoring only applications, 95% minimum.
3. For safety applications, 100% minimum.
4. Any analyzer installation with an availability less than 90% is unacceptable and the analyzer should be either removed from service or upgraded to an availability of 95% of greater, depending on its application.

What does this mean in terms of analyzer downtime? Assume that a minimum availability of 98% is desired for all analyzer installations. This means a total downtime of 14 hours per month per analyzer, of which, say, two hours constitutes calibration time. It is therefore evident that the method by which the maintenance data is obtained (data gathering) will have a significant effect on the accuracy with which availability can be calculated. For large operating companies with many process plants located in various regions of the world, it also means that there is justification for some form of standardization in the method of data gathering and documentation.

How near can one get to 98% availability for all analyzer installations. Is it feasible? This is a difficult question to answer since each type of analyzer may require a specific amount of attention to keep it operating properly, even with an experienced technician looking after it. A number of other factors are also involved:

- The maintenance procedures used.
- The extent and caliber of manufacturer support at the particular location.
- The availability of analyzer training programs.
- The ability to pinpoint major problem areas.
- The ability to solve these problems rapidly and effectively, thereby reducing required maintenance and downtime to a minimum.
- The ability to generate the data required to calculate availability, maintenance costs, and so on.

23.5.2 Information Requirements

If one is starting an analyzer documentation system from scratch, there is no easy answer to the question, "how is the necessary data obtained?" It will require a concerted effort on the part of site personnel and involve a number of different disciplines (e.g., process engineers, analyzer engineers, computer applications engineers, analyzer maintenance personnel). Even when the required data has been obtained, one is then faced with the problems of how to keep it up to date and how frequently such updating should be done.

The information required for analyzer performance evaluation may be divided conveniently into two types:

- Analyzer application data (see Table 23.1)
- Analyzer maintenance history data (see Table 23.2)

The key to any analyzer documentation system is the application data. Since the main objective of documenting analyzer installations is to calculate

TABLE 23.1 Analyzer Application Data

Plant location	Application/service
Process unit	Manufacturer
Analyzer type	Model number
Analyzer tag number	Year installed
Component tag number	Measurement range hi/lo
Process stream number	Engineering units
Measurement description	Potential savings (K$/yr)
Multi-stream analyzer yes/no	Date of current data
Analyzer computer yes/no	

TABLE 23.2 Analyzer Maintenance History Data

Plant location	Calibration frequency (days)
Process unit	Calibration time (hours)
Analyzer type	Calibration method
Analyzer tag number	Analyzer downtime (hr/month)
Application/service	Analyzer availability (percent)
Data of current entry	

Maintenance problems:
1. Sample handling system
2. Analyzer
3. Utilities, peripheral equipment
4. Spare parts
5. Vendor support at location
6. Other

the *actual* return on the investment (analyzer payback), the application data should include the *potential* savings that can result from use of the analyzer. Calculation of these savings usually assumes that the analyzer is operating at 100% availability, that is, it is operating, calibrated, and used by process operations whenever the process unit is operating.

The maintenance history data is concerned primarily with actual analyzer performance. Percent availability is the most important item here, since it will allow calculation of the *actual* savings accrued by use of the analyzer. Brief data should also be included covering the major problems (if any) with the analyzer installation, such as critical failure items, sampling problems, and so on. Such information will be an aid to analyzer maintenance personnel in improving the reliability and service factor of the installation and to analyzer engineers by feeding back information that will allow them to improve system design.

23.5.3 Data Base Management Systems

An analyzer documentation system of the type described can create major problems for large companies operating many plants:

- The effort and coordination required to gather the necessary information.
- The need to keep this information up to date.
- The resulting mass of paperwork generated.

With the advent of the computer and microprocessor, we have a powerful tool at our disposal in the form of computerized data base management systems (DBMS). These vary from small, stand-alone microprocessor-based systems to powerful systems run on a mainframe computer.

It must be emphasized that the computerized DBMS does not remove the need for gathering the required data, nor the effort required to enter the data into the data base. It does significantly reduce the effort of updating information and has the capability of providing a wide variety of searching, sorting, and report writing routines not available in manual systems, thus reducing the amount of paperwork involved.

With today's technology, large operating companies may have a terminal installed at each plant location, where plant personnel can address the DBMS to review their own data or data from another plant location. Maintenance of the system and data updating can be done either at a central location or at the individual plant location.

23.6 ANALYZER PERFORMANCE MONITORING

The DBMS described has the main function of documenting analyzer installations on a company-wide basis. Reports generated by the system can be used for a variety of purposes:

- Keeping management and engineering personnel up to date on current analyzer usage and performance.
- Highlighting major problems with analyzer installations and identifying critical failure items.
- Acting as a communication channel between different plant locations.

One must now address the question of how all the information entered into the DBMS is generated at the plant location. As mentioned earlier, this effort requires close coordination and supervision, together with some form of standardization in the method of approach. The data required for each analyzer installation will include the following:

- Calculation of percent availability.
- Calculation of MTBF.
- Calculation of MTTR.
- Analyzer downtime.
- Process downtime.
- Number of repairs carried out.
- The components repaired and their location (e.g., sampling system, analyzer, utilities).
- Cost of replacement parts.

If performance is to be monitored on a monthly basis, the data listed must be gathered at the end of each month and compared with the previous month's data. This all adds up to a mountain of paperwork.

The use of a stand-alone microprocessor-based data acquisition system can be a cost-effective and efficient way of reducing the effort involved. The analyzer technician can address the system through a terminal and enter all initial data required by the system through a keyboard. By making the system "user-friendly," the technician can be given some "ownership" of the system. Advantage can be taken of the computational ability of the microprocessor to calculate availability and other parameters and to offer software packages capable of searching, sorting, and writing reports. In addition, hard or floppy disks can be used to store data for historical recall.

Such a system can provide all the data necessary to evaluate analyzer performance and identify potential problem areas. It still requires a coordinated effort on the part of analyzer maintenance personnel to ensure accurate recording of all information and timely input into the system.

BIBLIOGRAPHY

American instruments abroad: Opportunities for trade. *Intech* (Nov 1984).

Black, J. W. Model estimates analyzer payouts. *Hcbn. Proc.* (Sept 1981).

Cropper, W. V. Effects of process monitors on refinery laboratory operations. *API Oper. Pract. Comm. Mtg.*, Houston, TX (May 1966).

Cropper, W. V. Instrumentation for quality assurance in refineries. *ASTM Comm. D-2 Mtg.*, Houston, TX (Dec 1966).

Davies, A. Assessing maintenance performance by Cusum monitoring. *Plnt. Eng.*, **22,** No. 4, 15–17 (1978).

Hough-Grassby, A. W. The evaluation of process analyzers. *Adv. in Instr.*, **34,** ISA (1979).

Jutila, J. M. On-stream analyzers. *InTech* (Oct 1980).

Robinson, C. User evaluation of on-line composition analyzers: A mixed bag. *InTech* (May 1985).

Schnelle, P. D. Jr., and Cole, T. L. Improving analyzer availability and accuracy using digital filters. *InTech* (Feb 1984).

24

The Role of Analyzers In Process Control Systems

24.1 INTRODUCTION

Since the first process analyzers were installed in the late 1950s, tremendous interest has been shown in the use of these analytical sensors in industrial process control systems. This is evident from the extensive coverage of the subject in the scientific literature.

In the early days, analyzer reliability unfortunately left much to be desired, confidence in analyzer performance was understandably low, and there was very little experience in either the users' or manufacturers' organizations. The net result was that if analyzers were used at all they were installed on an experimental basis, or in monitoring-only applications where the data generated by the analyzer was used as the basis for the *operator* to make a change to process operating conditions. This procedure is commonly referred to as "open-loop" control. There were a number of "closed-loop" control applications in which analyzers were used, but they were the exception rather than the rule.

After more than 25 years experience in analyzer technology, and much more in process control, the author is gratified to report that the situation is entirely different today. This has been brought about by improvements in measurement technology and materials design, but primarily by the introduction of the microprocessor and computer into the process control world. Process control computers and digital instrumentation systems have provided the capability for sophisticated and highly efficient advanced control systems, enabling industry to more closely control and optimize process plant operations. In many cases, these advanced control systems must have access to timely process stream quality data, provided by process analyzers, for them to function properly.

The role of analyzers in process control systems is therefore a real and important one. The main justification for installing process analyzers is their major contribution to efficient and effective industrial process control. This chapter summarizes the features of process analyzer systems that are important from a process control standpoint; they are also covered in detail elsewhere in this book. There is no attempt here to get involved in the detailed math of process control; there are many excellent textbooks on this subject to which the reader is referred. Rather, the purpose is to outline the main concepts associated with the analyzer system part of the control loop.

With regard to specific applications, it would be impossible to cover the thousands of applications that exist on a multi-industry basis. Therefore, the discussion here is limited to those *major* categories that are commonly applied in the petroleum and petrochemical industries with which the author is familiar. It is anticipated that the basic concepts outlined here will also apply to most process analyzer applications in other industries.

24.2 CONTROL REQUIREMENTS

The best way to illustrate the basic requirements for analyzers to be used in process control systems is to first discuss the hierarchy of process control, that is, the various levels of control at which the process analyzer can fit into the system.

24.2.1 Control Hierarchy

First, we must identify the differences between the two *modes* of control, manual and automatic, with which every instrument/control engineer is familiar. In manual control, the process operator is responsible for exerting control, the quality of which depends on the caliber of the operator. Obviously this is a low-cost solution to the control problem, but unacceptable from an efficiency viewpoint. Automatic control, on the other hand, is more costly (instrumentation is needed) but provides a consistently high quality of control and is the preferred solution.

In today's world of computer control and digital instrumentation systems, there are basically three levels of control, each of which has its own specific requirements for process analyzers.

(a) Regulatory Control (Level I)

This is the level at which the basic control of the process is carried out. The instrumentation system will contain and manipulate the elements necessary to perform this function (e.g., meters, controllers, valves, etc.). If analyzers are to be used at the regulatory level of control, certain requirements must be satisfied. The two most important of them are:

- Rapid response.
- High reliability and service factor.

Analyzer applications in Level I control systems are not very popular because analyzer failure is reflected directly to the final control element. Only selected analyzers (e.g., pH, density, conductivity), which are continuous analyzers with demonstrated high service factor and reliability, have been used. The majority of analyzers, except those capable of being installed in-situ (directly in the process stream), have response times that are too slow for Level I control.

(b) Supervisory Control (Level II)

Supervisory control of an industrial process is usually carried out in the process control computer, sometimes in the digital instrumentation system. Here the rules are somewhat different:

- A slower response time is acceptable.
- A lower analyzer service factor is acceptable.

This is because even in the event of an analyzer failure the basic regulatory control system will continue to function. Bear in mind that even if a lower service factor is *acceptable* in Level II control, it is not a reason to compromise the basic philosophy of keeping analyzer service factors as *high as realistically possible*. In other words, there is some leeway in Level II. Most process analyzers are applied at this level of control.

(c) Optimization Control (Level III)

Again, optimization of a process is carried out either in the process control computer or at an even higher level, since economic considerations such as profitability, supply and demand, feedstock and product costs, and so on, enter into the optimization computations.

At this level of control, response time is fairly slow, so there is a very real application for process analyzers here. Although this is a relatively new area of process control, it is expanding rapidly and is expected to show considerable development in the next few years.

One interesting development to come out of process optimization is the need for information on the absolute value, that is, the accuracy of the measurement, rather than its repeatability. In process control of the regulatory type, we have been interested mainly in maintaining the process at precisely stable conditions; the only reference to absolute values came in plant material balance and custody transfer operations.

With optimization control we are interested in moving the process from, say, one product purity to another, based on the economics of the moment. It is necessary to know where we started from, where we want to go to, and whether we have reached that point. Thus, the subject of analyzer accuracy, and therefore calibration, is becoming increasingly important.

24.2.2 Classification of Process Analyzers

From the standpoint of control engineering, process analyzers can be categorized as follows:

(a) Class I

The analyzer itself is continuous and the sensing element is mounted directly in the process stream, that is, there is no sample handling system. The analyzer output can thus be considered continuous and virtually instantaneous.

There are no dynamic elements in the analyzer interface, so this class of analyzer may be considered as identical to a conventional flow or pressure transmitter from a control engineering standpoint. The use of this class of analyzer in control results in the best loop controllability and performance. However, certain precautions must be taken (e.g., installation of redundant analyzers to prevent a dangerous situation occurring in the event of analyzer failure).

Examples of the use of Class I analyzers in regulatory control loops include furnace flue gas excess oxygen control (Chapter 7) and BTU control (Chapter 18). The subject of pH control is unique and is covered in more detail later in this chapter.

(b) Class II

The analyzer itself is continuous and gives a virtually instantaneous output. However, the process sample is incompatible with the analyzer and a sample handling system is required.

The dynamics of this class of analyzer system, which are entirely due to the sample handling system, have a major impact on the controllability/stability and performance of the control loop of which the analyzer is a part. Also, the dynamics can vary with changes in process operating conditions (e.g., variation in fast loop response time with changes in available pressure drop). As a design criterion, the sample handling system time constant (T_1 sec) should be insignificant compared with the major process time constant (T_2 sec); often $T_2 > 10T_1$ is set. Consequently, most analyzers falling into this class are used at the supervisory control level.

(c) Class III

The analyzer itself is discontinuous, that is, it requires a finite time to complete an analysis. In addition, the process sample is incompatible with the analyzer, so a sample handling system is required.

Here, there are two problems. First, the analyzer provides information only at discrete intervals (cycle time) and remains silent in between; there is thus a loss of information about the state of the process. Second, there is a time constant problem associated with the sample handling system.

The dynamics of this class of analyzer system greatly influence, and sometimes even dominate, the performance of the control loop of which the analyzer is a part. These analyzer systems are very difficult to apply in control and are only suitable for use at the supervisory control level. However, many of the most important analyzer applications, such as fractionator control, involve the use of the process GC, a Class III analyzer. Fractionator control is discussed later in this chapter.

24.2.3 Criteria for Analyzer Control

From this discussion, it can be seen that two major factors must be considered when using analyzers in process control:

- Analyzer system response time.
- Analyzer system reliability and service factor.

Both of these subjects are covered in other chapters.

Response time, as we have seen for continuous analyzers, is concerned solely with the sample handling system, covered in detail in Chapter 20. Generally, sample handling systems are designed with a maximum response time of 90 sec (60 sec in the transport system and 30 sec in the sample conditioning system). However, the dynamics of the process may dictate a faster response than the customary 90 sec; this must be decided by the analyzer engineer and the process control engineer. The response time for discontinuous analyzers is made up of cycle time and sample handling system response.

The reliability and service factor of an analyzer installation and its suitability for process control are largely determined by analyzer system design and installation, the caliber of maintenance personnel, and the method and frequency of calibration. The last of these has an impact on operator acceptability of the analyzer system (utilization). These subjects are covered in detail in Chapters 21, 22, and 23; they require no further elaboration here.

For discontinuous analyzers, such as GCs, cycle times of 4–5 min are common; the extremes are rapid cycle times of 60 sec or less and long ones of up to 8–10 min. Where a process control computer is used for supervisory control and there is a digital data link with an analyzer computer system, it is inefficient (and unjustified) to have the PCC gather analyzer data and execute advanced control every 5 sec, when new analyzer data is only available, say, every 5 min. Thus, analyzer data links, as discussed in detail in Chapter 19, either transmit data to the PCC only when new data are available, or set a flag to indicate to the PCC that new data are available in the analyzer computer (the traditional "read-me" flag).

Where process analyzers are used for composition (quality) control, they are usually required to give a high order of sensitivity. That is why composition control has generally been identified as a problem area. For instance, fractionator control systems are normally required to maintain product impurities to within $\pm 0.2\%$ or even 0.1% of the specification. We know it is virtually impossible to regulate flow to within less than $\pm 1\%$ in the unsteady state, but the composition controller is asked to perform 10 times better than this. Although we can *measure* process stream quality to $\pm 0.1\%$, it is unreasonable to expect it to be *controlled* to the same tolerance.

24.3 pH CONTROL

In the literature of industrial process control, pH is often cited as a difficult control problem, and in many cases rightly so. In addition to its usual properties as a composition control loop, the measurement is severely non-linear. As shown in Chapter 13, there is a logarithmic relationship between the potential of the glass electrode and the hydrogen ion concentration, as given by the Nernst equation. This very characteristic imposes exceptional demands on the rangeability of process equipment (e.g., control valves), which creates the major problem. This, and the other idiosyncrasies that make life hard for the control system, are discussed next.

24.3.1 Non-Linearity and Rangeability

The magnitude of the non-linearity problem in pH control is best illustrated by reference to the strong acid-strong base titration curve shown in Fig. 24.1. Note that at the neutral point (pH 7.0), the rate of change of pH with respect to increments of reagent addition becomes very large (that is, the process gain is high), whereas at pH 3 and pH 11, it is relatively small (process gain is low). Similar curves can be drawn for other ions (e.g., oxidation-reduction reactions).

In Fig. 24.2, pH and reagent demand are compared for the same strong acid-strong base titration. For instance, a strong acid of pH 6.0 may require only one drop of a strong base to produce a final pH of 7.0. If the same quantity of acid is at pH 5.0, 10 drops will be required. For an initial pH of 2.0, 10,000 drops will be required to arrive at a final pH of 7.0. Herein lies the main problem—metering accuracy and rangeability.

If, for example, the influent pH is 2.0, a 1% error in metering is equivalent to 100 drops of reagent. At neutrality, that magnitude error will take the pH to 4 or 10! In water treatment systems, where one is usually trying to control the effluent pH around neutrality (say 6.0 to 8.0), and the influent can be highly acidic or highly alkaline, it is virtually impossible to control within the specified limits using a conventional control system. Some way has to be found to greatly reduce these error effects.

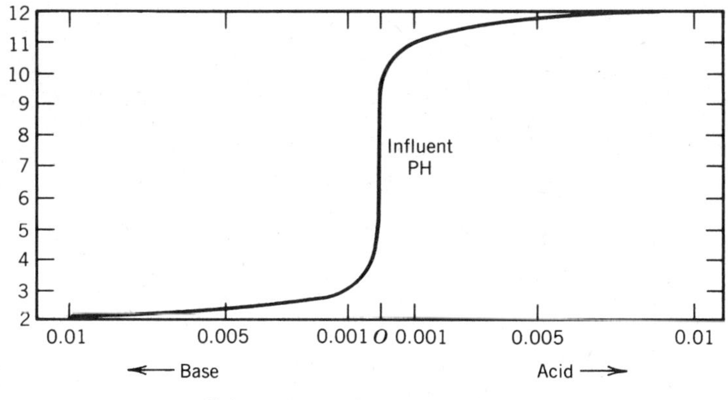

FIGURE 24.1 Strong acid–strong base titration curve.

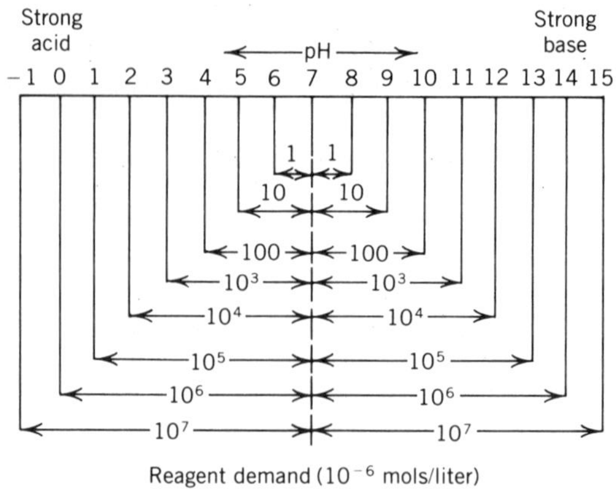

FIGURE 24.2 Relationship between pH and reagent demand.

Another factor that can create control problems is the presence of buffering agents in the process stream. As mentioned in Chapter 13, buffer solutions are resistant to change in pH by addition of an acid or base and consist, for instance, of weak acids and salts of weak acids. When weak acids and bases are present in waters in various mixtures and proportions, together with buffering agents such as metal ions, carbonates, acetates, phosphates, and so on, the shape of the titration curve may be completely different, as shown in Fig. 24.3. This situation frequently occurs in waste effluent treatment processes in a variety of industries (e.g., food, paper, chemicals, petrochemicals, metals).

In addition, the neutralization curves of certain waters show inflexion points. Such a condition usually arises from the formation of chemical intermediates in a reaction and is quite likely to occur in the chemical process

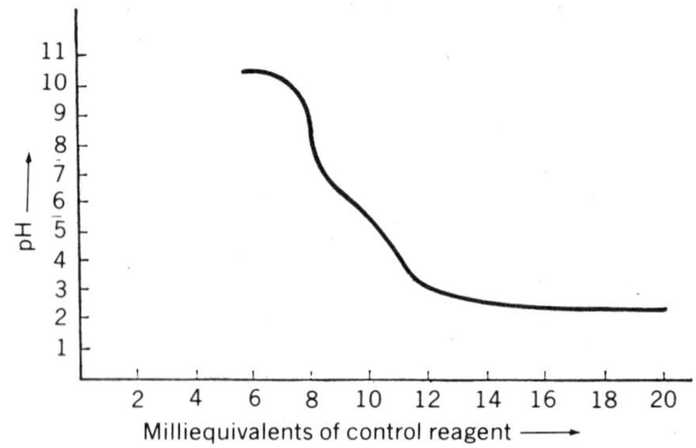

FIGURE 24.3 The effect of buffering on the pH curve.

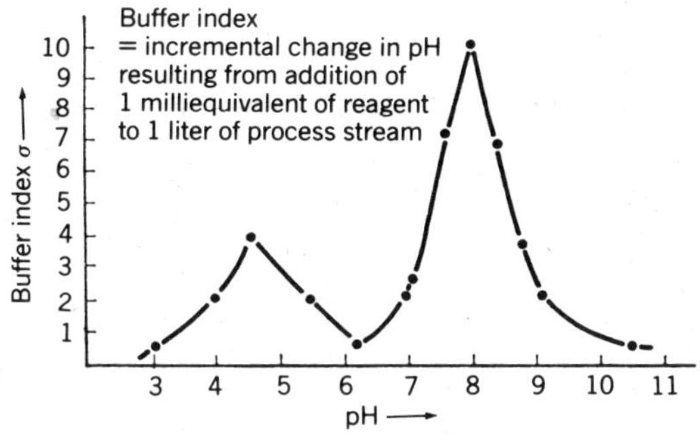

FIGURE 24.4 Typical buffer index curve.

industries. It is obviously impossible to control around an inflexion point since the pH controller would always be moving the process in the wrong direction by reagent addition. One must therefore be prepared to identify such a situation; the only way is to obtain the neutralization curve of the process sample at the earliest opportunity.

It is often difficult to get information on buffering effects directly from the neutralization curve. A good method is to construct a *buffer index* curve, as shown in Fig. 24.4. The buffer index (σ) is defined as the incremental change in pH resulting from the addition of 1 milliequivalent of reagent to 1 liter of process sample. This curve clearly shows the sensitive parts of the pH range for a given process stream.

The variation in slope of the neutralization curve gives rise to yet another problem, that of control asymmetry, as shown in Fig. 24.5. The variations in reagent addition shown are typical of the kind of situation that might occur in practice. If the setpoint were to be changed from pH 8.0 to pH 6.0, the effect would be reversed.

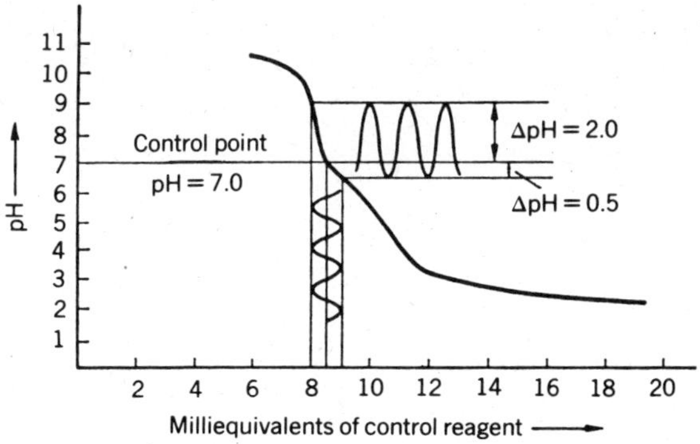

FIGURE 24.5 Control asymmetry.

24.3.2 The Influence of Process Equipment Design

Many control problems in the process industries can be solved merely by applying the correct, appropriately designed control system without having to make any changes to process equipment design. In pH control, the situation is different and somewhat unique in that process equipment design can have a significant effect on the success of any control system. Experience has shown that many neutralization processes are simply not controllable within the desired pH limits, mainly as a result of poor process equipment design.

The control of a wastewater neutralization process is a case in point, where the effluent is usually required to meet a stringent pH specification prior to disposal to a river or estuary. It is therefore essential, at an early stage in the facilities design, that the process engineer and control engineer design the complete system as one—the processing equipment and the control system.

To illustrate the limitations of controllability under imposed deviation limits, consider a stirred tank, as shown in Fig. 24.6, and subject the system to an instantaneous change in influent concentration. The change in concentration of the liquid in the tank at any time t after the load change is given by:

$$VdC = C_1Fdt - CFdt$$

or

$$\frac{dC}{C - C_1} = -\frac{F}{V}dt \qquad (24.1)$$

where V is the volume of liquid in the tank
F is the influent flowrate
C_1 is the concentration of the new influent
C is the concentration of the effluent after time t
C_0 is the concentration of the liquid in the tank just prior to the load change

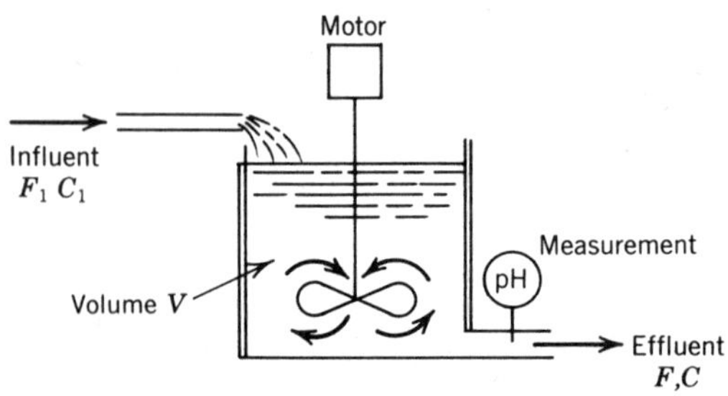

FIGURE 24.6 Controllability of a stirred tank.

Integration of (24.1) gives:

$$\int_{C_0}^{C} \frac{dC}{C - C_1} = -\int \frac{F}{V}\,dt \tag{24.2}$$

or

$$\ln \frac{(C - C_1)}{(C_0 - C_1)} = -\frac{Ft}{V} \tag{24.3}$$

Let $V/F = r$, the *retention capacity* of the tank. Then:

$$\ln \frac{(C - C_1)}{(C_0 - C_1)} = -\frac{t}{r} \tag{24.4}$$

or

$$(C - C_1) = (C_0 - C_1)e^{-t/r} \tag{24.5}$$

Rearrangement of (24.5) gives:

$$(C - C_0) = (C_1 - C_0)(1 - e^{-t/r}) \tag{24.6}$$

We have already seen that the pH setpoint and the influent pH are important factors in pH control. Equally important are the retention capacity (r) and the overall time lag of the system. With infinite retention capacity, $C - C_0 = 0$ and hypothetically no control is necessary. But as r decreases, the process deviation for any concentration change will increase. Trying to control pH in a pipeline is almost impossible because t/r is so large unless the influent pH variations are small. Therefore, neutralization reactions are usually carried out in a tank of adequate capacity fitted with an agitator.

To obtain values of r in equation (24.6) in terms of pH values, the concept of *acidity* can be adopted, where acidity values are given by $A = [H^+] - [OH^-]$.

At pH 7.0, $A = 0$ and values of A are positive below pH 7.0 and negative above pH 7.0. The same relationship can be arrived at by arbitrarily stating that below pH 7.0 the acidity values are the accepted hydrogen-ion concentrations and above pH 7.0 they are the accepted hydroxyl-ion concentrations but with a negative value. The relationship between pH and Acidity is therefore:

pH	0	1	2	3	4	5	6	7
acidity	1.0	0.1	0.01	10^{-3}	10^{-4}	10^{-5}	10^{-6}	0
pH	8	9	10	11	12	13	14	
acidity	-10^{-6}	-10^{-5}	-10^{-4}	-10^{-3}	-0.01	-0.1	-1.0	

Equation (24.5) then becomes:

$$(A - A_1) = (A_0 - A_1)e^{-t/r} \tag{24.7}$$

Equation (24.7) may be used for existing systems to calculate values of r from practical pH measurements.

24.3.3 The Importance of Mixing

Nearly all neutralization reactions take a finite time to reach completion, giving rise to what is known as *equilibrium lag*. It is therefore important that pH measurement electrodes are installed at a point where the reaction is complete.

Two types of mixing are important in pH control, intermixing and backmixing. Good intermixing is required to ensure that there are no areas of unreacted reagent or untreated water. Poor intermixing will result in a noisy signal from the effluent pH measurement system. Intermixing is usually accomplished upstream of the neutralization tank by promoting turbulence (e.g., suitable piping design incorporating right-angle bends, free-fall into the tank, etc.).

For tight pH control, backmixing is more important, its prime function being to allow sufficient time for the neutralization reaction to take place. For control purposes, backmixing is defined by the ratio of system deadtime to retention capacity for the tank. From experience, a ratio of 0.05 has been found to be adequate for good control.

The dead time is directly related to the rate of agitation. The agitator (stirrer) in the tank should be designed to turn the contents of the tank over (vertical pumping action) rather than to create a vortex effect, which will tend to increase the deadtime. The location of inlet and outlet connections on the tank can also have a considerable effect on deadtime. The preferred configuration is to have the inlet and outlet on opposite sides of the tank, with the inlet at high level and the outlet at low level, or vice versa. For below-ground tanks, it is common practice to install an inlet baffle so that both the inlet and outlet connections can be taken from the top of the tank.

The size of the tank will depend on the rate of reaction of the neutralization process and the influent flowrate. In general, the smallest feasible size is usually based on a time constant of about 3 min. The size would have to be increased if lime were used for neutralization instead of caustic, since lime can have a reaction time of anywhere from 5 to 20 min, depending on the type of lime. Another reason for increasing the tank size would be if a precipitate were formed by reaction; this could trap unreacted reagent or influent, thereby extending the reaction time because of diffusion effects.

The number of tanks required for effective neutralization depends entirely on the relative difficulty of the control problem. The following guidelines

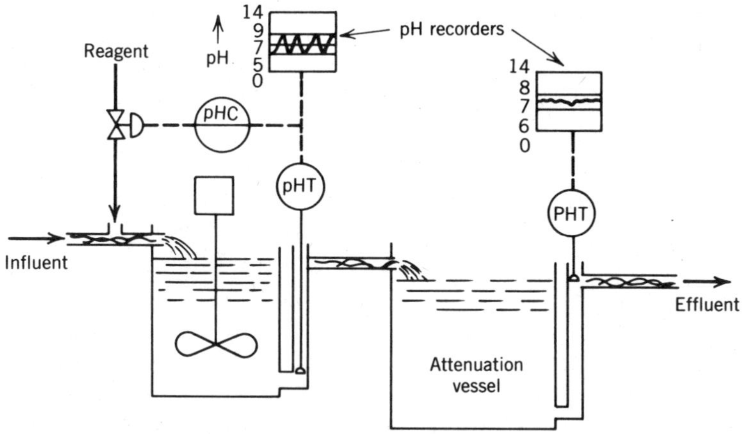

FIGURE 24.7 Typical two-tank system.

have been used successfully for strong acid-strong base neutralizations, where the ratio of deadtime to retention capacity of 0.05 is observed:

- One stirred tank for influent variations between pH 4.0 and 10.0.
- Two tanks, one stirred, one unstirred, for influent variation between pH 2.0 and 12.0.
- Three tanks, two stirred, one unstirred, where the influent pH can vary from less than 2.0 to more than 12.0.

The function of the unstirred tank is to dampen out the cyclic variations which can occur in the preceding tank; it is therefore an attenuation vessel and usually has a volume three to five times that of the preceding tank. A typical two-tank system is shown in Fig. 24.7, together with an indication of the extent of effluent pH control that can be achieved with such a system.

24.3.4 pH Control Systems

As we have seen, the success or failure of any pH control system depends on proper design of the process equipment; there is little one can do in the control area if the equipment is inadequate. Therefore the discussion on specific pH control instrumentation has been left to last.

In discussing control systems, the wastewater treatment process is again taken as an example, since this is probably the most difficult of all pH control problems to tackle. As mentioned earlier, the rangeability problem is the most difficult to overcome. Normally, influent wastewater flowrates do not vary appreciably (maybe over a 2 : 1 rangeability), but influent pH can vary by 4 pH units or more (i.e., 4 or more decades of acid or base concentration).

Realistically, a single reagent control valve can be expected to control adequately over a 50 : 1 range at most, or over 1.7 decades (log 50). Therefore a control valve delivering reagent to neutralize an influence of pH 2.0 will control erratically when the pH is greater than 3.7. It is necessary in practice to use two or more control valves on a sequential basis to achieve satisfactory control over a 4-decade range of concentration. If the influent pH is less than 2.0 or greater than 10.0, the situation is even worse. So the final control problem is severely compounded by the non-linearity of the measurement.

A variety of different schemes have been used for pH control over the years, with varying amounts of success (or failure). One that appears to have been the most successful is shown in Fig. 24.8 and will be used as an example. This control system has two elements, feedforward control and feedback control. In this example, it is required to neutralize an acidic influent stream by the addition of caustic (i.e., a strong acid-strong base neutralization).

The feedforward system is designed to control the effluent pH by correcting for changes in influent load. In the system shown, caustic reagent flow is manipulated in proportion to the total acid entering the neutralization tank. There are two components of influent load, flowrate and acid concentration (pH). As mentioned earlier, in most wastewater treatment processes the flow changes are gradual, whereas the changes in pH can be significant and rapid. Since a strong acid is completely ionized in solution, the flowrate of

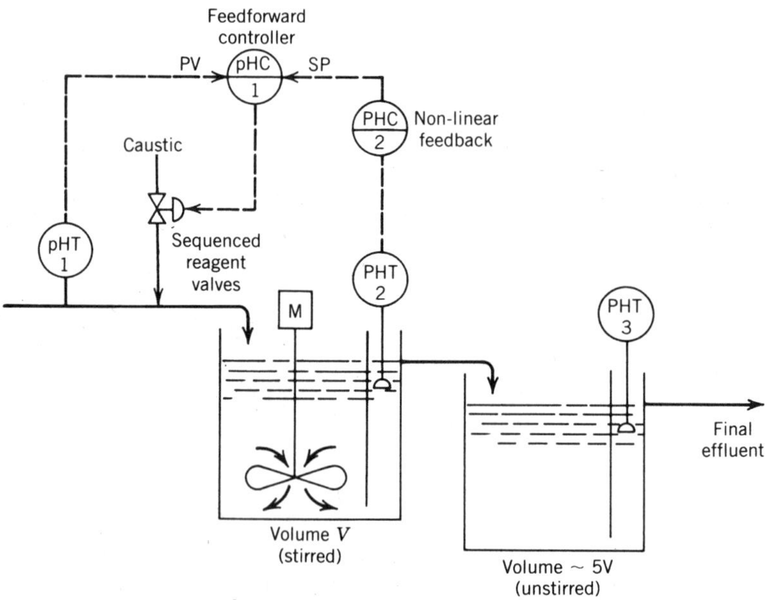

FIGURE 24.8 Feedforward-feedback pH control system. (Courtesy of The Foxboro Company)

caustic required for neutralization can be directly calculated from the measured influent pH. Provided that equal-percentage control valves are used for reagent addition, valve position can be calculated from influent pH by the following expression:

$$p = K(r - \text{pH})$$

where p is the valve position
K is the feedforward gain
r is a bias term (adjustable setpoint)

This equation can be solved by a simple proportional controller with influent pH as the measured variable. The bias term is adjusted to set the influent pH at which the control valve is delivering the minimum flow of caustic. This setpoint must be readjusted when the acid mixture changes. However, one cannot account for every possible variation in the influent stream and unfortunately there is no simple method of measuring total acid in the stream. Therefore, corrections for these variations, relative to influent pH, must be made using feedback control.

The feedback controller, which has effluent pH as the measured variable, is used to reset the setpoint r of the feedforward controller as necessary to achieve complete neutralization. This all seems straightforward, but there is a major problem. Feedback control of effluent pH is particularly difficult for three reasons, as mentioned earlier—deadtime, high process gain at the neutral point, and the extreme non-linearity of the measurement. For these reasons, conventional controllers have proved to be completely unsatisfactory for this application and it is necessary to use what is known as non-linear or adaptive feedback control.

The adaptive feedback controller is a PID controller with a non-linear control function that is complementary to that of the process stream neutralization curve. The error signal is characterized by this function, then passed to the proportional, integral, and derivative modes of control. The adjustable parameters in this controller are deadband width and gain within this deadband, with the gain outside the deadband equal to unity.

Although it is impossible to match the neutralization curve of the process stream exactly, this system is far superior to having a constant gain as with conventional controllers. The effect of this overall control system can be demonstrated by putting the adaptive controller on manual (no feedback control), then inhibiting the feedforward control (feedback control only). In either case the process is not controllable within desired limits, so it is necessary to have *both* elements of the control system for effective pH control.

As mentioned in Chapter 13, microprocessor-based pH controllers are now available that do a far better matching job on the neutralization curve than the analog controllers used previously.

The example control system described is relatively straightforward since the system only has to deal with a strong acid-strong base neutralization of an acid influent. The situation becomes far more complicated if the influent ranges from highly acid to highly alkaline. The presence of buffering agents, as explained earlier, can also complicate the situation. If all these problems are taken into account, one can develop a satisfactory system for pH control. The important point here is that to design an effective system one must have adequate information about the process stream and adequate process equipment with which to carry out the neutralization.

24.4 FRACTIONATOR CONTROL

24.4.1 Introduction

As discussed in Chapter 3, distillation is one of the most important separation processes used in the process industries, particularly in the petroleum and petrochemical industries. The process is, however, one of the major consumers of energy in a plant. In view of the large number of distillation towers (fractionators) that form part of the typical refinery or petrochemical plant, there are very substantial incentives for efficient fractionator control, both for energy conservation and in maximizing recovery of valuable products.

The fundamentals of the distillation process are summarized in Chapter 3 and require no further elaboration here. However, one important point of note is that although the incentives for improved control are high, fractionators are usually the most difficult to control and present a considerable challenge to the control engineer in designing an effective control system.

There are a number of reasons for his. First, product-stream purity specifications are usually very exacting and not easy to achieve. Second, today's process plants are highly integrated systems, where disturbances in upstream unit operations can propagate throughout the plant. Unit operations are also susceptible to changes in ambient conditions. Third, a fractionator is a multi-variable process unit, where control objectives are met simultaneously by several control loops that may interact severely, making control even more difficult. Therefore, although the basic principles of the distillation process are quite simple, the fractionator is relatively difficult to operate because of the many variables that can affect tower operation.

It is beyond the scope of this book to cover the subject of fractionator control in detail. There are many excellent textbooks on the subject, together with literature publications covering in-plant experience. The objective here is to outline the various control objectives and constraints in fractionator control, the types of control systems in common use, and to direct attention to the way in which on-line process analyzers may be used in these systems to contribute to the effectiveness of the control system in meeting the control objectives.

24.4.2 Control Objectives

It is important in the design of any control system to first identify the *operational objectives* of the unit, which can then be used as the basis for developing a set of *control objectives* for the unit. Operational objectives are usually based on safety and stability considerations, the economics of the operation, and ensuring that unit operational conditions are within the design limits of the equipment (i.e., the constraints). Based on these criteria, here are some typical control objectives:

1. Control tower pressure to ensure safe and stable operation.
2. Operating within equipment constraints.
3. Ensuring products meet desired specifications at all times and maximize recovery of the more valuable products.
4. Minimizing propagation of disturbances to downstream units.
5. Minimizing energy consumption while achieving the desired separation.

Note that the specific objectives for individual fractionators may vary depending on the particular unit configuration.

To achieve these control objectives, a number of important tower variables must be controlled by the manipulation of other variables. These are illustrated for a typical two-product fractionator in Fig. 24.9. Common controlled variables are:

- Tower pressure (P).
- Reflux drum level (L_R).

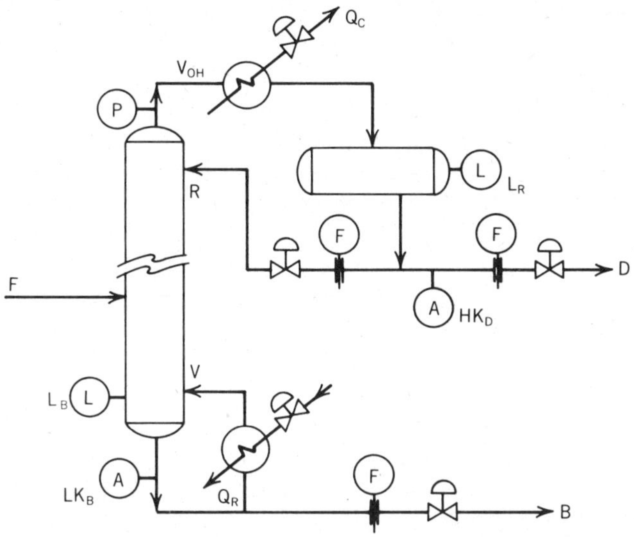

FIGURE 24.9 Control variables for a two-product practionator.

- Tower bottoms level (L_B).
- Distillate product composition (usually heavy key component concentration, HK_D)
- Bottoms product composition (usually light key component concentration, LK_B).

The commonly manipulated variables may include:

- Distillate flow (D).
- Bottoms flow (B).
- Reflux flow (R).
- Overhead condenser duty (Q_C).
- Reboiler duty (Q_R).

The control system must be designed to maintain the control objectives, even when the fractionator is subjected to disturbances such as changes in feedrate, feed composition, feed enthalpy (heat content), condenser cooling medium temperature, and reboiler heating medium temperature.

Since most fractionators receive feed from upstream units, feedrate and feed composition are highly dependent on upstream operation (or maloperation).

24.4.3 Factors Affecting Product Qualities

The primary goal of any fractionator control system is to maintain product qualities within specifications at all times. Two parameters are commonly used in distillation operations to describe the major variables that affect product qualities, *cutpoint* and *fractionation*.

Cutpoint is defined as the quantitative split of the feed (F) into distillate product (D) and bottoms product (B). It is an important parameter, since it determines not only how the feed is divided into the two product streams but also the purity of each product. At a constant tower-operating pressure, cutpoint can be defined as a temperature. An increase in cutpoint will result in more distillate production and less bottoms production. At the same time, the distillate purity will decrease and the bottoms purity will increase. A decrease in cutpoint will have the reverse effect. In other words, cutpoint changes increase the purity of one product and decrease the purity of the other and the product stream that increases in purity also decreases in volumetric flow.

Fractionation also has a major effect on product qualities. At constant feed/product ratios, fractionation is determined by either reflux/feed ratio (R/F) or reboiler duty/feed ratio (Q_R/F). Increasing either ratio will improve separation, resulting in an increase in the purity of *both* products. In other words, fractionation can be considered as the amount of energy required to achieve a given separation.

Either cutpoint or fractionation can be manipulated to control product quality. However, product quality is usually more sensitive to cutpoint changes; therefore this parameter is often used as the manipulated variable when only a single product stream quality needs to be controlled. When both product stream qualities need to be controlled, it is usually necessary to manipulate both cutpoint and fractionation.

24.4.4 Constraints in Fractionator Operation

Process constraints in fractionator operation are very closely related to the original equipment design. Control systems are designed to make use of the flexibility designed into the unit—to change operating conditions to meet the control objectives. However, when process conditions are changed too much from the original design values, or when too little excess capacity has been designed into the unit, what available flexibility there is can be used up rapidly. The fractionator is said to be under constraint and not all of the control objectives can be met. Some of the common constraints are described next.

(a) Energy Transfer Constraints

The energy balance of a fractionator is set mainly by the reboiler heat input and the overhead condenser heat removal.

Reboiler duty may reach a constraint for various reasons such as high bottoms temperature, high tower pressure, fouling, and low steam pressure (or low heating medium temperature). Lowering the tower pressure will move the tower away from a reboiler constraint (e.g., pressure minimization).

Condenser duty limitations can be caused by low tower pressure, fouling, build-up of non-condensibles, high coolant inlet temperature, and low overhead vapor temperature. In addition, weather changes can cause sudden upsets and, especially with air-cooled condensers, can have a drastic effect on tower control. Here, an *increase* in tower pressure will relieve a maximum condenser duty constraint. In fractionator pressure minimization applications it is usually found that the condenser duty is the most important constraint.

(b) Constraints Based on Tray Hydraulics

The trays in a fractionator are designed to operate within a range of vapor and liquid rates. Too high or too low vapor or liquid rates produce hydraulic imbalances that interfere with normal mass transfer, resulting in a greatly reduced tray efficiency in accomplishing a given separation. Common examples include flooding, entrainment, downcomer filling, weeping, and blowing.

(c) Feed and Product Flowrates

The maximum rate at which feed can be brought into the unit and products can be removed is determined by line pressure drops and control valve sizes. Therefore feed and product flowrates may also be considered as constraints on tower operation.

(d) Pressure Constraints

This is probably the most obvious constraint. All fractionators are protected by pressure relief valves set safely below the pressure at which equipment can be damaged by overpressuring. The tower operating pressure is usually kept well below this safety limit, imposing a pressure constraint on tower operation.

All of these constraints must be considered when designing the fractionator control system. A commonly used method for assessing the constraint situation is to construct a graph of boil-up rate vs tower pressure and express each constraint as a relationship with these two parameters. The region within which all constraints are satisfied represents the set of feasible operating conditions for the fractionator.

24.4.5 Fractionator Control Configurations

(a) Basic Regulatory Controls

There are two basic regulatory control schemes in common industrial use, material balance control (MBC) and energy balance control (EBC). Typical examples of each are shown in Fig. 24.10. Although EBC is the more traditional of the two, MBC has also gained acceptance. The choice of the quality controllers shown in Fig. 24.10 does not generally influence the pairing of the basic regulatory control loops. Some commonly used quality controls include conventional tower temperature control, temperature averaging control, temperature difference control, and analyzer control. Analyzer control is frequently used in conjunction with conventional temperature control, as will be explained later.

When comparing the two basic regulatory control configurations, one of the major considerations for selection should be the dynamic response of the systems to expected disturbances. These disturbances can take the form of material balance upsets and energy balance upsets.

The most common material balance upsets are changes in feedrate and feed composition. The two configurations respond differently to this type of disturbance. With MBC, large changes in cutpoint occur, reflected in significant variations in product qualities. With EBC, however, some self-regulation is inherent in the design, giving the system the ability to better handle feed disturbances.

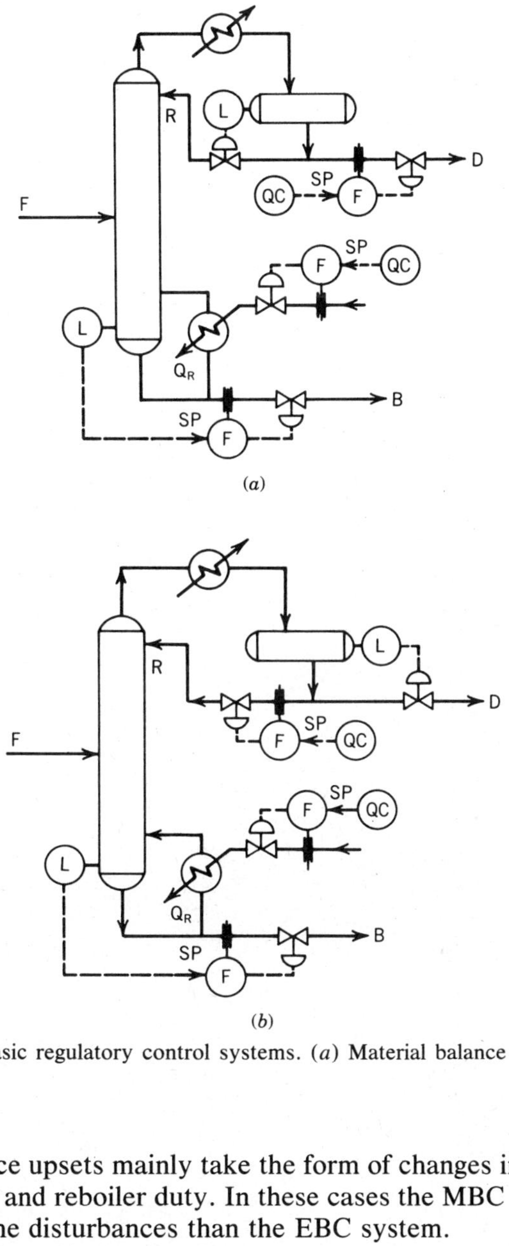

(b)

FIGURE 24.10 Basic regulatory control systems. (*a*) Material balance control; (*b*) energy balance control.

Energy balance upsets mainly take the form of changes in feed enthalpy, condenser duty, and reboiler duty. In these cases the MBC system is better able to handle the disturbances than the EBC system.

Which regulatory control system to use can be a difficult choice and depends on the particular fractionator and the types of disturbances likely to occur in practice. The control engineer must weigh the pros and cons of the situation and select a control system that gives the best compromise in meeting all of the control objectives.

(b) Advanced Controls

So far, we have limited our discussion to basic regulatory control configurations. There are also a number of commonly used advanced control systems that can be used in addition to the regulatory control system to significantly improve the overall control system response to disturbances. These include feedforward control of feedrate, feedforward control of feed composition (using an analyzer), internal reflux control, and multivariable control techniques.

If both feedrate and feed composition feedforward controls are implemented, the MBC configuration can be made effective in handling material balance disturbances. Internal reflux control is used in EBC configurations to increase the effectiveness of the control system in handling energy balance upsets.

In some fractionators where both product stream qualities must be controlled, the interaction between quality control loops can be severe. In these cases, the use of multivariable control techniques such as interactive or decoupling control (IAC) can effectively reduce these interactions.

24.4.6 Product Quality Control Using Process Analyzers

Controlling a tray temperature in the tower will usually provide adequate control of product quality in most fractionators. However, since the tray temperature/product quality correlation is only approximate, process operators will usually set the tray temperature setpoint at a value that ensures that on-spec product is made. Usually they are conservative in this respect, resulting in reduced recovery of valuable product. Product quality control can be markedly improved by the use of on-line analyzers, where the analyzer controller is the primary in a cascade control system, with the tray temperature controller as the secondary.

Several types of analyzers are used in fractionator control systems, such as RVP, IR, and boiling point. By far the most extensively used is the process gas chromatograph (GC), which has the advantage of being able to measure the concentration of one or more components in a multi-component process stream. For control purposes, the impurity concentration is usually measured (e.g., light key component in bottoms, heavy key component in distillate) since it can be measured with higher accuracy and sensitivity than pure component concentration. Distillation range or RVP is often the controlled quality in product stabilizer or debutanizer operations.

When considering analyzer control applications, it is important to recognize the "sampled-data" nature of the system. The inherent problems that arise because of this were discussed earlier in this chapter (Sections 24.2.2 and 24.2.3).

Two examples of fractionator control systems using on-line analyzers are shown in Fig. 24.11, both of which employ analyzer-to-tray temperature

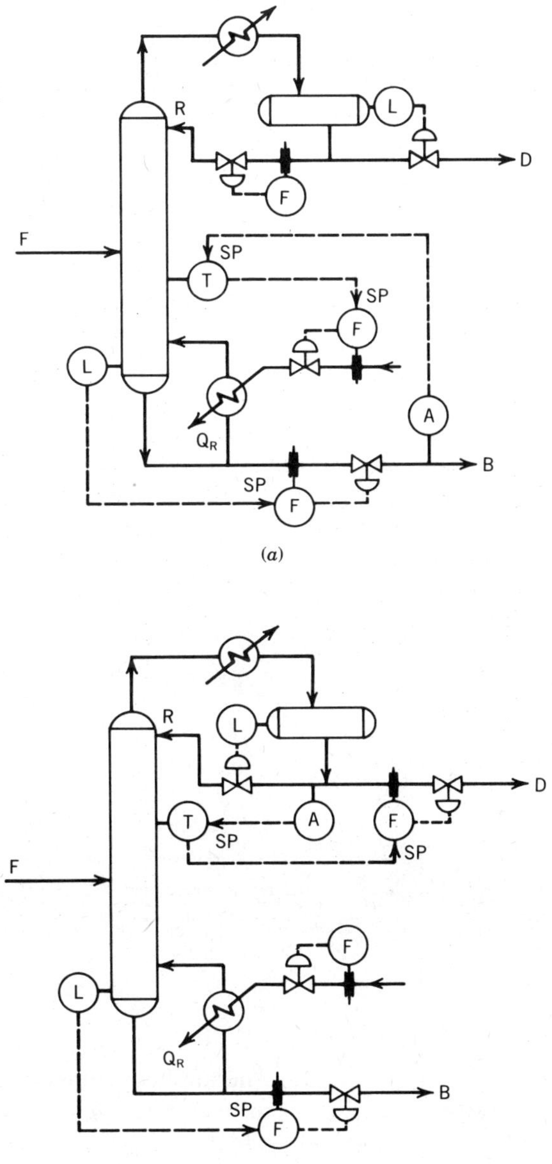

FIGURE 24.11 Fractionator control using on-line analyzers. (*a*) Bottoms quality control by energy balance; (*b*) distillate quality control by material balance.

cascade configurations. Fig. 24.11*a* shows an EBC system with control of bottoms product quality. Fig. 24.11*b* shows an MBC system with control of distillate product quality. Note that the secondary temperature loop of the cascade is provided to speed up the response of the control system to load disturbances and to provide backup regulatory control in the event of an analyzer failure. These systems are designed to control the quality of one product stream only; this is usually the more valuable stream.

Dual product quality control of a fractionator is inherently a multivariable control problem because of interactions between the quality control loops. In some towers the interactions can be quite severe, and control techniques such as IAC must be used to break the interactions. Nevertheless, it is

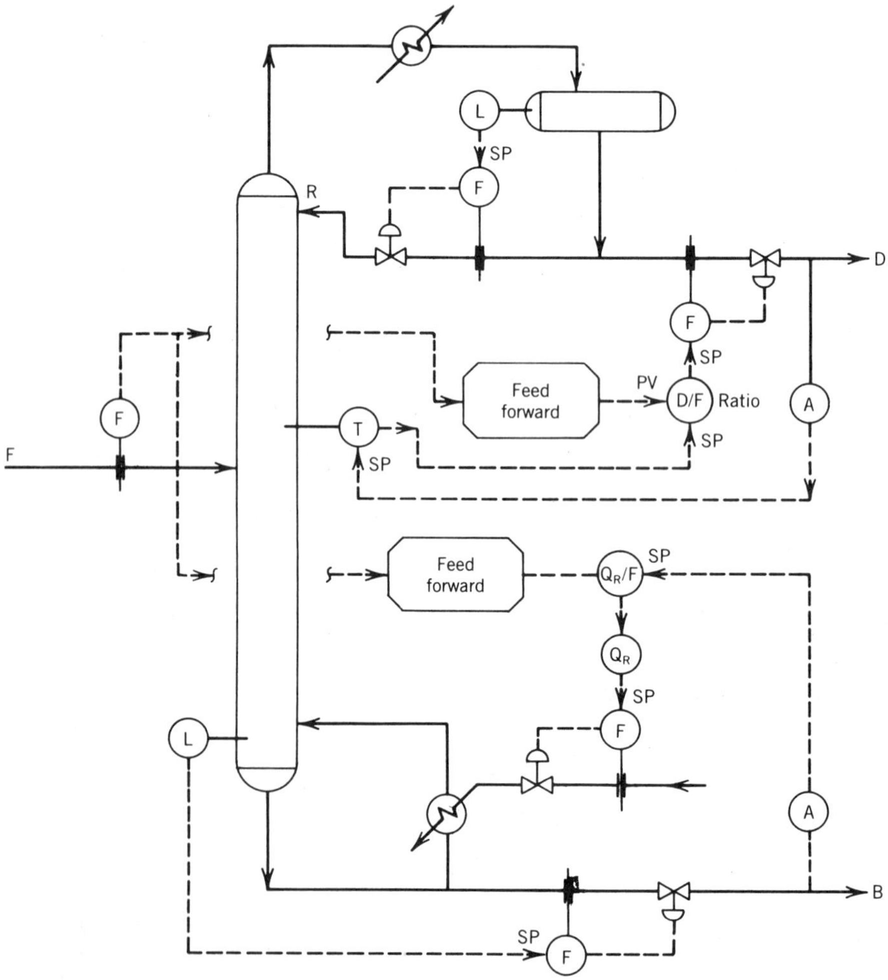

FIGURE 24.12 Dual analyzer control by material balance.

possible to achieve satisfactory quality control using two single quality control loops. Correct pairing of variables into regulatory control loops can also help to reduce interactions.

To obtain satisfactory dual quality control using two single loops, on-line analyzers are required. Usually only one analyzer loop is cascaded to the tray temperature controller, which adjusts tower cutpoint. The second analyzer loop is then used to manipulate the tower fractionation variable directly. For instance, if distillate product quality is the more important, the analyzer/tray temperature cascade should be on that stream.

Figure 24.12 shows a typical MBC configuration with dual analyzer control. The analyzer/tray temperature cascade adjusts the distillate/feed ratio (which represents the cutpoint). This ratio also introduces feedforward compensation for feedrate changes. The bottoms product is controlled by an analyzer controller, which resets the reboiler duty/feed ratio (which sets fractionation). If it is not possible to locate a tray in the tower where temperature correlates well with product quality, the distillate analyzer controller can manipulate the distillate/feed ratio directly. This has sometimes been done in superfractionators (e.g., ethylene/ethane splitters).

24.5 CRUDE DISTILLATION CONTROL

24.5.1 Control Objectives

As mentioned in Chapter 3, crude distillation is the primary separation process in any oil refinery; therefore, proper control of the crude distillation tower, or pipestill, is critical to refinery operations. The reader is referred to the typical atmospheric pipestill shown in Fig. 3.5, which will be used here as an example.

Pipestill operations are highly complex, since there are various modes of operation that depend on refinery economics (e.g., crude slate, supply and demand, etc.). The main objective of the pipestill control system is to allow automatic and smooth transition from one operating mode to another, simultaneously minimizing the time necessary to achieve product qualities within specifications.

The basic concepts of pipestill control are similar to those described for fractionators; however, the control problems are far more severe because of the complexity of the process itself. Here are some of the problems:

1. The multi-product drawoff configuration (sidestreams) results in severe interaction between control loops. Therefore, multivariable control techniques (e.g., IAC) are required to decouple these interactions.
2. The need to switch rapidly between various modes of operation puts a heavy burden on the control system.

3. Feed disturbances can have a drastic effect on pipestill control. Therefore, feedforward controls are frequently incorporated to eliminate these effects.

24.5.2 Process Analyzer Applications

Several different types of process analyzers are used in pipestill control systems, one of the more important being the boiling point analyzer. Some of the typical analyzer applications for both atmospheric and vacuum pipestills are described briefly next.

(a) Atmospheric Pipestills

Overhead Product Quality Control. A boiling point analyzer is used to measure the 95% point of the naphtha stream; the analyzer controller is usually cascaded to the pressure-compensated overhead temperature controller.

First Sidestream Quality Control. There are often two separate aspects to first sidestream quality control, using two different manipulated variables:

- A flash point or 5% boiling point analyzer can be used to manipulate stripping steam flowrate.
- A freeze point, cloud point, pour point, or 95% boiling point analyzer can be used to manipulate the sidestream drawoff rate to maximize yield to a particular limiting specification.

Feedforward and IAC controls are usually necessary to take account of feedrate changes and to decouple interactions.

Control of Other Sidestream Qualities. It is often required to maximize the yield of one sidestream at the expense of others. This is usually achieved by using a 95% boiling point analyzer to control the sidestream to be maximized at its endpoint specification by manipulating sidestream drawoff rate. Alternatively, a cloud point analyzer may be used, depending on the limiting specification.

If a lube oil feedstock is drawn as the bottom sidestream, a viscosity analyzer is used to manipulate drawoff rate.

Overhead Condensate pH Control. The overhead condensate from both the atmospheric pipestill and the vacuum pipestill is often acidic, and chemicals such as ammonia are injected into the overhead circuit to adjust the pH and avoid extensive corrosion of overhead equipment. This is an important application for pH control, where the pH controller adjusts the ammonia injection rate to keep the pH within acceptable limits.

(b) Vacuum Pipestills

Invariably the vacuum pipestill takes as its feed the bottoms stream from the atmospheric pipestill. The interactions between control loops in the VPS unit are even more severe than in the APS unit; in the VPS unit there are upward as well as downward interactions that must be decoupled.

Vacuum pipestills generally supply feed to a lube oil production unit or to a catalytic cracking unit. The bottoms stream is generally used for asphalt production. For lube oil production, each VPS sidestream is usually controlled to a viscosity specification by a viscosity analyzer, which is used to manipulate sidestream drawoff rate. Once again, feedforward and IAC controls are necessary here.

The bottoms (asphalt) stream is usually controlled to a viscosity specification (correlated with the lab penetration test). The viscometer is used to manipulate the bottom sidestream drawoff rate to control bottoms viscosity.

24.6 CONTROL OF IN-LINE BLENDING OPERATIONS

In-line blending systems are employed widely in the process industries to manufacture finished products by mixing a number of component streams together in proportion, to meet customer specifications. These systems are particularly important in the petroleum industry for the production of motor gasolines, middle distillates, and heavy fuel oils. In-line blending systems have the advantage that feedstocks can be blended directly, with little or no intermediate (spill) tankage. Increased supply flexibility can be created by reallocation of some finished product tankage to storage of components used for the in-line blending of products that can be immediately shipped. However, it is essential under such conditions that the blending control system be capable of delivering on-spec product at all times; there is no room for error. Therefore, the use of on-line analyzers for in-line blending quality control is almost mandatory.

Blending systems are basically flow ratio systems. A typical example is shown in Fig. 24.13 for two components blended to form a mixture of controlled composition, where one component is "wild" (i.e., its flowrate is uncontrolled). Of course, most blending systems are far more complex than this. A motor gasoline blending system is probably the most complex. Nevertheless, this example illustrates the principle.

The introduction of digital blending systems, using turbine meters or positive displacement meters that provide a pulse rate or frequency signal proportional to flowrate has led to a new dimension in in-line blending control. In the digital blending system, the volume delivered through each meter is compared on a digital basis with the volume desired for that stream, and the error is used to manipulate the control valve in the same way as the analog

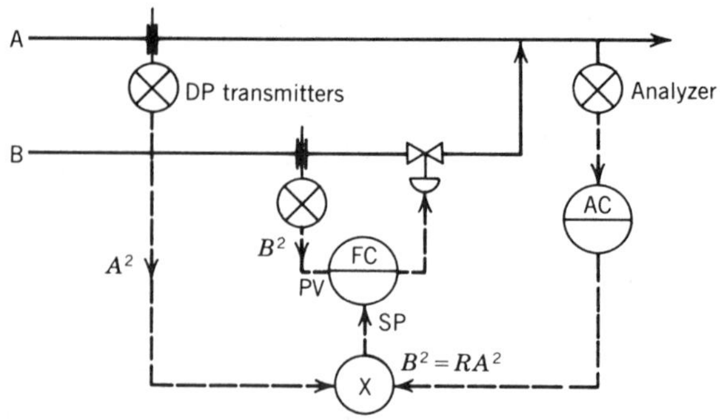

FIGURE 24.13 Flow ratio system for blending operation.

flow control system would. A master set station transmits a pulse rate, demanding a total desired flowrate, that paces the entire system. This pulse rate is multiplied by the ratio setting for each flow loop, producing a rate setpoint for each loop. A typical scheme is shown in Fig. 24.14. These systems are used extensively today.

Many different types of process analyzers are used in in-line blending systems, with the selection based on the primary specifications that must be

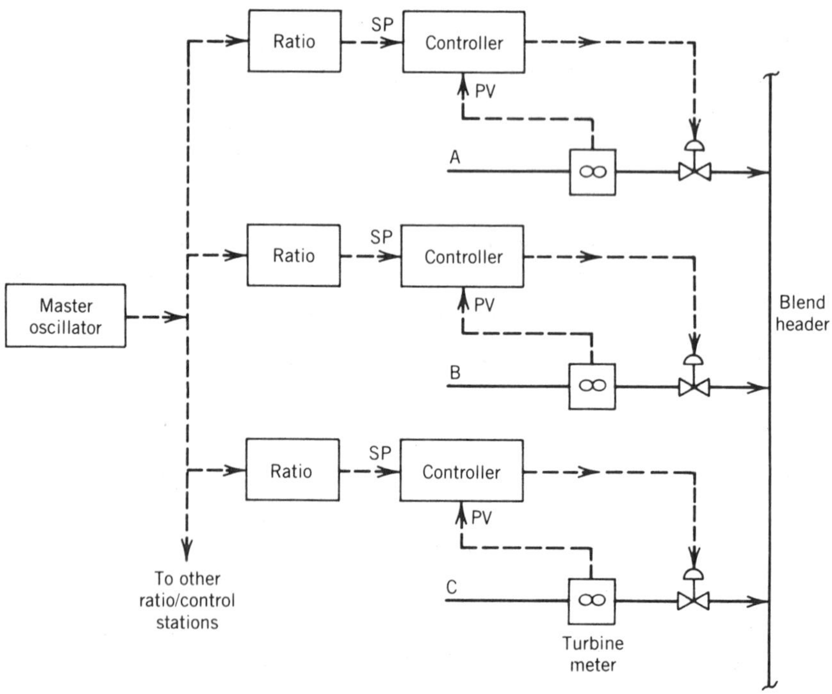

FIGURE 24.14 Typical digital blending control system.

met for the finished product. The use of on-line analyzers allows closer control of the optimum blend composition, reduces quality give-away by working closer to the specification limits, and can help to reduce the laboratory workload, which can be quite considerable in this kind of operation. Some examples of the types of analyzers used in these applications follow.

Motor Gasoline Blending		
Octane number	RVP (volatility)	Boiling point
Specific gravity	Lead content	
Middle Distillate Blending		
Cloud point	Pour point	Viscosity
Flash point	Boiling point	Sulfur
Specific gravity	Color	CFPP*
Heavy Fuel Oil Blending		
Specific gravity	Sulfur	Viscosity

* CFPP, Cold Filter Plugging Point (see Chapter 5)

There appears to be a consensus among analyzer experts that on-line analyzers in blending systems should be used not only for the benefit of the blending operation itself but for certifying the final product quality. The advantages here are the continuous analysis of limiting specifications and the better repeatability of the on-line analyzers compared with the corresponding laboratory tests.

In addition, possible sources of error such as unhomogeneous blends, bad sampling, and incorrect procedures in the laboratory can be avoided. In general, on-line analyzers in blending operations, if properly maintained and validated (see Chapters 22 and 23), can increase the confidence in the quality of production, can be the basis for reduced product quality margins, and can eventually lead to a reduced workload in the laboratory.

24.7 CONTROL OF CHEMICAL REACTION PROCESSES

24.7.1 Background

Chemical reaction processes are used extensively in the process industries, enabling several feedstocks to be combined to make one or more products of higher value. Many are catalytic processes, that is, the chemical reaction takes place in the presence of a catalyst, a substance that promotes the reaction without taking part in the reaction. Some reactions are exothermic (heat is evolved), others are endothermic (heat is absorbed). Some reactions are conducted on a batch basis, that is, the ingredients are added together in the reactor, the reaction is initiated and, when complete, the product is withdrawn. Others are continuous processes, which operate under conditions of constant feedrate, withdrawal of product, and removal or supply of heat.

Each type of chemical reaction has its own design and control requirements, the prime objective being to maximize the production of on-spec product(s). In catalytic processes, the catalyst may be packed in a fixed bed within the reactor, or it may be supported by the upward velocity of the reactant stream (which may be a gas or a liquid), in which case it is called a *fluidized bed* process.

Solid catalyst may also be dissolved or suspended in a liquid reaction medium. In some cases, the catalyst may be nothing more than a porous surface onto which the reactants are absorbed. The point to emphasize here is that the chemical reaction takes place *on the surface* of the catalyst.

Most catalysts can become deactivated by fouling of the surface with by-products or, more important, contaminated by impurities in the feedstock (poisons). Some of the very sophisticated catalysts used today are highly susceptible to poisoning by extremely low (ppb) levels of sulfur compounds (e.g., H_2S, mercaptans, etc.) and carbon monoxide. Feedstock poisons measurement is therefore one of the most important applications for on-line analyzers in these processes.

Most chemical processes in the petroleum/petrochemical industry are of the continuous type, such as the well-known catalytic cracking and reforming processes, although a few batch-type reaction processes are used, for example, in the production of lube oil additives. On the other hand, the pharmaceutical industry has used batch-type reaction processes for many years.

In view of the wide variety of chemical reaction processes used in industry today, it would be a formidable task to describe them all. It is the intention here to cover briefly the basic principles governing the conduct of chemical reactions, followed by a description of some of the more common processes used in the petroleum and petrochemical industries, with which the author is familiar. The emphasis is placed on how on-line analyzers are used in these processes to improve control.

24.7.2 Basic Principles of Chemical Reaction Processes

In most process plants, the reactor is at the front end of the process, so its production rate governs the overall plant production rate. This situation has some advantages, since the reactor is not usually subjected to rapid or unpredictable load upsets that other units experience. However, although many processes can tolerate a poorly designed control system, this is not the case for most chemical reaction processes.

Good control is essential here, otherwise disturbances that will be propagated throughout the rest of the plant can be generated. This can be particularly serious in plants with a high degree of heat integration (e.g., catalytic reforming). Often such disturbances can be sufficiently violent to ruin the product, destroy the catalyst, damage process equipment, and endanger life. In general, the success of any reactor control system very much depends on an intimate knowledge of the design of the reactor system, since the main control objective here is to maintain the optimum conditions of the reaction process established during the design.

The majority of chemical reactions are reversible, and an equilibrium can be reached where the rates of the forward and reverse reactions are equal. This equilibrium point determines the reactor *conversion—*, how much of the reactants are converted into products—and the operating conditions that will favor improved conversion. To illustrate, consider an exothermic reaction in which two reactants A and B are converted into a product C, represented by the following equation:

$$A + B \rightleftharpoons C \ (+ \text{ heat})$$

A certain ratio of product to reactant concentrations that will bring about an equilibrium can be reached. This ratio is known as the *equilibrium constant* (K) for the reaction, where K is given by:

$$K = \frac{[C]}{[A]\,[B]}$$

The equilibrium constant for any reversible reaction is a function of temperature and the type of catalyst used. This means that a certain fraction of the reactants must be withdrawn along with the product, limiting the conversion that can be achieved. However, there are ways in which the conversion can be improved:

1. A reduction in product concentration will allow a corresponding reduction in reactant concentration(s). Therefore, if the product can be removed from the reaction zone, for example by condensation, the conversion will improve.
2. An increase in reactant concentration will improve conversion (e.g., using one of the reactants in excess).
3. In reactions where the total moles (volume) of reactants is greater than the total moles (volume) of product, an increase in pressure will bring about an improvement in conversion.
4. In these examples, the evolution of heat will affect the equilibrium. Just as the separation of the product from the reactants will promote conversion, the removal of heat will have the same effect.

Some reactions may be carried out virtually to completion in a single pass, whereas others must be conducted at low conversion for various reasons, such as low reactions rates, reaction reversibility, and the occurrence of side reactions. In many continuous reaction processes, where conversion is incomplete, excess reagent is recycled.

Whereas endothermic reactions show considerable self-regulation and present no real temperature control problem, the reverse is true of exothermic reactions. To initiate such a reaction, heat must be applied. Once the reaction commences, heat must be removed, otherwise a reaction runaway can occur. This creates something of a challenge for the temperature control system, particularly in batch-type processes.

(a) Batch Reactions

Many of the commercial batch reactions used in industry today are polymerizations. The process is a sequenced operation, consisting of a number of steps, carried out in a stirred and jacketed reaction vessel of several thousand gallons capacity. Although there is considerable variation from one product to another, a typical sequence might be:

1. Charge the reactor with reactants and catalyst.
2. Heat the reactor to the desired operating temperature.
3. Allow the reaction to proceed to completion, with appropriate addition or removal of heat. This may take several hours.
4. Heat or cool to cure temperatures (if applicable).
5. Cool and empty the reactor.

Temperature control is most important during the early stages of the reaction process, where the conversion is usually at its highest. The major control objectives here are to maintain the optimum settings for operation at reaction temperature, to prevent overshoot when heating to temperature, and to conduct the reaction in the minimum time, thereby optimizing the production capacity of the unit.

Today's batch processes are usually conducted under the control of a computer system and/or programmable logic controller, whereby each step in the operating sequence is monitored to ensure it is completed and all parameters satisfied before proceeding to the next step.

At the completion of the reaction, heat will cease to be evolved. However, the temperature controller will tend to increase reactant flow to compensate for this, which is opposite to the action required at this point. Some logic must be incorporated to override temperature control on termination of the reaction.

Measurement of the endpoint of a batch reaction is a potentially important application for on-line process analyzers. Various parameters have been used, depending on the particular product being made. In the case of polymer production, molecular weight distribution is used extensively. Viscosity and refractive index measurement are also popular for certain processes. The endpoint of many reactions can be determined by measurement of pH (discussed in Section 24.3). Automatic titrimeters have also been used.

24.7.3 Types of Chemical Reaction Processes

(a) Catalytic Cracking

This is probably the most well-known of all the continuous reaction processes in the petroleum industry and produces unsaturated hydrocarbons from a heavy distillate feedstock originating from the vacuum pipestill. The vaporized feed is cracked at about 1000°F in the presence of a fluidized catalyst. Reaction products and catalyst are separated in cyclones at the top of the reactor and the catalyst is regenerated by steam stripping and coke burn-off in the regenerator. The two major control objectives are:

Heat Balance Control. To maintain heat transfer between reactor and regenerator in balance and avoid reactor temperature excursions. Heat balance is achieved by making the heat required in the reactor to vaporize and crack the feed equal to the heat released in the regenerator by combustion of coke. The controlled variable is the reactor temperature, and the manipulated variables are feedrate, preheat furnace coil outlet temperature, and catalyst circulation rate.

Carbon Balance Control. To maintain carbon production and carbon burning rate in balance for total feedrate and targeted reactor temperature. There are usually three modes of operation here.

1. *No CO burn,* where the controlled variable is the difference between the regenerator secondary cyclone outlet temperature and the regenerator bed temperature.
2. *Partial CO burn,* where the controlled variable is the percentage CO in the flue gas, monitored by an on-line analyzer.
3. *Full CO burn,* where the controlled variable is the percentage of O_2 in the flue gas, measured by an on-line analyzer.

(b) Catalytic Reforming

This is another continuous reaction process and is used to produce high-octane product from a low-octane feedstock. The reactions that take place in this process are described in Chapter 15. There are two types of process, fully regenerative and semi-regenerative. The process is a major source of hydrogen for some refineries.

The prime control objective in this process is octane maximization and so is an important application for an on-line octane analyzer. Usually, the octane number of reformate (stabilizer bottoms) is measured and used to adjust reactor inlet temperatures. The purity of hydrogen recycle gas is another important analyzer application.

(c) *Isomerization*

In this process, light virgin naphtha is upgraded in octane by conversion of n-paraffins to iso-paraffins. This is a fixed-bed process using a platinum-on-alumina catalyst in the presence of hydrogen, similar to hydrofining. The main control objective here is to maintain reactor outlet temperature constant by adjusting the reactor feed furnace coil outlet temperature. The hydrogen/oil treat ratio is an important parameter in this process. Recycle hydrogen purity is controlled using on-line analyzer measurement.

(d) *Polymerization*

This process uses a fixed-bed phosphoric acid catalyst to dimerize or trimerize propylene, butene, or pentene to C_7-C_{12} olefins. The primary control objective is to maintain high catalyst activity, which results in high conversion. The catalyst activity is determined by the hydration level (the water content of the reactor feed). Therefore, measurement of water in the plant feed is an important application for on-line analysis, since the amount of water injected into the reactor feed to produce the correct catalyst hydration is based on a knowledge of water content of the feed stream prior to water injection. This reaction is highly exothermic; therefore, reactor bed temperature monitoring is very important.

(e) *Hydrofining*

This is a very common process in any refinery and is used for sulfur removal. It is a fixed-bed catalytic process where hydrogen recycle gas purity is an important application for on-line analysis.

(f) *Butyl Rubber Production*

Butyl rubber is a copolymer of isobutylene and isoprene. The polymerization reaction is conducted at −140°F in methyl chloride as solvent and in the presence of aluminum chloride as catalyst. The reaction conversion is highly sensitive to feed impurities, which are measured by on-line process GC. The molecular weight distribution of the polymer is an important control parameter that can be measured by an on-line liquid chromatograph (see Chapter 1).

(g) *Steam Cracking*

This is a very important petrochemical process, used for the production of ethylene; the reactor in this case is a low-residence time furnace. Ethylene and other products are produced from a gas oil, naptha, or ethane feedstock.

One of the most important analyzer applications here is the monitoring of furnace effluent. The furnace effluent contains components ranging from hydrogen to tar, so there is a difficult sampling problem to overcome. The extensive use of the Fluid Data Inc. sampling probe, which provides a representative sample containing the light hydrocarbons only, has largely eliminated this problem. The on-line analysis of furnace effluent may be carried out by GC or mass spectrometer; both are used extensively.

Other chemical reaction processes of major importance in the petrochemical industry are those for the production of polyethylene (high pressure and low pressure, high density and low density), polypropylene, polyisobutylene, and polyvinyl chloride.

BIBLIOGRAPHY

Badavas, P. C. Feedforward methods for process control systems. *Chem. Eng.* (Oct 15, 1984).

Bailey, S. J. Energy, ecology crises spur on-line analysis. *Contr. Eng.* (Aug 1978).

Bailey, S. J. On-line analysis 1979: A microprocessor Market. *Contr. Eng.* (March 1979).

Bailey, S. J. Are on-line analyzers becoming analytical controllers? *Contr. Eng.* (Jan. 1982).

Bailey, S. J. On-line analyzers prevent plant operating pitfalls. *Contr. Eng.* (Aug 1983).

Callis, J. B. Optical imaging detectors in service of process analytical chemistry. *ISA Symp.*, Boston, MA (May 1985).

Carr-Brion, K. G. Analytical instruments for process control. *Meas. & Contr.* (June 1981).

Carter, R. J. Control systems for process plant. *Control* (England) (Oct–Dec 1964; 3 articles).

Deming, P. L. Process chromatographs with integral microprocessors upgrade composition-based control. *InTech* (July 1980).

Friedman, P. G., and Gavetti, B. J. Digital dynamic simulation for effective pH control system design. *18th Ann. Symp. ISA-AID,* San Francisco (May 1972).

Funk, G. L., Houston, B. F., and Stacy, G. D. Making energy conservation through automation. *Chem. Eng. Prog.* (June 1978).

Griffin, D. E. Fractionator controls can save energy. *Oil & Gas J.* (March 1981).

Griffin, D. E., Parsons, J. R., and Smith, D. E. The use of process analyzers for composition control of fractionators. ISA Joint Spring Conf., Houston (May 1978).

Griffin, D. E., and Webb, P. U. Process chromatographs and computers in optimizing control systems. *InTech* (July 1979).

Haggin, J. Faster, smaller integrated sensors in offing for process control. *Chem. & Eng. News* (June 1984).

Hall, J. (Ed.). On-line analyzers tackle new demands. *Instr. & Contr. Sys.* (Aug 1979).

Hoyle, D. L. The effects of process design on pH and PIon control. *18th ISA-AID Symp.,* San Francisco (May 1972).

Jutila, M. Multicomponent on-stream analyzers for process monitoring and control. *InTech* (July 1979).

Krigman, A. On-stream analyzers: Progress with new technologies. *InTech* (Oct 1983).

Krohn, D. A. Chemical analysis with fiber optics. *ISA Symp.,* Des Plaines, IL (May 1984).

McCoy, R. D. Adding capabilities to process chromatography with microprocessor-based programmers. *ISA Conf.,* Houston (May 1978).

Mix, T. J., Dweck, J. S., Weinberg, M., and Armstrong, R. C. Energy conservation in distillation. *Chem. Eng. Prog.* (April 1978).

Morrison, D. L., Snow, R. H., and Lamourex, J. P. Advances in process control. *Science,* **215** (Oct 1983).

On-stream analyzers: An update. *InTech* (July 1981).

Pusniak, T. J. Analyzers: The key to advanced control. *Proc. ISA Symp.* (1980).

Schaffer, R. D., and Casciano, R. M. pH control of acidic effluents. *Chem. Eng. Prog.* (May 1979).

Shinskey, F. G., High-performance pH control system. *Inst. Tech.* (June 1968).

Shinskey, F. G. *pH and pIon Control in Process and Waste Streams.* Wiley, New York (1973).

Shinskey, F. G. Adaptive pH controller monitors non-linear process. *Contr. Eng.* (Feb 1974).

Shinskey, F. G. *Process-Control Systems.* McGraw-Hill, New York (1979).

Shinskey, F. G., and Myron, T. T. Adaptive feedback applied to feedforward pH control. *25th Ann. ISA Conf.,* Philadelphia, PA (Oct 1970).

Smith, D. E., Stewart, W. S., and Griffin, D. E. Distill with composition control. *Hcbn. Proc.* (Feb 1978).

Stanton, B. D., and Sterling, M. A. A better way to control composition. *Hcbn. Proc.* (Nov 1979).

Topham, W. H. Simplified chromatographs for on-line control. *InTech* (Dec 1970).

Utterbach, V. C. On-line process analyzers. *Chem. Eng.* (June 21, 1976).

Weiss, M. D. Chemically-sensitive transducers: A new breed of miniature sensors. *InTech* (June 1985).

Williams, A. O., and Green, P. W. Applying analyzers in the process industries. *Hcbn. Proc.* (June 1980).

Williams, E. T. Feedforward for pH control. *Poll. Eng.* (March/April 1971).

Worthy, W. Analytical Chemistry. *Chem. & Eng. News* (July 31, 1978).

Index